To the Success of Our Hopeless Cause

To the Success of Our Hopeless Cause

THE MANY LIVES OF THE SOVIET
DISSIDENT MOVEMENT

BENJAMIN NATHANS

PRINCETON UNIVERSITY PRESS
PRINCETON & OXFORD

Copyright © 2024 by Benjamin Nathans

Princeton University Press is committed to the protection of copyright and the intellectual property our authors entrust to us. Copyright promotes the progress and integrity of knowledge. Thank you for supporting free speech and the global exchange of ideas by purchasing an authorized edition of this book. If you wish to reproduce or distribute any part of it in any form, please obtain permission.

Requests for permission to reproduce material from this work should be sent to permissions@press.princeton.edu

Published by Princeton University Press
41 William Street, Princeton, New Jersey 08540
99 Banbury Road, Oxford OX2 6JX

press.princeton.edu

All Rights Reserved

Library of Congress Cataloging-in-Publication Data

Names: Nathans, Benjamin, author.
Title: To the success of our hopeless cause : the many lives of the soviet dissident movement / Benjamin Nathans.
Other titles: History of the Soviet dissident movement
Description: Princeton : Princeton University Press, 2024. | Includes bibliographical references and index.
Identifiers: LCCN 2023046852 (print) | LCCN 2023046853 (ebook) | ISBN 9780691117034 (hardback) | ISBN 9780691255576 (ebook)
Subjects: LCSH: Dissenters—Soviet Union—History. | Soviet Union—Politics and government—1945–1991. | BISAC: HISTORY / Russia / General | POLITICAL SCIENCE / Political Ideologies / Communism, Post-Communism & Socialism
Classification: LCC DK277 .N38 2024 (print) | LCC DK277 (ebook) | DDC 947—dc23/eng/20240212
LC record available at https://lccn.loc.gov/2023046852
LC ebook record available at https://lccn.loc.gov/2023046853

British Library Cataloging-in-Publication Data is available

Editorial: Priya Nelson, Morgan Spehar, Emma Wagh
Jacket: Karl Spurzem
Production: Danielle Amatucci
Publicity: James Schneider (US), Kate Farquhar-Thomson (UK)
Copyeditor: Jane M. Lichty

This book has been composed in Arno Pro with Helvetica Neue LT Std

10 9 8 7 6 5 4 3 2 1

For Nancy, Gabe, Ilana, and Dora

The test of a first-rate intelligence is the ability to hold two opposed ideas in the mind at the same time, and still retain the ability to function. One should, for example, be able to see that things are hopeless and yet be determined to make them otherwise.

—F. SCOTT FITZGERALD, "THE CRACK-UP"

My intention was not to deal with the problem of truth, but with the problem of the truth-teller, or of truth-telling as an activity: . . . .who is able to tell the truth, about what, with what consequences, and with what relations to power.

—MICHEL FOUCAULT, *FEARLESS SPEECH*

> What's the matter, you dissentious rogues,
> That, rubbing the poor itch of your opinion,
> Make yourselves scabs?

—SHAKESPEARE, *CORIOLANUS*

CONTENTS

List of Illustrations	xi
Prologue: To Live like Free People	1

Part I. Stumbling Blocks — 21

1	Don Quixote in the Land of Soviets	23
2	Involuntary Protagonists	45
3	Transparency Meeting	75
4	The Court Is in Session	99

Part II. Movement of a New Type — 127

5	Rights Talk	129
6	Chain Reaction	147
7	The Dissident Repertoire	195
8	From Circle to Square	238

Part III. In Search of Form — 269

9	Leave the Politics to Us	271
10	Will the Dissident Movement Survive?	294

11	Recrimination and Reassessment	316
12	Taking the Initiative	336
13	The Inner Sanctum of Volpinism	373

Part IV. Disturbers of the Peace — 389

14	The Fifth Directorate	391
15	Fallen Idols	432
16	How to Conduct Yourself	452
17	Allies, Bystanders, Adversaries	487
18	Rights-Defenders among the Nations	510
19	Dissident Fictions	526

Part V. From the Other Shore — 545

20	The Kindness of Strangers	547
21	Adoptees at the Gate	572
22	Final Act	593

Epilogue: Breaking the Fourth Wall — 615

Chronology of the Soviet Dissident Movement — 641
Acknowledgments — 651
Notes — 655
Bibliography — 739
Illustration Credits — 771
Index — 775

LIST OF ILLUSTRATIONS AND TABLE

Figures

1.1 Alexander Volpin	30
2.1 Andrei Sinyavsky	46
2.2 KGB forensic analysis of typewriter patterns	49
2.3 Yuli Daniel	50
2.4 Andrei Sinyavsky and Yuli Daniel	52
2.5 Hélène Peltier	64
3.1 Larisa Bogoraz, Marina Domshlak-Gerchuk, Maria Rozanova, Andrei Sinyavsky, Yuri Gerchuk, and Yuli Daniel	76
3.2 Announcement of the SMOG event "We're Alive!"	83
3.3 Monument to Alexander Pushkin in central Moscow	87
3.4 Homemade banner for the first transparency meeting (December 5, 1965)	91
4.1 Yuli Daniel and Andrei Sinyavsky on trial	105
4.2 Chingiz Guseinov's pass for the Sinyavsky-Daniel trial	111
4.3 Cover page of the official (unpublished) stenogram of the Sinyavsky-Daniel trial	123
5.1 Postage stamps commemorating the fifteenth anniversary of the Stalin Constitution	136–137
6.1 Ludmilla Alexeyeva	170
6.2 Larisa Bogoraz and Pavel Litvinov	173

6.3 Anatoly Marchenko, Larisa Bogoraz, and their son, Pavel	181
6.4 Valeria Gerlin	183
6.5 Natalya Gorbanevskaya	190
6.6 Graph of founding of new samizdat periodicals, 1956–86	193
7.1 Petr Yakir and Sarra Yakir	209
7.2 Petro Hryhorenko (Petr Grigorenko)	215
7.3 Major-General Petr Grigorenko	221
7.4 Printed form for tallying and classifying anonymous "anti-Soviet" leaflets	227
8.1 Alexander Zinoviev's caricature of Leonid Brezhnev	247
8.2 "For your freedom and ours": banner held by demonstrators in Red Square on August 25, 1968	251
8.3 Thích Quang Duc's self-immolation in Saigon on June 11, 1963	257
8.4 Anatoly Yakobson	259
8.5 Family and friends of the Red Square protesters	261
8.6 Larisa Bogoraz-Brukhman	266
9.1 Andrei Sakharov	282
10.1 Gyuzel Makudinova, Andrei Amalrik, and Pavel Litvinov	300
10.2 Outside the courthouse where Red Square protesters were on trial	302–303
11.1 Vyacheslav Igrunov	326
12.1 Ivan Yakhimovich, Larisa Bogoraz, and Pavel Litvinov	339
12.2 Petr Grigorenko and Ivan Yakhimovich	339
12.3 Leonard Ternovsky	348
12.4 Petr Yakir and Viktor Krasin visiting the exiled Pavel Litvinov in Siberia	351

12.5	Postcard prepared by the International Committee for the Defense of the Rights of Man	368–369
13.1	Valery Chalidze	375
14.1	KGB diagram of distribution networks of the underground newspaper the *Bell*	400
14.2	Vera Lashkova	403
14.3	Involuntary detention of dissidents in Soviet psychiatric hospitals, 1960–1990	414
14.4	Natalya Gorbanevskaya and Venyamin Yofe	419
14.5	Alexander Solzhenitsyn	425
15.1	Petr Yakir and Viktor Krasin	433
15.2	Petr Yakir	435
15.3	Press conference with Viktor Krasin and Petr Yakir	444
15.4	Vladimir Dremlyuga	446
15.5	Sergei Kovalev	450
16.1	Arseny Roginsky	459
16.2	Andrei Sinyavsky, his wife, Maria Rozanova, and their son, Yegor	461
16.3	Vladimir Albrekht	464
16.4	Cover of Vladimir Albrekht, *How to Be a Witness*	465
16.5	Transcript of Sergei Kovalev's interrogation on February 21, 1975	472
16.6	Semyon Gluzman	475
16.7	Semyon Gluzman in Siberian exile	483
20.1	Andrei Tverdokhlebov	562
20.2	Valentin Turchin	566
21.1	Vyacheslav Sysoyev, "Human Rights"	584
22.1	Tatyana Osipova and Ivan Kovalev	604

E.1 Strategy-31 sign, Moscow, 2009 — 635
E.2 Constitution of the Russian Federation, Moscow, 2021 — 635

Table

1. Convictions under Articles 70 and 190 of the RSFSR, 1966–1980 — 401

Prologue

TO LIVE LIKE FREE PEOPLE

> Когда-нибудь некий историк
> Возьмёт и напишет про нас,
> И будет насмешливо горек
> Его непоспешный рассказ.
>
> Someday a certain historian
> Will undertake to write about us,
> And with mocking bitterness
> Will tell his unhurried tale.
>
> —ALEXANDER GALICH

In 1957, during the Sixth World Festival of Youth and Students in Moscow, a young Soviet art historian named Boris Shragin was assigned to accompany a group of foreign guests on a visit to the studio of Alexander Gerasimov, the president of the Soviet Academy of Arts. Gerasimov had made his name with fawning neo-classicist portraits of Joseph Stalin and Kliment Voroshilov and had used his position to crush "cosmopolitan" and "formalist" artists whose work strayed from the official aesthetic of socialist realism. Repulsed by the crude naturalism and pomposity of the paintings on display in Gerasimov's studio, Shragin turned to one of the foreigners, an art critic from Tokyo, and gesturing toward the much-praised master's works, quietly offered his own appraisal: "*merde*." The two men conversed in broken French and English until the foreigner asked, "When did your eyes open?" Shragin froze:

> I was unable to answer. And not only because I couldn't find the words. "Eyes open"—to what? Alexander Gerasimov had never been my ideal. "When did your eyes open?" I heard this question many times after I emigrated. Or, in even simpler form: "When did you become a dissident?" It's impossible to answer. Within the question itself lies a willful distortion of reality. Just as the ancient Sophists used to ask: "When did you stop beating your parents?" If you stopped, that means you used to beat them. If you didn't stop, that means you continue to beat them. But I didn't beat them. Ever.[1]

When it emerged in the 1960s in the midst of the Cold War, dissent in the USSR—unauthorized public gatherings, petitions in support of arrested intellectuals, wide circulation of uncensored, typewritten "samizdat" texts—caught everyone off guard. Soviet leaders assumed that, having forged the world's first socialist society under Stalin, the USSR had extinguished the class antagonisms that were seen as the ultimate source of crime and social deviance. Western observers, accustomed to regarding the Soviet Union as the quintessential Orwellian state in which independent thought, not to mention independent action, was virtually impossible, were similarly unprepared for public manifestations of contrarian points of view.

The two sides' attempts to explain these unexpected phenomena also displayed a curious similarity. Soviet officials (and many ordinary citizens) reflexively condemned public dissent as "anti-Soviet." Reluctant to contemplate the possibility that individuals born in the USSR and raised on its values would criticize the Soviet way of life, they routinely ascribed dissenting ideas to foreign influence, whether émigré organizations or Western intelligence services. Western commentators, for their part, were quick to cast the contrarians as surrogate soldiers of Western liberalism in the ongoing ideological struggle against Cold War totalitarianism.

The shock of foreign observers was not lost on the dissidents themselves. As one of them, Andrei Amalrik, put it in a 1969 samizdat essay that forecast the USSR's demise in the vicinity of the Orwellian year 1984, it was as if an ichthyologist had discovered talking fish. Suddenly there were natives inside the closely guarded Soviet aquarium who could not only speak, but speak their own minds. Smuggled texts penned by dissidents found their way to the West, where they appeared under titles evoking prophecy or the bearing of witness, such as Anatoly Marchenko's *My Testimony* (1969) and Andrei Sakharov's *Sakharov Speaks* (1974). The mere existence of people like Amalrik, Marchenko, and Sakharov seemed to take the "total" out of totalitarianism. It was not without irony, then, that Soviet dissidents soon became the darlings of those who continued to regard the USSR as irredeemably totalitarian. Indeed, it was precisely the regime's persecution of dissidents—arrests, harassment of family members, imprisonment in psychiatric hospitals, bogus trials, vicious press campaigns, harsh sentences to labor camps or exile—that now served as the most potent evidence of the Soviet system's abiding malevolence.

All but crushed by the KGB (Committee for State Security) in the early 1980s, Soviet dissidents surprised the world a second time during the heady days of the Mikhail Gorbachev era (1985–91). In a stunning reversal, their slogans—*glasnost* (transparency), rule of law, democratization—began to emanate from the Kremlin. To be sure, apart from Zviad Gamsakhurdia, the volatile and short-lived first president of post-Soviet Georgia, no Soviet dissident vaulted to power in the manner of Lech Wałęsa in Poland or Václav Havel in Czechoslovakia. But several of them, including Sakharov (released from internal exile), Ludmilla Alexeyeva (returned from exile in the United States), Revolt Pimenov (freed after multiple jail sentences and internal exile), and Sergei Kovalev (released after years in a labor camp and internal exile), were elected or appointed to positions of influence. Reborn as prophets of *perestroika* (Gorbachev's "restructuring"), dissidents and their ideas about human rights and civil society, long under siege in their native country, appeared poised to carry the day.

Which, needless to say, they have not. The dream of post-Soviet Russia abandoning its tradition of concentrated, personalized power in favor of laws and institutions has long since evaporated. So has the fantasy of

Moscow giving up its great power privileges to join the Pax Americana, or the European Union, or NATO, or any other alliance in which it would not be the dominant player. During the quarter-century following the USSR's dissolution, the afterlife of the Soviet dissident movement in Russia was a roller-coaster ride of praise, vilification, and willful forgetting. Praise for helping to discredit the Soviet order and ushering in a new era of freedom for Russians and other peoples of the former USSR. Vilification for helping to discredit the Soviet order and ushering in a new era of poverty, disarray, and loss of prestige. Forgetting, because the dissidents, in this version, were merely a sideshow of the superpower rivalry that propelled the Cold War. Or as a Russian colleague of mine put it, "Why are you wasting your time on those people?" Members of the Russian protest group Pussy Riot occasionally invoked the dissident legacy, but a 2013 public opinion poll by the respected Levada Center revealed that fewer than one in five Russians was able to identify any dissidents from the late Soviet era.[2] Contemporary protesters against Russia's assault on Ukraine have also invoked Soviet dissidents as role models. Before his death in February 2024 in a Siberian prison, the leading opposition figure Aleksei Navalny was reading works by Marchenko, Alexander Solzhenitsyn, and Natan Sharansky—and even corresponded with the latter from his cell. At the same time, President Vladimir Putin was busy liquidating the last surviving institutional traces of the dissident movement—the Moscow Helsinki Group, the Andrei Sakharov Center, and the Memorial Society—with barely a ripple of resistance by Russia's population. "History," former KGB director Vladimir Semichastnyi proclaimed in his memoirs, apropos Soviet dissidents, "has swallowed their names."[3]

My aim in this book is to retrieve those names—not to pluck them from history, but to give them a history they deserve. Rather than measure their significance solely by what they contributed to the unraveling of the Soviet order, or to post-Soviet Russia's short-lived experiment in political pluralism, I want to suggest that the story of Soviet dissent illuminates a deeper and more universal struggle between hopelessness and perseverance in the contemporary world. We ourselves are witnesses to a new wave of political authoritarianism along with subtler, technology-enabled forms of unfreedom; we face seemingly insurmountable threats of environmental degradation and expanding inequality. The story of the

Soviet dissident movement gives us a chance to reflect on the possibilities for public engagement under circumstances that appeared even more hopeless than our own. Among Soviet dissidents' many distinguishing qualities, perhaps the most remarkable is that, although they often mocked their own dreams of transforming the Soviet Union—their favorite toast, pronounced around countless kitchen tables, has become the title of my book—they succumbed neither to apathy nor to purely symbolic gestures, neither to what they called dogmatic pessimism nor to pathological optimism. That too demands explanation.

The chapters that follow explore why certain individuals "became dissidents" and how they eventually became a movement, the first movement for civil and human rights in the socialist world. My goal is neither to praise nor to vilify, but to inquire and to comprehend. "The moralist must praise heroism and condemn cruelty," the historian George Lefebvre wrote, "but the moralist does not explain events."[4]

Shragin's speechlessness notwithstanding, Soviet dissidents have had plenty to say about themselves and their movement. The talking fish, to borrow Amalrik's metaphor, proved to be prodigious writers, authoring over 150 book-length autobiographies, along with a slew of other texts.[5] It would be difficult, I think, to find another social movement in which such a high proportion of participants produced written accounts of their own lives.[6] For historians, of course, this is a gold mine—but a treacherous one. The remarkable outpouring of first-person accounts, before and after the USSR's collapse, has tended to foster an image of intrepid individualists fighting the tyranny of Soviet collectivism, thereby obscuring other identities as well as the collective life of the movement itself. Consciously or not, many dissident memoirs were aimed at readers on the Western side of the Iron Curtain, hungry for stories of heroic resistance to totalitarian domination. Western translators and publishers were not above excising passages that gestured in a different direction. "If you were to make little fishes talk," Boswell recorded someone sparring with Dr. Johnson, apropos the writing of fables, "they would talk like whales."[7]

To reach beyond autobiographical accounts by dissidents, I have drawn on the wealth of documentation that became accessible after the break-up of the Soviet Union. This includes records of interrogations by the KGB, inventories of materials seized during apartment searches, as well as internal KGB memoranda analyzing the movement's activities (based on information gathered via moles and listening devices) and recommending actions against it. In some cases, former dissidents were able to obtain photocopies of their KGB case files, often comprising multiple volumes, copies of which are now housed in various archives inside and outside the Russian Federation. During the two decades of relative openness in post-Soviet Russia, certain previously classified materials from within the Soviet government and the Communist Party also became accessible, including records of discussions of the dissident phenomenon by the Politburo, the Party's supreme executive organ.

In addition to shedding new light on Soviet policies toward internal dissent, official sources offer abundant insights into the dissident movement itself, in what we would now call "real time." Complementing them, at slightly greater remove, are materials generated by other close observers of the movement, including Western journalists stationed in Moscow, the human rights organization Amnesty International, and the Alexander Herzen Foundation (Amsterdam). I also draw on materials created in "real time" by dissidents themselves: unpublished diaries, private letters, and, above all, thousands of samizdat texts of every imaginable genre, copied and circulated within the USSR and occasionally smuggled abroad, published in the West, then smuggled back in, a practice known as *tamizdat*, or "publishing over there."[8]

Samizdat and tamizdat reached millions of Soviet citizens beyond the dissident movement thanks to decades of broadcasts by the Voice of America, the BBC, Radio Liberty, Deutsche Welle, and other Western shortwave radio programs collectively known as "the Voices." The fact that dissidents communicated with their fellow citizens as well as with the West via hand-typed text and radio broadcasts helps explain one of the striking lacunae in the movement's historical record, namely, the paucity of visual images of dissidents actually engaging in dissent. As the photographs and other images in this book suggest, the movement's extant visual record consists almost exclusively of portraiture of individuals and groups. In an age of iconic images of protest, whether of African-American civil rights demonstrators facing police dogs and fire hoses in Birmingham, self-immolating Buddhist monks in Saigon, or the "tank man" in Tiananmen Square in Beijing, one is hard-pressed to find images of Soviet dissent, including the historic August 25, 1968, demonstration on Red Square protesting the Soviet invasion of Czechoslovakia. Thousands of samizdat texts have been preserved and archived, but I know of no photographs of people engaged in the most widespread dissident activity of all—typing and reading samizdat. No doubt the Soviet intelligentsia's text-centrism helps account for this lacuna, along with the widespread fear of documenting activities that could lead to arrest and imprisonment. If a visual record of the Soviet dissident movement exists, it will likely be found some day in the Moscow archive of the former KGB, which was known to secretly photograph and film dissident protests, and perhaps other activities as well.

Most dissident texts about life in the USSR, especially the memoirs, highlight their authors' estrangement from official Soviet practices and ways of thinking—understandably, given the pervasive ideological rigidity that prevailed for much of the USSR's history. "There are times," the writer Andrei Sinyavsky urged, "when you have to defend yourself against reality." This was, among other things, a literary strategy designed to bypass the official dogma of socialist realism, to show that art could become a "means of vanquishing reality."[9] But it could also be a way of living, what the poet Joseph Brodsky called "the science of ignoring reality."[10]

Dissidents did not ignore Soviet realities; they called attention to them and sought to change them. They did so by appealing to Soviet and eventually international law—something many of their fellow citizens regarded as weightless fictions floating above the realities of the power-political. Dissidents regarded themselves as avatars of the storied nineteenth-century Russian intelligentsia, with its sense of moral calling on behalf of universal values, or as belated embodiments of the Western struggle for freedom, "reinventing the pursuit of liberty," as Ludmilla Alexeyeva put it. They imagined themselves, in other words, as something other than people of their own place and time.[11] Whatever talking fish might say, we should hardly expect them—paraphrasing the anthropologist Ruth Benedict—to mention the water in which they swim. My effort to understand Soviet dissidents has led me to a very different point of departure, namely, that we can begin to fathom the history they made only if we consider the history that made them.

Dissidents were Soviet people. This may seem a banal claim—akin to "Karl Marx was a German philosopher"—but is less banal than it first appears, especially if we interpret the movement they made in light of that claim.[12] Dissidents were children of the Soviet system, part of a generation that became the first adults in history who were not immigrants to socialism (voluntary or otherwise), but native born, the first indigenous speakers of its language. Shortly after seizing power, Vladimir Lenin cautioned his fellow Bolsheviks that they now faced the task of building socialism "not with abstract human material, or with human material specially prepared by us, but with the human material bequeathed to us by capitalism."[13] The dissident movement drew on human beings bequeathed by Soviet socialism, the system forged by Lenin and his successors. Dissidents were schooled in Soviet values, employed like everyone by the Soviet state, and scarred by the collective traumas of the Soviet experiment. They did not build the first socialist society; they inherited it. Their dissent was constructed largely out of ingredients taken from the Soviet world, especially from Soviet law. Sometimes the act of appropriation involved adapting or repurposing; sometimes it took the form of repudiation. "It doesn't matter what attitude people take in private or in conversation with friends," noted the philosopher Alexander Zinoviev. "What is important is that people are in constant contact with the powerful magnetic field of ideological influence ... they are willy-nilly particles in that field, and absorb from it a certain electric charge, standpoint, orientation, etc."[14] Rather than an opening of eyes to truth, an awakening or enlightenment, the theme that runs through this book is older and deeper: how orthodoxies generate their own heresies.

How do individuals and societies emerge from totalitarianism? The classic works on totalitarian regimes, by Franz Neumann, Hannah

Arendt, and Carl J. Friedrich and Zbigniew K. Brzezinski, pay far more attention to how such regimes come into being than to how they end.[15] These and many other scholars agreed that totalitarianism was a novel invention of the twentieth century, the age of mass participation in politics. The term "totalitarian" derived from the total control these new dictatorships sought over not just the behavior but the inner lives of their subjects. They explicitly rejected the principle of separation of powers among different branches of the state, instead favoring single-party rule, a centrally directed economy, a monopoly on communication, an all-encompassing and obligatory ideology, and mass terror. Dictatorships have come in many varieties, of course, including prior to the twentieth century. The distinguishing feature of totalitarian dictatorships was their insistence not just on the consent of the governed (a characteristic of all polities that locate sovereignty in "the people"), but on mass participation in state-driven projects of individual and societal transformation. Totalitarian regimes, in a nutshell, were participatory dictatorships.[16]

If consent in such systems was insufficient, dissent was intolerable. Ruthless in their suppression of internal non-conformity, totalitarian regimes, according to Western theorists, "would never change unless overthrown from outside."[17] Such an assumption was understandable, given that two of the three archetypal totalitarian states—Fascist Italy and Nazi Germany—were indeed destroyed by external force. For some contemporary observers, including Amalrik, that was how the third such state, the Soviet Union, was also likely to meet its end—at the hands of a resurgent China. But for many others, the likelihood of that scenario was diminished by the introduction of nuclear weapons into international relations. Unlike Benito Mussolini's Italy and Adolf Hitler's Third Reich, the USSR survived into an era when all-out war between great powers appeared literally a dead-end, leading inexorably to "mutual assured destruction" (or MAD, as the felicitous acronym put it). Under conditions of nuclear stalemate, historic precedents for the demise of totalitarian regimes offered little guidance.

Most people today regard Soviet totalitarianism as having unraveled in the 1980s as a result of economic stagnation, rising internal nationalisms, or Gorbachev's ill-starred attempts to fix those and other structural problems of the Soviet empire. They also imagine that it imploded quite suddenly, a distant echo of the pattern set by Nazi Germany and Fascist Italy despite the fact that the end of the USSR did not involve a military defeat or even a military attack. In fact, the Soviet Union's emergence from totalitarian rule began not in the 1980s but in the 1950s, under Stalin's successor, Nikita Khrushchev, and continued in zigs and zags over the course of the next three decades. Khrushchev's great wager, one he bequeathed to those who peacefully ousted him, was that socialism could forego the mass violence of the Stalin era while preserving the other fundamental components of the Soviet system, namely, monopoly rule by the Communist Party and the centrally planned economy.[18] For the post-Stalin leadership, removing state-sponsored terror, mass incarceration in the Gulag, and periodic Party purges from the repertoire of governing practices was first and foremost a step toward self-preservation. But it was also, as the historian Jörg Baberowski has noted, a "civilizing achievement that changed the lives of millions of people."[19] A formerly carnivorous regime responsible for murdering millions of its own citizens via execution, starvation, and forced labor entered—to borrow from the memoirist Nadezhda Mandelstam—its "vegetarian epoch."[20]

In contrast to the other totalitarian regimes born in the interwar era, and unlike nearly all of its revolutionary predecessors in Europe going back to the Anabaptists, the Roundheads, and the Jacobins, the state established by the Bolsheviks lasted long enough to encompass multiple generations.[21] Children and grandchildren of the revolutionaries, along with other Soviet citizens, grew up inside the state created by their parents and grandparents. Over the decades, they witnessed multiple leadership transitions, the metamorphosis of their country into a global superpower, and, by the early 1960s, its passage from a predominantly agrarian to a predominantly urban society. A seventy-four-year lifespan (1917–91), roughly that of a human being, might be regarded as rather brief for a sovereign state, but for a revolutionary polity it is exceptionally long. History is littered with short-lived revolutions. The second half of the USSR's history (1953–91), so often overshadowed by the dramatic upheavals and savage wars that marked its first half, affords us an unusual opportunity to trace the evolution of a society out of terror and mass violence into a condition without precedent, a set of arrangements for which contemporaries struggled to produce adequate terminology—as do we. "In the post-Stalin period," observed the physicist Yuri Orlov, born in 1924, "Soviet society became as different from that of the Stalin era as the sky from the earth."[22] In the third (1966) edition of her monumental study *The Origins of Totalitarianism*, originally published during Stalin's final years in power, Arendt registered "an authentic, though never unequivocal, process of detotalitarization," as a result of which "the Soviet Union can no longer be called totalitarian in the strict sense of the term."[23] Soviet leader Leonid Brezhnev called it "developed" or "mature" socialism. The computer scientist Valentin Turchin, borrowing the language of thermodynamics, described it as "steady-state totalitarianism," on the assumption that the USSR had channeled its revolutionary heat into stable patterns of governance and everyday life. Václav Havel preferred "post-totalitarian."[24] All these labels sought to make sense of the present by referencing what now felt like a completed chapter of the Soviet past.

What was "post-totalitarian" about the second half of Soviet history? Arbitrary arrests and mass violence virtually disappeared. Party leaders were referred to by their real names, rather than mythic *noms de guerre* such as "Lenin," "Trotsky," and "Stalin." The USSR no longer battled capitalism alone, having established an ensemble of socialist states around the globe. Soviet culture retreated from the ambition of merging individual selves into the greater collective, and instead sought to channel personal proclivities toward the larger social good.[25] An equally striking post-totalitarian symptom was the appearance of open dissent and the Kremlin's less than lethal response to it.[26] Solzhenitsyn called it "the return of breathing."[27] Turchin sensed a population no longer terrorized by a revolutionary regime but still fettered by "the inertia of fear."[28]

In the absence of regime change, totalitarianism's long afterlife was characterized by constant uncertainty over the boundaries of permissible behavior, and for dissidents and some other Soviet citizens, an abiding anxiety about a possible relapse to the Stalinist order. Despite the magnitude of the changes it brought, the post-totalitarian dispensation was never formally announced, much less codified, apart from Khrushchev's denunciation of certain of Stalin's crimes and cult of personality. To be sure, the official narrative had the USSR poised to enter the historically ordained transition from socialism to communism, in which the functions of the state would gradually devolve onto the Soviet population, the new "Soviet people" forged in the Stalinist furnace from which their country was now emerging. Communism, as Karl Marx and Friedrich Engels had prophesied and Lenin had affirmed, would mark the final stage of social evolution, the withering away of the state and the development of the human personality to its fullest potential. For most of the Soviet population, however, the official radiant future mattered less than "real existing socialism," which now meant a stable, largely peaceful (by Soviet and indeed contemporary American standards), seemingly perpetual socialism whose citizens could live in individual family apartments with their own telephones, refrigerators, and television sets and die in their beds. It was a socialism characterized not so much by increasing participation of ordinary citizens in the work of the state as by the state's retreat from the private lives of its citizens. Khrushchev and other Soviet leaders may have wanted to believe that the *homo sovieticus*, the new and improved version of humankind, had taken socialist values to heart. But the retreat from policing Soviet souls was driven as much by revolutionary exhaustion, and the USSR's external responsibilities as leader of the socialist world, as by confidence in the capacity of second-generation Soviet citizens to manage their own lives. The real story of the second half of Soviet history was not the transition to a communist future, but the struggle to exit a totalitarian past. Dissidents were the leading edge of that struggle. Three decades after the Soviet collapse, the story is not yet over.

"A specter is haunting Eastern Europe," Havel famously began his 1978 essay "The Power of the Powerless," "the specter of what in the West is called 'dissent.'" "So-called 'dissidents,'" the essay continues, "did not fall out of the sky." They are "a natural and inevitable consequence of the present historical phase of the system" they haunt.[29] It was a marvelous

The problem was not just that Soviet mass media had weaponized the word "dissident," sensing an opportunity to vilify domestic critics by branding them with a foreign epithet. The term itself implied a sort of religious sectarianism, a self-marginalization from established norms. That was certainly its earliest meaning in Russia, where the term "dissident" first appeared in the eighteenth century with reference to Protestant or Catholic minorities in Europe, before being extended in the nineteenth century to Christian sects in Russia itself.[33] In Latin, *dissident* means "sitting apart" (from *dis-sidere*)—quite the opposite of the dissident movement's civic engagement and attempted dialogue with the state. In colloquial Russian, moreover, "to sit" (*sidet*) means to "do time" in prison or a labor camp, suggesting that dissidents were people with criminal pasts or perhaps an appetite for martyrdom. As always in the Soviet setting, semantic ambiguity provided fertile soil for dark (and often untranslatable) humor. Among the endless cycle of jokes that feature the mythical Radio Armenia responding to listeners' questions, one goes as follows. Question: "Who are the dissidents?" Answer: "Who the *dissidents* are, we don't know. There are *dosidents* [those who have not yet done time], *sidents* [those currently doing time] and *postsidents* [those who have completed their time]."[34]

For all their unease, however, those labeled "dissidents" found the term inescapable—as do I.[35] I am nevertheless mindful that to introduce them as such could preempt my investigation into how certain Soviet people became dissidents before it has even begun. I have therefore taken a cue from Yelena Bonner. Rather than making explanatory shortcuts via the designation "dissident," I approach my protagonists with an eye to what animated them, what disturbed them, the kind of work they did, and the larger arc of their lives.

That the Western term "dissident" became inescapable, inside as well as outside the Soviet Union, speaks to the power of the Western gaze. An offspring of the Enlightenment, that gaze was refracted as it passed through the prism of the Cold War and continued to draw energy from Russia's, and the USSR's, historic fixation with the West. Since at least the eighteenth century, Europe had loomed before educated inhabitants of the tsarist empire as the image of Russia's own possible future, to be play on the opening lines of *The Communist Manifesto*, with its claim that capitalism breeds its own disrupters in the form of the industrial proletariat. While cleverly tweaking communism's founding document, however, Havel was also distancing himself from the terms "dissent" and "dissidents," casting the words as alien to the milieu they purported to describe. A purely indigenous phenomenon somehow got stuck with a strangely foreign name. In an essay called "On the Choice of Terms," Boris Shragin had anticipated this paradox. "The term 'dissidents' [Russian: *dissidenty*]," he noted, "came from the West. It was thought up by foreign journalists who observed a new phenomenon in the Soviet Union, wrote about it, and therefore needed to somehow give it a name. Like all such words invented by outsiders, the term 'dissidents' fits poorly with the consciousness of those to whom it is applied."[30] Among those to whom it was applied, the feeling of distaste was indeed remarkably widespread. "I've never cared for the term," Sakharov wrote in his memoirs. An American visitor to Moscow in the 1970s who casually referred to Sakharov as a dissident was promptly corrected by his wife, Yelena Bonner: "My husband is a *physicist*, not a dissident."[31] Solzhenitsyn, from his exile in Vermont, insisted in a private letter to President Ronald Reagan that he was neither a dissident nor an émigré.[32]

Some of those who were labeled "dissidents" preferred to be known as *inakomysliashchie*, or "other-thinkers" (literally, "people who think differently"), which certainly had the advantage of being recognizably Russian as well as appealing to the vanities of the intelligentsia. The term harkened back to the nineteenth century, when the intelligentsia had styled itself as "thinking Russia," a tiny island of enlightened consciousness in a sea of benighted peasants. By the 1960s, thanks in no small measure to the efforts of the Communist Party, tens of millions of peasants had learned to read, moved to cities, and, in some cases, received a higher education. "Thinking" was no longer a distinguishing mark; "thinking differently" was. But *inakomysliashchie* was a mouthful and difficult to render in foreign languages. Even more, it implied that the key quality was what one thought, when in fact it was what one said and did that mattered most. Others favored the terms *pravozashchitniki* ("rights-defenders"), or *zakonniki* ("legalists"), or *demokraty*.

emulated or loathed but impossible to ignore. The Bolshevik Revolution sought to drag the remains of the Romanov empire into a different future, a socialist society destined by the laws of history to serve as model for all other societies, starting precisely with the advanced (but now, according to Marx's stages of historical evolution, lagging behind the Soviet Union) capitalist countries of Europe and North America. Having installed themselves in the Kremlin, the Bolsheviks projected their role as vanguard party of the imperial Russian proletariat onto the new Soviet state, the presumed vanguard of history itself.

Such a daring leap into world-historical preeminence, not surprisingly, only heightened Moscow's sensitivity to what foreigners thought and said about the Soviet Union. Vanguard or not, the USSR had to make its way in a world still shaped by capitalism. The socialist construction project proceeded under watchful Western eyes, including those of foreign journalists with front-row seats in the Soviet capital. Beginning in the 1960s, they were the conduit by which the word "dissidents" was attached to protesters of all kinds in the USSR and, from there entered, with modifications, the Soviet lexicon. "We loved the dissident story," recalled James Jackson, Moscow correspondent for United Press International and the *Chicago Tribune* between 1969 and 1976, "because it was the only live story in town."[36] There were, needless to say, plenty of other stories, but few as captivating and accessible as that of the dissidents. Unlike most Soviet citizens, dissidents did not avoid contact with Westerners, especially journalists.[37] And no wonder: foreign correspondents became their lifeline to the West itself, to their most receptive audience and source of support.

In the West, as Zinoviev observed, the term "Soviet dissident" signified any citizen of the USSR who came into conflict with the Soviet system and was subject to repression as a result. This usage put nationalists, religious sectarians, neo-Leninists, terrorists, people desiring to emigrate, and writers who circulated their works in samizdat—regardless of content—all into a single pot. In the Soviet lexicon, by contrast, "dissident" tended to carry a narrower meaning, indicating those who conducted protests publicly in defense of civil and human rights.[38] The Soviet meaning is the one I have adopted in this book, with specific procedural and substantive elements that constitute what I call the "dissident repertoire."

Foreign correspondents did more than give dissidents their name. They trained a powerful international spotlight on a handful of individuals, sanctifying them and often shielding them, at least temporarily, from persecution—an effect of which dissidents (and would-be dissidents) were well aware. "They fashioned a halo around me," as one put it.[39] The spotlight trained on the chosen few left many others in the shadows, rendering a chiaroscuro of celebrity and obscurity within the dissident milieu. In their urge to tell person-centered stories of heroic resistance inside the Soviet behemoth, Western correspondents instinctively looked for (male) leaders in a movement that strenuously resisted internal hierarchies. The marquee names—Nobel laureates Sakharov and Solzhenitsyn—attracted enormous attention thanks to their formidable professional achievements and their trenchant critiques of the Soviet system, but they were not the movement's driving forces. By their own accounts, Sakharov arrived late, while Solzhenitsyn preferred to act alone, "preserv[ing] a jealous distance between himself and the other dissidents," as his biographer Michael Scammell observed.[40] The "many lives" in the subtitle of my book is meant to highlight the people who actually formed and propelled the movement, most of them scarcely known today, even in Russia. It also underscores the several near-deaths and resurrections through which the movement evolved as it faced suppression by the KGB as well as internal disagreements over organization and strategy.

If the bare plot and alleged leaders of the movement are well known, its inner drama is not. That drama has been obscured by two factors above all: dissidents' desire to protect one another from persecution by the Soviet state and the Western tendency to lionize rather than analyze Soviet dissent. As the USSR became the essential Cold War reference point for reflections on freedom and unfreedom, so too did Soviet dissidents. They were, one might say, useful to think with—and even more useful to polemicize with. One of the earliest studies of the subject was a 1966 report (updated in 1968) by the United States government titled *Aspects of Intellectual Ferment and Dissent in the Soviet Union*, a work whose gestation was carefully tracked by the KGB.[41] In his preface, Connecticut senator Thomas Dodd, a veteran of the Nuremberg Trials, had the following to say:

> In the battle for justice and liberty which is going on in the Soviet Union today, the intellectuals of the Soviet Union have given birth to a literature of freedom that ranks among the noblest and most heroic since the American Revolution. To write about freedom from the safety of the Western World requires little courage or commitment, for in the Western World freedom is an accomplished fact. But to write about freedom or to take a stand for freedom under the totalitarian Soviet regime requires the most sublime kind of courage, and a degree of commitment exceeding the understanding of those who have been brought up to regard freedom as their birthright.[42]

Soviet dissidents, in this telling, embodied the struggle for liberty in its purest form, a form no longer available to inhabitants of the West, where the perception of freedom as an "accomplished fact"—a perception fortified by the Cold War—rendered moot the idea of struggling for it, or even the capacity to fully comprehend and appreciate it.

Dodd's enthusiasm for dissidents as cognitively privileged creatures who could teach Westerners to re-embrace their own most cherished values, thus reclaiming their best selves, may strike today's reader as starry-eyed. Significant numbers of the senator's fellow Americans—those who identified with the civil rights or feminist movements, for example—hardly shared his impression of freedom as already fully achieved, especially if their notion of freedom extended beyond the presence or absence of restrictions imposed by the state. I hope nonetheless to show, in the pages that follow, that there is indeed much to learn from the Soviet dissident movement, including its inner drama, provided we are willing to peer beyond the familiar antinomies of heroism and cruelty, resistance and collusion. How did *this* movement emerge from *that* society? How did a diverse cohort of Soviet citizens find each other and collectively invent new forms of public engagement? How did they arrive, under the conditions of Soviet socialism, at the doctrine of universal human rights? They certainly did not start there. Many Western observers have taken the dissidents' invocation of rights and legal norms as a natural idiom for opponents of a Soviet-style regime. And no wonder: the rhetoric of rights, and especially of universal human rights, has seeped into the water in which we swim. To understand human rights as something less than self-evident, something other than finished goods imported from the West, we need to grapple with the history of rights, and those who fought for them, in non-democratic, non-Western settings. At some point, recalled Alexeyeva, "I realized that Westernizers were by no means Westerners."[43] Within the force field of the Cold War, and independent of the American policy of containment of the USSR, Soviet dissidents deployed a particular set of rights to promote their own version of containment of state power from within the Soviet order. In the process, they helped put human rights on the global map.

"The government was sole and absolute manager of the public business, but it was not master of individual citizens. Liberty survived in the midst of institutions already prepared for despotism. But it was a curious kind of liberty, not easily understood today." So wrote Alexis de Tocqueville, with a half-century's hindsight, about Louis XVI's France on the eve of the French Revolution.[44] The curious islands of liberty that formed inside the Soviet Union a half-century ago, during the second half of its history, are also not easily understood today. They belong to what the historian Quentin Skinner calls "liberty before liberalism."[45] Some involved "inner freedom," a retreat into cultural or religious oases of the spirit. Others fed on ironic detachment from official norms by people who chose to live "outside the system." Think of the philosopher employed as a boiler-room attendant, with ample time to read and reflect while "working." Still others involved "niche-making," the art of turning technical or disciplinary expertise into a form of insulation from ideology and politics. Perhaps the most commonly practiced form of liberty—which did not require being a member of the intelligentsia—derived from the unspoken agreement between the authorities and the population that in return for political docility, individual citizens could engage in various semi-legal or illegal economic activities.[46] As a much-cited Soviet saying put it, "They pretend to pay us, and we pretend to work."

Those who came to be known as dissidents pursued a different if equally curious form of liberty, one that demanded its own kind of make-believe. They pretended that the constitution of the USSR, with its guarantees of freedom of speech and assembly, was the law of the land. "They did something," Andrei Amalrik wrote, "simple to the point of genius: in an unfree country, they began to conduct themselves like free people."[47] This was the dissident drama inside the second act of Soviet history. Not surprisingly, it turned out to be anything but simple.

PART I
Stumbling Blocks

1

Don Quixote in the Land of Soviets

We must free ourselves from the influence of people with their deformed language and find a scientific expression for the concept of freedom. Only when we attain this will we be able to trust our own thoughts.

—ALEXANDER VOLPIN

In memoirs by participants in the Soviet dissident movement, an obscure name usually appears near the beginning of the story. No one claims that Alexander Volpin founded the movement. In good Soviet collectivist style, the movement had no founder, and, in any case, an eccentric mathematician known to walk the streets of Moscow in his house slippers would have been an unlikely candidate for that role. Volpin appears instead as the source of an idea that would come to serve as the movement's prime strategy as well as its central goal, an idea at once entirely consistent with the Soviet order and utterly foreign to many of those who grew up in it: that Soviet law should be taken seriously.

Andrei Amalrik described Volpin as "the first to understand that an effective method of opposition might be to demand that the authorities observe their own laws."[1] Ludmilla Alexeyeva, an editor and reform-minded Party member who befriended Volpin in the early 1960s, was amazed to hear him praise the 1936 Soviet Constitution (popularly known as the "Stalin Constitution") and castigate Soviet citizens for acting as if they had no rights. He would hold forth, according to Alexeyeva, on the proposition that Soviet laws "ought to be understood in exactly the way they are written and not as they are interpreted by the government, and the government ought to fulfill those laws to the letter." What would happen if Soviet citizens acted on the assumption that they have rights? "If one person did it, he would become a martyr; if two people did it, they would be labeled an enemy organization; if thousands of people did it, they would be a hostile movement; but if everyone did it, the state would have to become less oppressive."[2] Vladimir Bukovsky, who met Volpin at an unauthorized poetry reading in Moscow's Mayakovsky Square in 1961, described him as "the first person in our life who spoke seriously about Soviet laws. We laughed at him: 'What kind of laws can there be in this country? Who cares?'" That, Volpin replied, was the problem. "'Nobody cares. We ourselves are to blame for not demanding fulfillment of the laws.'"[3]

Demanding fulfillment of the laws required strictly abiding by the laws oneself. The Soviet dissident movement thus stands apart from one of the modern era's most distinctive forms of resistance to state power, the civil disobedience campaigns inspired by Henry David Thoreau, Lev Tolstoy, Mahatma Gandhi, and Rosa Parks. Civil disobedience, to quote the *Dictionary of the History of Ideas*, presupposes a "formal structure of law" and consists of "publicly announced defiance of specific laws, policies, or commands."[4] It was Volpin who developed the less well known but, in the Soviet context, equally provocative technique of radical *civil obedience*: engaging in practices formally protected by Soviet law, such as freedom of assembly or transparency of judicial proceedings, but often subject to extra-judicial or even judicial punishment by the state.

The history of imperial Russia and the early Soviet Union includes more than a few uprisings and other forms of disobedience, whether at the local level or against supreme authorities in Moscow and St. Petersburg. The Soviet Union would not have been born without a series of revolts against Tsar Nicholas II's regime and the Provisional Government that briefly took its place. Rarely, however, was resistance to perceived tyranny in Russia articulated and legitimated as a right, let alone a sacred right. More common was the poet Alexander Pushkin's famous verdict (in 1836) on Russian revolts as "senseless and merciless," a view that has since hardened into cliché. Even Lenin was inclined to view rebellions by workers and peasants as prone to "spontaneity," by which he meant lacking a historically informed purpose, unless they were guided by the higher "consciousness" of the Bolsheviks. The concept of *civil* disobedience never found much traction in imperial Russia or the Soviet Union, where civility was tainted by association with the bourgeoisie, and quotidian acts of unannounced disobedience were often necessary simply to survive.[5]

The absence of a tradition of engaging in civil disobedience, or of a legitimizing right to resist tyranny, may have helped open a space for Volpin's counter-intuitive strategy of civil obedience. But it hardly explains how he got there, how he injected new meaning into terms such as "transparency" (*glasnost*) and "rule of law" (*zakonnost*) that eventually became watchwords of Mikhail Gorbachev's fatal attempt to reform the Soviet system. The literary critic Yuri Aikhenvald, author of *Don Quixote on Russian Soil*, referred to Volpin as "the solitary knight at the beginning of our liberation movement."[6] This Don Quixote's Dulcinea, however, was not an ideal woman or even an ideal Russia, but an ideal language, a language free of ambiguity, capable not merely of flawlessly rendering thoughts but of making thought itself flawless.

The offspring of an affair between the translator Nadezhda Volpina and the immensely popular "village poet" Sergei Esenin, Alexander Volpin was a year and a half old when his father committed suicide in the Hotel Angleterre in Leningrad in 1925. Eight years later, he and his mother resettled in Moscow, where Volpin was to live for most of the next four decades. A bookish young Muscovite, Volpin constructed an inner life from the many available "leftovers of the past"—Pushkin, Tolstoy, Jeremy Bentham, Sigmund Freud—leavened by Bolshevik visions of a scientifically planned society of the future.[7] His adolescent mind associated the ubiquitous Soviet red star with Bentham's panopticon: the all-powerful Soviet state at the center of the star systematically observing and perfecting the character of the Soviet citizens gathered around it. He took pleasure in imagining a future of uninterrupted progress as the increasingly mechanized communist paradise expanded to the far reaches of the solar system.

At the same time, Volpin's adolescent preoccupations—as he recalled them a decade later, in diary entries from the 1940s—included suicide, immortality, Tolstoy's gospel of non-resistance to evil, and the search for a path to truth. To this list could be added, by the time of his fifteenth birthday, an obsession with dates, calendars, and odd word combinations. Rather than suffer the uncertainties of playing with other children in the courtyard of his apartment building, Volpin often sat at home and performed the elaborate calculations necessary to establish daily calendars far into the future.[8] His awkward relations with schoolmates, meticulously recorded in his notebooks, often left him distraught. After one particularly traumatic falling-out with a high school friend on April 15, 1939—"a day worthy of a monument in my life"—a despairing Volpin decided that his frictions with peers stemmed from his own inner struggle between thought and feeling. "I swore to myself that I would overcome my lack of will," he noted. "To hell with my useless heart. Only mind, only logic! I will cease to have enemies. . . . The era of dual power is over—the dictatorship of reason has begun."[9] His own psyche, Volpin concluded, was reenacting the revolutionary drama of October 1917, the replacement of uneasy power-sharing between the Provisional Government and the Councils (Soviets) of Workers' and Soldiers' Deputies by the Bolshevik "dictatorship of the proletariat."

As with many inhabitants of the Soviet Union in the 1930s, Volpin's awareness of the climate of repression was muted by the sense that life in the West—with its economic free-fall and rising fascist menace—was far worse.[10] Although a half-brother, Georgy Sergeyevich Esenin, disappeared following his arrest in 1937, the wave of political terror that engulfed Soviet society in the late thirties appears to have left Volpin and his immediate family relatively unscathed.[11] He was also spared the danger of military service when the Red Army declined to draft him in the summer of 1941, during the desperate months following Hitler's invasion. An army doctor informed by her colleagues that Volpin was "not of this world" warned Nadezhda Volpina that her son "would not last a minute" in the army. The two women agreed on a lifesaving diagnosis: "schizophrenic—unsuitable."[12]

The diagnosis did not prevent Volpin from enrolling at Moscow State University in August 1941. As Nazi air raids against the Soviet capital intensified, those faculty and students who had not been drafted—among them a shy senior named Andrei Sakharov—were evacuated to Central Asia.[13] In the transplanted department of mathematics, Volpin pursued his interest in the emancipation of reason from emotion and faith, a subject that led him to a number of key works on mathematical logic. These included Bertrand Russell and Alfred North Whitehead's seminal *Principia Mathematica*, which argued that all of mathematics could be derived from a priori principles of logic alone, and therefore that mathematical proofs, or truth-statements, need not depend on unproven assumptions, or belief-statements. Much of Volpin's work in the postwar years can be understood as an elaboration of the analytic philosophy championed by Russell, and in particular "ideal language" philosophy, whose mission was to create a formal system of communication free of the ambiguities of natural languages.

Friends and acquaintances from Volpin's student years recalled his devotion to the dictatorship of reason in the form of strict logic and legalism. One account has him deciding during the war that since butter contained more calories than bread, he would exchange his bread ration for butter—a decision that briefly landed him in the hospital.[14] Another describes an early run-in with the local branch of the Communist Youth League, which in 1943 adopted a resolution ordering the famously unkempt Volpin to take a bath. Since he was not a member, he considered himself outside the League's jurisdiction and refused to comply.[15] Yet another relates how Volpin once publicly challenged the head of the Communist Party bureau within the mathematics department at Moscow State University, Pyotr Ogibalov, after the latter demanded the expulsion of a group of students who had allegedly formed a secret organization. "What is it that makes you conclude the organization was secret?" asked Volpin, who was not in the group, at a meeting called to discuss the charges. "The fact," Ogibalov answered, "that I was unaware of its existence." "Forgive me," came Volpin's reply, "but until today I was unaware of your existence, but that has not led me to conclude that you exist secretly."[16] By his own account, during his student years Volpin felt no particular hostility toward Stalin or the Communist Party. He was simply an instinctive contrarian and would have been one "even under the best regime."[17]

If anything, Volpin's occasionally antagonistic behavior in the immediate postwar years was rather conventional. He began to cross out the names of candidates or otherwise disfigure election ballots, which featured one candidate per office—a safe gesture for those wishing to register anonymous dissatisfaction with the USSR's peculiar form of democracy. This was not "an expression of my relationship to the system," Volpin noted, but rather a piece of youthful non-conformism, fueled by the same spirit that once led him to salute a Communist official with his right hand while secretly making the obscene "fig" sign (thumb inserted between the middle and index fingers) with his left. Under Stalin, to be sure, even symbolic gestures such as these could have terrible consequences if discovered. When asked years later whether he had given any thought to the "social significance" of such behavior, Volpin's response was unequivocal: "That expression would probably have made me vomit. I simply didn't think in terms of that category of struggle with the regime."[18]

Circumstances, however, soon changed. After successfully defending his dissertation in the spring of 1949, Volpin was sent to the recently reincorporated city of Chernivtsi, in Soviet Ukraine, to teach mathematics at the local state university. Here he continued to write and, more significantly, to read aloud to acquaintances his non-conformist poems. Perhaps an eccentric mathematician from Moscow caught the eye of local authorities in a small provincial city. Or perhaps authorities in Moscow wished to arrest the non-conformist son of a famous poet in an out-of-the-way place in order to avoid attention.[19] Whatever the case, within weeks of arriving in Chernivtsi, Volpin was seized, sent on a plane back to Moscow, and deposited in the KGB's infamous Lubyanka prison. There he was re-arrested (possibly in order to transfer jurisdiction over the case from the Ukrainian to the Russian Republic) and charged with "systematically conducting anti-Soviet agitation, writing anti-Soviet poems and reading them to acquaintances."[20]

In a poem begun on the day of his arrest, Volpin struck a heroic pose, claiming that he feared "neither prison nor reprimands":

> What's the use here of "why" and "could it really be,"
> Everything is obvious without any "why's":
> Since I'd dispensed with all belief in human aims,
> Was it any wonder I was locked up in prison!
>
> . . . .
>
> I'm a spider, proficient in webs,
> Under interrogation I shall invent no lies at all. . . .
> I shall penetrate their protocols and their minds.[21]

Inside the Lubyanka, reality was somewhat different. Volpin was sufficiently apprehensive about the prospect of prison or labor camp that he faked a suicide attempt in order to initiate a psychiatric evaluation. At the time, he considered mental institutions "a salvation from a more awful punishment."[22] At the Serbsky Institute of Forensic Psychiatry in Moscow, doctors declared Volpin mentally incompetent, and in October 1949 he was transferred to the Leningrad Psychiatric Prison Hospital for an indefinite stay. A year later, he was labeled a "socially dangerous element" and exiled for five years (without trial) to the town of Karaganda, nearly two thousand miles east of Moscow, in Soviet Kazakhstan, where he found employment as a math teacher for adults. Scarcely two weeks after Stalin's death in March 1953, an amnesty was announced for over a million Soviet prisoners and exiles, Volpin among them. By April, he was back in Moscow.

Compared with the hundreds of thousands of other political prisoners, Volpin had suffered a relatively mild incarceration and exile. But they were enough to initiate a significant departure from his youthful posture as contrarian. In a poem called "Fronde" (French for "sling" or, figuratively, a group of hostile insurgents), he lamented the naiveté that had led to his arrest:

> On sunny days, behind locked doors,
> We indulged in careless chatter . . .
> How foolish—a *fronde* without a sling![23]

FIGURE 1.1. Alexander Volpin

It was not only his naiveté that came under scrutiny. The romantic nihilism he had cultivated as an adolescent now appeared to offer not liberation from faith but another form of it, "a risky self-deception," he wrote on New Year's Eve 1953. And yet negation remained "the deepest value of my identity," the axiom that made free thought possible.[24] The next eight years were a twilight zone of unemployment, during which Volpin survived on occasional translating and editing jobs. It was to be one of the most creative and fruitful periods of his life.

The Second World War cast a long shadow across the USSR.[25] Hitler's invasion had brought nearly a million square miles of Soviet territory and 40 percent of the Soviet population under Nazi control. Roughly twenty-seven million Soviet citizens—one in seven—lost their lives during the war, more than half of them civilians.[26] A significant portion of the USSR's infrastructure was destroyed: roads, railways, apartment buildings, factories, entire towns and cities. Desperate to rally the Soviet population against the Nazi onslaught, Stalin had loosened the Communist Party's grip over many arenas of military and civilian life, feeding hopes for an era of post-victory liberalization. Those hopes proved illusory. Or rather, they had to wait a decade until Khrushchev's momentous decision, in one of history's great Oedipal revolts, to denounce his predecessor (now safely dead) for executing thousands of innocent Bolsheviks, perverting the principles of Marxism-Leninism, and defiling the Party itself. Khrushchev's so-called Secret Speech to the Party's Central Committee in February 1956, which quickly became anything but secret, inaugurated the era of Soviet history known as the "Thaw."

As the metaphor implies, the Thaw set loose forces dormant or frozen beneath Stalin's dictatorship.[27] Millions of Gulag inmates returned to Soviet society, shrinking the USSR's prison and labor camp population to a fraction of its Stalin-era peak. In the USSR's cities—where by 1962 the majority of the Soviet population lived—tens of millions of individual family apartments went up, granting their inhabitants a novel experience: privacy. A youth subculture emerged that prized individual sincerity over the epic bombast of high Stalinism. The "living word" now took the form of lyric poetry, documentary prose, and films about the struggles of everyday life. The Soviet Union's extraordinary postwar scientific achievements—the design and testing of nuclear weapons and the launching into space of the first satellite (Sputnik, 1957), the first man (Yuri Gagarin, 1961), and the first woman (Valentina Tereshkova, 1963)—fostered a veritable cult of scientific knowledge and socialist progress.

The Thaw also opened the gates to select currents from outside the USSR, above all from the West. Not since Peter the Great carved a "window to Europe" at the beginning of the eighteenth century had so many Russians been exposed so suddenly to contemporary Western literature, art, music, and design. Translations of Ernest Hemingway, Erich Maria Remarque, and J. D. Salinger flooded Soviet bookstores; people waited for hours to view works by Pablo Picasso, to witness Glenn Gould's performances of Johann Sebastian Bach, or to walk through an (utterly atypical) "American kitchen" at the American National Exhibition outside Moscow.[28] Foreign cultural objects were not the only things showing up in Soviet cities; actual foreigners did too. In the summer of 1957, Moscow hosted the World Festival of Youth and Students, drawing thirty-four thousand visitors from over a hundred countries to engage in mostly supervised mingling with their Soviet counterparts. For many Soviet participants, the two weeks of direct contact with foreigners came to be associated with a euphoric "loosening of inhibitions" after decades of Stalinist isolation.[29]

One of the participants in the World Festival of Youth and Students was Volpin. Neither a student nor, at thirty-three, particularly youthful, Volpin found himself in a small crowd of people outside a hotel, where an enthusiastic visitor from France had just announced her desire to live in the Soviet Union. Volpin followed the woman into the hotel and, in an act of what he described as "Don Quixotism," attempted to explain the negative consequences of Soviet citizenship. After exiting the hotel, he was seized by watchful agents, who announced that he was being detained on suspicion of theft. While under interrogation the following day, Volpin learned that the charges had been changed to anti-Soviet agitation and propaganda, under Article 58 of the Russian Criminal Code. Insisting that his interrogator's interpretation was based not on the text of Article 58 but rather on "patriotic and ideological nonsense which might have its own independent value but had no juridical force," he demanded to speak to a prosecutor. Instead, a psychiatrist appeared, and after a sharp exchange it was determined that Volpin would be transferred to a mental hospital. The interrogator announced that "the slightest anti-Soviet move" by Volpin would result in indefinite detention. "I said that I didn't understand the meaning of the term 'anti-Soviet,'" Volpin wrote to friends a month later, "that I was not a Marxist and that perhaps his words meant that I had to keep silent in general." The interrogator, lumping Volpin with "philosophical idealists and other muddle-headed persons," warned him against "undermining the prestige of our state." Released after three weeks of confinement in a psychiatric hospital, Volpin described the incident as "nasty—but what a contrast to the bad old days."[30]

Like many of his fellow citizens during the Thaw, Volpin eagerly absorbed newly accessible works from the West, creatively adapting them to the Soviet world. In his case, those works tended to come from the fields of mathematics and philosophy. Some of them he translated himself, such as Stephen Kleene's *Introduction to Metamathematics*.[31] Of particular interest was the emerging meta-discipline of cybernetics. As a purported "science of control" over dynamic processes, cybernetics—from the ancient Greek word meaning "steersman"—aimed to translate a wide variety of phenomena into the precise language of mathematics and computer modeling. It applied concepts of control and feedback, entropy and order, signal and noise, to the flow of information regarding everything from machines to living organisms to human societies.[32]

Nowhere did the new approach to information elicit loftier hopes than in the USSR during the 1960s.[33] With its promise of rationalizing immense and complex processes, cybernetics appealed especially to those responsible for price setting, investments, production quotas, and consumption patterns across the Soviet Union's vast centrally planned economy. At the same time, a distinctive feature of Soviet cybernetics was its implicit rivalry with another would-be meta-discipline, namely, dialectical materialism, or *diamat*, as it was known to generations of Soviet college students. No discipline was beyond *diamat*'s reach. Mathematicians, linguists, historians, philosophers, biologists, jurists—all were vulnerable to charges of "formalism," "idealism," and other sins proscribed by the *diamat* catechism. Those who attempted to analyze human society via algorithms risked being charged with "detachment from life" and ignoring "the needs of the people."

Cyberneticists sought to isolate the methodological from the ideological. Whereas Alexander Voronsky, the editor of the literary journal *Red Virgin Soil*, had once praised the "life-giving spirit of the dialectic," by the postwar era that spirit had all but dried up.[34] "We were tired of the phraseology of official philosophy," recalled the linguist Vyacheslav Ivanov. "We wanted to deal with precisely described concepts and with notions defined through rigorously described operations." The mathematician Andrei Kolmogorov, a pioneer in the field of computer simulation, sought to make it impossible "to use vague phrases and present them as 'laws,' something that unfortunately people working in the humanities tend to do."[35]

Volpin's intellectual coming-of-age is inseparable from the rise of cybernetics and the cross-disciplinary goal of applying "exact methods" to the study of language, thought, and society. Many Soviet cyberneticists understood their approach, and the various modeling systems it spawned, as a form of insulation from official dogma. What distinguished Volpin's thinking during the Thaw was his insistence on applying "exact methods" to official ideology itself, that is, to the language of Marxism-Leninism.

An American participant in the World Festival of Youth and Students, twenty-one-year-old Sally Belfrage, happened to meet Volpin and recorded her impressions of him. Born in Hollywood to British writers, Belfrage had watched her father be grilled before the House Un-American Activities Committee at the height of the McCarthy era and then deported to England on suspicion of spying for the Soviets. After the festival, she stayed on in Moscow, working as a translator and gradually shedding her illusions about the USSR. Volpin appears in her notebooks as "an absent-minded professor" whose apartment was "chaos—littered with ancient cigarette ends and trash, completely buried in books." About his years in prison and exile, she reported, Volpin was "very philosophical and cheerful." "He gloats over the fact that, as a mathematician, they could never take away what counts for him—during those years he worked out calculations that have since been incorporated into his books and teaching, quite important ones." A self-described anarchist, he ridiculed the Soviet state's failure to wither away as Marx had predicted. When Belfrage asked what economic system he favored, Volpin replied, "'For Russia, the economic system is not essential; what is needed now are radical political changes,'" while declining to specify what those might be.[36] He was especially bitter about the discrimination that resulted from his being officially identified as Jewish in his passport: "He couldn't understand why he should be penalized for something he was so indifferent to." "'I am the enemy of any religion or nationalism,'" Volpin told Belfrage, "'or any kind of intellectual norms. Faith is evil; it makes men blind—faith in any kind of principle, I mean, not just enthusiasm. [Enthusiasm] is a fine thing.'" His dream, according to Belfrage, was to emigrate.[37]

During the following years, Volpin continued to consume works by contemporary Western thinkers and repurpose them for his own circumstances. He gathered his thoughts in a treatise initially titled "Why I Am Not a Communist," a nod to Bertrand Russell's iconoclastic essay "Why I Am Not a Christian."[38] By the time he arranged for a revised version of the treatise (together with some of his poems) to be smuggled abroad by a member of the visiting Yale Russian Chorus, he had dropped the allusion to Russell and renamed the work "A Free Philosophical Tractate," shifting his nod to Ludwig Wittgenstein's *Tractatus logico-philosophicus*, a Russian edition of which had been published in Moscow in 1958.[39] Volpin's "Tractate," published in New York in a bilingual edition in 1961, offers a first glimpse of the ideas from which the central dissident strategy of legalism would emerge.

The central concern of Wittgenstein's *Tractatus* is the relationship between language and reality. "Most of the questions and propositions of philosophers," it famously announced, "are based on our failure to understand the logic of our language."[40] In pursuit of semantic clarity, Wittgenstein sought to separate the sphere of values from the sphere of facts, a distinction that resonated with Volpin's own ambition to draw a firm line between emotion and reason or, in a later incarnation, between ideology and law. Equally important for Volpin was Wittgenstein's notion that logic was the only reliable generator of universal ethical imperatives, the first of which was intellectual honesty.

Formally, however, the two treatises could hardly have been more different. Wittgenstein's, like the field of logic it explores, was highly structured, with each statement numbered according to its rank within various thematic hierarchies.[41] Volpin's looked like his apartment. Explicitly repudiating the need for systematic presentation, he gave his "Free Philosophical Tractate" a telling subtitle: "An Instantaneous Exposition of My Philosophical Views," reflecting his claim (notwithstanding the work's gestation and revision) to have composed it in a single day.[42] Another key difference: Volpin's "Tractate" was not only published abroad but written with foreign readers explicitly in mind.[43]

These circumstances inspired Volpin's repeated apologies for the work's alleged lack of novelty. "Much that is written here is not new," he explained, adding that "if all this is familiar to everyone . . . I shall be very pleased. In that case please deposit it in the museum of Russian nonsense." Even the text's final self-justification—"Every student in Russia who has arrived at philosophical skepticism by his own thinking can consider himself a new Columbus"—was full of irony, insofar as the Russian saying "He thinks he just discovered America" was a common means of puncturing inflated claims of originality.[44] Reviewing the work in 1961, the British Sovietologist (and poet) Robert Conquest praised Volpin's courageous dedication to freedom of thought, but accepted at face value the claim of non-originality. "Much of it," Conquest claimed, "*is* reasonably familiar: it is its spontaneous rebirth from a barren soil that is so striking."[45]

Conquest may have been duped by the work's subtitle, for Volpin's "Tractate" was anything but spontaneous. The soil from which it arose was not so much barren as an intricate and novel blend of Soviet and Western elements. Even when Soviet elements served as antipodes for Volpin's thought, they were hardly barren. In fact, they were highly productive.

"A Free Philosophical Tractate" is above all an essay on the nature of knowledge. Consistent with Volpin's interests in the foundations of mathematics, it explores not so much truth itself as the conditions under which truth can be ascertained. For Volpin, those conditions require first of all the freedom that comes from rejecting all forms of faith. Marx's and Engels's definition (via G.W.F. Hegel) of freedom as "the recognition of necessity," he noted, "implies that, if I find myself in prison, I am not free until I have realized that I cannot walk out; but, as soon as I become aware of this, I shall immediately discover 'freedom.' Need I explain that such terminology is very convenient for the 'liberators of mankind'?" Addressing the "liberators" directly, he continued: "Demagogues, you who are merely interested in attaining your ends at the price of confusion in people's minds! You can do nothing but grunt like pigs. We must free ourselves from the influence of people with their deformed language and find a scientific expression for the concept of freedom. Only when we attain this will we be able to trust our own thoughts."[46]

Volpin's search for a scientific language was explicitly directed against the sacred cow of "realism," the notion that language and thought ought to orient themselves exclusively to "reality" and lived experience, or as Russians like to say, to "life itself." The "Tractate" refers obliquely to Volpin's own adolescent crisis, that fateful day in April 1939 when he pledged himself to reason over emotion, but tellingly recasts it as a "break with my belief in realism, [to which] I never returned again. Intuition usually makes us lean toward realism, but here we must not trust intuition until such time as it has been emancipated from language."[47] The primacy of metaphysical truths (ideally formulated in the language of mathematical logic) over the "real" world of emotion and experience was encapsulated in a phrase that appears again and again, mantra-like, in Volpin's writings: "Life is an old prostitute whom I refused to take as my governess."[48] Like the repeated retelling (and reworking) of his adolescent crisis, this phrase, with its suggestion of heroic struggle for intellectual autonomy from the fickle lessons of experience, forms a leitmotif in Volpin's fashioning of his life story.

The "Tractate" extrapolates, from mathematics to thought in general, the goal of self-emancipation from all forms of belief via the creation of an ideal, transparent language. It endorses anarchy as the system most likely to prevent beliefs of any kind from being imposed on individuals.[49] Having rejected Marx's notion that legal and ethical norms are determined by material factors, Volpin cleared a space for alternative sources of such norms. As a prerequisite for productive thinking about those sources, the "Tractate" called for a reform of the Russian language so as to make it conform more closely to the requirements of modal logic—the branch of logic that classifies propositions according to whether they are true, false, possible, impossible, or necessary. Volpin ridiculed what logicians call *ignoratio elenchi*—offering proof irrelevant to the proposition at hand—especially in the form of conspicuous quotations from Marx or Lenin ("a child's rattle") as a substitute for reasoned argument. It was not only ideologues who engaged in such practices. "This defect in our thinking is a paradise for poetry," Volpin noted, "which likes nothing better than this obscurantism. For this reason precisely, I have reacted with scorn during the past eight years to this genre of art which had earlier so fascinated me. . . . Yet to this day I love poetry, simply because a wedge is the best means for knocking out another wedge; and the former illusions, engendered by poetry, can best be destroyed with the aid of new poetry."[50] Volpin aimed to unsettle the rhetorical habits not just of the Communist Party, with its Bolshevik-speak, but of the Russian intelligentsia, with its faith in the transcendent value of the poetic Word.

To be sure, the call for a new, purified language, grounded in the idea that language structures consciousness and therefore thought, had a lengthy pedigree. For centuries, philologists (and, more recently, software engineers) have dreamed of constructing a language of perfect clarity built on principles of logical reasoning. Volpin no doubt drew on the impulses of the Russian avant-garde's "Promethean linguistics," as represented by the Futurist poet Velimir Khlebnikov and the ethnographer Nikolai Marr. Yet Volpin's imagined scientifically reformed Russian had little in common with Khlebnikov's anti-scientific "transrational language" (*zaum*) or Marr's "unified language," which claimed to restore the primordial proto-language from which all human tongues allegedly derived.[51] The ideal language, according to Volpin, would liberate its users from the layers of semantic ambiguity baked into natural languages. It would enable them, for the first time, to trust their own thoughts.

Volpin's singular contribution to the fertile cross-disciplinary debate taking shape during the Thaw involved a practical application of the utopian project of fashioning an ideal language. Instead of building such a language with the tools of logic (as Wittgenstein and, following his lead, analytical philosophers were attempting to do), or embarking on a wholesale reformation of the Russian language so as to rid it of ambiguity, Volpin sought to apply modal logic to two humanistic fields he considered most susceptible to "exact methods": jurisprudence and ethics. Ironically, the Soviet government, with its relentless insistence that intellectuals produce useful knowledge for the laboring masses, inadvertently fostered

Volpin's interest in finding practical applications for his rather abstruse ideas about ideal languages.[52] Many thinkers associated with cybernetics responded to such pressure by developing uncontroversial applications such as computer programs that could accurately translate one language into another, or simply by going through the motions of applied research. "I can't say that we intentionally deceived anyone," Vyacheslav Ivanov recalled, "but it is now impossible to overlook the fact that in those past discussions the practical utility of new methods was, if not strongly exaggerated, then at least strongly emphasized. . . . Everybody knew the rules of the game."[53]

The practical utility of a perfectly transparent language first came to Volpin's mind in connection with a specific and usually very unpleasant game, namely, the cat-and-mouse dialogues that inevitably occurred during interrogations by KGB officials. Interrogations provided rich material for thinking about language and ethics: when to tell the truth and when to remain silent; how to refuse to answer a question, even under pressure; how to avoid lying or setting a trap for oneself. In his quest for a language free of ambiguity, Volpin had concluded that the fundamental task of ethics was the eradication of lying.[54]

His interest in the language of face-to-face conversations between the individual and the personified state was, needless to say, more than academic. By 1963, Volpin had been incarcerated in mental hospitals four times (1949, 1957, 1959, 1963) and subject to numerous grillings by KGB officers and psychiatrists.[55] In his search for strategies to strengthen his own position vis-à-vis his interrogators, he stumbled upon a copy of the pre-revolutionary Russian Code of Criminal Procedure, published in 1903, among the books left to him by his maternal grandfather, a lawyer from the tsarist-era Pale of Jewish Settlement.[56] Volpin then acquired the current Soviet Code of Criminal Procedure, which had undergone major revisions in the late 1950s in response to the rampant abuse of procedural rules in the administration of justice under Stalin.[57] There he found a surprisingly dense web of protective measures designed to constrain the power of prosecutors and judicial investigators over defendants and witnesses. The revised Soviet code explicitly banned "leading questions"; it granted individuals under interrogation the right to write down their own responses (rather than have an official transcribe their words), to request explanation of terms used by their interrogators, and, in certain cases, to refuse to answer questions. The cat-and-mouse game, in other words, had rules, a kind of formal grammar governing speech between state and citizen. They were imperfect rules, to be sure, and sometimes ignored in practice, but they were designed to regulate verbal exchanges and the meaning of specific words. One could learn and master them.

Volpin's interest in Soviet legal codes initially centered around his own case. Having immersed himself in the fine points of Soviet law, he launched a formal appeal in December 1961 regarding his arrest and imprisonment in 1949 on charges of "anti-Soviet activity," for which he had been amnestied in 1953 without the charges themselves being repudiated (perhaps because the charges had never been confirmed by a court or judge). "I never considered as lawful the decision taken against me by the Ministry of State Security," Volpin's appeal announced. "With the present declaration I request that it be reviewed. The basis for this review lies in the violation of a series of regulations outlined in Soviet legislation," which he proceeded to lay out in great detail.[58] Volpin's claim, it should be noted, consisted not of a denial that he had committed a crime (an issue he declined to engage), but of the charge that in arresting and exiling him, the Soviet government had violated its own rules.

Volpin's appeal was probably not helped by the publication in New York earlier that year, under his own name, of "A Free Philosophical Tractate" together with a selection of his poems under the title *A Leaf of Spring*. To appreciate the audacity of this move, one should recall the ferocious Soviet campaign against Boris Pasternak following the 1957 publication in Italy of his novel *Doctor Zhivago* and the awarding to him of the Nobel Prize in Literature later that year. So enormous was the pressure brought to bear on Pasternak—public vilification as an "enemy of the people," expulsion from the Union of Soviet Writers, threats of criminal prosecution and exile—that he decided to formally decline the prize and withdraw from public life. He died a broken man two years later.[59]

The publication abroad of Volpin's explicitly anti-Marxist writings triggered a slew of public attacks on its author. At a 1962 meeting with representatives of the intelligentsia, Khrushchev pronounced Volpin "insane," noting that "our enemies" published his work and passed it off as emblematic of Soviet youth. Volpin's mental illness, Khrushchev added, was inherited from his father, since only an insane person would commit suicide.[60] The chair of the Central Committee's Ideological Commission, Leonid Ilichev, denounced *A Leaf of Spring* as "pretentious and illiterate" and its author as "mentally ill." Volpin's work was full of what Ilichev called "poisonous skepticism" and "hatred toward Soviet society and the Soviet people."[61] At a similar meeting a week later, the poet Yevgeny Yevtushenko described Volpin as "scum" and *A Leaf of Spring* as a "disgusting, dirty little book."[62] An article in the newspaper *Pravda* (Truth) and letters to the editor from two of Volpin's aunts— Sergei Esenin's sisters—characterized both Volpin and his work as "sick."[63] The campaign climaxed with the publication in the popular journal *Ogonek* (Spark) of a vitriolic essay titled "From the Biography of a Scoundrel." Its author, Ilya Shatunovsky, a prominent journalist who headed *Pravda*'s culture department, denounced Volpin as a slanderer of Soviet power, "which gave him everything in life," as an accomplice to treason (for having attempted to enter the grounds of the American embassy), and a speculator in foreign currency.[64]

Volpin's response to this assault was unusual. When his employers at the Institute of Scientific and Technical Information demanded that he perform the obligatory Soviet ritual of "self-criticism" by publicly repudiating the views expressed in *A Leaf of Spring*, he refused. When it was then suggested that he apply for permission to leave the country, he insisted that the government first formally acknowledge his right to emigrate.[65] In a letter to Khrushchev, he affirmed his "moral responsibility" for the contents of *A Leaf of Spring*, noting that nothing prevented even an anarchist like Volpin from being "a loyal citizen of the Soviet state, that is, abiding by its laws." "You have done more than anyone," he told Khrushchev, "to expose the lawlessness permitted under Stalin. That lawlessness was the basic cause of the viewpoints expressed in my book."[66]

The response to Volpin's letter was three months of forced confinement—without formal charges or a trial—in Moscow's Gannushkin Psychiatric Hospital. Following his release, Volpin proceeded to sue *Ogonek* for libel, in effect reversing the charges of slander leveled at him not only by the journal but by the Soviet government in 1949. In pretrial depositions, Volpin insisted that the court evaluate the truthfulness of Shatunovsky's assertion that *A Leaf of Spring* was "anti-Soviet." The juridical meaning of "Soviet power," he argued, referred exclusively to those institutions sanctioned by the Soviet Constitution—not to Marxism, not to the Communist Party, not to individual leaders.[67] His work had made no mention of Soviet institutions, and only a willful misreading by Shatunovsky, Volpin argued, could have produced the charge of "anti-Soviet" slander. To this Shatunovsky offered the following response: "From my Party-minded point of view, the conventional definition of 'slander' as a deliberate falsehood is irrelevant." Volpin noted that if this astonishingly frank statement had been made public, he would have dropped his lawsuit.[68]

In the end it was the court, not Volpin, that dropped the charges against Shatunovsky and *Ogonek*. The case nonetheless served as an important stimulus to Volpin's thinking about the value of Soviet law as a language in which dissenting positions could be articulated and defended. The distinction between "Soviet" and "Communist" was more than mere wordplay. It tapped into the historical fact that the October 1917 revolution had been carried out in the name of Soviet, not Bolshevik (that is, communist), power, and into the abiding distinction, in theory at least, between the institutions of the Soviet state and those of the Communist Party. By the 1960s, the adjective "Soviet" had come to signify not just a political agenda but an entire country and its population of some 250 million people. In an essay titled "What Is 'Soviet'?," Volpin noted: "We are all citizens of the USSR by virtue of having been born on its territory. But there is no law obliging all the citizens of the USSR to believe in communism or to build it, or to collaborate with the security organs, or to conform to some mythical ethos. The citizens of the USSR are obliged to observe the written laws, not ideological directives."[69]

Law as a transparent, formal language of specific behavioral obligations and prohibitions; ideological directives as a form of coerced belief: these were the categories, derived from modal logic and Wittgenstein's distinction between fact and value, that shaped Volpin's emerging concept of law-based dissent. His uncommon reaction to Khrushchev's forced resignation in October 1964 (due to "actions divorced from reality," as *Pravda* put it) and the Communist Party Central Committee's election of Alexei Kosygin as chairman of the Council of Ministers (in effect, head of state) illustrates this new way of thinking.[70] It was not, Volpin insisted, a matter of whether one approved of Khrushchev's policies or his erratic personality. His removal from power—Khrushchev was the first Soviet leader not to die in office—was illegal on procedural grounds. According to the Soviet Constitution, the chairman of the Council of Ministers was elected by the Supreme Soviet, not by the Communist Party, and therefore only the Supreme Soviet had the authority to replace him.[71] Volpin's view of the war in Vietnam, in the context of fervent Soviet denunciations of American imperialism, was similarly grounded in a stark proceduralism: what troubled Volpin most of all was that the American Congress, the only body authorized by the U.S. Constitution to declare war on a foreign state, had not done so.[72]

"The greatest paradox in the fate of Russia and the Russian Revolution," the philosopher Nikolai Berdyaev wrote, in exile from Stalin's Soviet Union, "is that liberal ideas, ideas of rights as well as of social reformism, appeared in Russia to be utopian."[73] There is indeed something paradoxical about Volpin's proceduralist utopia, built less on a vision of human dignity or human empathy (as with most rights-invoking movements) than on the ideal of semantic transparency. Volpin approached rights and the rule of law not through classic liberal ideas of contract and self-interest, with the right to private property at their core, but through ideal language philosophy and the "exact methods" of cybernetics. The dictatorship of reason he imagined as an adolescent, and repeatedly refashioned thereafter, never found expression in a perfectly unambiguous language. Instead, he settled for an imperfect but more pragmatic alternative: the already existing and (in theory) binding language of

Soviet law. Traces of the original utopian impulse, however, survived this compromise. One was Volpin's emphatically literal reading of Soviet law, as if it were already transparent rather than in need of interpretation. Another was his version of the Russian intelligentsia's belief in the transcendent power of the Word, a peculiar variant in which the Word resided in the Soviet Constitution and the Code of Criminal Procedure rather than in poetry and other works of artistic imagination. A third utopian impulse, perhaps the most ambitious of all, was his conviction that what Russia needed, after decades of violent revolutionary upheaval, was a "meta-revolution"—a revolution in the way revolutions are accomplished, a revolution in the minds of Soviet citizens, transforming their relationship to the language of law.

For much of Soviet history, those who sought to cultivate islands of personal freedom did so by publicly acting and speaking in accordance with official norms. Conformism in this sense was a protective shield under which Soviet citizens could privately think and behave in ways that, while not necessarily at odds with the Soviet order, were not aligned with it either. Volpin's philosophy of civil obedience recast both the content and purpose of conformism. It emphasized strict conformity to Soviet law rather than to exhortations by the Communist Party. Conformism was no longer a veil for contrarian practices, but a device for openly pressuring the Soviet government to similarly conform to the letter of its own law. This conformism required no compromises and no shame.

The Soviet Union, like every other complex, modern, variegated society, required at least the appearance of the rule of law in order to function. Volpin was groping his way toward a novel technique: to act as if the veneer were more than just a veneer, as if the language of law were transparent and binding on both the state and individual citizens. In his diary entry for New Year's Eve 1963, shortly after the dismissal of his lawsuit against Shatunovsky, Volpin wrote: "Let's wait and see. We will be patient and think things over. But we will make no concessions whatsoever, and when the time comes, we will issue a public manifesto of our rights."[74] Less than two years later, the time came.

2

Involuntary Protagonists

The more I observe, the more it seems to me that, as a rule, history is made by involuntary protagonists.

—MARK AZBEL

They seized him on September 8, 1965, near Nikitsky Gate, not far from Red Square, as he was waiting for a bus. The operation was so quick, so professional, that passers-by failed to notice. Catching him off guard with the simplest of ruses—calling out with mock familiarity his name and patronymic, "Andrei Donatovich?!"—they whisked him into a parked car from which there was no escape. No doubt the job was made easier by the fact that Andrei Donatovich Sinyavsky, a thirty-nine-year-old literary critic, put up no resistance. On the contrary: prison life, as he later wrote, "had fascinated me like a whirlpool." But the man now on his way to the Soviet Union's most notorious prison, deep inside KGB headquarters on Lubyanka Square, was not arrested for being the literary critic Andrei Sinyavsky. He was arrested for being the writer Abram Tertz. After nearly a decade of living a double life, this red-bearded, gnome-like figure, whose misaligned eyes peered out in slightly different directions as if hinting at his two personae, was now, as he put it, "heading for an eyeball to eyeball confrontation with myself."[1]

It had taken the KGB the better part of that decade to establish that Tertz—the pseudonymous author of *The Court Is in Session* (which

FIGURE 2.1. Andrei Sinyavsky arrest photos, September 8, 1965

Time magazine called "perhaps the most remarkable novel to have come out of Russia since the Revolution"), *Fantastic Stories,* and other works published to wide acclaim in the West beginning in 1959—was in fact Sinyavsky, a senior researcher at the Gorky Institute of World Literature.[2] By complete chance, the protracted effort to unmask Tertz had begun not far from its ultimate destination, with a KGB directive to the Gorky Institute to monitor foreign coverage of Soviet literary developments. Soviet security services had just been humiliated by the publication abroad of Boris Pasternak's novel *Doctor Zhivago* in 1957 (long on enthusiasm but short on superlatives, *Time* called it "the most remarkable Russian novel of the 20th century").[3] Even worse, the Swedish Academy decided a year later to award Pasternak the Nobel Prize in Literature. In its effort to unmask Tertz and to wage a more effective ideological war against the enemies of socialist realism, the KGB gave staff members of the Institute—among them, presumably, Sinyavsky himself—privileged access to foreign journals, including those in which his works were published in translation. There they found, among Tertz's many unorthodox ideas, this skewering of Marxism-Leninism's theory of history, from the essay *What Is Socialist Realism?*:

> If we ask a Westerner why the Great French Revolution was necessary, we will receive a great many different answers. One will reply that it happened to save France; another, that it took place to cast the nation into an abyss of ethical ordeals; a third, that it came to establish for the world the marvelous principles of Liberty, Equality, and Fraternity; a fourth will object that the French Revolution wasn't necessary at all. But ask any Soviet schoolchild—to say nothing of more highly educated people—and all of them will give you the precise and exhaustive reply: the Great French Revolution was needed to clear the way to, and hasten the arrival of, Communism.[4]

One early analysis, presented at a Gorky Institute meeting at which Sinyavsky himself was present, proposed that the mysterious author of these lines belonged to "White émigré circles" that were carrying on the decadent religious-philosophical tradition of pre-revolutionary, capitalist Russia, washed-up relics banished from the socialist paradise.[5] Western commentators, by contrast, proposed that Tertz, who took such obvious delight in deviating from the official dogma of socialist realism, came from a "literary underground" within paradise itself.[6]

A small army of KGB and associated personnel from Paris to Yalta—philologists, criminologists, handwriting experts, customs officials, and undercover agents—labored to discover the true identity of the subversive author(s). It took six years to identify Sinyavsky and his friend Yuli Daniel as likely suspects; in May 1964, the KGB formally opened a case against them under the code name "Epigones."[7] An agent who obtained the list of books and journals requested by Sinyavsky over the years at the Lenin Library found significant overlaps with the works mentioned in *What Is Socialist Realism?* Forensic analysis of texts by Tertz and Sinyavsky—comparing lexicon, style, and phraseology—revealed further similarities. Although contemporary rumors endowed the KGB with a registry of every typewriter in the Soviet Union, complete with individual keyprints as unique as fingerprints, allegedly allowing the government to trace any typewritten text to a specific machine (and its owner), this was not quite the case.[8] Rather, the KGB managed to secretly confiscate several pages of a text typed on Sinyavsky's "Optima" typewriter and to identify minute idiosyncrasies in the spacing and shapes of certain letters, matching them to those found in a samizdat typescript of one of Tertz's stories.[9] A listening device was planted in Sinyavsky's apartment and his meetings with foreigners in Moscow were secretly filmed. As late as July 1965, however, a pathbreaking volume of Pasternak's poetry with a scholarly introduction by Sinyavsky (the two writers had met shortly before Pasternak's death in 1960) was permitted publication in Moscow, suggesting that Soviet authorities had not yet definitively confirmed Tertz's identity.[10]

Within days of Sinyavsky's arrest, Yuli Daniel—poet, translator, and war veteran—was also brought to the Lubyanka. A KGB agent posing as a neighbor's relative had taken up residence in the communal apartment on Lenin Prospect where Daniel lived with his wife, Larisa Bogoraz, and had managed to secretly search their apartment.[11] Under the pseudonym Nikolai Arzhak (a name borrowed, like Abram Tertz, from songs about the criminal underworld), he too had sent works of fiction abroad, stories that probed the condition of post-Stalinist Soviet society in highly unconventional ways. *This Is Moscow Speaking* posits a time in the near future when the Party, having determined that the new Soviet Man had evolved to an unprecedented ethical level, decides to test the possible effects of the withering away of state and law, as famously predicted by Marx and Engels in their *Communist Manifesto*. The story begins with the announcement of "Public Murder Day," a single day on which certain categories of homicide will be de-criminalized, and follows the reaction of Soviet citizens to this unprecedented freedom. In another, *Atonement*, a man falsely accused by his peers of having been an informer jumps onstage during the intermission of a concert (or fantasizes about jumping onstage—the boundary separating fantasy and reality having faded). "Comrades!" he cries, "they are still repressing us! The prisons and camps have not been shut down! It's a lie! The press lies! It makes no difference whether we are in prison or the prison is in us!

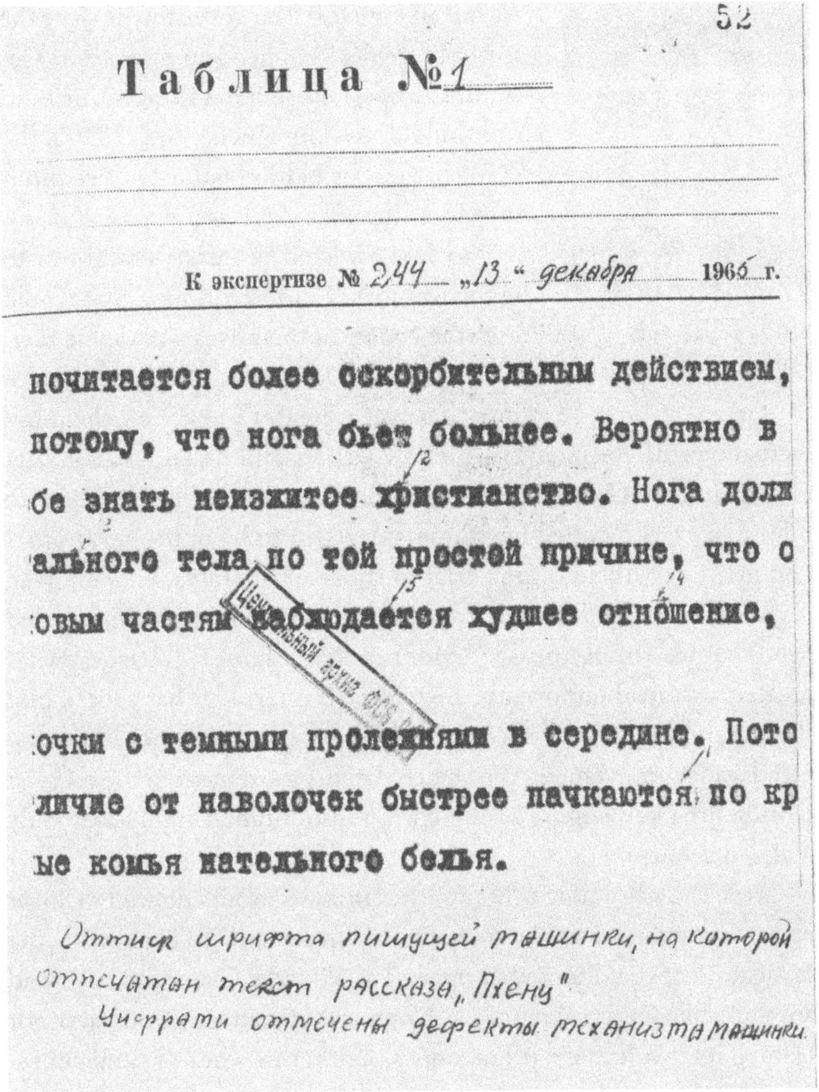

Таблица №1

К экспертизе № 244 „13" декабря 1965 г.

почитается более оскорбительным действием,
потому, что нога бьет больнее. Вероятно в
бе знать неизжитое христианство. Нога долж
ального тела по той простой причине, что о
овым частям наблюдается худшее отношение,

очки с темными пролежнями в середине. Пото
личие от наволочек быстрее пачкаются по кр
ые комья нательного белья.

*Оттиск шрифта пишущей машинки, на которой
отпечатан текст рассказа „Пхенц".
Цифрами отмечены дефекты механизма машинки.*

FIGURE 2.2. KGB forensic analysis of distinctive typewriter patterns in a samizdat text of the short story "Pkhents," by Abram Tertz (Andrei Sinyavsky) (1965)

FIGURE 2.3. Yuli Daniel

We are all inmates! The government lacks the power to free us! We need an operation! Cut out the camps, release them from within you! You think it was the Cheka, the NKVD, the KGB that put us behind bars? No: we did it ourselves. The state—it is we."[12]

The two arrested writers were about to be thrust into a Cold War drama that would reset the course of their lives and define their place in history. "No *cause célèbre* in modern times has had a greater impact on the intellectual world," declared their fellow writers and defenders Günther Grass, Graham Greene, François Mauriac, Arthur Miller, and Ignazio Silone in a 1966 collective statement.[13] Not to be outdone, another commentator likened their trial to those of Socrates, Galileo, Jean Calas (the executed French Protestant championed by Voltaire), and John Scopes (of Tennessee "Monkey Trial" fame).[14] At somewhat greater historical remove, the Sinyavsky-Daniel affair has settled into the annals of the twentieth century as the event that launched the Soviet dissident movement, the first civil rights movement in the socialist world and a key engine of the elevation of human rights to a truly global discourse.

Given the size and weight of such claims, we would do well to linger on that moment, to hold it up for closer inspection. To begin with, the instigators of the drama that opened at Nikitsky Gate did not intend anything like its actual outcome. Nor, for that matter, did Sinyavsky or Daniel, its involuntary protagonists, neither of whom would subsequently take an active part in the Soviet dissident movement.[15] How a trial meant to signal to the Soviet intelligentsia that there were strict limits to the "Thaw" was in effect hijacked and how a pair of writers who sought to defend works of the imagination from the clutches of Soviet law inspired a movement on behalf of the rule of that same law—these are the subjects of the present chapter. For in the end, the trial of Sinyavsky and Daniel became a test of the post-Stalinist, post-totalitarian version of the Soviet system, a system attempting to rescue itself from the lethal habit of using terror and mass violence as techniques of governance.

It was by no means obvious at first that Sinyavsky and Daniel, and they alone, would be put on trial. In October 1965, *Le Monde* informed its readers that *three* Soviet writers had been arrested: Sinyavsky, Daniel, and an unnamed individual, now known to be Andrei Remizov, a childhood friend of Sinyavsky's who had gone on to become head librarian at the All-Union State Library of Foreign Literature.[16] Under the pseudonym "Ivan Ivanovich Ivanov," Remizov had sent several plays and essays abroad, including "American Disturbers of the Russian Conscience" and "Is There Life on Mars?" His work was both more political and less stylistically adventurous than that of Sinyavsky and Daniel. Striving, as he put it to one of his interrogators, to "call things by their real names," Remizov called out the anti-semitism behind Stalin's "anti-cosmopolitan" campaign and the "petty tyranny" of Khrushchev's incipient cult of personality.[17] All three writers had smuggled their work to the West via the same person: Hélène Peltier, daughter of the French naval attaché in Moscow. Her father's status had made it possible for Peltier, starting in 1947, to become one of the very few foreigners to enroll at Moscow State University, in the Philological Faculty, where she met Sinyavsky in a required course on Marxism-Leninism. That status also gave her access to diplomatic mail privileges at the French embassy.[18]

Sinyavsky, Daniel, and Remizov were part of an expanding cohort of authors whose work—published or not—aroused the concern of Soviet leaders at the highest level. As Leonid Ilichev had noted nearly a year before Sinyavsky's and Daniel's arrest, taboo topics such as the Gulag and Stalin's tactical blunders during World War II were surfacing in a growing number of texts by Soviet authors. Even more alarming, "a certain portion of such manuscripts [were] turning up abroad," where the "bourgeois press" used them "to inflate its anti-Soviet propaganda."[19] True, relatively few Soviet citizens seemed to be aware of Tertz, Arzhak, and Ivanov, or other writers who had dared to follow Pasternak's lead by having their work published outside the socialist world. Soviet leaders,

FIGURE 2.4. Andrei Sinyavsky and Yuli Daniel carrying the lid of Boris Pasternak's coffin, 1960

however, understood the Cold War as a contest not just for Soviet but for global public opinion. "Not since the White emigration," fumed KGB chairman Vladimir Semichastnyi, "have anti-Soviet diatribes been published abroad on such an enormous scale, whereby a significant portion of such 'works' are by authors living on the territory of the USSR."[20]

The KGB had options, and its decision to single out Sinyavsky and Daniel from this larger cohort is therefore telling. During Stalin's reign, millions of Soviet citizens had occasion to wonder why a certain individual was arrested and others not, or why this person was exiled while that one was executed. The extra-judicial character of many of the state's punitive actions certainly helps explain their extreme unpredictability. So does their colossal number, which overwhelmed the security apparatus's resources. In some instances the sense of randomness may have been deliberate: the logic of terror as a technique of social control, after all, is to induce fear in vastly greater numbers of people than are directly targeted by a given action. After Stalin's death, however, terror was largely removed from the repertoire of Soviet governing practices, if only to keep the Soviet leadership from drowning itself in yet another bloodbath. Khrushchev's resuscitation of "socialist legality" aimed to restore the connection between punishment and crime. Henceforth punishment would be more or less predictable, thus finally giving a semblance of plausibility to the folk saying "They don't arrest people for nothing."[21]

Sinyavsky and Daniel—or, rather, Tertz and Arzhak—stood out as much for the uses made of their fictions in the West as for the actual content of their work. Alexander Volpin likely benefited from the relatively quiet reception of *A Leaf of Spring*, which cost the Soviet Union little in the way of negative publicity, and he was undoubtedly protected by his father's posthumous fame. Valery Tarsis, who had originally used a pseudonym and was targeted for arrest at around the same time as Sinyavsky and Daniel, appears to have been helped by outcries from abroad following his involuntary incarceration in a psychiatric hospital. In an October 1965 memorandum, KGB chairman Semichastnyi and procurator-general Roman Rudenko proposed permitting Tarsis to leave the Soviet Union and then barring his return. "Such a step," they noted, "would allow us to cut short various claims about his 'persecution' and ... to neutralize the activities [abroad] of assorted 'committees in defense of Tarsis,' which in the event of his arrest would doubtless associate his name with those of Sinyavsky and Daniel, which would hardly be politically favorable for us."[22]

That political and not legal considerations were uppermost in the mind of not only the KGB chairman but the country's highest judicial officer should not be cause for much surprise. From a purely tactical perspective, Sinyavsky and Daniel appeared to be ideal choices for arrest. Virtually unknown (and thus lacking the shield of fame), they could be held up as slanderers of their homeland before Western audiences, cowardly and deceitful figures who hid behind pseudonyms. The government could thus position itself to perform in a single stroke two of the most resonant Soviet political rituals: unmasking individuals who had disguised their identities and vilifying those who aired internal grievances to foreigners.

In principle, the same considerations should have applied to Remizov, who was keen to explain his "underground literary activity" and his "rejection of the policies of the Communist Party" to his interrogators. "To be honest," he told them, apparently frustrated with the pace of the investigation, "I can't understand why I haven't been arrested for my works. I did the same thing as Sinyavsky."[23] From the KGB's perspective, however, he had not. It was not just that Remizov's work had attracted little attention abroad. His frontal attacks on Stalin and Khrushchev actually worried Semichastnyi *less* than the kind of studied indirection found in *The Court Is in Session* and *This Is Moscow Speaking*:

> One can't speak of... growing dissatisfaction in the country with the existing order or of serious intentions to create an organized anti-Soviet underground. That is out of the question. However... we frequently encounter a loss of political vigilance, revolutionary fighting spirit, and class instinct, or simply a political dissoluteness among a certain portion of the intelligentsia and above all the creative intelligentsia. It seems that this latter circumstance deserves the greatest degree of attention, since it is taking on a rather widespread character, drawing a significant part of the intelligentsia and university students toward nihilism, rebelliousness, and an atmosphere of apoliticism, especially in the country's major cities.[24]

The most insidious texts, according to the KGB chairman, were those that not only "take an ironic stance vis-à-vis Soviet reality, but do so allegorically, as if to demonstrate the impossibility of speaking the truth or of openly criticizing shortcomings." Such meta-criticism could not be tolerated: authors must be punished not only for uttering the impermissible, but for implying that the impermissible could not be uttered. In these circumstances, disciplining the allegorists rather than an emphatic literalist like Remizov would send the desired message.[25]

For the time being, however, the Soviet government was sending no explicit messages at all. During the four months following their arrest in early September, not a single mention of Sinyavsky or Daniel, or for that matter Tertz or Arzhak, appeared in the Soviet press. As usual, rumors rushed in where facts feared to tread. Those rumors reveal what contemporaries imagined to be likely or plausible transgressions. Some had Sinyavsky and Daniel being arrested for speculating in icons or for distributing anti-Soviet materials sent from abroad. Others reported that Sinyavsky was being punished for his controversial introduction to the recently published volume of Pasternak's poetry or for writing subversive materials under the pseudonym Valery Tarsis (whose last name could easily be confused with Tertz). Still others assumed that the Sinyavsky in question must be the popular soccer commentator on Moscow's leading radio station.[26] But the Kremlin's silence opened the gates for more than rumors. By mid-October, the Voice of America, Radio Liberty, the BBC, Deutsche Welle, and other Western shortwave radio stations (the Voices), having repeatedly broadcast audio versions of *The Court Is in Session*, *Lyubimov*, and *This Is Moscow Speaking*, broke the news of their authors' arrest to Soviet listeners.[27] A new round of speculation began: Was this the beginning of a broader freeze on creative activity, marking the definitive end of Khrushchev's "Thaw"? Would Sinyavsky and Daniel simply disappear without a trace? Or would the authorities stage a show trial like those of the 1930s, in which the defendants confessed to fantastic crimes against the Soviet Union and begged the court not to spare them?

No one could be sure. Behind closed doors, Soviet authorities were planning a combination of what they called "administrative measures" and "explanatory and prophylactic work." These involved informing representatives of the Union of Soviet Writers and the Gorky Institute about the two cases; arranging for members of the Union to take part in "concluding measures" against Sinyavsky and Daniel once those measures "had been determined by the Procuracy, the KGB, and the judicial organs"; taking the "exceptional" step of preemptively confiscating the unpublished manuscript of the novel *The First Circle* by Alexander Solzhenitsyn, to whom foreign publishers had "penetrated," with the prospect of "noticeable damage to the country's political prestige"; "intensifying the ideological hardening of creative cadres" with the help of materials from investigations against individuals who had "contributed to the anti-Soviet press"; and, finally, "preparing materials for possible publication in the Soviet and foreign press."[28]

Behind a different set of closed doors, Sinyavsky and Daniel were separately undergoing a series of interrogations by their KGB captors. The two writers had anticipated these circumstances—in fact, at one point during the summer before their arrest, they had revealed to each other, in a conversation in Sinyavsky's apartment (which may well have been bugged by then), that they had often felt the desire to confess their clandestine literary activities to the KGB, but had concluded that such a move would place their friends and relatives in an impossible situation.[29] Initially they were determined to deny any connection to Tertz and Arzhak. Sinyavsky's first interrogation took place within hours of his being delivered from Nikitsky Gate. His apartment had already been searched, yielding among other things his manuscript titled "An Experiment in Self-Analysis," a samizdat translation into Russian of Arthur Koestler's novel *Darkness at Noon*, and a dozen bootleg tape recordings (*magnitizdat*) of Vladimir Vysotsky and other Soviet bards performing songs in the distinctive thieves' jargon brought back by former Gulag inmates. It was those songs that had introduced Sinyavsky to the character of Abram Tertz, the legendary Jewish thief from Odesa, as well as to the prison argot subsequently put to such eloquent use in Tertz's stories.[30]

Sinyavsky arrived at the Lubyanka knowing more, however, than just the vocabulary of Soviet camps and prisons. His own father had spent time in that world. Born into a gentry family, Donat Evgenievich Sinyavsky had followed the Russian tradition of the "repentant nobleman" by abandoning his inherited privileges and joining, in 1909, the pro-peasant Socialist Revolutionary (SR) Party. The revolutionary year 1917 pulled him still further to the left, into the more Bolshevik-friendly Left SR faction, on whose slate he was elected to the ill-fated Constituent Assembly. After the Bolsheviks and Left SRs disbanded the Assembly in January 1918, the elder Sinyavsky threw his hat in with the fledgling government. When the Left SRs quit the coalition later that year, he resigned from his party rather than join its armed resistance to Bolshevik power in the Russian civil war. In the 1930s, he had taken part in the forced collectivization campaigns—and nearly lost his life at the hands of enraged peasants. The Bolsheviks repaid his loyalty by arresting

Donat Sinyavsky in 1950 on charges of having engaged in espionage on behalf of the United States in the 1920s, while he helped distribute food from Herbert Hoover's American Relief Administration to some of the millions of famine victims in his native Volga region. The accusation of spying for foreigners, soon dropped for lack of evidence, was replaced by charges of "anti-Soviet agitation and propaganda" under Article 58-10.[31] It was all part of an operation, as Andrei Sinyavsky put it, to purge the USSR of "the last Mohicans of the revolution—former Mensheviks, anarchists, SRs who by some miracle had survived the 1920s and 1930s."[32] Officially "rehabilitated" after five years of prison and exile, his father emerged in 1956 physically and mentally broken. He died four years later.

When he entered the Lubyanka in 1965, therefore, the younger Sinyavsky had good reason to expect the worst. Indeed, it took his KGB interrogators, Colonel Vasilenko and Lieutenant-Colonel Kantov, barely forty-eight hours to get him to admit that he was Tertz.[33] There is no evidence, however, of physical coercion having been used, or even threatened, during these or subsequent interrogations. On the contrary, one month into his confinement, after nearly a dozen encounters with his handlers, Sinyavsky composed the following statement:

> Under the impression of stories and conjectures about the working methods of the organs of state security during the period of violations of the norms of socialist legality (in particular, the stories of my father), I at first did not testify sincerely. Being in solitary confinement while under investigation by the KGB, I had the opportunity to be personally persuaded that the administration and staff members of the investigative apparatus are far removed from those images that had formed in my mind. I was treated tactfully, with humane attention, and I did not encounter here any sort of intimidation or pressure, despite my incorrect behavior during the [initial] investigation. The many conversations I had with Colonel Vasilenko and Lieutenant-Colonel Kantov not only revealed the inaccuracy of my former understanding of the KGB's style of work, but convinced me that today's security personnel are highly civilized, proper, honest,

and principled people. This forced me to reflect upon and review my past, and to offer sincere testimony.[34]

One cannot discount the possibility that Sinyavsky composed this statement under pressure or with ulterior motives, such as persuading his interrogators that henceforth his responses to their questions would be truthful and complete (even if they were not). Looking back in 1984—now from Paris—on his time at the Lubyanka, Sinyavsky noted that "the KGB stopped beating people some time ago; these days, when they have the facts in hand they use dirty tricks, deceit, threats, but, above all, logic—logic!—to drive a prisoner onto the path of redemption."[35] Volpin was not the only one seeking to harness logic for strategic purposes.

After submitting his statement to his captors, Sinyavsky acknowledged for the first time that an unknown sum of royalties from his works had accumulated abroad and offered to give them to the Soviet government. He began providing the names of nearly two dozen people to whom he had read Tertz's stories prior to his arrest. Daniel—like Sinyavsky the son of a revolutionary, in his case Mark Mendelevich Meirovich, an alumnus of the Pale of Settlement who became the Yiddish writer Mark Daniel and had the good fortune to die of tuberculosis in 1940—did the same.[36]

One of the people Sinyavsky named was the thirty-three-year-old physicist and cybernetician Mark Azbel, whose own experience confirmed Sinyavsky's sense that the rules of engagement between citizen and state (the latter in the form of the security police) had changed. As one of Azbel's interrogators in Kharkiv told him, in a spirit of utter dejection after a fruitless fourteen-hour session, "If you'd landed with me ten years ago, you would have confessed that your grandmother had balls."[37] Unintentionally, this remark expressed a central flaw in Stalinist methods: the use of violence or torture against witnesses and defendants often produced wildly false testimony. By contrast, transcripts of the interrogations of Sinyavsky (thirty-three total), Yuli Daniel (nineteen total), and a dozen acquaintances suggest that the KGB's primary goal, in this case at least, was to collect evidence that could be used in the prosecution of the two writers—in a Soviet courtroom as well as in the wider court of public opinion.[38]

Within two weeks of their arrest, Sinyavsky and Daniel were informed that they would be charged according to the recently revised Criminal Code of the Russian Soviet Federated Socialist Republic (RSFSR), in the section "Especially Dangerous Crimes against the State," under Article 70, "Anti-Soviet Agitation and Propaganda." Article 70 required that the prosecution demonstrate at least one of two charges. The first was that their work had "the purpose of subverting or weakening Soviet power," without regard to whether the allegedly damaging passages were true or false. The second held that their work constituted "slanderous fabrications that defamed the Soviet state and social system," in which case—as Volpin had argued in his civil lawsuit against Ilya Shatunovsky—it would have to be shown not only that the allegedly damaging passages were false, but that the accused knew they were false. As in the Stalin era, terms such as "weakening Soviet power," "anti-Soviet," and "defamation" (when practiced against a state rather than an individual) continued to leave extraordinary latitude for interpretation by prosecutors and other authorities. The only significant reforms of the Stalinist law codes were procedural. The 1960 Russian Code of Criminal Procedure, for example, guaranteed the right to defense counsel and mandated that judicial decisions be based solely on evidence presented at trial.[39]

It had taken the KGB almost no time to get Sinyavsky and Daniel to confess to being Tertz and Arzhak. But even after multiple interrogations in which they were required to explicate the meaning of various passages in their works and why they had written them, both writers steadfastly denied that those works were anti-Soviet. They were helped by the fact that with one exception (Sinyavsky's *What Is Socialist Realism?*), these were unmistakably works of fiction with multiple characters expressing multiple points of view. As self-conscious heirs of the modernist movement, Sinyavsky and Daniel had studiously avoided any hint of an all-knowing, authoritative narrator. These facts alone, of course, hardly provided immunity: over the preceding decades, Soviet authorities had sent numerous writers of fiction and poetry to prison (or worse). But very few of them had gone through a formal trial, and those who had were typically charged with spying, sabotage, currency speculation, parasitism (that is, unemployment), or other crimes unrelated to their literary work—and often to reality. Under the new, post-Stalinist dispensation, courts were supposed to base their proceedings on evidence subject to inspection by both sides—even if it took the form of fictional texts.

Sinyavsky's and Daniel's interrogators thus found themselves trying to fasten criminal charges onto imaginative works swimming with irony, sarcasm, surrealism, and humor. In a typical encounter, Sinyavsky's interrogator attempted to build a trap using language lifted directly from Article 70:

> INTERROGATOR: The majority of [your] works contain slanderous fabrications that defame the Soviet state and social system. Explain what led you to write such works and illegally send them abroad.
>
> SINYAVSKY: I don't consider my works . . . to be slanderous or anti-Soviet. What led me to write them were artistic challenges and interests as well as certain literary problems that troubled me. In my works I resorted to the supernatural and the fantastic; I portrayed people who were experiencing various maniacal conditions, sometimes people with ill psyches. I made broad use of devices such as the grotesque, comic absurdities, illogic, and bold experiments with language. . . .
>
> INTERROGATOR: In *The Court Is in Session*, and in the collection published under the title *Fantastic Stories*, the Soviet Union is described as a society based on force, as an artificial system, imposed on the people, in which spiritual freedom is impossible. Is it really possible to call such works fantastic?
>
> SINYAVSKY: I don't agree with this evaluation of my works.[40]

It was an excellent question, which would surface again at the trial of the two writers: Was a text that depicted the USSR as fundamentally unfree recognizable as a departure from Soviet reality—that is, fantastic—rather than as a critique of that reality?

Daniel's interrogator was no more successful:

INTERROGATOR: For what reasons did you consider going to the KGB and confessing?

DANIEL: This desire appeared in me because of various circumstances. First, the situation inside the country that had brought to life a work like "This Is Moscow Speaking" changed. Second, the reaction to my works abroad turned out to be not at all what I had assumed. My works were called anti-Soviet and were evaluated almost exclusively from this perspective. . . .

INTERROGATOR: When did you come to the conclusion that your works were anti-Soviet?

DANIEL: I didn't conclude that and don't see them that way now. I wrote my works not against the Soviet system, but against violations of the Soviet system.[41]

When interrogators zeroed in on allegedly anti-Soviet pronouncements by this or that character, Sinyavsky and Daniel claimed that the character in question did not speak for the author. When asked to identify the "positive hero" of their stories—that is, according to the tenets of socialist realism, the implicit bearer of the author's point of view and the progressive march of history—they insisted that none of their protagonists was entirely positive. As Sinyavsky had written in his programmatic essay *What Is Socialist Realism?*, his protagonists were "neither for the Purpose nor against the Purpose"; they were "outside the Purpose," "creatures of different psychological dimensions."[42] When Daniel's interrogator asked whether he agreed with Lenin's position on literature, as articulated in the article "Party Organization and Party Literature," Daniel replied, "I have no objections to the arguments expressed in Lenin's article, but I'm not sure Lenin's demand that literature be Party-minded applies to works of fiction, since as far as I recall, in that article Lenin himself did not have works of fiction in mind." Apparently unaware that Lenin had made no such distinction between fiction and non-fiction, the interrogator switched to another topic. His clumsy invocation of Lenin offers a small but telling example of the very waning of ideological vigilance that so worried Ilichev and other Soviet leaders.

In Sinyavsky and Daniel they confronted writers whose stories featured not anti-Soviet heroes but something far more insidious: no heroes at all. A literature of non-heroes precisely captured the Party leadership's deepest fears about apathy among the postwar generation of Soviet youth.[43]

Having repeatedly failed to persuade the two prisoners to concede that their works were anti-Soviet, the KGB fell back on indirect evidence: What, other than anti-Soviet content, could explain why they had not even *tried* to get their stories published in the Soviet Union? But here too Sinyavsky and Daniel could claim, plausibly, that it was their radical stylistic experimentation that had led them to assume that no Soviet publisher would accept their work. Without mentioning the word "censorship," both could cite examples of other texts (submitted under their real names) languishing for years in Soviet publishing houses—and thus explain their turn to foreign venues as a result of impatience or artistic vanity, rather than anti-Soviet content.[44]

Another indirect argument involved the reception of the writers' work in the West. Insofar as Article 70 criminalized "subversion," "weakening," and "defamation" of the Soviet Union, the KGB sought to document these effects and found no shortage of supporting evidence. The *New York Times* had called *The Court Is in Session* a "devastating" portrait of "the essence of Soviet life." According to *Time* magazine, "Tertz has made his mark as a bitter, bedrock enemy of Communism." The introductions to Russian-language editions of Daniel's stories, published by the émigré Boris Filippov in Washington, D.C., described them as a rebellion by the "free human personality" against "socialist falsification," against "being turned into a simple cog in the Soviet communist machine."[45] When presented by their interrogators with such evidence, Sinyavsky and Daniel expressed their dismay while insisting that they had no control over the misreading of their work by anti-Soviet elements in the West. Sinyavsky's offer to meet with foreign journalists in order to clarify the actual motives and themes of his work went nowhere: having failed to extract confessions from the two writers, the KGB was not about to dilute what appeared to be proof positive of the damage they had caused to the Soviet Union's reputation abroad.[46] This was the kind of evidence that would serve well not only in a formal trial, but in the larger effort to galvanize the Soviet public.

It was also the kind of evidence that needed to be carefully controlled, since the foreign reviews selected by the KGB were hardly representative of the full spectrum of Western responses. "If we are to understand [Tertz]," the Polish émigré poet and diplomat Czesław Miłosz (a future Nobel laureate in literature) had written in his introduction to *What Is Socialist Realism?*, "we must abandon the division of people into Communists and anti-Communists." The *New Leader*, a liberal American journal of ideas, cautioned its readers against a "narrowly political" view of Tertz's stories, which were "more universal in significance than Russia alone or its communism."[47] For Sinyavsky's and Daniel's interrogators (and, one suspects, for *Time* magazine), such subtle efforts at nonalignment in the midst of the Cold War's ideological combat were beside the point. What mattered was that the two writers had violated the venerable Russian taboo against "taking garbage out of the hut," or, as their American adversaries would put it, airing dirty laundry in public. Not before the Soviet public, which would have been bad enough, but in front of *foreigners*. They had turned to the daughter of an officer in the French army—part of the NATO alliance—to smuggle their works abroad, thereby giving the West additional weapons for its psychological warfare against the Soviet Union. And for this they had earned royalties—in Western currencies, no less—while hiding behind false names. Even Soviet citizens who would later come to their defense were shocked by these revelations.[48]

Unsanctioned communication with foreigners was an essential ingredient of the Sinyavsky-Daniel affair, highlighting its geopolitical dimension and suggesting the possibility of treason. The role of Westerners as source, messenger, and/or audience for ideas marked as heretical in the Soviet setting hints at the international aspects of the Soviet dissident movement that will become visible in the chapters ahead. For now, I want to focus on a particular facet of the foreign presence in the story, a single but hardly singular moment in the centuries-long history of fraught relations between Russians and Westerners: the encounter between Sinyavsky and Hélène Peltier.

FIGURE 2.5. Hélène Peltier

In the available transcripts of Sinyavsky's numerous interrogations, as well as in those of the trial that followed, Peltier is barely visible. She appears as a bit player, a foreign student whose sole function was to serve as courier to the West of uncensored Soviet texts. Whatever their truth-extracting purposes, however, interrogations—and, for that matter, trials—can harbor powerful silences mutually beneficial to the opposing parties. Both Sinyavsky and the KGB knew, for instance, that in the late 1940s and early 1950s, as a dedicated member of the Communist Youth League and recipient of a coveted "Lenin Scholarship," Sinyavsky had served as an informer for the Soviet security services, secretly reporting on Peltier and her Soviet acquaintances at Moscow State University, among whom was Stalin's daughter, Svetlana Alliluyeva. They knew that Sinyavsky, having become friendly with Peltier, had been encouraged by his handlers to start an affair (despite the fact that he was already engaged to Inessa Markovna Gilman, who would become his first wife), presumably so that the security services could later blackmail her father.[49] Sinyavsky, furthermore, knew that Hélène

Peltier, or "Lenochka," as he came to call her, had in the end played a very different role: not as a stepping stone toward the manipulation of her father, but as a stumbling block for Sinyavsky himself, a catalyst of what he later called his "ideological breakdown," the beginning of his transformation into a dissident.[50]

Whether to save face in the wake of a mission gone bad, or because of an unspoken rule in the intelligence world (East and West) never to acknowledge the existence of one's informers, the KGB studiously left these facts in the shadows. Years later, however, memoirs published outside the Soviet Union by the two protagonists and several mutual friends shed considerable light on how the Soviet state inadvertently pushed a devoted communist off the beaten track. They also confirm that the Soviet leadership's extraordinary anxiety vis-à-vis foreign, especially Western, influence inside the Soviet Union was well founded. Not for nothing did Sinyavsky title the relevant chapter in his memoir, with a nod to Pierre Choderlos de Laclos's novel of French intrigue, "Dangerous Liaisons."

The encounter between Sinyavsky and Peltier began at the renowned Philological Faculty of Moscow State University in 1947 with a familiar conversation—familiar, that is, to anyone who has been an exchange student in a foreign country—about their respective languages.[51] Was it curiosity that led the twenty-two-year-old, "visibly intimidated" Sinyavsky to ask Peltier why she was studying Russian, or was he simply unable to think of a better opening line? Her answer—"because I love your literature and your country"—was to be expected from a diplomat's daughter but also happened to be true (Peltier went on to become a professor of Slavic literatures at the University of Toulouse, in France). To her query about why he was studying French, Sinyavsky responded, "Very simple: France is the birthplace of modern art." Their shared love of literature and art notwithstanding, each came to regard the other as a polar opposite. Peltier understood herself as the product of a conservative French Catholic milieu, "a Christian by birth and by conviction," raised in a military family, and Sinyavsky as an atheist, "a completely convinced communist" whose family fully embraced the "cult of the Revolution." To Sinyavsky, she was "a unique foreigner who had descended to earth on a diplomatic parachute," whose "invisible God I could not comprehend." He was not alone in this impression. His childhood friend and Moscow State University classmate Sergei Khmelnitsky, who introduced Sinyavsky to Daniel as well as to his future (second) wife, Maria Rozanova, described Peltier as a "charming creature" who had arrived "as if from another planet."[52] "Everyone," Sinyavsky noted, "seemed secretly in love with her," this "fearless Cinderella [who] entered our enchanted house." He and Khmelnitsky would visit Peltier at her residence in the stately Metropole Hotel in the heart of Moscow. When she returned from trips to France, she would invariably bring gifts for them, exquisite books from the Louvre and other Parisian museums. "Never," recalled Sinyavsky, "had I seen anything more beautiful in my life."[53]

This Cold War *ménage à trois*, apparently platonic, was too good for Stalin's secret police to leave alone. In 1948, Khmelnitsky and Sinyavsky were recruited by "the organs" to deliver weekly reports on Peltier and those with whom she interacted. As loyal Young Communists, neither was particularly disturbed by this assignment. It was simply assumed that "not a single foreigner in our country was to be left unwatched," and helping the security services was "a patriotic cause serving the flourishing of our Motherland."[54] Not that there were many foreigners to watch: during the final years of Stalin's reign, the USSR virtually sealed its borders to citizens from other countries, apart from the special category of German POWs, of whom hundreds of thousands remained in carefully guarded labor camps. In any event, Peltier's diplomatic upbringing and "European restraint" left Khmelnitsky and Sinyavsky with relatively little to report. Whatever he conveyed to his handlers, Sinyavsky himself formed an impression of Peltier as the incarnation of the Rousseauist soul of France: pure-hearted, almost child-like in her innocence, incapable of deception. At one point, when he used an earthy expression (*bliadstvo*, "whoring") and she asked him to explain what it meant, Sinyavsky wondered whether the French language "has any adequate equivalent": "I don't think two people can really curse each other in French: the language is too aesthetic."[55]

Peltier understood the Soviets better than Sinyavsky did the French. In contrast to his fairy-tale vision of France as a country incapable of cursing—a not uncommon inversion of the cliché of the wicked West that dominated Soviet public life—Peltier grasped the central paradox that had shaped Sinyavsky and his generation: "The fathers were the revolutionaries, not the sons." Among Sinyavsky's cohort of second-generation socialists, "not one . . . had formed his opinions himself or had had to fight against his milieu for his views, as is so often the case with Communists in the West." Indeed, their milieu and their lived experience had only affirmed their parents' teachings:

> Marxism offered a complete view of life and thus a rational basis for [Sinyavsky's] need of the absolute and of social justice. What attracted him was . . . above all the Marxist philosophy of history. Truth was historical, and Communism embodied it in our time. Wasn't this sufficiently proven by the Soviet Union's achievements in the war and by its place as a world power? Because it appealed to his patriotism this was perhaps the argument which influenced him most. I found the same attitude in many of my other fellow students. When they defended Communism, were they defending a doctrine or merely using Marxist language to praise their country? I could never be sure.[56]

Christianity, Sinyavsky informed Peltier, had been in decline since the Renaissance, "ever since it made personal salvation the one thing that mattered. Modern Christianity is individualist, Communism is concerned with the good of mankind, so its moral meaning is higher." Sinyavsky, according to Peltier, was "repelled by Western notions of the freedom of the individual as the highest good," regarding them as fostering selfishness. Western political life left him "indifferent." On the rare occasions when he spoke about it, "it was with the condescension of a Soviet citizen who looks on the multi-party system as a sign of weakness if not of decadence."[57] To a purpose-driven country like the USSR, Western pluralism appeared as little more than aimlessness.

Peltier provided one of the first instances in which Sinyavsky and his fellow students had to measure their received opinions against those from a strikingly different milieu. They were particularly unsettled by her presence in the required class on Marxism-Leninism:

Come on. How could a foreigner be so intent on fathoming a doctrine that we, in our heart of hearts, were already a little sick of, though it continued to play its obligatory role as a privileged possession, a consistent worldview accessible to us alone, a view, we were taught, that was very well constructed. Later on Hélène was almost the death of our poor teacher of Marxism; after passing the course with flying colors and answering all the thorniest questions, she modestly confessed at the end of her exam that, personally, she still retained her idealistic views. What?! To have an excellent grasp of Marxism and still remain an idealist?! I think they lowered her grade for that.[58]

"She showed us," recalled Khmelnitsky, "that our world was not absolute, and thereby recast our convictions, our inert faith, on a different, more modest scale."[59] Not that these adjustments were simple or painless. There were sharp disagreements, especially between Peltier and Sinyavsky. "Often I felt uncomfortable," she noted, "and was hurt by his hasty judgments and his habit of laying down the law. But the more violently he attacked 'individualism,' the more I felt that he was not arguing with me, or with Christianity, or with the West, but with himself. Just because my ideas were opposed to his, and I never concealed them, he used me to clarify his own thought and bring into the open the inner conflict which was later to become so sharp."[60] Part of Sinyavsky's inner conflict, no doubt, stemmed from his dual role as confidant and informer. But only part. According to Peltier, Sinyavsky was "too honest to accept the hackneyed 'movement of history' slogan as the answer to the everlasting question of whether the end justifies the means." Increasingly, the conflict took the form of a larger, haunting question: "Does one have the right, in the name of historical progress, to crush innocent individuals?" In this as in other respects, Sinyavsky had not yet been tested; the ethical dilemma "remained to some extent theoretical because he had not as yet come up against the facts in his personal life."[61]

That was about to change. The security services, evidently frustrated that reports on Peltier by Sinyavsky and Khmelnitsky had produced so little compromising material, decided to ramp up its approach. At roughly this time, Khmelnitsky, observing how close Peltier and Sinyavsky were becoming, intuited that Sinyavsky too must have been recruited as an informer. The two childhood friends broke a strict intelligence rule, revealing to each other their work as informers and discovering in the process that they were both reporting to the same agent. In the summer of 1948, the agent informed Sinyavsky that he was being "mobilized" to take the relationship with Peltier to a more "intimate" level.[62] It is not clear whether Sinyavsky shared this information with Khmelnitsky, who already had reason to be jealous. For the first time in his life, Sinyavsky was being told to harm, and possibly ruin, an innocent individual in order to safeguard his Motherland. The injunction to seduce Peltier, according to Sinyavsky, precipitated what "may have been the most serious crisis of my life," forcing him to confront a profound ethical choice. And choose he did: in a climactic encounter in the labyrinths of Moscow's Sokolniki Park, he confessed the entire plot to the unsuspecting Peltier.[63] To protect themselves from possible repercussions, they agreed to feign her rejection of his advances and to break off their friendship. This was the account that Sinyavsky gave to his handler and that Peltier gave to Khmelnitsky, so that he would "independently" relay it to the same handler.[64]

Sinyavsky's claim, more than three decades later, that after confessing to Peltier "it was emotionally impossible for me to return to the ranks of moral and political unity with Soviet society," can probably be set aside as hyperbole, or ironic play with a stock Soviet phrase, or perhaps as the compression into a single event of what was in fact a series of seismic tremors. The encounter with Peltier coincided with the official condemnation of formalism and abstraction in the arts by Stalin's director of cultural policy, Andrei Zhdanov. As Peltier observed, Sinyavsky was not one to be disturbed by revolutionary violence or run-of-the-mill Soviet propaganda, whose "stylistic devices" he could even appreciate as "an imaginative reconstruction of reality." But crude and deceitful attacks on Picasso (about whom he would later co-author a book), André Malraux, and other favorite modernists—*that* he could not abide.[65] His father's arrest in 1950 on charges of anti-Soviet activity produced a third tremor, all the sharper given Donat Sinyavsky's lifelong devotion to the revolutionary cause and Andrei's gnawing suspicion that he himself—having betrayed the secret police to protect a foreigner—ought to have been the one arrested.[66]

That same year, Peltier returned to France after four years of study in the USSR.[67] By the time she returned to the Soviet Union in the fall of 1953, now for graduate study, Stalin was dead and the country he had forged during the prior quarter-century was beginning to emerge from its deep ideological freeze. Over the next three years, Peltier observed her friends adapt to a new, uncertain landscape, an inescapable feature of which were the millions of prisoners returning from exile and the Gulag—among them Sinyavsky's father. She watched people absorb Khrushchev's sensational (though highly selective) exposé of Stalin's crimes. "I know how much the Soviet Union means to me," Sinyavsky told her at the time, "by the depth of my shame when I heard the report." These were no longer private tremors; they were collective shocks.

> Wounded in their national pride, all my friends reacted in the same, typically Russian way. Ashamed for their country—a country which had claimed to set an example of humanitarianism to the world while setting up an inhuman dictatorship and allowing thousands of innocent people, including some of its most gifted artists and writers, to be shot or slowly put to death in camps—they were even more ashamed of their own inertia, an inertia due to the fact that they had confused patriotism with passivity. Astonishingly, none of them blamed the government, they blamed themselves as, at best, accomplices; they felt that their silence had meant consent and that even genuine ignorance was inexcusable. . . . A new concept of patriotism was born, as well as a new will to action: something had to be done to prevent a recurrence of the catastrophe. But what in the world were they to do? Not one of the many intellectuals I knew gave so much as a thought to political action. They felt neither hatred nor any other emotion toward their rulers. . . . It was the strangest of paradoxes that, in so intensely political a climate . . . , political activity itself was regarded, even by sophisticated intellectuals, as an utterly irrelevant and useless way out of the country's problems.[68]

Some retreated into cynicism. Khrushchev's account of what had gone wrong—Stalin's "cult of personality" (which could also be translated as "cult of the individual," thereby emphasizing Stalin's sins against the collective)—quickly assumed the status of a new orthodoxy, in which blame for millions of deaths was placed on the shoulders of a single person who was conveniently already dead.[69] The endless repetition of this pseudo-explanation in the Soviet press and public discourse inspired the following *anekdot*, the compact genre featuring characteristically Soviet punch lines:

What is the cult of personality?
—That's when one person spits on everybody.

And what is the unmasking of the cult of personality?
—That's when everybody spits on one person.[70]

There were, to be sure, individuals who sought to widen the circle of responsibility to include Stalin's accomplices in the Party leadership, who happened to be the Soviet Union's current rulers. Sinyavsky's childhood friend Andrei Remizov was one such person. But Sinyavsky and Daniel were groping toward something different, "to invent new ways of life," as Peltier described it, based on the idea that *all* Soviet citizens shared responsibility for the atrocities of the Stalin era.[71] Never mind that the Communist Party had never gained the formal consent of a majority, let alone the totality, of Soviet citizens, or that so many Soviet citizens—including Communists—had themselves suffered persecution during the quarter-century of Stalin's rule. The fictional works that Sinyavsky and Daniel produced in the wake of Khrushchev's revelations offered no pure heroes, no pure victims, no characters untainted by the cultish atmosphere of the Stalin era. "I am guilty of sins of omission," announces the narrator of Daniel's *Atonement*. "I am guilty of the unaccomplished—of indifference, of cowardice. I am guilty of the same thing you are!"[72] When he declares, "The state, it is we," the effect is to collapse the traditional distinction between "us" (ordinary people) and "the bosses," to reimagine the Soviet "people's democracy" in the idiom of Louis XIV's absolutism. Not only his narrator, but Daniel himself endorsed such views: asked by his KGB interrogator in the Lubyanka to summarize the main idea of *Atonement*, Daniel stated that "I considered the entire Soviet people as a whole to blame for the appearance of the personality cult and the associated repressions."[73]

This is a distinctly twentieth-century Leviathan, in which the state not only acts in the name of the people, but mobilizes them, fashions them into participants and therefore collective bearers of the glory as well as the shame of the state's actions. As Hannah Arendt once observed, mass participation is what most distinguishes modern totalitarian systems from classic tyrannies.[74] Participatory dictatorship reaches far beyond the pre-revolutionary Russian tradition of *krugovaya poruka*, the collective liability (literally, "circular bail") of each peasant commune for the tax payments, military conscription, and in certain cases the crimes of its members. That responsibility was strictly local, within a face-to-face community, and was limited precisely to liabilities.

Daniel's notion of collective responsibility for state-sponsored violence also departed dramatically from the poet Anna Akhmatova's famous dictum, following the mass amnesties from the Gulag, that "two Russias are now eyeball to eyeball, those who were in prison, and those who put them there."[75] Akhmatova's stark division of the population into victims and perpetrators, and her unselfconscious identification of both as Russian, rather than Soviet, hint at her pre-revolutionary sensibilities (she was born in 1889). For the narrator of Daniel's *Atonement*, by contrast, it made no difference who had actually been in the camps. The reappearance of millions of former Gulag inmates in Soviet society—what the cultural historian Alexander Etkind has called "the return of the repressed"[76]—revealed a deeper truth to Daniel's narrator: "The prison is in us." For Daniel and Sinyavsky, both born in 1925 and formed by elite Soviet institutions, it was always the Soviet collective "we" who put "us" behind bars, "we" who "need an operation."[77] "So that prisons should vanish forever, we built new prisons," lamented Sinyavsky; "so that not one drop of blood be shed any more, we killed and killed and killed."[78]

"We will never permit such crimes to return": this is the phrase Peltier heard again and again from her Moscow friends following Khrushchev's

"Secret Speech."⁷⁹ But how? The fiction of collective responsibility flattened any attempt to identify, let alone render justice to, the specific perpetrators and victims of those crimes. The turn to self-liberation as a means to enact collective moral responsibility exemplified what Peltier described as the general lack of interest in political action on the part of Sinyavsky, Daniel, and their circle during Khrushchev's Thaw. Their Soviet intelligentsia upbringing had instilled in them both the sense of collective moral responsibility and the urge to work on the individual self as a means of escape from the internal prison.

Work on the self figured powerfully in Sinyavsky's unpublished (and uncirculated) essay "An Experiment in Self-Analysis," which the KGB confiscated during a search of his apartment in 1965 and preserved in its archive.⁸⁰ Regardless of their worldview or situation in life, Sinyavsky wrote, the decisive experience in the formation of his generation's collective consciousness was and always would be "the year thirty-seven," the zenith of mass arrests and executions under Stalin:

> This is where we come from. From this we learn to think. This is what our nostalgic memory longs for, like the parental home.... Why are we obsessed with it? Why, at the slightest recollection, do we prostrate ourselves, falling into ecstasy? Why do we ceaselessly argue and think about it? For what reason do we love—yes, love— the year thirty-seven?⁸¹

As if plunged into the grotesque world of Hieronymus Bosch, those who passed through 1937 experienced a magnified fear of moral disfigurement, of losing not one's life, but one's soul:

> There is no need to wonder at the unprecedented unanimity with which society branded the "vile traitors." There was no hypocrisy here. The normal reaction of a healthy person consisted precisely of such attempts to separate oneself, to create a protective wall, to exorcise one's self against the encroachment of the enemy. Each of us was potentially capable of becoming the enemy and, sensing this potential, we disowned [others] all the more intensely. The ubiquitous denunciations of that era served precisely this role, forming a magical

defensive circle.... In them there was not the least element of baseness or duplicity. On the contrary, it was honest, honorable people with a highly developed fastidiousness who were most drawn to denunciations.... In denunciations our age-old longing for justice found expression. To denounce meant to purify oneself and to receive at least some kind of psychological guarantee: that I am not like "him" and do not wish to have anything in common with "him." Naturally the state fostered in every way possible this act of ethical affirmation of the self. The morality of the pre-revolutionary past, which condemned denunciations, turned out to be a dangerous illusion, capable of destroying the righteous person's soul, transforming it into an accomplice in a horrifying crime.[82]

This diagnosis of a generation's collective psyche, needless to say, sheds considerable light on Sinyavsky's own activity as a KGB informer (unmentioned in this essay), perhaps in an attempt to justify it. But whereas Daniel's stories pushed readers to purge themselves of Stalinism's magical thinking and its catastrophic consequences, Sinyavsky preferred to linger, harnessing the Soviet grotesque for purposes of self-knowledge and artistic imagination. Having "paralyzed the will and reason," he noted, Stalin's sinister charisma "sharpened one's intuition, refined one's fantasy, [and thus] we should be grateful to it for this lesson in irrationality, taught in the language of dull newspaper clichés."[83]

Alexander Volpin was not looking for lessons in irrationality. For this supreme admirer of logic and literalism, the arrests of Sinyavsky and Daniel would serve as an opportunity to defend not the freedom of artistic imagination, but the transparent language of law and reason.

3

Transparency Meeting

> The word [*glasnost*] had no political meaning, and until Alek Esenin-Volpin pulled it out of ordinary usage, it generated no heat.
>
> —LUDMILLA ALEXEYEVA

In the fall of 1965, readers of Soviet newspapers did not find "dull clichés" regarding Andrei Sinyavsky and Yuli Daniel. They found nothing at all. But that hardly stopped word from spreading. As KGB chairman Vladimir Semichastnyi observed with displeasure shortly after their arrest, certain "isolated individuals" who sympathized with the two writers were already attempting to "portray their activities in an apolitical manner."[1] Among them were their friends Mark Azbel (a biologist), Ludmilla Alexeyeva (an editor at Nauka publishing house), Igor Golomshtok (an art historian who had co-authored with Sinyavsky a long essay on Picasso), and Nina and Alexander Voronel (playwright and physicist, respectively). Azbel and his wife, Naya, the Voronels, and Daniel and his wife, Larisa Bogoraz, had first met at intelligentsia gatherings outside Moscow during summers in the late 1950s. The Voronels lived five buildings away from Sinyavsky and his wife, Maria Rozanova, on Bread Lane, in Moscow's storied Arbat neighborhood. Alexeyeva had met both writers through acquaintances at Moscow State University. Along with a dozen other members of the Sinyavsky-Daniel social circle, they had taken part in numerous gatherings in cramped communal

FIGURE 3.1. Clockwise from upper left: Larisa Bogoraz, Marina Domshlak-Gerchuk, Maria Rozanova, Andrei Sinyavsky, Yuri Gerchuk, and Yuli Daniel

apartments and dachas where the two writers gave readings of their works and Golomshtok read his translations of George Orwell's *Animal Farm*, Arthur Koestler's *Darkness at Noon*, and the stories of Franz Kafka.[2]

Semichastnyi was correct when he labeled Azbel and others "isolated," in the sense that they did not speak for a larger group, and certainly not for an official collective. Unbeknownst to Azbel's circle of friends, for example, the Moscow literary critic Lev Kopelev, a Gulag veteran and Party member, wrote his own letter to the Central Committee in defense of Sinyavsky.[3] But attempts to gather support for the arrested writers among the liberal Moscow intelligentsia proved unsuccessful. Among the scientists and scholars whom Azbel approached that fall about signing a petition, most responded that, while the two writers might have written the truth about Soviet life, publishing their work abroad was "unpatriotic." Others resented even being asked. "This sort of decision," declared one Academician, "should not be forced; it has to be left up to each individual for himself. This is a moral issue, a matter of one's own conscience, on which one can speak only for oneself—not as part of a group."[4]

It is tempting to regard such phrases as an attempt to construct a noble façade around a response grounded in fear. In the uncertain atmosphere following Khrushchev's fall from power, it was impossible to know what consequences such a petition might have for those who signed it or whether it would help or harm Sinyavsky and Daniel—assuming it made any difference at all. Beyond the question of practical effects, however, the idea that protesting the arrest of a pair of writers was an issue "on which one can speak only for oneself—not as part of a group" carried its own significance. Many Soviet citizens had had more than enough experience of being coerced to sign collective letters for causes about which they had misgivings, whether endorsements of Party policies or denunciations of colleagues. They were tired of being mobilized. The relentless pressure to operate as part of a collective meant that the most elemental form of moral self-assertion was the decision *not* to participate in this or that joint action: not to sign a group letter, not to cast a ballot in an election, not to speak at a choreographed public meeting. In a climate of incessant collective affirmation and denunciation—hallmarks of a participatory dictatorship—apathy became a significant gesture. Soviet citizens, as one put it, looking back on the 1960s, "expressed defiance by *not* marching."[5] "Silence," wrote another, "had become the benchmark of a person's courage and decency."[6] Given the courage it required, "criticism via silence" (*kritika molchaniem*) should be understood not as a form of passive resistance, but as an act of individuation: the assertion of one self.[7]

Plenty of liberal intellectuals, to be sure, refused to support Azbel's petition on behalf of Sinyavsky and Daniel for simpler reasons: they were appalled by what the two writers had done. Rather than contribute to the collective struggle to push back the walls of censorship in the USSR, the authors of *This Is Moscow Speaking* and *The Court Is in Session* had dug their own private tunnels in order to publish their works abroad. Nina Voronel recalled typical reactions: "They ruined everything for us! We had gotten almost everything, we were on the verge of a real thaw, and they betrayed us! [Now] everything is going to be shut down." The intelligentsia was about to experience "real freedom . . . and then those bastards went and wrote that crap. They wanted to show off. Who needs this? They ruined everything, all the achievements of the Soviet intelligentsia have turned to dust!"[8]

This was the *liberal* reaction. For die-hard Soviet conservatives, it was treason pure and simple. Indeed, as Azbel, Alexeyeva, and the Voronels all subsequently acknowledged, they themselves would not have supported Sinyavsky and Daniel at the time had it not been for a single fact: their personal friendship with the writers. "The general political situation doesn't interest us," Alexander Voronel recalled thinking at the time. "We just need to rescue those guys!"[9] "I had made up my mind years before," Azbel noted, "that the guiding principle by which I had to live was, very simply, loyalty to my friends. It would be impossible to let caution affect my association with Yuli—to lock myself in a moral prison for fear of a future prison."[10]

Consciously or not, Azbel was echoing Daniel's *Atonement*, with its call for self-liberation from the fearful, magical thinking of the Stalin era. That Azbel and the Voronels even attempted to organize a petition in defense of persecuted writers already contrasted sharply with the absence, less than a decade earlier, of any such campaign in response to the public vilifying of Boris Pasternak.[11] But it hardly refutes Semichastnyi's impression of Sinyavsky's and Daniel's supporters as isolated. True, they were not quite "atomized," as Western observers at the time typically imagined the citizens of a totalitarian state. They were more like a compound molecule, a micro-community tightly bound by ties of adult friendship. "Let us join hands, friends," sounded the refrain of the popular song "Union of Friends" by the unofficial bard Bulat Okudzhava, "so that we don't fall one by one."[12] Soviet society was a loose assemblage of such face-to-face communities, whose strong internal bonds were both cause and effect of the weakness of broader social trust. The intense personal ties that led Sinyavsky's and Daniel's friends to mobilize on their behalf, despite manifold risks, also imposed limits on their ability to widen the network of support beyond the circle of friendship.

In this context it is of more than passing interest that the person who first drew significant public attention to the case of Sinyavsky and Daniel, and who successfully framed it in an utterly novel manner, was not a friend of either one.[13] Alexander Volpin had been waiting for an opportunity to issue "a public manifesto of our rights." With the arrest of the two writers, he found one. Although he too had published work abroad (under his own name) that was attacked as anti-Soviet slander, Volpin was emphatically uninterested in Sinyavsky's and Daniel's texts. In fact, he refused to read them, regarding them as distractions from the real issue, which for him had nothing to do with the all too familiar drama of Russian writers persecuted by the state, let alone with personal friendship. The real issue was to be found in a different text: Article 111 of the Soviet Constitution, guaranteeing that "examination of cases in all courts shall be open, in so far as exceptions are not provided for by law," and analogous language in the Code of Criminal Procedure of the Russian Soviet Federated Socialist Republic (RSFSR).[14] "Let them go ahead and convict those fellows," Volpin recalled thinking, "but let the words, like those expressed by Shatunovsky at my court case against him—'From our Party-minded point of view, the conventional definition of "slander" is irrelevant'—let this entire pseudo-argumentation be heard loud and clear. The more such occasions, the more quickly an end will be put to similar repressions."[15]

If this agenda struck many of Volpin's acquaintances as oddly minimalist—did anyone really need to be reminded of the authorities' frequent disregard for the law?—the means by which he proposed to realize it did not: a public "*glasnost* [transparency] meeting" demanding that the trial of Sinyavsky and Daniel be open to the public as per the Soviet Constitution.[16] The meeting itself would exemplify strict fidelity to the constitution, Article 125 of which, "in conformity with the interests of the toilers and in order to strengthen the socialist system," guaranteed "freedom of assembly and meetings." Together with fellow mathematician Valery Nikolsky, the journalist Yelena Stroeva, and the abstract artist Yuri Titov, Volpin set about drafting a "Civic Appeal," announcing the meeting's time, place, structure, and purpose, and inviting those who received a copy to take part.

The Civic Appeal would become, in effect, the founding document of the Soviet dissident movement. Like many founding documents, it evolved across multiple drafts, offering glimpses into the evolution of its authors' thinking during a period of intense ferment.[17] Successive iterations allow us to trace, for example, how the face-to-face community gradually receded as the presumed collective actor in the transparency meeting, opening up the possibility for broader, more impersonal, interest-driven networks and thus more expansive participation. They allow us to witness an unforeseen effect of the Kremlin's post-Stalin retreat from the use of mass terror as an instrument of social control. Sparked by the Communist Party's desire to end the self-destructive purges that characterized the Stalin era, the project of removing terror from the repertoire of governing practices was meant to help stabilize the Soviet state—which it did. But it also unintentionally opened up spaces for Soviet citizens to experiment with new forms of independent social action—"new ways of life," in Hélène Peltier's words—within the horizon of possibilities of a barely post-totalitarian Soviet society. The transparency meeting was one such experiment.

Originally, which is to say in the weeks following Sinyavsky's and Daniel's arrest, Volpin's plan was to recruit somewhere between five and twenty-five people to gather on October 31, either on Moscow's Nogin Square, across from the imposing building that housed the Central Committee of the Communist Party, or in front of one of the city's central courthouses. As he wrote in an early draft of the Civic Appeal, "The natural choice is a square or boulevard close to an institution on which depends the authorization of transparency."[18] If the initial meeting should fail to produce a response, participants would gather again at the same spot on November 7, the anniversary of the Bolshevik Revolution, with the hope of drawing greater attention to their cause.

Valery Tarsis, who like Sinyavsky, Daniel, and Volpin had gotten into trouble for publishing abroad (and who lived in the same apartment building as Volpin's mother), enthusiastically supported the plan for a transparency meeting but dismissed the idea of demonstrating in the name of "juridical formalities," let alone those contained in Stalin's constitution.[19] One had to fight on behalf of writers, in the name of Russian literature! Volpin countered that if the KGB had arrested a pair of veterinarians or tailors, the principle of judicial transparency would be the same. Tarsis began to protest, "Don't you see the obvious difference between..." but was cut off by an exasperated Volpin. "Valery Yakovlevich, not every obvious difference is relevant to the issue at hand. When it comes to procedural rights it is not I but the law that equalizes people of different professions."[20] Others dismissed as "idiotic" the idea of demanding transparency rather than freedom: "Where, even in the West, has anyone ever demonstrated on behalf of transparency?" Volpin's reply: "If in the West someone were to be tried the way they put people on trial here, then it's entirely possible that there would be precisely such demonstrations. In any case we have to confront our problems, not those of some other country."[21]

Yuri Aikhenvald, the literary critic who had befriended Volpin during their overlapping exiles in Karaganda, was initially more sympathetic. He urged that the meeting be silent, expressing the demand for an open trial exclusively via hand-held banners. This halfway house between "criticism via silence" and open protest was too constrained for Volpin: "Perhaps in 1948 it might have been necessary to proceed along such lines, but here we are in 1965." The earliest draft of the Civic Appeal gave participants the option to remain silent or to recite a single slogan, the same one that would appear on the banner: "We demand an open trial for Sinyavsky."[22] That Daniel's name did not even appear at this early stage of planning suggests how socially distant Volpin and his fellow organizers were from the two writers and their circle. The absence of phrases such as "socialist legality" and "struggle against the legacy of the personality cult" from any drafts of the "Appeal" similarly suggests how far its authors had moved from official rhetoric.

The organizers of the "Appeal" had little trouble arriving at a single, unambiguous, non-provocative, and perfectly legal slogan, which they hoped would discipline participants against the temptation to widen the meeting's purpose. No one was to criticize the Soviet government, make grand statements about Russian literature and artistic freedom, or demand that the arrested writers be released. Far less certain, however, was the composition of the imagined cast for the drama that was about to unfold. Initially, the idea was to organize a small number of people known and trusted by the drafters of the Civic Appeal. But it was not at all clear that five, let alone twenty-five such individuals could be found who were willing to serve as guinea pigs for the first-ever transparency meeting. Among Volpin's circle of friends (including his wife, Victoria) responses were overwhelmingly negative: "Have you lost your mind?" and "Have you forgotten where you live?" were two of the more common reactions.[23] No one could say for sure how the authorities would respond. Why should participants "give up so much," Alexeyeva wondered, "just for a few narcissistic moments?"[24] In one version of the "Appeal," Volpin countered such fears by pointing out that the right to assemble was guaranteed by Article 125 of the Soviet Constitution, and therefore participants faced "only a small risk." By his own account, however, that risk included not only "milder" forms of punishment such as dismissal from one's job, brief arrest, or placement in a psychiatric hospital, but "extra-judicial repression" and "administrative exile." Potential participants were unlikely to have taken comfort from one of Volpin's favorite sayings at the time: "Down with the instinct for self-preservation and the sense of proportion!"[25]

The idea that those who took part in the transparency meeting would face only mild punishment, or none at all, was not without foundation. During the preceding spring, in April 1965, advance word about a public demonstration organized by a youthful group of poets and artists known as SMOG—an acronym for Boldness, Thought, Form, Depth or, more to the point, The Youngest Society of Geniuses—had spread via hand-typed "appeals" distributed among high school and university students. "We are engaged in a desperate struggle with everyone and everything," declared their manifesto. "From members of the Communist Youth League to the Philistines, from the secret police to the bourgeoisie, from the untalented to the ignorant—everyone is against us."[26] From their ranks would subsequently emerge the writer Sasha Sokolov, the sociologist Boris Dubin, and the filmmaker Andrei Razumovsky, among other luminaries.[27] Having been denied permission by the Communist Youth League to register as an official student artistic association, SMOGists were taking their cause to the "adult" literary world (no one over twenty-five could join SMOG). On April 14, they demonstrated in front of the stately Central House of Writers with

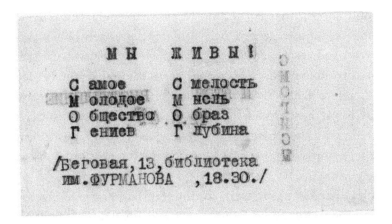

FIGURE 3.2. Announcing the fifth SMOG event: "We're Alive!" The text lists both meanings of the SMOG acronym, the location and time of the gathering, and (on the reverse) the names of seventeen SMOG members.

edgy banners reading "We will deprive Socialist Realism of its virginity!," "We will tear the buttons of censorship from Soviet literature's Stalinist uniform!," and "Art outside Politics!" Remarkably, only a few participants in the demonstration were punished, and even then for no more than five days of corrective labor.[28] Undeterred, the same loose conglomeration of students, now calling themselves the "Amazers," formed a "creative association" under the name "Avant-Garde," whose founding conference was to be held in mid-September 1965, with representatives arriving from Leningrad, Kyiv, and Minsk. The Amazers had applied to the Central Committee of the Communist Party for access to an assembly hall somewhere in Moscow and, unamazingly, had been turned down.[29]

Volpin learned of all this from Vladimir Bukovsky, the audacious expellee from Moscow State University who in the early 1960s was drawn to Volpin's legalist approach and who served as a bridge to a younger generation of non-conformists in and around SMOG.[30] It was precisely people such as Bukovsky and Yuliya Vishnevskaya and Vladimir Batshev (the latter two still in high school) who, having come of age after Khrushchev's "Secret Speech" and lacking the instinctive fear of unsanctioned demonstrations, proved to be more receptive to Volpin's ideas than were members of his own generation.[31] And it was precisely bohemians like Bukovsky and his friends whom Yuri Aikhenvald mistrusted, regarding them as troublemakers who would derail the transparency meeting and therefore had to be kept off the invitation list. For the majority of SMOGists, it must be said, the mistrust was mutual: they wanted nothing to do with their parents' generation, the Soviet Constitution, or "politics," preferring instead to keep SMOG's agenda, as one put it, "purely aesthetic."[32]

Volpin needed participants for the transparency meeting, and he needed them soon, before the KGB made any irrevocable decisions about the fate of Sinyavsky and Daniel. Stymied by squabbles over who should or should not be recruited, he rewrote the Civic Appeal, adopting a cautiously expandable system of invitations designed to harness but ultimately transcend the bonds of trust anchored in face-to-face communities. Recipients were now instructed to type two copies of the "Appeal" to pass on to "serious people whom one knows well." Only those able "to conduct themselves calmly" were to be invited. Each participant "should keep secret the name of the person from whom [the "Appeal"] was received as well as the names of those to whom he gives or attempts to give copies," except in cases where exposing an informer required otherwise. Having copied the "Appeal," recipients were to destroy the original. Anyone who received a copy from a stranger or casual acquaintance "may decline to accept it without coming under suspicion of indifference to the idea of the meeting."[33] The sign that this procedure—or "geometric progression," as Volpin, ever the mathematician, called it—had generated a sufficiently large "public" would be that "almost every participant will see many unknown faces."[34] To welcome such faces, to be willing to expose oneself to unknown risks together with them, to overcome the fear that among them might be informers or provocateurs: these were the symptoms of an expanding network of information exchange and social trust, the blueprint for which evolved across successive iterations of the Civic Appeal.

This was already too much for Aikhenvald, who accused Volpin of indulging in "utopian" fantasies. "Logic is one thing," Aikhenvald cautioned, "history—especially Russian history—is quite another." As Volpin later recalled, Aikhenvald launched into a disquisition on how the individual was powerless to resist history's tide, but Volpin was having none of it. "I've had it with silent gestures, with history!" he interrupted. "We're dealing with a juridical fact! Maybe what needs to change is history itself, so that a new humanity will live according to the legal code. Is that not so?" "Yes," came Aikhenvald's reply, "but this is not the way to do it."[35]

Volpin refused to budge. Indeed, by the time the Civic Appeal assumed its final, single-page form, he had abandoned all efforts to restrict the channels of its diffusion. Gone was the air of conspiracy, the requirement of special qualities on the part of would-be participants, and other virtuosi habits from Russia's revolutionary underground.[36] Gone, too, was the symbolic link to the revolution itself—though more by default than design, since the November 7 anniversary had slipped by while Volpin and others were debating the wording of the "Appeal." It was Valery Nikolsky who proposed that the transparency meeting be rescheduled on a different anniversary: December 5, Soviet Constitution Day (in 1965, a Sunday), marking the date in 1936 when the current constitution of the USSR had been ratified.

The final text of the "Appeal" read as follows:

CIVIC APPEAL

Several months ago the KGB arrested two citizens: the writers A. SINYAVSKY and Y. DANIEL. In this case there is reason to fear violations of the law on transparency of the judicial process. It is well known that any kind of lawlessness is possible behind closed doors, and that violation of the law on openness (Article 111 of the Constitution of the USSR and Article 18 of the Code of Criminal Procedure of the RSFSR) in and of itself constitutes a form of lawlessness. It is implausible that the creative work of writers could constitute a state crime.

In the past, lawlessness on the part of the authorities cost the lives and freedom of millions of Soviet citizens. The bloody past calls us to vigilance in the present. It is easier to sacrifice a single day of rest than to endure for years the consequences of an arbitrariness that was not checked in time.

Citizens have the means to struggle against judicial arbitrariness: "transparency meetings" during which those who gather recite a single slogan, "We demand an open trial for . . ." (followed by the names of the accused), or display corresponding banners. Any other phrases or slogans, going beyond the demand for strict observance of the law, are absolutely to be considered detrimental and possibly as provocations, and should be cut short by the meeting's participants themselves. During the meeting it is essential that order be strictly observed. At the first demand by the authorities to disperse, participants must disperse, having communicated to the authorities the meeting's aim.

You are invited to a transparency meeting that will take place on December 5 of this year at 6 o'clock in the evening in the central plaza of Pushkin Square, near the monument to the poet.

Invite two more citizens, using the text of this appeal.[37]

Actually, there was no "final" version of the text. Whether in the tenth century or the twentieth, there rarely is a final version when a text's reproduction depends on multiple rounds of copying (whether by hand or by typewriter) by multiple copyists, some of whom will inevitably be tempted to "correct" or "edit" perceived infelicities, while others are simply careless. Such were the realities of samizdat as a system of communication, as captured by one of its greatest practitioners, the poet and editor Natalya Gorbanevskaya: "If you are a reader of samizdat and you can type, even with only one finger, you become a publisher of samizdat."[38] In this specific sense, then, the process of "editing" the Civic Appeal did not end in mid-November; instead, it entered a second, more diffuse phase. Thus, for instance, even in the "final" version reproduced above, the phrase "state crime" at the end of the first paragraph had previously read "state secret," a legal category that could be invoked by the government to deny public access to a criminal trial. The anonymous typist who substituted "state crime" appears to have sought preemptively to defend Sinyavsky and Daniel against the most serious potential

FIGURE 3.3. The monument to Alexander Pushkin in central Moscow, with the office building of the newspaper *Izvestiia* in the background.

charges—in contrast to Volpin's focus on observing procedural norms. Another sympathizer, Yelena Stroeva, took it upon herself to substitute the phrase "near the monument to the poet [Pushkin]" for what Volpin had written: "across the street from the office of [the newspaper] *Izvestiia*." By November the strategy behind the meeting's location had shifted: rather than target "an institution on which depends the authorization of transparency" in Sinyavsky and Daniel's trial, the idea was to attract attention from an institution that could serve as a means of publicity for the transparency meeting. But Stroeva, uninspired by proximity to the office of a Soviet newspaper, could not resist invoking instead Russia's preeminent literary icon, "our everything," as the implied patron of a meeting on behalf of persecuted writers.[39]

Whatever its textual variants, by mid-November the Civic Appeal had irrevocably entered a new, unregulated mode of existence. "Done!" Volpin exulted to his wife, having launched the text on its geometrical progression. "We've burned our bridges."[40] He could now watch and wait as one advocate of judicial transparency became two, two became four, and four became . . .

The carefully planned algorithm dissolved almost instantly. Copies of the Civic Appeal were left on a stairway at Moscow State University. An unsuspecting student brought one to his seminar on the history of the Communist Party, where it became an object of heated discussion (the instructor who permitted the discussion was later reprimanded). Within days the entire university knew about the upcoming transparency meeting, and word quickly spread to other institutions of higher education in Moscow and beyond. On December 2, SMOG member Vladimir Batshev was arrested carrying thirty copies of the "Appeal."[41] Sinyavsky's former editor at the journal *Novyi mir* (*New World*), the poet Alexander Tvardovsky, incensed at the damage done to his prestigious journal by his contributor's double life as Abram Tertz, grumbled in his diary about "the Sinyavsky brouhaha" (*Siniavshchina*).[42] According to Bukovsky, copies of the "Appeal" traveled along paths previously carved out by the dissemination of samizdat. These "channels of trust," as he called them, "turned out to have been our single greatest achievement during the preceding decade, and thanks to them, by December practically everyone in Moscow knew about the planned meeting."[43] "Everyone," of course, was an intelligentsia term of art for "the intelligentsia," and even that was an exaggeration. But word spread more quickly and widely than Volpin had ever imagined.[44]

"Everyone," it goes without saying, also included the KGB, which learned of the planned transparency meeting within days of Volpin's dissemination of the "Appeal." Dozens of individuals suspected of distributing copies were brought in for "prophylactic" conversations with KGB officers or with administrators at their schools and universities. For their defiance during such conversations, Yulia Vishnevskaya and Bukovsky were forcibly confined in psychiatric hospitals on December 2. According to the head of the Moscow city branch of the Communist Party, these and other measures "made it possible to prevent a massive turnout on Pushkin Square on Constitution Day."[45]

We will never know how many people would have participated in the meeting on December 5. Some suspected the Civic Appeal was a KGB provocation designed to flush out troublemakers.[46] It is difficult even to gauge the number who actually did take part: eyewitnesses reported as few as fifty and as many as a thousand.[47] Part of the confusion was tied to the fact that Pushkin Square was a busy spot for pedestrians, some of whom may have happened upon the meeting by accident. But even among the people who were present by design, there were three functionally distinct groups: those who came as participants (including some who held banners); those who came as spectators (who were reluctant to take part in the meeting, but did not want to miss out or abandon friends who participated); and those who were assigned to break the meeting up (including plainclothes KGB agents, Communist Youth League activists, and *druzhinniki*, or civilian anti-crime squads). By most accounts, spectators easily outnumbered participants, especially if one considers that, in their retellings of the event, more than a few of the former retroactively placed themselves among the latter.[48]

As the crowd milled around Pushkin waiting for the meeting to begin, it was difficult to tell who was who.[49] Members of the Communist Youth League, for example, could be found among participants as well as among those who came to break up the meeting. However one calculated the numbers, the mere presence of so many people in an unsanctioned public gathering in Moscow was enough to impress those who took part. One of them was Alexander Ginzburg, a twenty-nine-year-old compiler of some of the earliest samizdat literary journals, featuring poetry by Bulat Okudzhava, Joseph Brodsky, and others. At the transparency meeting, he wrote in a letter a few days later, "I realized that not just I, but hundreds of people are concerned about the fate . . . of Andrei Sinyavsky and Yuli Daniel."[50] For Volpin, who had spent the preceding days staying in friends' apartments so as to avoid possible arrest, the crowd's size and composition mattered less than the fact that, to his delight, he recognized almost no one. The meeting had already achieved one of its goals, drawing together, despite barriers of fear and mistrust, Soviet citizens who were complete strangers.[51] As other eyewitnesses noted, the crowd appeared to be composed of small groups of friends roaming about the square in search of familiar faces in other groups. Oleg Vorobev arrived with fellow members of the "Idiots," a literary club at Moscow State University. "The six of us stuck closely together," he recalled; "everyone was going around in packs," taking stock of one another.[52]

At precisely 6:00 P.M., Volpin silently unfurled a homemade banner stowed under his coat: "Respect the Constitution (the Fundamental Law) of the USSR!"[53] Nikolsky held up another inscribed with the words "We demand an open trial for Sinyavsky and Daniel!" Other participants raised duplicate banners as well as a third reading "Free Bukovsky and others confined to psychiatric hospitals in connection with this transparency meeting!" A middle-aged man held up a copy of the constitution and called out the numbers of the articles pertaining to freedom of speech and assembly.[54] Alexeyeva, watching together with her friends Natalya Sadomskaya and Ada Nikolskaya (Valery Nikolsky's wife), recounted the scene that followed:

> Twenty or so young people run past me, surrounding Alek. The banners are still up over Alek's and Valera's heads. Another wave runs past, overpowering the first. The banners disappear before I

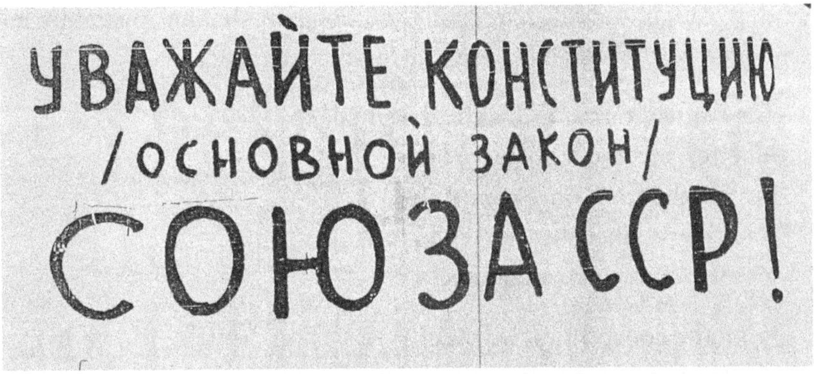

УВАЖАЙТЕ КОНСТИТУЦИЮ /ОСНОВНОЙ ЗАКОН/ СОЮЗА ССР!

FIGURE 3.4. Homemade banner for the first transparency meeting (December 5, 1965): "Respect the Constitution (the Fundamental Law) of the USSR!"

get a chance to read them. There isn't a word . . . , not a sound in front of me, not a sound behind me, nothing but crisp, tense silence. A bright flash, then another, and another. The square lights up, then instantly fills with the sound of camera shutters. Foreign reporters are photographing Alek, the banners, the demonstrators, the KGB, the spectators. The KGB is photographing Alek, the demonstrators, the banners, the spectators, the foreign reporters. In the glow of photoflashes, two men drag a third past the pedestal. He is slight; they are burly; he is in a short leather jacket; they are in heavy coats. A black Volga pulls up to the curb on Gorky Street. The man is thrown into the back seat. Another Volga pulls up. Another man is thrown in. Three minutes later, the demonstrators are gone. So are the spectators. The square is empty, except for the bronze Alexander Sergeyevich [Pushkin], arm crossed on his chest, his head bowed.[55]

One of those hustled into a waiting car was Yuri Galanskov, a veteran of the Mayakovsky Square poetry readings well known in student circles for his *Humanist Manifesto*. He had barely managed to address the crowd—contrary to Volpin's plan—with the words "Citizens of Free Russia!" before being taken away.[56] Another was Boris Sotin, a twenty-three-year-old student caught holding freshly minted doggerel that he apparently intended to read aloud, including the following verses:

> The constitution is brilliantly designed,
> A better one you'll nowhere find.
> But tell the truth once in a while
> And they're likely to put you on trial.
> It's much more practical to lie!
>
> From the constitution we glean
> The laws governing our routine,
> Granting us FREEDOM of expression
> And the right to public procession.
>
> But our people are so full of fear
> That of SOCIALIST LAW they do not wish to hear,
> Cowed by the personality cult
> And by Khrushchev's tendency to exult.
>
> Therefore our request we present:
> Apply SOCIALIST LAW without relent.
> Let IT wipe out as unfair
> The grievances that all of us bear.[57]

Altogether, twenty-eight people ranging in age from nineteen to fifty-four were taken to a nearby police station for questioning in connection with "disturbing public order" and "demagogic slogans." Among them were a dozen members of the Communist Youth League as well as seven individuals, including Volpin, whom the KGB described as "mentally ill."[58] It was police interrogators, however, who were driven crazy by Volpin's tautological responses to their questioning, a style his wife knowingly characterized as "absolutely precise and utterly devoid of information":

> INTERROGATOR: Alexander Sergeyevich, why did you come to the square?
> VOLPIN: In order to express that which I was trying to express.
> INTERROGATOR: But you were detained with a sign in your hands.

VOLPIN: I didn't detain myself; why did others detain me?
INTERROGATOR: No, but still, you were carrying a sign saying "Respect the Soviet Constitution," correct?
VOLPIN: Correct.
INTERROGATOR: Why did you make such a sign?
VOLPIN: So that people would respect the Soviet Constitution.
INTERROGATOR: What, do you think anybody doesn't respect it?
VOLPIN: That was not written [on the sign].
INTERROGATOR: Why [did you choose] this day?
VOLPIN: If I had come to the square on the First of May [Labor Day] with a sign reading "Respect the First of May," would that surprise you? Today is Soviet Constitution Day.[59]

The transparency meeting sent shock waves through Moscow and beyond. In an anonymous letter circulated in samizdat, the writer Varlam Shalamov—who had watched the meeting from a nearby side street—observed that an unsanctioned demonstration had not been seen since 1927.[60] A demonstration (sanctioned or not) demanding judicial transparency and respect for the Soviet Constitution had *never* been seen. Although the Soviet press—including *Izvestiia*, with its front-row seat—passed over the event in silence, millions of listeners to the BBC and Voice of America heard about it within a week.[61] A day after the meeting, the sinologist Vitaly Rubin, who together with his wife had watched it from a safe distance, recorded a single sentence in his diary: "Yesterday we saw what happened in Pushkin Square," an entry that left unclear whether its author declined to describe the event because he was frightened or unsure what to call it, or both.[62] According to his wife, Ina Akselrod-Rubina, "At the time, to take part struck us as insane."[63] People responded to the transparency meeting's utter novelty by assimilating it to familiar genres. Alexeyeva's husband, Nikolai Williams, overheard an expansive version in a Moscow bar: "Esenin has a son. He organized this demonstration of a thousand people to march on Gorky Street, with him marching in front of everyone with a banner; then he walked over to the KGB, threw a list of demonstrators on the table, and said, 'Here are the names of everyone who marched, but keep your hands off them. I answer for everyone.' He isn't afraid of anyone. And his name is Wolf."[64] As a vernacular rendition of the transparency meeting, this account is revealing in several respects. It translates a mathematical logician who advocates strict rule of law into the more culturally recognizable figure of a militant revolutionary with a recognizably Jewish name. It transforms a quiet, stationary gathering into a massive street march with a charismatic rebel at the fore. Creatively deploying the transparency theme, it has the folk hero defiantly deliver a list of marchers' names to the authorities. And perhaps in an attempt to account for the surprisingly mild treatment of those who were detained on December 5 and released after several hours, it invests the hero with the power to command the Soviet Union's most feared institution: "Keep your hands off them." Which, by all accounts, the KGB did.

In contrast to this nascent urban legend, Soviet authorities seem to have understood the transparency meeting as yet another display of political immaturity by the students of SMOG and their hangers-on.[65] Apart from those it had preemptively confined to psychiatric hospitals, the KGB left the job of punishing participants to the Communist Youth League, which dutifully arranged disciplinary hearings in which twelve students who had taken part in the meeting—on Volpin's side—were purged from its ranks for "apolitical actions":

> INTERROGATOR #1: Why did you come to the square?
> STUDENT: I was curious, and I believed that [the meeting] was about defending the constitution.
> INTERROGATOR #2: So were you in agreement with the contents of the leaflet [that is, the Civic Appeal] or not?
> STUDENT: Well, people occasionally write about how sometimes the Soviet criminal justice system commits violations.
> INTERROGATOR #3: You came to the square in defense of fairness. Don't we have organs that are capable of defending fairness?[66]

Remarkably, only one of the twelve (a graduate student, who evidently could no longer be trusted to instruct undergraduates) was expelled from the university. KGB chairman Semichastnyi, himself the former head of the League, diagnosed the transparency meeting as a symptom of the "low effectiveness of political-educational work, especially in institutions of higher education."[67]

A poem dedicated to the twelve purged students and circulated in samizdat suggested otherwise. For the author of "To the Descendants of the Decembrists," and presumably for more than a few of its readers, the transparency meeting exemplified precisely the heroic revolutionary tradition they had imbibed from countless Soviet textbooks:

> On that bitter, frosty day,
> It seemed freedom was near.
> You understood: you aren't students,
> You are descendants of the Decembrists.
>
> You understood that your wounds
> Will not cease to bleed,
> That Stalin will rise again
> If you do not go out to the square.
>
> Do not look to the West,
> It is for you to decide things for yourselves.
>
> Discard your books,
> It's time to look around.
> Look hard, without flinching.
> Today they take away your neighbor,
> And tomorrow you'll be the ones behind bars.
>
> . . . .
> I know they will condemn you
> And the rumors will fly.
> But Zasulich lives within you,
> You are the new Sazonovs.
>
> It is for you to remake Russia.
> It is for you to remake yourselves![68]

The Decembrists, aristocratic Russian army officers who returned from victory over Napoleon with grand plans to reform the Russian Empire, were the first to "go to the square." Not the one in Moscow subsequently named for Pushkin, but Senate Square in St. Petersburg, where their attempted coup d'état on December 14, 1825, was quickly crushed by the new tsar, Nicholas I—and with it the Decembrists' plans for Russia's first constitution. Half a century later, Vera Zasulich, a Populist who later converted to Marxism, secured herself a place in the Russian revolutionary pantheon by shooting the military governor of St. Petersburg, General Fedor Trepov, after the latter ordered the flogging of an imprisoned activist who had refused to remove his cap in Trepov's presence. Yegor Sazonov's claim to revolutionary fame, following his expulsion from Moscow University, stemmed from his assassination of Russia's minister of the interior, Vyacheslav Plehve, by tossing a bomb under his carriage as he made his way to Tsar Nicholas II's summer palace in the summer of 1904.

Semichastnyi's diagnosis notwithstanding, the revolutionary pedigree on display in "To the Descendants of the Decembrists" suggests that decades of political-educational work had been quite effective, though not entirely in the manner intended by the educators. Many of those who came to the square on December 5 were inspired less by Volpin's commitment to the rule of law than by the revolutionary romanticism they had absorbed from the Soviet curriculum, which drew heavily on the lore and allure of tsarist Russia's radical intelligentsia.[69] Others, especially the aesthetes in and around SMOG, were drawn to the transparency meeting as a form of street theater, a novel species of performance seemingly untainted by politics.

For Volpin, differences of generation and worldview among participants and spectators were not supposed to matter. The Civic Appeal intentionally invoked a different basis of solidarity, expansive but thin, familiar but unpracticed: the category of citizenship. It was a pair of Soviet "citizens," according to the "Appeal," who had been arrested; it was millions of Soviet "citizens" who had suffered from lawlessness in the past; it was "citizens" who now "have the means to struggle against judicial arbitrariness"; and it was two additional "citizens" whom each reader of the "Appeal" was asked to invite. Even the name of the "Appeal" drew on the idea of citizenship: in Russian, as in many languages, the words for "civic" (*grazhdanskoe*) and "citizen" (*grazhdanin*) are closely related. By grounding collective action in citizenship, rather than in friendship with the accused or solidarity among "creative intellectuals," Volpin was gesturing toward what the American sociologist Mark Granovetter would famously call, a few years later, "the strength of weak ties."[70] The micro-communities that constituted Soviet society, like micro-communities everywhere, were built on "strong ties," bonds of kinship and friendship.[71] In a variety of Western settings, Granovetter and others have shown that for purposes ranging from dissemination of ideas to finding a job to organizing protests, "weak ties"—those involving no more than passing or purely functional acquaintance—can play a crucial role in overcoming the parochialism of small, face-to-face communities. Volpin himself was an excellent example: his ties to Sinyavsky and Daniel were extremely weak, yet he played precisely the bridge-building role that, under the right conditions, allows multiple, distinct micro-communities to form a large-scale network. This was the logic behind his notion that the transparency meeting should lead "almost every participant" to see "many unknown faces."

In practice, the transparency meeting looked less like an assembly of citizens than a mélange of roving clusters of friends. Inside each cluster was a high degree of trust and intimacy, but between them was considerable wariness. This indeed was the predicament of Soviet society as a whole, which had only recently emerged from what the historian Geoffrey Hosking called the "catastrophic breakdown of social trust" produced by wave after wave of Stalinist witch-hunting. The Bolsheviks had never accepted the concept of loyal opposition. Under Stalin, "opponents" became "oppositionists," "oppositionists" became "deviationists," and "deviationists" became "enemies."[72] As one of Solzhenitsyn's characters remarked, Stalin "didn't trust his own brother."[73] Stalinism not only eviscerated social trust; it fostered new forms of "forced trust" based on mutual blackmail, or *kompromat*. It led, as Sinyavsky observed, to the desire to create a "magical defensive circle" around oneself and one's allies by denouncing perceived or imagined enemies.[74]

The transparency meeting was part of the process whereby social trust hesitantly emerged from the fearful shadows of Stalinism. Volpin had been searching for an ideal language that would allow him to trust his own thoughts and communicate without ambiguity. He and the other participants in the transparency meeting were attempting to enact among themselves the very thing they demanded of the Soviet state: openness. If only momentarily, they dared to connect as citizens with other citizens beyond their circles of intimate friendship—by coming to the square.

4

The Court Is in Session

> Many are simply confused: how could something like this happen at this time? Two relatively young men (both about 40), graduates of Soviet schools and then higher educational institutions, living side by side with us, who suddenly became the accomplices of our worst enemies. Why? To what end?
>
> —*IZVESTIIA*

Inside the Lubyanka prison, the two men whose arrest had spurred the transparency meeting were unaware that it had taken place. Interrogations of Andrei Sinyavsky and Yuli Daniel continued as the KGB laid the groundwork for their trial. There was, to be sure, never any doubt that they would be found guilty. In late December, Vladimir Semichastnyi and procurator-general Roman Rudenko informed the Central Committee of their plan to "convict the criminals"—they preferred this locution over "try the defendants"—under Article 70, for "writing and circulating slanderous fabrications that defame the Soviet state and social system."[1] The only question was when—and how.

For Semichastnyi and Rudenko, it was important that the two writers be convicted in a high-profile setting and that "detailed information be made available to the public" in order to "cut off analogous activities by isolated hostile figures." They intended to accomplish these aims by holding what they called an "open trial" in the Supreme Court of the Russian Republic, presided over by chief justice Lev Smirnov.[2]

Smirnov had an imposing résumé. Two decades earlier, he had assisted Rudenko as part of the Soviet prosecution team at the international military tribunals in Nuremberg and Tokyo; more recently, he had served as presiding judge in the principal trial of participants in the 1962 workers' protest in Novocherkassk, sentencing seven of them to death and an equal number to a decade or more of hard labor.[3] In 1964, he became president of the Association of Soviet Jurists. Another Rudenko protégé, Oleg Temushkin, was chosen to serve as prosecutor. Fulfilling the role of "public accusers"—ordinary citizens who, according to Soviet legal practice, represented Soviet society in the judicial process—were two carefully selected members of the Union of Soviet Writers, Arkady Vasilev and Zoia Kedrina. Semichastnyi and Rudenko proposed to fill the courtroom's roughly 150 seats with handpicked "representatives from among Party activists and the literary world." To reach a wider audience, "appropriate publications exposing the true character of Sinyavsky's and Daniel's 'literary activities'" were being prepared for placement in the Soviet press; additional materials were to be disseminated after the trial via print media and radio.[4]

Soviet citizens who learned of Sinyavsky's and Daniel's arrest, especially those with long memories, had good reason to assume that what awaited them was a classic Soviet show trial. The archetype (though hardly the first) of such events was the spectacular run of trials held in Moscow thirty years earlier, in which dozens of members of the Bolshevik pantheon, including intimate allies of Lenin, suddenly confessed to spying for Nazi Germany (or Great Britain, or Japan), plotting Stalin's assassination, sabotaging the Soviet economy, and other jaw-dropping crimes for which they implored the court to punish them without mercy. Broadcast live over Soviet radio, selectively filmed for newsreels, and reported at length in Soviet newspapers, these astounding courtroom performances introduced the term "show trial" into many of the world's languages. And no wonder: Soviet authorities themselves used the term (*pokazatel'nyi protsess*) without the slightest hint of derision. Just as the word *propaganda* in the Soviet setting referred to information meant for widespread dissemination, without implying its unreliability, so "show trial" was meant to signify not a sham procedure but a trial that "shows" or teaches something to a large audience: a morality play, in other words.[5]

The Moscow show trials of the 1930s were a resounding pedagogical success.[6] They taught huge numbers of Soviet citizens (and more than a few Communist sympathizers abroad) that spies were everywhere in their midst, that the USSR was surrounded by hostile powers bent on destroying the world's sole socialist state, and that Stalin alone could be trusted to lead the country forward. These lessons derived their force not least from spectacular confessions by the accused. Like any good teacher, chief prosecutor Andrei Vyshinsky had prepared for class well in advance, in his case by torturing defendants, threatening to ruin the lives of their spouses and children unless they agreed to confess to the surreal charges against them, and preparing interrogation transcripts before the interrogations themselves had occurred.[7]

And therein lay the problem for Sinyavsky's and Daniel's prosecutors. They too wanted a trial that would not just convict but instruct; they too sought to deploy Soviet mass media to deliver the desired lesson to as many pupils as possible. But it was by no means clear how to achieve these effects under the new, post-Stalin dispensation in which Stalin's heirs—those who literally survived him—had decided, if only for purposes of self-preservation, that torture and terror should no longer be regarded as legitimate tools of judicial investigation. During the final years of Stalin's rule, only ten to 20 percent of "political" trials—those involving charges of counter-revolutionary activity—had been conducted in regular courts. The rest were held in "special courts," including military tribunals designed to expedite convictions. After 1956, the proportions flipped; henceforth the vast majority of "political" trials would take place in regular courts.[8] Convictions were now supposed to be based strictly on verifiable evidence. The new dispensation also featured a Soviet audience substantially better educated than its parents. Thanks in no small measure to the Communist Party's efforts, tens of millions of illiterate, impoverished Soviet citizens had learned to read, moved to cities, and, in some cases, enrolled in institutions of higher education.

Semichastnyi and Rudenko, in sum, were attempting something of which neither they nor their generation had any experience: to mount a demonstrative trial in which the government retained only a limited ability to script the proceedings and control their reception.

In January 1966, the Central Committee of the Communist Party together with the KGB formed a coordinating group to manage all aspects of the trial and its presentation via the mass media to the Soviet public and the outside world. Among the group's members were Alexander Yakovlev, deputy director of the Department of Agitation and Propaganda and future architect of Mikhail Gorbachev's reforms, as well as Filipp Bobkov, deputy director of the KGB's counter-intelligence department who would soon be promoted to lead a new department devoted to suppressing internal dissent.[9] Having silently endured months of broadcasting on the affair by shortwave radio stations abroad, the Soviet press finally received permission to launch a coordinated assault on the two writers, starting on January 12 with an article in *Izvestiia* under the title "The Turncoats"—the first of more than a dozen published pre-trial attacks.[10] For Alexander Tvardovsky, Sinyavsky's former editor at the journal *Novyi mir*, this was a disastrous development. "It's strange to imagine how these people in the Central Committee," he confided to his diary in mid-January, "could fail to understand that, once arrested (and in all likelihood convicted), Sinyavsky—who deserves contempt and social ostracism—will only come out ahead, and not only in the opinion of 'the West'":

> With this noisy campaign, this "mobilization of public opinion," we will obtain results diametrically opposed to the calculations of the political heads who are organizing the affair. That is, first of all, we will demonstrate before our foreign opponents our vulnerability to "libelous statements," our anxiety that they threaten to "shake our foundations." Second, we will generate painful misunderstanding and confusion among our friends. Finally, here at home we will intensify the already heightened sense of alarm in people's minds—people who correctly grasp that the basic contours of this entire campaign match those of an all too memorable time. Strength and wisdom

should have been manifested in our concluding that there is no basis for repression, that punishment should have taken the form of supreme contempt toward [Sinyavsky] and his customers.[11]

Tvardovsky's remarkable prognosis was being confirmed even as the ink was drying on the pages of his diary. Whereas during the previous fall Mark Azbel and others had failed to attract support among the intelligentsia for a petition on behalf of Sinyavsky and Daniel, the vicious campaign in the Soviet press had the effect of rallying support even among those who found the pair's actions distasteful. "Brave guys," announced one Muscovite after reading *Izvestiia*'s account. "Too bad they'll shoot them."[12] By late January, a petition warning of a return to Stalinism had garnered the signatures of Kornei Chukovsky, Boris Slutsky, Viktor Nekrasov, Valentin Kataev, Konstantin Paustovsky, and other prominent writers—as noted in the diary of the literary scholar Raisa Orlova, who also meticulously recorded the names of those who had refused to sign, among them the writers Konstantin Simonov, Yevgeny Yevtushenko, Margarita Aliger, and Vasily Aksenov.[13] "Everyone," wrote Chukovsky in his diary, "is talking about the Sinyavsky affair."[14]

The joint KGB–Central Committee coordinating group determined that the trial would begin on February 10 and last three days. In order to control access to the courtroom in the face of swelling public interest, a novel system of numbered passes was introduced, with different colors for each session (morning and evening) to guard against counterfeits. "There were tickets to a trial!!!" marveled Yevtushenko, who used his Party connections to get one.[15] "How to get a ticket to the trial??" wondered the writer Chingiz Guseinov in his diary on February 9—"that's the problem of the day."[16] Friends of the defendants who tried and failed to secure tickets protested their exclusion from what they called "an open trial behind closed doors." "What do you want?" retorted one exasperated official. "That we should hold the trial in a stadium? Like in China?"[17] The coordinating group granted access to journalists from *Izvestiia* and half a dozen other prominent Soviet newspapers. All others were instructed to reproduce the official reports from TASS, the Soviet telegraphic agency. No cameras were to be allowed inside the courtroom. Soviet radio stations were similarly ordered to rely exclusively on TASS and articles appearing in the half dozen officially chosen newspapers. The Novosti Press Agency, together with the KGB, was responsible for preparing articles on the trial for publication abroad. In response to the crescendo of international protests against the arrests of Sinyavsky and Daniel, foreign journalists—even those from socialist countries—were barred from the courtroom.[18] Soviet citizens old enough to remember noted the contrast to the show trials of the 1930s, which a supremely confident Vyshinsky had opened to journalists from around the world.[19]

In addition to handpicking the judge, the prosecutor, the audience, the reporters who would convey the trial to the broader court of public opinion, and, in some sense, the defendants themselves, the Soviet government exercised de facto veto power over the choice of defense attorneys. Daniel's wife initially secured the services of Konstantin Simes, a jurist at the Institute for State and Law, while Sinyavsky's wife hired Vasily Samsonov, chairman of the Moscow Bar Association. Less than a month before the trial began, however, Simes was denied the necessary security clearance to work on a case involving politically sensitive materials, while Samsonov was pressured by the Moscow Party Committee to withdraw. Simes's wife, Dina Kaminskaya, a Moscow attorney who did have a security clearance, was also denied permission to defend Daniel—without explanation. In the end, both defendants were left with little choice but to accept a pair of lawyers whose goal, following an unspoken rule in Soviet political trials, was to minimize the sentences rather than to establish their clients' innocence.[20]

Notwithstanding these and many other tactical advantages over the defendants, the government's case lacked the key ingredient of the show trial, upon which Stalinist justice had become thoroughly dependent: confession by the accused, the "queen of proofs."[21] Was the prosecution hoping that the pressure of rigorous public interrogation would somehow cause Sinyavsky and Daniel to break down? Or did it assume that in this case confessions were unnecessary, because the pseudonymous texts by Tertz and Arzhak, not to mention their publication by the USSR's Cold War enemies, were already sufficiently damning? Despite

FIGURE 4.1. Yuli Daniel and Andrei Sinyavsky on trial. The sole woman pictured in the courtroom audience, with glasses, looking downward, is Maria Rozanova, Sinyavsky's wife.

their high standing, neither the prosecutor nor the judge appeared certain as to what the government's strategy should be. The legal structure of the case was complex. For many Soviet observers, the most egregious aspect of Sinyavsky's and Daniel's behavior was not so much the content of their work (with which relatively few were directly acquainted) as their decision to have it published under pseudonyms in the capitalist West, to lead Janus-faced lives inside the Soviet Union, and thus to flagrantly shirk writers' moral duty. The only problem was that, formally speaking, none of this was illegal under Soviet law. It was not even clear whether, from a strictly legal standpoint, works by Soviet authors published outside the USSR came under the purview of Soviet censorship (which itself operated outside any legal framework).

As Sinyavsky and Daniel were led into the courtroom at the opening of their trial on February 10, they got their first glimpse of the government's carefully assembled public. Except for their wives, they saw an audience consisting almost entirely of men, some of whom Sinyavsky undoubtedly recognized as fellow members of the Union of Soviet Writers. One member of the audience was surprised at the appearance of the two defendants, who showed "no signs of having been under investigation," that is, beaten or tortured while held in the Lubyanka. After requesting that everyone rise, the bailiff pronounced the familiar phrase that suddenly took on new meaning: "The court is in session!" Chief Justice Smirnov, according to one eye-witness, entered with "the purposeful look of a man accustomed to being in command."[22]

Except that he wasn't. The fifty-eight-year-old Smirnov was a seasoned jurist, but most of his experience came from an era when defendants charged with anti-Soviet activities were tried behind closed doors. It cannot have been easy for him, or for Temushkin, the prosecutor, to adjust to the new post-Stalin arrangements. Even if guilty verdicts for Sinyavsky and Daniel were a foregone conclusion, the unscripted nature of their trial, along with the highly literary quality of the principal evidence, lent a certain unpredictability to the proceedings. To be sure, much of what transpired during the trial could have been predicted on the basis of their interrogations the previous fall and winter. Over the course of the five months leading up to the trial, the two writers had repeatedly pointed out to their interrogators that in works of fiction one cannot automatically assume that ideas expressed by a character, or for that matter by the narrator (who may also be a character), represent the views of the author. But all of this was somehow lost on Smirnov and Temushkin, who despite having reviewed (and occasionally citing) the interrogation transcripts, fell into the same trap, as in the following exchange over Daniel's *This Is Moscow Speaking*:

> PROSECUTOR: I would like to ask Daniel to read the epigraph to Chapter 4 of the story.
> JUDGE: I don't see any need for uncensored swearing to be heard in this courtroom.
> PROSECUTOR: I would nonetheless like permission to read the epigraph, with cuts—leaving out the obscenity.
> JUDGE: Go ahead, but without the obscenity.
> PROSECUTOR: (*Reading.*) "I hate them so much I have spasms, screams get caught in my throat, I tremble. Oh, if only all these [*"whores"—omitted*] could be gathered and destroyed at once!" Daniel, how do you explain this epigraph?

DANIEL: It's an epigraph to the protagonist's thoughts. (*Laughter in the audience, at whom Daniel glances nervously.*)
PROSECUTOR: Whom do you hate? Whom do you want to destroy?
DANIEL: To whom are you speaking? To me or to my protagonist, or to someone else?
PROSECUTOR: Who is your positive hero? Who in the story expresses your point of view?
DANIEL: During our preliminary conversation I already told you that in my story there is no entirely positive hero. Nor does there have to be one.[23]

During his interrogations, Sinyavsky (unlike Daniel) had insisted that sending literary texts abroad did not violate Soviet law. When pressed in court by Temushkin to explain why he sent his manuscripts "illegally," Sinyavsky calmly reiterated that he had sent them "unofficially, not illegally"—an assertion Temushkin chose not to dispute.[24] Both writers, moreover, refused to concede that their works were anti-Soviet or that enthusiastic claims to this effect by Western reviewers proved that they were. "I have the strong impression," Sinyavsky declared to the court, "that bourgeois propaganda indulges in wishful thinking. The epithet 'anti-Soviet' is often used in the West for sensational purposes."[25] When Temushkin read aloud an aggressively anti-communist passage from émigré Boris Filippov's introduction to the American edition of *This Is Moscow Speaking* and then asked Daniel to comment, the latter replied, "I suggest you ask Filippov for an explanation. I am not responsible for what he writes."[26] Both defendants cited examples in which Western reviewers discussed their stories without reading anti-Soviet purposes into them, as well as instances in which officially approved Soviet fiction had been used by Western commentators to criticize Soviet society.[27] The diversity of Western readings of their work buttressed Sinyavsky's claim that literary texts defy definitive interpretation and that literary language by its very nature contained multiple meanings.

It did not take much effort to show that Sinyavsky's and Daniel's work departed from the dictates of socialist realism. Sinyavsky's literary-critical essay was quite explicit on that score. But the court seemed to have trouble digesting the idea that their work intentionally moved beyond the bounds of realism per se, socialist or otherwise:

> JUDGE: I want to clarify something. Imagine a communal apartment where Ivanova is arguing with Sidorova. If Ivanova were to write that a certain lady is ruining another lady's life, that would be an allusion, an allegory. But if she were to write that Sidorova was pouring garbage into her soup, then that would be grounds for legal investigation, as a kind of denunciation, a form of slander or something like that. You wrote directly about the Soviet government, not about ancient Babylon but about a specific government, claiming that it announced a "Public Murder Day." You name the date: August 10, 1961. Is that a device or outright slander?
>
> DANIEL: I will take your example. If Ivanova were to write that Sidorova literally flies around on a broomstick or turns herself into an animal, that would be a literary device, not slander. I chose a deliberately fantastic situation.
>
> JUDGE: Here is what Filippov wrote: "Is what Arzhak portrays in fact so unreal?" So it turns out, Daniel, that it's not a literary device after all, right?
>
> DANIEL: It is a literary device.
>
> PROSECUTOR: Daniel, are you really denying that "Public Murder Day," supposedly announced by the Soviet government, is in fact slander?
>
> DANIEL: I consider slander to be something that—at least in theory—you can make other people believe.
>
> JUDGE: I want to clarify this (*reads from the Criminal Code*): "Slander is the dissemination of deliberately false and defamatory information." That's the letter of the law.
>
> DANIEL: How then should we treat the fantastic?[28]

It was a remarkable scene: in Daniel's view, in order for statements to qualify as slander they had to be credible, whereas readers of his work would intuitively recognize it as not only fictional but deliberately implausible—at least as regards the post-Stalin era. Smirnov, by contrast, using Filippov as his foil, insisted that Daniel's work qualified as slander precisely because it *was* credible—that is, sufficiently close to Soviet reality. Reading Daniel's and Sinyavsky's work through a socialist-realist lens, Smirnov effectively articulated a position more damning of Soviet reality than that of the defendants.

From the court's point of view, utterances in works of fiction had to be either true or false. After Sinyavsky reminded the prosecutor, not for the first time, of the difference between author and protagonist, Smirnov steered the examination to *The Court Is in Session*:

> JUDGE: Let's talk about what comes from the author. "The filter under the sewage pipe" [*to catch documents flushed down toilets*]—what's that? The scene in the camp—isn't that the author [speaking]?
>
> SINYAVSKY: I'll explain the scene in the camp. The real historical events in the story are strictly confined to the period from the end of 1952 to the beginning of 1953—from the Doctors' Plot to Stalin's death. But a number of scenes only appear to refer to reality. This is an artistic work, not a political document. (*Laughter in the courtroom.*) Please allow me to speak. I portray a character who is a maniac . . .
>
> JUDGE: (*Interrupting.*) I want to clarify another aspect of your work that interests us. You write about mind readers, about filters under sewage pipes. Doesn't that mean that someone made the decision to install such devices? That's Article 70. Doesn't that contain a slander, [as per] Article 70? Doesn't that defame our society, our people, our system?
>
> SINYAVSKY: No, these characters are portrayed during a particular period, on the eve of Stalin's death. These characters are detectives. It's the era of the Doctors' Plot, with its atmosphere of arrests and suspicion. The epilogue is written in the voice of the storyteller, dated 1956, the year the story ends. The "I" in the story is conventional, it's neither Sinyavsky nor Tertz, it's the storyteller. The storyteller lives half in fear, half in exaltation. His

character reaches its logical end-point in Kolyma [*a corrective labor camp in the far north*]. This isn't reality but rather what haunts the storyteller in his moments of fear. This isn't a representation of the historical reality of 1956. (*Laughter in the courtroom.*) It's a literary device—it assumes an unreal situation.

JUDGE: But the abortion is real. Why did he do that?

SINYAVSKY: It is described but not shown. In works of fiction there is the concept of convention.

JUDGE: We'll talk about conventions later. You depicted the Russian people as drunkards.[29]

"The abortion is real": by this point the boundary separating fact from fiction had all but disappeared, as fictional characters rather than Sinyavsky and Daniel appeared to be under cross-examination. "In this court," Sinyavsky noted in his closing statement, "the literary image suddenly and horrifyingly loses its conventional [or conditional] character and is being perceived literally. As a result, the judicial proceedings merge into the literary text, as its natural sequel."[30] Having taken on the phantasmagoric qualities of Sinyavsky's *The Court Is in Session*, what started out as an unscripted trial appeared more and more to have a script after all, as if Smirnov and Temushkin were manifestations of life imitating a very Sinyavskian kind of art.

Sinyavsky was not alone in this impression. "Is it a dream or reality?" wondered Guseinov, who had managed at the last minute to secure passes to two of the trial's sessions.[31] He wasn't sure. The prosecutor's technique of repeating the same decontextualized quotations over and over had itself become a literary device, designed to erase the figurative nature of literary language, or, as Sinyavsky put it, "to make the metaphor real."[32] Delivering his final statement, according to Guseinov, "as if for history," Sinyavsky sought to place literature beyond the reach of the law: "In the prosecution's view, literature is a form of agitation and propaganda, and agitation is either Soviet or anti-Soviet. If it's not Soviet, then it's anti-Soviet. I do not agree with this. . . . In the depths of my soul I feel that one cannot approach literature with juridical formulas."[33]

FIGURE 4.2. Chingiz Guseinov's pass for the trial of Sinyavsky and Daniel, February 12, 1966

For the first time in Soviet history, according to Varlam Shalamov, defendants in a major political trial pleaded not guilty, refusing to deliver the "accursed confessions" that had become a staple of Soviet show trials. They rejected the ritualized apologies of the Stalin era and refused to turn against each other. "Sinyavsky and Daniel took up the battle after nearly fifty years of silence" and did so with cool deliberation. "Nothing would have been simpler than to close with a political speech," Shalamov noted, "recounting how allegedly since childhood they had despised [Soviet power] and were ready to die, etc." Instead, they calmly and firmly defended freedom of conscience, freedom of creativity, freedom of the personality, "inscribing their names in gold letters for all time."³⁴

On February 14, 1966—the tenth anniversary of the opening of the Twentieth Party Congress, where Khrushchev had performed his startling dethronement of Stalin—Sinyavsky and Daniel were found guilty under Article 70. "History loves to arrange such awkward coincidences,"

Tvardovsky wrote in his diary. "Ten years have passed, and so another era in our life recedes into the past, as if it had never been."[35] Sinyavsky received the maximum sentence of seven years in a corrective labor camp; Daniel, as a wounded veteran of the Great Patriotic War, and perhaps because he expressed regret for having unintentionally damaged the USSR's reputation abroad, received five years. The carefully selected audience thundered its approval. "Have you ever heard applause for a sentence passed down on you?" Sinyavsky wrote years later. "You will rejoice that there are people ... lower than you, if they dare celebrate so openly, with such pure human hearts, someone else's misfortunes."[36]

Outside the courtroom, however, reactions were noticeably more diverse. A dazed Guseinov recorded the spectrum of responses among his acquaintances: from "shoot the bastards" and "punish them severely" to "punish them lightly," "let the court of public opinion decide," "no trial," and "declare them innocent."[37] His bewilderment, captured in staccato diary entries in the days following the trial, echoed the broader cacophony of opinion:

> Thoughts and thoughts. Should there have been a criminal trial? Better a trial by public opinion, a moral trial. But without a court, how to deter others? Layer upon layer of hypocrisy, including mine. Heroism: To be silent? To vent on paper? To be pure before history and one's descendants? Yes, cast blame, debunk. But do it here, among us. The way [Sinyavsky and Daniel] did it was harmful. It intensified persecution, prohibitions; it messed things up.
>
> I lay out my views. In my diary. For myself. But I dissemble too: I consult my internal censor as I write. I want both to justify myself before my descendants and to secure myself in the present. I want to be a member of the Party out of conviction. The ideological struggle should not lead to physical extermination, to criminal prosecution.
>
> Last night I listened to the Voice of America. Letters to [Soviet premier Alexei] Kosygin from leaders of the Communist Parties of Sweden and, apparently, Denmark: "One should fight ideas with ideas, not with political measures." And: "The arrests [of Sinyavsky

and Daniel] did more for anti-Soviet propaganda than their works themselves."[38]

Opinions were shifting with extraordinary speed. "One has to write about what is happening every day now," Raisa Orlova confided to her diary. "About how people listened to [foreign] radio broadcasts, how they waited for the verdict, how all of us and everything we've written and will write—was condemned." The intelligentsia, she wrote, were in the position of people who have received a telegram: "Start worrying—details to follow."[39]

In the days and weeks following the verdict, KGB chairman Semichastnyi followed the details, reporting regularly to the Central Committee on the Soviet population's response to the trial. The roughly one thousand spectators who had sat in the courtroom during various phases of the trial, he boasted, had "unanimously approved the court's action." In an effort to guide the broader public response, the KGB organized a series of meetings at the Gorky Institute of World Literature (Sinyavsky's former place of employment), the Central House of Writers, and various universities, research institutes, and other institutions. These gatherings, Semichastnyi assured the Central Committee, had made it possible to neutralize various "rumors" stirred up by the Western press and radio.[40] He did not mention that at the Gorky Institute meeting, Svetlana Alliluyeva, Stalin's daughter and an employee of the Institute, had asked, "Why was there a trial at all? Why didn't they let people read [Sinyavsky's and Daniel's works], think, and come to their own understanding?"[41] Nor did he mention that, at a meeting on March 2 with Chief Justice Smirnov at the Central House of Writers, nearly all of the roughly fifty questions submitted in writing by audience members expressed support for Sinyavsky and Daniel or that Smirnov was interrupted several times by cries of "unconvincing!" from Tatyana Litvinova, the daughter of Maxim Litvinov, Stalin's commissar of foreign affairs in the 1930s.[42]

While their pedigrees enabled them to speak out in an unusually public manner, it wasn't only the children of famous Soviet leaders who argued that the entire trial had been a mistake. In a private letter to the

Central Committee, the celebrated author and six-time Stalin Prize recipient Konstantin Simonov noted that the government's handling of the case had mixed "old and new methods of education" (by which he presumably meant coercion and persuasion) and that it would have been better to let the "community of writers" judge the two offenders.[43] A group of sixty-three prominent writers and poets, led by Ilya Ehrenburg, Kornei Chukovsky, Viktor Shklovsky, Pavel Antokolsky, and Venyamin Kaverin, went even further, requesting in an open petition to the Russian Supreme Court that Sinyavsky and Daniel be placed on probation under the supervision of the petitioners, so that the convicted writers might learn from their mistakes.[44] In a letter to the Union of Soviet Writers, Tvardovsky described himself as part of the "majority" of Soviet writers who thought that Sinyavsky and Daniel "should have been condemned and subject to ostracism in a societal rather than criminal framework." Perhaps, he added, it was not too late to do so: "Instead of criminal punishment, there could be a more terrible but more just punishment, namely, stripping them of their Soviet citizenship and expelling them from the USSR." This would instantly deprive them of the aura of being "martyrs" or "prisoners of socialism" and would "cause the value of their wares to plummet in the anti-Soviet market."[45]

These and other proposals all had one thing in common: they sought to extract Sinyavsky and Daniel from the Soviet legal system by offering various extra-judicial alternatives. Some of those alternatives, such as stripping the two writers of their Soviet citizenship without trial and expelling them from the USSR, were inconsistent with Soviet law, though not with Soviet practice—the writer Valery Tarsis had been subjected to precisely this treatment just days before Sinyavsky and Daniel's trial.[46] But they were entirely consistent with a central idea of the Khrushchev era, namely, transferring more and more of the state's disciplinary functions to various lay collectives within Soviet society. Calls for a "societal" rather than judicial response to Sinyavsky's and Daniel's transgressions paralleled the revival of "comrades' courts," panels of ordinary citizens delegated by housing and work collectives to formally admonish their wayward peers.[47] Now that the Soviet population had officially been purged of class enemies—so the thinking went—it could be entrusted with ever greater levels of self-administration as part of the historic transition from socialism to communism, if not (yet) to the withering away of the state prophesied by Marx and Engels.

Semichastnyi had little patience for critics of the trial, dismissing Ehrenburg, Tvardovsky, Simonov, and others as concerned above all with not appearing "old-fashioned" or "orthodox" to their Western colleagues for fear of jeopardizing their chances of being invited abroad and offered favorable publishing contracts. "Deceived by bourgeois propaganda" and the illusion of "apoliticism," he concluded, they had failed to appreciate "the intensity of the battle on the ideological front."[48]

What verdict might the court of Soviet public opinion have delivered on Sinyavsky and Daniel? One can only speculate. To be sure, the idea of the public as a source of moral authority was hardly original to the Soviet Union or to the Russian intelligentsia.[49] But in the Soviet context, the notion that public opinion could be a source of authority distinct from that of the Communist Party, despite the latter's abiding claim to constitute "the mind, honor, and conscience of our era" (as Lenin's ubiquitous slogan put it), *was* new. So new, in fact, as to lack procedures for arriving at, let alone enforcing, public opinion's judgments. None of its proponents seems to have regarded the government's carefully selected "public accusers" at Sinyavsky and Daniel's trial as genuine representatives of public opinion, but neither were they prepared to suggest an alternative mechanism by which "the public" could render its verdict.

The Soviet government, unsurprisingly, preferred the more formal legal framework of Smirnov's court. As we have seen, it was also keenly interested in managing the trial's public reception. In his analysis of some four hundred letters sent to the Procuracy, the Russian Supreme Court, and several major newspapers, Semichastnyi confidently concluded that "the absolute majority of Soviet people" approved the punishment meted out to the two writers. In truth, neither the KGB nor anyone else possessed reliable information about the contours of public opinion—on this as on many other vital issues. Even at the local level, for example, in the Lithuanian Soviet Socialist Republic—where one month after the trial the KGB discovered graffiti in Vilnius declaring "Free Sinyavsky"—reports on public opinion rarely rose above the anecdotal.[50] Historians are hardly better equipped in this regard. Nonetheless, several features stand out among the admittedly spotty available data. First, letters by concerned citizens came from cities and towns across the entire Soviet Union, suggesting the affair's wide resonance. Whatever their position regarding the trial, moreover, a high proportion of letter writers were aware of protests on behalf of Sinyavsky and Daniel by left-leaning foreign luminaries such as Jean-Paul Sartre and Heinrich Böll as well as by representatives of communist parties in Western Europe, suggesting substantial access to non-Soviet sources of news. Indeed, many Soviet citizens had the impression, as Shalamov put it, that "the attention of the entire world is riveted" by the trial.[51] Finally, despite Semichastnyi's blithe claim that most letter writers endorsed the punishment meted out to Sinyavsky and Daniel by the Russian Supreme Court, the majority appear to have either questioned the wisdom of the trial itself or called for significantly *harsher* punishment on the basis that the actual crime was treason—a far graver offense than anti-Soviet slander. The consensus claimed by Semichastnyi was thus a fiction.[52] A collective letter from workers at a factory in Sverdlovsk, for instance, enthusiastically confirmed the West's "fuss" about a "purge" of the Soviet literary world: "Yes, gentlemen, our literary circles have purged themselves of two good-for-nothing double-dealers.... There is no place for them in our land." "It's embarrassing," wrote members of the Communist Youth League at a school in the western Urals, "to even call them human beings, citizens of our Motherland." A letter to the Procuracy demanded that the "anti-Soviet fleabags" be shot, in order that they no longer be able "to live on our holy soil." A certain Semenov, an artist from Irkutsk, surmised that the authorities, "taking into account the international climate," had refrained from imposing the full and appropriate measure of punishment. But Semenov wasn't worried: "In the camps, they'll top it off."[53]

Should we regard such letters as less genuine than those from Sinyavsky and Daniel's defenders? It could be argued that the use of stock phrases and doctrinal clichés betrays a kind of mimicry, a lack of authenticity, and that the predominance of collective letters among messages from those hostile to the two writers reduces their utility as barometers of genuine opinion. But unoriginal opinions expressed in stock phrases can be just as heartfelt as exceptional ideas conveyed in a distinctive voice; if anything, received wisdom tends to be the most tenacious kind. Standard phrases, of a different register to be sure, are also sprinkled across letters sent by Sinyavsky and Daniel's supporters. And whatever the significance of group letters, they too were found on both sides of the divide.

Taken at face value, then, the far from perfect evidence points to multiple fissures in public opinion. Among both supporters and detractors of the convicted writers, however, there was a relatively widespread preference for extra-legal frameworks for thinking about and responding to what Sinyavsky and Daniel had done. This preference, one might note, was shared by Daniel and especially by Sinyavsky, whose central argument revolved around the idea that literature and literary language operate outside law's jurisdiction. For their supporters, the arrests and harsh sentences suggested an ominous return to the Stalin era, a sign that the Party was preparing to reverse Khrushchev's campaign against the "remnants of the cult of personality." For Sinyavsky and Daniel's opponents, notably the writer Mikhail Sholokhov, freshly crowned Nobel laureate for his epic *Quiet Flows the Don*, the prospect of returning to the values of an earlier era triggered an undisguised nostalgia. "If these lackeys with black consciences had landed in the 1920s," Sholokhov announced to delegates at the Twenty-Third Party Congress in April 1966, "when people were tried not in accordance with strictly defined articles of the Criminal Code, but 'in accordance with the spirit of revolutionary justice,' then oh, these turncoats would have gotten a quite different punishment."[54] He was right. "In our day," recalled a survivor of Stalin's Gulag, entirely sympathetic to the convicted writers, "we called these 'kid sentences.'"[55]

Once again Aleksandr Volpin found himself an outlier. He shared the sense that the trial's significance could be measured via its relation to earlier periods of Soviet history, but that relation, in his view, was cause for neither alarm nor nostalgia. In a statement dated March 4 and circulated in samizdat, he noted that, for the first time, legal proceedings against Soviet citizens accused under Article 70 were public, and "the essential laws regarding transparency were fully observed." Even though the courtroom audience was stacked with Party loyalists, he wrote, "formally speaking, everything was done by the book. In questions of law I assign the highest significance to questions of formality." With all its flaws, "this trial is vastly superior to analogous trials in previous years. It used to be that writers and other thinking people would simply disappear for long years or forever; the world would learn no more about them than about the fate of a rabbit torn to pieces in the forest by a wolf."[56]

The trial's managed transparency had supplied Volpin with enough information to conclude that the prosecution had failed to demonstrate guilt as defined by Article 70. Specifically, the court had failed to show that statements made in works by Sinyavsky and Daniel were either "slanderous" (that is, knowingly untrue) or "anti-Soviet" in the sense of being opposed to the system of norms and laws outlined in the Soviet Constitution. Unlike Sinyavsky and Daniel, Volpin was less interested in the workings of literary language (which he distrusted) than in the language of the law. Literary conventions and devices, genres of the grotesque and fantastic were for him not so much privileged forms of verbal artistry beyond law's reach as simply irrelevant to the black-letter legal definition of anti-Soviet slander—as irrelevant, say, as whether the defendants were writers or veterinarians or tailors. The prosecution, he noted, had frequently described the two writers' works as "blasphemous," a term that appeared nowhere in the Criminal Code and that belonged "to the jurisprudence of the Inquisition, not the Soviet Union."[57] Unlike the prosecution, which insisted on a literal reading of literary texts for the purpose of tarring authors with their own fictions—as if metaphors could magically be made real—Volpin sought a literal, disenchanted reading of Soviet law itself.

Stalin's daughter, among others, had called for the Soviet public to be allowed to read Sinyavsky's and Daniel's works and to judge their content for themselves. Volpin, consistent with the demands of the *glasnost* meeting and his unrelenting focus on procedure, urged that the public be given access to a complete protocol of the *trial*. It is not entirely clear what he had in mind, since Soviet courts did not normally create stenographic records of trial proceedings, much less make them public. Nor is it clear what precisely Semichastnyi and Rudenko had intended when, in a memorandum to the Central Committee in December 1965, they had instructed that "after the trial the appropriate documents be given to the press and radio."[58] That, of course, was before the underwhelming performances by Smirnov and Temushkin, the eloquent defense of artistic freedom by Sinyavsky and Daniel, and their remarkable refusal to plead guilty.

As Volpin may well have known but chose not to reveal in his March appraisal, a complete transcript of the trial was already being prepared for publication—but not by the government and not through official media. Instead, Larisa Bogoraz along with Mark Azbel, Alexander Voronel, and other friends of the defendants had decided to take matters into their own hands. This time, in contrast to the show trials of the 1930s, the transcript would be created after, rather than before, the trial.[59] As Daniel's wife, Bogoraz had been allowed to attend all the trial's sessions, and each time she brought one of the dozen blank notebooks that Azbel and Voronel had purchased for her. Trained in linguistics, she captured in shorthand—"like a machine," as Voronel recalled—not only the words of the judge, prosecutor, witnesses, and defendants, but the grumbling and laughter of the audience. Between sessions, she would meet Azbel or Voronel in the women's bathroom—usually empty owing to the dearth of women in the carefully selected audience—to trade a filled-up notebook for a blank one. At the end of each day, the small band of conspirators would retire to the Voronels' apartment, where Bogoraz would spend hours relating to her friends what had happened. During these retellings she did not use or even mention the notebooks, since it was assumed that the apartment had been bugged. Indeed, part of the purpose of the retellings was to distract the KGB from the fact that the notebooks had been secreted away to a different address, where Nina Voronel and Leonid Nevler, friends of Bogoraz and Daniel, would spend the night decompressing Bogoraz's shorthand into a fully legible transcript.⁶⁰ From this round-the-clock operation—stenography by day, decoding by night—there would soon emerge a complete typed account of the proceedings.⁶¹

Bogoraz's was hardly the first attempt to create an unofficial record of a Soviet trial. In 1956, shortly after Khrushchev proclaimed a new era of "socialist legality," law professor R. D. Rakhunov visited a Moscow courtroom to observe judicial practices. When the presiding judge noticed Rakhunov taking notes, the following exchange ensued:

> JUDGE: Citizen! Did you come here to listen to the case or to work?
> RAKHUNOV: I came to listen to the case and to take notes.
> JUDGE: No. Taking notes is not permitted.
> RAKHUNOV: On what do you base that opinion?
> JUDGE: It is not permitted. (*Rakhunov continues to write.*) Citizen, I am addressing you: writing is not permitted. If you want to write, leave—there's a special room for writing.⁶²

In 1964, the journalist Frida Vigdorova, who was also a deputy of a Moscow district *soviet* (workers' council), quietly took notes at the trial of the twenty-three-year-old Leningrad poet Joseph Brodsky, who was charged with parasitism (that is, lack of employment). They contained the following immortal lines:

> JUDGE: What is your profession?
> BRODSKY: Poet. Poet and translator.
> JUDGE: And who told you that you were a poet? Who assigned you that rank?
> BRODSKY: No one. Who assigned me to the human race?⁶³

Within months of Brodsky's conviction, a furious Leonid Ilichev, chair of the Ideological Commission of the Central Committee, noted that Vigdorova had been distributing copies of her "so-called 'stenogram' of the trial" with the goal of "making those who spoke against Brodsky . . . look stupid, and to present him as the innocent victim of dim-witted bureaucrats."⁶⁴ He was right. Vigdorova's transcript gained Brodsky sympathetic attention from thousands of people inside and outside the

Soviet Union, far more than had ever read his poetry. The poet Anna Akhmatova, Brodsky's mentor, presciently observed of Soviet authorities, "What a biography they're fashioning for our red-haired friend! As if he'd gone out and hired someone to do it."[65]

It was a sign of how powerfully the sense of history-in-the-making hovered over the trial of Sinyavsky and Daniel that at least three other members of the audience, unbeknownst to each other, were also covertly transcribing the proceedings. The art historian Igor Golomshtok, called as a witness in the trial, took notes on the testimony of other witnesses; the Moscow writer Igor Vinogradov shared his notes with Tvardovsky within days of the trial's conclusion; Boris Vakhtin, a Leningrad sinologist and fiction writer, began circulating copies of his notes in May.[66] But none of them had been permitted to attend the entire trial; that distinction, and the claim to a complete transcript of the proceedings, belonged to Bogoraz.

Or so it seemed. We now know that the Russian Supreme Court, presumably at the behest of the KGB, created its own transcript, possibly via a recording device wired to microphones, visible in extant photographs of Sinyavsky and Daniel in the defendants' dock (see figure 4.1). It seems likely that excerpts from the official transcript were meant to be included among the "appropriate documents" that the KGB had planned to make public via the Soviet press and radio. Indeed, in the months following the trial there were rumors that the state publishing house "Political Literature" was planning to release a "Report on the Case of Sinyavsky and Daniel," featuring speeches and documents from the prosecution's side.[67] Whatever its intended purposes, the now accessible official transcript confirms the accuracy of Bogoraz's text and in fact reads like a lightly sanitized version of it.[68]

Even with careful editing, the official version of the trial would not have fared well against Bogoraz's. Nor, in the end, was it allowed to: apart from the final verdict, the KGB–Central Committee coordination group did not make public a single excerpt of the trial. Rather, it was the unofficial transcript that, within weeks of the sentencing, began to circulate anonymously in samizdat. And it was the unofficial text that, following in the footsteps of *Doctor Zhivago*, *The Court Is in Session*, and

Vigdorova's transcript of Brodsky's trial, made its way abroad, appearing in the summer of 1966 in a Russian émigré journal in Germany and as a Russian-language book in the United States, with prefaces by Hélène Peltier and Boris Filippov, the latter evidently flattered to see how prominently his Cold War readings of Sinyavsky's and Daniel's works had figured in the prosecution's case.[69]

This was just the beginning of the process by which the *glasnost* meeting's demand for judicial transparency was realized, as the Russian expression has it, "without official approval" (*iavochnym poriadkom*). Alexander Ginzburg, an editor of samizdat literary anthologies who had attended the meeting, decided to create a new kind of anthology combining Bogoraz's transcript with a wealth of related documents.[70] These included reviews of Sinyavsky's and Daniel's works and reactions to their arrest and trial in the Western press, Volpin's Civic Appeal, coverage of the trial in the Soviet press, letters to the editor by Soviet citizens (published as well as unpublished), and statements by writers inside and outside the USSR. The sheer volume of documentation, encompassing over four hundred typewritten pages, and the enormous risk involved in collecting and distributing such materials, was impressive enough. But it was in his editorial work, or rather in his editorial ethos, that Ginzburg displayed his greatest skill. The collection included unabridged materials both hostile to and supportive of the two writers—a purposeful pluralism of opinions entirely absent from official Soviet publications. Without disguising his sympathy for Sinyavsky and Daniel, Ginzburg nonetheless kept commentary to a bare minimum, resisting the temptation to tell readers what to think and instead letting the documents speak for themselves—which is to say, trusting readers to draw the appropriate conclusions. Texts were arranged chronologically, without regard to geography ("West" versus "East") or point of view ("anti-Soviet" versus "Soviet"), thereby making visible a transnational conversation as it unfolded across the Iron Curtain, with critics from abroad commenting on Sinyavsky's and Daniel's works, Soviet officials reacting to those critics, Sinyavsky and Daniel responding in court to Soviet officials as well as foreign critics, and seemingly everyone, from Manila to Moscow to Mexico City, weighing in on the trial. Rather than present

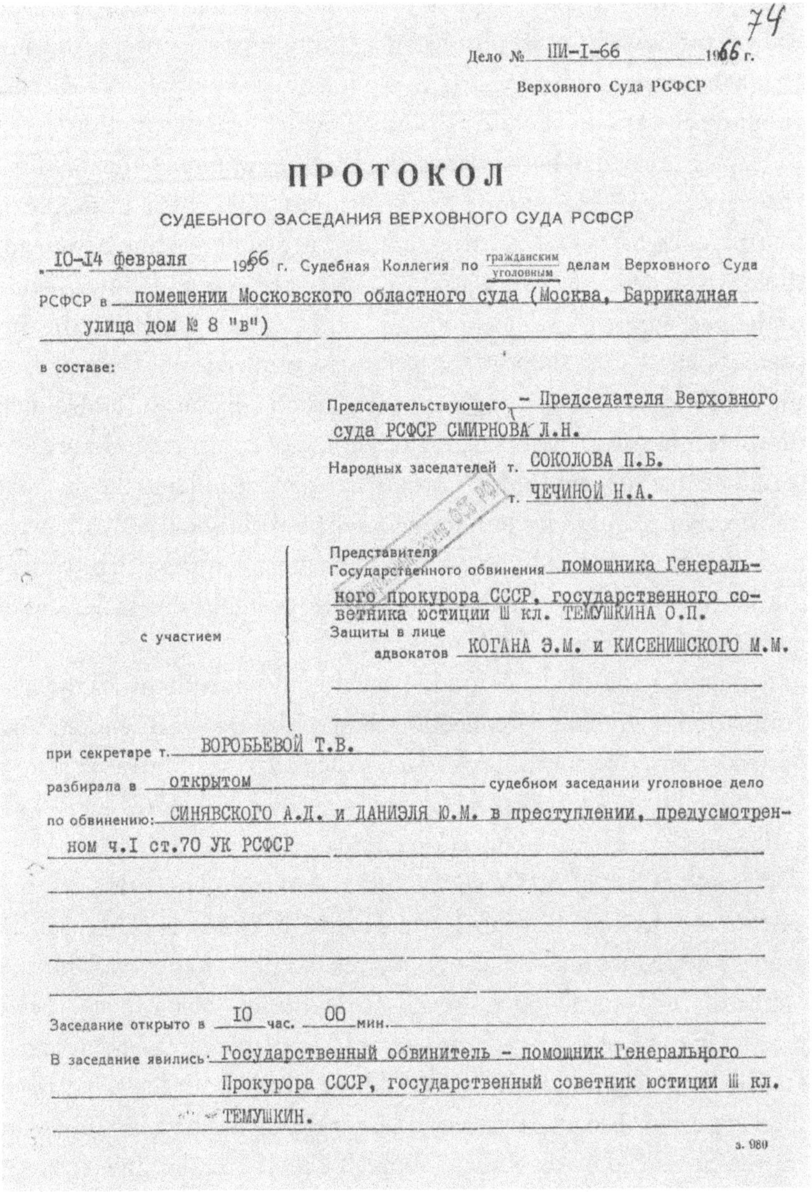

74

Дело № ШИ-I-66 1966 г.

Верховного Суда РСФСР

ПРОТОКОЛ

СУДЕБНОГО ЗАСЕДАНИЯ ВЕРХОВНОГО СУДА РСФСР

„10-14 февраля 1966 г. Судебная Коллегия по гражданским / уголовным делам Верховного Суда РСФСР в помещении Московского областного суда (Москва, Баррикадная улица дом № 8 "в")

в составе:

Председательствующего – Председателя Верховного суда РСФСР СМИРНОВА Л.Н.

Народных заседателей т. СОКОЛОВА П.Б.

т. ЧЕЧИНОЙ Н.А.

с участием

Представителя Государственного обвинения помощника Генерального прокурора СССР, государственного советника юстиции Ш кл. ТЕМУШКИНА О.П.

Защиты в лице адвокатов КОГАНА Э.М. и КИСЕНИШСКОГО М.М.

при секретаре т. ВОРОБЬЕВОЙ Т.В.

разбирала в открытом судебном заседании уголовное дело

по обвинению: СИНЯВСКОГО А.Д. и ДАНИЭЛЯ Ю.М. в преступлении, предусмотренном ч.I ст.70 УК РСФСР

Заседание открыто в 10 час. 00 мин.

В заседание явились: Государственный обвинитель – помощник Генерального Прокурора СССР, государственный советник юстиции Ш кл. ТЕМУШКИН.

з. 980

FIGURE 4.3. Cover page of the official (unpublished) stenogram of the Sinyavsky-Daniel trial the trial transcript as a single document, Ginzburg punctuated each day's sessions with Soviet press coverage, thus allowing readers to compare what was said in court with what Soviet newspapers reported about what was said in court. The juxtaposition was devastating.

Having learned of his activities, the KGB summoned Ginzburg for a "conversation."[71] He ignored their warnings. Taking the principle of transparency to its logical extreme, Ginzburg put his name on the anthology and delivered copies to the KGB as well as to Ehrenburg—one of the best-known Soviet writers and at the time a deputy to the Supreme Soviet—in the hope of having the case reconsidered.[72] Having seen the complete typescript, another writer asked Ginzburg: "How much do you think you'll get for this?" Both knew, of course, that the answer would be in years, not rubles.[73] An urban legend, similar to the one that transformed Volpin into the brazen Jewish revolutionary "Wolf" who marched into KGB headquarters, had Ginzburg attempting to blackmail the KGB with threats of making the anthology public unless Sinyavsky and Daniel were immediately released.[74]

Fantasies aside, the KGB found itself no more able to thwart the dissemination of Ginzburg's collection than it had, two years earlier, Vigdorova's transcript of Brodsky's trial. Dedicated to Vigdorova's memory (she died shortly after meeting Ginzburg, just weeks before Sinyavsky's and Daniel's arrest), the collection circulated in samizdat as *The White Book on the Case of A. Sinyavsky and Y. Daniel*.[75] The title's double entendre—invoking the genre ("white paper" or "white book") typically used by governments to make policy-related documents available for public discussion, as well as the Russian term for "blank spots" (*belye piatna*) in the historical record—cast into sharp relief the Soviet government's reluctance to release such documents itself.[76] The *White Book* soon appeared abroad in two dozen languages, including a 1967 Russian edition produced by the émigré People's Union of Labor in Germany, in a small format designed for "convenient dissemination"—that is, smuggling—into the USSR.[77] Radio Liberty broadcast detailed discussions of the book to listeners in the Soviet Union.[78] Despite having been almost entirely barred from Smirnov's "open" courtroom, Sinyavsky's and Daniel's supporters thus succeeded in turning a "demonstrative trial" to their own purposes, in effect inviting the world to attend and draw the appropriate conclusions.

If "socialist legality" was supposed to supplant terror as an instrument of governance under the post-Stalin dispensation, then the trial of Sinyavsky and Daniel did not bode well. Within a year, the Party's Central Committee acknowledged as much, noting that their conviction "had provoked a very unpleasant reaction."[79] At home, it failed to produce the intended deterrent effect—on the contrary, it galvanized a significant portion of the intelligentsia in support of the two writers. Sinyavsky and Daniel had shown, as one observer put it, that "even in the defendants' dock, you can preserve your dignity."[80] And not only the intelligentsia: when the two convicted writers arrived at separate corrective labor camps in the Mordovian Autonomous Region, to begin what Daniel, in his first letter home, called his "ethnographic expedition" into the Soviet netherworld, they were amazed to discover that they were already celebrities, admired by the other prisoners for having defied the authorities with their refusal to plead guilty.[81] In a confidential memorandum, minister of internal affairs Nikolai Shchelokov bitterly confirmed Akhmatova's and Tvardovsky's intuition: "Those scumbags will be transformed into martyrs for an idea, and we ourselves made this happen."[82] Three months after the trial, Soviet censors banned all mention of Sinyavsky and Daniel from the official mass media, but once summoned, the aura of martyrdom would not go away. According to an urban folk ditty that circulated at the time, "Yuli got five, Andrei got seven, altogether twelve they'll spend in prison."[83] Even high school students were taking up the cause. At the end of 1966, three students at Special School #17 were expelled from the Communist Youth League for distributing a homemade newspaper ("Organ of the Central Committee of the Union of Independent Youth") containing transcripts of Western radio broadcasts about the trial.[84] Years later, Sinyavsky's and Daniel's works, including samizdat editions of their final statements to the court, continued to turn up in apartment searches by the KGB. Questioned about where she had obtained copies of their statements, one subject of a search was unable to remember: "They were so widely read; I'm studying at the university, and practically everyone there has them."[85] Outside the USSR, as the pioneering historian and Party member Roy Medvedev noted, "the recent trial has provided our enemies with considerably more material for anti-Soviet purposes than all of Sinyavsky's and Daniel's writings taken together. In many countries where no one had ever heard either of Tertz-Sinyavsky, or of Arzhak-Daniel, all of their works are being urgently translated and prepared for publication. These writers could not have dreamed of a better advertisement."[86]

The clearest evidence of the state's central role in fashioning dissident biographies can be found in the contrasting fates of Sinyavsky and Daniel, on the one hand, and Andrei Remizov, the third pseudonymous writer to be caught availing himself of Hélène Peltier's diplomatic privileges, on the other. While the first two became involuntary protagonists in a Cold War drama thanks to the KGB's decision to publicly prosecute them, Remizov languished in obscurity, notwithstanding his burning desire to join them. It was precisely this apparent randomness that struck Mark Azbel, Sinyavsky's friend who was drawn into the investigation and forced to testify in court. "The nightmare of history," he concluded, "turns out to be that it gets made by people who haven't the least desire or are utterly unsuited to make it."[87]

The Soviet government used Sinyavsky and Daniel to make history, but, to paraphrase Marx, it proved unable to make history as it pleased. The two writers turned their trial into a platform from which to defend literature's autonomy against the imperious demands of both law and politics.[88] The heretical idea for which they became martyrs was that of an artistic imagination neither for nor against official doctrine but outside it. They too, however, proved unable to shape the impact of their trial as they pleased. Not because the Soviet government got the better of them inside the courtroom, but because Volpin, Bogoraz, Ginzburg, and several dozen other Soviet citizens in effect changed the topic of conversation *outside* the courtroom. What began as an unmasking of two writers accused of slandering the Soviet system transformed, even before Judge Smirnov pronounced the inevitable guilty verdict, into a public call for the Soviet state to abide by the rule of law.

PART II

Movement of a New Type

5

Rights Talk

Закон—что дышло: куда повернул, туда и вышло.

Law is like a wagon's tongue: whichever way you turn it, that's where it runs.

—RUSSIAN FOLK SAYING

Laws, of course, do not rule; people do. The question is whether people rule within a system of laws and, if so, what kind of laws and to what effect. It had taken undeniable courage for Khrushchev to condemn Stalin as "lawless" in his 1956 speech to delegates at the Twentieth Congress of the Communist Party. It required something beyond courage, however, to recognize the inconvenient fact that there had been plenty of laws under Stalin and that the concept of "socialist legality"—championed by Khrushchev as the antidote to Stalin's "cult of personality"—originated with none other than Andrei Vyshinsky, Stalin's procurator-general and the architect of the infamous show trials of the 1930s.

There has long been a tendency, particularly in the Anglo-Saxon world, to dismiss not just the rule but the role of law in autocratic and totalitarian states as little more than window dressing.[1] Both autocracy and totalitarianism, in fact, are defined largely by the absence of legal constraints on those who hold power.[2] "The problem with Soviet legal history," quipped the historian Martin Malia, who in the early 1960s smuggled works by Andrei Sinyavsky and Yuli Daniel into the USSR, "is that there's not enough of it."[3] His remark was meant to register the pervasiveness, among elites and masses alike, of extra-legal ways of doing things, as well as the Russian intelligentsia's traditional loathing of the bourgeois passion for contractual relationships. If not quite one of Rudyard Kipling's "lesser breeds without the Law," Bolsheviks, in this view, used laws and constitutions strictly as a façade, behind which operated the real, which is to say, informal and personal, mechanisms of power, what contemporaries knowingly referred to as *telefonnoe pravo*, "law by telephone." Implicit in this argument is an assumption of bad faith: that Soviet laws—or at least some Soviet laws, especially those pertaining to the rights of Soviet citizens—were not meant to be actionable, instead serving strictly ornamental purposes that allowed ruling elites to have their laws and ignore them too. It is a critique whose pedigree reaches back to the German sociologist Max Weber and his dismissal of Tsar Nicholas II's "pseudo-constitutionalism" following the revolution of 1905.[4] Or perhaps even further, as captured in an oft-quoted nineteenth-century saying according to which "the severity of Russian laws is compensated for by the non-obligatory nature of their fulfillment."[5]

The political theorist Judith Shklar turned this critique on its head. In *all* systems, she insisted, law as an instrument of social control is inescapably political; the question is rather which political ends it serves. In settings such as Hitler's Germany or Stalin's Soviet Union, according to Shklar, law was "insignificant," but not because it was subordinated to political interests. "Law as a political instrument," she argued, "can play its most significant part in societies in which open group conflicts are accepted and which are sufficiently stable to be able to absorb and settle them in terms of rules. As an instrument of terror, of coercive persuasion, ... it is all but useless."[6]

To dismiss the role of law in non-democratic societies, however, makes as little sense as to dismiss the role of economics in non-market societies.[7] Bad faith does not suffice either as a description or as an explanation of the relatively underinstitutionalized character of tsarist and Soviet law. Like most continental European states, imperial Russia employed an administrative legal system in which constituent units of the state, and the officials who ran them, were subject to different laws, courts, and judicial procedures from those that applied to the population at large. During the half-century before the Bolshevik Revolution, Russia took significant strides toward fashioning a uniform body of public law for all its diverse subjects, but efforts to regularize and enforce administrative law repeatedly came to naught.[8] Once in power, the Communist Party similarly operated in its own separate legal universe, retaining the prerogative to decide whether and how to discipline members suspected or accused of crimes, rather than turning them over to state prosecutors (apart from the period of severely diminished Party power during the era of Stalin's purges).[9] The Party's insulation from general laws persisted even as the USSR, reenacting the steps of its tsarist predecessor, gradually curtailed legal discrimination within the Soviet population based on social origin, profession, or ethnicity.

And what of the population at large? We know far less about the vernacular legal consciousness of ordinary Soviet citizens, their ideas about rights and duties, crime and punishment, than about legal debates among professional jurists and political elites. Laypeople, moreover, are most likely to express their views on law when they are accused of breaking it. The fact that much lawmaking in the USSR was not the result of open deliberation by popularly elected legislators, coupled with the persistence of informal, personal forms of power, kept most citizens at a considerable distance from official legal norms. Daily life taught people the art of when and how to get around the law, and the Soviet state, with its frequent recourse to "administrative measures" instead of legislation, was itself an expert at such detours. Under Stalin, a vast repertoire of what the historian Stephen Kotkin calls the "little tactics" of accommodation and circumvention became a matter of survival. Prominent among these was the ability to "speak Bolshevik," to express oneself in the idiom of revolutionary struggle and socialist values, whether from inner conviction or self-interest, or the many possible crisscrossings of the two.[10] By the time the USSR entered its post-totalitarian phase, speaking in official idioms had become a rote practice, something everyone simply took for granted, at least in public. Russia had always been a "service state," where one's relationship to the ruling power was defined largely by the kind of service one rendered. The Muscovite (1480–1721), Petrine (1721–1917), and Stalinist (1928–53) systems all exacted their characteristic forms of service from the populations they governed. The post-Stalin era gave birth to its own characteristic form: the lip-service state. In school, at work, in letters published by mass circulation newspapers, at public meetings and celebrations, millions of Soviet citizens routinely reproduced phrases and slogans originally designed to convey the fervor of a revolutionary crusade—in many cases with little or no interest in their meaning. It was not a matter of being for or against; the phrases were simply taken for granted. Pronouncing them was enough.[11]

The approach developed by Alexander Volpin and other "rights-defenders" was designed to disrupt the pervasive practice of lip-service by reanimating Soviet law, and in particular the rights enshrined in the Soviet Constitution. We will therefore need to take a close look at the supposedly hollowed-out, taken-for-granted material out of which they crafted their improbable strategy.

The constitutional rights that we associate with the French and American Revolutions emerged in the eighteenth century from the dialectical struggle between bourgeois rights and aristocratic and royal privileges.[12] Insofar as the idea of socialism emerged in the nineteenth century from the critical analysis of capitalism, it should come as no surprise that the Soviet conception of law developed via a series of critiques of bourgeois law. Marx argued that in bourgeois societies, legally enshrined equality among citizens (or at least among adult males) served to mask entrenched hierarchies of class. Bourgeois law was thus a smoke screen behind which profit-driven relationships of exploitation flourished. Since class antagonisms are the root of criminal behavior, the argument went, in a classless society crime would disappear, and law as an instrument of state coercion would no longer be necessary, destined to wither away along with the state itself as part of the transition to communism. Lenin—like Marx, trained in the law—continued in this vein, dismissing what he called "petty bourgeois constitutional illusions" and insisting that "a constitution is fictitious when law and reality diverge."[13] Divergences between law and reality, however, are everywhere the norm, if only because law's purpose is not to describe reality but to shape it via state-sanctioned codes of behavior. When we study how law functions in a given society, what should draw our keenest attention is the tension not between law and reality but between law in theory and law in practice.

At the heart of the Marxist-Leninist critique of the "fictitious" nature of bourgeois constitutions are the allegedly inalienable rights they proclaim—the façade of equality among individuals designed to mask the reality of class domination. For Marx, individual rights such as those constitutionalized by the American and French Revolutions served only to alienate their bearers from the human communities to which, as members of a fundamentally social species, they properly belonged, and without which they were unable to realize their full human potential.[14] Similar Bolshevik critiques of law as an alienating mode of social control led more than a few observers to characterize Lenin and his party as legal nihilists. Evidence to support this claim is not hard to find. Nikolai Bukharin and Evgeny Preobrazhensky's *The ABC of Communism* informed millions of readers in the 1920s that in the future "there will be no need for special Ministers of state, for police and prisons, for laws and decrees—nothing of the sort."[15] No less than the first president of the USSR Supreme Court, Petr Stuchka, agreed: "Communism means not the victory of socialist law, but the victory of socialism over any law, since with the abolition of classes and their antagonistic interests, law will die out altogether."[16] What Leon Trotsky famously called the "improved edition of humankind" forged by Soviet power was meant to be a higher breed without need of law. One struggles to find other historical instances of such a radical critique of law as an instrument of social order, particularly in a modern state. Even the early Christians, with their call for the transcendence of (Jewish) law by grace, did not repudiate legal frameworks entirely.

Marx could afford to leave unclear what if any role law would play during the revolutionary period of proletarian dictatorship, or for that matter during the socialist transition from capitalism to post-law communism.

The Bolsheviks, having captured what was left of the imperial Russian state in October 1917, could not. Even a cursory glance at Soviet legal history reveals that Bolshevik antinomianism quickly retreated before the exigencies of governing. Along with a wide range of laws, rights claims of various kinds—individual, collective, expressive, material—immediately found a place in the legal lexicon and never left.[17] The first Soviet Constitution, promulgated in July 1918, less than a year after the Bolshevik seizure of power, crafted legal rights for the explicit purpose of inverting rather than abolishing inherited social hierarchies. It granted the classic bourgeois freedoms of conscience, speech, assembly, and association exclusively to "toilers," whom it defined as those "who obtain their livelihood from productive and socially useful labor" as well as "soldiers of the Soviet army and navy" and the "rural proletariat" of poor peasants.[18]

The novelty of Soviet rights discourse, however, lay not only in using the selective granting of rights to help the modern meek (the proletariat) inherit the earth. From the outset, Soviet law parted ways with the notion that civil rights belong to what the liberal political philosopher Isaiah Berlin would later call "negative freedoms," that is, freedoms *from* interference by the state or other bodies.[19] Instead, Soviet law insisted that "real freedom of expression" and "real freedom of assembly" (in contrast to the fictitious freedoms advertised by bourgeois constitutions) *required* involvement by the state. As Lenin proclaimed at the Seventh Congress of the Bolshevik Party in March 1918, "For the bourgeoisie, what matters is a general proclamation of loud principles: 'All citizens have the right of assembly—but assembly in the open air, we won't give you any buildings.' But we say: 'Less talk and more substance.' We must seize palaces, and not just the Tauride Palace [where the Party congress was being held], but many others."[20]

Article 125 of the 1936 Soviet Constitution, trumpeted as "the most democratic constitution in the world," put it more formally:

> In conformity with the interests of the toilers, and in order to strengthen the socialist system, citizens of the USSR shall be guaranteed by law:

(a) freedom of speech;
(b) freedom of the press;
(c) freedom of assembly and meetings;
(d) freedom of street processions and demonstrations.

These rights of citizens shall be ensured by placing printing presses, stocks of paper, public buildings, streets, means of communication and other material conditions necessary for their realization at the disposal of the toilers and their organizations.[21]

Article 125 calls for careful parsing. Even as the Soviet state guaranteed the classic civil liberties to "citizens," it promised the material resources "necessary for their realization" only to "toilers" and required that liberties be exercised only in ways that conformed to the (unspecified) interests of those same "toilers." In the grammar of Soviet rights talk, no right could have practical value without an explicit commitment from the state to ensure the material prerequisites for its realization.[22] The state, in turn, expected the exercise of rights to serve the purposes of a socialist society. When a student at Moscow State University in the 1960s criticized Sinyavsky and Daniel for publishing their works abroad, a visiting American pointed out that they could scarcely have hoped to publish their works in the USSR. To which came the reply, "It's hard to argue that the [Soviet] state should subsidize something that slanders the state."[23]

Most of the rights enumerated in the 1936 constitution, including those in Article 125, are found in Section 10, "Fundamental Rights and Duties of Citizens." In the Soviet legal world, no rights were duty-free. The proximity of rights and duties reflected an implicit quid pro quo whereby the exercise of rights depends not only on material support by the state but on the fulfillment of duties by citizens.[24] The Stalin Constitution assumes a distinction between "having" a right, something any Soviet citizen could claim, and being granted (by the state) the material means to exercise it, which applies only to "toilers," people engaged in fulfilling the preeminent duty of socially useful labor.

The assertion that the 1936 constitution was the most democratic in the world rested on three claims, all of them ambiguous. First, prior to

FIGURE 5.1. Postage stamps commemorating the fifteenth anniversary of the Stalin Constitution. Left: Article 121—"Citizens of the USSR have the right to education." Right: Article 118—"Citizens of the USSR have the right to work."

its enactment, a draft of the constitution had been the subject of a yearlong "all-people's discussion" involving millions of Soviet citizens. Their comments, however—including some that were highly critical—had almost no impact on the final text. The mobilization of public opinion nonetheless allowed the USSR to contrast its constitution with that of an older rival, ratified in 1787 "by a convention in Philadelphia . . . consisting of members of the bourgeoisie and slave-owners who conducted

FIGURE 5.1. (*continued*)

their sessions behind closed doors."²⁵ Second, the Soviet Constitution introduced an unprecedented, comprehensive regime of social and economic rights, including the right to paid employment, leisure, material security in old age and in the event of illness or incapacity to work, and free education up through the university level. At the time, no other country in the world promised such a robust system of public welfare.²⁶ Discrimination in the allocation of social and economic rights based on gender, race, or nationality was forbidden. Here too, the constitution's reach exceeded its grasp, or rather strained the state's capacity to pay for this expansive menu of entitlements, as is known to occur with robust welfare states. For our purposes, what is most significant about the

USSR's pioneering social and economic rights is the way their logic—as entitlements granted by the state—structured the rights of speech, assembly, and association. The same language of material support guaranteed by the state applied to those rights as well. In this third sense too, then, there was a studied ambiguity, as an act of distributive justice had the inescapable effect of fostering dependence on the state.

The rights articulated in the 1936 constitution were thus extraordinarily "democratic" in the specific Soviet sense of prioritizing the material interests of the Soviet *demos*, the working class. In practice, of course, this same "democracy" denied workers, along with everyone else, any opportunity to participate in multi-party elections, to establish their own independent organizations, or to otherwise express political preferences distinct from those of the Communist Party.

Stalin's constitution would remain in effect for four decades, longer than all other Soviet constitutions (promulgated in 1918, 1924, and 1977) combined. This was the document Volpin had in mind when he held up a sign at the transparency meeting on Constitution Day (December 5) 1965, reading "Respect the Constitution!" Whatever ruptures were generated by Khrushchev's historic attack on Stalin's "cult of personality," Stalin's constitution remained the foundational text for all subsequent Soviet discussions of rights and other constitutional issues. And not only by jurists: the 1936 constitution was the subject of mandatory secondary-school classes, propaganda campaigns, and celebratory postage stamps well into the 1960s, exposing large portions of the Soviet population to the official rhetoric of rights.[27]

In one of his seventh-grade classes on the constitution, Sergei Kovalev—future biophysicist and dissident—was asked to name the relevant passages concerning "rights and duties of citizens of the USSR." In good Soviet fashion, the young Kovalev recited Article 125 from memory: "In conformity with the interests of the toilers, and for the purpose of strengthening the socialist order, citizens of the USSR shall be guaranteed by law. . . ." Prompted by the teacher to explain to the class what this meant, Kovalev chose an expansive interpretation: "The lawgiver believes that the existence of these freedoms corresponds to the interests of the workers and serves to strengthen order. That's why they're guaranteed in Article 125." The teacher corrected him: "These freedoms are indeed guaranteed by law, but only insofar as they correspond to the interests of the workers and serve to strengthen order. If they contradict those interests, they are not guaranteed by the constitution." Kovalev, who by his own account approached juridical formulas as one would a mathematical theorem, would not accept his teacher's restrictive interpretation, since it failed to indicate what was in the workers' interest, what served to strengthen the existing order, and most important—who decided. For this he was rewarded with a failing grade and a trip to the principal's office. "You're in the seventh grade," the principal reminded him. "Your job is to know the constitution, not to interpret it."[28]

It was not just future dissidents for whom the Soviet Constitution served as the point of departure for thinking about rights. Thanks to an official campaign in the early 1960s to solicit popular input for a planned new constitution (a plan quietly aborted after Khrushchev's removal in 1964), we can listen in on the rights talk of several thousand Soviet citizens who mailed their suggestions to the Central Committee of the Communist Party.[29] To be sure, the mere fact that these citizens responded to the campaign means they probably cannot stand for the Soviet population at large. But their letters have much to tell us about vernacular ways of thinking and talking about rights, as well as about how officially inscribed rights spoke to Soviet citizens. Written by men and women from a wide range of backgrounds who resided in cities, towns, and villages across the USSR, they articulate an array of political stances, including some that, had they been made openly, would have been reflexively condemned as "anti-Soviet" and potentially subject to criminal prosecution (for example, the demand to put an end to sham elections). In the aggregate, the letters display a conception of rights strikingly in sync with Stalin's constitution. Most construe rights as things the state bestows upon, rather than recognizes as inherent in, its citizens. Letter writers overwhelmingly assumed that rights are contingent on the performance of duties, part of an ongoing system of exchange between citizens and the state. "Prior to receiving rights and using them," urged Kyiv resident A. I. Avgustovsky, "citizens of the USSR should know and fulfill their duties."[30] Even rights classified in the Anglo-American tradition as "negative freedoms" or "immunities"—claims that in theory require only that the state refrain from certain activities (censoring speech, breaking up demonstrations, and so forth)—appeared in citizens' letters as requiring positive state action. As I. M. Abramovich, a high school history teacher from Murom, two hundred miles east of Moscow, put it, "Bourgeois constitutions merely proclaim the rights of the citizen, but the Soviet constitution materially guarantees their realization."[31]

Freedom of expression, central to the Sinyavsky-Daniel trial, emerged as a particularly controversial topic in letters about the planned constitution. Although a substantial minority of letter writers called for stricter constitutional limits on speech, especially for religious believers ("The constitution," urged A. S. Poluektov from the town of Oktiabr, "is supposed to protect the people from attacks by religion on citizens' consciousness"), the lion's share spoke in favor of stronger constitutional protection of public criticism of official policies.[32] One writer suggested that in order to make freedom of the press more than just a "formality," citizens should be allowed to publish opposing views, but at their own expense—as if it were inconceivable that the state would provide the material means for criticizing its own policies.[33] From Kolpino, on the outskirts of Leningrad, the pensioner D. K. Markov urged that those who prevent or retaliate against legitimate criticism be liable to legal action in court. Like a number of letter writers, Markov was unsure whether a new constitution could prevent a recurrence of the cult of personality:

> Criticism of our shortcomings, mistakes, defects, and survivals of capitalism is an inalienable, organic characteristic of the Soviet system. But this criticism is one-sided. It goes mainly from top to bottom. In 47 years of Soviet history I haven't once encountered an episode when, at a meeting or in the press someone criticized the Chairman of the Council of Ministers or the Chairman of the Presidium of the Supreme Soviet. People are more inclined to criticize them "around the corner" or "in secret." Along came Stalin. He promoted arbitrariness, made a

heap of mistakes, and where was the criticism? Just you try to criticize! Now we have comrade Khrushchev as Chairman of the Council of Ministers. I don't even want to compare him to Stalin; it's night and day. But comrade Khrushchev isn't anointed with the holy spirit either. He can make mistakes too. Down here it's more visible to us when he's mistaken, so why shouldn't we criticize openly and honestly? Why shouldn't we tell him, in a constructive way, when he's slipping up? But how? Where?[34]

While rights were invoked as constraints on the actions of misguided or abusive officials, not a single letter claimed that constitutional rights can (let alone should) place formal limits on the authority of the state itself. This is hardly surprising, given letter writers' rejection of the idea of separation of powers.[35] Lenin's achievement of "unified power" had marked a historic victory over "dual power," the tangled, unplanned power-sharing from February to October 1917 between the Provisional Government and the Soviets (Councils) of Workers', Soldiers', and Peasants' Deputies. "ALL power to the Soviets!," a letter writer from Moscow reminded the Central Committee, invoking Lenin's famous revolutionary slogan.[36] Even those who passionately defended civil liberties understood rights as emanating entirely from the state, rather than residing a priori in citizens or in human beings as such.

Soviet citizens' ideas about rights thus appear to have been consistent with the USSR's constitutional rhetoric. Both emphasized rights as gifts by the state to citizens in exchange for the duty of socially useful labor. Both treated the full spectrum of rights—civil and political as well as social and economic—as requiring the ongoing provision by the state of the necessary material preconditions for their fulfillment. And both understood rights as serving the larger project of building socialism. Close analysis of Soviet rights talk therefore undermines the Cold War cliché according to which the capitalist West championed "first-generation" civil and political rights while the socialist East promoted "second-generation" social and economic rights. It similarly unsettles what one scholar has dubbed the "Isaiah Berlin Wall" in the world of legal theory, the stark division of labor between negative and positive liberties, and instead highlights how, within the socialist camp, these two "generations" of rights shared a single logic.[37] The most distinctive formal quality of civil rights in the Soviet setting was the manner in which they were construed, by rulers and ruled alike, as depending on active state intervention to ensure that the material prerequisites for public expression and assembly were available to the entire population.

Khrushchev's abrupt removal from power in October 1964 put an end to his plan for a new constitution. Among his successors, the favored bulwark against the cult of personality nonetheless remained "socialist legality," as codified in reforms of the Soviet legal system in the late 1950s.[38] The trial of Sinyavsky and Daniel demonstrated that under the post-Stalin dispensation, the criminal justice system could operate without torturing defendants, even guaranteeing them the right to legal representation, if not necessarily of their own choosing.[39] The new system required prosecutors to argue on the basis of evidence openly presented in court and subject to scrutiny by the defense. It also set maximum sentences for a wide range of crimes, including "anti-Soviet agitation and propaganda" as outlined in Article 70 of the revised Criminal Code. For all its severity, Sinyavsky's sentence of seven years hard labor—the maximum—pales in comparison with what he might have received a generation earlier. Most of the 3.7 million Soviet citizens convicted of "counter-revolutionary agitation and propaganda" under Stalin were sentenced to far harsher punishments, including the death penalty.[40]

While it achieved dramatic success in tempering the arbitrary and often lethal violence of the Stalinist judicial apparatus, Khrushchev's version of "socialist legality" left a number of fundamental juridical questions unresolved. Well into the era of what his successor, Leonid Brezhnev, called "mature socialism," Soviet jurists debated whether the burden of proof in criminal cases was on the defendant (to demonstrate innocence) or the prosecution (to demonstrate guilt). Many legal experts considered a confession by the defendant as definitive proof of guilt, requiring no additional evidence. Acquittals—which rarely occurred—were taken to signify a failure of the judicial process. Leading jurists questioned whether the adversarial system itself—a bourgeois invention, after all—was suitable for a society that had eliminated class antagonisms. Roman Rudenko, the USSR's longest-serving procurator-general (1953–81), insisted in a 1961 memorandum that arguments for the presumption of innocence and the adversarial system "bear a certain scholastic character and help neither the practical judicial work of combating crime nor the elaboration of Soviet legislation concerning criminal law."[41]

Quite apart from the unsettled status of certain fundamental legal principles, the Soviet Union never fully abandoned the ideal of doing away with the law entirely. Among the "Leninist norms" that inspired visions of a communist future, one can find traces of the antinomianism of the 1920s. In 1961, for example, the Twenty-Second Congress of the Communist Party introduced the Moral Code of the Builder of Communism, a distillation of twelve supreme ethical values for the *homo sovieticus*. Beyond its short-term goal of fostering socially productive behaviors such as "conscientious labor for the good of society" (the second commandment) and "humane relations and mutual respect between individuals" (the sixth commandment), the Moral Code was meant to prepare the ground for the withering away of formal law.[42] At the Twenty-Second Congress it was announced, for example, that during the transition to communism, "the role of moral principles in social life grows, the sphere of activity of moral factors widens, and correspondingly the importance of the administrative regulation of relations between people decreases." According to one commentator, "In developed communist society, [moral norms] will be the only form of regulation of relations between people."[43] All this was consistent with the long-standing Soviet assumption that crime—and therefore the need for legally sanctioned punishment—would gradually disappear as a result of the elimination of private property and the class antagonisms that it produced.

In sum, Soviet legal culture was not so much feeble or insubstantial as riven by internal contradictions. The abiding ideal of a purely moral discipline, free of legal sanction, contradicted the inescapable need for formal law as a means of ordering a complex modern society. The Communist Party's persistent insulation from the Soviet judicial system—a structural inheritance from tsarist-era administrative law—collided with the declared principle of one law for all. Perhaps the most telling illustration of these unacknowledged contradictions can be found in the manner of Khrushchev's removal from the pinnacle of Soviet power. On the one hand, his bloodless fall—as he himself subsequently observed, not without pride—demonstrated his extraordinary success in ridding the Soviet system of pervasive political violence. Rather than subject Khrushchev to a sham trial followed within hours by execution in a military bunker—the fate, scarcely a decade earlier, of Khrushchev's former Politburo colleague Lavrenti Beria—his successors exiled their former boss to his dacha in the leafy outskirts of Moscow, where he lived out his remaining years, depressed but physically intact, with a generous state pension as well as a car and driver, even managing to smuggle his memoirs for publication abroad, following in the footsteps of Pasternak, Sinyavsky, and Daniel.[44] On the other hand, as Volpin noted, the entire procedure, while admirably non-violent, was nonetheless without legal foundation. The Party had taken yet another detour around the law.

Khrushchev's great civilizing achievement—the removal of mass terror and political violence from the repertoire of Soviet governing practices—proved so essential to the well-being of Soviet political elites that it became a permanent feature of "developed socialism." The post-Stalinist legal order, its flaws and contradictions notwithstanding, formed an indispensable part of that civilizing process. Never again—even when it was unraveling—would the USSR descend into the witch hunts and bloodbaths of the Stalin era. Even the Sinyavsky-Daniel trial, as the preceding chapters have shown, represented an attempt to accomplish state goals via juridical means, without recourse to torture, terror, false testimony, or withholding evidence.

When those goals were derailed by the unforeseen effects of the trial on Soviet public opinion, the state turned again to juridical means. Within months of the trial, on June 8, 1966, KGB chairman Semichastnyi and Procurator-General Rudenko wrote to the Central Committee that the current Criminal Code had proved inadequate for the struggle against "especially dangerous state crimes." A new law was needed in order to explicitly criminalize attempts at organizing another transparency meeting.[45] Their fears were not without foundation: as the KGB and other careful readers must have noted, Volpin's Civic Appeal referred to "transparency meetings" in the plural. Volpin had always conceived of the December 5, 1965, demonstration as reproducible across time and space. Indeed, an early draft of the Civic Appeal had stated that "in cases of extra-judicial repressions against participants in a transparency meeting, new meetings will be organized," including in cities other than Moscow.[46] The transparency meeting was not a unique event but a technique. A leaflet from an otherwise unknown entity (or perhaps an individual aspiring to become an entity) calling itself the "December 5 Movement" demanded to know what had happened to the "democratic constitution" promised by Khrushchev, urging readers to "defend the rule of law" by all legal means and to "demand the security of civil liberties" so as to avert a return of Stalinism.[47]

The Russian Supreme Soviet shared Semichastnyi's and Rudenko's concern, and in September 1966 its Presidium proposed adding Article 190 to the Criminal Code. Dispensing with the increasingly archaic language of "agitation and propaganda," the new article outlawed the organization of and participation in "group actions that violate public order," including "public transportation, state and public institutions." No reference was made to the political content of those activities. Indeed, the striking absence of the word "anti-Soviet" constituted an oblique acknowledgment that the December 1965 transparency meeting had successfully refrained from anti-Soviet content. Henceforth participants in such meetings could be held liable for the more politically neutral crime of disturbing public order. Article 190 also banned the "systematic dissemination in oral form of deliberately false statements slandering the Soviet state, as well as the preparation and dissemination of such statements in written, printed or other form," the final two words indirectly registering the growing presence of samizdat.[48] Unlike Article 70, under which Sinyavsky and Daniel had been sentenced, Article 190 did not require the prosecution to show that false statements "subverted or weakened Soviet power" or that the accused had intended them to do so—one of the more glaring weaknesses in the government's case against the two writers. To be sure, in keeping with the post-Stalin trend of diminishing severity of judicial punishment, Article 190 set the maximum sentence at three years detention or one year of corrective labor. But since intent to weaken or undermine Soviet power was no longer required to establish a defendant's guilt, the new law criminalized a potentially much wider circle of activities, targeting what Semichastnyi had called, during the hunt for Tertz and Arzhak, the insidious "apoliticism" evident among the country's urban intelligentsia.[49] Rumors began to circulate that the KGB planned to invoke the new law to arrest upwards of five thousand individuals about whom it had already prepared dossiers.[50]

The introduction of Article 190 revealed an important vulnerability in the law-based approach to dissent: rather than a fixed language of prohibitions and commandments, Soviet law could be altered at will to suit the fluctuating needs of the state. In its battle against dissidents, the Soviet government too was deploying a law-based strategy.

6

Chain Reaction

In a nuclear chain reaction, the splitting of nuclei under certain conditions produces a self-sustaining process of fission, in which an ever greater number of free particles break away, causing new nuclei to split. In this manner, the legal trials in recent years have called forth a self-sustaining process of self-liberating personalities, whose actions aim at splitting and crushing the forces of reaction, of bureaucracy, of hatred.

—GEORG BRUDERER

Article 190 went unmentioned in the Soviet press, but it did not go unnoticed. For what may have been the first time in Soviet history, a group of twenty prominent intellectuals, many of them at the pinnacle of their fields, publicly protested against a Soviet law in the form of an open letter to the Supreme Soviet. Among the signatories were the composer Dmitry Shostakovich, the filmmaker Mikhail Romm, the writers Venyamin Kaverin and Viktor Nekrasov, and nine members of the Academy of Sciences, including the 1958 Nobel laureate Igor Tamm, director of the Soviet thermonuclear bomb project, along with several of the physicists who had played key roles in that project—Vitaly Ginzburg, Andrei Sakharov, and Yakov Zeldovich. They were joined by historians, a biologist, and three Old Bolsheviks whose Communist Party pedigree predated that of the USSR's new leaders, Leonid Brezhnev and Alexei Kosygin. Without claiming any special juridical expertise (there were no lawyers among them), these self-described "Soviet citizens" proclaimed it their "duty" to convey to Soviet leaders—and since this was an open letter, to a wider audience of fellow citizens—their conviction that Article 190 "opens the way to the subjective and arbitrary interpretation of any statement as deliberately false and derogatory to the Soviet state and social system." Not only did the new law violate "Leninist principles of socialist democracy," it posed "a potential obstacle to the exercise of liberties guaranteed by the Constitution of the USSR."[1]

Despite the prominence of its signatories, the letter received no official reply. One signer, the physicist and Academician Mikhail Leontovich, was told privately by a member of the Central Committee that Article 190 was not intended to restrict the liberties of Soviet citizens. It took only weeks for that assurance to crumble.[2] Determined to silence the compilers (and any would-be imitators) of the widely discussed *White Book*, the KGB initiated a series of apartment searches and arrests in January 1967, targeting Alexander Ginzburg, Yuri Galanskov, Vera Lashkova, Aleksei Dobrovolsky, and Pavel Radzievsky. It had taken Alexander Volpin several months to organize a demonstration in response to the arrest of Andrei Sinyavsky and Yuli Daniel, but it took Viktor Khaustov, a furniture repairman, autodidact, and veteran of the Mayakovsky Square poetry readings, together with fellow "Mayak" participant Vladimir Bukovsky (who had helped spread word of Volpin's 1965 transparency meeting), only days to do the same for the compilers of the *White Book*. On January 22, they and a handful of protesters, standing silently at the Pushkin monument in the frigid evening air before a crowd of thirty to forty youthful onlookers, unfurled homemade signs calling for the release of those arrested as well as the revision of Articles 70 and 190. Volpin, ever the formalist, had opposed the slogans, since they strayed from the principle of demanding open trials and strict adherence to existing laws, but he attended the demonstration nonetheless.[3] Alerted in advance, undercover KGB agents and members of civilian patrol squads promptly seized the signs and broke up the gathering. Bukovsky (twenty-four years old), Khaustov (twenty-nine), Vadim Deloné (nineteen, a SMOGist recently expelled from the Lenin Pedagogical Institute), Yevgeny Kushev (also nineteen, a high school dropout and emerging samizdat poet), and Ilya Gabai (thirty-one, an editor employed at the Institute of the Peoples of Asia) were arrested, taken to Lefortovo Prison, and charged under the freshly minted Article 190, against which, for a few minutes at least, they had protested.

By now a pattern was emerging: arrests provoked demonstrations, demonstrators were arrested and put on trial. Trials begat protests and unofficial transcripts of the proceedings, which begat new rounds of demonstrations and arrests, and so on, forming what sociologists call "cycles of protest."[4] Contemporaries had a different name, inspired by the nuclear age: "chain reaction."[5] Whatever one called it, this was something new in the Soviet Union. During the Stalin era, a very different dynamic had prevailed: arrests and torture produced confessions (accurate or not) and naming of names, which led to more arrests and more naming of names. In this way, entire networks, real or imagined, could be crushed. There were no protests; on the contrary, those under arrest or merely under suspicion were usually shunned. Igor Golomshtok, arrested in 1966 for refusing to testify against Sinyavsky, noted that "unlike in Stalin's time, my acquaintances did not cross over to the other side of the street when they saw me coming. In fact, my circle of friends not only remained fully intact but expanded."[6] Larisa Bogoraz noticed the same thing after the arrest of her husband, Yuli Daniel. Three decades earlier, when her father was arrested on charges of participating in a Trotskyite conspiracy, her mother's friends stopped visiting their apartment. Her aunt announced at a work meeting that she had broken off all ties with her sister (Bogoraz's mother). Now, by contrast, "everything was just the opposite," she wrote. "Political repressions stopped being perceived as a private tragedy, and became a matter of public concern. This was something new and unexpected, not only for me, but for the KGB. Society began to resist that organization rather openly—the same organization that until recently had instilled a mystical fear in each individual person and in everyone collectively."[7]

When Stalin's officials arrested hundreds of thousands of Soviet citizens, convicted them in farcical trials (or without any trial at all), and executed them or sent them to the camps, most of the Soviet population believed the victims to be guilty. When Stalin's successors began to limit themselves, with occasional exceptions, to prosecuting people for what they had actually done, and to conduct trials with at least a minimum of due process, they started losing the confidence of Soviet intellectuals at an astonishing speed.[8] "Nowadays," observed Nadezhda Mandelstam, "the persecution of one intellectual only creates dozens more."[9]

Unlike the carefully calibrated chain reaction of nuclear fission, the metastasizing of dissenting activities during the second half of the 1960s was neither a controlled process nor an explosive one. At any given moment, neither the KGB nor participants in the various demonstrations, letter-writing campaigns, courthouse gatherings, and samizdat documentary collections were able to see, much less maneuver, beyond the immediate horizon of events. Soviet leaders were forced to recognize again and again that political trials failed to produce the desired pedagogical effect at home and were inflicting greater damage on Soviet prestige than the allegedly "anti-Soviet activities" of those put on trial. "What happened?" wondered the Ukrainian historian Valentyn Moroz. "Why are repressions not producing the usual [deterrent] effect?"[10]

For their part, protesters found their letters and petitions unanswered, their demonstrations subject to swift crackdowns by the KGB, and not a single political trial producing an acquittal. While Soviet émigrés such as Georg Bruderer, a leading figure in the People's Labor Alliance (known by its Russian acronym, NTS), based in West Germany and financed by the Central Intelligence Agency (CIA), remained stuck in the language of the Stalin era, indulging in fantasies of "splitting and crushing the forces of bureaucracy," Volpin and other advocates of strict constitutionalism continued to distance themselves from the legacy of revolutionary violence.[11] If anything, the rule of law could potentially strengthen bureaucracy, in the form of legislation, regulation, judicial procedures, and officials and courts to carry them out.

Although they neither aimed at nor produced a political explosion, the cumulative effect of arrests, trials, and protests was something historic: a movement for civil rights, the first such movement in a socialist country. Political trials in the Khrushchev era had not had this effect. During the late 1950s and early 1960s, trials of underground revolutionary groups—whose members usually numbered in single digits—remained isolated events, the various coteries unknown to each other and to all but a small circle of friends. Nor had protests by Baptists, Crimean Tatars, and other religious and ethnic minorities resulted from or given rise to connections with analogous groups or with the outside world. What made the dissident chain reaction possible in the setting of mature socialism, and how did it give rise to a movement whose ambition was not to overthrow but to set limits on Soviet state power?

Dina Kaminskaya, one of the attorneys whom Bogoraz had sought out to defend her husband, Daniel, in court, was the daughter of a Russian-Jewish lawyer elected (like Sinyavsky's father) to the ill-fated 1918 Constituent Assembly, in his case on the ticket of the Constitutional Democratic Party. Whatever his attitude toward the Bolsheviks, Isaak Kaminsky raised his daughter, in her words, "as a completely Soviet child," part of "an entire generation that, with very few exceptions, consciously or unconsciously deprived itself of the right to doubt."[12] As a high school senior in 1937, Kaminskaya and her classmates rapturously recited the following verse by the Soviet poet Pavel Kogan:

> In our days one can be certain
> That children in ages to come
> Will weep in their beds at night
> Over the era of the Bolsheviks.
> And they will lament to their dear ones
> That they were not born in those times.[13]

Khrushchev's "Secret Speech," but even more, his hounding of Boris Pasternak, set the first stumbling blocks in Kaminskaya's path. Disgusted by the poet's persecution, she engaged in "criticism by silence," she wrote, since it "simply did not occur to me, or to many others, that it was possible to participate freely in public life. From birth until early adulthood, I had been conditioned to associate public life with participation in official demonstrations, meetings and gatherings organized by the authorities."[14] The participatory dictatorship was the only organized public life she knew.

In Moscow in the 1960s, Kaminskaya was one of a handful of attorneys willing to defend those who organized their own meetings and demonstrations—a job that risked major career damage. Despite possessing the necessary security clearance to work with politically sensitive materials, Kaminskaya had been denied permission in 1965 to serve as Daniel's defense attorney. Approached by Nina Bukovskaya in the aftermath of her son Vladimir's arrest at the January 22, 1967, demonstration on Pushkin Square, this time Kaminskaya was cleared by the KGB to act as defense lawyer, perhaps because, thanks to the new approach taken by Article 190, "anti-Soviet" charges were absent from the indictment. Several months later, after the prosecution had completed its criminal investigation, the forty-eight-year-old Kaminskaya got her first chance to interview her young client in the KGB's Lefortovo Prison. This was standard Soviet judicial procedure: defense attorneys entered the fray only after the state had finished assembling what it considered the relevant evidence and witnesses. To his mother's dismay, however, Vladimir Bukovsky wanted nothing to do with defense lawyers, whom he regarded as little more than stool pigeons for the KGB.[15] A defense attorney's typical strategy in court was to announce his or her distance from the client's "incorrect" views (thereby establishing credibility before the judge), to gather character references (preferably demonstrating working-class origins, or personal hardship, or some form of service during the war), and to plead for a mild sentence following the client's contrite acknowledgment of guilt. Many defense attorneys performed an additional role. "Clients who approached attorneys," admitted Kaminskaya, "were often looking for us to serve as intermediaries who could pass on bribes to the judge."[16]

Instead of interviewing Bukovsky, however, Kaminskaya found herself being interrogated by him. Was she a member of the Communist Party, like his parents? She wasn't. Did she have a security clearance? She did. How could that be?[17] This wasn't simply a matter of a client not trusting his attorney. The supremely self-confident Bukovsky, who had been expelled from Moscow State University after one semester in the physics department and subsequently schooled in Soviet law by a mathematician (Volpin), was convinced of his own superior legal knowledge. "A [KGB] investigator's main weapon," Bukovsky explained, "is the

Soviet person's illiteracy in matters of law." He had therefore spent the months of pre-trial detention "devouring the criminal codes [including the Russian Code of Criminal Procedure] as if they were detective novels. I memorized them . . . and discovered to my surprise just how many rights I actually have." "I crushed [my interrogators] with laws," Bukovsky bragged, "I pinned them down with articles, stunned them with paragraphs. . . . I had the KGB by the throat."[18] The vaunted Russian tradition of inducing schoolchildren to commit long, intricate texts to memory turned out to have applications beyond reciting Pushkin to ecstatic grandparents.

And therein lay a problem. Bukovsky was convinced that he had fully mastered Soviet law, bombarding Kaminskaya, as he had his interrogators, with articles memorized verbatim from the Code of Criminal Procedure, articles that KGB investigators had allegedly violated—and in the process missing the most salient legal issue of all. Committing the Code's 420 articles to memory, Kaminskaya dryly observed, "did not make him a lawyer."[19]

As Bukovsky's trial approached, the professional lawyer and the jailhouse lawyer found themselves on a tactical collision course. Both knew that a guilty verdict was preordained. KGB investigator Voronin had said as much to Bukovsky's fellow defendant Vadim Deloné: "Your parents spent big money on a lawyer for you, and all for nothing. In cases like this, the defense decides nothing. We decide everything."[20] With this in mind, and inspired by Sinyavsky's and Daniel's widely circulated closing statements, Bukovsky was determined to use his trial as a tribune from which to "unmask the KGB" (unmasking was a standard Bolshevik ritual vis-à-vis "enemies of the people") and force his interrogators to admit, "honestly and truthfully, tell the whole world, that there is no freedom to demonstrate in the USSR, that you forbid everything you don't like." He would prove, "as if I were addressing the entire country," that Article 190 was unconstitutional and then, as an apostle of Volpin and martyr for the rule of law, be sent off to the camps.[21]

Kaminskaya had other plans, namely, to defend Bukovsky within the existing law. She rejected the tactic of trying to prove the unconstitutionality of Article 190 on straightforward procedural grounds. In the

Soviet Union, as in many legal systems at the time, there was no principle of judicial review, and no court was authorized to declare a law unconstitutional or even to criticize it. A lawyer could not ask a judge to do what was legally impossible. Instead, Kaminskaya would insist that Article 190 *was* constitutional and that it therefore should be interpreted strictly within the framework established by Article 125 of the constitution, guaranteeing "freedom of street processions and demonstrations." While Article 190 prohibited "group actions that grossly violate public order," it did not prohibit demonstrations per se—in fact, the word "demonstration" appeared neither in the article itself nor in the relevant legal commentaries. Bukovsky had organized a peaceful demonstration, to which Article 190 was simply irrelevant. Kaminskaya was going to do something virtually unheard of in the annals of Soviet jurisprudence: acknowledge her client's behavior precisely as described by the indictment and then insist that it was perfectly legal. She was going to ask for an acquittal.

The KGB investigator whose job it was to observe meetings between Kaminskaya and her client—like judicial review, attorney-client privilege was not part of the Soviet legal universe—was skeptical. "You and I both know that Bukovsky's fate has already been decided," he told Kaminskaya. "How can he be acquitted when he's an enemy? No amount of persuasion has any effect on him. He has an inflexible character. I keep thinking: where did we screw up, how did we lose such a person? After all, he's the sort you'd want with you on a reconnaissance mission; he'd never let you down." The investigator even hoped his teenage son would develop the same "human qualities" as Bukovsky.[22]

In certain ways, the trial of Bukovsky and his fellow demonstrators echoed that of Sinyavsky and Daniel (and Joseph Brodsky before them), as if Soviet officials were stuck in a conditioned response to dissenting behavior that, by their own account, was repeatedly making them look like "dim-witted bureaucrats" who "turned scumbags into martyrs." Once again, a special system of admission tickets was established to ensure that the Moscow City Court was stacked with a "public" hostile to the accused, while friends and foreign journalists were kept outside the building. Once again the prosecution tied itself in knots, this time by failing to carefully distinguish between charges of disturbing public order and anti-Soviet activity. Just as Sinyavsky and Daniel had insisted on the multiplicity of meanings contained in literary texts, those examined in connection with the January 22 demonstration defended themselves by defending pluralism:

> JUDGE: Do you consider that all views should be freely expressed?
> KHAUSTOV: Yes.
> JUDGE: No matter what they are?
> KHAUSTOV: Yes, because for every opinion there is a counter-opinion.
> JUDGE: Do your friends hold these same views?
> KHAUSTOV: Their views don't always coincide with mine.[23]

When asked by the judge what had "united" him with the demonstrators, Volpin, summoned as witness, replied, "I didn't say that something united us. Though there was of course a uniting factor—namely, that we are all human beings. This ought to unite me and you, and you and the accused."[24] Once again, Soviet authorities mistakenly assumed they could control the impression made by the trial at home and abroad. And once again, a handful of sympathetic observers in the courtroom managed surreptitiously to create a transcript of the entire affair, circulating it in samizdat within the USSR and arranging for its publication in the West.[25]

Amid these structural parallels, however, the individual events constituting successive phases of the chain reaction were strikingly different, their effects not merely additive but transformative. To begin with, the trial of Bukovsky, Khaustov, and others was not about metaphors, works of fiction, or authorial intent. The texts invoked were relatively straightforward declarative statements calling for the release of prisoners and the revocation of certain legal statutes. Unlike Sinyavsky and Daniel, the defendants did not attempt to dodge overtly political issues, let alone to claim an "apolitical" status for the January 1967 demonstration:

> BUKOVSKY: Beginning in 1961, I started thinking about whether the democratic freedoms guaranteed by the constitution are in fact a reality in the Soviet Union.
> JUDGE: Why in 1961? What happened?

BUKOVSKY: In that year my friends Osipov, Kuznetsov, and Bokshtein were convicted of producing a hand-written journal. I consider this unjust, and since then I've become opposed to the atmosphere of unfreedom and repression which exists in our country.

JUDGE: Today we are discussing not the actions of your friends but your own actions. Stick to the point.

BUKOVSKY: I am sticking to the point. The [January 1967] arrest of Galanskov and others is analogous to this case. In our country there are numerous such cases—such as the case of Sinyavsky and Daniel.

JUDGE: We're not talking about Sinyavsky and Daniel—we're talking about you.

BUKOVSKY: I request, based on Article 243 of the Code of Criminal Procedure of the RSFSR, that the trial protocol indicate that the judge is not letting me have my say. As an opponent of all forms of totalitarianism, I have made it my life's goal to combat the anti-democratic laws that lead to political inequality in our country. In particular, I oppose the anti-constitutional decree recently ratified by the Supreme Soviet, because it contradicts the basic democratic freedoms guaranteed by the Soviet Constitution. The demonstration I organized is, in my opinion, a form of protest and struggle aimed at abolishing anti-democratic laws. In conformity with Article 125 of the Soviet Constitution, this right is granted to citizens. The civilian patrol squads were the only ones who violated public order on the square, and I don't understand how we can be tried for such actions.

PROSECUTOR: Bukovsky, judging by your statements, you do not repent of anything, even though in the course of the seven months you've just spent in solitary confinement, you could have reconsidered your views of Soviet laws.

BUKOVSKY: I consider that a most contemptible argument coming from the mouth of a prosecutor.

PROSECUTOR: You have a very poor understanding of Soviet laws. As Article 125 of the Soviet Constitution declares, the freedoms you mentioned are guaranteed for the purpose of strengthening the socialist order and in the interests of the toilers.

BUKOVSKY: So who decides what is in the interests of the toilers and what isn't? That's sophistry.

PROSECUTOR: I don't understand you. You have no special training—what right do you have to judge Soviet laws?

BUKOVSKY: If I'm going to be judged according to Soviet laws, that means I have the right to judge them.[26]

Soviet citizens—including Bukovsky's ashen-faced mother, who was present in the courtroom—were not accustomed to hearing words like these flung at representatives of the state by a fully sober individual.[27] Undeterred, Bukovsky continued to goad the judge into interrupting his final statement and thereby violating judicial procedure, so that he could expose the court itself as breaking Soviet laws.[28]

While her client used the trial to broadcast his criticisms of the Soviet system, Kaminskaya stuck to strictly juridical arguments. The prosecution, she argued, had failed to demonstrate any negative effects on public transportation, pedestrian access to the square, or other tangible aspects of public order, instead retreating to the hypothetical: the unsanctioned gathering "could have attracted a large crowd," which in turn "could have led to large disturbances."[29] Pushkin Square, according to the prosecutor, was "a space sacred to everyone." By foregoing the established mechanisms "for citizens expressing dissatisfaction with various aspects of reality," the protesters "violated public order." Their demonstration, the prosecutor insisted, was "rude" because it was "impudent"; it was impudent because their slogans "criticized existing laws and actions of the organs of state security, thereby undermining their authority."[30] Kaminskaya's defense was straightforward: slogans themselves could cause a breach of public order only if they were illegal, but neither the indictment nor the prosecutor had made, much less demonstrated, such a claim. "The fact that Bukovsky organized the demonstration," Kaminskaya concluded, "has been acknowledged by him and is disputed by no one. The main point is that my client's behavior does not constitute a crime.... We cannot invent laws and we cannot hold people criminally responsible where there is no legal basis for doing so. Criticism of laws or of the actions of the organs of state security is not in and of itself a criminal offense according to Article 190. The actions of my client therefore do not constitute a crime. I ask the court to find him innocent."[31]

A decade and a half later, after being forced to leave the Soviet Union, the former defense attorney felt the need to defend her strategy. Why had she embarked on a path with virtually no chance of reaching its stated goal—Bukovsky's acquittal—and with a very real chance of lengthening his punishment, not to mention crippling her career? Arguing for acquittal was "not simply a position adopted for an admittedly hopeless cause," Kaminskaya insisted, but "part of the struggle against arbitrariness and lawlessness. This was my contribution to justice, my place in the struggle for the observance of law."[32] In this manner, she was able to align her professional ethos with the "conscious heroism" of her young client, who had chosen a far riskier role. Bukovsky, whose run-ins with the KGB began in the tenth grade, when he was expelled for distributing a hand-written satirical magazine (tellingly titled "The Martyr"), similarly acknowledged having had "not the slightest hope" that the January 22 demonstration would achieve its stated goals, namely, the revision of Article 190 and the release of Alexander Ginzburg, Yuri Galanskov, and other purveyors of samizdat. He too was driven by an unstated aim: to precipitate a visible confrontation with the Soviet state. Indeed, he had momentarily panicked when, having unfurled the banners on Pushkin Square, it briefly seemed that "nothing would happen." What sort of demonstration would it be if the authorities didn't react? "The whole plan had been that they would confiscate the banners and take us away. Not that we would take the banners back home. I assumed I wouldn't be going back home at all."[33] He needn't have worried: the authorities were hardly inclined to let nothing happen. Within minutes they performed their expected role, seizing banners, roughing up demonstrators, and arresting the apparent ringleaders. For Bukovsky, the demonstration's deeper purpose and the means of achieving it were thus one and the same: to create a moment of "transparency," as he put it, "so that no one could say later, 'I didn't know.' Each of us wanted to have the right to say to our descendants, 'I did everything I could. I was a citizen, I fought for the rule of law and never went against my conscience.'" It was "a struggle between the *living* and the *dead*, the *authentic* and the *artificial*."[34]

And, one might add, between children and parents. It was not only in postwar Germany that adults who emerged from the mass violence of the 1930s and 1940s, when asked by a younger generation about their actions, or more often their inaction, responded with "I didn't know." Khrushchev's speech at the Twentieth Party Congress in 1956 had pulled back the curtain on a portion of Stalin-era atrocities, but the sole named perpetrators—Stalin and his fellow Georgian, Lavrenti Beria—could not possibly have carried out alone the mass crimes attributed to them (in which Khrushchev himself had played a not insignificant role). In his "Ballad on Disbelief," a samizdat poem that made the rounds in the mid-1960s, Bukovsky's co-defendant Vadim Deloné asked:

> Was it only Beria's monsters
> Was it only the over-anxious Stalin
> Who made the woodchips fly?[35]

The Soviet debate over the traumas of the Stalin era was in some ways even more fraught than the roughly contemporaneous (West) German effort to come to terms with the Nazi past.[36] Stalin, after all, had ruled for twice as long as Hitler. His reforging of Russia, and, more broadly, of the expansive Soviet empire, was deeper and more enduring than anything witnessed in Hitler's Germany and the territories under the Third Reich's control. The lion's share of Nazi blood-letting had occurred outside German territory, against non-Germans, while the USSR's more intimate violence took place largely among Soviet citizens. Nazism was emphatically discredited by its total defeat (including the suicide of its leader) in the deadliest war in human history, even as the USSR's victory in that war served as the ultimate vindication of the Soviet system.[37] Nazism quickly became the twentieth century's benchmark of absolute evil, a status that enormously simplified the job of repudiating it while doing nothing to explain it. By contrast, it was—and remains—no simple matter to disentangle Stalin's crimes from Stalin's triumph. Most important, while the postwar "de-Nazification" campaign was directed by foreign armies of occupation before passing into the hesitant hands of Germans themselves, the Soviet "struggle against the cult of personality and its consequences"—or what in the West was labeled, via dubious analogy, "de-Stalinization"—was directed by none other than Stalin's former underlings, who denounced some of the deceased dictator's means but none of his ends.[38] Imagine, as the British novelist Robert Harris brilliantly does, a victorious post-Hitler Nazi empire run by Albert Speer or Rudolf Hess, who at some point decides to repudiate certain of the Führer's "mistakes," and you begin to get a sense of the post-Stalin predicament faced by Stalin's heirs.[39]

Nonetheless, within the confines of the Kremlin's campaign to enshrine a patriotic narrative of the recent past, a limited public conversation on moral responsibility during the Stalin era had begun. Among its participants was Konstantin Bukovsky, born in 1908 to a peasant family in Tambov Province. After joining the Communist Party in his early twenties, he had helped carry out the forced collectivization of the peasantry and later, as a professional journalist, was elevated—allegedly by Stalin himself—to the ranks of the Union of Soviet Writers. "Humble" and "honest in his own way" but also "difficult and despotic," according to his son Vladimir, in the 1930s Bukovsky *père* lived the Soviet dream of upward social mobility for children of peasants and workers.[40] A generation later, he was unwilling to see that dream tarnished by anti-Stalinists, especially those who themselves had profited from Stalin's policies. In an essay published in the conservative journal *Oktiabr* (October) just months before his son's January 1967 demonstration on Pushkin Square, Konstantin Bukovsky pounced on Ilya Ehrenburg's memoir, *People, Years, Life*, which had begun appearing serially in *Oktiabr*'s arch-rival, the liberal journal *Novyi mir*, some five years earlier. Ehrenburg described having harbored private doubts about Stalin's show trials and purges, as well as his own "criticism via silence." For Konstantin Bukovsky, this was nothing more than an attempt to trim one's sails to the winds set loose by Khrushchev's "Thaw." As against

Ehrenburg's alleged "grand conspiracy of silence" during the 1930s, Bukovsky insisted that "we didn't have any doubts whatsoever" and "we did not lie" by praising Stalin while harboring doubts about him. Ehrenburg and other "repentant intellectuals," according to Bukovsky, were fabricating versions of their younger selves in order to retrospectively foster the illusion of moral distance from Stalin's excesses.[41]

For Vladimir Bukovsky it was bad enough that, in Stalin's time, his father (and mother, also a journalist) had helped produce "the very propaganda that had so basely deceived me and that was used to justify murders or cover them up."[42] Now, more than a decade after Stalin's death and Khrushchev's "Secret Speech," his father continued to defend that propaganda with the claim that an unspecified but all-embracing "we" had believed in its truth, as if it had been impossible not to believe. By publicly dissenting from official policy, Vladimir Bukovsky and other activists sought to ensure that *their* children would never have occasion to ask, "'Why were you silent? How could you have allowed all that to happen?'"[43] Alongside the manifest issues of moral responsibility and the rule of law, a family drama was playing itself out, driven by a youthful desire to escape parents' moral contamination and achieve an integrity capable of withstanding the scrutiny of future generations.

After the Sinyavsky-Daniel trial, the writer Chingiz Guseinov had captured an inner tension between, as he confided in his diary, "justifying myself before my descendants" and "securing myself in the present."[44] Bukovsky and his fellow demonstrators were prepared to forego the latter. Though now considerably lower than during the Stalin era, the price of justifying oneself was hardly negligible: three years in a labor camp for Bukovsky and his co-organizer Viktor Khaustov, one-year suspended sentences for Vadim Deloné and Yevgeny Kushev, who pleaded guilty and repented—temporarily, it turned out—for their actions. The newspaper *Vechernaya Moskva* (Evening Moscow), tagging Bukovsky as a "hooligan," falsely informed readers that he had pleaded guilty and declined to mention the cause around which the January 22 demonstration had been held.[45] An incensed twenty-seven-year-old physics graduate student and instructor at Moscow's Institute of Precision Chemical Technology, Pavel Litvinov, pushed back by releasing into samizdat a corrective letter to the editor. Although unacquainted with Bukovsky and the other defendants, he was in a position to know, having surreptitiously helped compile a transcript of their trial. Less than a year earlier, Litvinov had been too nervous to attempt to gain entrance to the courtroom where Sinyavsky and Daniel were being tried and instead had read about it in Alexander Ginzburg's compilation. After the two writers were sent to labor camps, however, Litvinov visited them (together with Bogoraz and her teenage son, Alexander Daniel) in Mordovia, where they were being separately held.[46] Now he was preparing to disseminate a sequel, under the planned title *The White Book of the January Trial*.[47]

Litvinov, one of the "free particles" set off by the chain reaction, was the grandson of Maxim Litvinov, Stalin's commissar of foreign affairs until 1939 and later ambassador to the United States. From his bedroom window in the legendary House of Government, where elite Bolshevik families resided, young Pavel could see the Kremlin. Like millions of Soviet children, he idolized Stalin and enthusiastically joined the Pioneers and the Communist Youth League. Russian literary classics, canonized in the Soviet curriculum, planted in him an instinctive sympathy for "the little guy"—the poor and the weak who suffered injustice in capitalist countries. In school he learned that the Russian Revolution and the Soviet victory over Nazi Germany were the leading edge of the struggle by oppressed peoples everywhere. But Khrushchev's speech in 1956 and the Sinyavsky-Daniel trial a decade later redirected Litvinov's Soviet-bred instincts against the Kremlin's treatment of a new generation of "little guys," some of whom were becoming his friends.[48]

Litvinov's exalted Soviet pedigree may well have nourished a belief that the KGB would think twice before arresting him. It didn't look good when revolutions devoured their children, as the Bolshevik Revolution had, with gusto, in the 1930s. What would the world think if, on the eve of its golden anniversary in October 1967, the revolution began feasting on its grandchildren? Four weeks after the conclusion of Bukovsky's trial, on September 26, the KGB summoned Litvinov to the Lubyanka and warned him that dissemination of the transcript would result in his arrest. There was still time, he was told, to mend his ways. Such "prophylactic" conversations were often recounted at gatherings of dissident friends around kitchen tables. Litvinov, however, took things a step further. After leaving the KGB's imposing headquarters on Dzerzhinsky Square, he typed a more or less verbatim rendition of the exchange that had just taken place and proceeded to send copies to *Izvestiia* and several other Soviet newspapers as well as to the *Morning Star*, *L'Humanité*, and *L'Unità*—the leading British, French, and Italian communist newspapers—with the request that they publish it in full. If KGB officer Gostev's aim in summoning Litvinov for a chat had been to discreetly nip dissident activity in the bud, then Litvinov in effect turned the tables by publicly exposing the KGB's threats before they were carried out:

> GOSTEV: Pavel Mikhailovich, we have evidence that you and a group of people are planning to prepare and distribute a transcript of the recent criminal trial of Bukovsky and others. We warn you that, if you do this, you will be held criminally responsible.
>
> LITVINOV: I know our laws well and can't imagine which law could be violated by the compiling of such a document.
>
> GOSTEV: There is such a law—Article 190. Have a look at the Criminal Code and read it.
>
> LITVINOV: I know that article extremely well. Incidentally, investigations under Article 190 are not within the competence of the KGB. I can recite it from memory. It has to do with slanderous fabrications that defame Soviet society and the Soviet state. What kind of slander can there be in the transcript of a trial held in a Soviet court?
>
> GOSTEV: But your transcript is going to tendentiously distort the facts and slander the conduct of the trial. This will be proven by the organs within whose competence such matters fall.
>
> LITVINOV: How can you know that in advance? And more generally, instead of conducting this senseless conversation and initiating a new case, you yourselves ought to publish a

stenogram of the trial and cut short the rumors that are making the rounds in Moscow.

GOSTEV: Why should we publish it? It was just an ordinary case of violating public order. All the information about the case appeared in *Evening Moscow* on September 4. There you can find everything you need to know about the trial.

LITVINOV: To begin with, very little information was given there: a reader who hadn't heard anything previously about the case simply wouldn't understand what it was about. Secondly, the information was false and slanderous. The editor of *Evening Moscow*, or whoever provided the information, should be charged with slander. . . .

GOSTEV: Pavel Mikhailovich, the information was completely accurate. Keep that in mind.

LITVINOV: It said Bukovsky pleaded guilty, but I looked into this case and know for sure that he did *not* plead guilty.

GOSTEV: What does it matter how he pleaded? The court declared him guilty—so he was guilty. *Evening Moscow* printed everything Soviet people need to know about this case and it's completely truthful. We're warning you: even if it's not you but your friends or someone else who assembles that transcript, the person held responsible will be you. Imagine if the whole world learns that the grandson of the great diplomat Litvinov is involved in things like this—it would be a stain on his memory.[49]

Remarkably, much of the world did learn. Not thanks to the communist newspapers to which Litvinov sent the transcript; none of them published it. But three months later, on December 29, the *International Herald Tribune* did, along with an accompanying article under the headline "Soviet Intellectuals Playing Russian Roulette with the KGB." The story was picked up that same day by the Voice of America and the BBC, which broadcast a theatrical reading of the dialogue in Russian—now featuring "Dr. Litvinov, eminent physicist" and, as the radio announcers never tired of noting, grandson of the former Soviet foreign minister.[50]

It was hardly news to Soviet citizens that such "prophylactic" conversations took place; they were becoming standard KGB practice.[51] But that Litvinov would dare to expose the KGB by telling the world what was said—*that*, according to his friend Andrei Amalrik, "made a tremendous impression on me and, I think, not just on me."[52] In the spirit of transparency, Litvinov had even included his name and address in the document. Within days of its broadcast, he began to receive letters and telegrams from shortwave listeners in Riga, Kyiv, Leningrad, Moscow, Odesa, Tashkent, Volgograd, Yaroslavl, and more than a dozen other cities. Over sixty such documents arrived; others were likely intercepted by the KGB. "You could consider them," wrote one of Litvinov's friends, tongue only slightly in cheek, "the first independent research into Russian public opinion since November 7, 1917."[53] Some were anonymous and almost comically brief. "Privyet!" (Hello!) read one in full. "Happy New Year!" read another. Others praised Litvinov's courage and wished him success in his struggle, as one listener put it, "against obscurantists and Bourbons" (signed "with respect, from a stranger and a patriot"). A resident of Leningrad asked for a copy of Litvinov's original letter to the newspapers, the text of Bukovsky's final statement at his trial, and information about the fate of Ginzburg and Galanskov, in order to "form my own opinion on this matter." Most letter writers, however, had already formed their own opinion, including one who opened his message with "Why are you disgracing your grandfather's memory, you lousy kike?" It occurred to Litvinov, though evidently not to the letter writer, that he was a "lousy kike" precisely because his grandfather was a "kike."[54] Another letter, from a self-described "ordinary Soviet woman, not a Party member, who has known hunger and cold," is worth quoting at length:

> You, to whom Soviet power has given everything, to whom since infancy all roads have been open, you, who have always been able to go wherever you wished, who could choose whatever university you wanted, who never had to worry about rations, who received an apartment from the Moscow city council without having to wait in line—without a moment's hesitation you write a libel against Soviet

power and give it to filthy journalists for a filthy broadcast on the "Voice of America." You and those who think like you, smug and vacuous individuals whose upbringing cost the Soviet people so much, you seek acute sensations because you are spiritually impotent. Simple human feelings, *elementary* feelings like conscience, honor, and love of Motherland, have atrophied in you because you practice verbal fornication, a sort of tickling of deadened nerves. It's you who gave birth to the dandies [*stiliagi*] and other scum, you who with your deadened taste think up such infernal mixtures as the "Bloody Mary" to drink at your parties, where you perfect your plans for your next dirty trick, so as to pass for a "hero," a "personality of genius," etc. In your vile letter you even gave your address, to protect yourself while making sure that your gesture appeared like an act of bravery. You're not fooling anybody! You spat on your grandfather's dead eyes and defiled the New Year holiday for Soviet people who accidentally heard the "Voice of America" that evening. They should have propped you up in Revolution Square so that passers-by would spit in your face, you traitor.[55]

Notwithstanding the priceless disclaimer about accidentally listening to the Voice of America (a popular pastime among Soviet retirees), this letter captures as well as any the popular resentment against intellectuals such as Litvinov who dared to drape dirty laundry over the Iron Curtain.[56] Entirely bypassing the substance of Litvinov's transcript, the author targeted what she believed to be the immoral character of the Litvinov and his friends, their lack of gratitude to the Soviet state for bestowing on them the privileges of the intelligentsia, and their vain lust for the status of "outstanding personality," with its attendant fame at home and abroad.

For the time being, Litvinov appeared to be winning or at least surviving his game of Russian roulette. And perhaps enjoying it, too: he was known to ask the KGB agents who now constantly followed him in their black Volgas for a lift when he needed to get somewhere in a hurry.[57] On at least one occasion, when a group of agents began taking photographs of dissidents assembled for a trial of one of their friends,

Litvinov promptly borrowed someone's camera and started photographing them in return.[58] "We feared everything," he recalled, "and we feared nothing."[59] Samizdat copies of the transcript of Bukovsky et al.'s trial began to circulate, and the Soviet government's only response was to censor any mention of Maxim Litvinov from the media, either to posthumously punish the grandfather for his grandson's sins or, more likely, to minimize embarrassment to itself.[60] Less than two weeks after the Voices broadcast Pavel Litvinov's exchange with KGB officer Gostev, the chain reaction moved to its next phase. After nearly a year in pre-trial detention (well past the nine-month legal maximum), Alexander Ginzburg, Yuri Galanskov, Vera Lashkova, and Aleksei Dobrovolsky went on trial for anti-Soviet agitation and propaganda under Article 70. What became known as the "Trial of the Four" unfolded in the same Moscow courtroom where Bukovsky and other demonstrators had been tried, and a similarly handpicked audience ensured that sympathy inside the courtroom tilted, rather boisterously, toward the prosecution, while outside a crowd of several hundred supporters of the accused held daily vigils. Over a hundred of them had collectively petitioned in advance of the trial, insisting on the right to attend—itself evidence of rising coordination. Their demand that "laws regarding transparency in legal proceedings be observed" went unanswered.[61] By now, the practice of secretly creating transcripts of political trials had become sufficiently common that it hardly mattered when the KGB threatened relatives of the defendants with reprisals or confiscated notes from individuals as they exited the courtroom. Enough information was retained by enough witnesses to allow Litvinov, undeterred, to begin work on the next "White Book."[62]

One of the signers of a petition demanding the right to attend the "Trial of the Four" was Ludmilla Alexeyeva. "The first petition," she recalled, "was the hardest to sign," because unlike standing outside a courthouse or observing a demonstration, it created a permanent written record. Intellectuals defined themselves as people who took the written word seriously, and they knew the Soviet government did too. What would happen, Alexeyeva wondered, if she signed? Would she lose her job? Would she ruin prospects for her two young sons? What would happen if she *didn't* sign? Would her sons ask, "Why didn't you sign? Aren't you friends with those people?" "What goes on psychologically when these protests are signed by members of the intelligentsia," wrote another signer of similar petitions, "has never been explained anywhere." A thousand moral dramas played out among family and friends, as news spread of who had signed and who had not:

> Not to sign a petition in defense of an unjustly condemned man when a friend's signature was there was demeaning. Among the intelligentsia all were equally "opposed" to the official illegalities and therefore petitions and letters appealed directly to the conscience. But to sign meant to doom one's self and one's family to unknown and uncertain consequences, to future sacrifices whose scope was impossible to foresee. For the present, nothing might happen—the government might pretend not to notice. But everyone knows that the regime has a powerful memory. It can recall one's crime tomorrow, next year, or even in twenty years.[63]

To sign such a letter felt like "submitting a denunciation of yourself." When one signer heard her name (and address) mentioned in a Radio Liberty broadcast, she hid in a closet.[64] To seek additional signatures from friends and colleagues, according to the linguist Yuri Glazov, was to walk around carrying "a bomb." "You could tell what kind of people they were from how they reacted." One colleague signed, then changed his mind and asked to have his name removed, then changed his mind again and asked that his name be reinstated. As the computer scientist Valentin Turchin noted, "The highly placed say that, well, it's fine for the 'simple folk' [to sign]—they have nothing to lose. The 'simple folk' would say, it's fine for the highly placed—nobody will touch them. One person happens to be finishing a dissertation, another doesn't want to let his boss down, a third is worried about jeopardizing a trip abroad." "Reputations and long-standing friendships," Glazov recalled, "were destroyed."[65]

For Andrei Amalrik, the deeper drama was not between signers and non-signers, but within each individual signer. Putting one's name on an open petition marked "a step toward internal liberation—and for many that step was decisive. This or that particular signature might have no significance whatsoever for the country's political situation, but for the signer himself, it could become a kind of catharsis, a rupture with the system of double-think in which the 'Soviet person' had been raised since childhood."[66] The effect was striking enough to give rise, sometime in 1968, to a new Russian word: *podpisant*, a person who has signed.[67]

The first petition signed by Alexeyeva in anticipation of the "Trial of the Four" declared that the constitutionally guaranteed openness of judicial proceedings "is at the basis of all other judicial guarantees, and for that reason we cannot tolerate even isolated violations of the principle of transparency."[68] Judicial transparency seemed as fundamental to the rule of law as intellectual honesty was to ethics. Two decades later, however, in her autobiographical account of signing this first petition, Alexeyeva invoked something more visceral than the principle of open courtrooms. By arresting Ginzburg, Galanskov, and others, the government was trying to strangle samizdat itself, the medium that made the chain reaction possible. Western media coverage (including the Voices), unable to resist a seemingly ageless Russian archetype, cast "the Four" as persecuted writers. But to their Soviet supporters they were something different: *samizdatchiki*, compilers and distributors of uncensored texts of every imaginable provenance and genre, whether official (but not meant for public circulation) or unofficial, foreign or Soviet, fictional or documentary. By the late 1960s, according to Alexeyeva, consumption of samizdat had expanded so dramatically that "the intelligentsia could no longer imagine life without it."[69] Curators of the forbidden, *samizdatchiki* fed the Soviet intelligentsia's insatiable textual appetite. As one zealous reader put it (quoting the American scientist John Robinson Pierce), "From slaves of misinformation we shall become masters of information."[70]

The attraction of samizdat as an autonomous medium was so pervasive as to generate both alarm and humor. In a February 1969 memorandum to the Central Committee, KGB chairman Yuri Andropov reported his agency's discovery of the "preparation and dissemination of 'samizdat'" in Moscow, Leningrad, Kyiv, Odesa, Novosibirsk, Gorky, Riga, Minsk, Kharkiv, Sverdlovsk, Karaganda, Yuzhno-Sakhalinsk, Obninsk, and other Soviet cities and regions. Western radio broadcasts, he warned,

FIGURE 6.1. Ludmilla Alexeyeva

"are acquainting a significant number of Soviet citizens with the contents" of samizdat texts.[71] One popular anecdote of the time features a telephone conversation between two parties who assume the KGB is listening in:

"Did you finish the pie?"
"We'll finish it up tonight."
"When you're done, pass it on to Volodia."

In another, a woman brings Tolstoy's *War and Peace* to a typist to copy. "What for?" exclaims the bewildered typist. "You can buy this in a bookstore." "I know," answers the woman, "but I want my son to read it."[72]

Samizdat provided not just new things to read, but new modes of reading. There was binge reading: staying up all night pouring through a sheath of onion-skin papers because you'd been given twenty-four hours to consume a novel that Volodia was expecting the next day, and because, quite apart from Volodia's expectations, you didn't want that particular novel in your apartment for any longer than necessary.[73] There was slow-motion reading: for the privilege of access to a samizdat text, you might be obliged to return not just the original but multiple copies to the lender. This meant reading while simultaneously pounding out a fresh version of the text on a typewriter, as a thick raft of onion-skin sheets alternating with carbon paper slowly wound its way around the platen, line by line, three, six, or as many as twelve deep. "Your shoulders would hurt like a lumberjack's," recalled one typist.[74] Experienced samizdat readers claimed to be able to tell how many layers had been between any given sheet and the typewriter's ink ribbon. There was group reading: for texts whose supply could not keep up with demand, friends would gather and form an assembly line around the kitchen table, passing each successive page from reader to reader, something impossible to do with a book. And there was site-specific reading: certain texts were simply too valuable, too fragile, or too dangerous to be lent out. To read Trotsky, you went to this person's apartment; to read Orwell, to that person's.

However and wherever it was read, samizdat delivered the added frisson of the forbidden. Its shabby appearance—frayed edges, wrinkles, ink smudges, and traces of human sweat—only accentuated its authenticity.[75] Samizdat turned reading into an act of transgression. Having liberated themselves from the Aesopian language of writers who continued to struggle with internal and external censors, samizdat readers could imagine themselves belonging to the world's edgiest and most secretive book club. Who were the other members, and who had held the very same onion-skin sheets that you were now holding? How many retypings separated you from the author?

On January 11, 1968, four days into the "Trial of the Four" but before the widely expected guilty verdicts were pronounced, Pavel Litvinov and Larisa Bogoraz handed typed copies of a document to foreign correspondents who were mingling among the roughly two hundred protesters assembled outside the courthouse. By this time, Litvinov was meeting almost daily with journalists from Western countries.[76] In 1966, at the Sinyavsky-Daniel trial, friends and supporters of the defendants had mostly kept their distance from foreign journalists, instantly recognizable by their non-Soviet attire. KGB officers were keeping a watchful eye on those who lingered outside the courtroom, Alexeyeva recalled, and, in any case, she "didn't want to be quoted in a Western newspaper." But the problem wasn't merely that the journalists were Western. It was that they were outside the bonds of friendship that had moved Alexeyeva and others to show their support for the accused: "We are here because our friends are on trial. It's our problem; it's our grief. For reporters, this is just a political thriller. I don't want my life to be the subject of someone's curiosity."[77] Two years later, at the "Trial of the Four," such reticence had all but vanished. Circles of friendship were expanding, and foreign correspondents in Moscow had proved to be exceptionally valuable friends. Outside the courthouse where Ginzburg and others were being tried, numerous Western journalists, including Anatole Shub (*Washington Post*) and Karel van het Reve (*Het Parool*, a leading Dutch newspaper) easily established contacts with the defendants' supporters, despite KGB officials monitoring the crowd.[78]

Bogoraz and Litvinov were among those supporters. The appeal they handed to Shub and van het Reve was titled "To the Global Public." It excoriated the trial as "a violation of the most important principles of Soviet law," including the impartiality of judges, open access to the proceedings, and the right of witnesses to remain in the courtroom after testifying. It urged people around the world "in whom conscience is alive and who have sufficient courage" to demand punishment of those responsible for the violations as well as a new trial of the accused in the presence of international observers. Speaking directly to Soviet citizens, Bogoraz and Litvinov declared the trial "a stain on the honor of our state and on the conscience of every one of us. You yourselves elected this court and these judges—demand that they be deprived of the posts which they have abused. Today it is not only the fate of the accused which is at stake—their trial is no better than the celebrated trials of the 1930s, which involved us all in so much shame and so much bloodshed that we still have not recovered from them."[79] The appeal was being given to

FIGURE 6.2. Larisa Bogoraz and Pavel Litvinov, ca. 1968

"the Western progressive press" with the request that it be "published and broadcast by radio," rather than to Soviet newspapers, because, as the authors put it, "that is hopeless." At the bottom of the final page stood their names and addresses.

By assigning ultimate responsibility for the impending miscarriage of justice to the entire Soviet population, Bogoraz and Litvinov were in effect attempting to activate the claim articulated by the hero of Yuli Daniel's short story *Atonement*: "The state, it is we." Just as Volpin insisted on behaving *as if* the Soviet Union were a state governed by law, Bogoraz and Litvinov treated mass participation by Soviet citizens in public life—including the election of judges—*as if* it rendered citizens responsible for the state's actions.[80] In Russian, the same word, *vybory*, means both "election" and "choices." Soviet ballots, however, listed only one candidate for each office, an individual selected by the Communist Party. Casting a vote meant taking part in a public ritual that skeptics dubbed "*vybory bez vyborov*," an "election without choices." By the late 1960s, rituals of mass participation—especially elections—had become largely performative and indeed festive.[81] With candidates routinely receiving 98 percent (or more) of the votes cast, elections fostered the illusion of unanimity among Soviet people, an illusion constantly reinforced in official speeches and the mass media.

In a multi-party democracy with a robust press, even the appearance of popular unanimity is nearly impossible to achieve. The distinctive official culture of late Soviet socialism—in which, as a widely used slogan put it, "The People and the Party are One"—made possible Bogoraz and Litvinov's remarkable assertion that Soviet citizens were collectively responsible for the state's actions, including the actions of its judges. This was either a brilliant tactical move, designed to repurpose a Soviet ritual so as to foster genuine civic responsibility, or an astonishing case of literal-mindedness, taking the contrived unanimity of Soviet political rituals at face value. Or both.

No less remarkable was the call to "the global public," an entity as elusive as today's oft-invoked "international community."[82] In the USSR, however, the term had a distinguished pedigree, thanks in no small measure to the legacy of Soviet internationalism on which Bogoraz, Litvinov, and millions of Soviet citizens were raised. For much of their existence, the Bolsheviks had regarded existing borders as bourgeois conventions subject to adjustment or erasure as part of history's tectonic shift from sovereign nation-states to a global workers' state. History, it seemed, had ordained Moscow as the guiding force behind that shift. Moscow, for its part, was constantly reminding Soviet citizens that the USSR, as a typical *Pravda* headline announced, stood "at the center of attention of the global public."[83] What was noteworthy in Bogoraz and Litvinov's appeal was the notion that world public opinion could be deployed to *contain* Moscow's power, for example, by inviting foreigners to monitor the USSR's administration of justice to its own citizens. Even more than most other great powers, the Soviet Union treated relations with the outside world as a crown prerogative, the jealously guarded domain of the Kremlin alone.[84] More than a few supporters of judicial transparency and of the accused *samizdatchiki* balked at the idea of allowing foreigners to encroach upon the Soviet Union's sovereignty. This was not just "taking garbage out of the hut"—it was inviting strangers to inspect the hut from inside.

Most remarkable of all was what happened to the appeal once Bogoraz and Litvinov handed it to the foreign correspondents covering the "Trial of the Four." Not only did it reach its improbable addressee, "the global public," it did so not in months (as with Litvinov's transcript of his conversation inside the Lubyanka the previous fall), or weeks, or even days, but within *hours*.[85] By the afternoon of January 11 the text had been transmitted to news desks in London, and by evening it was being broadcast globally by the BBC, including its Russian-language service. In our era of social media it is difficult to recapture the effect, especially for Soviet listeners, of this stunning global dissemination of unauthorized information in what we would now call real time, that is, within the time horizon of the events themselves. "Where is the iron curtain?" asked one dumbfounded Soviet listener.[86] Another, echoing Bukovsky, noted, "We can remain silent this time, too, but we will not be able to lay the blame on our ignorance or naiveté, because I already know about the appeal by Larisa Daniel and Pavel Litvinov. I heard it in a BBC radio broadcast."[87] Even Litvinov's family members relied on the Voices to keep up with him. When Tatyana Litvinova was asked how her dissident nephew was doing, she pleaded ignorance—her shortwave radio wasn't working well.[88] As a popular ditty put it, "What happens in Russia currently // We will learn from the BBC."[89]

The conviction and sentencing of the four *samizdatchiki* triggered an outpouring of responses by Soviet citizens. In articles and letters to the editor published in Soviet newspapers, a surprising share of indignation was reserved for the fact that, as the *Literary Gazette* put it, "bourgeois propaganda has persistently passed off as Soviet *writers* people who have *never* published a single line *anywhere* (if we're talking about our country)." It was bad enough that Sinyavsky and Daniel had published their works while hiding behind pseudonyms. Those currently on trial weren't even authors to begin with. "All week long," complained a reader from Kharkiv, "the hoarse, mercenary 'Voice of America' has been breathlessly screaming over the airwaves: 'A Soviet court has unjustly convicted four Soviet writers! Demand the immediate release of the four writers!'"[90] Some blamed Ginzburg and Galanskov themselves for the imposture. Posing as "suffering littérateurs," argued *Komsomolskaya pravda*, these "enterprising seekers of glory" had "raised yet another fuss about 'writers falling victim to arbitrary rule,'" engaging in "unpardonable speculation in our country's long tradition of respect for literature, of trust in the person of the writer." "What right do they have," demanded a reader from Orel, "to call themselves writers? Literature is a calling for people who are strong, talented, and pure as crystal." A satirical article under the headline "How to Become Famous" described the American philanthropic organization "See-Eye-A" as offering Parker ballpoint pens, typewriters, and "copying" (carbon) paper to those in search of "quick success and global renown." In March 1968, Leonid Brezhnev himself condemned the "self-advertising" of those who had been praised by the forces of imperialism, "morally unstable and politically immature people ready to announce themselves as loudly as possible."[91] The problem, in a nutshell, was vanity.

In truth, it was Western media that reflexively described as "writers" individuals who claimed no such status for themselves, certainly not in the exalted Russian sense of the word. In Russia, a writer was someone who, while perhaps not as pure as crystal (Dostoevsky, after all, gambled in casinos, while Pushkin gamboled in bordellos), created works that constituted the collective conscience, or what Alexander Solzhenitsyn famously called a "second government." In the Soviet setting, where Stalin referred to them as "engineers of human souls," writers' lofty status was institutionalized via membership in the state-sponsored Union of Soviet Writers. By contrast, although Galanskov was an occasional and not overly gifted poet, both he and Ginzburg became known principally as compilers of other people's writings. Their hallmark, in fact, lay in their editorial minimalism, which is to say, in how *little* they wrote.

Beyond the official press, in the world of samizdat, reactions were no less vigorous. Within two weeks of the BBC's broadcast of Bogoraz and

Litvinov's appeal, over a thousand people residing in two dozen Soviet cities signed petitions in defense of the *samizdatchiki*—roughly ten times the number of names that had appeared on petitions in support of Sinyavsky and Daniel. "We have public opinion!" exulted one contemporary.[92] In many cases, individual signers—as one noted in her memoirs—had never met most of the other signatories.[93] Quite apart from its speed and scale, the most remarkable aspect of the petition campaign was its emphatic focus on law and judicial procedure. In 1966, the main thrust of letters in support of Sinyavsky and Daniel had been to propose various extra-judicial ways of handling their case, whether via the "court of public opinion" or the tutelage of fellow writers. Now, scarcely two years later, there was strong evidence that Volpin's strategy of demanding strict and transparent adherence to constitutional norms was becoming the dominant mode of intelligentsia protest, a non-violent, seemingly non-ideological, and apparently legal method of taming the Soviet state's abuse of power. Even Solzhenitsyn, who would subsequently dismiss the USSR's "deaf-and-dumb judicial carcass" along with Volpin's strategy, initially saw fit to attack Soviet censorship as "unsanctioned by the constitution and therefore illegal."[94]

Many of the petitions responding to the conviction of Ginzburg, Galanskov, Dobrovolsky, and Lashkova demanded that, as one put it, "all the material related to this trial be made available for detailed public scrutiny in accordance with existing law." Some called for a new trial to be broadcast live on Soviet radio and TV. Another petition enumerated no fewer than eighteen procedural violations of the Soviet Code of Criminal Procedure that had allegedly occurred during the trial. Yet another, bearing seventy-nine signatures (many accompanied by addresses), insisted that "the accused acted within the framework of our constitution and laws" and that the *samizdatchiki* be retried in strict accordance with the law. The single largest petition, with 224 signatures from individuals in seven cities ranging from Kaliningrad (formerly Königsberg) on the Baltic Sea across four thousand miles to Magadan on the Sea of Okhotsk, condemned what it called the "flagrant violation of legal norms" and insisted that "the persons guilty of organizing this trial and of bringing Soviet law into disrepute should receive the punishment they deserve."⁹⁵ A letter from 139 intellectuals in the Soviet Ukrainian Republic recounted a similar chain reaction of political trials in Kyiv, Lviv, and other Ukrainian cities, virtually unmentioned in the Soviet press. There too, trials of dissenting writers begat protests and demonstrations, inspiring the journalist Viacheslav Chornovil to compile a samizdat documentary collection roughly analogous to Ginzburg's *White Book*, for which he was arrested and tried, sparking new protests. The two chain reactions—each unknown to the other—intersected for the first time in the Gulag, where Chornovil and other Ukrainian political prisoners met Sinyavsky and Daniel, after which their relatives and supporters in Kyiv and Moscow also established contact.⁹⁶ Just as the spaces outside courthouses became arenas where supporters of the accused met foreign journalists (both groups having been excluded from the courtroom), so punitive labor camps and psychiatric hospitals fostered contacts among political prisoners (having been extruded from Soviet society) as well as their relatives from across the USSR, people who otherwise would have continued to reside on separate islands in Soviet society. "The great truth of the political camps today," wrote Daniel, "is the remarkable diversity of the people gathered there."⁹⁷ These were the zones where the chain reaction multiplied, linking individual rights-defenders with Crimean Tatars, Ukrainian autonomists, evangelical Christians, Lithuanian Catholics, Zionists, and Russian nationalists.⁹⁸

Among the many connections forged in the camps, one of the most unusual was that between Yuli Daniel and Anatoly Marchenko. It was unusual because Marchenko was neither a political prisoner nor a member of the intelligentsia. Born to illiterate parents in the Siberian town of Barabinsk, he dropped out of school in the eighth grade and was first arrested in 1958, as a twenty-year-old oil driller, for brawling with exiled Chechens in a workers' hostel, for which he was sentenced to two years in a corrective labor camp in Kazakhstan. After escaping, he and a friend were caught attempting to cross the Soviet border into Iran, resulting in a six-year sentence at a strict-regime camp in Mordovia. It was there that he met the newly arrived Daniel—writer, translator, poet, son of a poet, proficient in half a dozen languages. Marchenko's inherited assumptions about intellectuals like Daniel—that they "got paid for doing nothing" and lived a life of ease—were still strong when they met.[99] "What makes you think Daniel's going to stand at a machine or handle a shovel?" asked one of Marchenko's fellow camp prisoners. "He'll find himself a cozy little nook here too, the Jews always get away with it everywhere."[100] More than a few dissidents, Jewish or not, returned the suspicion, viewing ordinary Soviet folk as little more than submissive if occasionally unruly livestock. "O fellow citizens, cows and oxen!" one of Volpin's early poems declared, "to what have the Bolsheviks reduced you."[101] According to Bukovsky, his fellow citizens had been reduced to a silent crowd that "drags itself along the boulevards, through the underground crosswalks of the metro. . . . They all raise their hands at meetings, vote in elections, and—most important of all—they do not protest."[102] This was the participatory dictatorship.

While the chain reaction readily crossed lines of ethnicity, drawing dissident Russians, Ukrainians, Lithuanians, Jews, Crimean Tatars, and others into an emerging communication network, differences of class proved more difficult to overcome. Officially, of course, there were no class antagonisms to overcome, insofar as the abolition of private property and the passing of the pre-revolutionary generations (along with the lethal campaigns against "class enemies") had transformed the USSR into the modern world's first classless society. In practice, however, while disparities of income and status were far more muted than in the West, differences in cultural capital, and specifically in levels of education, continued to act as an enduring social classifier within the Soviet populace. Not for nothing did the Lithuanian poet Tomas Venclova remark about the tight-knit dissident milieu, "It's a small world—if you don't count workers and peasants."[103]

Having gotten to know Daniel during the several months they overlapped in Mordovian Camp 11, Marchenko began to enter that world. Upon his release in November 1966, he made his way to Moscow, phone numbers of Daniel's friends in hand, including that of Daniel's estranged wife, Larisa Bogoraz (they remained legally married in order to preserve rights of visitation and correspondence). To his new

Muscovite acquaintances, Marchenko was an emissary from a foreign country. To be sure, thanks to a series of mass amnesties beginning shortly after Stalin's death in 1953, millions of prisoners had returned from the Gulag. Hundreds, possibly thousands of them had produced written accounts of camp life.[104] But only a handful of such accounts had circulated inside the Soviet Union. Solzhenitsyn's novella *One Day in the Life of Ivan Denisovich* created a sensation when it was published in 1962, with Khrushchev's personal approval, in the journal *Novyi mir*. Yevgenia Ginzburg's memoir *Journey into the Whirlwind* and Varlam Shalamov's *Kolyma Tales* were blocked from publication in the USSR and available only via samizdat. Most other Gulag texts took up residence in desk drawers. But more to the point, these and all other extant accounts of life in the camps were about Stalin's Gulag. Khrushchev had publicly announced in 1959 that there were no more political prisoners in the Soviet Union; the Gulag as an administrative unit had been shuttered in 1960.[105] Most Soviet citizens—including Bogoraz, whose father and stepmother had returned from the camps in one of Khrushchev's mass amnesties—assumed that the system of confinement and forced labor was a thing of the past. "We believed that there were no longer political prisoners," Bogoraz recounted, "that is, people sentenced for 'politics,' or for nothing at all—for their 'tongue' (telling jokes) or for laziness or as family members of 'enemies of the people.' Now, we assumed, such people had been freed. The political camps were either empty or filled with criminals and gangsters. This is how I and many of my friends and acquaintances imagined things."[106]

If dissidents were concerned above all with preventing a return of Stalinism—"The bloody past calls us to vigilance in the present," as Volpin's Civic Appeal put it—then Marchenko's message was that in certain respects Stalinism had never left. Notwithstanding the prisoner amnesties, the reforming Twentieth and Twenty-Second Party Congresses, and the Thaw, the Gulag remained, in reduced form, an arena of near starvation, unconstrained violence, and sexual brutality, not only for thieves and gangsters but for untold numbers of arrested Baptists and sectarians, Ukrainian and Baltic nationalists, members of underground neo-Leninist

FIGURE 6.3. Anatoly Marchenko, Larisa Bogoraz, and their son, Pavel, ca. 1972

groups, and other dissenters. Encouraged by Bogoraz and others, Marchenko composed a graphic, book-length account of his experience in the contemporary Gulag under the title *My Testimony*. Circulating widely in samizdat—one reader commented that Solzhenitsyn's artfully understated narrative about the peasant Ivan Denisovich "paled by comparison"— *My Testimony* was smuggled abroad and translated into half a dozen languages.[107]

For all his sudden prominence as a working-class dissident, however, Marchenko was the exception that proved Venclova's rule: the chain reaction, and the dissident milieu that fostered it, remained an overwhelmingly intelligentsia phenomenon. By writing a book (the first of several, all published outside the Soviet Union), Marchenko began his own migration toward the intelligentsia. By the time he and Bogoraz married in 1972, he had arrived.

Not one of the letters, petitions, and appeals to members of the Politburo, the Supreme Soviet, the Council of Ministers, the KGB, the Procuracy, or the Supreme Court produced a written reply. There were, however, responses of another kind. Many of the roughly one thousand individuals who signed protests against the trial of the *samizdatchiki* were called in for "prophylactic conversations" with the KGB and warned to cease their "anti-social activities." Copies of the protests circulated among local Communist Party organizations, with instructions to expel, with or without the required hearing, any Party members whose names appeared among the signers. Alexander Ginzburg's attorney, Boris Zolotukhin, was stripped of his Party membership and fired from the Moscow Collegium of Lawyers for his "un-Party, un-Soviet" defense of his client. Copies of protest documents were similarly used to assemble blacklists of writers and scholars, which were then passed on to publishers and journal editors. Hundreds of signers were expelled from universities, were fired from their jobs, had their salaries halved, or otherwise had their career and future threatened.[108]

Several of these retaliatory gestures ignited their own chain reactions. On April 4, 1968, a closed meeting was held in the Siberian Branch of the Soviet Academy of Sciences in Novosibirsk to question Raissa Berg, a fifty-five-year-old population geneticist, about her signature on a letter in support of the four *samizdatchiki*. What disturbed her colleagues most was the fact that the "Letter of the 46," as it came to be known (indicating the number of signatures), had been published abroad and then broadcast by the Voice of America. "That letter," announced one fellow scientist, "provides grist for our enemies' mills. It confuses young people." "You should conduct yourself in such a way," scolded another, "so that our behavior strengthens the impression that we do everything correctly here [in Novosibirsk]. Suppose a transcript of this meeting were to wind up abroad?" Which is precisely what happened.[109] Berg was promptly fired.

Two weeks later, a literature teacher at Moscow's High School 421, Valeria Gerlin, was the subject of a similar disciplinary hearing by her local teachers' union. Gerlin had grown up in the family of a high-ranking NKVD (People's Commissariat of Internal Affairs) official in a well-appointed apartment on Lubyanka Square. Her father's Socialist Revolutionary past caught up with him in 1937, shortly before she turned eight. Following his arrest and extra-judicial execution, Gerlin's mother was exiled for eight years to Kazakhstan as the spouse of an enemy of the people. Gerlin herself—like her future husband and fellow teacher, Yuri Aikhenvald, Volpin's comrade in exile—was arrested in 1949, convicted *in absentia*, and sent to the camps. After Stalin's death, she was rehabilitated, along with her mother.[110] "I know what violation

FIGURE 6.4. Valeria Gerlin

of laws means," she told the forty-two colleagues assembled to judge her. "I know how important observing the letter of the law is for the honor of our state. History, for us, is not to be found in textbooks, in lifeless pages. For us, history is in our blood. We must feel personally responsible for it and before it. We cannot be indifferent to violations of the law."[111]

It is impossible to know whether the teachers who spoke out against Gerlin and who voted to terminate her employment did so out of inner conviction or as a form of lip-service to a decision rendered in advance by Party officials, or out of some combination of the two. Of the thirty-five who voted against her, most appear to have remained silent.[112] Those who did express themselves deserve close attention, for even if they were engaged in some version of double-speak, voicing opinions they did not necessarily or fully hold, their words reveal the kinds of criticisms of law-based dissent that contemporaries considered plausible and persuasive. Thus one teacher, L. P. Semerova, complained that Gerlin "says it's not about *who* was tried, but *how* they were tried. How can a Soviet person talk like that? We encourage people to speak about shortcomings, but why on earth sign letters? Why collect signatures? Why make this known to the whole world?" Another teacher, A. V. Novozhilova, dismissed Gerlin's criticism of restrictions placed on access to the court where "the Four" were tried: "One can't allow them to make their anti-Soviet statements in court in the presence of Soviet people. You mustn't force our people to be subjected to the influence of such slander." Besides, she added, there might have been other reasons for restricting access to the courtroom, and "we don't need to know all the reasons. We have to trust our organs [of state security], not be suspicious of them." For Novozhilova, the real problem was that Gerlin had simultaneously come to the defense of anti-Soviet people *and* taught literature to high school students. "Such two-facedness is impermissible: a person who wavers or doubts cannot be a transmitter of our ideology, cannot be an educator, cannot work in our school." Epshtein, a geography teacher, agreed: "We have to be entirely convinced, we cannot have any wavering—otherwise, what kind of teachers are we! How can we dare to emphasize political questions in class, if we lack lofty conviction and crystal purity in our worldview?" So did Ingerev, the school director: "You have no right to sow doubt in the minds of sixteen-year-olds! I fought at the front, I know how they condemned war criminals. You think I was interested in laws then?"[113] Whether they were sixteen-year-old students or sixty-year-old pensioners, Soviet people needed to be protected from ideas that could cause them to waver.

Gerlin claimed to have been driven by a sense of "civic unease" fed by the memory of Stalin's terror. Eidlin, a history teacher, wasn't buying it: "Enough already about 37. It's been condemned. There were other things besides the year 37. Like Soviet power." Another teacher rejected Bogoraz and Litvinov's argument that, having elected their judges, Soviet citizens should feel collectively accountable for the administration of justice. Opting for an ethic of solidarity over an ethic of responsibility, he noted, "We all voted for that judge, we trust our judge. A person who doesn't trust cannot be a Soviet teacher." This prompted a heated exchange between Gerlin and Georgy Vasilievich Gasilov, head of the regional school board:

> GERLIN: I consider each one of us to be obliged (and not simply to have the right) to write about facts that raise concerns regarding purity of the law. The concept of "citizen" imposes this responsibility, and not only on occasions when you are ordered to feel responsible.
>
> GASILOV: Stop trying to pull the wool over our eyes! Lega-a-a-lity. Who needs your (*using the familiar "tvoya"*) legality and your game of open courts! It's all talk!
>
> GERLIN: Georgy Vasilievich, are you really unable or unwilling to understand that a person who considers legality a game is in fact advocating lawlessness?! You yourself have had occasion to see the consequences of violations of law.
>
> GASILOV: You [*ty*] have no business holding forth on that!
>
> GERLIN: Our relationship is not such that we should address each other with *ty*. (*Murmurs in the auditorium.*) Legality is everyone's business. My unease, your unease, the concern and unease of everyone here about observing the laws is what guarantees their observance.... Our laws are sufficiently severe to punish criminals without recourse to dubious methods. We are asking not for mercy but for justice. Although it's no crime to request "mercy for the fallen."
>
> GASILOV: That's what liberal ideas lead to!
>
> GERLIN: Shame on you, Georgy Vasilievich! (*The auditorium buzzes.*) You've taken to speaking in the language of the Third Section!
>
> GASILOV: You see, she talked here about not having a grievance against Soviet power—but who can believe her! There's bitterness in her. Let's be open about this; there's no need to hide our sins. Who among us never has grievances against Soviet power? It happens with everyone. This person didn't get an apartment, that person wasn't paid a bonus, and so on.

> Everyone has some issue. But what do we do? Well, we come home, we grumble a little in the kitchen, but that's it. We take our grievance no further than the kitchen. But she—she writes it up!¹¹⁴

For Gerlin, the disciplinary hearing reenacted her experience under Stalin, indeed was a symptom of creeping re-Stalinization. "I am already familiar with everything that is happening to me," she announced to the assembled teachers. Then, as now, people believed what they were told by "higher authorities" and stopped saying hello to her.¹¹⁵

In the most crucial respect, however, what happened to Gerlin was not a reenactment. She appealed her case to a Moscow court, and, for unknown reasons, both she and her husband were reinstated as teachers.¹¹⁶ The fact that a thousand people across two dozen Soviet cities had taken part in a public act of protest against the trial of the four *samizdatchiki* and that uncensored documents pertaining to both the trial and the protests were now reaching thousands more via samizdat and possibly millions via the Voices—all this made Volpin's claim three years earlier in the Civic Appeal, that "citizens have the means to struggle against judicial arbitrariness," seem less fanciful. By early 1968, the chain reaction of interrogations, apartment searches, arrests, trials, demonstrations, transcripts, broadcasts, and protests had expanded far beyond the circles of friends with which it began, achieving what appeared to be a kind of critical mass. The sense that "so many events were occurring every day," that "hundreds of [judicial] sentences were just hanging in the air," as participants put it, presented a new challenge: how to keep up with the quickening pace and expanding reach of events, with what felt like the acceleration of history itself.¹¹⁷

Having included his home address on Alexei Tolstoy Street in the "Appeal to the Global Public," Litvinov found his two-room, centrally located apartment—a luxury by Soviet standards—becoming a destination for visitors from Moscow and other cities.¹¹⁸ Since he lacked a telephone (he assumed it would be bugged if he had one), Litvinov decided to establish Tuesdays as "visiting days" when guests could drop in during the evening to catch up on the latest news, read the latest samizdat, listen to the Voices, and, not least, share a drink.[119] There was no formal gate-keeping, but one had to at least know someone who knew Litvinov to gain entrance—not an especially imposing barrier, given that "half of Moscow" knew the eminently likeable Litvinov.[120] The KGB had no trouble placing a mole.[121] Tuesday evenings now featured twenty to fifty people crammed into Litvinov's apartment, engaged in fervent discussions—"the eternal requirement of thinking people in Russia."[122] Gone were the days when such conversations were conducted, if at all, in whispers. A neighbor, perhaps feeling the eternal requirement of a good night's sleep, reported to the authorities that Litvinov was "systematically holding collective drinking parties and listening to foreign radio stations at night, thereby violating the rules of socialist communal living."[123]

Petr Grigorenko and Viktor Krasin decided to be systematic too, setting their own visiting days on Monday and Friday evenings, respectively.[124] Petr Yakir announced that he didn't need to set a visiting day: at his place near the Auto Factory metro station, "every day was visiting day." Indeed, on pretty much any evening, other-thinkers (and fellow drinkers) arrived at apartment 75 from all across the Soviet Union—from Odesa, Vilnius, Kalinin, Obninsk, Melitopol, and other cities.[125] There they immersed themselves in an "unbounded sea of talk" that surged well into the night, or at least until the well-lubricated host kicked everyone out. "One of the most striking results of the past twelve years," wrote one contemporary, referring to the period since Khrushchev's "Secret Speech" in 1956, "is that we've taken a liking to conversation." The "terrible muteness" of the Stalin era had receded. "When we meet each other, we have lively exchanges of news, rumors, and predictions, we share emotional reactions to events in the political and cultural life of our society. Our speech has become freer; we've become bolder, more trusting."[126] "I don't think we'll ever drink ourselves to death," one frequenter of such gatherings observed. "We'll talk ourselves to death."[127]

According to the KGB mole, the talk at Litvinov's apartment was about "new forms of activism" that could "promote the conditions for the rule of law" without resorting to "Bolshevik methods."[128] That spring, the thirty-two-year-old poet Natalya Gorbanevskaya took up an idea that was making the rounds at the various visiting days: to start a news bulletin that would gather, organize, and disseminate information on state repression of dissent, information that was tendentiously reported by or, more often, entirely absent from the Soviet mass media. A dozen years earlier, in 1956, as a first-year student in the philology department of Moscow University, Gorbanevskaya had hit a stumbling block not unlike the one that had previously tripped up one of the instructors there, Andrei Sinyavsky: a call to collaborate with the KGB. Two of her friends, incensed at the USSR's suppression of uprisings in Poland and Hungary that year, had distributed leaflets around the university declaring that "we have indeed surpassed tsarist Russia—we've become the international gendarme of the entire world" and mocking Khrushchev as an "idiot."[129] One day, while attending a lecture, Gorbanevskaya was summoned to the dean's office and from there taken to the Lubyanka. Forced to don prison garb and placed in a holding cell, she held out for a day and a half before deciding—based on what she later came to regard as "shameful self-deception"—that as an upstanding member of the Communist Youth League, she should tell her interrogators the truth. Names, dates, locations, who said what, who wrote what: everything came out in a flood of honesty. Released three days later, she became the sole witness for the prosecution in a trial that resulted in multi-year prison terms for her two friends. To make matters worse, two other students had steadfastly refused to cooperate with the investigation.

Gorbanevskaya fell into a deep depression, horrified that she had become, as she put it, a "traitor"—not to her country, but to her friends. As a young woman, according to one of her lifelong friends, the writer Ludmila Ulitskaya, Gorbanevskaya was "categorical and intolerant."[130] Now she turned those qualities on herself. She began to hear voices at night and to feel nauseated whenever she touched a book or a piece of paper. It became impossible to sleep or work:

Preserve, and save, and forgive me,
a sinner, us sinners.
In a minefield, a thistle wound its roots
round a raspberry bush, before the explosion.[131]

When she decided to be baptized in 1967, she was told that ritual immersion would lift away all her sins, "but still I didn't forgive myself for this one."[132] Gorbanevskaya's studies took her to Tartu, in Soviet Estonia, where she worked with the semiotician Yuri Lotman, and Leningrad, where she became friends with a young conservatory student, the pianist Vladimir Ashkenazy. Her poems, lyrical and devoid of political intonation, appeared in some of the earliest samizdat journals edited by Galanskov and Ginzburg. She was among the first to sign petitions supporting them after their conviction in January 1968—a gesture, perhaps, of atonement for her earlier betrayal. At the various "visiting days," she took part in debates, along with Litvinov, Krasin, Yakir, and others, about how to create a news bulletin that would track current victims of state repression. "Our menfolk, you know," she recalled to an interviewer years later, "they were, well . . . inspired by the idea. But it was the women who did the work."[133]

There were of course precedents for such an undertaking. As every Soviet schoolchild knew, in the 1850s and 1860s the aristocratic socialists Alexander Herzen and Nikolai Ogarev had published an uncensored weekly journal called the *Bell* (*Kolokol*), reaching thousands of readers including high government officials and even Tsar Alexander II himself ("Tell Herzen to stop cursing me," Alexander once reportedly joked, "or I'll cancel my subscription").[134] But Herzen and Ogarev had produced their periodical from the safety of exile in London and Geneva, smuggling copies across imperial Russia's porous borders. With its masthead bearing the slogan *Vivos voco!* (I call the living!), the *Bell* featured works of literature and political tracts urging readers to join the revolutionary struggle against serfdom and tyranny. More recently, but virtually unknown to Soviet schoolchildren or anybody else, Baptists in the USSR had begun producing samizdat newsletters, starting in 1965 with the hand-written (and initially hand-reproduced) *Fraternal Leaflet* (*Bratskii listok*), whose masthead urged readers to "stand firm in the one Spirit, striving together as one for the faith of the gospel" (Philippians 1:27). The anti-religious campaigns unleashed by Khrushchev, along with the attempt to forcibly merge Soviet Baptists and evangelical Christians into a single organization supervised by the state, had triggered dozens

FIGURE 6.5. Natalya Gorbanevskaya

of arrests and church closings followed by trials, arrests of protesters gathered outside courtrooms, and the dissemination of trial transcripts via samizdat and tamizdat.[135] Crimean Tatars deported by Stalin en masse in 1944 had similarly begun producing "information bulletins" in the mid-1960s in support of their campaign to return to their ancestral peninsula.

Gorbanevskaya's news bulletin shunned its predecessors' impassioned advocacy, to say nothing of the ideological pathos of contemporary Soviet mass media. Unlike the *Bell* and the *Fraternal Leaflet*, it did not summon its readers. "Under no circumstances should you press on people's emotions," Gorbanevskaya insisted, summarizing her editorial ethos. "Give them information alone." It was a technique honed in the crucible of crafting group protest letters, where finding formulations acceptable to the maximum number of signers required focusing more on facts than on interpretation, let alone exhortation. Leaving out value judgments allowed readers to imagine an interpretive consensus whether one existed or not.

The news bulletin did not attempt "to encapsulate ultimate truth; it merely gathered evidence."[136] Or as one contemporary put it, "Truth is beyond our reach; we'll settle for facts."[137] Another predecessor, Roy Medvedev's monthly samizdat journal of political commentary, was devoted to "the development of Marxist theory."[138] Informally known as the *Months*, it was deliberately limited to five copies per issue, for circulation among several dozen carefully chosen intellectuals inside and outside the Party. By contrast, Gorbanevskaya's bulletin did not identify with any school or theory beyond the notion of universal human rights. Article 19 of the Universal Declaration of Human Rights appeared on the cover page of each issue, serving as the bulletin's minimalist credo and protective armor: "Everyone has the right to freedom of opinion and expression; this right includes the freedom to hold opinions without interference and to seek, receive and impart information and ideas through any media and regardless of state borders."[139]

In honor of the twentieth anniversary of the Universal Declaration of Human Rights, a group of newly decolonized countries in Africa and Asia had successfully lobbied the United Nations to designate 1968 as "International Year of Human Rights." The Kremlin fully endorsed this move in Soviet media, seeing yet another opportunity to excoriate the capitalist camp for its legacy of racist colonialism.[140] Evidently aware of this campaign, Gorbanevskaya decided to name the bulletin *Human Rights Year in the USSR*, with the anodyne subtitle *Chronicle of Current Events* borrowed from a popular Russian-language news program broadcast by the BBC.[141] From the outset, then, the bulletin announced itself using foreign rather than Soviet or Russian idioms. As the name suggested, it was not initially meant to continue beyond 1968.[142] Gorbanevskaya released the first issue in samizdat on May 1 but backdated it to April 30, to avoid possible association with International Workers Day, an official Soviet holiday.[143] Over the course of the next two years, the name changed multiple times—*Human Rights Year Continues, The Movement for the Defense of Human Rights in the Soviet Union Continues, The Struggle for Human Rights in the USSR Continues*—before settling down to the maximally minimalist *Chronicle of Current Events*, known to its readers (and the KGB) simply as the *Chronicle*.[144] By then, it was clear that Gorbanevskaya and those who worked with her intended to keep publishing issues indefinitely.[145]

Readers of the *Chronicle* could visualize, for the first time, the itinerary of the chain reaction, beginning with an overview of the "Trial of the Four" (without a word of criticism against Aleksei Dobrovolsky, who had turned state's evidence against the other three defendants), an account of the protests that followed, and a list of protesters singled out for punishment, including Gorbanevskaya herself, who was involuntarily confined to a psychiatric hospital for several weeks. Various thematic sections became a recurring feature: "Arrests, Searches, Interrogations," "Extra-Judicial Persecution," "In Prisons and Camps," "News of Samizdat," and others. While readers may initially have regarded the *Chronicle* as a kind of in-house newsletter of law-based metropolitan dissent as conceived by Volpin, within a few years new sections were introduced to cover persecution of religious believers across the Russian Republic, Tatars seeking to return to Crimea, as well as activists on behalf of Ukrainian cultural autonomy and Jewish emigration. Decades later, from her exile in Paris, Gorbanevskaya proudly observed that while each of these groups eventually (or, in some cases, already) produced a bulletin about its own members, "we wrote about everyone."[146] If the Communist Party was fond of proclaiming that "no one is forgotten, nothing is forgotten"—with reference to the suffering endured by the Soviet population during the Great Patriotic War—then the *Chronicle*, Gorbanevskaya insisted, "is the real fulfillment of this slogan."[147]

The birth of the *Chronicle* presaged an eruption of samizdat periodicals. If the *Chronicle*'s mission was to apply the principle of transparency on behalf of victims of state persecution, *Crime and Punishment*, a bulletin launched less than a year later, aimed to do the same to perpetrators, including names and, where possible, addresses. The second issue (in 1969), for example, identified Valentin Astrov as having denounced, in the 1930s, innocent communists connected to Nikolai Bukharin, many of whom, like Bukharin, were subsequently executed. "Astrov now

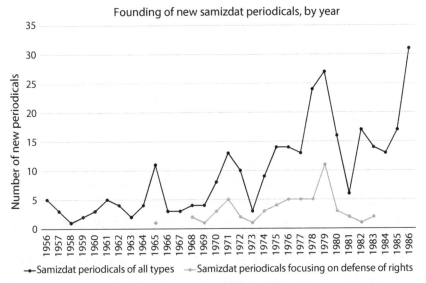

FIGURE 6.6. Graph of quantitative growth of samizdat periodicals, 1956–86

lives in Moscow," *Crime and Punishment* informed its readers, "and can often be seen in the Lenin Library."[148] The physicist Valery Chalidze founded his bi-monthly journal *Problems of Society* as a forum for uncensored analysis of Soviet law and human rights (Volpin was a frequent contributor), as well as for reviews of texts already circulating in samizdat, thereby "assisting the development of discussions within samizdat."[149] During the decade starting in 1968, the average number of new samizdat periodicals appearing each year in the USSR nearly tripled relative to the preceding decade, though, unlike the *Chronicle*, many were short-lived.[150]

Starting in 1969, Volpin's "transparency meeting" also became a regular, annual occurrence, honoring Soviet Constitution Day (December 5) without necessarily calling attention to specific arrests and trials, of which the chain reaction had now generated too many to associate with a single day.[151] To reduce the risk of harassment or arrest, participants would simply remove their hats as they silently gathered on Pushkin Square, as if to allow Russia's national poet, hat similarly in hand, to speak for them with words engraved at the base of his statue:

> I shall be loved, and the people will long remember
> that my lyre was tuned to goodness,
> that in this cruel age I celebrated Freedom
> and asked mercy for the fallen.[152]

All these activities—weekly visiting days, monthly uncensored news bulletins, annual demonstrations on Constitution Day—were meant to regularize and systematize dissident work. Taken together, they signaled a move away from the spontaneous trajectories of the chain reaction, which neither the Soviet government nor the activists had been able to control. The chain reaction itself was enabled by the emergence of the new communicative network of samizdat, whose dramatic growth owed much to the long reach of shortwave radio broadcasts emanating from the USSR's Cold War rivals. Extraordinary speech-acts by dissenting individuals such as Litvinov and Bogoraz, who not only spurned the camouflage of pseudonyms but openly announced when and where they could be found, produced an incalculable effect on those readers and listeners for whom "criticism via silence" had come to feel, by comparison, morally inadequate as well as wholly ineffective. The mode of activism pioneered by Volpin, emphasizing strict observance of constitutional norms and judicial procedure, also helped fuel the chain reaction by making it possible to articulate dissenting views without appearing to engage in conspiratorial or illegal behavior. "Defense of rights" promised an ideologically neutral and thus apolitical mode of civic engagement. The chain reaction was giving birth to a movement—a movement of a new type.

7

The Dissident Repertoire

На тыщу академиков и член-корреспондентов,
На весь на образованный культурный легион
Нашлась лишь эта горсточка больных интеллигентов,
Вслух высказать, что думает здоровый миллион!

Among the thousand academicians and academicians-in-waiting,
Among the entire legion of the cultured and educated
Was found this mere handful of feverish intellectuals,
To say out loud what the healthy million were thinking!

—YULI KIM

During Stalin's final years in power, and continuing into Khrushchev's Thaw, no worldview, no way of thinking, no set of texts contributed more to oppositional activity in the Soviet Union than those of Marxism-Leninism.[1] Even as it continued to fortify the Communist Party's monopoly on power, serving as a potent instrument of legitimation and social control, the USSR's official ideology also inspired countless clandestine "unions," "central committees," and "parties" dedicated to revolutionary struggle against the "bureaucracy" or "new class" that had allegedly hijacked the Revolution. There was the Union of Struggle for the Revolutionary Cause, the Communist Party of Youth, the Union of Patriots, the Union of Communards, and dozens, possibly hundreds, of similar groups. With membership typically in the single digits, most of them were as ephemeral as they were exuberant. Even when they were conspiring in the same town or indeed in the same university, the secrecy of such groups often kept them unaware of each other's existence—until they met behind prison bars. In most cases, historians discovered them only decades later, in the archives of the former Soviet Procuracy or the KGB.²

None of this is especially surprising. By the early post–World War II period, the demographic cohort most likely to rebel against authority—namely, youth—had been molded by Soviet institutions in which the revolutionary teachings of Marxism-Leninism had achieved the status of sacred canon. "For young people educated in Soviet schools," Vyacheslav Igrunov recalled, "there was no way out other than to begin one's political self-definition with Marxism."³ As we saw in the aftermath of the December 1965 transparency meeting, the revolutionary romanticism nourished by the official Soviet curriculum could not always be contained within approved channels, especially once the Soviet project entered its post-Bolshevik middle age. The dilemmas of postrevolutionary life—in particular, the gap between fulsome propaganda and flawed realities—were efficiently captured, as usual, via *anekdot*. A Moscow kindergarten teacher describes to her pupils how in the Soviet Union, children have the best toys, the sweetest candies, the nicest schools, and so on, when suddenly little Vova [Vladimir] bursts into tears. "Vovochka," the surprised teacher asks, "what's the matter?" To which Vova replies, between sniffles, "I want to live in the Soviet Union!"⁴

If Vova went on to join an underground student group, he would upgrade his desire for the best toys into a yearning for a "return to Leninist principles," or what one might call "doctrine-based reformism."⁵ In November 1956, for example, Viktor Trofimov, Ivan Potapov, and other founders of the Union of Leninist Communists (all of them members of the Communist Party's Youth League) distributed leaflets calling to "increase people's involvement in politics and restore Leninist freedom." "Where is the freedom that Lenin won for us?" they asked. "Let us destroy the chains of political bureaucratization under the banner of Leninism!"⁶ As Vladimir Bukovsky observed of those who gathered with him for poetry readings in Moscow's Mayakovsky Square in the early 1960s, "Taking the classics of Marxism-Leninism as their point of departure, and appealing to those works, people tried to force the authorities to abide by their own shining principles."[7]

From a tactical standpoint, this approach anticipates that of the rights-defenders, who tried to make the Soviet state abide by its own constitution and judicial procedures. Both appealed to a version of Soviet orthodoxy in order to transform the conduct of the Soviet state, an approach the anthropologist James C. Scott calls "strategic use of hegemonic values."[8] But it mattered which version was in play: the Soviet Constitution and the Criminal Code consisted of compact, already legally binding injunctions and prohibitions, while Marxism-Leninism filled dozens of volumes of texts spanning nearly a century and comprising letters, essays, articles, manifestos, and dense economic treatises, few of which were specifically designed to guide the behavior of state institutions. There was no personality cult hovering over the anonymous articles of the Soviet Constitution and the Code of Criminal Procedure. "You know as well as I do that our constitution is bad," Petr Grigorenko, a former major-general in the Soviet Army, once wrote to a fellow dissident. "It doesn't suit us at all. But it suits the authorities even less."[9] Unlike the neo-Leninist underground, moreover, rights-defenders had no desire to revive Bolshevik revolutionary traditions, with their ideological discipline, armed conspiracy, and *noms de guerre*. It was not simply that they shunned the use of violence. They did not seek to capture the state; theirs was a mission of containment by law.

Quite a few rights-defenders, to be sure, began their search for first principles with the idea of a return to Leninism. "I am a Leninist," Alexander Ginzburg, the future compiler of the *White Book* on the Sinyavsky-Daniel trial, told a friend in 1959. "I wish everything were the way it was under Lenin. What I am fighting is the remnants of Stalinism."[10] As a university graduate in the mid-1950s, Ludmilla Alexeyeva knew that tsarist Russia had been "an impoverished, unjustly governed country." Her grandparents told her so, and so did Pushkin and Tolstoy.[11] But she was plagued by a sense that "society was not becoming more just." "Life was not getting better. Could it be that we had been going in the wrong direction?" The only way to find out was "to start at the source": "I had to read Lenin from cover to cover." Thanks to the USSR's enormous, heavily subsidized print-runs of Lenin's writings, she had no trouble acquiring the Founding Father's collected works. For Alexeyeva, reading was a contact sport; her underlining and marginal notes "made each page look like a battlefield." But the result was not what she expected. Lenin emerged as a "gambler" who invoked class struggle and other impersonal forces, "Marxian abstractions invisible to the naked eye." His world was "orderly but barren." "It had no place for real people. I could not detect an iota of concern for soldiers, workers, or peasants. Nor could I detect even a glimmer of doubt. He was convinced that he knew precisely what needed to be done at any given moment, and he was capable of vicious attacks on anyone who disagreed with him on even the most minute of points." By the time Alexeyeva had read up to 1917, "I realized that I had lost all respect for Lenin."[12]

Equally troubling, the Founding Father was repeatedly discovered contradicting himself. When Grigorenko sat down in the early 1960s to "seek proof that the present Party-state leadership had deviated from Leninism," Lenin's writings themselves now appeared "contradictory" or flat-out wrong. "Returning to Leninist principles" turned out to be impossible because, while Lenin emerged as a supreme tactician, he lacked principles. He condoned extreme violence and his language itself was violent. He defined dictatorship of the proletariat, Grigorenko discovered, as "power based not on law but on force."[13] Poring over a memo in which Lenin had written that it was time to "rough somebody up" or "put someone up against the wall," the young writer Vladimir Voinovich had assumed that "perhaps Lenin was serious about the 'roughing up' part, but as for the wall, or 'execute as many as possible'—that was somehow meant figuratively. At some point I understood that no, it was not meant figuratively at all."[14] Reflecting on his own experience, the historian Aron Gurevich described this process as losing one's "Marxist-Leninist virginity."[15]

Not every Vova took leave of Leninism, of course, and not all those who took leave of Leninism turned to legalism. Some made their way to the Church, or the Nation, or Culture, or various combinations of the three—or to nothing in particular. Nonetheless, by the late 1960s rights-based legalism had effectively eclipsed revolutionary Leninism as the lingua franca of opposition in the Soviet Union, the dominant idiom of the chain reaction whose itinerary we followed in chapter 6. "For a while," observed Anatoly Chernyaev, the reform-minded head of the International Department of the Communist Party's Central Committee, in a diary entry, "one part of the intelligentsia expressed its alienation from the authorities and all of so-called public life through nostalgia for our revolutionary past and the revolutionary purity of the youth.... But this wave has passed. They grew tired and realized that this is just helpless nostalgia for a past one cannot bring back." Quoting the Italian communist politician Enrico Berlinguer, Chernyaev bitterly concluded, "The momentum of the October Revolution has dried up."[16] Leninism as a living mobilizational ideology had run its revolutionary course and was now the stale dogma of the lip-service state.[17]

The turn away from Leninism, however, hardly meant a rejection of socialism, which for most Soviet citizens, including most dissidents, had become an accepted way of life, something taken for granted and only loosely related to Leninism, or, for that matter, to law and rights. During the quarter-century of Stalin's rule, through extraordinary exertion and sacrifice, the Soviet people had built the world's first socialist system, whose vitality was conclusively demonstrated in the minds of many contemporaries by its emphatic victory over the Nazi empire. Khrushchev sought to revive revolutionary élan and push the USSR to the final stage of history, the transition from socialism to communism, during which the state apparatus would finally wither away. But few of his fellow citizens were in a hurry to follow him. Even fewer understood where he was going. Some suspected that he didn't understand either. Removed from office in 1964 in part for repeatedly sowing uncertainty, Khrushchev was replaced by a man who personified stability and predictability. Socialism, or as Leonid Brezhnev liked to call it, "mature socialism," settled into its new role as the USSR's natural order of things.

Soviet dissidents commonly understood themselves as reviving a tradition wider and older than Leninism, namely, that of the nineteenth-century

Russian intelligentsia. The pre-revolutionary "social movement" (*obshchestvennoe dvizhenie*) was a touchstone for collective engagement by intellectuals in public life.[18] More than a few historians have endorsed this alleged pedigree.[19] Having barely survived Stalin's terror, so the argument goes, the intelligentsia spirit of direct engagement with the "accursed questions" once again stirred courageous individuals not only to speak truth to power, but to live that truth in their own lives, inspired by engagé archetypes such as Alexander Herzen, Nikolai Chernyshevsky, Lev Tolstoy, and other pre-revolutionary writers. Thousands of outstanding members of the intelligentsia had perished in Stalin's prisons and camps, but lifelines could nonetheless be established between those who came before and those who came after. Varlam Shalamov likened Alexander Volpin to Petr Chaadaev, whom the government of Tsar Nicholas I had declared insane and placed under house arrest in 1836 in retaliation for the publication of his unsparing *Philosophical Letters*.[20] Natan Eidelman's hugely popular historical studies of the Decembrists were often read as coded portraits of contemporary dissidents.[21] "I felt an instinctive bond with the Decembrists," wrote Alexeyeva, describing herself in the 1950s. "Just a few years earlier, their country had won the war with Napoleon. It is citizens who win wars. But the war was over . . . and the regime no longer had any need for citizens, for citizens have a way of being a nuisance. They demand reforms; they demand the rule of law."[22] Foreign observers, too, found analogies to the nineteenth-century Russian intelligentsia irresistible. "Sakharov is my man, as he is, I am sure, yours," Isaiah Berlin wrote to Arthur Schlesinger, "the authentic noble liberal voice, very like Herzen's."[23]

The fact that rights-defenders and their admirers understood the dissident movement as reviving the moral and intellectual habits of the Old Regime intelligentsia tells us much about the latter's enduring appeal. It was the intelligentsia, after all, that had produced impassioned critiques of serfdom and autocracy and other forms of oppression, captured in works of literature that catapulted Russia into Europe's cultural pantheon and that have remained Russia's pride ever since. What nobler and more obvious lineage could dissidents have chosen for themselves?

Closer inspection, however, reveals a host of complications. To begin with, dissidents in the post-Stalin era, in contrast to their pre-revolutionary predecessors, emerged predominantly from scientific, mathematical, and technical fields. While well versed in Russian literature, their ways of thinking were nonetheless stamped by different standards of rigor and universalism. "My position," announced the physicist Valery Chalidze, who worked closely with Volpin, "does not depend on which country I live in."[24] Compared to their forerunners in the tsarist era, with their party congresses held abroad, their executive committees, and their active recruitment in imperial Russia's universities, Soviet dissidents remained a comparatively small and informal conglomeration of activists.

The tsarist intelligentsia, moreover, was imbued with a sense of moral duty (typically tinged with guilt) toward Russia's impoverished and illiterate masses, a duty that expressed itself in various forms of "going to the people" with a mission to civilize them. Soviet dissidents, to be sure, developed their own civilizing mission. "Critically minded people," according to Chalidze, "who do not hide their criticisms and who honor the principle of loyalty [to the existing legal system], provide the public with an unexpected example of law-abiding freedom of thought." This was exceptionally important for "developing a culture of thinking in a country where, for half a century, the governing ideology has claimed sole access to correct ideas" and where, in any case, "the majority of the population . . . has a low level of legal culture."[25] But many dissidents felt profoundly ambivalent toward a society that, despite having become better educated and more urbanized, was also more dependent than ever on the state. They did not share their intelligentsia predecessors' belief in the privileges of backwardness, the idea that Russia could leverage Europe's experience to leapfrog ahead in the process of self-modernization.[26] Nor did they share the pre-revolutionary intelligentsia's confidence in the "radiant future" or any other version of inevitable historical progress. When the American journalist Lincoln Steffens visited Soviet Russia in 1919, he famously proclaimed, "I have seen the future, and it works" (Soviet Russia at the time was in the throes of civil war, epidemic, and incipient famine, which would soon claim over ten million lives, most of them civilians). A half-century later, Soviet dissidents concluded that the future had indeed arrived, and it was not working.[27] They measured their society not as the late imperial Russian intelligentsia had, in relation to an imagined future purified by revolution, but against a specific, recent, and still traumatic past: Stalinism.

To further complicate matters, dissidents faced a competing claimant for the patrimony of the pre-revolutionary intelligentsia, namely, the Communist Party, the self-proclaimed "mind, honor, and conscience of our era." As the historian Richard Pipes noted in a prescient analysis on the eve of the Sinyavsky-Daniel affair, that competition forced the successors of the old imperial intelligentsia "to fight on entirely new ground." The sharpest divide between Soviet intellectuals and their pre-revolutionary predecessors lay in their respective attitude toward politics. "The old intelligentsia had a thorough commitment to politics," Pipes noted. "The modern one seems to shun and even despise it. The whole burden of the 'liberal' literature of the past several years is to assert the writer's right to an apolitical existence."[28] Politics, in the sense of public contestation over the common good, was now monopolized by the Communist Party, and therefore no longer available as a means for ordinary citizens to influence the behavior of the state. Instead, following in Volpin's footsteps, rights-defenders placed their hopes in the law, an arena of which the pre-revolutionary intelligentsia had been largely dismissive, regarding it as either an instrument of oppression by the autocratic state or a symptom of petty-minded bourgeois contractualism.

For Soviet rights-defenders to assume the mantle of the pre-revolutionary intelligentsia was therefore no simple matter. While they may have understood themselves as avatars of a tradition stretching back over a century, dissidents inevitably appropriated that tradition in a selective manner, highlighting certain elements, leaving others in the shadows, and transforming still others beyond recognition. Let us take Isaiah Berlin's man as an example, a person widely regarded as the living embodiment of the old intelligentsia spirit, Andrei Sakharov. In his memoirs, Sakharov recalls reading in the early 1930s, as a young boy in Moscow, Tolstoy's celebrated essay "I Cannot Be Silent," an impassioned denunciation of capital punishment written two and a half decades earlier, during the wave of judicial executions engineered by prime minister Petr Stolypin following the stalled revolution of 1905. So intensely did Sakharov identify with the "power of thought and feeling" in Tolstoy's jeremiad that he misremembered it as having been published in a collection of essays co-edited by his grandfather, Ivan Sakharov, as if to fortify his own connection to Tolstoy's—and his grandfather's—sense of "internal moral duty" and "civic courage."[29]

But it was not just the essay's editor, date, and venue of publication that got reworked in Sakharov's mind. So did its central argument. Tolstoy's despair at the military tribunals that were methodically sentencing thousands of suspected revolutionaries to the gallows ("Stolypin's neckties," as they were called) led him to denounce above all the judicial character of the executions. "More horrifying than anything," he wrote, "is the fact that this is done not in a fit of passion or emotion that smothers reason, as is the case in a fight, a war, or even during a robbery, but on the contrary, according to the demands of reason, of calculation that smothers feeling."[30] Like Tolstoy's essay, the phrase "the demands of reason" evidently stuck in Sakharov's mind and would resurface in the conclusion of his 1975 Nobel Peace Prize lecture, delivered in Oslo by his wife, Yelena Bonner. Human life may be little more than a fleeting episode in the history of the cosmos, Sakharov proclaimed, "a flash in the dark" within the endless expansion and contraction of the universe. Yet even if "we have emerged for a single moment from the black nothingness of unconscious material existence," he urged, we must nonetheless "carry out the demands of Reason and create a life worthy of ourselves and of the Goal we only dimly perceive."[31] Reason alone can protect humanity from the catastrophe of nuclear war; only "under the control of Reason" is progress possible and sustainable. Turning Tolstoy on his head—and echoing Volpin's commitment to the emancipation of reason from emotion and faith—Sakharov repurposed "the demands of reason" from heartless calculation to supreme ethical imperative.

Soviet dissidents looked to the pre-revolutionary intelligentsia for a usable past, and they found what they were looking for. They certainly appeared to many contemporary observers as heirs to that past. But in both substance (as a source of social and moral thought) and form (what the sociologist Charles Tilly called "repertoires of contention"), the Soviet dissident movement bears only a distant resemblance to the prerevolutionary "social movement."[32] We will have to look elsewhere to understand how the chain reaction evolved into a movement of a new type.

How do social movements come into being? What drives their evolution? Can we identify what actually moves, and why, and with what consequences?[33] The persistence of the totalitarian model in Western thinking about the USSR did not help in the search for answers to such questions, or even to recognize them as relevant to the Soviet case. The totalitarian model, for all its value as a means of identifying a new species of governance—a regime of a new type, forged in the crucible of total war and mass politics in early twentieth-century Italy, Germany, and Russia—had at least one crucial flaw: it did not offer ways to think about, much less explain, how totalitarian states and the populations they govern change over time.[34] The history of Italian Fascism and German Nazism spanned less than a single generation under a single leader. As sole survivor from the interwar pantheon of totalitarianism, the USSR became the world's first multi-generational experiment in that novel form of state power.

All of which helps explain why the study of social movements and their younger, more fashionable cousins, social networks, has barely begun to animate scholarly research on the Soviet Union and other socialist countries.[35] As the preceding chapters have shown, the image of an atomized, totalitarian society failed to register the strong ties formed among adult friends in the post-Stalinist USSR—bonds whose intensity seems to have inhibited the development of weak ties. All movements begin as networks, but some kinds of networks are better at becoming movements than others. What characteristics account for such differences, and what factors make networks more or less vulnerable to decay or suppression? Social movements, for their part, exhibit a variety of organizational arrangements, indeed they often experiment with structure and form. Do they practice, in their own milieu, the kind of arrangements they seek to promote in the larger society? Do they depend for vital support on forces (financial, institutional, or otherwise) outside the movement itself? If so, how do those forces shape movement goals and activities?

Just as the post-Stalin Soviet leadership was testing the removal of mass terror from the repertoire of governance, so certain Soviet citizens were abandoning Leninist models of revolutionary resistance and instead groping toward new repertoires of contention. Indeed, it was the Kremlin's renunciation of mass terror that enabled the search for "new ways of life," as Hélène Peltier put it, along with new forms of independent social action.

Legalism—getting the Soviet state to observe its own laws—was the grand strategy of Volpin and other rights-defenders. But that strategy dictated neither specific tactics (apart from avoiding obviously illegal activity) nor a specific organizational form. Even as rights-defenders sought to induce the Soviet state to abide by formal laws and procedures, they proved notably resistant to formalizing their own activities, whether by establishing an organization, defining distinct work roles, or codifying membership or leadership. They shunned such measures not simply because they might have made them more vulnerable to repression by the KGB, but because when it came to their own movement, they were anti-formalists. Given the long-established preeminence of personalities over institutions in Russian and Soviet political culture, we should not be surprised to discover similar preferences within the dissident milieu itself.

If one of the hallmarks of the chain reaction set off by the trial of Andrei Sinyavsky and Yuli Daniel was precisely the *reactive* character of the events that constituted it, then the year 1968 marks a turning point in the search for new forms of collective action in the post-Stalin era. That search, I have suggested, was made possible by the Soviet leadership's own quest, in the aftermath of the Stalin era's mass violence, to step away from state-sponsored terror, a quest that unintentionally opened up spaces for experimenting with new ways of living.

Now was the chance to become less reactive and more deliberate, or, as Major-General Grigorenko put it, "to go on the offensive."[36] But how?

"Visiting days" (the weekly gatherings of dissidents at designated apartments) and samizdat periodicals helped regularize the flow of critical information barred from the Soviet mass media, but they hardly determined what should be done with that information. Early in 1968, a group of activists began meeting to discuss future strategies. They included Andrei Amalrik, Larisa Bogoraz, Ilya Gabai, Natalya Gorbanevskaya, Viktor Krasin, Pavel Litvinov, and Petr Yakir.[37] Since their own apartments were presumed to be bugged, the gatherings took place on the outskirts of Moscow. Like Grigorenko and Amalrik, Krasin—whose father had perished in the Kolyma hard-labor camps, and who himself had spent four years in those camps before returning to Moscow and earning a degree in economics—was convinced that, as he put it in a subsequent interrogation, "the separate streams of which the movement currently consists should merge into a single current."[38] Krasin had been the moving force behind protest letters that garnered hundreds of signatures following the "Trial of the Four." He was organized, intensely focused on results, and openly contemptuous of the Soviet regime—all qualities that distinguished him from Litvinov. At meetings, Krasin always sought to come to some sort of conclusion, or consensus, identifying where people agreed and where they didn't.[39] Now, he and Grigorenko proposed, a "legal and transparent opposition" could emerge under the leadership of a committee whose goal would be to foster the "broad democratization of the Soviet social structure." The committee itself, Krasin argued, should consist of twenty to thirty individuals from various cities, each of whom should have a designated replacement in case of arrest. They would develop a plan of action, starting with the coordinated publication of all available materials regarding trials of dissidents, public protests, the condition of political prisoners, and the struggle by "small nations" (Soviet minorities such as the Chechens and the Crimean Tatars) and religious believers for equal rights. The proposed "Committee to Defend Civil Rights in the USSR" would also be responsible for coordinating ties to foreign organizations advocating civil rights around the world.[40]

Yakir, master of ceremonies at the Auto Factory's perpetual "visiting days," was largely indifferent to Krasin's and Grigorenko's proposal.

"With or without a committee," Krasin recalled him saying, "I don't give a damn. I've been taking action and I will continue to take action and it doesn't matter whether or not you pin to my last name the label 'committee member.'"[41] Yakir was right: his last name was already well known in the Soviet Union. His father, Yona Yakir, had distinguished himself as a charismatic Red Army commander in the Russian civil war, carrying out brutally successful campaigns against the Don Cossacks, the guerrilla forces of the peasant-anarchist Nestor Makhno, and the White army led by the tsarist general Nikolai Yudenich. By all accounts a brilliant orator and military innovator, Yona Yakir was a fervent Bolshevik and close friend of Khrushchev, with whom he served on the Central Committee of the Communist Party starting in 1934. Swept into the vortex of Stalin's terror in May 1937, Yakir was accused of participating in a Trotskyite military conspiracy and of spying for Nazi Germany, Japan, and Poland. After several days under torture, he was executed by firing squad in the basement of the Lubyanka at the age of forty. According to Khrushchev, when Stalin learned of Yakir's last words before he was shot—"Long live Stalin!"—his response was "What a bastard. What a Judas."[42]

In keeping with Soviet practice, Yakir's closest relatives were also treated as criminals: his brother, Morris, was executed in 1937, his sister, Isabella, was sentenced to ten years in hard-labor camps, and her husband, Simon, was executed sometime before the end of 1939. Yakir's wife, Sarra, was sentenced to sixteen years in the camps. Their son, Petr, fourteen years old at the time of his father's execution, spent the next seventeen years in prisons and camps. There he met and married Valentina Savenkova, a fellow prisoner. By the time of his release and rehabilitation in 1955, at age thirty-two, he had spent the majority of his life in the Gulag. After returning to Moscow, he joined the Party and found powerful patrons in the Soviet Army, high-ranking officers eager to restore the good name of their executed former commander. Khrushchev, guilt-ridden over having believed Stalin's surreal charges against a close friend and war hero, declared in a speech to the Twenty-Second Party Congress in 1961, "I knew Comrade Yakir very well. This year his son came to me. He asked about his father. What could I say to him?" The question hung in the air, unanswered, apart from Khrushchev's repeated assertion that Vyacheslav Molotov, Lazar Kaganovich, and Kliment Voroshilov were responsible for the father's execution. Shortly thereafter, Khrushchev arranged for the son to be given a research position at the Soviet Academy of Sciences, in the Institute of History.[43] Several admiring biographies and a commemorative postage stamp began the process of restoring Yona Yakir's reputation, while Petr Yakir toured the country giving public lectures about his father, whose charisma and oratorical skill he evidently inherited.[44] By the mid-1960s, he had become the de facto leader of a group of sons and daughters of prominent Bolsheviks killed in Stalin's purges, among them the children of Nikolai Bukharin, Karl Radek, and Nikolai Vavilov.[45] Within the dissident milieu, Yakir possessed a unique combination of moral authority and social capital: the son of a military hero wrongly executed by Stalin, an outspoken survivor of the Gulag, and someone with access to high Soviet officials, even after Khrushchev's fall from power in October 1964. A Westerner who frequently stopped by Yakir's apartment in the late 1960s recalled that "he, more than anyone else in Moscow, was the cement for the activist part of the movement." Young people in particular "were attracted to him simply by his sheer force of personality." The fearsome mystique of the Gulag saturated his apartment. Taken together, a Russian visitor observed, its three inhabitants—Yakir, his wife, and his mother—had spent a half-century behind bars and barbed wire. Imprisoned in his early teens, Yakir was a native speaker of the "thieves' dialect" that inspired a dozen dictionaries (translating camp argot into standard Russian) and captivated writers and musicians such as Shalamov, Sinyavsky, Alexander Solzhenitsyn, and Vladimir Vysotsky with its pungency. Gulag songs, accompanied by guitar, rang out from the apartment late into the night. Loquacious and combustible, especially with alcohol in the mix, Yakir "dared [people] to trust each other" and "speak their minds."[46]

But he was not one for committees. Neither were Bogoraz, Litvinov, Alexeyeva, or Gabai, who objected on principle to Krasin's and Grigorenko's call to harness the chain reaction into a movement with a leadership structure. For them, the moral force of protest depended on its being

FIGURE 7.1. Petr Yakir and his mother, Sarra Yakir

propelled exclusively by individual free will, unmediated by organizations and hierarchies.[47] The only acceptable form of obligation was that which emerged from within the individual conscience. The means deployed by the dissident movement were just as important as its ends. For Litvinov, in fact, the means—getting people to stop going through the motions dictated by the lip-service state, to abandon double-think and double-speak and instead act according to their conscience—*were* the ends.[48] Litvinov was also mindful that protesting carried different risks for different people, depending on how well known or connected they were, and that therefore the price of following one's conscience varied widely from individual to individual.

Beyond questions of conscience and risk stood an inbred aversion to the kind of badgering practiced by the Party-state, with its incessant reminding of Soviet citizens that they should contribute to the collective mission of building communism.[49] In the dissident movement, there should be no mobilizing, no recruiting, no external pressure to sign letters or attend meetings, no assigning of roles. "I never collected signatures," Litvinov reported years later. "I resisted the whole 'look, I signed, so you should sign too' argument."[50] After a decade and a half in the Communist Party, Alexeyeva recalled, "I didn't want to be anyone's subordinate or anyone's superior. I wanted to reserve the right to pick the people with whom I worked."[51] It was not just a matter of rejecting the oppressive hierarchy and structure of the Communist Party; it was about rejecting hierarchy and structure as such, in favor of what activists in our own time call horizontality. "The main thing in resisting a police state," the literary critic Raisa Orlova wrote in her diary, "is to live according to your own laws, not someone else's."[52] Activism would emerge spontaneously from within each individual, allowing the chain reaction to chart its own course. Only that kind of movement, Litvinov believed, could produce the transformation of consciousness imagined by Volpin—a revolution in the way revolutions are made. A movement of this new type required human personalities of a new type, people who would resist the descent into hierarchy.

Chalidze recognized that "free spirits," tired of being told what to think and how to behave, were keen to "seek their own paths of protest." But by spurning any subordination to collective requirements, "even those connected to some sort of tactical effectiveness," they were denying the very concept of "tactical optimization" in favor of "intellectual individualism."[53]

Positions were clarified, but there was no meeting of minds between the advocates of spontaneous, conscience-driven action and those who favored conscious organization and strategizing. The only thing actually decided at the spring 1968 gatherings outside Moscow, thanks to the initiative of Gorbanevskaya, was to launch the *Chronicle of Current Events*.

Not long after Bogoraz and Litvinov issued their appeal "To the Global Public" in January 1968, the British poet Stephen Spender, one of the contributors to the ex-communist manifesto *The God That Failed*, responded with a statement of support signed by more than a dozen renowned artists and intellectuals, including W. H. Auden, Cecil Day-Lewis, Stuart Hampshire, Mary McCarthy, Yehudi Menuhin, Henry Moore,

Sonia Brownell (widow of George Orwell), Bertrand Russell, and Igor Stravinsky.[54] Their one-sentence telegram—"We, a group of friends representing no organization, support your statement, admire your courage, think of you and will help in any way possible"—failed to arrive at 8 Alexei Tolstoy Street (Litvinov's home address indicated on the appeal), but its content nonetheless reached Litvinov and Bogoraz, and many other Soviet citizens, courtesy of BBC radio.[55] In a subsequent on-air interview, the eighty-five-year-old Stravinsky recounted how his teacher, Nikolai Rimsky-Korsakov, had similarly suffered under tsarist censorship. The effect on Litvinov, by his own account, was "colossal": "The Soviet regime had made every effort to separate the two Russias of before and after the revolution, and to create distinct worlds inside and outside state borders. Broken links in time and space were suddenly reconnected."[56]

In a private response to Spender, smuggled out by Karel van het Reve, Litvinov elaborated his views on the rapidly forming protest movement in the USSR. The movement was "diffuse and amorphous" and had not yet developed sufficiently clear goals beyond the struggle for human rights and the humanization of Soviet society. Nonetheless, he wrote,

> if a majority of citizens in our country would recognize that the individual human self, and first of all their own self, represents an independent value, and not just a means for fulfilling this or that distant, vague task, this would make it possible to create a healthier society without the horrors, violence, and bloodletting that have accompanied our long revolution, beginning in 1917. Such an appreciation of the human self does not exclude, of course, various metaphysical ideals—on the contrary, it makes them even more significant. All this may appear as truisms in the West, but we have arrived at them with our own minds and with great difficulties.[57]

Perhaps in the future, Litvinov speculated, the protest movement might "take on more concrete organizational forms," but this would "undoubtedly increase our difficulties." What was most urgently needed was the creation of an international committee whose task it would be to support the democratic movement in the USSR. Its members would include well-known progressive writers, scholars, artists, and activists from countries around the world, and it would steer clear of anti-communist or anti-Soviet sentiment. "It would even be helpful," he added, "to include people who have been persecuted in their own countries for their pro-communist or independent views, as for example in Greece," where the composer Mikis Theodorakis had recently been arrested, along with other left-wing artists, by the military junta—a development Litvinov regarded as a mirror image of the arrests of Sinyavsky and Daniel.

The proposed international committee "could take on all the practical work" of maintaining contacts with the democratic movement in the Soviet Union. It would help defend members of that movement subject to persecution for their social or literary activities, serve as a clearinghouse for links to other international organizations, keep the "global public" informed about the situation in the USSR, and help coordinate the publication of all manner of uncensored manuscripts abroad—in the process protecting authors against copyright infringement and loss of royalties, and perhaps even enabling a modest income for the legions of (mostly female) typists who, at great personal risk and little or no pay, enabled the diffusion of samizdat.[58]

An astonishing amount of what Litvinov proposed in this letter came to fruition. In 1969, van het Reve together with the Dutch historian Jan Willem Bezemer and the British Sovietologist Peter Reddaway established the Alexander Herzen Foundation in Amsterdam, dedicated to publishing—in Russian—important samizdat works from the Soviet Union.[59] By producing definitive versions of texts explicitly authorized by their creators, the Foundation inaugurated an important shift, at least outside the USSR, away from the unregulated, uncompensated forms of (re)production characteristic of samizdat and tamizdat, in favor of securing copyrights and royalties for authors.[60] Three years later, the British Slavicist Michael Scammell founded the journal *Index on Censorship*, a quarterly devoted to censored works from the Soviet Union and other non-democratic countries. Excerpts from Litvinov's letter to Spender appeared in one of its early issues.[61]

No less remarkable was the division of labor outlined in Litvinov's letter. It is tempting, in light of the recent surge of interest in transnational histories (including of protest movements), to interpret the letter as calling for the globalization of the emerging Soviet civil rights movement.[62] While Litvinov was indeed inspired by the sense of "reconnected links in time and space," what his letter in fact proposes is the outsourcing of the movement's formal organizational work to entities beyond the Iron Curtain. This was not simply a pragmatic gesture, driven by fear that founding a formal organization inside the USSR would effectively serve up its members on a platter to the KGB. Consistent with his critique of Krasin's proposed "Committee to Defend Civil Rights in the USSR," Litvinov sought primarily to shield dissidents from all forms of hierarchy and institutional pressure, allowing them to continue functioning as free radicals, powered solely by the dictates of conscience. All this was meant to unfold within a milieu that Litvinov's friend Vladimir Bukovsky likened to a beehive:

> Outwardly, the entire complex life of the hive appears to follow some sort of charter. But inside there are no commands or regulations, and every bee, obeying an inner impulse, either flies after honey, builds honeycombs, or guards the entrance. To a casual observer it appears that the queen bee is running the entire operation, but if you kill her, the bees immediately bring up another one without abandoning their work. One of the bees reports on newly discovered fields, rich with nectar, and right away dozens of its neighbors fly off to gather the nectar. How did they decide who flies, and who stays?

In this "amazing conductorless orchestra" there was "no assigning of roles, no recruiting and no agitating."[63] Or so it seemed to Bukovsky, who struck more than a few fellow dissidents as exhibiting the personality of a soloist. Or perhaps a conductor.[64]

Even in the absence of explicit commands and regulations, bee hives and orchestras are capable of maintaining internal hierarchies. In the case of the Soviet dissident movement, the most obvious division of labor was between men and women. The overwhelming majority of samizdat texts were composed by men; women disproportionately took on the work of typing and retyping them.[65] Women were also the leading force in organizing assistance for the families of arrested dissidents, the majority of whom were men.⁶⁶ What stands out, however, especially in comparison with similar movements in other settings, is precisely the prominence of women at the epicenter of dissident activity—a phenomenon to which the foreign journalists (all men) who assiduously covered the movement were seemingly blind. Not only was Gorbanevskaya the moving force behind the founding of the *Chronicle of Current Events*; over the course of the journal's fifteen-year lifespan, five of its ten editors and twenty of its forty-nine compilers were women.⁶⁷ Among the roughly one thousand signers of open letters and petitions in the spring of 1968, somewhere between a third and a half were women.⁶⁸ The most important lawyers for rights-defenders standing trial were women. Women exercised considerably more influence in the Soviet dissident movement than among analogous movements in Soviet-bloc countries in Eastern Europe, or among the various nationalist movements (Russian, Ukrainian, Lithuanian, and others) within the USSR.⁶⁹ If anything, their influence recalls that of their counterparts in the revolutionary movement of late imperial Russia.⁷⁰

The first imperative for a social movement of a new type was to break out of the conspiratorial mindset of the underground. A century earlier, Fedor Dostoevsky had plumbed that mindset in works such as *Notes from Underground* and *The Devils*. Lenin had repurposed it by fashioning a "party of a new type," a vanguard of professional revolutionaries dedicated not to terrorism or to winning elections but to forging the consciousness of Russia's still embryonic working class, transforming it into a fighting force capable of smashing the autocracy. During the first half of Soviet history, political opposition outside the Communist Party, to the extent it existed at all, by and large replicated the Bolshevik model of subterranean conspiracy. In the post-Stalin era, rejection of the underground went hand in hand with the waning of neo-Leninism and the turn to legalism and transparency.⁷¹ The Bolshevik "party of a new type" had forged a regime of a new type; as it moved into its post-totalitarian guise, that regime in turn created the conditions for an oppositional movement of a new type.

Some dissidents, for example, Grigorenko, experienced this transformation in their own biographies. Born to an impoverished peasant family in the Ukrainian province of Zaporizhzhia in 1907, Grigorenko

FIGURE 7.2. Petro Hryhorenko (Petr Grigorenko) in the 1920s

(Hryhorenko) was the first in his village to join the Communist Youth League. From there his career followed an exemplary Bolshevik trajectory: attendance at a local "workers' faculty," study at the Kharkiv Polytechnical Institute, and then training at the elite Military-Engineering Academy in Moscow. By his late twenties, Grigorenko was in charge of a battalion responsible for demolishing cathedrals and building bridges in the Belorussian Soviet Republic, before being recalled to Moscow in 1937 for post-graduate training at the Higher Military Academy of the Soviet General Staff. His career illustrated the new forms of upward social mobility offered by the Bolshevik state to millions of Soviet workers and peasants, many of whom eagerly filled positions previously occupied by the victims of Stalin's purges. Well on his way toward the commanding heights of the Soviet system, Grigorenko dropped the Ukrainian "Petro" in favor of the Russian "Petr."

Even in the midst of this rapid ascent, however, Grigorenko—an imposing figure, over six feet tall with a shaved head and booming voice—displayed an unusual readiness to take risks. After his brother Ivan was arrested in 1938, Grigorenko arranged a meeting with Andrei Vyshinsky, procurator-general of the USSR and architect of the Moscow show trials, to complain about abuses of power by the NKVD in Zaporizhzhia. Three years later, in June 1941, during a conversation with a fellow officer after the surprise Nazi attack on the USSR, he criticized Stalin's lack of preparedness for war.[72] Either of these incidents (and there were others) could have ended Grigorenko's career—or his life. None of them, however, involved the slightest doubt about the Party or the mission to build socialism. On the contrary, they were the result of Grigorenko's steadfast belief in the Party's crusade—a belief that, by his own account, had convinced him during the Great Terror that there was in fact a fifth column inside the USSR and that "the lion's share of those arrested were indeed 'enemies.'"[73]

That belief did not waver during the war. As a commanding officer, Grigorenko was injured while fighting the Japanese in Mongolia and severely wounded during combat against the Germans in Ukraine, where the Communist Party representative assigned to his battalion was the thirty-eight-year-old Brezhnev.[74] Nor did it waver after the war, despite the long delay in his promotion to major-general (1959), possibly in retaliation for various episodes of outspokenness. By then, Grigorenko was a senior instructor at the Frunze Military Academy in Moscow and chairman of its cybernetics department. His outspokenness continued, whether in the form of publicly criticizing a colleague's antisemitism or condemning the return of the "cult of personality" around the person who had famously denounced it—Khrushchev. For the latter remark, Grigorenko was fired from the Frunze Academy and forbidden to defend his doctoral dissertation. Still a firm believer in the communist cause, in 1963 he helped found the clandestine Union for the Struggle for the Rebirth of Leninism, whose handful of members chiefly occupied themselves with distributing anonymous leaflets describing the Party's betrayal of Lenin, including the recent deadly assaults on protesting workers in Novocherkassk, Temir-Tau, and Tbilisi.

One day in the fall of 1963, Major-General Grigorenko stood nervously outside the venerable Hammer and Sickle metallurgical complex on the outskirts of Moscow, handing typed leaflets criticizing Party policies to workers who were streaming out at the end of their shift. With one exception, neither they nor he felt comfortable enough to strike up a conversation. Those workers who took leaflets (most declined) did so without stopping as they exited the factory gate. Keen to expand his "experiment" to a broader public, but wary of the extreme risk to which he had exposed himself at the Hammer and Sickle, Grigorenko decided to try a different method. Now dressed in full military uniform—trying, as it were, to visually inhabit his official identity—he sat down on a bench in Moscow's Paveletsky train station, took out a single copy of "A Reply to Our Opponents" (the Union for the Struggle for the Rebirth of Leninism's most elaborate brochure to date), and silently pretended to read it. During the commotion of passengers entering and exiting a newly arrived train, he left the brochure on the bench and moved to the opposite end of the hall to observe what would happen next. Within a short time, a young man and woman sat down together and, taking note of the brochure, began to glance through its pages:

> They started a heated conversation about something, stood up, and headed in my direction. I was reading *Ogonek* and pretended not to notice them.
>
> "Comrade General," the young man addressed me, "isn't this yours?" He held out the brochure.
>
> "No, not mine," I replied firmly.
>
> "But . . . we . . . we thought we saw you reading it," he said with some awkwardness.
>
> "Yes, I was reading it. But it's not mine. I read it and left it where it had been before I arrived."
>
> They walked away in a state of confusion. Then the young woman (practically a girl) tore herself away from her companion and, thoroughly embarrassed, ran up to me.
>
> "Comrade General! Perhaps you have another copy? We're heading in different directions, and we'd both like to take a copy with us."

> "I'm sorry, my dear girl," I smiled sympathetically. "Honestly, there aren't any more!"
>
> "Well, in that case, excuse me"—and off she went.[75]

Such experiments convinced Grigorenko that "the people want the truth." But they also taught him that anonymous leaflets produced no tangible effect, no human connection, indeed they inhibited human contact and led him to lie and dissimulate to the very people he was hoping to reach.

It didn't take long for the KGB to catch up with the heralds of a new Leninism. Grigorenko was arrested on February 1, 1964, and immediately brought to the Lubyanka for questioning by KGB chairman Vladimir Semichastnyi, Lieutenant-Colonel Georgy Kantov, and two other deputies:

> "Well, what exactly have you gotten yourself into?" asked Semichastnyi, turning to me.
>
> "I don't understand your question."
>
> "What is there to understand? You probably think we don't know anything. Georgy Petrovich, show him please." Kantov pushed several leaflets over toward me that appeared to have been picked up off the streets or torn from walls.
>
> "Do you really intend to deny your participation in this undertaking?" Semichastnyi again addressed himself to me.
>
> "No! I intend to deny the KGB's right to take part in reviewing this question."
>
> "How so?" he exclaimed with surprise.
>
> "Very simple. I have a conflict with *my* Party. I insist on my legal right as a member of the Party. And insofar as people are trying to use unlawful, un-Party methods to interfere with that right, I am intensifying the struggle. Perhaps I overstepped the bounds permitted by Party statutes. For this the Party can punish me. According to Party rules it can punish up to the most extreme level—expulsion from the Party. But what do the police have to do with this? This is a purely Party matter."

There was an awkward silence, which Zakharov [one of the deputies] interrupted.

"Petr Grigorevich, it is unforgivable for you to speak this way. You declare yourself a Leninist, but Lenin said that the Cheka is above all an organ of the Party."

"This doesn't apply to you. First, you're not the Cheka, and the KGB is part of the Council of Ministers of the USSR [that is, an organ of the state, not the Party]. Second, Lenin said not only what you mentioned, but also emphasized that if the Cheka were left in the same form and with the same rights [as in the immediate post-revolutionary years], it would degenerate into a typical counter-intelligence agency. And that's precisely what we witnessed in Stalin's time."

"You're getting way off topic," Semichastnyi said, collecting himself. "That's all theory, and you're not here to lead a theoretical discussion. You created an underground organization whose goal was to topple the Soviet government. Fighting against that is the task of the organs of state security, not of Party commissions."

"That's an exaggeration. I didn't create an organization with the aim of violently overthrowing the existing order. I created an organization for the dissemination of undistorted Leninism, for the unmasking of its falsifiers."

"If it was only a matter of propagating Leninism, why were you hiding in the underground? Preach within the system of Party political education and at meetings."

"You know better than I that that's impossible. The fact that Leninism has to be preached from the underground demonstrates better than anything that the current Party leadership has deviated from Leninist positions and thereby lost the right to leadership of the Party, and has given to Communist-Leninists the right to struggle against that leadership."[76]

The Communist Party was indeed an island with its own laws and its own procedures. Grigorenko had acted as a Party member, in the belief that he was a better custodian of Leninist values than the current leadership. Semichastnyi must have sensed the danger of this more-Catholic-than-the-pope argument. Rather than risk putting a Soviet major-general and avowed Leninist on trial, he decided to have Grigorenko declared mentally incompetent and confined to a different island with its own laws and its own procedures—the Serbsky Institute of Forensic Psychiatry.

At the Serbsky Institute, Grigorenko befriended Vladimir Bukovsky, who had also been declared mentally incompetent. Once again a zone of incarceration inadvertently became the seedbed of precisely the kind of human contacts that Grigorenko had failed to find in the underground. Prisons, camps, and psychiatric wards had introduced Alexander Volpin to Yuri Aikhenvald; Anatoly Marchenko to Yuli Daniel and Valery Ronkin; Alexander Ginzburg to Aleksei Dobrovolsky. After they were released (in 1965), the twenty-two-year-old Bukovsky began introducing the fifty-eight-year-old army general to his dissident friends. It was a disorienting experience for Grigorenko, and not just because of differences of age or occupation. "I suspected the presence among them of some sort of organization," Grigorenko recalled. "I didn't perceive their community of individual personalities, nor did I understand that the only way to join that community was by my own volition and my own work. Trapped in my old psychology, I was waiting for instructions. In my soul I even took offense that there was a demonstration [in January 1967] and no one had even mentioned it to me."[77]

It was the "old psychology" that drove Grigorenko to support the idea of forming a civil rights organization in Moscow to coordinate various dissident initiatives and groups across the territory of the USSR. As a military man accustomed to discipline, structure, and clear chains of command, Grigorenko—whose "authoritative expression, the result of a long-standing habit of giving orders," initially led some of Bukovsky's friends to take him for "a typical Stalinist"—found the shapeless dissident milieu disorienting.[78] "After losing his general's stars," according to Alexeyeva, Grigorenko craved hierarchy and action, "something to take the place of battle plans, firefights, air support, and matériel."[79] The emerging dissident movement, by contrast, appeared to him as a kind of "guerrilla warfare." For their part, his fellow activists referred to him,

FIGURE 7.3. Major-General Petr Grigorenko

with affectionate irony, as "the General."[80] No one but his wife called him by his first name.

Grigorenko's struggle to adapt to the new milieu surfaced in a series of letters delivered in 1969 to Litvinov, then in Siberian exile. "The strength of our movement," Grigorenko wrote, attempting to speak Litvinov's language, "lies in the initiative of its participants":

> We are not an organization—not in form, not in our essence. For me, the absence of an organization in the typical Russian sense is not a tactical move, deployed for purposes of self-protection. It is based on conviction. I am organically unable to accept the very idea of such an organization. The most I can endorse is an association of independent personalities who share certain goals. I can discuss with them any topic under the sun, I will respectfully listen to any opinions and proposals, but I will act in accordance with my duty and my conscience.[81]

This is the perspective that Grigorenko chose to retain in his autobiography a dozen years later:

> What organizational form should we give this movement? I reflected on this for a long time and then firmly decided: NONE AT ALL. First, as soon as even the tiniest, weakest groups attempt to formally unite, the KGB will immediately liquidate them. Second, I don't want to join any party whatsoever. I've had it with parties. Every party is a coffin for a living cause. A party means a struggle for power; it substitutes bureaucratic intrigues for living communication with people. Arguments over charters and programs sink living causes. No! You simply have to work and to love people, that is, fight against what you yourself oppose. Only on this basis of bringing people together can there be a genuine, non-organizational unity, a spiritual brotherhood. It seems to me that in a totalitarian society this kind of unity can develop spontaneously, encompassing the majority of society and thereby keeping tyrannical elements from seizing power. It can create a set of social relations different from what we have now.[82]

The almost comical notion of a former major-general "firmly deciding" that the movement should be spontaneous suggests that Grigorenko's struggle was not yet over. The instinct to give and receive orders battled with the dream of a voluntary community, in which there would be "no organization and no leaders, since 'everyone was an INDIVIDUAL PERSONALITY.'" "No one organizes their work or teaches them how to act or recruits them into a movement," Grigorenko observed. "Everyone who feels himself to be an individual personality—who wants to be an individual personality, who refuses to destroy himself by submitting to the arbitrariness of the ruling powers—inserts himself into the movement. On his own initiative, answering an inner calling, he does what he considers necessary for the defense of the individual personality."[83]

For his new friend Bukovsky, this was not a dream. It was the chain reaction, the conductorless orchestra.[84] The result, Grigorenko agreed, was nothing short of "a miracle."[85]

But there was also a less miraculous side. In another letter to Litvinov, Grigorenko confessed that duty and conscience did not always give clear instructions. "I absolutely need to have certainty that we are acting correctly," he implored. "I don't see any authoritative judge around here. I think people simply trust me and therefore approve of the path we're taking. I need word from you."[86] And there was more. The movement's lack of leadership was starting to breed "the typical Russian disease" that had infected "all oppositional tendencies in the past: FACTIONALISM." It was "the worst form of that disease, the atmosphere of family circles. More and more one hears the expression 'our people [*nashi*],' 'your people,' and 'my people.'" "Many of those who used to visit you on Tuesdays," he told Litvinov, "and who have not joined any narrow family circle, are already complaining openly to me about the unpleasant shift in atmosphere: 'At Pavel's, I felt I was among my own, but now I show up and sense that I'm an outsider.'"[87]

Even members of the original family circle, close pre-trial friends of Sinyavsky and Daniel, were beginning to feel like outsiders. "We already sensed who 'our people' were," recalled Alexander Voronel. "Immediately after the trial the inventory of friends began to change. From a tight group there emerged a movement ... consisting not only of friends, in fact not primarily of friends but rather of 'friends in the struggle,' which is to say, comrades in arms. Among old friends this sometimes produced a bitter feeling of having been cast aside."[88] Three years after his arrest, a large crowd gathered to celebrate Yuli Daniel's forty-third birthday *in absentia* at his apartment on Lenin Boulevard (the honoree was midway through his sentence at a labor camp in Mordovia). Several of Daniel's friends who had stood vigil outside the courthouse where he and Sinyavsky were tried had a decidedly mixed reaction to what they called the "new people"—including Litvinov and Grigorenko, neither of whom had been present at the trial. "We felt lost in an enormous crowd of people we barely knew," recalled Voronel, "for whom Yulik was not a living person, but an abstract symbol, even an idol. I remember how Tolya Yakobson called out, 'Brothers, old friends, let's head to the kitchen and just reminisce about Yulik! What we have here is a congress of the Democratic Movement.'" A "congress of the Democratic

Movement" was by no means a compliment. Among those "new people," several struck Voronel as having walked straight out of Dostoevsky's *Devils*.[89] Writing to Litvinov, Grigorenko—who despite considerable self-Russification appears to have retained a number of negative stereotypes about the USSR's dominant nationality—sensed the formation of a "typical Russian atmosphere," in which certain individuals "played at being rebels," "had a bit to drink and 'discussed global problems.'" "The truth," he confided, "is that in our movement there have begun to appear our very own 'devils.' How to survive them I don't know. It's not easy to kill off the Russian totalitarian spirit in Russian people. God forbid these kinds of 'democrats' should ever get power in their hands."[90]

One of the central topics of political philosophy is how the few manage to rule the many. Here we are concerned with a different problem: What happens when the few *become* the many, when strong ties of intimate friendship are threatened with dilution by weak, purely functional ties? Can a band of brothers (and sisters) learn to work with others? For Volpin, that was the whole point: "to see many unknown faces" at a protest meant that what had begun as a circle of friends was expanding into a network and perhaps into a bona fide movement. The chain reaction's capacity to overcome inherited deficits of social trust, connecting diverse people who otherwise led separate lives, was, for many, cause for celebration. A guest at Litvinov's visiting days recalled his encounters there: "We were astonished to see how different we were. Different social origins, diametrically opposed activities as youths (in school and at university), radically different characters. And yet our paths had come together here."[91] Amalrik, too, was "constantly amazed at how people of such contrasting fates and temperaments could come together."[92] Even Major-General Grigorenko, for all his misgivings about *arriviste* devils, marveled at the new sociability: "The Soviet corporative system was set up so that a person worked only among people of his particular social group, lived only among them, shopped in stores only with them, and socialized with no one else."[93] Inside the Serbsky Institute, Grigorenko met political prisoners "from different places, different strata, different professions":

[They had] acted in isolation ... but when circumstances brought them together, they immediately understood each other. It was clear to me that if they had met outside [*na vole*] they also would have found a common interest. But they didn't meet there. Why? Above all, because the conditions of life do not allow people to recognize likeminded individuals. In ordinary life, the lie reigns supreme, people don't dare express their disagreements with the ruling elite for fear of repression. So they use leaflets to search for the like-minded.... But in the underground you won't find them. In the underground you meet only rats.[94]

The dissident movement fostered the mingling of individuals from different cities and towns, different nationalities, different belief systems: Russians, Ukrainians, Lithuanians, Crimean Tatars, Jews, Catholics, evangelical Christians, Russian Orthodox, atheists, neo-Leninists, social democrats, and those for whom all variants of Marxism were anathema. In certain ways, the movement replicated the staggering diversity of the Soviet Union itself. In addition to its appeal as an ostensibly non-ideological means of containing the power of the Soviet state (thereby hindering a relapse into Stalinism), the defense of constitutional rights made it possible for these varied constituencies to celebrate their heterogeneity without fully confronting its implications. Below the surface, however, the expansion of the trusted few into the unknown many was fraught with tension.

We are approaching a persistent puzzle in the history of the dissident movement: the nearly unanimous claim by rights-defenders, even as they were being lionized by Westerners for their political courage and condemned by Soviet authorities as political renegades, that they and their activities were not political at all and that they in fact shunned politics entirely. As with the resistance to formal organizations and divisions of labor, here too one is tempted to discount the rhetoric of political abstinence as little more than a tactical gesture designed to reduce the threat of repression by the KGB.[95] To be sure, dissidents referred to trials of fellow activists as "political trials" whose inevitable guilty verdicts turned those activists into "political prisoners." But it was the state, in their view, not the defendants, that made trials and prisoners "political" by subordinating legal norms to considerations of power and ideology. It was "exceptionally important," Litvinov wrote, "to emphasize the *extra-political* character" of the dissident movement itself.[96] How else to account for the extraordinary range of worldviews held by participants in a single movement? And even more important, how else to prevent that diversity of opinion from eventually tearing the movement apart?

Face-to-face communities built on ties of friendship do not require formal rules. Indeed, they typically shun them in favor of unwritten ones. During the period of the chain reaction (roughly from 1966 to 1968), when asked in courtrooms or over kitchen tables why they took part in public demonstrations or signed their names on open letters of protest, participants invariably invoked the same categorical imperative. "I simply could not desert my friend at such a moment"; "It was in defense of my friends; I couldn't stay out of it"; "Friendship has always been the most important thing in life for me—this is why I went"; "It feels terrible to be at liberty when your friends are in prison."[97] What would happen when those arrested, or those who went to the square to defend them, were not friends or even acquaintances? Could impersonal ties lead to analogous acts of solidarity, or was it necessary first to expand the circle of friendship itself—assuming such an expansion were possible? Would the presence of friends among the signatories of a protest letter be sufficient to inspire additional signatures, even at the risk of reprisals against signatories and their family members? Or would people react the way friends of Alexeyeva's did, when asked to sign a petition on behalf of Ginzburg, Galanskov, Dobrovolsky, and Lashkova: "We wrote letters in defense of Daniel and Sinyavsky because they are our friends. We were defending our friends. That's all. Getting into this fight would mean getting into politics. We don't want to do that; we are private people"?[98]

This was the challenge to the emerging movement: how to expand the bonds of friendship into a wider, thinner, more public solidarity

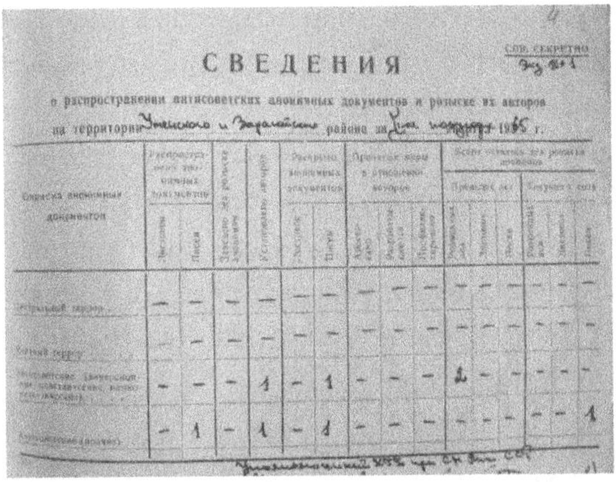

FIGURE 7.4. A printed form, labeled "top secret," for tallying and classifying anonymous "anti-Soviet" leaflets (1965)

willing to defend the minimal and broadly shared demand that the Soviet state abide by its own laws. What repertoire of contention would such a movement adopt? Would it impose on itself an analogous demand to abide by impersonal rules of conduct?

In the decade after Stalin's death, the dissemination of oppositional leaflets—single, often hand-written sheets of paper denouncing this or that aspect of the Soviet order—was so common that the KGB printed forms to help agents record and aggregate data on location, content, and suspected authorship. In 1961–62 alone, more than a million such leaflets were discovered.[99] Known as "*anonimki*," these typically unsigned works were produced by their authors in the dozens or hundreds, stuffed in mailboxes in the entryways of apartment buildings, left on benches inside metro stations or parks, or dropped from balconies of department stores and other populated spaces. A leaflet was precisely what Volpin had insisted his 1965 Civic Appeal was *not*, and what Grigorenko was determined to abandon along with the rest of the underground repertoire.

The problem with anonymous leaflets as a communicative technique, as Grigorenko discovered, was that they failed to spark dialogue, much less form communities of the like-minded. Unlike samizdat texts, which typically traveled via networks of friends, leaflets reached readers anonymously, untethered from social networks and active selection by readers. Unlike consumers of samizdat, readers of leaflets did not reproduce them and pass them on to others. Rights activists associated leaflets with graffiti and anonymous letters—common vehicles for seditious attacks against the authorities, typically addressed to the working class.[100] Leaflets often deployed highly personal (and abusive) terms meant to "call evil by its name" and perhaps thereby ritually neutralize its power, as in the following examples:

> Down with Khrushchev's clique!
> Down with the gluttony of administrators!
> Down with the gang of gendarmes!
> Bait those dogs, beat them, destroy them!
> Don't stand on ceremony!
> Demand higher pay!
> Hail a second 1917!

> Banish all Jews from Moscow.
> All Ukrainians, too.
> Reduce the population of Moscow.
> Supply all the stores in Moscow with all foodstuffs and manufactured goods.
> Provide housing for the entire population of Moscow.
> Give medals to those who have worked in production for at least twenty years.
> If you want to live, do this.

> You can't hide the truth. History will slam the Trotskyite Khrushchev and his clique in the stocks. Khrushchev's Trotskyite ears will stick out no matter how hard he tries to hide them behind the screen of Leninism.

> That bullshitter Khrushchev said at the Twentieth Party Congress that in 1957 the workday would be reduced and workers' vacation times would be increased, but all of that is just on paper, and all your bullshitting has become repulsive to all the working people of the USSR.[101]

Available data suggest a precipitous decline (by more than 50 percent) in the use of anonymous leaflets during the first half of the 1960s, paralleling the decline in neo-Leninism as the dominant ideological idiom of opposition. By the late 1970s, *anonimki* were largely the preserve of individuals or groups threatening terrorist or other violent attacks.[102]

The shift from leaflets to samizdat represented a change not merely in technique of dissemination but in constituency. The social networks within which samizdat spread were overwhelmingly those of the intelligentsia, with its consensus in favor of non-violence and legality as the only acceptable exit ramps from Stalinism. Within this consensus, however, vigorous debates emerged over the efficacy of demonstrations versus group letters, transparency versus secrecy, openly demanding freedom of speech versus simply enacting it via samizdat. Carefully orchestrated demonstrations, for example—even small ones—had proved their ability to attract attention well beyond the events themselves (which were often quite brief), to spark conversations about constitutional rights and the rule of law, and, not least, to breathe new life into the iconic act of civic courage known as "going out to the square." Going out to the square meant leaving the privacy of one's apartment, the world of family and friends, and laying claim to a quintessentially public space, whether Leningrad's Senate Square, Moscow's Pushkin and Red Squares, Kyiv's Kalinin Square (today's Maidan), Yerevan's Lenin Square (today's Republic Square), or elsewhere. Going out to the square meant demonstrating the capacity for civilized protest, notwithstanding the stereotype of Russians as collectively prone to alcohol-infused rampaging (Pushkin's "senseless and merciless Russian riot"). "Among us," a friend had warned Volpin on the eve of the first transparency meeting, "people with real composure are a rarity. The moment someone is offended, he'll forget all your instructions, start screaming 'Bitches!' and stir up trouble."[103] So far, rights-defenders' demonstrations had managed to avoid that trap.

As with leaflets, however, there was a certain random, not to say accidental, quality to the audience for demonstrations. To Litvinov's aunt, Tatyana Litvinova, who was present at the February 1967 protest against the arrest of Ginzburg, Galanskov, Dobrovolsky, and Lashkova, the whole thing seemed "childish, unwise, as pointless as it is innocuous":

"To whom are they addressing their demands? Nobody knows."[104] Bukovsky, as we saw in chapter 6, was trying to ensure that "no one could say later, 'I didn't know.'" Vadim Deloné, arrested along with Bukovsky, compared their demonstration to the spontaneous gatherings to celebrate the successful launches of Soviet astronauts. "I don't see why the question should arise, 'Whom are you addressing?' To whom is the slogan 'Hurrah, our guys are in outer space' addressed? Just like in our case—to everyone."[105] Older activists tended to be skeptical about the value of going to the square. "When you are over forty," Alexeyeva admitted, "the act of shouting slogans in the streets loses its allure." It didn't help to express slogans in written form, on homemade banners: aside from looking amateurish, they were easily torn up, seized, or otherwise made to disappear. Furthermore, "what we had to say could not be reduced to slogans."[106] Having attended several of the annual December 5 (Constitution Day) meetings on Pushkin Square, the forty-something Sakharov agreed: "I'd never been all that enthusiastic about these demonstrations, which smacked of 'revolutionary' party rallies." Nor did he like having dirty snow dumped on his head by plainclothes KGB disrupters. There was always the possibility of violent confrontations.[107]

For most middle-aged, text-centric members of the intelligentsia, a carefully composed group letter circulated via samizdat—epistolary dissidence—was the preferred genre.[108] Open letters and other samizdat texts were designed to be individually reproduced on a scale that rendered confiscations futile. Such letters could make a deep impression, not only for their publicly critical stance, but because they often featured signatures by individuals from multiple cities and multiple professions—combinations not typically seen in group letters to the editor that appeared in Soviet newspapers, which generally came from a single workplace or work-defined collective. The impact of an open letter depended in no small measure on the status of the individual signers, rather than—as with most demonstrations—their quantity. Numbers mattered, to be sure, but the logic of open letters was that of moral witness and persuasion. "A Word said openly and out loud," remarked Lyudmila Polikovskaya, an editor and journalist who also took part in compiling the *Chronicle of Current Events*, "is already a Deed."[109] Words said openly and out loud initially meant words designed for the eye, typed and retyped in samizdat. Better informed and more literate than the vast majority of leaflets, open letters stood a greater chance of finding their way to foreign journalists, and from there to the Voices, which in turn broadcast them to millions of Soviet ears, a process known as *radizdat* ("radio publishing"). Without listening to the Voice of America, noted the Dutch correspondent Karel van het Reve, "you will not be able to participate in conversations with cultured people."[110]

Here too, however, despite explicitly identified addressees— procurator-general Roman Rudenko, chairman of the Council of Ministers Alexei Kosygin, general secretary of the Communist Party Leonid Brezhnev, members of the Russian Supreme Court, or the Central Committee, or the Presidium of the Supreme Soviet—the familiar nagging question reappeared. "To whom are we appealing?" the former Bolshevik believer Raisa Orlova asked in her diary. "To whom do we send letters and telegrams? To those for whom Yevgenia Semyonovna [Ginzburg, author of the samizdat Gulag memoir *Journey into the Whirlwind*] washed floors in Kolyma? Are we waiting until this belief, too, will be crushed?"[111] The geneticist Raissa Berg was not waiting. "We pretend to believe," she told Sakharov, "that there is someone to whom we could appeal for help, when in fact we know perfectly well that we are addressing our grievances about the evil spirit to . . . the evil spirit." When Berg signed letters of protest, it was simply because "it's better at least to do something than to do nothing at all."[112]

Sakharov was less pessimistic. As a preeminent weapons scientist he had had repeated contacts with Soviet leaders, who, to be sure, often struck him as boorish and unsophisticated but rarely as evil. Even after he lost his elite access and his letters and telegrams ceased to elicit responses from the Party leadership, Sakharov insisted that open letters on public issues "are a useful means of promoting discussion" with the possibility of "educat[ing] the public at large, and just might stimulate significant changes, however belated, in the policy and practice of top government officials."[113] The biologist Sergei Kovalev called this "transparency from below."[114] As regards the samizdat readership, this was no doubt true. In official circles, however, ideas promoted in samizdat texts were reflexively stigmatized because of their provenance, usually dismissed as "anti-Soviet" or "immature" with little if any discussion of their content.[115] A prime example was the open letter addressed to Brezhnev, Kosygin, and Nikolai Podgorny on freedom of information and the airing of opinions, co-authored by Sakharov, Valentin Turchin, and Roy Medvedev.[116] In his memorandum about the letter to the Central Committee, KGB chairman Yuri Andropov merely noted that the authors "are known for their politically harmful works" and listed names of individuals who were "aware of the letter's existence" or whose signatures Sakharov had sought out. About the letter's central argument—that without the free flow of information, the Soviet Union could not remain globally competitive in science and technology—there was not a word.[117] Nor did the letter's authors receive a response.

Most open letters were signed by fewer than a dozen people. Those bearing a hundred or more signatures, such as appeared after the "Trial of the Four," were exceptional. As a military strategist, Grigorenko considered letters with a small numbers of signers to be of limited utility, regardless of the signers' status. "Calling on [the state] to respect our demands is easier than forcing it to do so," he wrote to Reshat Dzhemilev, unofficial Moscow representative of the quarter-million Crimean Tatars deported en masse to Central Asia on Stalin's orders in 1944. Dzhemilev had recently met with the new KGB chairman, Andropov, to discuss the future of his exiled people, as a result of which the charge of collaborating with the Nazis during the war was officially rescinded in 1967 (but not the exile from Crimea).[118] To gain real leverage, Grigorenko wanted to organize "a mass collection of signatures" demanding that the USSR honor its own constitution as well as the Universal Declaration of Human Rights. In tiny Czechoslovakia, the writer Pavel Kohout's open letter to the Party leadership, "Socialism, Alliance, Sovereignty, Freedom," had garnered a million signatures in support of political reforms. If that feat could be reproduced in the Soviet Union, Grigorenko noted, "it would be an extraordinary event [and] our demands would begin to be respected." Such a feat had in fact been achieved in the USSR, he reminded Dzhemilev, in the form of the Stockholm Appeal of 1950.[119]

Initiated by the French Communist physicist Frédéric Joliot-Curie via the World Peace Council, the Stockholm Appeal called for a global ban on atomic weapons. Ostensibly signed by over 100 million Soviet citizens, the appeal went on to gain the support of another 150 million individuals worldwide.[120] In the Czechoslovak case, however, Kohout's letter was published in several major newspapers before circulating among the population at large. The Stockholm Appeal was similarly published in *Pravda*, and Stalin deployed considerable resources to generate the astounding (and likely fraudulent) tally of signatures.[121] No such resources were at the dissident movement's disposal, and Grigorenko's proposal to gather signatures by canvassing door-to-door failed to gain the support of his fellow dissidents.[122]

Grigorenko had good reason to turn to Dzhemilev for help. Among dissenting groups in the USSR, only national and religious minorities had succeeded in producing petitions with signatures in the thousands. Crimean Tatars were the most successful of all. Their appeal to the Twenty-Third Congress of the Communist Party in 1966, demanding a return to their ethnic homeland after two decades of exile, bore upwards of 120,000 signatures. By 1968, it was estimated that various Tatar appeals over the previous decade had gained some 3 million signatures— meaning the average adult Crimean Tatar had endorsed more than ten appeals.[123] No wonder, then, that Grigorenko sought to persuade Dzhemilev that "your people should play a very important role" in gathering support for a petition demanding the observance of civil and human rights in the USSR. If Crimean Tatars could muster 300,000 signatures for such a petition, Grigorenko speculated, "then we could collect roughly another 700,000. And with a million you can start banging on the table."[124] As Volpin had argued, if isolated individuals acted as if they had rights, they became martyrs; if everyone did so, however, the state "would become less oppressive."[125] But as Alexeyeva recognized, at least in hindsight, it all seemed "too logical to be applicable to real life." The Soviet system, in which a façade of unanimity replaced public opinion polls and competitive elections, made it impossible to know what proportion of Soviet citizens sympathized with the demand that the Soviet government abide by its own laws. The bard Yuli

Kim (Petr Yakir's son-in-law) was merely guessing, or perhaps wishing, when he sang in 1968 of "the healthy million" who silently agreed with what "a handful of feverish intellectuals were saying out loud."[126] Rights-defenders never came close to collecting 7,000, let alone 700,000 signatures. More to the point, they never tried: mass petitions were not part of their repertoire of contention, if only because most dissidents did not regard the Soviet masses as likely allies. The dissident repertoire also excluded a host of other techniques commonly used by twentieth-century non-violent movements in other settings, such as strikes and labor stoppages, pickets, civil disobedience, boycotts, street theater, and public processions.[127]

More than anything else, more than human beings coming out to the square, what "moved" in the dissident movement were texts. Its central activity, the only one that achieved truly mass proportions, and the only one that required a new word to describe it, was samizdat—the unregulated dissemination and consumption of uncensored texts produced and reproduced on thousands of typewriters in apartments across the USSR (and occasionally amplified by the Voices). Soviet people, as their government never tired of proclaiming, may have read more than any other people in the world.[128] What is certain is that they typed and retyped more than any other people in the world. Every document reproduced and circulated via samizdat became a kind of chain letter, addressed to an indeterminate, risk-sharing, infinitely expandable readership.

The primacy of samizdat over other techniques of contestation—what one might call the triumph of the textual—was all but overdetermined. The cardinal role of intellectuals in the dissident movement was bound to foster an emphasis on the written word. Texts in turn tended to attract intellectuals more than other social groups to the movement. The fact that samizdat was an ownerless technology of free speech fit well with the leaderless structure of the dissident movement. The strict legalism of the rights-defenders mitigated against the use of strikes and labor stoppages—both considered extra-legal actions in the Soviet system, in which work was not just a constitutional right but a constitutional duty—as well as any form of civil disobedience, which by definition involved the deliberate flouting of established law.[129] By contrast, Soviet law not only refrained from banning specific topics or texts (unlike the tsarist government, which published lists of banned books and topics), but made no explicit reference to the existence of censorship itself, thereby leaving the production and consumption of samizdat formally within the realm of lawful practice—unless "anti-Soviet" content could be demonstrated.[130]

The significance of samizdat lies not just in the way it performed an end-run around Soviet censorship. It also performed an end-run around the specific logic of Soviet civil rights, according to which, in order to make civil rights "real" (in contrast to the "fictitious" bourgeois variety), the Soviet government was obliged to provide the material prerequisites for realizing freedom of speech and the press. "Freedom of the press," as the American journalist A. J. Liebling famously put it, "is guaranteed only to those who own one."[131] Samizdat in effect declared: there is no need to transform the negative liberty of an uncensored press into the positive right of state-sponsored entitlement to material resources of publicity. There is no need to reconfigure the right to free speech by way of analogy to social and economic rights to education, housing, health care, and employment. Dissidents had their own non-state means of making the right to free speech real: typewriters, onion-skin paper, and carbon paper. Thanks to Western shortwave radio broadcasts, moreover, do-it-yourself samizdat texts received a massive boost subsidized by other states—not as an entitlement but as part of the Cold War's battle for hearts and minds.

The beehive of samizdat production, reproduction, dissemination, and consumption was the distinguishing feature of the Soviet dissident repertoire. Its distinguishing paradox was hiding in plain sight. For despite their constant emphasis on legalism, among themselves dissidents displayed a fierce resistance to law-based procedures, or indeed formal rules of any kind. The sprawling corpus of dissident memoirs contains no mention of this paradox or of the underlying tension between the laws of the state and individual conscience ("one's own laws," as Orlova wrote). Nor does it appear in contemporary sources such as diaries, letters, or essays circulated in samizdat. What one finds instead, some three years into the self-perpetuating cycle of arrests, protests, and trials, is an increasingly divisive debate over forms of activism. For Litvinov, Gorbanevskaya, Bogoraz, and others, the dissident chain reaction was the ideal modus operandi. Spontaneous, conscience-driven, non-hierarchical, it allowed participants to decide freely for themselves how, when, and with whom to defend the rights of fellow citizens. Put differently, it allowed individuals to realize within themselves the values of autonomy and agency that constituted both ends and means of the movement.

Critics of spontaneous, conscience-driven, non-hierarchical activism did not oppose it because it was at odds with the movement's external emphasis on strict legal and procedural formalities. That tension appears to have gone not just unmentioned but unnoticed. Rather, they opposed it because they considered it ineffective. "There was no organization in the movement," Chalidze complained, "no one was obliged to do anything. Nobody was required to submit to the opinion of the majority, in fact the opinion of the majority was usually unknown. Everyone did what he considered acceptable, either on his own or with others who wished to do the same thing. . . . This is not what is usually called organized behavior."[132] According to Krasin, Yakir, and eventually Grigorenko, rights-defenders needed a leadership structure and a formal division of responsibilities in order to rise from transient expressions of moral outrage to coordinated, sustained, targeted action. In a word, they needed to become an organization.

There are good reasons to regard this divide as belonging to a deep pattern in modern Russian and Soviet culture, the struggle between forces of "spontaneity" and "consciousness." Spontaneity stood for instinct and intuition; consciousness expressed the need for thought, deliberation, and strategy. The dichotomy extends at least as far back as the rupture between Mensheviks and Bolsheviks in 1903 and was kept alive thereafter in countless Soviet novels whose protagonists struggled to harness their desires to the higher goal of building socialism.[133] It would be a mistake, however, to regard Litvinov, Gorbanevskaya, and other advocates of unregulated, conscience-driven activism as somehow analogous to pre-revolutionary Russian workers unaware of the larger political—indeed, world-historical—significance of their strikes and protests. Opponents of hierarchy and formal divisions of labor in the dissident movement were acting not simply on instinct but on a consciously worked-out synthesis of self-liberating ends and means. Advocates of structure and organization, for their part, were driven in varying degrees by their own impulse to lead. Both groups were committed to the strategy of persuading the Soviet state to abide by its own laws and procedures even as they clashed over whether the dissident movement should have laws and procedures of its own.

Tensions between spontaneity and consciousness found expression not just within the dissident movement but within individual dissidents. One such figure was Volpin, an advocate of both anarchism and the rule of law. Another was Grigorenko. "We're like photons," Grigorenko wrote to Litvinov, referring to weightless particles of light. "When we're not moving, we have no mass. For participants in a progressive movement, to stop moving is fatal."[134] With their form and repertoire still undecided, Soviet dissidents were about to discover the full truth of that statement.

8

From Circle to Square

> А на тридцать третьем году
> я попала, но не в беду,
> а в историю.
>
> In my thirty-third year I fell
> Not into misfortune
> But into history.
>
> —NATALYA GORBANEVSKAYA

Whether under socialism or capitalism, civil and political rights inscribed in constitutions don't implement themselves. To gain force, they have typically required political mobilization in the form of popular resistance to officials jealously clinging to the prerogatives of state power. "Until people learn to demand what belongs to them as a right," insisted Vladimir Bukovsky, "no revolution will liberate them. And when they learn, there will be no need for a revolution."[1]

The dissident debate over how to make rights granted by the Soviet Constitution real, through which techniques and what kind of movement, developed under the tense circumstances of the chain reaction, whose hallmark was its unpredictability. Who would be arrested next, when would protests re-erupt, and where was it all leading? Within the emerging movement, a fault line was growing between those who wanted the spontaneous chain reaction to continue, attracting more and more individuals willing to follow their moral intuition, and those who saw a need for greater coordination, leadership, and strategic direction.

A further source of tension, but also of hope, was emerging from the outer tier of the Soviet empire. It began in June 1967, with a skirmish inside the Czechoslovak Writers' Union: a small faction of writers demanded that literary production be freed from supervision by the Communist Party. The response came swiftly, with the expulsion of three of the troublemakers from the Party and the transfer of the Union's weekly literary journal to the Ministry of Culture, where it would be supervised even more closely than before. Within months, the skirmish found an echo inside the upper echelons of the Party, as the first secretary of the regional Communist Party of Slovakia, Alexander Dubček, unexpectedly challenged the hard-line policies of Party leader (and president of Czechoslovakia) Antonín Novotný. Dubček was not alone in his opposition, and by January 1968 he had replaced Novotný as head of the Czechoslovak Communist Party. Soon thereafter, Novotný relinquished the presidency to the popular General Ludvík Svoboda, who had distinguished himself as a soldier in World War I (fighting the Germans and Austrians), in the Russian civil war (fighting the Bolsheviks as a member of the renowned Czechoslovak Legion in Siberia), and World War II (fighting the Germans in an underground resistance group). During an anti-communist uprising by Czech workers in 1948, Defense Minister Svoboda had famously declared, "The army will not march against the people." He was later purged and briefly imprisoned.

By March 1968, the push for reform had spread beyond the confines of the Party. Responding to widespread pressure from below, Dubček announced an "Action Program" that included the near total abolition of censorship, a shift toward a mixed socialist and market economy, the prospect of multi-party elections, and, not least, the reorganization of the country as a federation of autonomous Czech and Slovak republics. Reforms were meant to foster European-style democratic socialism, or, as Dubček famously put it, "socialism with a human face." It was not a face that leaders of other Soviet-bloc countries wanted their citizens to see. Bulgarian authorities responded by blocking tourist travel to Czechoslovakia even as Czechs and Slovaks gained new freedom of movement; East Germany began jamming radio broadcasts from Prague even as the Czechoslovak government ceased jamming broadcasts by the Voice of America and other Western shortwave radio stations. By May, Soviet premier Alexei Kosygin had concluded that developments in Czechoslovakia were "completely abnormal and dangerous."[2]

In the eyes of Soviet rights activists, the Prague Spring had opened a conversation in Czechoslovakia on burning issues of the day, a conversation that was only beginning to emerge in the USSR and, for the time being, exclusively in samizdat. There, in fact, lay the key difference: in Czechoslovakia, the push for reforms came substantially from within the Communist Party, indeed from the new Party leadership. Some of the most dramatic proposals—including Party member Ludvík Vaculík's manifesto "Two Thousand Words," calling for a more democratic and humane socialism—appeared in official newspapers readily available at kiosks across the country. At least one of those newspapers, *Rudé právo* (Red truth), could also be purchased at kiosks in Moscow and other major Soviet cities, where selected articles, among them "Two Thousand Words," were translated and disseminated via samizdat—including by Natalya Gorbanevskaya, an accomplished translator of Czech and Polish.[3] Anxiety over such ideological contamination from abroad caused Leonid Brezhnev to raise the issue during bilateral talks with Czechoslovak leaders in May: "Your newspapers are read also by Soviet citizens," he warned, "and your radio broadcasts attract listeners in our country as well, which means that all this propaganda affects us just as much as it does you."[4]

The Prague Spring was more about democracy and decentralization than civil rights or the rule of law.[5] But that hardly diminished its appeal for reform-minded Soviet citizens, for whom experiments with modifying socialism in Czechoslovakia and Yugoslavia were more plausible and more compelling as possible models for the USSR's future development than the capitalist democracies of Western Europe or the United States. An anonymous samizdat poem making the rounds in spring 1968 cryptically referred to "our Soviet Dubček," causing consternation in the

KGB.⁶ Certain rights activists expressed solidarity with the Prague reformers "the Russian way": "After the traditional toast to those who could not be with us [because they were in the camps], we drank to Comrade Dubček, Comrade Mlynář, Comrade Černík. Some admirers of the Prague Spring aimed to honor each of the dozen or so comrades on Czechoslovakia's politburo."⁷ When Petr Grigorenko and a friend met with a pair of Czech visitors in Moscow to learn of the latest developments in Prague, the friend joked, "If only you could figure out how to occupy us!"⁸

Rumors, however, were trending in the opposite direction, namely, of a possible Soviet military occupation to put an end to the Czechoslovak experiment, which the Kremlin was now ominously characterizing as a plot by fascist counter-revolutionaries acting on instructions from Washington and Bonn. Among dissidents, euphoria at what was being born in Prague alternated with dread at the thought that Soviet troops might strangle it. "Our mood went up and down," Pavel Litvinov recalled. "'They're about to send in troops.' 'No, they're not sending in troops.'"⁹ On July 22, Anatoly Marchenko, whose revelatory account of Soviet forced labor camps, *My Testimony*, was soon to be published in the West, circulated an open letter addressed to the editors of half a dozen communist newspapers in Czechoslovakia, the USSR, and the West, as well as to the BBC. Soviet leaders, the letter announced, were frightened by what was happening in Czechoslovakia, not because it threatened socialism, but because it could "discredit the principles and methods of leadership that currently dominate the socialist world." If democratic socialism were to succeed in Czechoslovakia, "then perhaps our workers, peasants, and intelligentsia will wish to have freedom of speech in practice and not only on paper."¹⁰

A week earlier, Soviet and other socialist leaders had met in Warsaw and delivered a virtual ultimatum, threatening to use "all means" for the struggle against "anti-socialist" forces in Czechoslovakia. "It's a pity," Marchenko's letter taunted, that Soviet leaders "didn't specify, in concrete terms," what those means were. "Kolyma? Norilsk? The Hong Weibing [the Red Guards carrying out the Cultural Revolution in China]? 'Open' courts? Political concentration camps and prisons? Or merely the usual censorship and extra-judicial reprisals, such as firing people from their jobs?" Noting the much-trumpeted "unanimous" support by the Soviet population for the official warnings sent to Prague, Marchenko reminded his readers that "all the so-called mistakes and excesses in the history of our country were accompanied by stormy, sustained applause, rising to an ovation." Obedience, it turned out, "was not the highest civic virtue":

> I am ashamed of my country, which once again is playing the disgraceful role of gendarme of Europe. I would also be ashamed of my people if I believed that they really did unanimously support the Central Committee's policy regarding Czechoslovakia. But I am sure, in fact, that that is not the case, that my letter is not the only one. It's just that such letters don't get published here. Unanimity in this case is a fiction, artificially created by means of violations of precisely that freedom of speech that is being realized in Czechoslovakia.

Even if it turned out that he was alone in his opinion, Marchenko concluded, he would not change it, because "it was given to me by my conscience. And conscience, it seems to me, is more reliable than the general line of the Party, with its constant fluctuations and excesses."[11]

As intended, the letter struck a nerve. Marchenko was arrested a week later, on July 29, ostensibly for visiting Moscow in violation of the internal passport rules governing former camp inmates.[12] Justice was to be swift: a trial was set for August 21. Ludmilla Alexeyeva, Larisa Bogoraz, Yuri Gerchuk, Natalya Gorbanevskaya, Viktor Krasin, Petr Grigorenko, Pavel Litvinov, and Anatoly Yakobson immediately circulated an open letter addressed simply to "Citizens!" "If you do not want to put up with this tyranny in the future," they pleaded, "if you understand that Marchenko has been fighting for all of us and that all of us must fight for Marchenko, then we ask you to defend him in the way that you yourselves deem necessary.... Collective responsibility based on fear and servility must give way to collective responsibility based on humanity and civil courage."

Signers included their phone numbers and addresses, inviting readers "to write or telephone any of us if you wish to have additional information."[13] Bogoraz's cousin Irina Belogorodskaya, the daughter of an NKVD agent, accidentally left behind a bag containing signed copies of the letter in a taxi, after which the KGB operatives who had been tailing her promptly arrested her for disseminating anti-Soviet slander. The letter's signatories assumed they would be next. Litvinov began drafting a new letter protesting Belogorodskaya's arrest.[14] Another cycle of the chain reaction was set in motion.

Litvinov and his wife, Maya Kopeleva (daughter of Raisa Orlova and Lev Kopelev), stayed at Bogoraz's apartment on the night of August 20. Marchenko's trial was set to begin the next morning, with Dina Kaminskaya as Marchenko's defense attorney. There were plans for the usual gathering of supporters outside the courthouse. At 7 A.M. Bogoraz was awakened by a telephone call. A sobbing Gorbanevskaya could barely get out the words: "Soviet tanks are in Prague."[15] Litvinov immediately switched on the radio. The Voice of America and the BBC were jammed; Radio Moscow was reporting news of "fraternal assistance" to Czechoslovakia. They set off for the courthouse, where they met several dozen friends and supporters of Marchenko and an equal number of KGB agents, who began photographing all those present. Given that morning's news, the situation was tense; Marchenko's supporters sensed that the KGB wanted to provoke an altercation. To their amazement, however, the courtroom doors were open, and everyone was able to enter and find a seat. Litvinov speculated that, given the explosive news of the Soviet military action, the KGB had instructed its officials to contain the crowd inside the courtroom rather than risk a public spectacle. The Soviet invasion also explained the strange absence of foreign journalists—they now had a bigger story to report.[16] Alexander Volpin arrived late; he hadn't yet heard the news. When someone whispered it to him, he replied in a loud voice, "What, they sent troops to Czechoslovakia?! Those thugs!" and was promptly expelled from the courtroom.[17]

Many reform-minded Soviet intellectuals were similarly aghast. Those who were old enough remembered the phrase "fascist tanks in Prague," referring to Hitler's capture of the city in March 1939. Alexander Tvardovsky, who as a war correspondent had covered the liberation of the Czech capital by the Soviet Army, penned this verse in his diary within days of the Soviet invasion:

> What shall we do with you, oath of mine
> Where to find the words to relate
> How in forty-five they greeted us in Prague
> And how they greet us now in sixty-eight.[18]

The poet Yevgeny Yevtushenko, too young to have witnessed the liberation, released the following lines in samizdat:

> Let them simply put the truthful words,
> Without sobbing, on my tombstone:
> "A Russian writer. Crushed
> By Russian tanks in Prague."[19]

"Nobody took the Soviet side," recalled the physicist Yuri Orlov of his acquaintances in Moscow, Leningrad, Novosibirsk, and Yerevan, "nobody liked the current Soviet government, and—and nothing. Debates did not move beyond the kitchen walls."[20] Well before the invasion, in fact before Marchenko's letter and others warning of possible Soviet military action against Czechoslovakia, Bogoraz, Gorbanevskaya, and Litvinov had begun contemplating what they would do in the event of such an attack. The usual textual repertoire of group letters, petitions, and international appeals—to say nothing of poetry—seemed an inadequate response to an invasion of a small, progressive, socialist ally by half a million troops, most of them Soviet.[21] Nobody bothered to take notes at Marchenko's trial on August 21; there was no samizdat transcript recounting the inexorable farce of his conviction and sentencing to a year in a strict-regime camp in Perm (a sentence that would subsequently be extended by two years for "circulating fabrications defaming the Soviet system").[22] The talk now, at least among advocates of spontaneous, individual, conscience-driven action, was of something more dramatic. There would need to be a demonstration, a spectacle.

Two days later, Bogoraz happened upon a spectacle. On August 23, President Svoboda flew to Moscow to meet his country's occupiers. Local cynics, playing on Svoboda's last name (which means "freedom" in both Czech and Russian), joked that once again Hegel had been proved correct: "Freedom is the recognition of necessity."[23] As Bogoraz was walking from the Lenin Library, she came upon a large crowd along the boulevard where Svoboda's cavalcade was scheduled to make its way to the Kremlin:

> As usual during such ceremonies, Muscovites driven from their workplaces were lining the route of the honored guest, chewing sandwiches, licking ice cream cones, buying glasses of lemonade, and clasping in their free hand the appropriate little flags they had been given for the occasion. The idea was to wave the flags at the limousine as it passed by and then quickly disperse. On this occasion the crowd was getting tired of waiting. The appointed hour had come and gone, and then another hour, and another. The person they were supposed to greet still hadn't arrived, but neither their bosses nor the police would let them go. Another sandwich, another glass of lemonade . . . I was literally shaking at the sight of this chewing mass, this *people* [*narod*] who had just voted "to support and approve" [the invasion]. Here they are, waving flags of the Czechoslovak Soviet Socialist Republic—and no one's hand is burning from holding that flag. What are they, cretins? Unfeeling cattle?

At last the open-air limousine approached, moving slowly along the boulevard with its passenger, Svoboda, standing erect and staring straight ahead. His face reminded Bogoraz of the tragic masks worn by actors in the ancient theater.

> And *they*: they put on smiles, called out "hurrah!" and jabbed their flags in the air. I was so ashamed on behalf of this elderly Czech man . . . for whom I am sure it was difficult to bear the tragedy and humiliation of his people. I was ready to push my way to the front and scream I don't know what. But I didn't. I persuaded myself that I would produce nothing more than a pathetic peep that no one would hear, and even Svoboda wouldn't notice.[24]

It was a double humiliation: first the attack against a peaceful ally and then, closer to home, over and over again, the relentless lie that the

Soviet people unanimously endorsed the invasion. Newspaper headlines declared "The Will of the Soviet People Is Unified and Unshakeable," "All Soviet People Firmly and Decisively Support the Action to Defend the Cause of Socialism and Peace," "The Soviet People Fulfills Its International Duty," or simply "We Approve!"[25] As the sociologist Yuri Levada, the preeminent analyst of public opinion in post-Soviet Russia, wrote in his memoirs, "No one was supposed to know that somebody disagreed."[26] Even worse was the willingness of so many Soviet citizens to actually endorse the invasion. It was a matter of feeling "insulted by your inferiors," Alexeyeva wrote, "humiliated so deeply that your entire being rebels, you want to stand up and do something—anything—to separate yourself from that thing called 'the masses.'"[27]

Major-General Grigorenko, populist instincts still intact, was less contemptuous of that thing called the masses. His proposal to organize a "mass collection of signatures" condemning the invasion, however, fell on deaf ears among the dissidents.[28] What were the "healthy million" actually thinking about the invasion? Alexeyeva and Bogoraz may have persuaded themselves that popular approval was coerced, bought, or otherwise insincere—which hardly diminished their fury at the "cretins"—but there is now good evidence that most of the Soviet population, however well or poorly informed, was perfectly willing to go along with the invasion, as most populations are wont to do in the first flush of their country's military engagements abroad.[29] Many patriotic Soviet citizens resented what they perceived as a lack of gratitude from the Czechoslovaks for their liberation by the Red Army, which suffered significant casualties in the process of driving Nazi forces out of Eastern Europe.[30]

Gorbanevskaya, who had Czech friends and a long-standing love of Czech literature, was determined to puncture the unbearable lie of Soviet propaganda. "If even one person doesn't approve of the 'fraternal assistance,'" she reasoned, "then the approval stops being unanimous." The "whole nation minus one" was no longer the whole nation. This was the "arithmetic of atonement" that made it possible to "cleanse one's conscience" and "atone for part of the historical guilt of one's people."[31] It was an arithmetic shaped by the legacy of participatory dictatorship, now in the form of a lip-service state that demanded explicit consent not from a majority of the governed but from everyone,

FIGURE 8.1. Alexander Zinoviev's private caricature of Brezhnev, "Мы за" (We're for). The image plays on Brezhnev's fondness for awarding himself medals, his labored speech when reading even the simplest of texts, and the resemblance—at least in Zinoviev's mind— between Brezhnev's famous eyebrows and Stalin's famous mustache.

the entire collective. The only escape from shared responsibility for the state's actions was public disassociation from those actions. "We were defending ourselves," Gorbanevskaya wrote, "from the unavoidable feeling—had we remained silent—of participating in a crime." In her case it was also a chance to atone for another, much more personal guilt, stemming from her betrayal of fellow students who had protested the Soviet invasion of Hungary in 1956. She hinted at this years later when she insisted that, far from an act of self-sacrifice, her decision at age thirty-three to publicly protest the Soviet invasion of Czechoslovakia was "almost egotistical."[32]

Andrei Amalrik learned of the invasion at Marchenko's trial. Afterward, Litvinov mentioned to him the idea of staging a protest of some kind. Amalrik reacted negatively, pointing out that any such protest stood zero chance of reversing the occupation and would be suicidal for participants, who could expect long prison sentences—or worse. Litvinov was trespassing once again on issues of foreign relations, an arena the Kremlin regarded as its exclusive prerogative. "Besides," Amalrik added in his usual unsentimental manner, "as history teaches, the Czechs themselves will work out a compromise."[33] Why sacrifice yourself in the name of moral principles when the victims were open to accommodation? The physicist Valery Chalidze, who was preparing to launch a samizdat periodical analyzing the problems of Soviet society, agreed: a demonstration would inflict damage on the nascent dissident movement far out of proportion to any conceivable gain.[34]

It was precisely this sort of cost/benefit analysis that Litvinov, Bogoraz, and Gorbanevskaya rejected. No one, Gorbanevskaya wrote, was under the illusion that "the Kremlin gates would fly open and the entire Politburo would rush out to express its gratitude, sobbing, 'guys, thank you for opening our eyes; we're such idiots, we didn't understand what we were doing.'"[35] Whether out of frustration with negative reactions to their planned protest or from a desire to prevent the plan from becoming too widely known, they stopped talking with other people about it after settling on Sunday, August 25, at noon, in Red Square. Proponents of conscience-driven voluntarism were helped by the fact that Grigorenko, Krasin, and soon Amalrik—advocates of a more organized strategy for the movement—were out of town ("thank God," Litvinov later recalled thinking), it being August, a time when many Muscovites had left the city for dachas in the surrounding countryside.[36]

It was important to avoid placing moral pressure on anyone to participate in an action that was sure to lead to arrest and imprisonment. Each person needed to follow his or her own conscience: "Whoever feels like it, whoever wants to come, let them come." Litvinov had made a point of never urging people to sign group letters and was firmly against the "herd instinct," including the familiar idea that "'so-and-so is going, so we're going too.'"[37] As a result, it was unclear who knew about the protest and who, if anyone, would take part in it. There was never a meeting of organizers or an announcement via samizdat; news traveled by word of mouth (telephoning was out of the question). At one point Bogoraz told Litvinov that if he wanted to cancel the whole thing, that was fine; she would protest alone. But then they learned that Konstantin Babitsky, a researcher in linguistics at the Soviet Academy of Sciences, had told someone he was going, and since they had no way to reach him, they decided they needed to proceed with the plan. On the evening of August 24, Litvinov and his wife, Maya Kopeleva, were at her parents' apartment, where the popular bard Alexander Galich was trying out a new song called "Petersburg Romance," about the famed Decembrist revolt of 1825 in Senate Square, in Russia's imperial capital:

> We repeat their whispers,
> We repeat their steps:
> Experience has saved no one,
> So far, from disaster!
>
> . . . .
>
> Just so, no simpler
> Our epoch tests us:
> Will you go out to the square?!
> Dare you go out to the square?!
> Dare you go out to the square?!
> Will you go out to the square
> When the hour strikes?[38]

Galich seemed to be glancing at Litvinov, who felt an overwhelming urge to tell those present about the demonstration planned for the following day. Beneath the question "Dare you go out to the square?," of course, was another: "Are you willing to go to prison?" Litvinov kept his mouth shut.[39]

The plan was to show up at Lobnoe Mesto, the circular platform on Red Square adjacent to St. Basil's Cathedral and the Kremlin, at noon.[40] This was a new location for dissident protests; the transparency meetings always took place at Pushkin Square, half a mile away. Since vehicular traffic was banned from Red Square (apart from high officials being whisked into and out of the Kremlin), Bogoraz and Litvinov hoped they might avoid the accusation of disrupting public order, especially on a Sunday. It was going to be a "sit-down demonstration," an idea borrowed from the American civil rights movement, which had received extensive coverage in the Soviet press. By sitting, protesters might at least be less likely to be charged with hooliganism.[41]

For sheer historical resonance, nothing could match Red Square, the epicenter of the Soviet universe. Here was Lenin's mausoleum, patrolled round the clock by goose-stepping honor guards; here annual parades commemorated the USSR's greatest triumphs, the Bolshevik Revolution and the victory over Nazi Germany. Red Square was the preeminent space of the Soviet Union's official public sphere, an arena designed less for citizens to articulate and exchange ideas and interests than to participate in the representation of state power. Choosing this space to challenge that power was therefore itself an acute form of dissent. According to KGB surveillance records, Litvinov had alerted foreign journalists about the planned demonstration, urging them to take photographs of the protesters "as representatives of the progressive part of the Soviet intelligentsia" opposed to the occupation of "freedom-loving Czechoslovakia." The journalists, according to the report, had in turn advised Litvinov to stage the protest with St. Basil's Cathedral in the background, with its iconic multi-colored onion domes, so that Western newspaper readers would instantly grasp the event's extraordinary setting.[42] No one had ever seen a photograph of a dissident protest in the USSR, let alone on Red Square.

They wouldn't see this one either. Somewhere between Litvinov or Bogoraz and the Dutch journalist Henk Wolzak, whose Russian was shaky at best, information about the timing of the protest got garbled. Foreign journalists arrived at eleven, rather than twelve. Tired of waiting, all but one departed before protesters arrived at noon.[43] As Litvinov

FIGURE 8.2. "For your freedom and ours," one of the banners held by demonstrators in Red Square on August 25, 1968

made his way to the square, tailed by a KGB operative, he had no idea who besides Bogoraz would join him.[44] Gorbanevskaya arrived with her three-month-old son, Yosif, in a baby carriage, similarly uncertain whether anyone else would show up. Hidden under the sleeping Yosif was a small Czechoslovak flag (presumably a leftover from the crowds marshaled to welcome Svoboda two days earlier) and homemade cloth banners reading "Hands off the ČSSR," "Long live free and independent Czechoslovakia" (in Czech), and "For your freedom and ours." The last banner, Gorbanevskaya later noted, reproduced "the slogan of Russian democrats who came to understand in the course of the previous century that a people that oppresses others cannot itself be free."[45]

Five other individuals arrived as well. In addition to Babitsky, there was Vadim Deloné, a young poet and grandson of a prominent Russian mathematician, who had learned of the demonstration only hours before; Vladimir Dremlyuga, who had been expelled from Leningrad State University and had attended several "visiting days" at Litvinov's apartment (he brought two homemade banners of his own: "Down with the occupiers" and "Freedom for Dubček"); Tatyana Bayeva, a friend of Irina Yakir (Petr Yakir's daughter) and, like her, the daughter of a Gulag survivor; and Viktor Fainberg, who had just graduated from Leningrad State University with an undergraduate thesis on Holden Caulfield, the narrator of *The Catcher in the Rye* who rebels against everything "phony."

Gorbanevskaya handed banners to her fellow protesters and the eight sat down silently at the chosen spot, holding the banners over their heads. Curious onlookers began to approach the group, but before they could get close enough to read the slogans, a group of roughly ten plainclothes officials—presumably police or KGB employees—rushed in, seized the banners and the flag, and began pummeling those seated, yelling "They're all yids!," "Beat the anti-Soviets!," "Stomping on, that's what they need!," and "The whore's got herself a child—now she comes to Red Square!" According to an eye-witness, someone—it wasn't clear whether it was a bystander or one of the plainclothes officials—called out "You hooligans, what are you sitting here for?" Gorbanevskaya replied, "In twenty years or so you too will understand and then you'll be ashamed."[46] Litvinov, who like other protesters offered no resistance to the attackers, was bleeding profusely from his head; four of Fainberg's front teeth were knocked out. An official grabbed a camera from the lone foreign journalist who had waited long enough to catch the protest and ripped out the film. Within minutes, three cars pulled up to transport the protesters to a nearby police station. A female plainclothes officer handed Gorbanevskaya her baby and pushed her into a car. Lowering the window, Gorbanevskaya shouted to the crowd of stunned onlookers, "Long live free Czechoslovakia!," at which point the official slapped her mouth. Undeterred, Gorbanevskaya called out again, and this time the woman punched her in the face. "How dare you hit me," she cried. "Who hit you?" came the reply. "Nobody hit you."[47]

The entire episode was over by a quarter past noon. At four o'clock, as the protesters sat in jail waiting to be interrogated, the first BBC report on the incident was broadcast to listeners in the Soviet Union and around the world.[48]

Much could be said about the biographies of the Red Square protesters, but for our purposes the more salient point is this: apart from Bogoraz, Gorbanevskaya, and Litvinov, they barely knew each other. Several had never met prior to the event whose aura would trail them for the rest of their lives. Once again Volpin's goal of seeing unfamiliar faces at a protest gathering had been met. But more than that, the demonstration on Red Square represented a nearly perfect consummation of individual, conscience-driven voluntarism: no one had been pressured to take part, no roles had been assigned, there were no leaders and no followers, no wrangling over an elaborate group statement, just eight individuals who had chosen to follow their own moral intuition in full awareness of the mortal risk that entailed. Three days later, Gorbanevskaya smuggled a letter out of the USSR, which appeared on August 29 in the *New York Times* and the *Times* of London. "My comrades and I," she wrote, "were able, if only for a moment, to break through the flood of unbridled lying and cowardly silence, to show that not all the citizens of our country approve of the violence carried out in the name of the Soviet people. We hope that the people of Czechoslovakia have learned or will learn about this. The belief that Czechs and Slovaks, when thinking about Soviet people, will think not only of the occupiers, but also of us, gives us strength and courage."[49] In multiple ways, it was a symbolic inversion of the mass rally organized by the Kremlin to welcome the humiliated Svoboda: the same Czechoslovak flag, but now accompanied by messages of Soviet shame and solidarity with the Prague reformers, messages borne by a handful of individual citizens who had come entirely on their own initiative, and whose reward was not sandwiches, lemonade, and ice cream, but prison.

The August 1968 demonstration in Red Square became the most celebrated fifteen minutes in the history of the Soviet dissident movement. It captured what U.S. senator Thomas Dodd, who like the Soviet judge Lev Smirnov had served on the Allied prosecution team at the Nuremberg Trials, had in mind when he invoked the "sublime courage" of those who "take a stand for freedom under the totalitarian Soviet regime."[50]

Dodd wasn't the only one in awe. Joan Baez recorded a song called "Natalya" celebrating Gorbanevskaya; Adrienne Rich dedicated the poem "For a Sister" to her. Ivy Litvinov, widow of Maxim Litvinov, sent a letter to a friend in her native England referring proudly (but anonymously) to "the hero" in the family—her imprisoned grandson Pavel.[51] In the following years, film companies in West Germany and Japan produced admiring docu-dramas based on the August 25 demonstration and the subsequent trial of the protesters. When KGB chairman Yuri

Andropov learned of the West German production, he conveyed a message to Bonn that broadcasting the film on television—where it could be viewed by much of the population of Communist East Germany—would damage Soviet–West German relations.[52] Writing from exile in West Germany in the early 1980s, Raisa Orlova spoke for many dissident Soviet intellectuals inside and outside the country: "If humanity does not perish, this event will take pride of place in textbooks on twentieth-century Russian history."[53] Amalrik insisted (also with hindsight and from exile), "It was historically necessary that someone say an emphatic 'no' to Soviet imperialism. Perhaps in the final accounting that decisive 'no' from seven people in Red Square [Amalrik was either unaware that Bayeva had participated in the demonstration or wished to shield her from persecution] will prove weightier than the indifferent 'yes' from 70 million people at 'meetings of the toilers.'"[54] The demonstration on Red Square completely overshadowed several dozen other protest actions against the invasion of Czechoslovakia, many of them by lone individuals and altogether involving some 160 people in Moscow and other cities, virtually unknown to contemporaries. The August 25 demonstration has been reenacted *in situ* on several of its post-Soviet anniversaries, once with Gorbanevskaya herself taking part. More than any other single event, it has contributed to the canonization of the Soviet dissident movement in general and seven individual protesters in particular.[55]

It didn't start out this way. Initial reactions to the Red Square protest among the Moscow intelligentsia, including those horrified by the Soviet assault on the Prague Spring, were overwhelmingly negative. Amalrik lashed out at Gorbanevskaya for committing such a "tactical mistake," and especially for allowing a crucial figure like Litvinov to attend (Amalrik suspected that Litvinov took part only so as not to abandon his friends Bogoraz and Gorbanevskaya, that is, out of precisely the kind of motive that Litvinov repudiated). The demonstration was also "terribly organized," with wholly inadequate attention to media coverage. "If I hadn't by chance returned to Moscow," Amalrik fumed several weeks later, "no one would have learned right away that there had been a demonstration on Red Square or who was arrested."⁵⁶ Like Volpin's "transparency meeting," which Alexeyeva had initially dismissed as a pointless risk for the sake of "a few narcissistic moments," the demonstration on Red Square struck many as "senseless" and "unnecessary," hopelessly devoid of practical effects.⁵⁷ This was certainly the view taken by Grigorenko as well as by Chalidze, who rejected "the almost fatalistic indifference to the question of efficacy (for that is what the principle 'I cannot remain silent' is from the tactical point of view)."⁵⁸ According to Gorbanevskaya, Krasin, the foremost advocate of a more strategic approach by the dissident movement, told her, "If I had been in Moscow [at the time], I would have canceled your demonstration. How could we lose such people? Larisa, our irreplaceable connection to the camps. Pavel, the remarkable compiler of books about political trials." Krasin recalled putting it somewhat differently: it was "necessary to produce victims in carefully calibrated doses." Either way, such remarks only confirmed Gorbanevskaya's worst suspicions regarding calls for leadership and a division of labor within the movement. Krasin "spoke this way," she lamented, "as if dealing with functionaries, when he was talking about people who considered him a close friend."⁵⁹ Was he blind to the difference between weak ties and strong ties? Krasin seemed to take for granted not only that Gorbanevskaya had directed the demonstration, but that he had the authority to overrule her, as if the dissident movement were built on a hierarchy of assigned roles rather than on freely made decisions powered by moral intuition. It was not the last time he would think that way.

Others disapproved of the demonstration for different reasons. "At a time when the Hong Weibing [Mao Zedong's Red Guards] are taking their revenge on intellectuals and the term 'intellectual' has become a curse word," Kornei Chukovsky wrote in his diary, "it is important to remain in the ranks of the intelligentsia and not to abandon them—for prison. We need our intelligentsia here for the sake of the day-to-day intelligentsia cause. It would hardly have helped for Chekhov or Constance Garnett [Chekhov's English translator] to serve time in prison."⁶⁰

Earlier that year, the signing of group letters criticizing the arrest and trial of protesters, complete with signatories' addresses, had already been ridiculed as "self-denunciation," needlessly exposing participants to extra-judicial reprisals including termination of employment, university enrollment, and Party membership.[61] It was all fine and good to reject underground revolutionary traditions of conspiracy and *noms de guerre* in favor of transparency, but why make the KGB's job easier? Allegedly driven by despair, or an appetite for martyrdom, the demonstration in Red Square inspired a new term of derision among nominally sympathetic observers: "self-imprisonment." The Russian *samosazhanie* puns on the nearly identical *samosozhzhenie*, meaning "self-immolation," the term used to describe the suicide of South Vietnamese Buddhist monk Thích Quang Duc, who in 1963 famously set himself on fire not far from the Presidential Palace in Saigon, in protest against the anti-Buddhist policies of the American-installed dictator (and fervent Catholic) Ngo Dinh Diem. A photograph of that event had appeared on the front pages of many major newspapers, leading President John F. Kennedy to declare that "no news picture in history has generated so much emotion around the world as that one."[62] For Soviet intellectuals appalled by the Soviet invasion of Czechoslovakia but queasy about the demonstration on Red Square, describing the protest with a term that invoked religious zealotry—or perhaps a stereotypical Asiatic fanaticism—served to shield them from guilt over their own inaction.

What all these ironic terms had in common—self-denunciation, self-imprisonment, self-immolation—was precisely the *self*. The Ukrainian mathematician and Marxist dissident Leonid Plyushch called it "the intellectual's original sin . . . of belief in the power of personal protest," a mirror image of "the romantic reaction to Soviet reality that took the form of monarchism, Slavophilism, and nationalism."[63] This claim was repurposed into an impassioned defense of the Red Square demonstrators by Anatoly Yakobson, a charismatic high school teacher, poet, and literary critic. Yakobson deeply regretted not having been with them that day. His wife, Maya Ulanovskaya, a survivor of Stalin's Gulag, had decided not to relay to him a message from Bogoraz about the upcom-

FIGURE 8.3. Thích Quảng Đức's self-immolation near the Presidential Palace in Saigon on June 11, 1963

ing demonstration, convinced that he would be unable to resist taking part and that she and their young son might never see him again. In an open samizdat letter dated September 18, Yakobson urged people not to judge the protesters "according to the usual political standards, where every act is supposed to produce a direct, materially measurable result, some tangible utility." The demonstration was "a manifestation not of political struggle (for which conditions do not exist), but of MORAL struggle":

> Human dignity does not permit one to make peace with evil, even if one is powerless to stop that evil. This doesn't mean that all those who sympathize with the demonstrators must follow their footsteps and go to the square. Nor does it mean that every moment is appropriate for a demonstration. But it does mean that everyone of the same mind as the heroes of August 25, using one's own powers

of reason, should choose the moment and form of protest. There is no single method for all, just one universally understood fact: "prudent" silence can turn into madness—the restoration of Stalinism. Ever since the trial of Sinyavsky and Daniel . . . , not a single act of arbitrariness and violence by the government has failed to be met by PUBLIC protest and rebuke. This is a precious tradition, the beginning of people's SELF-LIBERATION from humiliating fear, from participation in evil. Let us remember Herzen's words: "Nowhere do I see free people, and I cry out—halt! Let us begin by liberating ourselves!"[64]

The letter brilliantly staged the protest as something far greater than the repudiation of a specific act of aggression by the Soviet state. Those fifteen minutes on August 25 could now be seen as part of an ongoing psychic battle taking place within every conscientious Soviet citizen. Yakobson's version of work on the self, moreover, did not proceed from the specific need to repent for past sins that had drawn Gorbanevskaya to the square or from the particular disgust with public conformism that had driven Bogoraz. Instead, it positioned the desire for self-liberation ("for your freedom *and ours*") as part of a broader recovery of personal dignity from the depredations of the Stalin era. That recovery was now entwined with the mission of the dissident movement.

But there was more. Yakobson's letter drew attention to the fact that, as the intervening weeks had shown, the demonstration in Red Square had also produced effects out in the world. Thanks to the BBC and other Western news agencies, anyone who wanted "to know the truth" had learned of this spectacular event: "the people of Czechoslovakia found out, all of humanity found out." This was not simply about the power of publicity and public opinion. Again invoking Alexander Herzen, Yakobson reminded readers that the exiled Russian aristocrat's famous jeremiad against tsarist Russia's assault on the Polish independence movement in 1863—an obvious parallel to the crushing of the Prague Spring—had "single-handedly rescued the honor of Russia's democratic movement." Similarly, Yakobson continued, the handful of demonstrators on Red Square "rescued the honor of the Soviet people." It was a

FIGURE 8.4. Anatoly Yakobson

striking formulation, calling to mind, among other things, Jesus redeeming the sins of others (and then being crucified). Was the assumption that the eight demonstrators had redeemed the Soviet masses—Bogoraz's "cretins"—from their shameful approval of the assault on Czechoslovakia? Or was this another instance in which dissidents understood themselves as "saying out loud what the healthy million were thinking," at a time when "prudent silence" was no longer honorable?

All these issues were in play in the Moscow courtroom where, beginning on October 11, five of the eight protesters stood trial. Bayeva had managed to convince the police that she was a bystander rather than a participant; she was released without charges. Four missing front teeth and a badly swollen face made Fainberg ill-suited for the courtroom, since they gave credibility to the claim that it was not the protesters who should be prosecuted for disturbing the peace but those who attacked them. To forestall such a scenario, Fainberg was declared mentally unfit and dispatched to a psychiatric prison, where over the course of the next five years he would repeatedly be beaten and forcibly injected with antipsychotic drugs, driving him to the brink of suicide. Gorbanevskaya, who authorities may not yet have realized was editor of the samizdat *Chronicle of Current Events*, was released, either because no one wanted to put a nursing mother on trial or because she was likely to be as outspoken in court as she had been on Red Square—or both.

The remaining five defendants were charged under Article 190 with engaging in "group activities that violate public order" and making "deliberately false statements that slander the Soviet state." Once again, a trial of dissenters produced preordained guilty verdicts along with a public relations debacle for the Soviet government, complete with a covertly assembled verbatim transcript edited by none other than Gorbanevskaya. Widely circulated in the USSR via samizdat, the transcript and associated documents were published abroad under the title *Red Square at Noon*, an obvious nod to *Darkness at Noon*, Arthur Koestler's best-selling novel based on Stalin's show trials. In contrast to Koestler's protagonist Nikolai Rubashov, however, none of the defendants pleaded guilty. Their lawyers, among whom were Dina Kaminskaya and Sofia Kallistratova, adopted the now familiar strategy of arguing not that their clients hadn't done what the prosecution accused them of doing, but that what the prosecution accused them of doing was not a crime under Soviet law. For its part, the prosecution struggled to prove that the demonstrators had interfered in any way with pedestrian or vehicular traffic in Red Square, instead retreating to claims that their actions *could have* interfered or that publicly expressing disagreement with Soviet policies itself constituted a violation of public order.[65] When the prosecutor asked Bogoraz why she didn't send a letter to the government expressing her views, she noted: "I have had occasion to address the government before in other connections and not a single one of my letters has received a reply."[66] In response to the charge of anti-Soviet slander, Deloné pointed out that "we presented no facts in the texts of the banners, merely our attitude" toward the military action in Czechoslovakia. The banners "therefore cannot be false, let alone deliberately false." Notwithstanding the fact that the banners constituted the sole evidence

FIGURE 8.5. Family and friends of the Red Square protesters outside the courthouse where they are being tried, October 1968. The man on the upper left, looking down, is Pavel Litvinov's father, Mikhail Maximovich Litvinov. Just below him, in the fedora, is Larisa Bogoraz's father, Yosif Aronovich Bogoraz. To his right, in the striped knitted hat, is Tatyana Bayeva, one of the eight protesters, who convinced police that she was a passerby and was released. To her left, in the center of the photograph, with short dark hair and a scarf, looking up, is Yekaterina Velikanova. Directly behind her, to the right, is Alexander Daniel, her future husband and the son of Larisa Bogoraz and Yuli Daniel. The woman in the headscarf and glasses is Maya Kopeleva, Pavel Litvinov's wife. Anatoly Yakobson is at the upper right. Looking straight at the camera is a Soviet militiaman.

for the charge of making "deliberately false statements that slander the Soviet state," the prosecutor stated that "it is unnecessary to analyze the banners and demonstrate their deliberately false and slanderous nature."[67]

To a remarkable degree, the defendants' final statements echoed Yakobson's view of the protest as a quest for personal dignity and honor. "All my conscious life," Dremlyuga told the court just before his sentencing, "I have wanted to be a citizen, that is, a person who proudly and calmly speaks his mind. For ten minutes I was a citizen." Deloné insisted that his "five minutes of freedom" were worth years in prison.[68] It fell to Bogoraz, who declined to be represented by a lawyer, to articulate most fully the inner struggle that brought her to Red Square on August 25:

> I found myself facing a choice: to protest or to keep silent. For me, keeping silent meant associating myself with the approval of actions of which I do not approve. To keep silent meant to lie. I do not consider my way of acting the only *correct* decision, but for me it was the only *possible* decision. Knowing that I was not speaking "in favor" was not enough for me. For me it mattered that my voice "against" would be absent. It was precisely the demonstrations, the radio, the reports in the press about universal support that aroused me to say: I am against this, I do not agree. Had I not done that, I would have considered myself responsible for these actions of the government, just as all adult citizens of our country bear responsibility for all the actions of our government, just as our entire people bears responsibility for the Stalin-Beria camps, for the death sentences, for—

The prosecutor cut her off. "The defendant is not entitled to speak about the actions of the Soviet government or the Soviet people."[69]

The defendant, of course, had already spoken about both in her appeal "To the Global Public," co-authored earlier that year with Litvinov. Now, however, at the trial's climax, in a closing statement that, according to a century-old Russian tradition, was likely to serve as her final testament before she disappeared into prison or exile, Bogoraz chose a more intimate register to describe what led her to speak out. Her account rests on two paradoxical ideas. First: that it is possible to lie even while remaining silent. The conventional definition of lying refers to making a statement one knows to be false or, more subtly, creating an impression one knows to be false. In both cases, one has to make or create something, that is, exercise agency. Lying in this sense is distinct from merely not telling the truth. For Bogoraz, however, the Soviet system imposed a different ethical framework. The constant assertion that on all essential matters Soviet public opinion was unanimous (an achievement allegedly made possible by the forging of a society free of class conflict) and that public opinion and official policy were identical effectively transformed silence into an act that created the impression of consent. It didn't matter what your actual opinion was. The pervasive façade of unanimity—sustained by centrally coordinated mass media, a standardized, country-wide school curriculum, and the ubiquitous and infallible Party line—created a presumption of consent. Because consent was the default setting, you didn't need to exercise agency in order to foster the impression that you too consented. You could do nothing and say nothing but, according to Bogoraz, still be lying.

The second paradox was that even as Bogoraz felt compelled to shatter the myth of the Soviet population's unanimous support for the actions of its government, she insisted with equal fervor on the collective responsibility of all (adult) members of that population for those same actions. A skeptical Chalidze called this the "totalitarian view of responsibility"—a version that ignores gradations of culpability among individuals.[70] Totalitarian responsibility can be heard as a distant echo of the principle of "collective liability," which endured in tsarist Russia long after it was abandoned by peasant communities in other countries. Collective liability was powerfully reinforced in the Soviet participatory dictatorship forged under Stalin, a symptom of a society in which state capacity and the reach of its law-enforcement agencies were still limited. If, according to Yakobson, a small band of dissidents could "rescue the honor of the Soviet people," that meant that honor, too, was something held in common, along with sin, guilt, and responsibility—the constituent elements of a moral economy. But the question remains: How could

Bogoraz construe responsibility as universal if support was not? How can there be collective moral responsibility in a system where decision-making power is highly concentrated in the hands of an unelected elite?

It has been argued that Bogoraz's claim of collective responsibility was a tactical device, a useful fiction designed to induce Soviet citizens to behave *as if* they were responsible for the political life of their country, just as dissidents were determined to behave *as if* they were free citizens and *as if* the Soviet state honored the rule of law—even though everyone understood that none of these things was true. Since useful fictions—including the "as if" variety—are not all that different from useful lies, this interpretation would require us to dismiss Bogoraz's claim to an extreme aversion to lying. The tactical argument relies on the apparent implausibility of holding ordinary Soviet citizens responsible for, say, the brutal collectivization of the peasantry, or the mass deportations of ethnic minorities, or the horrors of the Gulag—that is, it assumes that Bogoraz could not possibly have meant *that*. The intensity of her vilification of popular support for the assault on Czechoslovakia, however, and the fact that she chose supremely risky means to advance the claim of collective responsibility—first at the Red Square demonstration, then in her closing statement at her trial—suggests that something deeper than tactics was at work. I am inclined in this case to take her at her word, to accept her self-proclaimed aversion to lying (including via silence) and to read that aversion as symptomatic of an overwhelming desire for transparency, for the absolute alignment of public face and private self. The roots of that desire, I believe, can be found in her biography.

Bogoraz's parents lived the Soviet dream: both came from poor Jewish families in the tsarist Pale of Settlement, and both found new lives and new careers thanks to the Bolshevik Revolution. Her mother, Maria Samuilovna Brukhman, was rescued by the Red Army during the 1920 war with Poland; after joining the Communist Party, she found work as a state censor in the town of Vinnitsa and then as an instructor at the University of Kharkiv. Her father, Yosif Aronovich Bogoraz, also a Party member, worked at the State Planning Commission of the Ukrainian Soviet Socialist Republic and taught political economy at the elite Institute of Red Professors. Neither held a university degree, a fact that hardly hindered their careers given the dearth of educated white-collar workers in the young Soviet state. Among Yosif Bogoraz's cousins was the prominent ethnographer Vladimir Tan-Bogoraz (born Natan Bogoraz). On her mother's side, Larisa had an uncle in the NKVD. Both parents had previously been married (a fact about which her mother never spoke). They were not legally married to each other—an unexceptional arrangement at a time of widespread experimentation with alternatives to traditional marriage and family life. They lived well in Kharkiv, Soviet Ukraine's first capital, where Larisa was born in 1929, with access to a state-owned dacha and regular shipments of groceries while millions of Ukrainian peasants were dying from the famine triggered by Stalin's forced collectivization of agriculture. They raised their daughter as a fervent Soviet patriot, an eager member of the Pioneers and the Communist Youth League for whom stock phrases such as "enemy of the people," "fifth column," "when you chop wood, the chips fly," and "if the enemy does not surrender, he is destroyed" rolled easily off the tongue. "Public activism," Bogoraz later recalled, "was apparently in my blood. There was a feeling of being connected to society, of responsibility for what was happening in society. This appeared very early [in me]. In what direction that feeling was steered, how and by whom it was used—that's another question."[71]

Bogoraz's family also lived the Soviet nightmare, starting with her father's arrest as an alleged Trotskyite in 1936. Her mother, refusing to defend or even discuss Larisa's father and what had happened to him, decided that henceforth Larisa would be registered at school under the last name Brukhman, in the hopes of concealing her kinship with an enemy of the people. Only years later did Larisa learn that Yosif Bogoraz had been sentenced to Vorkuta, north of the Arctic Circle, one of the deadliest outposts of the Gulag. During a shattering visit there in 1947, she began to grasp that he was not an enemy of the people. In the meantime, he had married a fellow camp survivor. By then, Maria Brukhman, frightened by the state-sponsored wave of anti-semitism spreading across the USSR after the war, had re-registered her daughter under the name Bogoraz, the better to conceal her Jewish ancestry.

FIGURE 8.6. Larisa Bogoraz-Brukhman

Multiple imposed changes of public identity, requiring a certain dissemblance by the young Bogoraz, coupled with her discovery of her father's innocence after years of assuming that her mother's silence signaled his guilt—one can imagine how all this might have nourished an unusually strong urge to live free of hypocrisy and dissimulation. In the years that followed, she stopped regarding political repression as "a personal tragedy" and instead began to understand her family's misfortune as "a fact of social fate."[72] The desire for transparency didn't arise overnight, or even over the course of a few years; it accumulated in fits and starts, filling the spaces left vacant by the erosion of childhood illusions.[73] Bogoraz was discovering, as Hannah Arendt put it in an essay on moral responsibility under totalitarian rule, that "it is better to be at odds with the whole world than ... with yourself." In words that were meant to illuminate motives for resistance under Nazi rule but that could apply to dissidents like Bogoraz, Arendt insisted that protesters acted

> not because the world would be better (not because of political responsibility) and not because they were worried about the salvation of their soul, but because they wanted to go on living with them-

selves. The presupposition for this kind of judging is the habit of examining and living together with yourself. We call that silent dialogue in which you speak with yourself Thinking, but it is not technical, not the privilege of the educated and sophisticated. No one was easier to lure into the new morality-trap than the intellectuals, because they can produce ideas for everything.[74]

The Leningrad literary scholar Lidiya Ginzburg described the silent dialogue this way: "In the twentieth century . . . a new conversation began—about how to survive and go on without losing one's human visage."[75]

The demonstration on August 25, 1968, illuminates one of the great enigmas of the Soviet dissident movement, namely, its persistent claim that it was apolitical. The eight protesters that day—three women and five men—had no illusions of influencing Soviet foreign policy. Their protest was an act conceived beyond the realm of political calculus, an act of care for the self designed to show that there are always moral alternatives, even under conditions of seemingly eternal Soviet power. Learning to conduct yourself like a free person meant living with an authentic version of yourself; doing so in an unfree country required a willingness to live, indefinitely, at odds with your own world.

The demonstration marked, ironically, the end of the three-year chain reaction of arrests, protests, trials, protests about trials, new arrests, and so on. There were no massive letter-signing campaigns on behalf of the protesters in Red Square and only a handful of detentions of suspected accomplices. The demonstration also put an end to dissident debates between those who favored spontaneous, conscience-driven forms of engagement and those eager for greater organization, strategic planning, and division of labor. It wasn't that the latter side won the argument; rather, the arrests of Bogoraz, Litvinov, and eventually Gorbanevskaya removed three of the most forceful proponents of voluntarism and minimal structure. For all its morally sublime qualities, the Red Square protest revealed the enormous price of acting on conscience alone. From now on, the debate was not whether to organize the movement, but how.

PART III
In Search of Form

9

Leave the Politics to Us

There is a need to create ideals even when you can't see any way to achieve them, because if there are no ideals, there can be no hope and one would be left completely in the dark, in a hopeless blind alley.

—ANDREI SAKHAROV

It took only a few days in the KGB's Lefortovo jail for Larisa Bogoraz to develop "prison hearing," the ability to decipher the sounds of prisoners tapping out messages on the walls, a technique dating back to the tsarist era. The system of coded spelling and abbreviations "seemed to be in my blood."[1] During the long days and nights, a song by Vladimir Vysotsky, passed around on audiotapes years before, floated into her mind:

> We have no need for plots and stories—
> Try as you will, we know all their features.
> As for me, I regard the best book in the world
> To be our Code of Criminal Procedure.
>
> If I'm having trouble sleeping,
> Or a hangover after going on a bend,
> I open the Code at any page
> And I can't help it—I read to the end.[2]

Like Vladimir Bukovsky, Bogoraz decided to use the period of pre-trial detention to immerse herself in the Code of Criminal Procedure, a copy of which was available in the prison library. Where Bukovsky was surprised to learn how many rights he had, Bogoraz was shocked at how many crimes carried the death penalty. Prison guards occasionally interrupted her reading, curious to get a glimpse of one of the Red Square protesters mentioned in a new underground song by Yuli Kim. "The Drunken General Secretary's Monologue" featured a bushy-eyebrowed Leonid Brezhnev as narrator:

> My eyebrows are thirsty for blood—
> sing along, no back-talk!
> Sing to our health,
> until I start frowning.
>
> Grab a mandolin, Yura [Andropov],
> confiscate it from [Alexander] Galich.
> Where are you, you idiot censor?
> Come on, sing like we used to.
> Eh, c'mon! Let's go, one more time,
> again, again, and again,
> and Pashka [Pavel Litvinov],
> and Natashka [Natalya Gorbanevskaya]
> and Larisa Bogoraz![3]

One guard, evidently pleased to have a freshly minted celebrity on his watch, opened the hatch for food delivery, winked at Bogoraz, and quietly sang the last three lines. On another occasion, he whispered "Belogorodskaya's in 45, and your Litvinov is in 43"—indicating cells along the same corridor as Bogoraz's. A prison administrator stopped by to strike up a conversation, trying to convince Bogoraz that she and Litvinov had taken the wrong approach. "Why did you protest openly? And with just a handful of people, what could you possibly change? You should have established an organization."[4]

Bogoraz knew better than to share her thoughts with a prison official. But as previous chapters have shown, these questions were coming up in conversations beyond the prison walls as well. Why did dissidents sign open letters, in effect handing themselves over to the KGB? Why did they appeal to the Soviet Constitution when the Kremlin seemed perfectly willing to ignore it?

For signers of protest letters in the spring of 1968 who had had their salaries cut, or been fired from their jobs, dismissed from institutes and universities, expelled from the Party, denied permission to defend their dissertations, or blacklisted from future publication, the question was now whether to withdraw their signatures in the hopes of restoring their former status. By Stalinist standards, of course, the punishments were mostly mild; signers were rarely arrested, let alone tortured, exiled, or shot. The reprisals took place without even the pretext of a legal framework. A telephone call from a member of the Central Committee or the KGB was all it took.

The classic Russian method of recanting was to claim inebriation when signing. But there were other options. A scientist in Kyiv explained that a group letter had been brought to him by a beautiful post-doctoral fellow in mathematical physics, and "I couldn't turn her down"—a phrase that enjoyed a long parodic afterlife among dissidents.[5] To many, however, the dissident movement itself appeared to be entering its afterlife. The Red Square demonstrators were exiled to Siberia or locked up in prisons. The *Chronicle of Current Events*, under the anonymous editorship of Natalya Gorbanevskaya (still at liberty, but not for long), provided a stream of information about extra-judicial reprisals against signers of protest letters.[6] Stories of recantations also began to spread.

Unlike institutions, social movements require motion: they cannot survive, much less thrive, standing still. Even more, they require periodic successes, victories large or small that convey not just motion but momentum. What did Soviet dissidents have to show in this regard in 1968? They had helped galvanize an uncensored, samizdat-reading counter-public, distinct from the official Soviet public sphere and significantly (if selectively) amplified by the Voices. They had brought fundamental issues of constitutional rights and the rule of law to the attention of that counter-public. They had embarrassed the Soviet government—or, rather, they had facilitated the Soviet government embarrassing itself—before the West and a significant portion of the Soviet intelligentsia. They had also, via chain reaction, significantly expanded their own numbers. Gone were the days when, as one put it, "the entire movement could fit on Yakir's couch."[7] The growth in numbers, however, was now threatened by official reprisals against the movement's outer tier of supporters, the roughly one thousand signers of open letters, as well as its most visibly committed inner core, those who had protested in Red Square against the invasion of Czechoslovakia.

Among those who had suffered extra-judicial reprisals in the spring of 1968 was Andrei Sakharov, one of the USSR's most highly decorated scientists. Despite working at a top-secret nuclear weapons research facility in the Soviet interior, several hundred miles east of Moscow, Sakharov was not beyond the reach of Western shortwave radio broadcasts, and therefore of the chain reaction triggered by the trial of Andrei Sinyavsky and Yuli Daniel. By early 1968, he was regularly listening to the Voices, including to reports on the Prague Spring and dissident activity in the USSR. Having learned about the letter-writing campaign on behalf of Alexander Ginzburg, Yuri Galanskov, and other arrested *samizdatchiki*, Sakharov sent a private note to Brezhnev, whom he knew personally from the latter's job (between 1956 and 1960) as the Central Committee's overseer of the Soviet weapons industry. The letter endorsed the free flow of information as essential to the USSR's modernization. Brezhnev, who admired Sakharov's intellect, did not respond, but Sakharov learned from colleagues that the Soviet official in charge of secret weapons research, Yefim Slavsky, had been notified.[8] "Sakharov is a good scientist," Slavsky reported at a Party meeting several weeks later. "He's accomplished a great deal and we've rewarded him well. But as a politician he's muddleheaded, and we'll be taking measures."[9] Shortly thereafter, Sakharov's salary of a thousand rubles a month—sky-high by Soviet standards—was cut in half, and he was demoted from his position as department chair at the Academy of Sciences' Institute of Physics.

Sakharov had certainly accomplished a great deal for the Soviet Union. Shortly after receiving his PhD in theoretical physics at age twenty-seven, he was drafted into the Soviet atomic weapons research program, where he contributed to the USSR's breaking (in 1949) of the American atomic monopoly. He then went on to design a hydrogen bomb and presided over its successful testing in 1953, shortly after the Americans had first tested theirs.[10] For Sakharov, these were heady days. "A thermonuclear reaction," he wrote, "the mysterious source of the energy of the sun and stars, the sustenance of life on earth but also the potential instrument of its destruction—was now within my grasp. It was taking shape at my very desk."[11] These were also milestones for Sakharov's country. Together with the victory over Nazi Germany and the launching of the Sputnik satellite, the development of thermonuclear weapons signaled the USSR's meteoric rise from backward agrarian kingdom to socialist superpower. Sakharov's career made a stunning ascent too, starting with his induction in 1953, at the exceptionally young age of thirty-two, into the Soviet Academy of Sciences, followed by the awarding of the Stalin Prize, the Lenin Prize, and the Hero of Socialist Labor Prize (three times). Most of these awards (and the hundreds of thousands of rubles that accompanied them) were bestowed secretly because of the classified nature of Sakharov's contributions to Soviet nuclear weapons technology. Having helped turn the USSR into a superpower, Sakharov was now among the most richly rewarded scientists in Soviet history. KGB chief Yuri Andropov referred to him as a "golden head"; Khrushchev described Sakharov as "sparkling like a gem among all the scientists."[12] Sakharov and his family enjoyed a luxurious apartment in Moscow, a summer house in the coveted Zhukovka region not far from the capital, a personal driver, and access to goods and services unavailable to ordinary Soviet citizens. His office at the top-secret weapons research complex known as "the Installation" (the Soviet counterpart to the American facility in Los Alamos) included a special high-frequency phone with a direct line to the Kremlin. The town of Sarov, where the facility was located, disappeared from Soviet maps. Officially, it was referred to as Arzamas-16, after the nineteenth-century literary society that counted Pushkin among its members. Unofficially, the scientists who worked there called it "Los Arzamas." The Kremlin considered Sakharov's research so critical to the USSR's military strength that it forbade him to fly or use public telephones.[13]

Of all the varied itineraries to the dissident movement, none seem more unlikely than that of Sakharov. What could have been the stumbling block for someone who enjoyed the highest educational and material privileges the Soviet system bestowed, who had never been arrested (nor had any members of his immediate family), and whom the KGB had never coerced into betraying friends or colleagues?[14] Several explanations have been offered. Given Sakharov's lifelong concern about the danger of nuclear war, it has been argued that a sense of guilt over the staggering destructive powers he helped create—mirroring the experience of the American nuclear physicist Robert Oppenheimer, director of the Los Alamos National Laboratory—drove Sakharov to become a dissident in an attempt to redeem the sins of his youth. This, incidentally, was the KGB's favored explanation, perhaps because it avoided any criticism of the Soviet system.[15] Another view has it that, despite being born and raised *in* the Soviet Union, Sakharov was never quite *of* the Soviet Union, never became a "Soviet person," having been home-schooled as a child in the cocoon-like atmosphere of a pre-revolutionary intelligentsia family (his father was a physics teacher and amateur pianist who occasionally accompanied silent movies) and subsequently absorbed into the rarefied atmosphere of quantum physics and cosmology. Still another approach maintains that it was the scientific mindset itself—its insistence on verifiable truths, its suspicion of dogmas, its cosmopolitanism—that put Sakharov on a collision course with the Soviet regime, despite the support it lavished on him.[16]

Neither alone nor in their various combinations, however, can these arguments satisfactorily account for Sakharov's movement toward public dissent. At no point, for example, did he express guilt at having helped create nuclear weapons. Unlike the United States, the Soviet Union never actually deployed such weapons against an enemy, with the resulting hundreds of thousands of civilian deaths and mass irradiation. On the contrary, Sakharov remained convinced that by helping to break the American nuclear monopoly he had made a vital contribution to the balance of power and therefore to world peace.[17] As for his never having absorbed Soviet values, Sakharov, like Alexander Volpin, struck many of his colleagues not as un-Soviet but rather as un-worldly. One even dubbed him "the Martian" (just as Volpin was known to some of his friends as "the computer"), a nickname Sakharov tellingly took as "flattering."[18] By Sakharov's own account, however, he considered himself "part of the [Soviet] state."[19] Even a brief perusal of his writings in the 1960s, including his early statements of dissent from official policies, reveals a characteristically Soviet lexicon, sprinkled with phrases such as "monopolist capital," "ruling classes," "the socialist elevation of the moral significance of labor," and references to Marx, Engels, and Lenin. In 1968, Sakharov described his views as "deeply socialist."[20]

As with Soviet dissidents generally, we will begin to understand Sakharov only when we see him not as a Western liberal caught on the wrong side of the Iron Curtain, or as a holdover from the nineteenth-century Russian intelligentsia, or as someone from another moral or cognitive planet, but as a person immersed in Soviet norms, in constant dialog with them, shaped by the internal tensions and possibilities they contained. Every orthodoxy houses the seeds of its own potential disruption, and Soviet socialism was no exception.

Sakharov's views on social and moral questions were profoundly shaped by his experience as a scientist. As his biographer Jay Bergman has persuasively argued, Sakharov didn't merely draw on scientific methods as a way of tackling social problems. He regarded the scientific community itself, the republic of science, as the model of an ideal society: rational, evidence-based, progress-driven. And managed, one might add, by meritocratic elites. In this he was certainly not alone. For Valentin Turchin, the physicist turned computer scientist, quantum mechanics—or, as he put it, "the penetration of physics into the world of elementary particles"—had shown that the laws of nature could henceforth be expressed only in terms of probabilities rather than mathematical certainties. This was a death sentence for any kind of determinism, and therefore for the notion that individuals cannot influence the course of history.[21]

Any attempt to draw a line from scientific worldview to dissent, however—a connection that scientists might call "reproducible" and historians would call "generalizable"—faces daunting obstacles. To begin with, while it is true that members of the USSR's scientific-technical intelligentsia constituted a striking proportion of participants in the chain reaction of dissent, making up close to half of the roughly one thousand signers of open letters in support of Ginzburg and Galanskov, they also constituted a roughly equivalent proportion of the intelligentsia as a whole, that is, of the social milieu within which the chain reaction largely took place.[22] Dissident scientists, moreover, constituted only a small fraction of the USSR's scientific-technical intelligentsia, the majority of whom were silently sympathetic, indifferent, irritated by, or flat-out opposed to protests, letter-writing campaigns, and other dissenting activities. One cannot credibly claim, therefore, that scientific ways of thinking, or the high status of science in Soviet society, or the cosmopolitan framework of scientific activity, by themselves, fostered an actively critical stance toward the Soviet state. They seem to have had no such effect on the majority of scientists.

Becoming a dissident invariably involved a personal stumbling block connected to Soviet power: an unjust punishment, a morally compromising demand, an egregious gap between theory (or propaganda) and reality. Scientists were no more likely to experience such an obstacle than other members of the intelligentsia. If they did, however, and if they decided not simply to make peace with it but to respond critically, then approaches drawn from scientific thought and practice could come to the fore. This was certainly the case with Sakharov. By his own account, the initial stumble was tied not to nuclear weapons per se, but to the kind of people who controlled their use. While still in his thirties, Sakharov was sufficiently eminent in the world of nuclear weapons research to have met quite a few leading Soviet officials, including Stalin's secret police chief Lavrenti Beria, the Party's ideological gendarme Mikhail Suslov, and future premier Nikita Khrushchev. Such encounters often produced troubling impressions. One particularly disturbing encounter took place in November 1955, following the USSR's successful testing of a new thermonuclear bomb. At a celebratory banquet just hours after the detonation, Sakharov, as the bomb's principal designer, was given the honor of making the first toast in the presence of Marshal Mitrofan Nedelin, the deputy minister of defense:

> Glass in hand, I rose, and said something like: "May all our devices explode as successfully as today's, but always over test sites and never

over cities." The table fell silent, as if I had said something indecent. Everyone froze. Nedelin grinned and, rising from his seat with glass in hand, said: "Permit me to tell a parable. An old man wearing only a shirt is praying before an icon: 'Guide me and give me firmness, guide me and give me firmness.' An old woman, who is lying on the stove, calls over: 'Just pray for firmness, old man, I can guide it in myself.' Let's drink to firmness."

My whole body tensed, and I think I turned pale—normally I blush. For a few seconds no one in the room spoke; then everybody began talking loudly. I drank my brandy in silence and kept my mouth shut for the rest of the evening. Many years have passed, but to this day I retain the sensation of having been lashed by a whip. . . . Nedelin considered it necessary to rebuff my pacifist deviation, to put me and anyone else who might share these ideas in our place. The point of his tale (half indecent, half blasphemous, which added to its unpleasantness) was clear. We—the inventors, scientists, engineers, and workers—had created a terrifying weapon, the most terrifying in the history of humankind. But we would have no say whatsoever over its use. That decision would be made by them—those at the pinnacle of power, of the Party and military hierarchy.[23]

The offspring of peasants and workers, "they" often spoke in a way that struck Sakharov as betraying a certain moral crassness or "indecency," as he put it (even as Sakharov himself committed a Soviet indecency by ignoring, or being oblivious to, the invisible line separating scientists from political power). Were people like Nedelin, Suslov, Khrushchev, and Brezhnev, lacking university educations, qualified to decide matters of life and death on a global scale?

From its inception, the Bolshevik state had assumed a parental role vis-à-vis the Soviet masses: rewarding them, punishing them, and above all educating them at every available opportunity. The Soviet legal system was designed to teach (or "show," as in "show trials") as much as to administer justice; Soviet newspapers, radio, and television were designed to instruct as much as to inform. "Study, study, study!" Lenin had urged.

Those who failed to learn their lessons were charged with "political immaturity." To be sure, before the Second World War, most European states considered it their duty to instruct their masses in moral and other matters, but in the postwar order Western governments gradually retreated from paternalism. In the USSR, where Stalin had played the role of Great Teacher with exceptional brutality, his successors retreated from the use of terror but maintained their position as teachers. In the meantime, however, the students had grown up. The Communist Party had spread literacy to millions of workers and peasants, moved them into single-family apartments, and provided higher education to tens of thousands. By the 1960s, it was no longer clear that teachers such as Khrushchev and Brezhnev were wiser than their brightest pupils.

Sakharov developed serious concerns about the radioactive fallout produced by above-ground testing of nuclear weapons, the standard postwar method. Such tests spewed massive quantities of toxic particles into the atmosphere, where they could eventually (Sakharov sketched out projections over the next five thousand years) induce higher rates of cancer and genetic mutations in unsuspecting populations around the world. Above-ground nuclear testing now appeared to him as a "crime against humanity."[24] Rather than hunkering down within the fortress of his own expertise in particle physics, he began to explore biology and meteorology in order to assess the wider implications of his own work. In this he resembled Volpin, trained in mathematical logic but determined to apply it to language, law, and other domains. Sakharov's quest was partly stymied by the fact that Soviet data on air pollution were classified.[25] His concern about the long-term effects of atmospheric nuclear testing on human health—his stumbling block, as it were—nonetheless set him on a collision course with the Kremlin, culminating in what must have felt like an eerily familiar banquet scene.

In 1958, the Soviet Union, the United States, and the United Kingdom agreed to a voluntary moratorium on atmospheric testing of nuclear weapons while they began the process of crafting an international treaty formally banning such tests. As is often the case with arms control negotiations, fears of appearing politically weak, disagreements over methods of verification, and the persistent desire of both East and West to prevail in the arms race all impeded progress. By the summer of 1961, Khrushchev was aggravated by frequent shifts in the Anglo-American position. Perhaps wishing to test the mettle of the youthful new American president, John F. Kennedy, the Soviet leader decided that the USSR would no longer honor the moratorium. On July 9, he announced the news to a gathering of several dozen nuclear physicists in the Kremlin, including Sakharov. Once again ignoring social norms (as well as the possibility that Khrushchev was engaged in a tactical bluff vis-à-vis the West), Sakharov expressed his conviction that the Soviet Union stood little to gain by resuming atmospheric testing. When Khrushchev failed to respond to his remarks, Sakharov passed a hand-written note to him reiterating his point. The Soviet premier read the note, glanced at Sakharov, and silently slipped the paper into his jacket pocket.

He did not remain silent for long. At the banquet that followed, Khrushchev turned what began as a toast celebrating the assembly of distinguished scientists into a thirty-minute dressing-down of Sakharov. As his face became redder and his voice louder, Khrushchev echoed the drubbing delivered six years earlier by Deputy Minister of Defense Nedelin:

> Sakharov has moved from technical matters into politics. Here he's poking his nose where it doesn't belong. You can be a good scientist without understanding a thing about political affairs. Politics, after all—it's like the old joke about two Jews traveling on a train. One asks the other: "So, where are you going?" "I'm going to Zhitomir." "What a sly fox," thinks the first Jew. "I know he's really going to Zhitomir, but he said it in such a way as to make me think he's going to Zhmerinka." Leave the politics to us—we're the specialists. You make your bombs and test them, and we won't interfere with you; we'll even help you. We have to conduct our policies from a position of strength. . . . Sakharov, don't try to tell us politicians what to do or how to conduct ourselves. I'd be a willow in the wind and not Chairman of the Council of Ministers if I listened to people like Sakharov![26]

Nedelin had punctured what he took to be a naive, inflated intellectual with a joke about sex; Khrushchev used a joke about Jews. Functionally,

FIGURE 9.1. Andrei Sakharov

the two jokes were equivalent, as were the reactions in the room as recalled by Sakharov: "Everyone sat frozen, some averting their gaze, others maintaining set expressions.... No one looked in my direction."[27]

Sakharov was not content to make and test bombs. In 1964, at a meeting of the Soviet Academy of Sciences, he spoke out against a motion to grant full membership in the Academy to Nikolai Nuzhdin, an associate of Trofim Lysenko, the anti-Mendelian biologist responsible for turning the theory of environmentally acquired inheritance into Soviet scientific orthodoxy, in the process gutting the field of genetics. To criticize Lysenko and Nuzhdin for "the shameful backwardness of Soviet biology," "the dissemination of pseudoscientific views," as well as "the defamation, firing, arrest, even death, of many genuine scientists" was shocking enough. To do so as a physicist, publicly, with Lysenko sitting in the audience, was unprecedented.[28]

Because of these and other unorthodox gestures, Sakharov began to be approached by various figures seeking to harness his prestige to this or that reformist cause. These included the collective letter against the introduction of Article 190 (banning "group activities that violate public order"), and the campaign to block the perceived re-Stalinization of the Communist Party. Alerted by an anonymous announcement typed on onion-skin paper and left in the mailbox of his Moscow apartment building, Sakharov attended the second annual December 5 transparency meeting in Pushkin Square in 1966. The announcement, composed by Volpin, invited participants to assemble at the monument shortly before 6 p.m. and, at the stroke of the hour, to remove their hats and observe a minute of silence "as a sign of respect for the Constitution and support for political prisoners."[29] By Volpin's criteria, Sakharov's participation was a success: the physicist didn't recognize a single face among the several dozen people gathered in the square. It's unlikely that anyone recognized him either, given the secrecy shrouding his career. Nor did he join any of the conversations that sprang up after the minute of silence.

It cannot have escaped the KGB's notice (plainclothes officers filmed the demonstration on Pushkin Square, using infra-red cameras) that Sakharov's interventions were, for the time being at least, measured and discreet.[30] He kept his distance from other demonstrators and from Western journalists. When he requested leniency for arrested writers and *samizdatchiki*, he did so by calling KGB director Andropov on the high-frequency telephone in his office at the Installation or by sending a private note to Brezhnev.[31] Such behavior was correctly interpreted as a sign that Sakharov wanted to maintain his access to political elites, or, as he later wrote, "deep down I still felt that the government I criticized was *my* government."[32] The government in turn sought to steer his growing interest in world affairs in more palatable directions, hoping thereby to keep him under its wing. In 1967, a deputy of Soviet premier Alexei Kosygin reached out to Sakharov to commission an essay on the future of science and technology, for limited circulation within the *nomenklatura*. The field of "futurology" was developing rapidly in the United States, the deputy explained, and Premier Kosygin wanted Soviet scientists to produce their own long-range analyses to assist the

Kremlin in its perpetual planning of the Soviet economy. The resulting essay, surveying past and possible future developments in astrophysics and cosmology under the title "The Symmetry of the Universe," appeared in the restricted-access journal *Future of Science*.[33] Sakharov brought together two contrasting arenas: on the one hand, "the world of elementary particles, vanishingly minute sizes, and tiny fractions of seconds," and, on the other, "the world of galaxies, unimaginably great distances, and billions of years."[34] Writing about topics beyond the confines of his own work, he later noted, "had a profound psychological effect on me." For years, Sakharov's research, responding to the needs of the Soviet state, had largely been limited to designing nuclear weapons. To be sure, the physics of atomic and thermonuclear explosions offered Sakharov "a theoretician's paradise," as well as the chance to combine theory with the practical task of constructing an actual device, a task at which he also excelled. "The Symmetry of the Universe," however, pushed him to think more expansively about links between micro and macro scales of change over time.[35]

Sakharov's appetite for wide-angle thinking continued to develop in an article he drafted at the suggestion and with the collaboration of a journalist who went by the name Ernst Henri, one of several aliases of this shadowy figure whose career included working for Soviet counterintelligence, a fact of which Sakharov was then ignorant. Targeted for publication in *Literaturnaya gazeta* (Literary gazette), the essay dealt with the role of the intelligentsia in the nuclear age. The topic struck the journal's editors as sufficiently sensitive to require approval from higher-ups. Sakharov dutifully shared the manuscript with Suslov, a Politburo member, whose catchall verdict found the article "interesting but unsuitable for publication, since its ideas might be interpreted incorrectly." The text nonetheless found its way, without Sakharov's permission, into the samizdat periodical edited by Roy Medvedev, where it circulated among several dozen reform-minded members of the Soviet political and scientific elite.[36]

Sakharov continued to ponder the human impact of scientific knowledge and specifically of nuclear technology—or, as he put it, questions of "war and peace, the environment, and freedom of opinion."[37] His elite standing gave him access to restricted libraries housing a wide range of books and articles on these topics, including works by Rachel Carson, Albert Einstein, and Bertrand Russell. Spurred once again by the suggestion of an acquaintance who later turned out to have connections to the KGB (the physicist Yuri Zhivlyuk), Sakharov set out to organize his thoughts in the form of a broad, programmatic essay. It is likely that, as with the overture from Kosygin's deputy a year earlier, the KGB was attempting to subtly draw Sakharov away from the chain reaction of public protests and petitions. KGB chairman Andropov had already complained to the Central Committee that Sakharov, "at the request of antisocial individuals," had signed "politically immature documents, which were then sent to various [Soviet] government and state bodies."[38] If Sakharov's golden head was losing interest in designing weapons of mass destruction, perhaps it could be kept occupied with harmless labor of a grandly speculative nature.[39]

If that was indeed the KGB's plan, it was about to backfire spectacularly. Following Zhivlyuk's suggestion, Sakharov began work on a quintessentially Russian topic: the role of the intelligentsia in the contemporary world. There were already enough treatises on this subject to fill a warehouse, but Sakharov, working after hours at the Installation, was entering new terrain. What began as a statement on the role of the intelligentsia evolved into a series of reflections on the application of scientific knowledge to governance—and a belated repudiation of Khrushchev's injunction to "leave the politics to us—we're the specialists."[40] While announcing that his views were "profoundly socialist," Sakharov insisted that the most important contemporary challenges—environmental devastation, overpopulation, and above all the threat of universal nuclear annihilation resulting from the antagonism of the world's two superpowers—were inescapably global in nature and could be effectively confronted only through cooperation by scientific and technical elites across the Cold War divide. Echoing his earlier essay on the symmetry of the universe, he proposed that the capitalist and socialist worlds were destined to become more alike, each adopting the most effective features of the other in an evolutionary process of convergence. This, he argued, was the only viable survival strategy for humankind.

Titled *Reflections on Progress, Peaceful Coexistence, and Intellectual Freedom*, Sakharov's essay contained only a few genuinely new ideas.[41] The impossibility of winning a nuclear war, looming environmental and population catastrophes, convergence—all these were staples of contemporary debates in the West, though less so in the Soviet Union. "Peaceful coexistence" with the capitalist bloc had been Moscow's official foreign policy mantra since 1956. Even Sakharov's most daring observations—on affinities between Stalinism and fascism, official violations of the civil rights of Soviet citizens, overweening censorship by the Communist Party—would have been familiar to any dedicated reader of samizdat.

What *is* new is the way the essay connects these issues and, in particular, links the imperative of intellectual freedom to effective governance. A year earlier, Alexander Solzhenitsyn had issued an eloquent appeal to the Fourth Congress of the Union of Soviet Writers, demanding an end to censorship of works of literature so that writers could express themselves freely about society's most urgent moral problems.[42] Innokenty Volodin, a character in Solzhenitsyn's recently completed Gulag novel *The First Circle* (blocked from publication in the USSR), put it even more boldly. "A writer is a teacher of the people," Volodin tells his brother-in-law (an aspiring writer); "a greater writer is, so to speak, a second government."[43] Sakharov, however, was thinking about the first government, the Soviet government, *his* government, the one that actually set and carried out policies. He had in mind something far more comprehensive than freedom for artistic expression, namely, the free flow of information within the USSR as well as between it and the rest of the world—a necessary precondition for the scientific management of the most complex present and future challenges. *Reflections* not only mentions the particular cases of Sinyavsky, Daniel, Ginzburg, Bukovsky, and other producers or purveyors of uncensored texts, it links them to the imperative of unfettered access to information, and therefore to what Sakharov called "the scientific method of directing policy," a method "based on deep study of facts, theories, and views, presupposing unprejudiced, dispassionate, open discussion and conclusions."[44] Once again, the nuclear physicist was connecting elementary particles to larger processes. He was doing so, moreover, not in the name of preventing a return to the Stalinist past—the leitmotif of the dissident movement to date—but in order to imagine a more desirable future. In this sense *Reflections* bore the traces of Sakharov's exercises in futurology that preceded it. Instead of adopting the dissident movement's reflexive insistence on the non-political nature of demands for the observance of rights already enshrined in the Soviet Constitution, Sakharov was making the case that those rights, and especially the right to free exchange of information, were of paramount importance for the process of political decision-making, and thus for confronting the cardinal challenges facing the USSR and the world. In a nutshell, human rights were meant to function not simply as a constraint on *raison d'état*, but as one of its constituent elements.

Reflections makes that crucial argument, it must be said, in a clumsy, roundabout manner. The essay's vague invocation of the 1948 Universal Declaration of Human Rights, for example, fails to mention Article 19 ("everyone has the right . . . to receive and impart information and ideas through any media and regardless of frontiers"). Like many Soviet citizens, Sakharov seems to have been unaware that the USSR had abstained from the Universal Declaration and that, in any event, the Declaration did not constitute a legally binding document even for those countries that had ratified it. Moreover, *Reflections* endorses the distinctly Soviet approach to civil liberties as requiring state intervention (the state "must make available the material resources for freedom of thought") rather than the do-it-yourself ethos of samizdat, according to which the state need do nothing more than refrain from censorship.[45]

Nonetheless, *Reflections* contains in embryo an idea that would subsequently form Sakharov's most important contribution to the dissident movement, and perhaps to the world: that human rights not only were about protecting individuals from the depredations of repressive governments, but could reshape the way governments behave generally—including toward each other. *Reflections* offered itself not as a finished program but as a text designed to promote "open, frank discussion under conditions of transparency," to make it possible "to express not only true, but also dubious ideas" without fear of punishment.

As Masha Gessen notes, *Reflections* gave its readers an example of "thinking in public..., an invitation to think—and to argue with the author." "It is essential," Sakharov wrote, "that we get to know ourselves better." Soviet society had "started out on the path of self-cleansing from the foulness of 'Stalinism.' Drop by drop, we are squeezing the slave out of ourselves."[46]

Given the fate of the considerably less audacious essay he had co-authored with Ernst Henri the year before, Sakharov was under no illusions that *Reflections* would be published in the Soviet Union. Even before the drafting was finished, he began to make arrangements with Medvedev to solicit comments from a small circle of Moscow writers, intellectuals, and scientists, fully aware that this could lead to wider circulation in samizdat and possibly beyond the borders of the USSR. In the spirit of transparency, Sakharov also sent a copy to Brezhnev. Well before this, however, the KGB had begun to discover hand-typed copies of the essay in apartment searches in and beyond Moscow. Andropov already had a copy in his office safe. He sent a stern message to Sakharov via his supervisor, the physicist Yuli Khariton, director of the Installation: withdraw the manuscript from circulation.

In the world of Soviet official print-culture, this would have been a plausible request: the authorities did indeed recall and destroy books featuring a name, a passage, or an image that had fallen out of favor. They also reissued books with offending pages or photographs excised, as if they had never existed (this is how the entry on Lavrenti Beria in the relevant volume of the *Great Soviet Encyclopedia* was replaced by an article of the exact same length on the Bering Straits). If Andropov had not yet fully grasped the radically different system of samizdat, in which readers, rather than the state, owned the means of reproduction, then Sakharov's reply to his demand let him know: it was too late.[47] One of the samizdat copies of *Reflections* ended up in the hands of Andrei Amalrik, who passed it on to Karel van het Reve, the Moscow correspondent for the Dutch newspaper *Het Parool*. Van het Reve immediately grasped the historic nature of the text and, in addition to arranging for an abridged translation to appear in *Het Parool*, gave a copy to his colleague Raymond Anderson, the Moscow correspondent of the *New York Times*. On

July 22, 1968, the *Times* dedicated three entire pages to Sakharov's unabridged text, introduced on the front page with the headline "Russian Physicist's Plan: U.S.-Soviet Cooperation."[48] By then, the Voices had begun broadcasting excerpts to listeners in the Soviet Union. Over the course of the next year and a half, more than eighteen million copies were published around the world, briefly putting Sakharov in third place for global book printings, just behind Mao Zedong and Lenin and slightly ahead of Agatha Christie. He would never set foot in his office in the Installation again.[49]

Because it was unpublished in the USSR, we cannot know how many Soviet citizens read or listened to Sakharov's *Reflections*. Within the dissident milieu, judging by the number of memoirs that mention it, as well as the number of responses to it that soon began circulating in samizdat, the number of readers and listeners was quite high. The essay also found a substantial readership thanks in part to a covert CIA program of book distribution behind the Iron Curtain, of which *Reflections* became a "staple" text.[50] It made the rounds at Moscow State University, where students discretely left behind copies for each other in lecture halls. Andrei Grachev, a Soviet diplomat traveling in Western Europe at the time, bought a copy of *Reflections* and smuggled it into the Soviet Union upon his return. A few years later, so did a young provincial Communist Party official named Mikhail Gorbachev. Yefim Slavsky, head of the government ministry that oversaw the production of nuclear weapons, received calls from local Party secretaries across the USSR complaining about "counterrevolutionary propaganda" by one of his employees (that is, Sakharov).[51]

Publicly, the Kremlin maintained a studied silence regarding both Sakharov and his essay—a stance that led some Western analysts to conclude that Soviet leaders could not agree on how to respond or that they had deliberately leaked Sakharov's text as a trial balloon for a planned rapprochement with the United States, to counterbalance an increasingly belligerent China.[52] Careful readers of the Soviet press, however, noted a pair of articles in *Izvestiia* and *Pravda* condemning the idea of convergence as a smoke screen for the restoration of capitalism—

without mentioning Sakharov or *Reflections* by name.[53] Internally, the Soviet leadership showed no signs of dissension, instead coalescing around three lines of attack. First, they agreed that the notion of capitalism somehow peacefully merging with socialism into a single world government was, as Slavsky put it, "utopian nonsense," a symptom of Sakharov's political immaturity regarding the fundamental ideological and geopolitical divide of the current era. In a one-on-one meeting—"the only serious discussion I've ever had with anyone in authority about *Reflections* or any other statement of mine on public issues," Sakharov later noted—Slavsky insisted, "Capitalism can't be made humane. Their social programs and employee stock plans aren't steps toward socialism.... We'll never give up the advantages of our system, and capitalists aren't interested in your convergence either." Second, as Andropov wrote in a memorandum to the Central Committee, Sakharov's essay essentially called for "a transfer of power from politicians to the technical intelligentsia," and thus dared to "question the leading role of the Party in the task of building communism."[54] "You pit the intelligentsia against the leadership," Slavsky argued in a similar vein, "but aren't we, who manage the country, the real intelligentsia of the nation?"[55]

The third criticism was personal. For the seventy-year-old Slavsky, whose Jewish father had been drafted as a young boy into the tsar's army, and who himself had worked in the Donbass mines, fought side by side with Petr Yakir's father (Yona Yakir) in the civil war, and had risen to the upper echelon of the Soviet political elite, Sakharov was above all a hypocrite.[56] "You criticize the leaders' privileges," Slavsky scolded, "but you've enjoyed the same privileges yourself.... You have no moral right to judge our generation—Stalin's generation—for its mistakes, for its brutality; you're now enjoying the fruits of our labor and our sacrifices."[57] Sakharov had indeed been showered with extraordinary privileges, but Slavsky's rebuke could be (and was) applied to virtually anyone of a younger generation—anyone who had been housed, fed, and educated by the Soviet state, or spared the ravages of war, but nonetheless had the temerity to criticize the system that Stalin's newly elevated peasants and workers (including Slavsky) had built.

Stripping Sakharov of the moral right to judge became a convenient way to avoid engaging with the central points advanced in his *Reflections*.

Outside ruling circles, the reception of Sakharov's ideas was, in its own way, also highly personal. Amalrik spoke for many: "The thing about Sakharov's article that most shook me up was not *what* had been written but *who* had written it." Many of the article's individual points had already been made in various samizdat works, but never by someone with Sakharov's extraordinary record of scientific achievement and high status as an Academician. Public dissent by a figure like Sakharov signaled to Amalrik the presence of "deep geological faults within the core of the Soviet establishment." Unlike Slavsky, moreover, Amalrik was struck not so much by the elite privileges Sakharov enjoyed but, on the contrary, by his willingness to sacrifice them in order to freely speak his mind. In contrast to so many other scientists and writers, Sakharov had managed not to become entangled in the web of official favors—a fact that appeared to Amalrik as "a miracle."[58] Solzhenitsyn agreed: "Weighed down by our common sins, and the sins of each of us as individuals, [Sakharov] has left behind the abundant material comforts with which he was provided . . . and stepped out in front of the jaws of almighty violence."[59]

The quasi-religious tone was not an accident. Russian culture is hardly unique in bestowing special moral authority on those who have suffered. But it is surely at the upper end of the scale. Within the dissident milieu, Petr Yakir and Viktor Krasin rose to prominence in part because, as survivors of the savagery of Stalin's Gulag, they could speak with exceptional credibility about the threat of neo-Stalinism. While a substantial portion of the Soviet population regarded those who returned from the camps with suspicion ("They don't arrest people for nothing," went a popular saying, even though under Lenin and Stalin they certainly did), for others, such individuals were endowed with an almost mystical aura of righteousness.[60] Neither Yakir nor Krasin, however, nor any other veteran of the Gulag had gone willingly to the camps. Sakharov, by contrast, had voluntarily sacrificed his career along with his dacha, sky-high salary, and other elite privileges. Precisely because he engaged in open dissent from an exceptionally lofty perch, he stood to fall farther than others.

If Andropov and the KGB regarded Sakharov's views as "politically immature," so, in essence, did many Soviet readers. In the July 1969 inaugural issue of the samizdat journal *Problems of Society*, Sakharov's friend and fellow physicist Andrei Tverdokhlebov summed up the reception of his essay to date: "Readers of *Reflections*, as far as I am aware, unanimously share the view that the author is naive when it comes to political questions." The crushing of the Prague Spring, just weeks after Sakharov's optimistic call for convergence between socialism and capitalism, did not help in this regard. "The 'best outcome,' which Sakharov presents as a program of action for all mankind," noted Tverdokhlebov, was almost universally regarded as "unrealistic."[61] Some readers took for granted that Soviet socialism was incapable of reform and therefore of converging with another system. Others, such as Leonid Plyushch, regarded convergence as possible, but feared that each system would adopt not the best but the worst features of the other—the USSR sliding into "state capitalism in its most inhuman form," the West becoming "less democratic, [with] greater concentrations of capital, and merging monopolies with the state."[62] Still others, including Solzhenitsyn, opposed convergence as such, on the grounds that it would erase national identities—including Russia's—that served as "the wealth of mankind, its collective personalities . . . bearing within themselves a special facet of divine intention."[63]

Sakharov had anticipated the charge of naiveté, though not its full extent. "I can already hear the howls about revisionism and blunting the class-based approach," he wrote toward the end of *Reflections*, "the smirks about political naiveté and immaturity."[64] The essay had emerged, after all, from a series of commissioned exercises in futurology, and Sakharov made clear that he regarded his ideas as starting points for discussion rather than as demonstrated conclusions. Forecasting history's path, however—let alone its timing—is a notoriously risky business. In 1961, Khrushchev made the mistake of announcing an actual year (1980) when the USSR would complete its transition from socialism to communism, by which time the state would presumably have withered away, Soviet citizens would have mastered the art of give-and-take (according to their abilities and needs), and everyone would live happily ever after. Seven years later, in his *Reflections*, Sakharov set the completion of the process of convergence in the year 2000, by which time all "national contradictions" would be resolved and, far from withering away, a single world government would administer the entire planet.

No wonder, then, that Sakharov's Soviet readers pounced on his predictions while largely overlooking his argument that freedom of information was indispensable not only for protecting individuals from the states that govern them, but for optimizing policy-making and global security. They overlooked that argument partly because of its sheer novelty, along with its overshadowing by the notion of convergence between capitalism and socialism. Many of those readers, including the dissidents among them, had built a firewall in their minds between the defense of rights and the rule of law, on the one hand, and the realm of politics, on the other. Sakharov, almost universally (at least in the USSR) criticized as naive or politically immature, was the first dissident to treat the defense of rights as a manifestly political project. His *Reflections* launched a new samizdat conversation whose subject was not just how to prevent a return to the Stalinist past, but how to solve the most acute problems of the Soviet and global order.

10

Will the Dissident Movement Survive?

> CRITIC: What sort of person is he? Has he published?
> SON: No, as a formalist he doesn't get published.
> LADY: What does he live on?
> SON: Friends help him out. He has lots of friends, about five hundred, or even a thousand.
> LADY (*inspired*): A thousand?! That's almost a million!
> SON (*modestly*): Yes, almost.
> CRITIC (*critically*): Still, a thousand is less than a million.
> SON (*displeased*): Yes, a bit less.
>
> —ANDREI AMALRIK

Soviet dissidents, this book has argued, were not born but made, and made with Soviet ingredients. Their making typically began with a stumbling block, an ethical dilemma or cognitive dissonance rather than outright repression, set in their path by the Soviet system, triggering a reevaluation or repurposing of the norms by which that system initially formed them. If there were a case to be made for a born dissident, however, or at least a born contrarian, then Andrei Alekseyevich Amalrik would surely serve as Exhibit A.

A dissident among the dissidents, Amalrik was known for his brash, fearless, and abrasive personality. The phrase "humble as Amalrik" was a running joke among his friends.[1] Family lore included a story that captured his tendency, even as a small child, to question authority. It was set during the Second World War, during one of the first German air raids on the Soviet capital. As sirens warned Muscovites of the impending attack, a terrified crowd, including Zoya Amalrik with her three-year-old son in her arms, was being driven by police into one of the city's deep subway stations. As his mother recalled, there was probably no need to forcibly herd people into the shelter, since they were already running in that direction. That, at any rate, was Andrei's sentiment: as they huddled on the subway platform underneath the city, waiting for the bombing to end, he exclaimed over and over, "They drove us like pigs. Like pigs."[2]

Meanwhile, his father, a Red Army officer, made the mistake of criticizing Stalin for failing to prepare the Soviet Union for the German assault, a remark that earned him a sentence of eight years in the Gulag. Having served two of those years, Aleksei Amalrik was pardoned and sent back to the front, where he was gravely wounded by a land mine.[3] He went on to become a popularizer of archeology and history, as well as author of unpublished poems and fairy tales starring his young son, Andrei. One poem, penned in 1949, touched on the *Handbook for Parents* by the renowned Soviet psychologist Anton Makarenko:

"Andrei: Raiser of Children"

Makarenko wrote a handbook for parents,
Which little Andryusha snatched from the shelf.
On this you can count, my dear friends, have no doubt:
Andrei will be raising himself.[4]

What delighted his father did not go over so well at school. "Andrei is well read and capable," one early report card indicated, "but displayed a slipshod attitude toward his studies. He frequently skipped school. In class he was inattentive, distracted by secondary issues and talked out of turn. He has a very high opinion of himself and was often nasty to his teachers." Looking back three decades later, Amalrik more or less agreed: "I was a very bad pupil and often played hooky. I still consider my best school days to have been the ones I skipped."[5]

Things got worse in high school. In the spring of 1955, a letter to Amalrik's parents informed them, "Because of his systematic violation of student rules (missing 232 classes this year), bad behavior in class and a disrespectful attitude toward the school and its teachers, your son is being expelled."[6] The pattern continued in Moscow State University's history department, to which Amalrik managed to gain admission in 1959, presumably with his historian father's help. Within a month of enrolling, he received a reprimand for "rude and tactless behavior toward an instructor." In Amalrik's improbable telling, he was simply unaware that students were supposed to stand up when responding to an instructor's question. Apparently, he was similarly unaware that when a conversation ends with the instructor yelling "I know better than you," something has almost certainly gone wrong. While awaiting his punishment in the vice-dean's office, Amalrik was quietly advised by the secretary to apologize. He declined, noting that his classes were often boring and that certain instructors and administrators seemed intent on "offending" students and treated them "rudely." He also complained about the constant need to show one's pass to gain entrance to university buildings, as well as the shabby condition of classroom furniture. "Lack of respect—that's what several instructors accused me of," Amalrik wrote in an appeal to annul his reprimand, "whereby they cited not the content of what I say, but certain elusive nuances in my tone, my facial expression, etc." In any case, he argued, "truly authoritative people, people who have earned respect, do not ascribe any significance to lack of respect from some first- or second-year student."[7]

By his own account, Amalrik was shy and withdrawn, disinclined to fit in with other students or engage in mandatory public-service projects. "I came to the history department to study history," he announced, "not for public service." Just as he had refused to join the Pioneers in elementary school, so he declined to join the Communist Youth League, in contrast to most of his classmates at Moscow State University. This, he believed, had been misinterpreted as "a symptom of arrogance" and "anti-Soviet" tendencies.[8] A character trait had been turned into a political stance.

Amalrik threw himself into a research project on the role of Vikings in the origins of the Russian state. According to the controversial "Norman Theory," Viking explorers from Scandinavia (Norsemen or, in Russian parlance, Normans) had established Kyivan Rus in the ninth century. Debates on the subject, tinged with nationalist indignation at the idea that foreigners had laid the foundations of Russian (not to mention Belorussian and Ukrainian) history, had been going on for centuries. It didn't help that the scholars who first proposed the theory were German. Writing in 1947, the émigré historian Nicholas Riasanovsky noted that "no question in the entire field of Russian history has drawn more attention in historical literature, and has created more controversies than the problem of the origins of the Russian state." That same year, indirectly confirming Riasanovsky's assessment, the Soviet journal *Bolshevik* officially condemned the Norman Theory for "denying the ability of the Slavic peoples to create an independent state by their own efforts."[9]

Such was the minefield into which Amalrik stepped with his research on the Normans and Kyivan Rus. Tellingly, what attracted him to the Norman Theory was not the theory itself, or the evidence used to support it, but rather his own "anti-anti-Normanism," his dislike of the arguments deployed by the theory's critics, with their "narrowly nationalistic historical concepts dating from the era of the cult of personality." There was "nothing offensive to Slavs" in the Norman Theory, Amalrik insisted: "If every people were to begin all their undertakings from scratch, we would not have gotten past the neolithic era."[10] Offensive or not, Amalrik's instructor, the medievalist Anatoly Sakharov (no relation to the physicist) refused to accept the paper (which encompassed roughly 150 pages) because it was too strident, insufficiently Marxist, and submitted well past the deadline. Sakharov suggested instead that Amalrik write a different paper the following semester. True to form, Amalrik rejected the offer, insisting that he wanted a detailed critical assessment of his work by an expert in the field. He began to approach other members of the history faculty. One by one, they either declined or returned the paper with minimal comments.

By this time, the history department had already initiated Amalrik's expulsion from the university. With both of his parents experiencing severe health problems, Amalrik was now missing more classes and deadlines than ever. His mother died from a brain tumor in January 1961; a heart attack and a stroke left his father half-paralyzed and barely able to speak.[11] Amalrik unleashed a barrage of letters criticizing his expulsion, but more important, he now extended his quest for an expert assessment of his seminar paper (on which he continued to work) beyond Soviet borders. In March, he visited the Danish embassy in Moscow to request the mailing address of professor Adolf Stender-Petersen, a Slavic philologist at Aarhus University whose work on early Russian texts Amalrik had used in his research. Soviet police picked him up for questioning afterward, warning him to stay away from foreign embassies. Within weeks, Amalrik received Stender-Petersen's address and proceeded to send him a letter describing his work as well as his desire "to receive a serious and objective assessment." The professor responded, indicating that he was delighted to help. In fact, he had grown up in the Russian Empire and had studied at the University of St. Petersburg before emigrating in 1915. He considered himself "neither an émigré, nor a refugee, nor a defector, but simply a Danish citizen who retains the warmest recollections of the Russian scholarly world, and who relates to Soviet scholarship, especially that of the past five years, with enormous sympathy," notwithstanding his diverging point of view. "You can rest assured," Stender-Petersen added, "that I detest with my entire being any kind of nationalism, and absolutely do not consider myself a despised bourgeois."[12]

Undeterred by the police warning, Amalrik returned to the Danish embassy, magnum opus in hand.[13] Rather than forwarding it to Stender-Petersen, however, embassy officials, apparently fearing a provocation, decided first to clear the text with the Soviet Ministry of Foreign Affairs, which passed it on to the KGB. "So instead of the Danish professor," Amalrik recalled, "I had to have a discussion with a Soviet investigator."[14] KGB officials "helped me understand the incorrectness of the path I had chosen. They promised to assist me with my research (and helped my father receive his pension, for which I am very grateful), and advised me to seek re-admission to the university."[15] The university, however, was not interested. In a thirty-page letter to the Central Committee of the Communist Party, titled "Am I Worthy of the Attention of the Central Committee?" and laying out in detail the entire saga of his expulsion, Amalrik cast his re-admission as a hopeless cause. "I have written this letter," he concluded, "in the interest of future students who have not yet enrolled in the history department. I would not want my story to be repeated by them. I appeal above all to common sense: is it rational to declare as enemies people who have done nothing hostile?"[16]

The seminar paper on the founding of Kyivan Rus never reached the Danish professor. By now, Amalrik was a person of interest to the KGB, especially as he seemed to be spending more and more time in the company of foreigners. He befriended a number of visiting Americans, including the historians Richard Wortman and, somewhat later, Richard Pipes, who noted "a childish impudence about him: he did not hate communism, he just laughed at it and mercilessly twitted his police interrogators."[17] Amalrik also began to act as an intermediary between Moscow's unofficial art scene (in which his wife, Gyuzel Makudinova, participated) and prospective buyers among foreign journalists and diplomats stationed in the Soviet capital. Despite speaking no language other than Russian, he was one of the few Muscovites who was not afraid to invite Western correspondents to his apartment.[18] Amalrik, as one friend put it, "knew how to socialize with foreigners." According to Boris Shragin, the art historian, this was because Amalrik lacked "the inferiority complex vis-à-vis the West and the resulting absurd arrogance and eagerness to present a more attractive version of oneself... which constitute a characteristic feature of the Russian national mentality."[19] Amalrik had a different explanation, inspired by Alexander Volpin: Soviet law did not prohibit associating with foreigners, therefore "there was nothing criminal about it." Since his time at Moscow State University, he had "sought friends and acquaintances among foreigners." "The main thing for me was to find some almost metaphysical exit from the world that surrounded us. We were supposed to believe that the Soviet world was a closed sphere, that it was the universe. Those of us who were poking holes in that sphere, however small, were able to breathe a different air, sometimes actually a noxious air, but at least not the rarefied air of totalitarianism."[20]

For the KGB, Amalrik's contacts with foreigners were an asset to be monitored and exploited. In 1963, he was detained by agents who demanded—under threat of arrest—that he provide testimony against

FIGURE 10.1. Gyuzel Makudinova and Andrei Amalrik, with Pavel Litvinov behind them

an employee of the American embassy. He refused. They also tried, unsuccessfully, to pressure his disabled father into submitting a report alleging Amalrik's failure to care for him.[21] Several months later, after his communal apartment neighbors grew suspicious of the stream of foreign guests, Amalrik was visited by a pair of agents intent on taking him in for interrogation:

"Have you got a summons?" I asked.

At first they misunderstood and thought I was asking whether they had an automobile. When they realized what I meant, the senior lieutenant was baffled.

"What do we need a summons for?" he said. "We're just inviting you for a talk."

"I won't go without a summons," I said.

"Yes you will [using the informal *ty*]."

"Take me by force, if you want," I said. "Otherwise I won't go."

"Get your stuff and let's go!" Captain Kiselev suddenly barked. Until then he had played the role of a silent extra, but my stubbornness angered him. "What, you're not a Soviet person, you don't obey the authorities?!"

"What do you mean?" I replied. "I obey Soviet power, but not you." This Jesuitical answer perplexed him.

"Let's go—otherwise it'll get worse. Let's go, we just want to have a friendly talk with you." The senior lieutenant continued to argue and threaten me for a long time, but I had made up my mind not to go. Not because I considered this the most sensible thing to do (perhaps it wasn't sensible at all), but simply from a feeling of contrariness, a desire not to give in.[22]

Let us allow for a dose of retrospective bravado in this account (as U.S. secretary of state Dean Acheson was fond of noting, one rarely reads a memorandum of a conversation in which the author lost the argument). It nonetheless illustrates how a born contrarian such as Amalrik—who five years earlier, during his battle with the history department at Moscow State University, had declared that he "didn't assign too much significance to any sort of formalities"—learned to deploy Volpin's strategy of legal formalism.[23]

The KGB arrested Amalrik in May 1965 on charges of leading a "parasitic and anti-social way of life," the same charge used against the poet Joseph Brodsky the previous year. His main source of income had come from arranging sales of unofficial works of art to foreigners. At his trial a new charge was added, alleging that a series of samizdat plays written by Amalrik were "anti-Soviet and pornographic."[24] He was sentenced to two and a half years of internal exile on a collective farm in Siberia. Aleksei Amalrik, deprived of his sole caregiver and convinced that he would never see his only child again, died later that year. In the meantime, Andrei Amalrik's lawyer appealed the verdict to the Supreme Court of the Russian Republic, which overturned it—a rare exception

FIGURE 10.2. Outside the courthouse where Red Square protesters were on trial, October 1968. Above, a plainclothes (and fashionably dressed) KGB officer, conspicuously taking photographs of the crowd of supporters of the defendants. On the next page, the same KGB officer (now with his back to the camera), speaking with his colleagues. To his right, attempting to enter the conversation, is Andrei Amalrik, in glasses and beret. To the left of the group, listening in, is the poet Ilya Gabai, with heavy beard and glasses.

to the norms of Soviet jurisprudence. By July 1966, Amalrik was back in Moscow—just in time to be swept into the chain reaction.[25]

When Amalrik joined gatherings outside courthouses where rights activists were on trial, he would chat with KGB officers while others kept their distance.[26] He produced and consumed works of samizdat. He argued for greater organization and a clear division of labor within the dissident movement, half-jokingly designating himself the movement's

FIGURE 10.2. (*continued*)

"communications officer" because of his frequent contacts with foreigners (such official-sounding titles appeared comical in the free-form dissident milieu). But Amalrik refused to participate in demonstrations or sign group letters of protest, preferring to formulate his own ideas and present them in his own name, a practice that earned him the nickname "lone wolf." "I have an intuitive dislike for collective actions," he explained, "a highly developed individualism with which a Soviet upbringing is at war. I have always been repelled by the need to 'rally around the flag.' All group actions based on the imitation of some people by others—whether they're reasonable or not—contain an element of psychosis."[27]

Having helped transmit Andrei Sakharov's *Reflections* to the West, Amalrik decided in the spring of 1969, at the age of thirty, to offer his own prognosis of the Soviet system and the dissident movement to which it had given rise. It was an indirect reply to Sakharov, formed among other things by a starkly contrasting life experience. *Reflections*, in Amalrik's view, reflected "Sakharov's position in the establishment, his lack of experience of persecution, his education in a scientific milieu, his career in science, and his faith in people's innate nobility." By contrast, Amalrik was shaped by "social alienation, the experience of exile, poetic intuition, a skeptical attitude toward the social role of science, and the consciousness of human imperfection."[28] Unlike Sakharov, he made no assumption of goodwill on the part of the USSR's rulers.[29] As suggested by its incendiary title, *Will the Soviet Union Survive until 1984?*, Amalrik's counter-exercise in futurology departed as radically from Sakharov's as it did from Khrushchev's.

If Sakharov appeared to many as a naive dreamer descending from the Olympian heights of the Soviet Academy of Sciences, Amalrik was the opposite: a guerrilla activist and merciless puncturer of wishful thinking of every variety. Where Sakharov derived his categories of analysis from astrophysics, Amalrik thought like a historian—a historian, moreover, who took a class-based approach to his subject matter, though not exactly in the Marxist mode. Today, Amalrik's *Will the Soviet Union Survive until 1984?* is usually described as a prescient warning of the Soviet collapse by a dissident who posed what was then an utterly counter-intuitive question, to which history has in the meantime provided the answer.[30] That is accurate, though with qualifications large (the actual cause of collapse) and small (its precise timing).[31] But Amalrik posed another question that remains unanswered and very much alive today: Does the growth of an educated, consumption-oriented middle class necessarily foster democratization and the rule of law? Was the dissident movement, the middle class's leading edge, saying out loud, as Yuli Kim claimed, what millions of others were thinking?[32] Amalrik's answer, the first attempt at an analysis of the dissident movement by a participant-observer, was as unexpected as the title of his essay.

By the late 1960s, the USSR appeared nearly imperishable, especially to its own citizens. After surviving the ultimate stress-test—Hitler's Operation Barbarossa—the victorious Soviet Union had settled into its new role as one of two global superpowers, its armed forces second to none, its ruling Communist Party now a mass organization, and Moscow the capital not just of the world's largest country but of an expanding socialist world—"an eternal state," as one contemporary described it.[33]

The dissident movement, according to Amalrik, arose from samizdat networks of exchange and was now entering a new "political" phase. This reflected not the emergence of a coherent political ideology but increasing persecution of the movement's participants by the Soviet government. The movement itself had "no clearly defined program" beyond a shared commitment to "the rule of law, founded on respect for the basic rights of man." Some members gravitated to "genuine Marxism-Leninism" (a return to the original principles of the Russian Revolution), others to "Christian ideology" (blending a Russia-centric Slavophilism with Christian moral principles), and still others to "liberal ideology" (by which was meant Western-style democratic socialism). These worldviews, however, were "largely amorphous," according to Amalrik—that is, they were assumed rather than articulated in any comprehensive or persuasive way.[34]

Ideology, therefore, was a poor guide to the dissident movement's character and future potential. Instead, Amalrik turned to the movement's social composition, ingeniously mining the protest letters of spring 1968 for data on the profession and social class of some seven hundred signatories. His findings were striking: while there were artists, engineers, professors, scientists, teachers, doctors, lawyers, students, and workers among them, nearly all signers came from a single social group, the intelligentsia, which constituted roughly 1 percent of the Soviet population. The single largest contingent, moreover—nearly half of all signers—worked at institutions of higher education. It didn't much matter, Amalrik argued, whether they were Marxist-Leninists or Christians or liberals. What mattered was that "the kind of thinking fostered by scholarly work has a more speculative than pragmatic character," and therefore "intellectuals seem to me to be those least capable of purposeful action. They are eager to engage in 'reflections'"—a not-so-subtle jab at Sakharov's essay—"but act extremely indecisively." Soviet workers might currently be more politically conservative and passive than scholars, but it was not difficult to imagine them going on strike under the right conditions. By contrast, Amalrik could not imagine a strike in any academic research institute.[35]

In a specifically Soviet way, the intelligentsia constituted a "middle" class, situated between the ruling elite and the masses of peasants and workers. As a potential source of recruitment or at least support for the dissident movement, however, the Soviet middle class struck Amalrik as unpromising. To begin with, for decades the Soviet state had systematically removed from society "the most independent-minded and active of its members," whether via exile, forced emigration, imprisonment, or physical annihilation. The result was "an imprint of grayness and mediocrity on all strata of society." Even those members of the middle class who recognized the need for systemic reform tended to doubt that it was possible, or that they could contribute to it in any way: "There's nothing to be done—you can't break down a wall by banging your head against it," as the popular saying put it. They were stuck in a Nash equilibrium, according to which each individual perceives himself or herself as having nothing to gain by changing only their own behavior so long as others do not change theirs—and the system thus perpetuates itself indefinitely. The middle class subscribed to "a cult of its own impotence" with respect to the state, a cult that conveniently absolved it of the moral responsibility to act. Finally, Amalrik noted, "in any country, the stratum of society least inclined toward change or any sort of independent actions is that composed of state employees"; since in the USSR everyone was an employee of the state, "we all have the psychology of government workers."[36]

And what of the government itself? A similar anemia, Amalrik argued, plagued the Soviet leadership. Its cadres too had been subjected, under Stalin, to a process of "unnatural selection," resulting in "the retention of those who were most obedient and eager to carry out orders." This mechanism had fostered increasingly weak and indecisive leaders, men (and they were virtually all men) "incapable of generating new ideas," indeed "regarding any novel thought as an attack on their rights":

> Evidently we have already reached that full stop where the concept of power is no longer connected to a doctrine, the personality of a leader, or a tradition, but only to power as such.... Self-preservation is the

regime's only goal. It wants neither to "restore Stalinism" nor to "persecute representatives of the intelligentsia" nor to "render fraternal assistance" to those who have not requested it [for example, Czechoslovakia]. The only thing it wants is for everything to be as it was before: for the authorities to be recognized, the intelligentsia to keep quiet, and the system not to be wrecked by dangerous and unaccustomed reforms. Its motto is: Don't touch us, and we won't touch you.[37]

"Bloody Stalinist dynamism" had given way to relative stability followed by "stagnation." Economic reforms had repeatedly stalled because of fears of short-term instability and loss of control by the state. Similarly, attempts to strengthen the rule of law had collided with the leadership's inability to wean itself from the use of extra-judicial punishments. Amalrik estimated that over 15 percent of the signatories of recent petitions regarding political trials had been subject to such punishment, whether fired from their job or stripped of their Party membership.[38]

The 1960s, to be sure, had brought greater security, a higher standard of living, and more personal freedom than any previous decade in Soviet history. This had caused many people mistakenly to conclude that a process of liberalization had set in. For Amalrik, however, liberalization was driven not by any sort of official plan but rather by "the growing decrepitude of the regime [which is] simply growing old and can no longer suppress everyone and everything with the same strength and vigor as before."[39] Foreign observers, Americans in particular, "grasp with hungry impatience every trivial fact that testifies to the Soviet system's 'liberalization.'" They were inclined to believe that "the gradual improvement in the standard of living, as well as the spread of the Western way of life, will gradually transform Soviet society—as if foreign tourists, jazz records, and mini-skirts will make a 'humane socialism' possible. Perhaps we will indeed have 'socialism' with bare knees, but hardly one with a human face."[40]

The anemia of the Soviet middle class thus put the success of the dissident movement in grave doubt, while any apparent liberalization on the part of the government was better understood as a symptom of senescence rather than reform. As for the USSR's workers and peasants, it was difficult for anyone, even the country's leaders, to know what attitudes prevailed among them. One certainly couldn't rely on the state-controlled mass media for this purpose, and even the secret reports assembled by the KGB were probably of questionable accuracy. This didn't stop Amalrik from forming his own impressions, however, including those drawn from his year in exile on a Siberian collective farm. The dominant mood was "passive discontent," directed not against the government—"the majority do not think about it, or they feel that there is no alternative"—but rather against particular aspects of the Soviet system. Workers resented their factory managers, peasants resented the directors of their collective farms, and everybody resented inequalities in wealth, restrictions on place of residence, and the deficit of consumer goods. But this hardly gave grounds for hope on the part of rights-defending dissidents:

> Whether due to its historical traditions or some other reason, the Russian people find the idea of self-government, of equality before the law and personal freedom—and the responsibility that goes with these—almost completely incomprehensible. Even in the idea of pragmatic freedom, the average Russian sees not the possibility of securing a good life for himself but the danger that some clever person will do well at his expense. The majority of Russians understand the word "freedom" as a synonym for "disorder," for the ability to engage with impunity in anti-social or dangerous activities. As for respecting the rights of the human personality as such, the idea simply arouses bewilderment. One can respect strength, authority, even intellect or education, but to the popular mind, the idea that the human personality itself represents any kind of value is preposterous.[41]

Even the strongly rooted Russian idea of justice offered little basis on which the dissident movement could seek or cultivate allies. In fact, it contained "the most destructive aspect of Russian psychology," the desire that "nobody should live better than I do."[42]

Much of the Soviet population, Amalrik maintained, suffered from "acute social disorientation" induced by rapid migration to cities, a process triggered by Stalin's campaign to forcibly collectivize the countryside. Millions of recent urban transplants were ill at ease in their new environment, uncertain of their social standing and unmoored from traditional peasant beliefs. Christian morality, including basic concepts of right and wrong, had been "driven out of popular consciousness" and replaced with "purely opportunistic moral criteria."[43] The result, in Amalrik's estimation, was "a country without beliefs, without tradition, without culture, and without the ability to do business." While the amoral masses took pride in the USSR's enormous size and strength, "the basic theme of the cultured minority has been the description of its own weakness and alienation."[44]

Will the Soviet Union Survive until 1984? thus presented a bleak tableau: an entrenched but decrepit regime, an anemic middle class, disoriented and resentful peasants and workers. Against this background, one could of course "hope that the emerging dissident movement will, despite persecution, become influential, will work out a sufficiently concrete program, will find the necessary structure and attract many supporters." But Amalrik was not optimistic. The Soviet middle class, the movement's social base, was "too weak and too beset by internal contradictions to allow the movement to engage in a real face-to-face struggle with the regime or, in the event of the regime's self-destruction or its collapse as a result of mass disorders, to become a force capable of reorganizing society in a new way." In this regard, Amalrik's book could just as well have borne the title *Will the Dissident Movement Survive until 1984?* A more likely outcome, he argued, was the rise of a virulent Russian nationalism "with its characteristic cult of strength and expansionist ambitions" and its need for nationally defined enemies, whether external (the Chinese) or internal (the Jews). With ethnic Russians constituting just over half the Soviet population, such a turn was bound to lead to mass violence, against which "the horrors of the Russian revolutions of 1905–7 and 1917–20 would look like idylls." Even more likely, according to Amalrik, was the outbreak of war between the Soviet Union and the People's Republic of China, whose fraternal socialist alliance had recently deteriorated into a series of alarming military skirmishes along their 2,500-mile border.[45]

Whether through internal decay or external assault by a newly radicalized China, Amalrik concluded, the "great Eastern Slavic empire, created by Teutons, Byzantines, and Mongols, has entered the last decades of its existence. Just as the adoption of Christianity postponed the fall of the Roman Empire but did not rescue it from inevitable doom, so Marxist doctrine has delayed the break-up of the Russian Empire—the third Rome—but does not possess the power to prevent it." With more than a glimmer of *Schadenfreude*, Amalrik set a coda on his disastrous experience a decade earlier at Moscow State University where, "for reasons beyond my control, I was forced to interrupt my research on the origin of the Russian state." Now, by way of compensation, he informed his readers, "I hope as a historian to be rewarded a hundred-fold by becoming a witness to that state's end."[46]

Amalrik would not live long enough to reap his intended reward. But *Will the Soviet Union Survive until 1984?* was an instant sensation inside and outside the USSR, catapulting its author to fame and fortune along with infamy and prison. Years later, readers often remembered where and when they first encountered it. In the city of Ulyanovsk, Lenin's birthplace, Gulag survivor Irina Verblovskaya spent a sleepless night binge reading Amalrik's prophecy of the Soviet apocalypse, a night she vividly recalled four decades later.[47] Alexander Solzhenitsyn was reported to have had a similar experience while holed up at the dacha of the cellist Mstislav Rostropovich in Zhukovka, not far from Moscow.[48] Amalrik, Solzhenitsyn wrote, "doesn't chew the common cud, doesn't wrap himself in quotations from the Fathers of the Vanguard Doctrine, but dares to give an independent analysis of the contemporary structure of society and forecast the future, things that may in fact happen to our country." The Soviet soccer star Sergei Bondarenko recalled taking part in a group reading: "Passing the [sheets of] text to each other, reading over each other's shoulder. We were absolute fans." Radio Liberty broadcast *Will the Soviet Union Survive until 1984?* in its entirety to Soviet listeners.[49]

What Amalrik's fans seem to have appreciated above all was his *sangfroid*. Everything was subjected to clear-eyed, critical assessment: tsarist Russia, the Soviet Union, the KGB, the intelligentsia, the dissident movement, peasants and workers, the Russian people, Western public opinion, foreign journalists in Moscow. Anatole Shub, the *Washington Post*'s Moscow correspondent who befriended Amalrik and reported on his essay before it began to circulate (and whose coverage of the dissident movement got him expelled from the Soviet Union in 1969), called him "the coolest political mind I encountered in Russia."[50] Amalrik didn't just flout the USSR's unwritten taboos; he acted as if they didn't exist, thus modeling, according to fellow dissident Boris Shragin, genuine "inner freedom," the kind that overcomes the most powerful censor of all, the one inside the writer.[51]

Amalrik also ignored taboos regarding samizdat authorship. Insofar as publication of one's work in capitalist countries could increase the likelihood of official harassment or arrest (especially if royalties were involved, raising the specter of treason for hire), many writers of samizdat were hesitant about their texts making their way to the West. Some took pains to distance themselves in advance from that possibility, either because they genuinely didn't want it or in order to be able to deny involvement and blame publication abroad on the uncontrollable itineraries of samizdat texts. The Russian orthodox priest Sergei Zheludkov, for example, concluded his response to Sakharov's *Reflections* with a plea that readers help prevent the text from traveling abroad (it was published in Paris later that year).[52]

Amalrik approached the question of publication abroad completely differently. "I wouldn't want to create the impression," he wrote to his Dutch friend Karel van het Reve, that *Will the Soviet Union Survive until 1984?* "wound up in the West via 'samizdat.' I intentionally sent it there myself. That way, if I am accused in some future trial of 'receiving instructions from abroad,' I can look the prosecutor straight in the eye and respond, 'On the contrary—I gave them.'" "There" meant the Alexander Herzen Foundation's publishing house in Amsterdam, which van het Reve had just co-founded. As we saw in chapter 7, it was Pavel Litvinov who, in a letter to Stephen Spender, had launched the idea of setting up a publisher of samizdat texts abroad, in order to create definitive, authorized Russian-language versions of those texts and then to negotiate translations with commercial presses around the world, thus retaining both copyright and royalties for their authors. Amalrik was the first to put the new arrangements to use. In a series of impatient letters, he badgered van het Reve—jokingly addressed as "Dear and Most Respected Treasurer"—for the latest news on contracts, sales, and royalties. Was the price of the Russian-language edition perhaps too high? If the KGB were to arrest him, could van het Reve distribute to the press copies of a photograph of Amalrik with Litvinov and Larisa Bogoraz?[53]

The news from Amsterdam was spectacularly good. Packaged as a slender paperback, Amalrik's long essay was being translated into over a dozen languages and was already receiving wide press coverage. The American diplomat George F. Kennan, architect of the Cold War policy of containment and briefly ambassador to the Soviet Union, publicly expressed his hope that Amalrik's book would be published in English—which it was, prompting the Book of the Month Club in the United States to distribute copies to its three hundred thousand members. Henry Kissinger, the U.S. national security adviser, made sure that President Richard Nixon received a copy. Letters from foreign readers began to arrive at Amalrik's apartment in Moscow. So did pilgrims from the Russian provinces who had listened to *Will the Soviet Union Survive until 1984?* courtesy of the Voices, and who now wished to meet the author. "I am very satisfied," Amalrik reported to van het Reve. "Reality has surpassed all my expectations."[54]

Not all of Amalrik's readers were fans, of course. Nadezhda Mandelstam found his central argument preposterous: "You think the Soviet Union won't last another fifteen years? I think it will last another thousand!" The mathematician and Russian nationalist Igor Shafarevich, while describing Amalrik's book as "one of the most striking and intelligent works of post-revolutionary Russian thought," nonetheless registered "feelings of despair and inner protest at [Amalrik's] verdict on Russia." Viktor Sokirko, a graduate student in engineering, accused Amalrik of being overly pessimistic: "The main thing for you is not so much active struggle toward the goal [of a democratic Russia] but rather the stoic defense of your personal inner freedom, your personal conscience and independence from the regime." Seemingly everyone was reading or listening to Amalrik's book, "a samizdat best-seller" and a coveted item on the tamizdat black market.[55] There was even a joke about how to respond in court if a prosecutor asked whether you had read *Will the Soviet Union Survive until 1984?*: "Of course it will survive. How could there be any doubt?"[56]

Initially, van het Reve arranged for various travelers to the USSR to personally deliver royalty payments from Western publishers to Amalrik and his wife, who lived in a communal apartment where they shared a toilet, bathtub, kitchen, and telephone with six other families. "The communal apartment," his wife once remarked to a foreign visitor, "is our country in miniature."[57] Within a year, royalty payments—now many thousands of dollars—had become too large for such informal methods. But there was more: consistent with the dissident movement's emphasis on strict adherence to the law, Amalrik now wanted to receive his royalties officially, via wire transfers from Western banks to the Soviet bank where he had a savings account. He was informed that authorities in Moscow had secretly blocked that move on the grounds that, having published his book abroad without permission, hence "illegally," he was not entitled to receive royalties. At best, he could receive the funds as a "gift," which meant they would be subject to a 38 percent tax—far higher than the tax on foreign royalties. When he turned for help to the USSR's All-Union Office for the Defense of Authors' Rights, Amalrik was told that publication of works abroad was fully legal but that the office could not assist him in his endeavor to defend his rights.[58]

Those who knew Amalrik were not surprised by what came next: a blistering letter to the editors of the *Times* of London, the *New York Times*, the *Washington Post*, *Le Monde*, and *Het Parool* denouncing the Soviet government for failing to observe its own laws. Amalrik declared that he had hoped "to prove that a Soviet citizen, like the citizen of any other country, has the right to publish abroad books not issued in his own country, to do so under his own name, to negotiate himself with the publishers the terms of publication, and to enjoy all the author's rights that flow from this." The payments from abroad, Amalrik insisted, "are not a gift, but money I have earned." If the Soviet Union refused to allow such arrangements, he would instruct his publishers to deposit his royalties in the West rather than send them through the USSR's "unreliable official bodies." His letter was intended to "publicly shame the Soviet government for its stinginess and pettiness." "While Stalin would have shot me for publishing my books abroad," Amalrik noted, "his pathetic heirs are merely capable of trying to appropriate part of my money. This only confirms my view of the degeneration and decrepitude of this regime."[59]

Leaving aside the question whether Stalin's heirs should be seen as pathetic, or rather as measured in comparison with their murderous predecessor, the fact that Amalrik had not been arrested aroused suspicions in more than a few contemporaries. The title of his book alone, as one observed, was "bold to the point of suicidal."[60] Why would the KGB allow such a provocateur to remain at large? And why, come to think of it, had his conviction in 1965 for social parasitism been overturned and his Siberian exile cut short? In the Cold War hothouse of psychological warfare and double agents, a rumor began to circulate that Amalrik was working for the KGB—possibly in return for having his term of exile reduced—and that *Will the Soviet Union Survive until 1984?* was a hoax. It all made a certain kind of paranoid sense: rather than shattering Soviet taboos, the book was actually intended to shatter the dissident movement by having an insider expose its weaknesses and further tarnish it with his unpatriotic views and slandering of the Russian people. At the same time, it would lull the West into complacency (and reduced defense budgets) by falsely presenting the USSR as poised to collapse on its own or as a result of a Chinese assault.[61]

It is quite possible that this rumor was fueled by Amalrik's withering verdict on Russia, which did in fact upset an older generation of Russian émigrés.[62] But émigrés were not the only ones who assumed that Amalrik's book must have been sponsored by the KGB. More than a few dissidents came to the same conclusion.[63] To dispel the idea that he sought to undermine—rather than critically appraise—the movement in which he himself was a participant, Amalrik enlisted Petr Yakir to draft an open letter vouching for his bona fides. Since the arrest and exile of Litvinov following the August 1968 protest in Red Square, Yakir had emerged as one of the movement's most prominent activists, invested with the additional moral authority of a survivor of Stalin's Gulag and son of one of Stalin's most prominent victims. "I appreciate the precision, honesty, and impartiality of your positions," Yakir announced, in the pages of the *Chronicle of Current Events* as well as the *Times* of London, even as he took his distance from Amalrik's critical assessment of "the Russian character" and the future of the dissident movement. While the movement's "social base" to date was indeed quite narrow, Yakir noted, "the ideas it has proclaimed have started to be widely disseminated around the country, and this is the beginning of an irreversible process of self-liberation."[64]

Yakir's own trajectory would soon put the lie to that last claim. But he was hardly the only one who thought this way. "Twenty years ago Sinyavsky and Daniel would have been executed," a Moscow State University student told a visiting American. "Today, they only get a few years. Ten years from now their works will be published. Isn't that progress?"[65] He was right, of course, though, like Amalrik, slightly premature in his prediction: it would take not one decade but two before Sinyavsky's and Daniel's stories were published in the USSR. Amalrik had warned against precisely this way of thinking, which he paraphrased as "the situation now is better than it was ten years ago; therefore in ten years it will be even better."[66] For his fellow dissidents, he had a sobering message: don't mistake movement for progress.

According to the American writer F. Scott Fitzgerald, the test of a first-rate intelligence is "the ability to hold two opposed ideas in the mind at the same time, and still retain the ability to function. One should, for example, be able to see that things are hopeless and yet be determined to make them otherwise."[67] At one time or another, every Soviet dissident faced this test. Sinyavsky feared that "democracy as a social and government system has no future whatsoever in Russia" but insisted that "it is our calling to remain proponents of freedom, because 'freedom,' like other 'useless' categories—for example, art, goodness, and human thought—is a value in itself and does not depend on historical or political trends."[68] "Struggle does not always lead to victory," admitted the dissident sociologist Boris Kagarlitsky, "but without struggle not only can there be no victory, there cannot even be elementary self-respect."[69] Amalrik, a born dissident, was determined to live like a free person in a country he was certain would remain unfree.

11

Recrimination and Reassessment

This personal responsibility for one's country, personal responsibility for one's history—it's a strange business.

—VYACHESLAV IGRUNOV

In their starkly different ways, Andrei Sakharov's *Reflections on Progress, Peaceful Coexistence, and Intellectual Freedom* and Andrei Amalrik's *Will the Soviet Union Survive until 1984?* unsettled Cold War orthodoxies. While Soviet mass media maintained a strict silence about both works, their authors became household names among the Soviet intelligentsia, thanks to samizdat, tamizdat, the CIA, and the Voices.[1] Supercharged attention from the West—or, to put it a bit more theoretically, the selective gaze of First World mass media, with its peculiar culture of celebrity—helped make Sakharov and Amalrik internationally famous, even as it ignored many other participants in the emerging debate over the strategies and goals of the dissident movement. The itineraries of samizdat texts were not just unregulated but in many ways capricious—including whether those texts made their way to the West (most did not) and what happened to them once they arrived. To trace the evolution of rights-based dissent past the chain reaction, we will need to widen the spotlight beyond the handful of individuals and texts that achieved renown outside the USSR.

The sense of defeat triggered by Moscow's smothering of the Prague Spring, together with the KGB's campaign of extra-judicial reprisals against signers of protest letters, made 1968 the psychological equivalent of a failed revolution. And like many failed revolutions, it was followed by a period of recrimination and reassessment, the first substantial stock-taking—for some, a post-mortem—of the dissident movement. Along with conversations in countless kitchens, a wide-ranging samizdat debate dissected the movement's goals and tactics, successes and failures, featuring authors both within and beyond the circle of activists. By now, the existence of a "dissident" or "democratic" movement (*demokraticheskoe dvizhenie*, or *demdvizhenie* for short) and "movement people" (*dvizhentsy*) was widely taken for granted, whatever one thought of its prospects. A group calling itself the "Democratic Movement of the Soviet Union" even managed to circulate a seventy-six-page manifesto and have it published abroad (and read by members of the Politburo), before it became evident that the group consisted of a single anonymous author. It was a symbol, perhaps, of the abiding asymmetry between the movement's small numbers and its outsized ability to command attention at home and abroad.[2]

One of the key participants in that conversation, unknown to the outside world but not to Soviet dissidents, was Alexander Malinovsky. In 1964, two years into his training at the Institute of Eastern Languages at Moscow University, where he specialized in Hindi, Malinovsky had announced at a Communist Youth League meeting that, while his political views remained those of a communist, he had decided to be baptized in the Russian Orthodox Church because he "believed in the immortality of the soul." A certain familial trait may have been at work, a longing for eternity: his grandfather, also named Alexander Malinovsky but better known under the pseudonym Alexander Bogdanov, was one of Lenin's earliest collaborators, a leading revolutionary theorist, psychiatrist, science fiction writer, and, more to the point, a proponent of periodic blood transfusions as a path to eternal rejuvenation. The younger Malinovsky's great-uncle, Anatoly Lunacharsky, another Bolshevik luminary, had promoted socialism as an ersatz religion and served as the first Soviet commissar of enlightenment. Pedigree notwithstanding, after declaring his metaphysical beliefs the younger Malinovsky was expelled from the Communist Youth League and soon thereafter from Moscow State University. He made his way into dissident circles and became a close observer of the emerging movement.[3]

Malinovsky's long samizdat essay "Considerations concerning the Liberal Campaign of 1968" offers the most comprehensive critique of the dissident movement produced by a contemporary.[4] Like Sakharov, Malinovsky regarded nuclear war and environmental devastation as existential challenges for the USSR and the entire world.[5] Like Amalrik, he considered the Soviet government incapable of meeting those and other existential challenges, which he estimated would lead to its collapse within five to fifteen years. Like Amalrik, he regarded the dissident movement in its current form as unlikely to offer a viable solution to the USSR's looming crisis. Unlike Amalrik, however, Malinovsky—whose samizdat works appeared under the pseudonym A. Mikhailov—located the movement's principal weaknesses not in its social composition (too many intellectuals, with their preference for cogitation rather than action) but in its worldview.

There were two fundamental shortcomings, according to Malinovsky, both fixable. The first was the strategy of demanding that the Soviet state observe its own laws. "It is strange," Malinovsky noted, "when grown-ups write open letters to the Central Committee, complaining about injustices committed, as everyone knows, on orders from that very same Central Committee." In theory, criticizing the regime's failure to honor its own laws would be a useful technique for unmasking official phraseology "if anyone besides Young Pioneers took that phraseology seriously anymore." The majority of the Soviet population was already "quite familiar with the regime's hypocrisy and has gotten used to it—in fact we take it as a given." It was also strange to see dissidents engage in loud moralizing aimed at a government famous for its deafness and intolerance of criticism. This was not naiveté but pretend naiveté (*naivnichan'e*), "pseudo-Don-Quixotism." Liberal romantics loved heroic gestures—"Look, we have nothing to hide, we're not keeping any secrets"—but protest letters with signatories' names and addresses accomplished nothing besides helping the KGB retaliate against the most active "other-thinkers." If the strategy of legalism and transparency was supposed to protect its practitioners from reprisals while strengthening respect for the rule of law, Malinovsky argued, then it had failed to achieve those goals. The dissident movement took for granted that when there is injustice, an honest person must not remain silent and must protest. But "when you're dealing with bandits in a dark alley," he countered, "a formal, open, verbal protest is not appropriate, and may even be harmful." Such methods inspired neither respect for nor trust in the dissident movement's political judgment.

The second problem was the ethos of individual moral purity and "living like free people." Malinovsky's critique of the August 1968 demonstration on Red Square is worth quoting at length:

> Larisa Bogoraz stated that she and her comrades were fully aware of how politically senseless their protest on Red Square was. In her words, it was an act not of political but of moral struggle. This was a very characteristic statement, perfectly reflecting the romanticizing mystification of the 1968 campaign. The liberal romantics were indeed guided in their actions more by personal-ethical motives than by social ones. They saw and understood the mounting injustice. They experienced a psychological inability to remain silent and go on with their normal routines while people were being thrown in jail who were close to them personally or in spirit, while "eternal and inalienable rights" to spiritual freedom (or, if you prefer, "rights guaranteed by the constitution") were being trampled. In the end, they barely thought about the realities of society; they were simply incapable of thinking about them. They were busy saving their own soul, the purity of their individual conscience and honor as they understood them, selflessly and uselessly delivering themselves up for slaughter. The only thing that excuses them in this undertaking is the mystical, completely unfounded, irrational conviction that, in cleansing their conscience, they somehow, in some secret, inscrutable way, were helping the world.

Dissidents were justifiably accused of being "poseurs" who created elaborate moral spectacles that ended in labor camps. If the liberation of society was to be "something other than a fiction or a surrogate," it had to "deliver some sort of results, express itself in a pragmatic approach to questions that demand solutions." Otherwise, it amounted to nothing more than "a carousel of 'spiritual' life," spinning round and round but stuck in one place. "Seriousness, even a crushed and disfigured seriousness," Malinovsky insisted, "stands a better chance of achieving genuine liberation than any spiritual games."

Spiritual games, lack of seriousness, a merry-go-round, irrational thinking, behaving like Young Pioneers: according to Malinovsky, dissidents were childish. Or rather, they were adults disingenuously attempting to adopt a child-like innocence. "They pretend to be naïve," he wrote, "good little boys [sic] who see things from the perspective of the Communist Youth League, who don't understand what is happening and are outraged by 'isolated acts of injustice.' No matter how much poison-tipped irony you inject into this stance, it still deprives you of the possibility of a rigorously scientific assessment of society, of articulating genuinely persuasive positions, of appealing to the interests of real social groups." It was not only in the USSR, of course, that people reacted to protesters bearing an unfamiliar rhetoric and lifestyle by comparing them to children. It was the 1960s, after all. But Soviet dissidents did not in fact project the personae of children, let alone flower children; most were in their thirties and forties and many were already parents.

Malinovsky had additional analogies in his arsenal, none of them favorable. One invoked the "classic pattern" of the nineteenth-century Russian intelligentsia: "the individual personality with a developed moral consciousness versus an all-powerful, soulless society." Prerevolutionary intellectuals had engaged in "archaic, irrational worship" of "the people"; dissidents exhibited "an equally irrational contempt and pronounced chilliness" toward "the population." Dissidents were also mirror-images of Bolsheviks:[6]

> Old ideas of which [the dissidents] are sick and tired simply get replaced by their opposites. Did the Bolsheviks teach that the radiant goal—communism—justifies any methods, even the most grotesque? Look where that got us! From this [the dissidents] conclude

that, on the contrary, no end justifies any means other than ideal means (which have no need for justification). And in general there's no need to worry about ends, just about the virtue of one's immediate actions. [The dissidents] think they are liberating themselves from the Bolshevik spirit, when in fact they are embodying it once again. One group [Bolsheviks] were fanatics of the radiant goal, the other [dissidents] of clean hands and clean methods. This is precisely how ... sectarian and aristocratic thinking comes into being. It's the same principle everywhere: negativism, an abundance of noble emotions, and the complete absence of a constructive and simple approach to reality. All these teachings are very effective at getting a well-meaning, respectable person to take his leave of reality, to neither hear nor see it, to sacrifice himself.

Malinovsky didn't seem to notice, or care, that his various analogies were mutually contradictory: that children generally do not have a "developed moral consciousness," that worshipping "the people" is quite different from deploring them, or that rejecting an ends/means calculus puts one fundamentally at odds with Bolshevik thinking. His was a jeremiad by a self-identified communist (or, by the time his essay appeared, a democrat socialist), a staunch critic of the Soviet regime who regarded the dissident movement as the last hope for rescuing the USSR from its impending doom—provided the movement rethink its ends and means.

The first step, according to Malinovsky, was to recognize the need to articulate a better alternative to the existing Soviet order. Dissidents not only had failed to produce any such alternative, but had failed to concern themselves with even the immediate consequences of their actions. They did this, moreover, "not by mistake, but on principle." They (apart from Sakharov) had "lost faith in the old values, in the scientific approach to society, in socialism, in service to the people." True, they made pronouncements about "the people" and "the interests of society," but such expressions appeared to be merely part of their strategy of using official phraseology to mask their true goals. In practice, those goals concerned the creative intelligentsia alone, with its narrow "guild" interests.[7] Dissidents had repeatedly come to the defense of a handful of arrested "freethinking intellectuals" with demands to review their trial, to release them from prison, to stop arresting them, and to publish their work. "Only a very narrow circle of people," Malinovsky maintained, "could regard such demands as their own, as of existential importance." There was no doubt that "in the long run, liberalization would have a favorable impact on the entire society," but when expressed in such abstract terms, it could interest only the creative intelligentsia. How could the average Soviet citizen relate to such an agenda? The dissidents' social isolation "was therefore not a consequence of underdevelopment and backwardness on the part of the Soviet people, as it would be fashionable to suggest, but flowed from the very essence of their approach." Their failure to convert the chain reaction into a mass protest movement was due above all to the "narrow, negative, and purely defensive character of their position."

Life was not easy, Malinovsky argued, for average Soviet citizens. They endured two to three hours a day on public transportation, spent a similar amount of time waiting in lines at stores, and "barely make ends meet—knowing full well that some of their fellow citizens live ten times better." At their jobs they were required to tremble in front of the boss, women were stretched by competing demands of work and family, health care was abysmal, alcoholism and crime on the rise. Dissidents had ignored all this, so why should one expect average Soviet citizens to support them? For participants in the various transparency meetings, Yuli Daniel and Andrei Sinyavsky and Alexander Ginzburg and Vera Lashkova "were worth going to jail for—but millions of insignificant, crippled, and downtrodden inhabitants were not." It was all well and good for Alexander Volpin to declare that the defense of constitutional rights was just as important for veterinarians and tailors as for writers. But no one was going to the square on behalf of veterinarians and tailors, let alone miners or collectivized peasants. "One must try to understand these regular people," Malinovsky urged, "who booed the protesters at their trials, who were ready to tear them to pieces in Red Square [at the August 1968 demonstration], and who are certain that liberals are parasites and traitors." Dissidents needed to reach out to them, "to understand their interests—as they themselves understand them—and make them their own, include them in their demands."

At the same time, one needed to approach social activism scientifically, armed with a theoretical model that would serve as the basis for engagement with the world. When dissidents invoked scientific knowledge at all, "they had in mind not substantive macro-sociology (of the Marxist type), but jurisprudence, a purely formalistic discipline belonging to the superstructure." To think sociologically was to recognize workers and peasants and other classes as agents of history. To think legalistically (by invoking, for example, the "rule of law") was to pretend that class distinctions didn't matter. It was impossible to achieve genuine social democracy simply by grafting the juridical forms of "classical bourgeois democracy" onto the existing Soviet order. The dissident movement, Malinovsky wrote, needed to ground itself in social-democratic theory. This would make it possible to formulate a unified program with ten or twelve points, none of which Malinovsky named.

As should by now be evident, "Considerations concerning the Liberal Campaign of 1968" had more to say about what the dissident movement should stop doing than about what it should do. It urged an immediate end to "senseless self-imprisonment" and "useless" petitions, protests, and appeals—useless, that is, because their tangible benefits failed to match the human cost of carrying them out. The custom of putting one's name, address, and phone number at the conclusion of a samizdat text should also be curtailed, unless the author was already known to the police, and, even then, the decision should be a matter of calculation, not ethics. There was no reason to invoke honesty or transparency in this context. "The goal of samizdat," Malinovsky noted, "is the exchange of opinion and information, the discussion and dissemination of ideas, not absurd self-advertising that reaches the employees of the KGB." Because the "actions of the state against other-thinkers are a form of terrorism, progressive forces have not only the right but the duty to defend themselves," insofar as they bore responsibility for the "free and purposeful development of thought in society." Dissidents should abandon the pretense of loyalty to the Soviet Constitution and should stop directing grand appeals to "the global public," concentrating instead on the daily work of forging bonds with a variety of social groups in their own country.

The significance of the dissident movement had so far been mostly symbolic, according to Malinovsky. Trapped in moral and legal argumentation, it was approaching a dead-end. Of much greater significance than the movement itself was the "passive interest" it had aroused in "the thinking part of society." In place of its previous, misguided tactics, it was now time for the movement to accept its inescapable status as an illegal opposition and embrace the tasks of "conspiracy, agitation, and propaganda." Malinovsky did not elaborate on these terms either, whether because of uncertainty about what they would actually entail or out of fear of being too explicit, except to note rather enigmatically that "of course, this in no way signifies a revolutionary stance," by which he presumably meant the use of violence to usurp the Soviet government. A revolution of a new type was nonetheless central to his plan. "What we need now more than anything else," Malinovsky urged, "is a moral revolution, a revolution that would become permanent, that would not retreat into aristocratic isolationism, into the sectarian twilight of 'spiritual life' and the 'purely moral' sphere of private life, but would spread across the entire society." Such a revolution would "rouse society from its slumber," enabling it "at last to reach a higher level of consciousness and come face-to-face with reality."

Rights-defenders could be forgiven for wondering whether Malinovsky's "moral revolution" was all that different from Volpin's "meta-revolution." True, Volpin had in mind the dissemination of a specifically legal consciousness among the Soviet population, while the content of Malinovsky's revolution remained unspecified. But both operated at the level of consciousness, that is, on the superstructural rather than the macro-sociological level so ardently promoted by Malinovsky. It was strange, moreover, to urge dissident intellectuals to get to know how other social groups understood the world—groups that Malinovsky characterized as either unthinking or asleep. If the Soviet state was indeed run by bandits—a curious choice of term by someone aspiring to sociological analysis—then why should their victims eschew revolutionary methods? And what odd bandits these were, with their incessant talk of socialist legality, people's democracy, and the radiant future.

Notwithstanding its many contradictions, Malinovsky's critique hit multiple nerves. Like Sakharov's *Reflections* and Amalrik's *Will the Soviet Union Survive*, it was widely debated within the world of samizdat. "All my friends were outraged," reported Leonid Plyushch, paraphrasing their reaction: "He sits in his little corner, keeping mum, not doing a damn thing, and proudly lectures us, daring to toss around accusations, recommending theoretical works without even showing his face."[8] "Considerations concerning the Liberal Campaign of 1968" cast dissidents as children and singled out Larisa Bogoraz (a woman) for committing "hysterical acts" and for "a psychological inability to carry on with normal routines"—by which Malinovsky presumably meant normal routines of double-speak and dissimulation.

In contrast to Soviet mass media and internal government memoranda, however, Malinovsky offered not just attacks but arguments, based on close analysis of the dissident movement's practices. Whereas official denunciations sought to cripple the movement, Malinovsky's criticisms aimed to help it survive and flourish. In fact, his essay along with those by Sakharov and Amalrik galvanized an intense debate over the movement's past and possible futures. Prior to 1968, face-to-face conversations among several dozen Moscow activists had pitted advocates of voluntarism and equality, such as Bogoraz, Pavel Litvinov, and Natalya Gorbanevskaya, against those favoring greater organization and hierarchy, such as Petr Grigorenko, Viktor Krasin, and Petr Yakir. Now, having spilled over into samizdat, the debate became textualized, which is to say dispersible, and therefore open to a substantially larger number of participants—authors as well as readers—extending well beyond the core of Moscow activists.

One of the main points of contention, not surprisingly, was what had gone wrong in 1968. Why had the movement lost momentum? For Vyacheslav Igrunov, a student at the Odesa Institute of Economics, an enthusiastic reader of Amalrik and Malinovsky, and a prodigious collector of samizdat, it was not enough to blame the KGB's extra-judicial reprisals against signers of protest letters. The real problem, he argued in his samizdat essay "On the Problematics of the Social Movement" (signed with the pseudonym "Ego"), was the absence of a theoretical framework for dissent, something that could lend resilience to victims of reprisals,

FIGURE 11.1. Vyacheslav Igrunov

inspiring them to continue the struggle rather than retreat from it. The struggle itself had devoted too much energy to securing formal rights to free expression instead of "putting to use at least that piece of freedom that we [already] have inside us." The rapid retreat of so many movement people in 1968 testified to their deficit of inner freedom. "Spiritual and cultural life," Igrunov insisted, "can exist with or without the state's permission." Its value lay precisely "in the process of self-discovery, of opening up one's inner world."[9]

Anatoly Marchenko, author of the samizdat memoir *My Testimony*, considered Igrunov a hypocrite. "What kind of person," he asked, "writes about inner freedom and then signs with a pseudonym?"[10] According to Venyamin Kozharinov (writing under the pseudonym Andrei Slavin), however, it was precisely the heroic disdain for elementary measures of self-protection (as exhibited by Marchenko and others) that had led the dissident movement to its current "stagnation." Activists who adopted a devil-may-care attitude when it came to using pseudonyms or hiding samizdat texts from the KGB, who "flaunt their lawfulness, their contempt for 'behind-the-curtains maneuvers,'" who refused "to play hide and seek" with the KGB, suffered from what Kozharinov called a "complex of doom and hopelessness." People who "tickled Gulliver's heels" in this way were guilty of "criminal negligence with respect to our entire movement." In some cases these "princely lords" of inner freedom even acquired their own fiefdom, a circle of admirers of their reckless heroism. When the prince was imprisoned or sent into exile, his admirers invariably fell apart and abandoned dissident activity. Certain dissident strivers, Kozharinov lamented, displayed symptoms of what Lenin had called "leaderism."[11] The movement needed to be built around ideas, not personalities.

For Volpin, the rule of law was still the cardinal concept, and Malinovsky had entirely missed its significance. Without the legal enforcement of civil rights, Volpin insisted, "nothing serious can be accomplished" and "there can be no freedom," internal or otherwise. Respect for law remained "the only basis on which, in our era of universal skepticism, we can build consensus without infringing on each other's convictions." The setbacks of spring 1968 were a symptom of insufficient solidarity with victims of rights violations. The key mistake was the failure to respond to the KGB's retaliation against signers of open letters—the firings, blacklistings, expulsions, salary reductions, and other forms of extra-judicial punishment. "Our movement," Volpin wrote, should have "answered with the doubling and tripling of petitions against these repressions" and should have "ridiculed those who renounced their words out of cowardice and who called on others to do the same." This was the only way to sustain the chain reaction, to foster people's willingness to perform "public acts of civic engagement" and to ensure that "new people would eagerly join us." Notwithstanding its tactical mistakes, the movement had successfully put an end to "the former mockery with which, until 1965, many liberals responded to their friends' efforts to elicit a movement for the rule of law in our country."[12]

Most participants in the post-1968 reassessment of the dissident movement did not share Malinovsky's contempt for the pursuit of inner freedom. Anatoly Yakobson, the impassioned defender of self-liberation as embodied by the eight protesters in Red Square in August 1968, expressed his own contempt for Malinovsky's insistence on grand theory as the basis for action. Rights-defenders "knew of no recipes for the salvation of humanity. They sought to defend people. They protested against individual acts of arbitrariness and coercion. They said and continue to say what they thought necessary—and this is an honorable position. Sacrificing oneself does not mean provoking others to recklessness."[13]

Vladimir Zelinsky's essay "Ideocratic Consciousness and the Individual Person" offered a deeper defense of the dissident movement's cultivation of inner freedom. A researcher in the Institute of Philosophy at the Soviet Academy of Sciences, Zelinsky had specialized in the work of Martin Heidegger before being baptized in the Russian Orthodox Church and abandoning a promising academic career.[14] Inner freedom, he asserted (writing under the pseudonym Dmitry Nelidov), was a precondition for human dignity, the "fundamental concern" of the dissident movement. Only against the background of the USSR's "dull, cold, clumsy tyranny" and the "spiritual illness" of Soviet society could one grasp the movement's full meaning. Dissidents were engaged in a search for healing, a quest for "self-liberation from conditioned, almost inborn ideological reflexes and imposed mechanisms of behavior." Theirs was a struggle against "the degenerate, reflexive herd mentality that inevitably arises when people are bred on 'ideological farms' and in 'incubators of opinion.'" In comparison with this core mission, the "external narrowness" of the movement's striving for the rule of law and the "notorious similarity of its rhetoric to the phraseology of the [Soviet] state" were of only secondary importance.[15]

The Soviet state, according to Zelinsky, maintained an implicit social contract with its citizens: "I will speak to you about your freedom and you will act as if there is no freedom." In general, citizens abided by this arrangement, this "elementary etiquette of double-think," since the penalty for not doing so was severe:

> And so it went. Every day, words about freedom and democracy were repeated with the expectation that they would be understood within the framework of previously conditioned reflexes. And when certain people attempted to understand these words differently, having first overcome those reflexes within themselves and then acting as if those reflexes didn't exist at all, they were doing something impermissible, in violation of all decency. Regardless of anyone's subjective intentions, they shattered the framework of double-think in which such phrases were usually pronounced. And because of this, everything changed fundamentally. Suddenly all our magnificent constitutional forms, all our triumphal words and rituals started looking fake, like props. The entire

structure of our rusty ideological machine was exposed, a machine that had gotten used to conditioned reflexes and sclerotic beliefs.[16]

The dissident movement in effect practiced what the Formalist literary critic Viktor Shklovsky, in an influential essay published in 1917, had called "defamiliarization" or "estrangement."[17] Just as literature and art had the capacity to reanimate our experience of everyday life and objects by "making them strange," pulling them out of habitual, automatized modes of thought and feeling, so dissidents were changing the meaning of official Soviet rights-talk by removing it from ideologically conditioned (or, as Zelinsky put it, "ideocratic") consciousness and routinized lip-service. The act of displacement, or de-automatization, made it possible to become aware of one's own "ideocratic consciousness," to step outside it, to repudiate it. "It is precisely from this simple awareness (perceived not cynically, but spiritually)," Zelinsky argued, that the dissident movement had begun. Inner freedom began with mindfulness.

"Not cynically, but spiritually": this parenthetical distinction was crucial. The spiritually inclined sought to reanimate the official discourse of rights and the rule of law by linking it to human dignity and freedom. "Behind any public protest," Zelinsky wrote, "is an individual's attempt to break through his own social adaptation, through his own double-think, and find himself. When young people enter into hopeless political conspiracies for which they are subsequently sentenced to ten or fifteen years, they do this because they want to possess their own identities among the surrounding masks."[18] Self-unmasking was a means of achieving transparency of the self, or, as Lidiya Ginzburg put it, of retaining one's "human visage."[19]

And what about the cynics? They liberated themselves from "ideocratic consciousness" by concluding that *all* talk of rights and the rule of law—along with the radiant communist future—was semantically hollow, ideologically conditioned, a discursive game, and, in any event, impossible to realize within the eternal Soviet order. They kept their masks on and performed lip-service in order to make ordinary life possible in a system that demanded public conformity. Malinovsky had in effect claimed that nearly the entire adult Soviet population belonged to the category of cynics, having long since ceased to take official propaganda seriously. Zelinsky was not far from this position. Most of educated society, he admitted, did not share the dissident movement's "moral pathos" and remained "indifferent to the idea of rights as such. The layer of cynical double-think and the associated ironic indifference are still too dense."[20]

And what of those outside the intelligentsia, the vast majority of the Soviet population? Malinovsky's main criticism of the dissident movement targeted its lack of interest in building alliances with ordinary Soviet citizens, listening to their grievances and finding ways to address them. Evidence of such indifference—or worse—was certainly not hard to find. Ludmilla Alexeyeva expressed an overwhelming urge to separate herself "from that thing called 'the masses'"; Bogoraz raged at the "unfeeling cattle" who publicly supported the Soviet invasion of Czechoslovakia.[21] Perhaps the most extreme example appeared in the samizdat essay "The Intelligentsia and the Democratic Movement," whose pseudonymous author ("K. Volnyi") announced that "the role of the intelligentsia is and always has been the moral education and re-education of the people." Volnyi's assessment of this civilizing mission, beginning with the Bolshevik Revolution, was grim:

> The people fled to the villages, leaving only a small group of citizens committed to freedom and morality.... We used to idealize the people. Now we have lost respect for them, they who once seemed a pure sufferer. The people have betrayed and slaughtered their intelligentsia—their only protector. Immeasurable is the guilt of the people toward the intelligentsia, repeating the fate of the crucified Jesus! We refuse to idolize the animalistic and egoistic people. We serve not the various barbaric imperfections, but great ideals—Kindness, Truth, and Beauty.[22]

Even a self-proclaimed populist such as Grigorenko could not resist casting dissidents as spiritual virtuosi, worldly ascetics seeking to better the world. "Whoever wants to struggle against arbitrariness," he declared, "must destroy in himself the fear of arbitrariness. He must take up his cross and climb Golgotha. Let people see him, and in them the desire to take part in this procession will awaken. The people will see those who march, and they will approach them." Igrunov took a similar if less overtly Christological stance. "Movement people need to be present in all spheres of life," he wrote. "If only by means of our irreproachable conduct, we need to set an ethical example for those around us." Perhaps Amalrik put it best: by conducting themselves like free people, dissidents had begun "to change the country's moral atmosphere and its governing tradition."[23]

There is more going on here than simply setting a good example. Confronted with their apparent failure to inspire a large following, some dissidents took solace in a notion that was very much part of Russia's governing tradition, reaching back through the doctrine of socialist realism to the politically charged writers of the nineteenth century. This was the idea that certain individuals, whether in life or fiction, could achieve transpersonal significance and become "guides to living" (as Nikolai Chernyshevsky famously wrote in the 1860s) for others.[24] When asked by the CBS News correspondent William Cole what he hoped to achieve with his activism and how many supporters the dissident movement had, Vladimir Bukovsky responded that the essence of the struggle was against "the fear that has gripped the people since the time of Stalin and thanks to which the system of dictatorship, of pressure, of oppression continues to exist":

> In that struggle, great importance attaches to personal example—the example which we give people. I personally did what I considered right, spoke out on those occasions when I wanted to, and I am alive, I am now sitting here and not in prison. I'm alive, I can get about, I can live. For me and for many other people, that is very important—it shows that it is possible to fight and that it is necessary. It is difficult for me to estimate the number of people who follow us, who think the same way as we do. We are by no means in a position to know all the people of that kind. The fact is that I cannot know in advance who—upon hearing what I say today or what I said in court or to other people, or who has read Solzhenitsyn, Marchenko, or someone else—who else will take up the same kind of activity as ours. I can never know this.[25]

The essential unknowability of the dissident movement to its own virtuosi, the invisibility of the networks of samizdat beyond one's immediate friends, did not inhibit Bukovsky's activism. The first installment of his memoirs explained how exemplary individuals could acquire transpersonal significance:

> In an extreme situation, there is a qualitative difference between the behavior of a single individual and that of a human crowd. In such a situation, a people, nation, class, party, or simply a crowd are unable to go beyond a certain limit: the instinct of self-preservation proves too strong. They can sacrifice a part in the hope of saving the rest, they can split up into little groups and seek salvation that way. This is what destroys them.
>
> To act alone is an enormous responsibility. With his back to the wall, the individual understands: "I am the people, I am the nation, I am the party, I am the class—and there is nothing else." He cannot sacrifice a part of himself, and cannot divide himself into parts and stay alive. There is nowhere for him to retreat, and the instinct of self-preservation pushes him to extremes—he comes to prefer physical death to spiritual death.
>
> It's an extraordinary thing: in defending his wholeness he simultaneously defends his people, his class or party. It is precisely such individuals who win the right for their communities to live, even, perhaps, if they are not thinking of this at the time.
>
> "Why me of all people?"—each member of a crowd asks himself. "I can't accomplish anything by myself."
>
> And all are lost.
>
> "If not me, then who?"—the individual with his back to the wall asks himself.
>
> And he saves everyone.[26]

It was indeed an extraordinary, almost mystical thing: how resistance by a single individual, or a handful of virtuosi, could somehow transform the moral status of the collective to which they belonged, or, in Yakobson's words, how a handful of demonstrators on Red Square could "rescue the honor of the Soviet people," just as ten righteous men would have been enough to save the entire city of Sodom.[27]

All these cases imply a kind of moral economy. Not the one of traditional peasant society, which demands that economic activity be governed by the imperatives of subsistence and equity, but a metaphysical economy in which guilt and honor, like debts and credits, can flow between an individual and a group.[28] Even if the meaning of Bukovsky's phrase "he saves everyone" remains unclear—saves how, from what, with what consequences?—his account captures an important figuration of the dissident as an outstanding personality whose power derives from the assertion of inner freedom in a setting of unfreedom. A personality of a new type.

The preceding discussion has by no means exhausted the variety of viewpoints on the dissident movement that found expression in samizdat during the months and years following the turning point of 1968. The sheer diversity of opinion may have constituted the most significant and unsettling message of all: it was now clear that the movement's minimal consensus around the demand that the Soviet state abide by its own constitution was yielding to a plethora of views regarding both tactics and goals. The only remaining near-universal points of agreement were the taboo against the use of violence and the appeal of samizdat, the seemingly unstoppable hive of communication that provided an alternative to the official Soviet public sphere. If the Kremlin had so far rejected dissidents' attempts to initiate a dialogue between state and society, even with a prized figure like Sakharov, samizdat had nonetheless succeeded beyond anyone's imagination in fostering uncensored dialogue within the intelligentsia (with a considerable assist from the Voices). The dense, multi-layered post-1968 debate over the future of the dissident movement bore eloquent witness to that development.

The sheer vitality of samizdat struck some observers as grounds for abandoning the hitherto unsuccessful demand that the state honor the free speech rights inscribed in the Soviet Constitution. "We're not going to make a fuss. We're not going to fight for freedom of the press," announced one commentator. "We're going to put it into practice."[29] Igrunov, who established in Odesa what may have been the first samizdat lending library, was convinced that the dissemination of dissident literature "can now play a more important role in changing public consciousness than letters of protest, trials, and demonstrations." The movement needed to brace itself for "long years of slow growth, years of working behind the scenes, without external effect. The creation of a new culture, the total transformation of our selves, is a slow process, but without it, no serious and lasting results are possible."[30]

What was the prerequisite of meaningful change: the transformation of Soviet selves or the enforcement of Soviet citizens' constitutional rights? Inner freedom or external freedom? The primacy of defending rights had been grounded in the imperative, following Khrushchev's removal from power, of preventing a return of Stalinist lawlessness. "The bloody past," Volpin's Civic Appeal had declared in 1965, "calls us to vigilance in the present." As the Stalin era receded, however, and as it began to appear that what Brezhnev's Kremlin prized above all was stability, preventing a relapse into Stalinism became less and less persuasive as a rallying cry.

Of course, seeking one kind of freedom did not preclude striving for the other. Volpin, the godfather of the rights-defense strategy, advocated the dissemination of legal consciousness—a state of mind—as part of the dissident movement's "meta-revolution." But like every movement, Soviet dissidents had to set priorities. During the formative years of the chain reaction, disagreements among activists focused largely on repertoires of contention: demonstrations versus open letters, organized assignment of roles versus spontaneous, conscience-driven voluntarism. Ideological differences between, say, a neo-Leninist like Grigorenko and a post-Leninist such as Alexeyeva had been conveniently submerged under the lingua franca of constitutional rights and judicial procedure. There had even been a certain exhilaration in the diversity of political views within the movement—a welcome contrast to the stifling Soviet myth of unanimity and a signal that support for the rule of law did not require political uniformity.

Once the chain reaction exhausted itself, however, and once critical assessments of the movement began circulating—including some that read like eulogies—it was all but inevitable that disagreements would rise to the surface. Should dissidents establish their own formal organization, putting into practice among themselves their call for transparent, rule-governed activity? Or should they move their operations underground, deploying covert methods of struggle? Should the focus be on expanding the ranks of the movement and, if so, among what sorts of people—potential reformers within the Soviet government, self-identified supporters in the West, the Soviet population at large? The historian Roy Medvedev and the physicist Valentin Turchin advised focusing on reformers within the Communist Party and avoiding the creation of non-Party organizations.[31] Medvedev in particular warned against the movement's "anarchic character," with religious believers working alongside Leninists: "It is easy for thoroughly dubious people to insert themselves into a movement open to everybody."[32] The fear of being compromised by "devils" of one sort or another—undisciplined, unscrupulous, hungry for fame and/or martyrdom—would not go away. Some dissidents harbored a similar fear regarding the movement's links to supporters abroad or to foreign journalists in Moscow, whose motives and funding sources were nearly impossible to scrutinize. Inveighing against "hypnosis by the West," Malinovsky warned that such links "especially compromise our engagement with the masses, who understand the need for reforms but who are frightened that the people who embody the democratic movement are not quite 'ours.'"[33]

In a conversation with a fellow dissident, the ever pragmatic Kozharinov suggested that, despite the crushing of the Prague Spring and the KGB's campaign of reprisals against signers of protest letters, now was the time to search for new people to join the movement. "Search how? Where?" came the bitter reply. "People who wanted to join us, joined us. Others ran away. What else do you propose?"[34] During the following years, Soviet dissidents would have to decide what they were: a social movement, a conspiracy, or a band of virtuosi.

12

Taking the Initiative

Even the appeals of our best scientists and writers bounce back like peas off a wall.

—ALEXANDER SOLZHENITSYN

Jan Jahimowicz was an ideal communist. Born in 1931, the tenth child of Polish working-class parents in Daugavpils (also known as Dvinsk or Dinaburg) in southeast Latvia, not far from what had recently become the border with Poland, he was nine years old when Latvia was annexed by the Soviet Union in accordance with the secret protocols of the Hitler-Stalin Mutual Non-Aggression Pact. A year later, Hitler's army launched a surprise attack, incorporating Latvia and other Baltic territories into the Nazi empire's eastern borderland, the Reichskommissariat Ostland. One of Jahimowicz's brothers enlisted in the Latvian division of the Waffen-SS; another joined its nemesis, the Red Army. By 1944, Latvia was back in Soviet hands, soon to be reincorporated into the USSR as the Latvian Soviet Socialist Republic. Jan Jahimowicz, now a teenager, became Ivan Yakhimovich. Russian became his main language, and in 1951 he joined the Communist Youth League. After graduating from Latvian State University, he volunteered for the Virgin Lands campaign to accelerate agricultural production in northern Kazakhstan and western Siberia. There he met his future wife, a fellow volunteer. After returning home, he began to teach Russian language and literature at a middle school before becoming the director of the collective farm Jaunā Gvarde (Young Guard), where he helped build socialism in Soviet Latvia. By the time he joined the Communist Party in 1961, life had taught him the supreme value of "friendship of the peoples."[1]

Yakhimovich never stopped studying. While running the collective farm, he took correspondence courses at the Latvian Agricultural Academy and continued to read the classics of Marxism-Leninism. In 1963, he spoke out against Khrushchev's erratic agricultural policies, for which he was expelled from the Party. After Khrushchev's removal in October 1964—due in part to his erratic agricultural policies—the Party promptly reinstated Yakhimovich. That same year, he and Jaunā Gvarde were singled out for praise in the newspaper *Komsomolskaya pravda*.[2] In his voluminous private journals—part diary, part scrapbook— Yakhimovich interspersed lengthy passages copied verbatim from Lenin, Marx, Hegel, August Comte, Benjamin Disraeli, Che Guevera, and John F. Kennedy with clippings from the Soviet press on international affairs and socialist economic development. From the Bolshevik revolutionary Anna Kalygina he culled this maxim: "One is a human being only when one is a fighter."[3]

On his collective farm, five hundred miles west of Moscow, Yakhimovich kept track of events in the Soviet capital. A journal entry from February 1966 pronounced the trial of Andrei Sinyavsky and Yuli Daniel a "mistake": "They should have been tried by the court of public opinion, in front of students, in front of workers. They should have been unmasked as hypocrites and double-dealers, and thereby disarmed. You don't like the Soviet Union, you're not happy with Soviet power? Here's your passport, go off to some capitalist paradise. Don't stink up things here, gentlemen!"[4] News of the ensuing chain reaction of protests, arrests, and trials only deepened Yakhimovich's conviction that the Kremlin was mishandling dissent. In January 1968, having heard Larisa Bogoraz and Pavel Litvinov's "To the Global Public" on the BBC, he sent a letter to Mikhail Suslov, secretary of the Central Committee, requesting a review of the trials of *samizdatchiki* in order to avoid the "enormous damage such trials cause to our Party and to the communist cause in our country—and not only there." It was Lenin, he reminded Suslov, who insisted that "we need full and truthful information." On a more practical level, he noted,

> I live in the provinces, where for every house with electricity there are ten without it; where in the winter the buses can't get through; where it takes weeks for the mail to arrive. If [despite all this] significant quantities of information have been reaching us, you can imagine what seeds are being sown across the country. There is only one way to destroy samizdat: by expanding democratic rights rather than curtailing them. By observing the Constitution rather than violating it. By putting into practice the [Universal] Declaration of Human Rights, insofar as Vyshinsky signed it on behalf of our country. Remember: Leninism—Yes! Stalinism—No![5]

On the eve of the fiftieth anniversary of the founding of the Soviet Army, and in advance of the upcoming international congress of communist parties, Yakhimovich considered it his "communist duty" to request amnesty for Sinyavsky, Daniel, and Vladimir Bukovsky and a fresh evaluation of the case of Alexander Ginzburg, Yuri Galanskov, Aleksei Dobrovolsky, and Vera Lashkova. Political prisoners, he argued, should not be deprived of their liberty; they should be re-educated by means of persuasion, clear explanations, and agitational programs at their place of work.[6]

The Party responded to Yakhimovich's performance of his communist duty by expelling him again. Shortly thereafter he was removed from his position as director of the collective farm. His wife, a high school teacher, was fired. A KGB officer told Yakhimovich he'd been duped: the appeal by Bogoraz and Litvinov was a BBC fabrication. Who was telling the truth, Yakhimovich wondered, the KGB or the BBC? To find out, he traveled to Moscow in search of Bogoraz and Litvinov, who befriended him, as did Petr Grigorenko, a fellow Leninist. On July 28, Grigorenko and Yakhimovich joined forces to hand-deliver to the Czechoslovak embassy a statement of support for the Communist Party and the people of Czechoslovakia. Yakhimovich had already returned to Latvia when Soviet troops launched their assault on Prague. After the BBC informed listeners on the evening of August 25 about the protesters on Red Square earlier that day, he immediately dispatched a

FIGURE 12.1. From the left, Ivan Yakhimovich (Jan Jahimowicz), Larisa Bogoraz, Pavel Litvinov, spring 1968

FIGURE 12.2. Petr Grigorenko and Ivan Yakhimovich near the Czechoslovak embassy in Moscow, July 28, 1968

note to Litvinov: "I am full of pride and admiration, and if I had been in Moscow, I would have been with you on Red Square."[7] The KGB decided to search the one-room apartment shared by Yakhimovich, his wife, their three daughters, ages five, six, and seven, and his mother-in-law, on the unlikely pretext that he was a suspect in a local bank robbery. In fact, Yakhimovich was as devoted a communist as ever. As he wrote (again) to Suslov, this time in an open letter, "To occupy a country where the leading role belongs to the Communist Party, without the latter's consent and against its will—such an act is inconsistent with the moral values of the Soviet people, a peace-loving, selfless people who prize the friendship and trust of others." Ringing with classically Soviet rhetoric, Yakhimovich's letter condemned Stalinism as "a left deviation, outright revisionism, unsocialism." "Communists of the entire world," he urged, "stop the Stalinists before it's too late!"[8]

On January 16, 1969, a Czech student named Jan Palach set himself on fire in Wenceslas Square, in the heart of Prague, to protest the Soviet occupation. He died three days later. In the ensuing weeks, half a dozen of his countrymen performed similar self-immolations. Yakhimovich and Grigorenko joined forces again, this time issuing a samizdat appeal "To the Citizens of the Soviet Union!" in which they demanded an immediate withdrawal of Soviet troops from Czechoslovakia. "We all bear a portion of guilt for the death of Jan Palach," they declared, sounding the dissident theme of collective responsibility. "By endorsing the introduction of troops, by justifying them, or simply by virtue of our silence, we enable the continued burning of human torches in the public squares of Prague and other cities." On March 24, KGB officers returned for another search of Yakhimovich's apartment. As they rifled through his papers, he drafted what he anticipated would be his final, frantic appeal, noting, "I am forced to speak about myself, because soon, a river of lies and hypocrisy may spill forth from a courtroom. I am forced to speak about myself, because my fate is the fate of my people, my honor is their honor."[9] Meanwhile, his daughters waited in the garden outside the window, singing the "The Internationale":

> We will tear down to its foundations
> the entire world of violence,

and then we will belong to ourselves.
We will build a new world:
Those who were nothing
will become everything.[10]

Their father was arrested and taken to prison later that day.[11]

The KGB now had a problem: how to put a devout, Lenin-quoting communist on trial. To avoid that unappealing scenario, it was arranged to have Yakhimovich declared mentally ill. The doctors who examined him could not help noting his "outstanding mastery of the classics of Marxism-Leninism." They flagged another quality prized by the Communist Party: "He considers his duty to society as standing considerably above his duty to his family." Diagnosed with "paranoid development typical of a psychopathic personality," Yakhimovich found himself committed indefinitely to a psychiatric prison in the Latvian capital, Riga.[12]

Yakhimovich's case, like that of Grigorenko, offers the remarkable spectacle of a fervent Marxist-Leninist colliding with an avowedly Marxist-Leninist regime. It also illustrates how a devoted communist could enter the orbit of the dissident movement, embracing its rhetoric of constitutionalism and human rights. The most important thing about Yakhimovich, however, was not his clash with the Kremlin or his burgeoning friendship with dissidents, but what came next. His arrest unintentionally reignited the dissident debate that had begun a year earlier, in the spring of 1968, over whether to establish a formal organization, a party, or some other entity that could provide structure and leadership to the dissident movement. That debate had been interrupted by the Soviet invasion of Czechoslovakia and the ensuing demonstration on Red Square, which effectively removed from the movement both Bogoraz and Litvinov, two of the strongest proponents of a purely conscience-driven approach that outsourced formal organizational work to sympathetic allies in the West. During the intervening year, as we have seen, the debate had expanded well beyond the ranks of core activists in Moscow. In broader samizdat circles, dominant opinion favored the establishment of an organization—the question now was what kind, and how. With Yakhimovich's arrest, the search for organizational form began in earnest. Not surprisingly for a movement consisting mostly of intellectuals, there was no shortage of ideas, but Soviet history offered few if any models of what coordinated legal dissent might actually look like. Even the typically commanding Grigorenko was unsure. Responding to advice from Litvinov (writing from Siberian exile) not to "rush ahead," Grigorenko lamented: "As if any of us knows where 'ahead' and 'behind' are."[13] By necessity, then, the search for form consisted of a series of experiments.

Dissidents advocating the rule of law faced a conundrum: Soviet law did not recognize the category of an organization independent of the state and the Communist Party. It was not that the law expressly forbade voluntary citizens' associations; on the contrary, Article 126 of the Soviet Constitution granted "the right to unite in social organizations: trade unions, cooperative associations, youth organizations, sport and defense organizations, cultural, technical, and scientific societies." But that right, like other Soviet civil rights, was prefaced by the catchall phrase "in conformity with the interests of the toilers" and, in addition, was followed by an even more restrictive clause designating the Communist Party as "the leading core of all organizations of toilers."[14] In the Soviet system, the state exercised a monopoly not just over the legitimate use of violence (a defining feature of all modern states), but over the legitimate organization of public activity as well. Under Khrushchev, there had been much talk of devolving certain state functions onto groups of citizen-volunteers, whether in the form of "comrades' courts" to arbitrate everyday disputes or "people's patrols" to help maintain public order. But none of these entities—alleged harbingers of the "withering away of the state"—were intended to question or challenge the state, nor did they function that way. On the contrary, both their membership and their activities were subject to strict supervision by the Party.[15]

Thus when Grigorenko, at a meeting of dissidents at his apartment in May 1969, presented a plan to establish the Committee in Defense of Ivan Yakhimovich, everyone understood the gravity of what was being proposed. Gathering signatures in support of an arrested fellow activist was risky enough, as the previous year's firings, expulsions, and other punitive measures against hundreds of "signers" had shown. Presenting such signatures under the banner of an unsanctioned organization or committee marked a quantum leap both legally and psychologically. It was tantamount to a public declaration of independence in a country whose citizens, as Andrei Amalrik observed, "had been bred by the Soviet regime to fear the word 'organization.'"[16]

Grigorenko invited one of Alexander Volpin's disciples, the physicist Boris Tsukerman, to explain the legal aspects of forming a committee. Tsukerman had made a name for himself as a one-man cottage industry of samizdat briefs exposing the illegal actions of every conceivable branch of the Soviet government, including the security services.[17] As a KGB officer once complained, "Tsukerman poked a lot of sticks in our wheels." But his argument that Grigorenko's proposed committee should cast itself as a trade union—one of the "social organizations" recognized by Soviet law—found few sympathizers among those gathered in Grigorenko's apartment. For starters, they came from a variety of trades and professions and could hardly make the case that they belonged together in a single union. It also didn't help that Tsukerman was just as stubborn and relentless with his fellow dissidents as he was with the Soviet government. "He was a pain in the neck," recalled Amalrik. "If you interrupted him, he would just resume with the same word where he left off."[18]

Legal formalities aside, others present that day wondered why a committee should be formed in defense of just one person, especially a devout communist such as Yakhimovich. In theory, rights-defenders were not supposed to care about the political views of those whom they defended, but anti-communist Gulag survivors such as Viktor Krasin balked when it came to risking arrest for the sake of an avowed Marxist-Leninist, especially one unknown personally by most of those assembled in Grigorenko's apartment.[19] Maya Ulanovskaya, another Gulag survivor, pleaded with supporters of Grigorenko's proposal—among them her husband, Anatoly Yakobson—to reconsider. If they formed a committee, they would almost certainly go to prison, this time on charges of organized conspiracy. "You haven't been there," she implored.

"You haven't been in a cell with inmates awaiting execution. It's horrible.... It's non-stop horror, day in and day out. You shouldn't walk straight into their jaws." Hadn't Russia had enough of unions, societies, committees, and parties? Was the purpose of gaining greater freedom from the Communist Party really to found yet another organization?[20]

Grigorenko, who had hoped to galvanize support for a committee to defend comrade Yakhimovich, was furious at this "hysterical" outburst. His fury was still alive a decade later, when he suggested in his memoirs that Ulanovskaya had denounced his proposal at the behest of the KGB, which supposedly exploited her fear that her husband would be arrested (a fear that had already led her not to inform him of the planned demonstration in Red Square on August 25, 1968).[21] Unsupported by evidence, the insinuation against Ulanovskaya—like that against Amalrik—serves as a reminder of how, alongside the deep bonds of trust among activists, anxieties of infiltration by the KGB, or what dissidents called "informer mania," never disappeared.[22] And with good reason.

Nearly all of those present that day voted against forming a committee. Several weeks later, at a second meeting with a slightly different roster of participants, the majority again opposed forming any sort of organization. It took the KGB's intervention to inadvertently flip the consensus. Yuri Andropov was well aware of what had transpired at the meeting at Grigorenko's apartment, thanks to either a listening device or a mole. He also knew that Grigorenko, whom he described as "the ideologue of the so-called 'anti-Stalinist democratic movement,'" continued to push for the creation of a "society for the defense of human rights" with headquarters in Moscow and branches in Leningrad and Kyiv, whose job would be to monitor the penal system and "ensure transparency of judicial proceedings." Refusing to heed warnings from KGB officials, Grigorenko released statement after statement condemning the Soviet invasion of Czechoslovakia and demanding that exiled Crimean Tatars be allowed to return to their homeland. Some were published by the *Washington Post* and the London *Sunday Times*. An informant reported Grigorenko saying, "If I were certain that my self-immolation could in some way help [Czechoslovakia], I would do it."[23]

It is hard to know how serious Grigorenko was about martyring himself or how concerned Andropov was about Grigorenko's intentions.[24] It certainly wouldn't have looked good if a former major-general of the Soviet Army set himself on fire—possibly on Red Square—to protest the Soviet invasion of Czechoslovakia. But that was all hypothetical. What was already happening, Andropov noted, was that foreign radio stations and newspapers "are constantly informing world public opinion about Grigorenko's hostile activities, advertising him as 'the conscience of Russia,' 'the avant-garde of a great historical movement,' 'a well-known fighter for rights and freedom.'" That didn't look good either. The publicity was helping recruit young Soviet citizens to engage in what Grigorenko called "partisan warfare" against the Party, leading the father of one such citizen to request that the KGB put an end to Grigorenko's "harmful influence."[25]

Grigorenko was not the "ideologue" of the dissident movement. As a Leninist, he was on the movement's fringe, and, in any case, rights activists prided themselves on having neither an ideology nor a leader. Even though Andropov recognized that the former major-general had so far failed to persuade his fellow dissidents to establish a committee or organization, he was not about to take chances. When Grigorenko flew on May 3 to Tashkent, the capital of Soviet Uzbekistan, to attend the trial of Crimean Tatar activists, the KGB was waiting. Two thousand miles from the eyes and ears of Moscow's foreign press corps, they arrested him.

If Andropov thought he could disrupt the plan to create a dissident organization by jailing its most forceful advocate, he was about to be disappointed.[26] Within days, Petr Yakir's apartment was packed with activists debating how to respond to the major-general's arrest. The discussion quickly moved past the idea of forming the Committee in Defense of Grigorenko, especially once news arrived that Ilya Gabai, a poet and prodigious signer of protest letters, as well as Viktor Kuznetsov, a graphic artist who had taken part in a controversial discussion of the sources of cynicism in Soviet society, had also been arrested. "We have to respond more sharply," Krasin insisted, "in a more categorical form." It was time to take advantage of the indirect effect of previous punitive measures against "signers": "cowards and hesitaters have withdrawn" from the movement, he argued, freeing the remaining activists to proceed more boldly. The KGB seemed to be launching a new wave of arrests, and since Soviet institutions had turned a deaf ear to all prior letters of protest, Krasin proposed that this time a letter be addressed to the United Nations' Committee for the Defense of Human Rights and to UN secretary-general U Thant.[27]

Krasin had brought with him a list of rights violations, which the assembled group proceeded to expand and edit, including the major dissident trials that had formed the chain reaction, the persecution and trials of religious believers, along with Ukrainians, individuals from the Baltic republics, Crimean Tatars, and Jews seeking the right to emigrate. The list also mentioned the recent arrests of Yakhimovich, Grigorenko, Gabai, and Kuznetsov, as well as the alarming practice of confining dissidents indefinitely in psychiatric hospitals. All of these, Krasin argued, violated Article 19 of the United Nation's Universal Declaration of Human Rights, on the freedom to hold opinions and to seek, receive, and impart information and ideas. In dozens of trials, moreover, there had been gross violations of procedural norms. The idea was to request that the United Nations "examine the question of the violation of basic civil rights in the Soviet Union."[28]

Those present expressed their support for Krasin's initiative and agreed to reconvene the following evening to finalize the text of the letter. As the guests were leaving his apartment, Yakir pulled Krasin aside: now was the time, he indicated, to form an organization for the defense of rights. The two began to discuss who its members should be, focusing on "authoritative people," experienced individuals "of middling age" (that is, not university students) who were seasoned participants in the movement. They made sure to include in the list people from cities other than Moscow, to emphasize the geographic reach of the planned organization. After coming up with roughly a dozen prospective members, they drew up a second list of prospective "supporters," people who could co-sign the letter and be prepared to replace members who were arrested in the event of a KGB crackdown.[29] As for the organization itself, they decided to call it an "initiative group" (*initsiativnaia gruppa*), a generic Russian term for a voluntary association of citizens, or at least a group presenting itself as such.[30] In this case, it would be the Initiative Group for the Defense of Civil Rights in the USSR—the first civil rights organization in the socialist world.

The next evening, thirty to forty people showed up at Yakir's apartment to finalize the text of the letter.[31] Yakir and Krasin arrived an hour and a half late with stunning news: that afternoon, they had hand-delivered a copy of the letter to Frank Starr of the *Chicago Tribune*, Anthony Collings of the Associated Press, and Lars Nelson, a reporter for Reuters, with the request that it be transmitted to the United Nations and published as widely as possible. At the bottom of the letter, dated May 20, 1969, stood the name Initiative Group for the Defense of Civil Rights in the USSR followed by the names of fifteen "members" and thirty-nine "supporters." Apart from Yakir and Krasin, none of those listed had heard of the Initiative Group, much less agreed to join or support it. None had given their approval of the final version of the letter; some had not even attended the previous night's discussion and had little knowledge of the letter's contents.

Leonard Ternovsky, a thirty-five-year-old physician, stared at the letter in disbelief. "I was shaking," he recalled, "my heart was pounding; thoughts were jumping around randomly in my head."

> What?! It was bold and noble, of course, but was such an impertinent challenge justified? Did the founders [of the Initiative Group] not understand that they'll all be rounded up in three days, or at most a week? For decades in the USSR there has been an unwritten but iron rule: any self-initiated organization not overseen by the authorities must be instantly and mercilessly crushed. And here, an openly oppositional group with a list of names. Come right in—take them away.

Then he noticed his own name listed among the supporters.

> But I had never laid eyes on this appeal! I hadn't even heard anything about it! What to do? Protest? Demand that my name be removed? But didn't I essentially agree with the letter's content? And besides,

FIGURE 12.3. Physician Leonard Ternovsky

I was told that it had already been given to the foreign journalists. Was it even conceivable for me to compromise such an important undertaking? Should I be silent? Make my peace with this scandalous outrage? With this arrogant manipulation of me and my name? ... The spirit of *The Devils* hung over the room.[32]

It was indeed a scandal of Dostoevskian proportions, made even more awkward by the suspicion that Yakir's apartment was bugged. No one wanted a rift inside the movement to burst open while the KGB was listening. No one, moreover, wanted to seem like a coward by asking that his or her name be removed from the letter. "That would have been embarrassing," the biologist Sergei Kovalev recalled, "especially vis-à-vis Yakir and Krasin," survivors of Stalin's Gulag whose boldness put them at even greater risk than others. Kovalev rebuked Krasin, telling him that this was no way to operate, but, like Ternovsky, he was essentially in agreement with both the letter and the idea of establishing the Initiative Group. Krasin apologized, explaining that the offer by foreign journalists to meet had come up very suddenly that morning. Had he and Yakir turned it down, it was unclear when the opportunity would arise again. Determined to transmit the letter to the West before the KGB learned of their plan, they went ahead without getting permission from other signatories.[33]

That was a lie.[34] It was Krasin who at Yakir's urging had called Starr to arrange the meeting that day. Krasin had acquired excellent English by listening to the BBC's "English by Radio" program, and he kept a detailed inventory of names and phone numbers of Western journalists, most of whom did not speak Russian well.[35] He was what Russians call a *delovoi chelovek*, a person who gets things done.[36] A major collector and distributor of samizdat, he had amassed an enormous archive, taking photographs and using negatives to print documents in bulk, rather than engaging in the usual laborious process of manually retyping on onion-skin paper.[37] According to his friend Venyamin Kozharinov, Krasin was frustrated by "the mood in the democratic movement, the fact that there were few decisive, business-like people, and a lot of chatterboxes, people who love to attach themselves to causes without demonstrating any initiative. The most they could manage was to sign protest letters." Dissidents had wasted nearly a year debating whether or not to form an organization. Now that the "cowards and hesitaters had withdrawn," as Krasin put it, all that was needed was a nudge from someone with a vision of how to propel the movement forward.[38]

But there was more. In a subsequent search of his apartment, the KGB discovered cassette tapes on which Krasin, already mindful of his place in history, had dictated notes for a planned memoir of the dissident movement. In one recording, he said that after the arrests of Litvinov, Bogoraz, and Grigorenko "it became necessary for me and Petr Yakir to assume what in Bolshevik terminology is called the leading role in the movement." The "leading role" is precisely what the Communist Party had assigned itself in Soviet history. When asked by a KGB interrogator about this statement, Krasin demurred:

> The way I expressed that was not quite right and moreover inaccurate. In the movement for civil rights there was no leadership and no leaders. It was, if one can put it this way, a self-organizing system. Of course some people worked harder than others. It would be more correct to say that after the arrest of Litvinov, Bogoraz, and Grigorenko, the most active participants in the movement were myself and Petr Yakir. Yakir and I were not leaders, but merely demonstrated more initiative in comparison with others. There was greater regularity and constancy in our actions.[39]

That too was a lie, or at least a gross misrepresentation, although wholly understandable given whom Krasin was talking to. When Krasin and Yakir visited the exiled Bogoraz and Litvinov in Siberia in 1969, Krasin had bragged about now being the de facto leader of the dissident movement.[40]

A decade and a half later, in the course of another interrogation, this time not in Moscow but in New York, and not by a KGB officer but by his second wife and fellow dissident, Nadezhda Yemelkina, Krasin put it this way:

> I was never known for my humility, and in the rights movement, as I increasingly found myself in the center of events, my ambition grew immeasurably. People respected me because of my past in the camps, because of my activism in the movement; they trusted me. This allowed me to exercise influence over them and over the course of our actions. At the end of the day I really did believe that it was my calling to be the "leader" [*vozhd*]. I constantly abused people's trust—as in the episode when I gave foreign correspondents that first letter to the United Nations.[41]

The problem of "leaderism" was not unique to Krasin. Yakir's nearly two decades in Stalin's Gulag, along with his high Bolshevik lineage, his charisma, and his oratorical skills made him an even more revered figure. According to Kovalev, after Grigorenko's arrest, Yakir "became the most popular dissident in the country."[42] That status was enabled in no small measure by the spotlight wielded by Starr and other foreign correspon-

FIGURE 12.4. Petr Yakir (left) and Viktor Krasin (right) visiting the exiled Pavel Litvinov (center) in Siberia, near the Manchurian border, 1969

dents. At the same meeting on May 20 where Yakir and Krasin handed Starr the first letter of the Initiative Group for the Defense of Civil Rights in the USSR (which soon appeared in the *New York Times*, the London *Observer*, the *Frankfurter Allgemeine Zeitung*, and dozens of other newspapers, as well as in broadcasts by the BBC and the Voice of America), the American journalist peppered Yakir with questions about his life. Within days, an article appeared on the front page of the *Chicago Tribune* under the headline "Pyotr Yakir Is New Russian Voice of Intellectual Dissent." With the arrest of Litvinov and Grigorenko,

Starr announced (Bogoraz, a woman, went unmentioned), "the tacitly understood leadership of Moscow's intellectual opposition has passed to [Petr] Yakir, a burly affable man of 46 with bushy black hair who has borne the full tragedy of Stalinist repression and describes himself as a pathological optimist." Starr singled out Yakir for his "fearlessness."[43] The *Washington Post*'s Anatole Shub called Yakir an "authentic hero."[44]

Such assessments were beamed back to Russia by the Voices, with the result that, according to Ludmilla Alexeyeva, "with every passing day an increasing number of people perceived Yakir as the leader of the opposition." Letters from Soviet citizens in Kyiv, Chișinău (Kishinev), and other cities began to arrive at his apartment.[45] Soviet citizens themselves began showing up too. "They came with their samizdat and their news dispatches" for the *Chronicle of Current Events*, Alexeyeva noted. For Yakir, "encouraged by skin-deep stories written by Western reporters" and "egged on by provincial truth-seekers" who learned about him via those stories, "heroic stances had become his identity. He lived for them, trying to forget that beneath his pamphleteering surged the total, paralyzing fear of a privileged fourteen-year-old boy taken away from his mother."[46] That was a fear the KGB did not forget.

The manner of the Initiative Group's founding seemed to confirm the darkest premonitions of the voluntarists, figures such as Bogoraz, Litvinov, and Natalya Gorbanevskaya, the leading proponents of antihierarchical, conscience-driven dissent (and currently exiled or under arrest for demonstrating on Red Square). Gorbanevskaya regarded Krasin and Yakir's preemptive strike as the Initiative Group's "original sin," cleansed only by its subsequent good work.[47] Striving for power and influence within the movement, Krasin and Yakir had deceived several dozen individuals into publicly associating themselves with an unauthorized organization, thereby putting them in danger of imprisonment, exile, or worse. They then lied to those individuals about what they had done. True, Krasin apologized in response to Kovalev's rebuke, but as he subsequently confessed to his wife, "I paid no attention to it."[48] How could a movement devoted to transparency, the rule of law, and living like free people possibly proceed on such a Machiavellian foundation?

The truth was not so simple. To begin with, not everyone who discovered his or her name on the Initiative Group's founding document was outraged. When describing how Krasin and Yakir made him a member, the Russian Orthodox writer Anatoly Levitin-Krasnov chose the same phrase dissidents used in praise of samizdat: they did it *iavochnym poriadkom*, "without prior approval," which is to say, on their own initiative. "I did not express any objections to my inclusion in the Initiative Group," Levitin-Krasnov told a KGB interrogator.[49] Even Ternovsky made peace with what happened on May 20, sympathetically comparing the Initiative Group to an illegitimate child. "So what?" he reflected three decades later. "Despite all the persecutions, which began literally the next day, the group survived, got on its feet, and grew to become part of our contemporary history."[50]

The issuing of unauthorized statements, however, continued. In subsequent years, Krasin would sometimes collect members' signatures on an appeal and then, dissatisfied with the document's content or tone, "write a new text, more severe" and hand it to foreign journalists in the name of the Initiative Group. For some, this was infuriating. "Maybe you could at least inform us," seethed one signer, "when you change our decisions." By contrast, some of Krasin's friends gave him carte blanche, orally or in writing, to sign statements on their behalf on those occasions when communication became too risky. Active dissidents, after all, were frequently tailed by the KGB, their apartments and telephones bugged, their mail intercepted and read. Those who lived in cities other than Moscow often communicated with their counterparts in the Soviet capital via notes hand-delivered by trusted couriers. The perennially impatient Krasin manipulated this system, adjusting the calculus of risk to accommodate his own preferences:

> Over time, signing on behalf of other people became like a reflex. Why call X, Y, or Z to arrange a meeting to discuss a text? That took so much time. A day or two later, when the document was published [abroad], people would learn that they had signed it without even knowing its content. In this manner, one of the most important achievements of the rights-defense movement was nullified, namely,

democratic procedure, in the course of which each person takes responsibility for his own decision as dictated by conscience. Instead of free choice there was once again the humiliating old stereotype, whereby the few arbitrarily and irresponsibly decide for everyone. One of my friends told me, "You're a Bolshevik in reverse."[51]

Krasin was the most ethically flexible member of the Initiative Group. But he was hardly alone. Yakir was notorious not so much for cunning as for carelessness when it came to securing signers' permission, altering agreed-upon texts, and using the telephone to discuss sensitive topics.[52] Even Alexeyeva and Gorbanevskaya, fierce critics of "leaderism" and undemocratic decision-making, could succumb to the latter. Once, having decided to write an international appeal on behalf of the imprisoned Anatoly Marchenko, Alexeyeva, who was not formally affiliated with the Initiative Group (she was, as she put it, disinclined "to join anything called 'group'"), turned for help to Gorbanevskaya, who was. Gorbanevskaya agreed, but there was a problem: "Where would you find the Initiative Group in the middle of June?" "It's such a clear-cut case," Gorbanevskaya reasoned. "No one will object if we issue something." The appeal went out in the name of the Initiative Group and was soon picked up by the Voices. Now it was Gorbanevskaya's and Alexeyeva's turn to be scolded by irate members who learned of the appeal only when it was broadcast from the West.[53]

Along with their extraordinary courage, one cannot help note the Initiative Group's strikingly informal, not to say haphazard, approach to actions that entailed enormous personal risk. As I argued in chapter 7, informality should not be understood primarily as a strategy designed to shield activists from KGB reprisals. Indeed, in the case of the Initiative Group, informal practices had no such effect. Instead, the persistence of improvisational behaviors, even among people seeking to promote strict observance of the law, reflects what the social scientists Alena Ledeneva and Svetlana Barsukova call the "embeddedness of informality" across Russian political culture. Historically, they observe, informal practices and networks in Russia have been "not an option, but a necessity." During the Soviet era, they became a habit and a preference in daily life, often more effective and more humane than the formal institutions of the state.⁵⁴

The Initiative Group is especially interesting in this regard. At first glance, its founding appeared to mark the victory of those dissidents who favored greater organization and the formal assigning of roles (that is, the division of labor) within the movement. That victory flowed less from winning the contest against egalitarians than from the latter's arrest and exile. Subsequent Machiavellian maneuvering by Krasin and Yakir, however, reflected the persistence of informal ways of doing things even among the contest's winners. A year after its founding, when the Initiative Group finally got around to announcing its goals and methods, the resulting statement sounded more like a declaration of victory by the anti-organizational wing of the movement. "The Initiative Group has neither a program nor a charter nor any sort of organizational structure," it announced. "We are not bound by any formal obligations. Each of us can choose not to participate in the composition of any document that emanates from the Initiative Group and not to sign it. Each of us is free to act as he sees fit when speaking in his own name. Each of us can freely leave the group."⁵⁵

Insiders, of course, understood the subtext: the problem of issuing appeals in the name of the Initiative Group without members' prior approval.⁵⁶ More broadly, the statement sheds light on how dissidents were coming to understand the elusive concept of "free association," the kind mentioned in the Universal Declaration of Human Rights as well as the Soviet Constitution, both of which the statement cited: an association of individuals who not only come together freely, without compulsion or hindrance by the state, but who continue to be free agents vis-à-vis each other. The statement emphasizes the absence of constraints on members by other members or by the group itself, without mentioning any duties or responsibilities within the organization, apart from "the feeling of personal responsibility for everything that happens in our country." The Initiative Group imposed no political litmus test on its members, who included "believers and non-believers, optimists and skeptics, people of communist and non-communist views," all of whom "understand social progress primarily in terms of the progress of freedom." "We do not propose any positive steps in the arena of state governance," the group announced, in the by now familiar dissident refrain. "We merely say: do not violate your own laws."[57]

The Initiative Group was thus an experiment: Could the spontaneous, conscience-driven energy of the chain reaction survive within the framework of an organization? Apparently, it could. Over the course of the following decade, members managed to overcome the group's flawed birth, issuing some forty collective appeals publicizing rights violations by the Soviet government, including the punitive use of psychiatric confinement, the persecution of Christians, and discrimination against Crimean Tatars, Bashkirs, and other ethnic minorities. Most appeals, however, concerned the arrest and imprisonment of rights activists, including members of the Initiative Group itself, who began to be targeted within weeks of the organization's founding.[58] Within six months, half of its members, including Gorbanevskaya and Krasin, were either in prison or in psychiatric institutions, a clear indication of the seriousness with which the KGB approached the group.

Amalrik, a close observer of the Initiative Group, captured its dual impact on reform-minded Soviet citizens. On the one hand, it successfully "broke the psychological barrier" against forming independent citizens associations, opening a path for other such organizations in the years that followed. In this sense it represented the leading edge of the process of overcoming the inertia of fear inherited from the Stalin era. On the other hand, Amalrik observed, by focusing on "the defense of those who were in prison because they had defended others who were imprisoned before them," the Initiative Group gave the impression of "moving within an increasingly narrow circle, turning the dissident movement in on itself."[59] Defending the defenders is of course vital for any rights-based movement; without them nothing can be accomplished. Having come under siege by the KGB, however, the Initiative Group, whose goal was to publicize "violations of civil rights" in the USSR, quickly found itself focusing on the persecution of its own members. Even when it managed to look beyond them, moreover, it tended to focus on groups perceived by much of the Soviet population as marginal: non-Russians, religious believers, Czechs and Slovaks.

The Initiative Group was experimental in another sense as well: its appeals were addressed almost exclusively to entities outside the Soviet Union, starting with the United Nations.[60] Simultaneously, its criticism of the Soviet government expanded from violations of civil rights enshrined in the Soviet Constitution to violations of international human rights, especially as codified in the 1948 Universal Declaration of Human Rights. The substance of what dissidents found objectionable didn't change; what shifted was the framework within which they articulated their objections, the body of norms and laws against which the Kremlin's behavior could be cast as illegitimate or illegal.

This was a subtle but significant shift. A landmark study of human rights by the historian Samuel Moyn goes so far as to describe it as "unintentionally fateful for world history."[61] Calling attention to the Soviet government's violation of its own laws made the Kremlin look hypocritical and its legal system farcical. But these were essentially domestic concerns. Calling attention to the USSR's violation of international human rights norms, by contrast, threatened to position the Kremlin as an outlaw vis-à-vis global society, deviating from the "common standard of achievement for all peoples and all nations," as the Universal Declaration described itself.[62] It lay the groundwork for grievances on the part of entities abroad that regarded themselves as stakeholders in that common standard.

"Puzzlingly," Moyn writes, "there is no clear-cut answer to why the founders [of the Initiative Group] chose in 1969 to allude to human rights rather than mainly domestic protections, like the earliest dissidents."[63] Indeed, neither in the group's numerous appeals nor in the memoirs of its founders can one find anything resembling an explanation for the shift from civil to human rights. Once again we are confronted with improvisation and informality rather than an articulated strategy. Two people—Krasin and Yakir—had chosen the name Initiative Group for the Defense of Civil Rights, and two people—Gorbanevskaya and Alexeyeva—subsequently took it upon themselves to attach (without explanation) the name Initiative Group for the Defense of Human Rights to the appeal they issued without consulting any other members.[64] It was the latter name that stuck, despite the fact that no general discussion of which kind of rights the group would defend (and hence the group's name) ever seems to have taken place. Nor is it clear whether members initially grasped the difference between rights belonging to citizens of a particular state and rights belonging to all human beings simply by virtue of being human—or whether they thought it mattered very much. The Initiative Group counted among its ranks accomplished physicists, mathematicians, and engineers (as well as an economist, a linguist, and a translator), but not a single lawyer, which may help explain their casual indifference to the distinction between civil and human rights—a distinction that carried important consequences regarding jurisdiction, procedure, and enforceability.[65] Nearly a year after its founding, the group was still using what for lawyers would have been nonsensical phrases such as "the defense of the civil rights of man."[66] "Violation of civil rights in any country," Krasin insisted during a KGB interrogation, "concerns all people of goodwill, and moreover concerns the UN, which includes a special Commission on Human Rights." The Initiative Group, he declared, could also invoke the Geneva Convention to protest the treatment of Soviet political prisoners—apparently unaware that the Convention, though ratified by and legally binding on the USSR, pertained exclusively to the rights of prisoners taken during wartime.[67]

Such confusion, caused above all by a dearth of reliable information, was pervasive, reaching well beyond the Initiative Group. Many dissidents, and probably many Soviet citizens, were under the impression that the Universal Declaration of Human Rights was a legally binding treaty subject to enforcement by the United Nations, rather than a nonbinding declaration of norms.[68] Even more important, a widespread misconception held that the USSR had ratified the Universal Declaration after it was approved by the General Assembly of the United Nations in December 1948.[69] True, Stalin's deputy foreign minister at the time, Andrei Vyshinsky, had been a central player in negotiations over the text of the Declaration, ensuring that it mentioned economic and social rights (Article 22), including the right to work (Article 23), arenas in which the Soviet Union was a pioneer.[70] In the final vote by the General Assembly, however, the USSR and other Soviet-bloc states (along with Saudi Arabia and South Africa) had abstained.[71] Whatever appeal the Universal Declaration may initially have had for the Kremlin—as an opportunity to showcase the Soviet Union's system of "real" rights— was subsequently overshadowed by Stalin's concern that the document would be used by the Kremlin's Cold War adversaries as an excuse to interfere in the USSR's domestic affairs. Whether Stalin, Vyshinsky, or any other Soviet leader ever imagined that Soviet citizens themselves might use the Universal Declaration to challenge Soviet policies— which would be difficult to characterize as interference in domestic affairs—remains unknown.

The Universal Declaration did indeed get swept up into the Cold War, each side using it to embarrass the other over its human rights failings, of which there was no shortage. Despite having abstained in the final vote, the Kremlin used every opportunity to draw attention to American violations of the Declaration, especially in the form of state-sanctioned discrimination and violence against Black citizens.[72] At the same time, Soviet authorities, guided by deputy foreign minister Andrei Gromyko, studiously avoided making the text of the Declaration available to the Soviet population. That policy began to change only after Stalin's death, as his successors grew more confident of the USSR's global stature and more willing to selectively relax their country's isolation from the outside world. Soviet mass media now referred to the Declaration as a "progressive document" that, while marred by compromises imposed by bourgeois states, nonetheless shared some of the principles of the USSR's supremely progressive constitution.[73] The complete text of the Declaration was published in classified journals, including one to which Volpin gained access at a restricted library and then proceeded to copy and circulate via samizdat. Freed from Stalin's paranoia and presiding over what was now an internationally recognized superpower, Leonid Brezhnev publicly endorsed a pair of international covenants designed to anchor the Universal Declaration, one covering civil and political rights, the other covering economic, social, and cultural rights. Both were adopted by the United Nations in 1966 and signed by Moscow two years later, nearly a decade ahead of the United States.[74]

In sum, whether because of a dearth of information, mixed signals sent by the Soviet government, or the absence of professional lawyers among members of the Initiative Group, considerable confusion reigned over the Universal Declaration and the USSR's obligations to it. Did the Soviet dissident movement thus stumble onto international human rights by accident? Moyn's solution to the puzzle invokes larger, indeed global forces. Loss of faith in other projects for collective redemption, he argues—chief among them socialism and world government—cleared the ground for human rights to achieve a "breakthrough" in the 1970s as the next and perhaps final utopia.[75] Whether they were aware of it or not, Soviet dissidents were thus part of a global response to the "crisis of utopianism," a mass flight of moral idealism away from the sinking ships launched by Karl Marx and Woodrow Wilson, toward the transcendent idea of inalienable rights for all human beings everywhere.[76]

The fact that members of the Initiative Group repeatedly sent appeals to the United Nations—the successor to Wilson's League of Nations and the leading embodiment of the ideal of global government—should lay to rest the claim that they had given up on that ideal. With only a handful of exceptions, dissidents did not reject socialism, either.[77] However convenient it was for the KGB to paint dissidents as anti-Soviet or anti-communist—a charge often uncritically embraced by dissidents' allies in the West—one rarely finds criticism of collective ownership of property or centralized economic planning in dissident texts, let alone arguments asserting an indispensable link between private property (or capitalism) and liberty. One of the most trenchant analyses of the topic can be found in the samizdat essay "Is a Non-totalitarian Type of Socialism Possible?" by the physicist Yuri Orlov.[78] Even Orlov, after subjecting Soviet-style socialism to a withering critique, answered the essay's guiding question affirmatively. He further argued that the safest and most realistic means to reform the Soviet system would require the decentralization of the command economy while preserving state ownership of its critical sectors. To this end, he called for the creation of an "ethical anti-totalitarian" (not anti-socialist) approach.

Dissidents did not criticize the USSR's elaborate system of social and economic rights, from free child care, medical care, and education (up through the post-secondary level) to subsidized housing and vacations. In fact, most showed little to no interest in the economy or how economic structures might inflect legal arrangements, including civil and political rights. "I knew of no opponents of socialism in my country," recalled Alexeyeva, who knew as many dissidents as anybody.[79] Paradoxically, the most common position among Soviet dissidents seems to have combined an unspoken acceptance of socialism as a system of coordinated, egalitarian public welfare (the "base," to use Marxist terminology) with a determination to detach the sphere of culture, politics, and ideas (the "superstructure") from Soviet dogma. Late Soviet socialism had become demagnetized: no longer a regulative ideal, but simply taken for granted.[80]

Moyn's larger point—that human rights in the second half of the twentieth century were conceived as an alternative to politics[81]—certainly applies to Soviet dissidents, who never tired of insisting that their movement was about law and conscience rather than any particular set of political arrangements. To understand the Initiative Group's turn to human rights, however, requires that we attend to a more immediate motive, one hidden in plain sight. The group's inaugural appeal informed readers: "We are turning to the UN because our protests and complaints, sent over the course of several years to the highest Soviet state and judicial authorities, have received no reply. The hope that our voice might be heard, that the authorities would cease the lawlessness which we have constantly pointed out—this hope has been exhausted."[82] Rather than a loss of faith in socialism, of which there is not a hint in the texts produced by the Initiative Group, it was their own government's refusal to engage in even a minimal, low-level dialogue that drove members to search for alternative interlocutors abroad. "The authorities did not respond," complained Alexeyeva. "It made no sense to keep writing to them. Such letters had become a dead genre."[83] As Alexander Solzhenitsyn put it, in his characteristically vivid manner, "Even the appeals of our best scientists and writers bounce back like peas off a wall."[84] There were rumors that the authorities, including the KGB, weren't even bothering to read protest letters anymore.[85] In the context of the Cold War, the United Nations—the embodiment of aspirations for global governance, a non-partisan international organization of which the Soviet Union was a founding member—appeared the best alternative.

It was the pragmatic turn to foreign entities, starting with the United Nations, that facilitated the Initiative Group's invocation of human rights—not the other way around, not a loss of faith in socialism and certainly not a loss of faith in the ideal of world government. Other factors may have been at work as well. Starting with Khrushchev's Thaw, Soviet legal scholars increasingly emphasized what they called the "rights of the person" or the "rights of the individual" (*prava lichnosti*) as part of "developed socialism."[86] The concept of *lichnost*—the human "personality" or "self"—had a distinguished pedigree in Russia, reaching back to the nineteenth-century intelligentsia with its cult of "consciousness" and the "developed personality," inspired by German Romanticism's ideal of self-perfection and wholeness.[87] Soviet legal scholars tended to take for granted that a fully developed personality was possible only within the framework of a human collective in which private property (and therefore selfishness and exploitation) had been eliminated—which is to say, within the framework of socialism. It is unclear, however, whether dissidents were familiar with Soviet legal scholarship. Volpin and his disciples tended to read law literally, without consulting commentaries and interpretations by professional jurists. Nonetheless, the Initiative Group shared some of the official scholarship's rhetoric, announcing, for example, that "at the foundation of the normal life of a society lies the recognition of the absolute value of the human personality [*lichnost*]."[88] Once the value of the human personality was cast in universal rather than system-specific terms, its surest legal protection presumably lay in human, rather than civil, rights.

In a sense, the improvised search for interlocutors abroad had begun in 1968 with Litvinov and Bogoraz's appeal "To the Global Public." But the global public was a hazy entity with no address. The United Nations, by contrast, was a global institution that had its own Committee for the Defense of Human Rights along with a "sacred duty" to defend those rights. Or so the Initiative Group imagined.[89] Here too there were signs of confusion and wishful thinking, starting with the fact that the UN entity was actually called the Commission on Human Rights, with not a word about defending them. More to the point, according to its own self-limiting by-laws, the Commission had "no power to take any action in regard to any complaints concerning human rights," opting instead merely to forward such complaints to the government of the country from which they came and to keep confidential records of which countries were accused of violating which articles of the Universal Declaration. It was, as the UN official responsible for overseeing this ritual of bureaucratic impotence famously remarked, "the most elaborate waste paper basket ever invented."[90]

Unaware that the Commission on Human Rights refused to accept petitions from unofficial entities, the Initiative Group continued to appeal not only to the Commission but to the UN secretary-general, U Thant, as well as to his successor (as of January 1972), Kurt Waldheim. The new appeals tracked a mounting despair:

> 26 September 1969: "We lack information regarding the Commission's stance regarding our [previous] petition. Silence on the part of international legal organizations unties the hands of those set on further repressions. We request that you inform us of the measures you have taken."
>
> May 1970: "We have appealed five times to the UN Commission on Human Rights. So far the Commission has not reacted to our statements. Perhaps there are reasons for this which are unknown to us."
>
> 5 January 1972: "We are appealing to you [Mr. Waldheim] just as we appealed to your predecessor in the high post of Secretary-General of the United Nations—unfortunately, without receiving a reply."[91]

Even when publicized by Western mass media, the Initiative Group's messages to the United Nations—like earlier letters to the Soviet government—went unanswered. The group began to look elsewhere, reaching out to the World Health Organization and the World Psychiatric Association (on the use of psychiatric incarceration against dissidents), as well as to Pope Paul VI and the Council of Orthodox Churches (on persecution of Christians). These letters, too, went unanswered.[92]

By the fifth anniversary of its founding, all but a handful of the group's original members had been arrested.[93] Those who remained at large continued to generate statements documenting rights violations, mostly against fellow dissidents, while retreating from the search for institutional allies in the West. Their appeals increasingly took the form of "open letters" and "announcements" addressed to no one.

Even as the Initiative Group struggled to open a conversation with the United Nations and other institutions abroad, it found itself being wooed by obscure Western non-governmental organizations (NGOs) eager to partner with the USSR's first independent rights association. What were these foreign entities? Whom did they represent? What were their ultimate goals? All this was a complete mystery to members of the Initiative Group, whose lack of information about the formalities of human rights was more than matched by the lack of information about the various citizens' groups in the West that took it upon themselves to advance those rights. Frustration with Cold War gridlock fostered a new citizen-to-citizen diplomacy in Europe and the United States, powered by "policy entrepreneurs" and other transnational actors.[94] Apart from a handful of exceptions such as Amnesty International and the International League for the Rights of Man, the history of citizens' associations focusing on human rights in the post–World War II era remains largely unresearched. Such groups proliferated as the United Nations' impotence in the human rights arena became clear. Their typical format was a small, advocacy-oriented NGO, rather than a broad social movement. The doctrine of human rights, moreover, was embraced across much of the political spectrum, from left-liberal to reactionary, religious to secular.[95]

None of this was apparent to Soviet dissidents. Just as authors of samizdat texts in the Soviet Union had little say over where, when, and by whom their work might be published abroad, so were they unable to control which foreign groups took up their cause, or attempted to contact them, and for what ultimate purpose. The same *glasnost* ethos that led dissidents to include their home address or telephone number (or both) in samizdat documents—so as to register their break from the tradition of underground conspiracy and signal their conviction that their activities did not violate Soviet law—placed their contact information at the disposal of a global public. Christian youth groups from the United Kingdom, Flemish human rights activists, and Italian neo-Fascist organizations, inspired by the novelty and heroism of protest at the epicenter of communism, dispatched youthful emissaries to Moscow on tourist visas, hoping to establish contact with known Soviet activists, to deliver assistance of various kinds, and, in several cases, to disseminate their own leaflets at prominent sites in the Soviet capital. Such episodes, which typically caught Soviet dissidents unaware, invariably ended badly for both them and their would-be allies.[96]

Shortly after the Initiative Group's founding, Yakir, Krasin, and Gorbanevskaya received a congratulatory telegram from an organization identifying itself as Norwegian Supporters of SMOG, being the nonconformist Moscow student group that had helped populate the first transparency meeting on Pushkin Square in December 1965. The Norwegian organization had counterparts in Denmark and Sweden, all founded in the second half of the 1960s. Seeking to promote the "liberation of persecuted intellectuals in the USSR who are struggling against the infringement of human rights," the three groups had taken it upon themselves to send petitions to Soviet officials in defense of Sinyavsky, Daniel, Ginzburg, and other imprisoned dissidents. They also mounted demonstrations in front of Soviet embassies, offices of Intourist (the Soviet agency responsible for coordinating visits by foreigners to the USSR), and Soviet art exhibitions that traveled to Scandinavia. When Lev Smirnov, the presiding judge at the trial of Sinyavsky and Daniel, visited Norway in June 1971, protesters disrupted his public appearances, causing him to cut his trip short.[97]

NGOs in other West European countries similarly applauded the founding of the Initiative Group. Whether by telegram, telephone, or letter, the International Committee for the Defense of the Rights of Man (Paris), the Foundation for Spiritual Freedom (Bern), the Flemish Committee for Eastern Europe (Antwerp), and European Civilization

(Rome) all reached out to express their support.[98] Taking advantage of the Soviet Union's growing accessibility to foreign tourists, some followed up by sending representatives to Moscow to meet dissidents in person. When a mysterious visitor from Belgium showed up unannounced one evening at a gathering in the apartment of Lyudmila Ginzburg, bearing stacks of leaflets demanding the release of her son Alexander, the assembled dissidents informed him that "no one needs leaflets—all they do is arouse their accidental readers in vain" and suggested he toss them in the Moscow River.[99] Visiting members of Europa Civiltà invited Yakir to join them as they handcuffed themselves to a mezzanine banister at GUM, the massive department store across Red Square from the Kremlin, and threw leaflets to bewildered shoppers below. To Soviet onlookers, the self-administered handcuffs could only have reinforced the image of protesters as bent on "self-imprisonment." Yakir, explaining that "in general, we're against leaflets," preferred to observe the action from a distance, along with several invited foreign journalists. He was, however, willing to accept a four-thousand-ruble gift from the organization, to assist the families of Soviet political prisoners. Other foreign groups carried out similar operations—including the handcuffs—in some of Moscow's train stations and in Leningrad. Most ended with the arrest and expulsion of the foreign demonstrators, a flash of attention in the Western mass media, and utter silence on the Soviet side, apart from coverage in the *Chronicle of Current Events*.[100]

During one of his many evening gatherings in the summer of 1969, Yakir opened the door to his apartment to find a pair of strangers who had just arrived from Paris. In broken Russian, the two young men introduced themselves as representatives of the International Committee for the Defense of the Rights of Man, founded earlier that year. They proposed that the Initiative Group formally affiliate itself with the Committee for purposes of information-sharing and collaborative work.[101] That process had in fact already begun: the Committee had reprinted the Initiative Group's first appeal to the United Nations, adding photographs of the recently arrested Grigorenko, Marchenko, and Galanskov. The Parisian pair brought dozens of copies to distribute to Soviet citizens as leaflets.¹⁰² Later that summer, representatives of the Flemish Committee for Eastern Europe arrived in Moscow and got in touch with several dissidents to announce their plan for organizing a public discussion of Andrei Sakharov's *Reflections on Progress, Peaceful Coexistence, and Intellectual Freedom*. From their hotel, they sent invitations to a hundred prominent Soviet writers, scholars, and artists, as well as to the editorial boards of *Pravda*, *Izvestiia*, and other newspapers. For logistical support they reached out to the Soviet Academy of Sciences, the Ministry of Culture, and the Ministry of Foreign Affairs. Within days, they were expelled from the country.

The Initiative Group struggled to find a common language with these audacious young supporters from the West. This was true in the literal sense, as nervous conversations unfolded in halting Russian or English, but also figuratively, in the sense of a language of protest. For dissidents, dispersing leaflets among random passersby was a low-tech, inefficient means of communication compared to the "boomerang" method of using Western shortwave radio stations to broadcast samizdat texts to thousands or even millions of curious Soviet listeners. Few Soviet citizens were inclined to publicly pick up leaflets dropped by strangers, especially if those strangers had engaged in the undignified and culturally alien spectacle of handcuffing themselves to banisters. The content of some of the leaflets was incongruous too, even for dissidents. A postcard-size handout printed by the International Committee for the Defense of the Rights of Man, for example, featured a photograph of Vladimir Bukovsky and the following message:

> Remember Bukovsky!
> Fight on his behalf and together with him for Truth, Freedom, and Justice! Demand his liberation!
> If you are able, provide him with moral and material support.
> Follow his example!¹⁰³

This fusillade of moral exhortations was a far cry from the restrained, non-prescriptive, just-the-facts idiom of the *Chronicle of Current Events*. If anything, it is likely to have reminded Soviet readers of the Communist Party's mobilizational rhetoric (*Forward to the victory of communism!*,

FIGURE 12.5. Front and reverse side of a postcard publicizing the case of Vladimir Bukovsky, prepared by the International Committee for the Defense of the Rights of Man (Paris)

Strengthen the union of hammer and sickle!, *Young Pioneer, you are responsible for everything!*, etc.). Even among committed dissidents, few were inclined to replicate Bukovsky's relentless defiance: "'If not me, then who?'—the individual with his back to the wall asks himself. And he saves everyone."[104] His stance had already garnered him six years of confinement in mental hospitals and labor camps and, in January 1972, a twelve-year sentence for sending abroad evidence on the use of punitive psychiatry against dissenters. There was something incongruous, moreover, about a call to martyrdom at the hands of the Soviet state issued by a committee operating from the safety of Paris.

Incongruity, in fact, was the hallmark of the initial encounters between Soviet dissidents and their would-be foreign allies. At one of Krasin's "visiting days," a young Swede showed up with a stack of sixteen printed booklets, each containing Sakharov's *Reflections*, a commentary on it by Estonian intellectuals, the Universal Declaration of Human Rights, and the program of the People's Labor Alliance (NTS), the West leaflets.[102] Later that summer, representatives of the Flemish Committee for Eastern Europe arrived in Moscow and got in touch with several dissidents to announce their plan for organizing a public discussion of Andrei Sakharov's *Reflections on Progress, Peaceful Coexistence, and Intellectual Freedom*. From their hotel, they sent invitations to a hundred prominent Soviet writers, scholars, and artists, as well as to the editorial boards of *Pravda*, *Izvestiia*, and other newspapers. For logistical support they reached out to the Soviet Academy of Sciences, the Ministry of Culture, and the Ministry of Foreign Affairs. Within days, they were expelled from the country.

The Initiative Group struggled to find a common language with these audacious young supporters from the West. This was true in the literal sense, as nervous conversations unfolded in halting Russian or English, but also figuratively, in the sense of a language of protest. For dissidents, dispersing leaflets among random passersby was a low-tech, inefficient means of communication compared to the "boomerang" method of using Western shortwave radio stations to broadcast samizdat texts to thousands or even millions of curious Soviet listeners. Few Soviet citizens were inclined to publicly pick up leaflets dropped by strangers, especially if those strangers had engaged in the undignified and culturally alien spectacle of handcuffing themselves to banisters. The content of some of the leaflets was incongruous too, even for dissidents. A postcard-size handout printed by the International Committee for the Defense of the Rights of Man, for example, featured a photograph of Vladimir Bukovsky and the following message:

Remember Bukovsky!
Fight on his behalf and together with him for Truth, Freedom, and Justice! Demand his liberation!
If you are able, provide him with moral and material support.
Follow his example![103]

This fusillade of moral exhortations was a far cry from the restrained, non-prescriptive, just-the-facts idiom of the *Chronicle of Current Events*. If anything, it is likely to have reminded Soviet readers of the Communist Party's mobilizational rhetoric (*Forward to the victory of communism!*,

FIGURE 12.5. Front and reverse side of a postcard publicizing the case of Vladimir Bukovsky, prepared by the International Committee for the Defense of the Rights of Man (Paris)

Strengthen the union of hammer and sickle!, *Young Pioneer, you are responsible for everything!*, etc.). Even among committed dissidents, few were inclined to replicate Bukovsky's relentless defiance: "'If not me, then who?'—the individual with his back to the wall asks himself. And he saves everyone."[104] His stance had already garnered him six years of confinement in mental hospitals and labor camps and, in January 1972, a twelve-year sentence for sending abroad evidence on the use of punitive psychiatry against dissenters. There was something incongruous, moreover, about a call to martyrdom at the hands of the Soviet state issued by a committee operating from the safety of Paris.

Incongruity, in fact, was the hallmark of the initial encounters between Soviet dissidents and their would-be foreign allies. At one of Krasin's "visiting days," a young Swede showed up with a stack of sixteen printed booklets, each containing Sakharov's *Reflections*, a commentary on it by Estonian intellectuals, the Universal Declaration of Human Rights, and the program of the People's Labor Alliance (NTS), the West

Помните о Буковском!

Большую часть своей жизни он провел в психотюрьмах, тюрьмах и лагерях за то, что отстаивал Правду, Свободу и Справедливость. 5.1.1972 г. он снова осужден на 12 лет — тюрьмы, лагеря и ссылки — за передачу на Запад документов, свидетельствующих об использовании КГБ психобольниц, как мест заключения инакомыслящих. В последнем слове на суде он высказал только сожаление, что за год, два месяца и три дня на свободе сделал так мало.

    ...Я выбираю Свободу!
    Но не из боя, а в бой,
    Я выбираю свободу
    Быть просто самим собой...

    ...Я выбираю Свободу!
    Я пью с ней нынче на ты!
    Я выбираю свободу
    Норильска и Воркуты!

    Где вновь огородной тяпкой
    Над всходами пляшет кнут,
    Где пулею, или тряпкой
    Однажды мне рот заткнут...

    ...Я выбираю Свободу!
    Пускай груба и ряба,
    А вы, валяйте, по капле
    «Выдавливайте раба»!..

                    (А. Галич)

Помните о Буковском!
Боритесь за него и с ним за Правду, Свободу и Справедливость! Требуйте его освобождения!

Если у вас есть возможность, окажите ему моральную и материальную поддержку. Следуйте его примеру!

        Comité international pour la défence
        des droits de l'homme,
        26, rue St.-Placide, Paris 6

FIGURE 12.5. (*continued*)

German–based organization of Russian émigrés dating from the interwar period. A self-described "revolutionary movement," NTS was not squeamish about its goals:

> For Russia!
> We Will Bring Death to Tyrants!
> We Will Bring Freedom to the Toilers!
>
> Friend! You are surrounded by the lie of communist propaganda.
> Only from the underground can you hear truth.
> Listen to the underground radio station "Free Russia."
> Listen to NTS—the voice of revolution.[105]

As soon as the Swede left, Krasin ripped out and burned the NTS program from all but one of the booklets.[106] It was dangerous enough to meet with foreigners or keep samizdat (of which Krasin possessed one of Moscow's largest collections) in one's apartment. Getting caught with NTS materials risked tarnishing the dissident movement by association with a long-standing bête noire of Soviet propaganda, which repeatedly—and accurately—described NTS as having been supported first by the Nazis and more recently by the CIA.[107] Soviet propaganda aside, NTS's conspiratorial methods and advocacy of armed insurrection, not to mention its language, made it an anathema in dissident circles. Less well known but similarly tainted was the anti-communist Europa Civiltà, which bore a fascist legacy, including paramilitary training camps on the outskirts of Rome.[108]

Dissidents now faced a host of troubling questions. Were these obscure human rights organizations fronts for the CIA or other spy agencies?[109] Or perhaps an elaborate means of entrapment by the KGB? Why were they so keen to establish contacts with the Initiative Group? Who financed their activities? Such concerns came to a head when a lawyer from Paris, representing the International Committee for the Defense of the Rights of Man, appeared in Moscow in an attempt to formalize a partnership with the Initiative Group. The Committee proposed to officially represent the Initiative Group's interests in the West and in return to have exclusive rights to publish its documents. There were forms to sign. "Why do we need a formal agreement?" Krasin asked. The lawyer explained that "in the West there were certain traditions that required that if any organization represented the interests of another organization, it had to be officially authorized to do so, otherwise no one would want to negotiate with it."[110]

Western traditions notwithstanding, the half-dozen members of the Initiative Group who assembled at Yakir's apartment that evening unanimously declined this and all other partnerships with foreign NGOs. When the International Committee for the Defense of the Rights of Man came back with a more modest proposal for a joint declaration condemning the persecution of "other-thinkers" in the USSR, to be submitted to the United Nations office in Moscow and then to foreign correspondents, that too was rebuffed.[111] Many members of the Initiative Group were convinced that establishing such contacts would discredit them in the eyes of the Soviet public, which, as Venyamin Kozharinov noted, "is constantly bombarded with the notion that the Soviet democratic movement is a creature of American and other intelligence organs."[112] Members of the Initiative Group were part of that Soviet public; they too had been primed since childhood to view foreign organizations with suspicion. They were aware, moreover, that so-called citizens' associations in the USSR were actually controlled and financed, directly or indirectly, by the Communist Party.[113]

Soviet propaganda insisted that the same was true in the West. Self-proclaimed NGOs were actually fronts managed by governments or by the bourgeois ruling class, which amounted to the same thing, since governments were themselves instruments of class rule. The first Soviet handbook on international NGOs, published in 1967, listed over four hundred entities, noting that capitalist governments were eagerly weaponizing them:

> The appearance of these kinds of organizations is linked to the effort by the dominant classes of capitalist society to exploit the growth in mass activism for their own interests and to confine it within a certain framework. As a result, many international NGOs de facto carry out the directives of bourgeois parties and are kept on retainer by the

Western powers and various monopolies. Regular subsidies from these sources constitute the basic means of [the NGOs'] financing, although the overwhelming majority of such organizations seek to highlight their alleged independence from the ruling bosses of the capitalist world and to underscore in every possible way their "apolitical" nature.[114]

The political neutrality trumpeted by the International Committee for the Defense of the Rights of Man was powerfully appealing to Soviet dissidents. It mirrored the Initiative Group's conviction that defending civil and human rights was a cause separate from politics. Was it possible that similar assertions by Western NGOs were a smoke screen designed to conceal the agendas of hidden sponsors?

Western NGOs claimed to operate not just independently of their own governments, but outside the force field of the Cold War. From what position could one make such claims? How could one be sure that mysterious groups such as Europa Civiltà or the International Committee for the Defense of the Rights of Man represented, as Kozharinov put it, "the Western public" or, for that matter, anyone besides themselves? Part of the uncertainty, of course, stemmed from sheer lack of information due to Soviet censorship and the relative obscurity of the NGOs that approached the Initiative Group. It didn't help that most of their emissaries to Moscow were barely out of college.

But there was a deeper source of uncertainty. As the Initiative Group attempted to inhabit the role of loyal opposition, a public voice that was independent but not anti-Soviet, critical but not treasonous, it did so without the benefit of a script or precedent. There were no indigenous post-revolutionary models for what a legal Soviet NGO might look like or how it should function. The Soviet system rejected the very category of non-governmental organization, just as it rejected the principle of checks and balances, or separation of powers, not just among institutions of government but in relations between government and society. The Initiative Group was seeking to operate in a space that did not exist, an imagined crevice in what the Kremlin insisted was solid ground.

13

The Inner Sanctum of Volpinism

> You want us to make a revolution? Don't be silly.
> —VALERY CHALIDZE

The tension between advocates of conscience-driven activism and proponents of organization and leadership propelled one of the great inner dramas of the dissident movement. Implicated in that tension was the dissidents' unacknowledged dilemma: a movement determined to promote strict proceduralism and rule of law in the USSR resisted instituting those same principles in its own work. The arrest of the most prominent representatives of the voluntarist wing at their demonstration on Red Square in August 1968, followed by the founding of the first dissident association less than a year later, seemed to mark a victory for the forces of organization. As chapter 12 showed, however, the modus operandi of the Initiative Group for the Defense of Civil Rights looked more like a triumph of informality over proceduralism—or spontaneity over consciousness—than the other way around.

In the ongoing search for new forms of civic engagement, Valery Chalidze took a different approach. Where the Initiative Group shunned rules and procedures in favor of informality in the conduct of its business, Chalidze dreamed of raising formalism to new heights. Trained as a physicist, he worked at Moscow's All-Union Research Institute of Plastics, focusing on synthetic polymers. His father, an ethnic Georgian, was killed in the Second World War; Chalidze was raised largely by his mother, an architect and descendent of Polish political activists exiled to Siberia under the tsars. He was married to Vera Slonim, Pavel Litvinov's cousin and, like him, a grandchild of Maxim Litvinov.[1] At a scientific conference in the late 1950s, Chalidze chanced to meet Alexander Volpin, who was keen to share his ideas about how to persuade the Soviet state to obey its own laws. "The principle of fidelity to Soviet law," Chalidze later wrote, "came quite naturally to me."[2] Those who knew him agreed: austere and formal (his nickname among dissidents was "the Prince"), Chalidze, according to a British journalist who got to know him, had "an exaggeratedly pedantic and literal mind" and was accustomed to "fight to the last micrometer of principle and legality on every issue."[3] As Andrei Sakharov more gently put it, Chalidze's "quick, analytical intelligence was made to order for juridical 'games.'"[4]

Chalidze was not one to go to the square. He did not attend the annual December 5 gatherings at the Pushkin monument. In fact, he didn't much like leaving his apartment, where antique Georgian sabers hung from the walls and collections of precious stones and dried scorpions were kept under glass.[5] Amid piles of books and samizdat documents, he received visitors and edited his bi-monthly samizdat journal *Problems of Society*, a kind of scholarly counterpart to the more widely read *Chronicle of Current Events*, featuring in-depth analyses of Soviet constitutional rights, international human rights, and other legal topics, most of them by physicists and mathematicians such as Volpin, Boris Tsukerman, and Chalidze himself.[6] In the spirit of transparency, each issue listed Chalidze's home address in the Arbat neighborhood as well as his telephone number.

In the fall of 1970, Chalidze approached Sakharov with an idea: to conduct the kind of research and analysis featured in *Problems of Society* under the aegis of a Human Rights Committee that would offer expert advice to the Soviet government on an array of legal issues. The goal would be to study theoretical and practical aspects of rights and to clarify areas of tension between Soviet and international human

FIGURE 13.1. Valery Chalidze

rights law. The Kremlin, to be sure, was not asking for advice. "If the government ignores us," Chalidze anticipated, "we will just go on making more recommendations," which would be shared with the Soviet public, thereby advancing Volpin's goal of cultivating popular legal consciousness. Most Soviet citizens, according to Chalidze, "see little distinction between the law as it is written and as the authorities choose to interpret it." Citizens should learn to understand the law in its literal sense, proceeding from the principle that everything not expressly forbidden by law is permitted.[7]

For its part, the Human Rights Committee would present itself not as a "voluntary association" but as a "creative association." Chalidze's claim to this status was itself rather creative. According to Article 482 of the 1964 Civil Code of the Russian Soviet Federated Socialist Republic, copyright over a work authored by two or more individuals (a "collective work") belonged jointly to the authors, regardless of whether the work in question was an indivisible whole or consisted of separate parts. "Relations among co-authors," stipulated the section of Article 482 that caught Chalidze's eye, "can be determined by their mutual agreement." Not a word about official approval. The Human Rights Committee would constitute just such an "authors' collective," jointly producing analytical papers by "mutual agreement" and therefore not requiring official registration.[8] To forestall accusations of engaging in political activity, only persons would be permitted to join the Committee who "are not members of political parties or other organizations claiming a role in governing the state, who are not members of organizations whose principles permit participation in orthodox or oppositional political activity, and who do not intend to use their participation in the Committee for political ends."[9]

Sakharov was evidently not persuaded by such "juridical games" but was supportive enough of the larger purposes of the proposed Committee to accept Chalidze's invitation to join. A third physicist, Andrei Tverdokhlebov, also came on board. The son of a former Soviet diplomat and deputy minister of culture, Tverdokhlebov had become active in dissident circles after signing open letters in the 1968 protest campaign. To increase the Human Rights Committee's prestige and visibility, the popular actor, playwright, and bard Alexander Galich was made a corresponding member, as was Alexander Solzhenitsyn—who just weeks earlier had been awarded the Nobel Prize in Literature. It was the only time Solzhenitsyn allowed himself to be publicly linked to the dissident movement. Volpin and Tsukerman agreed to serve as expert legal consultants. Chalidze drafted the Committee's governing principles and by-laws, Sakharov and Tverdokhlebov signed them, and on November 11, 1970, the three held a press conference in Chalidze's packed apartment, surrounded by fellow dissidents and foreign journalists.

Chalidze was a master at presenting the Human Rights Committee in the most benign possible light, as a kind of think tank. Its functions, he insisted, were strictly "academic and advisory":

> The Committee's task is not to expose or demand, but rather to study and recommend, which implies that our activities must take into ac-

count both tradition and practical difficulties. However indignant individual Committee members may be over particular violations of human rights, the Committee as such must base its conclusions on a careful consideration of the practicality of defending, at the given moment, the right attacked. The Committee must be prepared to wait for years for the realization of its good intentions and yet this delay must not lead to a sense of frustration or futility. The task of the Committee is first to promote the growth of legal consciousness in society; we can hope for serious progress in the defense of human rights only after an advance in the legal consciousness of people including those in power.[10]

The idea was to cultivate "methods which would have less risk of bringing about reprisals," he told Harrison Salisbury of the *New York Times*, "so it became clear to me that the chief principle of my future work would be the defense of political prisoners—before they became political prisoners."[11]

Soviet media made no mention of the founding of the Human Rights Committee, but countless Soviet citizens learned thanks to the Voices.[12] News of what soon came to be known as the "Sakharov Committee" unleashed a flood of rumors. As with Petr Yakir and the Initiative Group, letters from cities across the Soviet Union began arriving, most of them addressed to Sakharov. The KGB reported to Brezhnev and the Central Committee that "politically immature Soviet citizens" were offering "to establish contact, to collaborate, and to provide various kinds of assistance, including financial support." The West, meanwhile, was allegedly using the Committee "to ignite an anti-Soviet campaign" and "to portray Sakharov as the leader of an opposition that purportedly exists in the USSR."[13]

Sakharov did not consider himself the leader of anything, though he was well aware of the prestige and visibility that his presence bestowed on the Committee.[14] Weekly meetings were held on Thursdays at Chalidze's apartment, where the host insisted on parliamentary procedure even when only he, Sakharov, and Tverdokhlebov were present, as was usually the case. Each meeting produced a written protocol of the discussion, which Chalidze then included in the following issue of *Problems of Society* (where the Committee's by-laws and governing principles also appeared), in an effort to model the kind of transparent, rule-bound behavior that dissidents demanded from the Soviet government. A starker contrast to the improvisatory style of the Initiative Group would be difficult to imagine.

Members of the Initiative Group were not impressed. Anatoly Levitin-Krasnov dismissed the Human Rights Committee's work as "academic discourse by learned liberals" and "a step backward in the development of the Russian Democratic Movement." Another member was even blunter: "So what? Our physicist comrades will stand outside a courthouse in the freezing cold, watching the drunken face of the law, listening to an obscene interpretation of the constitution, getting punched in the head a few times ... and then they will give up trying to advise the KGB on matters of jurisprudence." But it wasn't only the Initiative Group that dismissed Chalidze's highly formalized approach. Even die-hard supporters of "legalism," according to Leonid Plyushch, found the idea of "assisting the lawless guardians of the law" simply "laughable." Only by deploying the law in an adversarial manner, exposing the regime's illegal actions to the whole world, did one stand a chance of changing the Kremlin's conduct. "Laws are our weapon," Plyushch insisted, "but not our illusion."[15]

Some who observed the Human Rights Committee up close also found themselves laughing. One thing was immediately apparent: these were not Litvinov's or Yakir's "visiting days," where people called each other *Yura* and *Pasha* and *Lara* in freewheeling, well-lubricated conversations. Here one was addressed with full first name and patronymic, and the beverage of choice was tea. Sakharov's soon-to-be second wife, the physician Yelena Bonner (his first wife, Klavdia Vikhireva, had died of cancer in 1969), dubbed the trio of physicists "Ve-Che-Ka," a play on the name of the KGB's predecessor, the All-Russian Extraordinary Commission (widely known as the Cheka). In this case, the three letters signified the Thursday meetings' core ingredients: Volpin, tea, and cookies.[16] When Ludmilla Alexeyeva first read the Committee's by-laws and charter, she recalled, "I started to giggle." They were written "in impeccable, almost comical legalese. I laughed at every sentence, imagining the look on the face of some KGB operative the morning Valery's *Principles and Regulations* were placed on his desk. My God, it seemed they were going to take minutes at their meetings! That was the opposite of the accepted strategy: we tried not to put much on paper, mostly as a way of keeping our KGB dossiers as thin as possible. The committee even had an administrative structure."[17]

Curious to observe these highly formalized rituals, Alexeyeva requested permission to attend a Thursday meeting in order to present information on political prisoners' right to send and receive mail. "The Committee doesn't hear petitions from the public," Chalidze replied, suggesting that she submit a written statement instead. Alexeyeva prevailed. In early 1971, she met with the three scientists, who sat stiffly around a table in Chalidze's apartment while she stood and delivered her remarks. For weeks thereafter, "I entertained my friends with the story of my wondrous journey into that inner sanctum of Alek-Yesenin-Volpinism."[18] Even Sakharov was known to refer to the group as the "Pickwick Committee," with reference to Charles Dickens's comic novel.[19]

All this laughter is worth taking seriously. As is often the case, the humor was precipitated by juxtaposing incongruous elements. Dissidents, especially the legalists among them, were well accustomed to using the highly formal, not to say stilted, language of Soviet law in their dealings with the government. Indeed, it was Volpin's great discovery that such language could be deployed not only as an instrument of state power but—counter-intuitively—as a means to place limits on that power. The by-laws and minutes of the Human Rights Committee conformed as closely as possible to the language of the Soviet Civil Code and, in that respect, were the opposite of funny. What made them funny—in the minds of dissidents—was their deployment *within* the movement, as a way of governing relations among dissidents themselves. Like Andrei Amalrik declaring himself the movement's "communications officer," the laughter-inducing incongruity was between the officious world of Soviet law and government ("them") and the persistently informal, intimate dissident milieu ("us"). In this sense, the

Human Rights Committee was the exception that proved the rule: dissidents were by and large allergic to formal structures.

Sakharov confessed that he "didn't understand very well the juridical aspects of the problems studied by the Committee" and was occasionally annoyed by Chalidze's "overly speculative" approach. Solzhenitsyn immediately regretted having agreed to join as a corresponding member and took no part in the Committee's work or in Chalidze's "supremely boring samizdat law journal." "It was a strange committee," he recalled in 1975 from his exile in Vermont, designed "to advise the cannibals (if they requested) on the rights of those they were about to tear to pieces."[20] The Politburo, to no one's surprise, was not interested in the Committee's advice. "Whom does Sakharov want to defend," Mikhail Suslov asked dismissively, "and from whom?"[21] Chalidze, Volpin, and Tsukerman nonetheless proceeded to produce a slew of pioneering briefs on important issues, including the right to defense counsel in Soviet courts; the presumption of innocence in Soviet jurisprudence; discrimination on the basis of political opinion; unwarranted curbs on the free exchange of information; the legal definition of the terms "parasite" and "political prisoner"; the legal rights of persons declared mentally ill, of homosexuals, and of so-called resettled peoples such as the Crimean Tatars; and legal restrictions on the right to leave, return to, and choose one's place of domicile within the USSR.[22] Some briefs highlighted contradictions between Soviet law and international norms, for example, between the Soviet requirement that parents raise their children in the spirit of "a builder of Communism" and the United Nations Convention against Discrimination in Education (ratified by the USSR in 1962).

Many of these articles appeared in Chalidze's *Problems of Society*. Dense, detailed, and often arcane, they did not elicit the kind of samizdat discussion that debates on the future of the dissident movement had aroused two years earlier. Nor did they elicit a response from Soviet lawyers or jurists, for whom such topics were broached strictly within their own professional journals—or not at all.[23]

If the work of the Human Rights Committee largely failed to reach Soviet readers, Soviet readers—and listeners to the Voices—were nonetheless finding their way to the Committee and specifically to Sakharov. Chalidze's determination to preserve the strictly academic nature of the Committee and to shield it from petitions from the public was no match for the simmering reservoir of grievances within Soviet society. As rumors about the "Sakharov Committee" spread, Sakharov—like Yakir and the Initiative Group—found himself besieged by hundreds of letters from across the Soviet Union seeking his advice and assistance. Many of them were sent to Sakharov at the Soviet Academy of Sciences, possibly on the mistaken assumption that the Human Rights Committee was sponsored by that august institution. Some were addressed to Sakharov as "Chairman of the Legal Advice Bureau 'Defense of Human Rights,'" others to the "Chairman of the Committee to Guarantee Human Rights in the USSR."[24] The KGB carefully tracked this "rising torrent of correspondence" from "politically immature Soviet citizens . . . who consider Sakharov a 'defender of the insulted and the injured.'" Some letter writers volunteered to collaborate with or join the Committee; others offered to send money; still others submitted ideas on how to improve the socialist system, requesting that their proposals be included in the upcoming Twenty-Fourth Congress of the Communist Party. In a memorandum to Brezhnev, Yuri Andropov noted that "most of the letters contain complaints about the actions of local authorities, court sentences, etc. Sakharov . . . believes that this kind of activity gives him the reputation of an active, independent political figure and will force the leaders of the Soviet state to consider his views on issues of foreign and domestic policy."[25]

In fact, this activity filled Sakharov with despair. "All these letters, requests, and complaints," he lamented. "What were we supposed to say, how were we supposed to respond? That we're a committee for studying, not for defending? That would amount to mockery. . . . The impossibility of helping all these people, the idea that I had raised false hopes among them, tormented me."[26] Sometimes strangers would show up unannounced at Sakharov and Bonner's apartment on Chkalov Street to plead their case. The largest group of supplicants were people trying to leave the Soviet Union; next were those with employment grievances (illegal dismissals, mistreatment by bosses, and so forth), followed by elderly and/or disabled individuals seeking government assistance and relatives of convicted criminals reporting judicial corruption or prison abuse. Hundreds of letters arrived from labor camps, jails, and psychiatric hospitals, all based on the assumption that the Human Rights Committee was less think tank than legal aid bureau.[27] To manage the flood of correspondence, Sakharov was eventually able to secure the assistance of Sofia Kallistratova, a lawyer who had risked her career defending dissidents. "Only in rare instances," he admitted, "were we able to render practical assistance. But at least the letters were answered"—unlike those sent by dissidents to the Soviet government and the United Nations.[28]

Another category of visitors also attempted to contact Sakharov and Chalidze: representatives of various Western human rights NGOs, including some who had tried and failed to establish ties with the Initiative Group.[29] Unlike the latter, the Human Rights Committee was eager for what Chalidze called "creative partnerships." As elite scientists, he and other members of the Committee were perhaps already accustomed to thinking of their work in transnational terms. Chalidze was looking not for foreign organizations with whom to lodge complaints against the Soviet government, but for entities analogous to the Committee: scholarly, non-partisan, independent. Rather than accept offers that came to him, he sought out appropriate partners himself, in part by consulting the official Soviet handbook of international NGOs—the one cautioning readers that such entities "de facto carry out the directives of bourgeois parties and are kept on retainer by the Western powers and various monopolies."[30] Two in particular stood out: the International League for the Rights of Man, formed in 1941 in New York by Roger Baldwin (co-founder of the American Civil Liberties Union), with member organizations in Great Britain, Germany, Japan, Israel, and over a dozen other countries, and the International Institute of Human Rights, established in 1969 in Strasbourg by René Cassin, the French jurist who was awarded the Nobel Peace Prize for his leading role in drafting the Universal Declaration of Human Rights two decades earlier.

Baldwin, whose initial enthusiasm for the Bolshevik Revolution had yielded by the 1950s to a staunch anti-communism, had to be persuaded that Chalidze and other members of the Committee would not face reprisals by establishing links with a foreign entity such as the International League. As Chalidze noted, "traditional ways of thinking" assumed that since the Human Rights Committee was not controlled by the Soviet government, it must be an underground organization opposed to the ruling power. "You want us to make a revolution?" an exasperated Chalidze asked *Newsweek*'s Moscow bureau chief Jay Axelbank during an interview shortly after the Committee's founding. "Don't be silly."[31] One purpose of the Committee's existence was to demonstrate the possibility, under the post-Stalinist dispensation, for independent but fully legal action. Too many people in the West as well as in the USSR took for granted that certain activities should be avoided when in fact no Soviet law forbade them. "As strange as it may sound to the public," Chalidze announced in the *Chronicle of Current Events*, "for me the prospect of criminal proceedings [against the Human Rights Committee] is very appealing, since they would make it possible, in an open court dispute, to examine the state of law and practice as regards the right of association."[32]

Establishing ties with independent citizens' groups abroad was one of those allegedly not-forbidden-therefore-permitted activities. With Chalidze's assurance in hand, the International League announced its formal acceptance of the Human Rights Committee as a member group in June 1971. John Carey, the International League's chairman, noted that "both organizations, the Committee and the League, are entirely non-political and non-governmental and their only concern is the furthering of human rights throughout the world." Chalidze and Carey began conferring weekly by telephone, supplemented by a stream of United Nations material on human rights sent by Carey to Chalidze's apartment. A month later, the Human Rights Committee formally established ties with Cassin's Institute in Strasbourg (also via telephone), which set about translating and publishing Chalidze's *Problems of Society* in French and English.[33] In the meantime, the International League produced a Russian-language book of briefs prepared by the Committee, in pocket-size format for travelers to bring undetected into the USSR.[34]

What counted as "non-political" or "non-governmental" depended, of course, on whether one was in Moscow or New York. The KGB noted that the International League's mission included fostering "the implementation of the guarantees of political liberties not only in relations among states, but especially their observance within states." Andropov, in an August 1971 note to Suslov, insisted that "in reality, the activity of the 'League' often assumes, under the influence of reactionary circles, a sharply anti-Soviet, anti-communist character." His concern only grew as it became clear, from the KGB's bugging of conversations in Chalidze's apartment and phone calls with human rights activists in New York and Strasbourg, that the Committee intended to expand its contacts with human rights NGOs, including Amnesty International, the International Federation for Human Rights, the Association for International Law, and others. All of this, according to Andropov, demonstrated that the Committee was attempting "to raise its authority in world public opinion and strengthen its position inside [the USSR]." Chalidze's increasing contact with foreign visitors to Moscow "attests to the fact that hostile propaganda centers are taking a noticeably greater interest in the 'Committee.'"[35]

The dissident movement's relationship with the West was indeed shifting. Three years earlier, in his August 1968 letter to Stephen Spender, Pavel Litvinov had proposed a division of labor between Soviet dissidents and their allies beyond the Iron Curtain. In order to preserve their purely conscience-driven, non-hierarchical activism, dissidents would outsource formal organizational work to a committee abroad consisting of progressive representatives of the "global public"—prominent writers, scholars, and artists such as Spender and the other signers of the statement supporting Litvinov and Larisa Bogoraz earlier that year. The human rights struggle was going global, but in highly asymmetric fashion, as activists on each side of the Iron Curtain deployed methods geared to their own local conditions.

Chalidze's Human Rights Committee abandoned that model. Rather than assigning different functions to Soviet dissidents and their allies in the West, the Committee sought to translate the form and methods of Western NGOs and apply them in the Soviet Union. The Committee's affiliation with multiple human rights NGOs abroad was both symptom and facilitator of that process. "For the first time in half a century,"

Chalidze announced at a Committee meeting in summer 1971, "a kind of non-governmental public association, controlled by neither the Party nor the government, has managed not only to survive for more than half a year, but to establish contacts abroad."[36] Having adopted the idea of human rights as a tool "to limit the state's authority over the individual," Chalidze and other dissidents were now embracing the NGO as an organizational form just as it was emerging as the preeminent global mechanism of human rights advocacy.[37]

It is tempting to see dissidents as taking the first steps toward building independent institutions of civil society under the shadow of the Communist Party's monopoly on public life. We could then place the Human Rights Committee as well as the Initiative Group within the venerable history of Montesquieu's *corps intermédiaires*, those voluntary citizens' associations designed to mediate between society and the state, creating a new framework for civic agency and therefore for civic dignity. So appealing is this narrative that during the global heyday of the civil society paradigm in the late 1980s and 1990s, a number of former Soviet dissidents retrospectively adopted it as their own.[38] Closer examination in no way diminishes the pathbreaking and courageous nature of dissidents' work, but it suggests a different framework of significance. Rather than an effort to build bridges between Soviet society and the state, the Human Rights Committee and the Initiative Group should be understood as a response to the failure of that strategy. The formation of dissident organizations instead signaled the beginning of a different kind of mediation: between Soviet dissidents and their potential allies in the West.

Dissidents' inability to open a dialogue with the Soviet government was not for want of trying. The KGB's campaign of intimidation and reprisals against signers of open letters, along with the failure of such letters to achieve their goal of fair and open trials for arrested activists, drove many sympathizers away. "People started saying, 'What's the point of risking our jobs or even arrest when it does no good?'" one dissident observed.[39] Protest letters, according to Viktor Krasin, experienced the equivalent of a currency devaluation, a "crisis of the genre" suggesting that it had outlived its usefulness.[40] What was once known as the Dem-Dvizhenie (short for "Democratic Movement") was now mocked as the DymDvizhenie (movement of smoke).[41]

All this was brought home to Sakharov in the spring of 1972. With the fiftieth anniversary of the founding of the Soviet Union (December 30, 1922) approaching, Sakharov took it upon himself to collect signatures for two petitions to the Presidium of the Supreme Soviet: one calling for abolition of the death penalty (a cause championed by his paternal grandfather a decade before the Bolshevik Revolution), the other requesting amnesty for all Soviet political prisoners. The first dozen signatures came quickly, from dissident friends of Sakharov and Bonner. Trouble started, ironically, with members of the Human Rights Committee. Chalidze hesitated, concerned that his name would appear alongside those of certain individuals whom he disliked. Solzhenitsyn announced that signing might interfere with his work on other matters for which he felt responsible; as he let it be known elsewhere, "I don't sign collective letters."[42] One of Sakharov's fellow academicians refused point-blank: "What's the matter with you? If the authorities want to declare an amnesty for political prisoners, receiving a collective letter like this will only offend them and they'll change their minds!" The physicist and future Nobel laureate Petr Kapitsa insisted that there were more important things than "a few political prisoners." The population explosion, for example, "threatens millions with death by starvation." The microbiologist Alexander Imshenetsky also refused: "Don't try to pull me into anti-Soviet ventures. I am not offended by the Soviet government—it sent me abroad thirty-six times." At one point, Sakharov and Bonner went door-to-door seeking signatures in an elite dacha colony for writers and scientists, only to learn that KGB agents were following in their footsteps with visits of their own.[43]

The two petitions, bearing just over fifty signatures each, were sent to the Supreme Soviet in October. There was no response. On the eve of the anniversary, Sakharov gave copies to foreign correspondents, after which they were broadcast by the Voices. Acknowledging his disappointment, Sakharov nonetheless maintained that "under the new circumstances, each signature carried a great deal of weight. The petitions expressed the opinions of many, and perhaps caused radio listeners to reflect."[44]

Just how many was anyone's guess. What is clear, at least in hindsight, is that the new circumstances involved more than just a precipitous drop in the number of committed allies of the dissident movement. An incipient social movement was undergoing a process of condensation, reformatting itself into a cluster of small, elite, Western-oriented NGOs. It is telling, for example, that when Vyacheslav Igrunov, proprietor of a samizdat lending library in Odesa, suggested to Chalidze that the Human Rights Committee establish a branch in that city, he was rebuffed.[45] Similarly, the Initiative Group was uninterested in gathering signatures for its appeals to the United Nations and other international bodies beyond its dozen or so members, as if the organization itself sufficed and there was no need to attract additional supporters. For Venyamin Kozharinov, whose name appeared on several Initiative Group documents, such an approach could only lead to "stagnation." The Initiative Group included a handful of members from outside Moscow (for "demonstrative" purposes), but was uninterested in using extant samizdat networks to develop branches in other cities. "We haven't taken [local groups] seriously," Kozharinov complained in a 1970 samizdat essay. "We haven't established regular, mandatory lines of communication [with them], we didn't care about maintaining their integrity as part of a working collective. In general, we just gave them a pat on the shoulder. We should see in every person a potential ally. Our personal qualities—our faith, our conviction in the rightness of our cause, our capacity to bring people over to our side—these alone will determine whether we will be a small group or have the support of the majority." Kozharinov detected in dissidents what he called "a condition of hypnosis vis-à-vis the West and its voluntary associations, its way of life—and that seriously harms us." It was a mistake, he thought, to act as if Western-style rule of law and human rights were a "panacea for all our ills." Such views were especially damaging to relations with "the masses, who understand the need for change, but who are afraid that those identified with the democratic movement are not entirely 'our people'; they have a 'Western odor.'"[46]

Neither the Initiative Group nor the Human Rights Committee was designed for communication with Soviet citizens. Indeed, dissidents' most efficient means to reach their fellow citizens was increasingly via foreign journalists and foreign radio stations. It was not long before the expectations and interests of those journalists and radio stations began to subtly inform statements issued by dissidents like Krasin, for whom connections with foreigners were more important than expanding the movement.[47]

Nowhere, in fact, did human rights activism take the form of a broad social movement.[48] Whether it was the International League for the Rights of Man in New York, or the International Institute of Human Rights in Strasbourg, or Amnesty International in London—to which we will turn in part 5—small NGOs run by highly educated, media savvy, globally oriented professionals constituted the emerging pattern in the West. In this sense, the Initiative Group and the Human Rights Committee—notwithstanding their differences—were symptoms of a subtle adaptation to international norms defined not just by ideas about human rights and the rule of law, but by a specific mode of self-organization and the repertoire of contention that went with it. The Soviet case, to be sure, exhibited a number of distinctive features. Dissident organizations emerged as an alternative to a stalled social movement. Among their leading activists there were almost no professional lawyers. Beneath the formal organizations, moreover, networks of samizdat continued to expand—the one truly mass phenomenon in the dissident world. Finally, unlike most human rights activists in the West, those in the USSR focused on their own country and were subject to repression by their government. But the salient fact remains: the dissident movement, having given up on establishing broad support in Soviet society, was reinventing itself as a series of NGOs. Once again, "the entire movement could fit on Yakir's couch."[49]

PART IV
Disturbers of the Peace

14

The Fifth Directorate

> Things were simple in Stalin's time: you walloped the intellectuals, did away with anybody who thought for himself. Now it's much harder.
>
> —KORNEI CHUKOVSKY

For Stalin's successors, the salient fact about the Cold War was that, in the places that mattered most to Moscow, it stayed cold. Despite lingering American talk of a "rollback" of Moscow's sphere of influence, by the 1960s it was becoming clear that neither NATO nor the United States was prepared to undertake a military assault against the USSR. Hitler's Operation Barbarossa, the most recent such attack, had assembled the most lethal military force in history, only to be driven back by a crushing Soviet counter-offensive. While it would take until the late 1960s for the Soviet Union to match (and then surpass) America's nuclear stockpile, even an inferior arsenal of such weapons exercised a powerful deterrent. The decision by Western powers not to deploy military force in response to Soviet suppression of protest movements in Hungary in 1956 and Czechoslovakia in 1968 reinforced the impression that its enemies regarded the Soviet empire as too strong to take by force of arms.

Armed resistance to Soviet power within the USSR, too, had all but disappeared by the early 1960s. After a decade of post–World War II guerrilla activity by tens of thousands of "forest brothers" in the annexed

Soviet republics of Lithuania, Latvia, and Estonia (with support from Western intelligence services), as well as by militant nationalists in western Ukraine, Soviet authorities could confidently regard their western borderlands as pacified. "The regime conveyed unequivocally," writes the historian Amir Weiner, "that resisting Soviet power by force was futile."[1]

But the Cold War was still a war and therefore, in Carl von Clausewitz's classic formulation, a continuation of politics by other means. American intelligence operatives called it "psychological warfare." The State Department preferred the more genteel "struggle for hearts and minds." Most of the relevant techniques had been developed during the two world wars with the goal of weakening morale on the enemy's home front. They included dropping propaganda leaflets from high-altitude balloons and other airborne devices as well as shortwave radio broadcasts to targeted populations behind enemy lines. As late as the 1960s, the KGB identified thousands of uncrewed devices per year sent from abroad into Soviet airspace, each bearing thousands of leaflets designed to discredit communist rule. Untold millions of Soviet citizens were listening to the Voice of America, Radio Liberty, the BBC, Deutsche Welle, Kol Yisrael, and other shortwave radio stations. Some of these tactics received covert funding from the CIA, as did a secret program that smuggled roughly ten million books and magazines to readers behind the Iron Curtain, a "Marshall Plan for the mind."[2]

Seen from Moscow, the American struggle for hearts and minds was self-evidently a class struggle organized by bourgeois imperialists to eat away at the socialist bloc. By corroding communist ideals, it aimed to achieve from within what could not be obtained by force from without: the dissolution of the Soviet system. The Soviet government at times spent more money jamming Western shortwave radio programs than it did on the USSR's own domestic and international broadcasting.[3] According to an internal KGB handbook, psychological warfare—or, as Soviet officials preferred to call it, "ideological sabotage"—consisted of "a complex of propaganda and provocation measures undertaken by capitalist states by means of the mass information media and by intelligence measures with the object of undermining the ideological foundations of the socialist social order, and undermining the moral and political unity of the peoples of the socialist countries by spreading all manner of lies and slander."[4] A pair of internal KGB reports in 1964 by counter-intelligence officers Filipp Bobkov and Vladimir Strunnikov described such tactics as having become the central modus operandi of Western anti-communism. Insofar as the USSR had successfully eliminated class antagonisms, the imperialists now lacked "a social base in which to cultivate subversive actions against our country" and therefore were increasingly relying on "the covert insinuation among Soviet people of an alien ideology in order to gradually and systematically evoke feelings of dissatisfaction with the socialist order." Western promotion of abstract art was part of this strategy, as were efforts to "artificially provoke discussions among young people of the problem of fathers and children" (as the "generation gap" was called in Russia).[5]

It was symptomatic of the West's ideological bankruptcy that, in their determination to chip away at the "moral and political unity" of the Soviet people, the imperialist powers gladly exploited *any* social tensions, whether between "the intelligentsia and the working class, workers and peasants, the city and the countryside" or by "playing on religious feelings or nationalist prejudices."[6] Precisely for this purpose, Bobkov would go on to claim, Western intelligence agencies planted the omnibus term "dissident" into global discourse, to unite disparate groups under a single label.[7] Disguising its true intent behind a screen of "apoliticism and the defense of 'spiritual values,'" psychological warfare reflected an "extreme degree of degradation, of alienation from all progressive ideas and political traditions."[8] According to Bobkov, the mask of apoliticism proved especially appealing to a portion of Soviet youth disoriented by the ideological zig-zags of Khrushchev's Thaw, many of them suffering from an "adolescent spirit of protest and romantic nihilism."[9] Unlike agents recruited in the classic manner by Western handlers working abroad, moreover, those who carried out "ideological sabotage," according to the KGB, gave the appearance of operating on their own. It was therefore difficult to prove that such informal agents of influence were linked to Western intelligence organizations.[10]

It fell to Yuri Andropov, former Soviet ambassador to Hungary, member of the Central Committee, and, as of May 1967, successor to Vladimir

Semichastnyi at the helm of the KGB, to organize the struggle against this new, more insidious variety of capitalist aggression. Within weeks of taking office, Andropov established a counter-intelligence unit within the KGB targeting the agents, unwitting or otherwise, of Western psychological warfare. Known as the Fifth Directorate, it would become the Kremlin's instrument of choice against what it regarded as the USSR's fifth column. Its mission was to gather intelligence on and neutralize the emerging dissident movement, among other targets.[11]

For Andropov, the idea of the Fifth Directorate had its roots in an episode predating the chain reaction, namely, the 1956 anti-communist uprising in Hungary. From his window in the Soviet embassy in Budapest, Ambassador Andropov had watched in horror as, in the space of a single week in October, a peaceful student protest metastasized into a mass armed insurgency that toppled the communist government and threatened to unhitch Hungary from the Soviet bloc. Despite the successful crushing of the uprising by Soviet troops, Andropov famously developed a "Hungarian complex"—a mortal fear of small, ostensibly peaceful groups sparking movements to undermine communist rule with direct (in the Hungarian case) or indirect encouragement by the West.[12]

Bobkov, a veteran of the renowned SMERSH (Death to Spies) counter-intelligence unit created by Stalin during World War II, was the man Andropov would eventually choose to head the Fifth Directorate, a post Bobkov held from 1969 to 1983.[13] Like Andropov, Bobkov believed that the failure of the Hungarian uprising had not altered the West's strategic goal of subverting Soviet power, but had instead inspired a shift toward subtler, more gradual, and thus more insidious tactics. Rather than attacking with conventional methods, "ideological sabotage" acted "like cancerous tumors, gradually and imperceptibly taking over the entire organism" from within.[14] Under Bobkov's command, the Fifth Directorate became the nerve center of what he called "political counter-intelligence," with subdivisions to monitor foreign tourists, journalists, and exchange students; Soviet university students and faculty; religious minorities, "sectarians," and "Zionists"; "nationalist and chauvinist" groups; "centers of ideological sabotage" abroad; and, by the early 1980s, hippies, punks, devotees of Hare Krishna, and other youth groups beyond the Communist

Youth League's shrinking control. Its full-time staff more than doubled during Bobkov's tenure, from 201 to 424 supervising officers overseeing some twenty-five thousand agents and informants across the USSR, including a young KGB lieutenant named Vladimir Putin, who worked in the Fifth Directorate's Leningrad branch from 1976 to 1979.[15] In addition to its central office in the KGB's headquarters on Lubyanka Square in Moscow, each Soviet republican and regional unit of the KGB had its own local branch of the Fifth Directorate.[16] According to the intelligence historian Christopher Andrew, "More of the KGB's elite corps of illegals (deep-cover personnel posing as foreign nationals) were used to penetrate dissident movements within the Soviet bloc than were ever deployed against the United States or other Western targets during the Cold War."[17] The journalist Ernst Henri (aka Semyon Nikolaevich Rostovsky, aka Leonid Abramovich Khentov), who had sought to distract Andrei Sakharov from the chain reaction of dissent by encouraging him to write about the role of the intelligentsia in the nuclear age, was one such "foreign national."

Combating those construed as internal ideological opponents, of course, was hardly a new activity for the Soviet security apparatus. Within the NKVD, the KGB's predecessor under Stalin, the Secret Political Division (SPO) had performed analogous work during the Great Terror. What was new in the Fifth Directorate was the targeted, evidence-based nature of its operations and their coordination with broader government policies. Whatever one might say about the Fifth Directorate's efforts to crush the dissident movement, it was hardly a rogue operation. Andropov became a candidate member of the Politburo in 1967, the same year he took over as head of the KGB, and a full member in 1973. In Leonid Brezhnev he found a boss whose highest goal was to safeguard the USSR's hard-won social stability, economic prosperity, and international prestige, and who by nature was a seeker of consensus. Brezhnev's main concern vis-à-vis dissidents was that they not disrupt détente with the West or compromise the Soviet Union's image abroad. Like many Soviet citizens, he seems to have been genuinely perplexed by the dissident phenomenon: How, given all its historic achievements, could Soviet society produce such antagonists from within its ranks? Whatever the explanation, both Brezhnev and Andropov preferred to handle troublemakers with as little noise as possible, so as not to arouse attention. Brezhnev left the details to Andropov.[18]

The KGB had plenty of experience combating monarchists, religious believers, fascists, black-marketeers, neo-Leninists, and nationalists, including those prepared to use force. What it had not previously encountered were devout constitutionalists.[19]

The Soviet dissident movement was made possible by the post-Stalinist retreat from state-sponsored terror in favor of more predictable governance. That retreat had saved the Communist Party from cannibalizing itself and helped usher in an era of unprecedented social stability. Its impact on the KGB's ability to suppress dissent, however, was decidedly mixed. On the one hand, virtually every dissident trial produced the juridical outcome desired by the state, that is, conviction of the accused. The Kremlin could point to violations of known laws, adequate defense counsel for the accused, and nearly all the visible trappings of a bona fide legal system. On the other hand, dissident trials typically failed to produce the political effect desired by the state, namely, the discrediting of the dissident movement. Defendants' refusal to plead guilty, much less confess, spoiled the intended pedagogical impact of most trials.[20] They turned the accused—in the eyes of a significant minority within the Soviet Union and the overwhelming majority of observers abroad—into heroes. "How much longer," wondered Central Committee member Anatoly Chernyaev, "will we keep shitting in our own pockets?"[21]

The defense strategy pioneered by Dina Kaminskaya in the 1967 trial of Vladimir Bukovsky—acknowledging her client's behavior as described in the indictment but insisting it did not violate Soviet laws—was becoming the new norm in dissident trials. The attorney Boris Zolotukhin was expelled from the Party for adopting this approach in his 1968 defense of Alexander Ginzburg. In a 1970 memorandum to the Central Committee, Andropov cited the "incorrect behavior" of several lawyers who in recent trials "took the path of denying that a crime had been committed." "This group of attorneys," Andropov warned, "is now taking on practically all analogous cases that arise in various regions of the country. Often they act in direct collusion with anti-social elements, briefing them about the materials of the preliminary investigation and jointly developing a strategy for defendants and witnesses during the investigation and trial."[22] The Central Committee got the message: three weeks later, Kaminskaya, Sofia Kallistratova, and two other attorneys were barred from participating in trials involving Article 190. The Moscow Procuracy, Judiciary, and Bar Association, along with the KGB, were instructed to "improve the coordination of their activities when organizing and carrying out public-political trials of great significance" and to "increase attorneys' personal responsibility when appearing in court."[23] In practice, coordination meant barring any defense attorney deemed "unreliable" from handling political cases and forbidding lawyers registered in Moscow from participating in trials in other cities—while the KGB had a free hand in deciding where any given trial should take place.[24]

For the KGB, defense attorneys' strategy of denying *corpus delicti* was one facet of a larger problem: the dissident movement's claim that its existence and its activities were consistent with Soviet law and in fact protected by the Soviet Constitution. The most elaborate argument along these lines, as we saw in chapter 13, came from Valery Chalidze's Human Rights Committee. But Andropov's predecessor, Semichastnyi, had already sensed the danger in Alexander Volpin's transparency meeting on December 5, 1965. "On a formal level," he noted, "these actions do not constitute a crime, but if we do not resolutely stop these antics, a situation may arise when it will become necessary to resort to criminal prosecution, which is hardly justified."[25] Within days of establishing the Fifth Directorate, in a speech to graduates of the KGB's training academy, Andropov warned that "enemies" were seeking "to undermine Soviet society using means and methods which at first glance do not appear to match what we understand as hostile activity." Their goal, he said, was "to stop short of violating statutes of the criminal code, to stop short of violating our Soviet laws, to act within their framework, and nonetheless to act in a hostile manner."[26] This was precisely what Andropov saw happening during the "Trial of the Four" in 1968. Supporters of the defendants, he wrote to the Central Committee, "are trying to legalize in our country the conduct of their anti-Soviet work, to achieve impunity for their hostile actions."[27] The specter of a legal opposition arose repeatedly, eventually finding its way into the KGB's classified history of itself, which served as a textbook for new officers. "Characteristic of the tactics of anti-Soviet elements," the textbook stated with remarkable frankness, is the "conduct of hostile activity at 'the brink of legality,' so that, without formally violating Soviet laws, they could commit hostile acts against the USSR under the guise of 'friendly criticism of socialism,' 'perfecting its democracy,' 'extending freedom,' and so on."[28]

Putting dissidents on trial who had not formally violated Soviet laws did not advance the KGB's image of having ascended from the dungeons of Stalinist terror into a worthy professionalism. More to the point, trials—which because of the requirements of "socialist legality" required considerable time and effort to prepare—often did greater reputational damage to the Soviet state, inside and outside the country, than the alleged anti-Soviet activity of the accused, much of which remained unknown until its authors were arrested and displayed in the defendants' dock. For the KGB, this was a source of considerable frustration. Stalin's show trials, with their perfunctory attention to matters of evidence and judicial procedure, had been a resounding pedagogical success. By the 1960s, trials of dissidents that appeared to operate within the bounds of due process were being received by a substantial portion of the intelligentsia—not to mention public opinion in the West—with skepticism or worse.

Andropov did not have the luxury of simply ignoring people like Andrei Amalrik or Natalya Gorbanevskaya. One of the peculiarities of the Soviet system, and more generally of authoritarian governments that claim comprehensive control over public life, is that whatever behaviors go unpunished are assumed to have the leadership's tacit approval. Neutrality is not considered part of the state's repertoire. Hence the concern often expressed in official Soviet circles that failure to suppress dissent could confuse, anger, or embolden ordinary Soviet people, leaving them in a state of ideological disorientation. Protest by dissidents, Andropov informed the Central Committee, "arouses in the absolute majority a feeling of outrage, but in certain circles of the intelligentsia and among young people, it sometimes inspires sentiments such as this: 'Act boldly, openly, attract foreign journalists, obtain support in the bourgeois press, and nobody will dare touch you.'"[29] It was imperative for the KGB to disprove such claims.

When it came to battling dissidents, the KGB had its own version of network theory. Instead of identifying strong ties and weak ties, it searched for leaders whose influence radiated outward, structuring groups under their alleged control. Dissidents may have imagined themselves as a conductorless orchestra, but the KGB's instinct—likely conditioned by the pervasiveness of patron-client relationships in the Soviet system—was to hunt for conductors. When the KGB pictured the dissemination of the samizdat journal *Kolokol* (Bell), for example, it deployed a hub-and-spoke model (figure 14.1). According to this model, the most efficient way to disrupt distribution was to remove the hubs—in this case, Valery Ronkin (the large circle pictured center-left) and Sergei Khakhayev (the large circle pictured center-right). For a time, the Fifth Directorate attempted to suppress all samizdat in this manner, which, given its radically decentralized means of reproduction, ended up resembling a gigantic game of whack-a-mole. "A lot of literature, leaflets, and letters are being disseminated, on the political and national rights of man, freedom of expression, of the press, democracy and other questions," complained the Kyiv-based Politburo member Petro Shelest in his diary. "But our security agencies can't find the 'root' from which all this comes. Moscow is scrambling too, and is unable to do anything to put an end to this phenomenon."[30]

Samizdat was irrepressible because there was no root "from which all this comes." There were thousands of roots. Every copy of every samizdat text had the capacity to spawn thousands more (or millions, if one counted dissemination via the Voices). No text could count as definitively suppressed so long as a single copy remained at large. In contrast to the Internet, there were no central platforms or servers from which the dissemination of texts could be blocked. Every typewriter was a virtual private network. Whereas the *Kolokol* network, as pictured by the KGB, encompassed twenty-eight people in roughly four different locations,

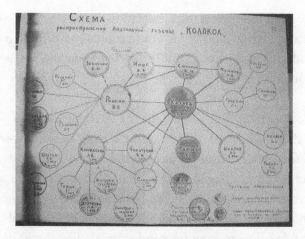

FIGURE 14.1. KGB diagram of distribution networks of the underground Leningrad newspaper the *Bell*, ca. 1962

the *Chronicle of Current Events* involved hundreds of contributors and distributors and thousands of readers in cities and towns across the USSR's eleven time zones.[31] As part of an ongoing investigation known as Case 24, the Fifth Directorate spent years hunting for the hubs, that is, the *Chronicle*'s compilers and editors. Dozens of apartments were searched and hundreds of suspects were brought in for interrogation. So extensive were the networks uncovered in Case 24 that rumors began to circulate that the KGB itself had sponsored the *Chronicle* precisely in order to flush out dissidents and their supporters.[32] The *Chronicle* reported as many of the searches and interrogations as it could, in its usual dispassionate style.[33]

Frustrated by the unintended fallout from political trials and the Sisyphean labor of attempting to disrupt the dissemination of samizdat, the Fifth Directorate turned to other methods. Available data for individual years are spotty, making it difficult to pinpoint the timing of the shift. In the Russian Republic, between 1966 and 1980, the number of convictions for anti-Soviet activity under Article 190 declined by roughly a third, while the number of convictions under Article 70 dropped by nearly two-thirds. There is no evidence that the volume of

Table 1: Convictions under Articles 70 and 190 of the Criminal Code of the RSFSR, 1966–1980

	Total number of people convicted	
Years	Article 70	Article 190
1966–70	295	384
1971–75	276	527
1976–80	112	285

Sources: Kozlov and Mironenko, Kramola, 36; Petrov, "Podrazdeleniia KGB," 166.

Notes: This trend may apply to other Soviet republics. According to Ludmilla Alexeyeva, Ukraine was the "testing ground" of new methods, which were then "extended to other republics." Alexeyeva, Soviet Dissent, 55. See also Bellezza, The Shore of Expectations, 217–312. For Lithuania, see Oleszczuk, Political Justice in the USSR (96–100), which notes that in the 1970s and 1980s the KGB was particularly reluctant to put Roman Catholic priests on trial because of the attendant negative publicity.

activity deemed "anti-Soviet" actually decreased during these years; what changed was the way the KGB responded to it. The Fifth Directorate increasingly deployed a trio of stealthier tactics designed to disarm dissidents with minimal fuss and publicity. Those tactics were "prophylactic measures," involuntary emigration, and forced confinement in psychiatric prisons and hospitals.

None of the stealth tactics was new. Some originated in the 1950s, part of the reforms that followed Stalin's death; others traced their roots back to the nineteenth century. But they all had one thing in common: in contrast to criminal trials, they allowed the KGB to operate entirely outside the bounds of the justice system. In this sense, the bravura courtroom performances by dissident defendants that helped propel the chain reaction, as well as the public protests that drew so much attention to dissident trials, proved Pyrrhic victories. Volpin's ultimate goal, after all, was to nudge the USSR toward the rule of law. In the short run, the transparency promoted by the dissident movement was moving the Soviet state not to abide by its own laws and judicial procedures, but to work around them.

The Russian word *profilaktika*, like the English "prophylaxis," derives from Greek and means "to guard against in advance." Its most common use is in the field of medicine, where it refers to techniques for preventing disease or infection. In the KGB lexicon, it signified a range of methods for preventing crimes before they happened. Formally inaugurated by Khrushchev in 1959, *profilaktika* was meant to reduce the mass incarceration that had characterized the Stalin era. By addressing minor transgressions, especially by young people, the KGB sought to nip deviant behavior in the bud, setting errant Soviet citizens on the path to productive and patriotic lives. Soviet leaders were also keen to inoculate "politically immature" individuals against contamination by Western influences, to which the USSR was increasingly exposed as it emerged from decades of isolation.[34]

The KGB's medicine took many forms. The most common was a "conversation" between a KGB officer and his wayward subject, either alone or in the presence of parents, teachers, neighbors, or work colleagues. Conversations were not interrogations; their primary purpose was not to extract information for an ongoing investigation but rather to chastise, to warn, as well as to instruct and encourage. As one official put it, "We lay out their mistaken views and point out the harm they can bring to the Soviet state.... [We demonstrate] the great concern that the Communist Party and the Soviet government show for the welfare of the Soviet person and for his moral and spiritual level."[35]

That, more or less, was the experience of the twenty-one-year-old Vera Lashkova when she was first "invited" to the Lubyanka for a conversation. While her parents were away in Smolensk, she had used their Moscow apartment to host a revolving cohort of poets and *samizdatchiki*, sometimes up to a hundred, passing around a bottle of wine and reciting someone's verses. Among her guests, it seems, was an informant who heard tell of a plan to distribute anti-Stalinist leaflets in Red Square. Lashkova soon found herself seated in the office of KGB officer Ivan Abramov as he addressed her in what she described as "a very fatherly way." "He didn't try to bend or break me," she recalled, "or to find out who said what. They already knew enough." Instead, to her amazement, Abramov—who would later succeed Bobkov as head of the Fifth Directorate—proceeded to ask about her family, and about whether anyone had been repressed under Stalin (one of her grandfathers had

FIGURE 14.2. Vera Lashkova

been). Lashkova was accustomed to thinking of herself and her friends as "us" and the KGB as "them." She was not inclined to respond to overtures from "them." But her parents had raised her to be polite: "It was difficult for me to maintain a stance of 'I don't know you, I won't talk to you.'" At one point in their conversation, Abramov turned to her and said, "Vera, you're very young, you're talented, you need to study. Think about what you've gotten yourself into." Gently putting his hand on her shoulder, he declared, "No, Vera, we will not give you up to the enemy!"[36]

She forced herself not to burst out laughing. The phrase "No, Vera, we will not give you up to the enemy!" stuck in her memory, an unintended piece of comic theater and a reminder of the chasm separating her and her friends from the KGB. The next time Lashkova appeared at the Lubyanka, it was after being arrested, along with Alexander Ginzburg, Yuri Galanskov, and Aleksei Dobrovolsky. A year later, in the

"Trial of the Four," she was convicted of helping to disseminate anti-Soviet slander via samizdat.[37]

Abramov's quasi-paternal concern for Lashkova's future was not necessarily typical of prophylactic conversations, especially when the subjects were male. More often than not, those conversations included threats of arrest, expulsion from school or university, or the loss of a job. Sometimes they involved pressure to become an informant.[38] Whatever their tone, such encounters were becoming the KGB's preferred mode of interacting with actual and potential dissidents. In the early 1960s, for every individual arrested for anti-Soviet agitation and propaganda, approximately five were subject to prophylaxis. In 1967, when the Fifth Directorate started systematically gathering country-wide data on the KGB's prophylactic work, that ratio began to skew dramatically. Over the next seven years (1967–74), for every individual arrested for anti-Soviet agitation and propaganda, ninety-six were subject to prophylaxis, producing an annual average of more than eleven thousand "conversations." Out of a total of 121,406 "conversations" conducted by the KGB during that period apropos all types of potentially criminal activity, 81,365 (two-thirds) had to do with suspected anti-Soviet agitation and propaganda.[39] As the historian Nikita Petrov notes, "prophylaxis was the basic working method" of the Fifth Directorate.[40]

We do not know what effects the tens of thousands of encounters with the KGB had, apart from the fact that the vast majority of those summoned were not subsequently arrested or prosecuted. It is impossible to distinguish the impact of "conversations" from that of harsher prophylactic methods such as termination of employment, expulsion from school, denial of university scholarships, blacklisting from publication, and various forms of harassment of spouses and children—none of which appear in the statistics assembled by the KGB. The campaign against the roughly one thousand signers of open letters in 1968 and 1969, however, suggests the powerful deterrence such punitive measures could produce.

All this punishing, it should be emphasized, took place outside any judicial framework. While the KGB was legally obliged to get permission from the Procuracy (at the union or republican level) before arresting someone, no such permission was required for prophylactic measures. Ex post facto notification sufficed.[41] The Fifth Directorate was thus empowered to decide by itself who was likely to commit a crime, who should be summoned for a "conversation," and, if necessary, whose education or career should be derailed. The Presidium of the Supreme Soviet formalized this power in 1972 via an unpublished administrative decree granting certain directorates within the KGB (including the Fifth Directorate) the authority to require subjects of prophylaxis to sign a document confirming that they had been warned about the impermissibility of their anti-Soviet behavior. In the event of a future arrest, the signed document could be used against the defendant in court.[42]

The Fifth Directorate was now effectively in the business of punishing a range of activities that it recognized were not forbidden by Soviet law. According to Vasily Mitrokhin, a KGB archivist who defected to Great Britain in 1992 with voluminous notes on internal KGB documents, those activities included

> listening to, recording, or recounting the content of foreign radio broadcasts; expressing negative views of Soviet reality and governing institutions; praising the Western way of life; assessing the Soviet way of life in a demagogic or unprincipled manner; posing provocative questions; telling political jokes; showing excessive interest in classified library and archival materials; collecting documents and statistical data or results of sociological research concerning negative phenomena in Soviet society; collecting biographical information about people who had been unjustly repressed in the past; gathering signatures for appeals, letters, and complaints; organizing gatherings in which negative aspects of literature, art, and the domestic and foreign policies of the Communist Party were discussed; contact with foreigners and the acquisition of books and articles from them; acquisition of polygraphic or photocopying devices; expressing approval of the policies of states hostile to the USSR; condemnation of communists, representatives of authority, or leaders of local organs.[43]

In neither their severity nor their extent nor their capriciousness did the extra-judicial activities of the KGB come close to the wanton violence of the NKVD under Stalin. People no longer returned from interrogations with their fingernails gone. Suspects did not disappear for unknown or invented reasons, never to be seen again. Nonetheless, across the lifespan of the Fifth Directorate (1967–89), prophylactic activity intruded on and in some cases disrupted the lives of hundreds of thousands of Soviet citizens. Overall, KGB threats and warnings seem to have produced the desired deterrent effect.

There was another, even more telling contrast between secret police practices of the Stalin and post-Stalin eras. NKVD interrogators of the 1930s and 1940s were interested not only in what a suspect had done and with whom, but why, based on what class origins and driven by what beliefs. Bolshevik zeal demanded a thorough mining of the inner lives of individual suspects. Motives could not be established without scrutinizing biographies. Depth probes were pivotal for the quintessential Bolshevik ritual of unmasking. Only after rendering their enemies transparent could the NKVD's "metaphysical police" (as the émigré novelist Vladimir Nabokov called them) consider the possibility of their redemption via conversion to the revolutionary cause.[44] "Our investigation has an educational role," an interrogator told Gleb Bonch-Osmolovsky, arrested in Leningrad in 1933 on charges of counter-revolutionary activity. "We will find out who you really are." Only by confessing in full "will you free yourself psychologically from the various impediments to the unremitting revolutionary self-criticism you need." After deciding that he had "no right to distrust the investigator," Bonch-Osmolovsky dutifully poured out his innermost doubts and hopes. He emerged from the encounter convinced that "the whole thing was an exercise in evaluating me, not an inquiry into a crime. At no point during the investigation were concrete crimes ever attributed to me."[45]

KGB officials in the 1960s and 1970s, by contrast, displayed a remarkable lack of interest in the inner lives of their dissident interlocutors.[46] For the most part born after the revolution and formed by Soviet institutions, dissidents had grown up in a country that had eliminated internal class enemies and, as of 1936, claimed to have built socialism. Under the post-Stalin dispensation, KGB interrogators no longer grilled suspects on their social origins, nor did prosecutors. Internal Soviet government memoranda also stopped mentioning class background—as if class had simply evaporated as a way of accounting for deviant or antisocial behavior. Nor were KGB interrogators in the habit of asking about political beliefs or worldviews or the content of samizdat materials confiscated in apartment searches. Stalin's interrogators had constantly demanded to know suspects' views on this or that topic, and how they arrived at such views, in the hope of unmasking counter-revolutionary or bourgeois forms of consciousness. They were known to repeat their questions over and over, for hours on end, until they got the answer they wanted.[47] Brezhnev's interrogators did not pose such questions—not because they would have violated new standards of professionalism, but because they simply weren't interested. When Semyon Gluzman, a dissident psychiatrist arrested in Kyiv in 1972, was told by his interrogator, "You still have time: recant, give us the information we want, and you won't get ten years," he was stunned. "Do you actually think," Gluzman asked, "that arresting people and investigating them are enough to change their beliefs?" "Who cares about your beliefs?" the interrogator shot back. "That's not what this case is about."[48]

Lack of curiosity about political views and inner lives increasingly characterized not just the way Bobkov's agents in the Fifth Directorate talked to dissidents, but the way the regime talked to itself about dissidents. By the 1960s, a series of administrative orders prohibited reproducing anti-Soviet statements in legal documents. At most, officials might reference them in the subjunctive, as, for example, "The person under investigation spoke as if there were no democracy in the USSR." At a time when foreign observers were busily cataloguing the political views of Soviet dissidents, deciding who qualified as a liberal, a social democrat, a monarchist, a nationalist, and so on, the Soviet government, including the Fifth Directorate, typically limited its in-house characterizations to such nondescript categories as "anti-Soviet," "politically immature," "politically harmful," or "reformist." The historian Vladimir Kozlov has interpreted this practice as reflecting a ritualized taboo, according to which dissident critiques were unrepeatable heresy.[49] The prosecutor and judge in the trial of Andrei Sinyavsky and Yuli

Daniel had indeed described the defendants' works as "blasphemy." Censorship within the state and Party bureaucracy certainly demanded the excision of vitriolic or obscene comments about specific Soviet leaders (in this regard there was nothing unusual about Soviet bureaucratic practices), of which there was no shortage among the thousands of homemade leaflets and defaced election ballots discovered by the KGB. Lack of engagement with the content of samizdat texts and the inner lives of their arrested authors and readers, however, suggests something else: the KGB's own ideological apathy.

Memoirs of former KGB officials, including Bobkov—all published after the collapse of the Soviet Union—are full of contradictions on this point. As part of the widespread veneration of Andropov, they highlight the KGB's professionalism, including its ability to see beyond the blinders of Marxism-Leninism.[50] Much of this, of course, stems from a retrospective desire not to appear as having been naive enforcers of a now discredited ideology. But it also reveals the decidedly un-Bolshevik assumption that ideological fervor was no longer a necessary component of, and perhaps even a hindrance to, professional intelligence work. In any case, the notion of transcending ideology in the name of sober professionalism offers a clue as to why the KGB demonstrated so little interest in dissidents' ideas.

At the same time, KGB memoirists insist—again, in hindsight—that ideological commitment constituted the "core" of the Soviet system.[51] Two decades after the Soviet collapse, Bobkov posed what he called "the main question, the painful question that tortures me: Why did the nation that defeated fascism and liberated not only itself but a whole series of European countries from German occupiers—why did this nation lose the Cold War?" His answer: the decline of ideological fervor or, as he rephrased it, perhaps in sync with the rekindled romance between church and state in the Putin era, the decline of "faith" in the Soviet system.[52]

Additional factors may have contributed to the KGB's relative indifference to the beliefs and biographies of Soviet dissidents (even as, following Anna Akhmatova's observation, the KGB unwittingly helped remake those biographies). In public as well as in private, Soviet officials took for granted that dissent was inspired, directed, and financed by the West as part of its psychological warfare against the socialist bloc. "Intelligence agencies in the countries of our main enemy," Bobkov wrote in a confidential 1972 memorandum, "are making efforts to consolidate anti-Soviet elements within our country and to set in motion their subversive activities. The leading role in inspiring these activities continues to be played by the 'Radio Liberty Committee,' which carries out orders from the CIA."[53] Radio Liberty—originally named Radio Liberation, until the U.S. government decided in 1959 that liberalization was a more realistic goal vis-à-vis the USSR—was indeed covertly financed and directed by the CIA, a fact that became public in 1967 thanks to an exposé in the San Francisco–based magazine *Ramparts*. But Bobkov didn't need *Ramparts* to reveal the CIA's involvement. From 1966 to 1985, the KGB had a mole inside Radio Liberty's headquarters in Munich, the ostensible defector Oleg Tumanov. Radio Liberty's mission, according to Tumanov's messages, was "to foment mistrust in the policy of the Soviet government, stimulate doubts in the Soviet state system, undermine belief in communist ideals, and enable the formation in our country of a so-called democratic movement."[54] All of the basic dissident activities, Bobkov maintained, were "methods of underground organization promoted by the enemy," from the "illegal preparation and dissemination of various kinds of hostile materials intended to ideologically manipulate [Soviet] citizens" (that is, samizdat) to the "establishment of organizational ties with foreign anti-Soviet centers" (that is, Western human rights NGOs).[55] Andropov went further: not just Western intelligence services and NGOs but journalists and diplomats were all part of a master plan to foster dissent in the USSR. In a speech titled "Communist Conviction Is the Great Strength of the Builders of the New World," published on the front page of *Pravda*, he warned that "'dissidence' has become a distinct profession that is generously paid in hard currency and other handouts, which in essence hardly differs from the way the imperialist special services pay off their agents."[56]

If dissidents were mercenaries employed by the USSR's enemies, then KGB interrogators had little reason to probe their beliefs. Neither, for that matter, did prosecutors, insofar as the post-Stalin reforms of Soviet law de-emphasized defendants' criminal intent and focused instead on their impact on a vaguely defined "public order."[57] Dissidents' beliefs, including what Semichastnyi, Andropov's predecessor as head of the KGB, called their insidious "apoliticism," mattered only in a negative sense, as spoilers of the ubiquitous claims of Soviet society's "moral and political unity." From the bottom to the top of the system, Soviet officials went to considerable lengths to promote the façade of unanimity and therefore to pathologize dissent. During their Tenth Party Congress, in 1921, the Bolsheviks formally banned factions within the Party—a ban that was never lifted. Open dissent inside the Party was henceforth understood as "objectively" serving the interests of domestic and foreign enemies.[58] Soviet elections, as we have seen, presented voters with a single candidate for each office and thus became mass rituals of acclamation rather than acts of choosing among different legitimate options. At the pinnacle of power, the Politburo developed a simple procedure to promote the appearance of unity: after discussing and voting on a given issue, members would vote a second time, only now everyone would endorse the majority position, so that a unanimous result could be recorded.

Another way to foster an imagined communist unity was to excise from the Soviet ecumene anyone who publicly dissented. Punishment by exile was a common practice of modern European empires, whether British convicts sent by the tens of thousands to colonies in America and Australia or Russian peasants and political prisoners banished by the hundreds of thousands to Siberia.[59] In the twentieth century, the Soviet Union became an outlier in this regard, retaining the practice of internal exile (usually to Siberia or Central Asia) at a time when most other states abandoned it. In the post-Stalin era, hundreds of Soviet citizens were punished this way, including Joseph Brodsky, Andrei Amalrik, Larisa Bogoraz, Pavel Litvinov, Nadezhda Yemelkina, Andrei Sakharov, Vladimir Bukovsky, and Yuri Orlov, to name just a few. Such cases were typically the result of judicial sentences, often generating negative international publicity for the Soviet government.[60]

After prophylaxis, the second most common extra-judicial punishment practiced by the KGB consisted of a specific form of exile: involuntary emigration from the Soviet Union accompanied by loss of citizenship and a de facto ban on returning. It might seem odd to use emigration as punishment, given the significant number of Soviet citizens, especially Jews, who actively sought to leave the USSR starting in the late 1960s. By 1985, over a quarter million would succeed in doing so, thanks in part to foreign incentives dangled before the Kremlin—American grain, technology, financial credits, and arms-control agreements.[61] Over one hundred known dissidents were pressured to leave the USSR during the same period, including a significant fraction of the roughly one thousand "signers" in 1968–69, but mere drops in the river of eagerly departing Soviet Jews.[62] That, in a way, was the point: to get rid of dissidents with as little notice as possible. No arrests, no trials, no protests—just a "conversation" in which the Fifth Directorate presented the subject with the options of emigration or prison. Whether or not they were Jewish, dissidents who "chose" emigration received a visa to Israel, which helped reinforce popular stereotypes that people who agitated for rights must be Jews. Inside the dissident milieu, offers of exit visas from the Fifth Directorate sometimes proved toxic, fueling speculation that certain individuals had cut a deal with the KGB in order to escape from the USSR, thereby abandoning the cause of civil and human rights. Indeed, there were increasing suspicions that certain people were taking up that cause, as the art historian Igor Golomshtok put it, "in order to accumulate political capital prior to their departure abroad."[63] In the West, meanwhile, the distinction between "dissidents" and "refuseniks" (Soviet Jews who had applied and been refused permission to emigrate) was frequently lost. That could only have pleased the KGB.

The third method of extra-judicial punishment involved the confinement of dissidents in psychiatric hospitals and prisons. One might think of this as the Fifth Directorate's version of psychological warfare, only in this version the weapons of choice were not ideas transmitted via smuggled texts and shortwave radio broadcasts, but tranquilizers and anti-psychotic drugs. The goal was no longer to change minds but to disable them.

As the cases of Volpin and others illustrated, in the late Stalin era, questionable diagnoses of mental illness occasionally served as a merciful alternative to prison, the Gulag, or conscription in the Soviet Army. To be sure, confinement in a psychiatric ward with people suffering genuine mental illness, including some convicted of violent crimes, could prove a harrowing experience, as could the abiding stigma of having been labeled "schizophrenic," "paranoid," or "delusional." Prior to the 1960s, however, such confinement rarely involved the application of mind-numbing drugs, and at least in Volpin's case, the periods of confinement were comparatively brief: a year in 1949–50 (for "anti-Soviet" poems), three weeks in 1957 (for discouraging a French visitor to the World Festival of Youth and Students in Moscow from applying for Soviet citizenship), a year in 1959–60 (for failure to inform on an acquaintance who had allegedly engaged in currency speculation), and four months in 1962–63 (following the publication in New York of *A Leaf of Spring*).[64] Prison and labor camp sentences for "anti-Soviet" activity, by contrast, typically ranged between three and seven years. There too, one's fellow inmates might include violent criminals.

The length of psychiatric internments was not set by judicial sentence. Court-mandated confinement to a psychiatric institution typically involved a closed-door trial *in absentia*, with the defendant declared guilty but not responsible for his or her actions (*nevmeniaemyi*, or "nonimputable") due to insanity. In many cases, however—including three of the four aforementioned instances involving Volpin—there was neither a trial nor a formal finding of guilt and therefore no court mandate. Under the Soviet system, medical personnel and prison administrators could initiate a psychiatric examination and, having found a suspect to be nonimputable, could confine him or her without any involvement by lawyers or judges.[65] Such administratively mandated (as opposed to court-mandated) confinement effectively placed the suspect at the whim of individual psychiatrists, who in turn were subject to influence by, and were sometimes employees of, the Fifth Directorate.[66] As usual, a Soviet *anekdot* summed up the situation:

How has socialist legality changed since Stalin's death?

They've stopped executing people before their sentences are announced. Now they put them in psychiatric hospitals before the investigation begins.[67]

Like coerced emigration, psychiatric internment gave the KGB multiple advantages over its targets. Officials could avoid the nuisance and publicity of criminal trials, confining suspects indefinitely without the right of appeal. Just as emigration inevitably reinforced the suspicion that all dissidents were Jews, psychiatric confinement nourished the widespread intuition that there was something unhinged about people who dared openly to disagree with the Communist Party. "In truth," Bukovsky conceded, "one has to be decidedly 'different' to become a dissenter in the USSR."[68]

For the period between 1960 and 1991, 674 cases of involuntary psychiatric confinement of individuals alleged to have committed various anti-Soviet acts have been documented.[69] The majority were administratively mandated; that is, they bypassed any opportunity to legally challenge the finding of non-imputability. Cases occurred in over one hundred Soviet cities and towns, the majority (60 percent) somewhere other than Moscow, Leningrad, and Kyiv. Over 90 percent of those confined were men. As the graph in figure 14.3 suggests, recourse to psychiatric internment of dissidents rose sharply in the late 1960s, following the chain reaction of arrests, trials, and public protests. In response to these developments, Andropov secured approval from the Politburo in 1970 for a significant expansion of the network of psychiatric hospitals and prisons, apparently anticipating more frequent KGB recourse to this method of punishment.[70]

In the overwhelming majority of cases, dissidents subjected to psychiatric incarceration were diagnosed as schizophrenic. Schizophrenia, literally "split mind," has been used to describe a wide range of symptoms involving the uncoupling of reliable (or "normal") associations among thought, emotion, perception, and behavior. For most of its history, the term has been dogged by extreme imprecision. Even today, there is no

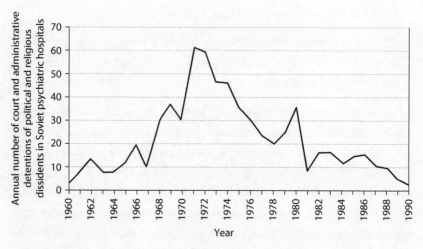

FIGURE 14.3. Involuntary detention of dissidents in Soviet psychiatric hospitals, 1960 to 1990

objective diagnostic test to confirm its presence. Soviet psychiatrists in the post-Stalin era, led by Andrei Snezhnevsky, director (1962–87) of the Institute of Psychiatry of the Soviet Academy of Medical Sciences, developed a particularly elastic understanding of the concept, including a variant called "slow progressive" or "sluggish" schizophrenia in which the characteristic symptoms of being unable to distinguish reality from unreality (delusions, hallucinations, and other psychoses) were allegedly hidden, latent, or otherwise imperceptible.[71] "The absence of symptoms of an illness cannot prove the absence of the illness itself," insisted a psychiatrist at Moscow's Serbsky Institute of Forensic Psychiatry, having examined twenty-year-old Olga Yofe following her arrest on charges of disseminating anti-Soviet materials.[72] Like the doctrine of socialist realism, the official aesthetic of Soviet literature and the arts, the concept of "slow progressive schizophrenia" was supposed to capture the reality of patients' mental lives not merely as they appear at the moment, but more deeply, in light of their future developmental trajectory.

Apart from the plight of Soviet Jews seeking to emigrate, no aspect of human rights in the USSR evoked more visceral outrage in the West than the use of involuntary psychiatric incarceration against "otherthinkers."[73] Part of that outrage was fueled by the Western vision of the

Cold War as a battle between freedom and totalitarianism, with dissidents subjected to straitjackets and mind-numbing drugs as the ultimate expression of "total" power over individual human beings. Another part drew on the specific Western backlash against psychiatry per se, beginning in the 1960s and inspired by figures ranging from the psychiatrist Thomas Szasz to the novelist Ken Kesey, the sociologist Erving Goffman, and the philosopher Michel Foucault. From different vantage points, they portrayed the psychiatric profession as a particularly insidious form of social control that used the idea of mental illness as a weapon against individuals who refused to conform to certain social norms.[74] In the Soviet Union there was no such backlash, neither within nor outside the psychiatric profession. What was distinctive about the Soviet setting, in this as in so many other arenas, was the overwhelming presence of the state. In nearly every case in which a dissident was diagnosed with mental illness, it was the state that initially identified the alleged symptoms, rather than family members or others close to the individual concerned, as is normally the case. Indeed, family members usually insisted that their incarcerated relative was mentally healthy.[75]

Soviet psychiatry mirrored Soviet jurisprudence. Just as defendants in courtroom trials bore the burden of proving their innocence, so arrested dissidents bore the burden of proving their mental health when examined by psychiatrists. For their part, Soviet psychiatrists resembled judges and lawyers in their lack of professional autonomy from the state, including the KGB. Like the handful of defense attorneys who dared to argue that their dissident clients had not broken any Soviet law, psychiatrists who questioned diagnoses of dissidents as mentally ill faced disbarment from their profession or worse, with little or no protest by colleagues.

It did not seem to matter that psychiatric explanations for dissenting behavior, which invoked dysfunctional families of origin, personality disorders, and biological illnesses of the mind, diverged markedly from that of the Fifth Directorate, which emphasized the malicious influence of the West's "psychological warfare" on "politically immature" Soviet citizens. The two approaches found common ground in diagnostic phrases that appear repeatedly in psychiatric reports on arrested nonconformists: "paranoid reformist delusional ideas," "over-estimation of his own personality," "poor adaptation to the social environment." As Andrei Snezhnevsky's *Handbook of Psychiatry* put it, "Affective tension and delusional interpretation by the patient of words and actions of individuals around him may, during his struggle for his 'rights,' become the cause of socially dangerous acts."[76] One could hardly sum up more concisely the pathologizing of dissent by the Soviet psychiatric establishment.

The Fifth Directorate had at its disposal a varied tool kit with which to manage dissidents and potential dissidents. Arrests and trials continued, but with declining frequency. Bobkov and other officials increasingly relied on extra-judicial methods of prophylaxis, forced emigration, and psychiatric confinement—the first ensnaring hundreds of thousands of Soviet citizens over the lifespan of the dissident movement, the second and third applied far more selectively against hundreds. The hybrid approach was both flexible and unpredictable—perhaps by design. Even today, with access to a significant if still limited store of internal KGB documents, it is difficult to discern a pattern or logic governing the choice of one or another method. Individual cases, however, offer glimpses of the KGB's modus operandi and of how Soviet leaders at the highest levels participated in deciding dissidents' fates.

The poet Natalya Gorbanevskaya was one of many signers of protest letters during the chain reaction who faced extra-judicial retaliation by the Fifth Directorate. In February 1968, during a brief pregnancy-related hospitalization, she was forcibly transferred from a maternity clinic to a psychiatric hospital, where she was diagnosed with "slow progressive" schizophrenia.[77] If this episode was intended to frighten her away from further activism, it had the opposite effect. After her release, Gorbanevskaya produced a caustic samizdat essay titled "Free Medical Care," recounting her experience of involuntary psychiatric confinement. In April, she organized the first issue of the *Chronicle of Current Events*, followed by more protest letters and culminating with the August 25 demonstration on Red Square.

Two weeks later Gorbanevskaya was required to undergo a new psychiatric evaluation, which now revealed "deep psychopathy: the possibility of slow progressive schizophrenia is not excluded." Placed under her mother's guardianship, she remained on probation, subject to immediate confinement in a mental institution whenever the authorities deemed it necessary. This too failed to deter her: a samizdat transcript of the trial of her fellow Red Square demonstrators appeared under Gorbanevskaya's name in 1969, and in the spring of that year she became a founding member of the Initiative Group for the Defense of Civil Rights. By December, Bobkov and Andropov had had enough: Gorbanevskaya was arrested and charged with slandering the Soviet system under Article 190 and once again subject to psychiatric examination, which now firmly concluded that she was "suffering from a chronic mental illness in the form of schizophrenia." Among the evidence cited was Gorbanevskaya's statement, regarding her protests, that she "acted thus so as not to be ashamed in the future before her children." "She does not renounce her actions," the report added, and "thinks she has done nothing illegal."[78]

Gorbanevskaya was tried *in absentia* on July 7, 1970. All but one of her friends who attempted to enter the courtroom were refused admission; Volpin and others watched through an open window, to the great annoyance of the presiding judge. Valery Chalidze, the sole friend who managed to gain entrance, produced a samizdat transcript which showed that Gorbanevskaya's lawyer, Sofia Kallistratova, had pulled no punches. The psychiatric report, Kallistratova pointed out to the judge, listed "not a single symptom of disordered mental activity," even as the indictment failed to show that Gorbanevskaya's actions had violated Soviet laws. To no avail: pronounced guilty but not responsible for her actions, Gorbanevskaya was sentenced to indefinite confinement in a mental hospital.[79]

Every day, she was forced to take haloperidol, an anti-psychotic drug used to treat schizophrenia and other disorders. Her hands soon began to tremble uncontrollably, she found it impossible to read or write coherently, and she slept poorly. When she complained, the attending psychiatrist doubled the dose. Another psychiatrist told her that the natural punishment for anti-Soviet writers was to execute them. After two years of confinement, Gorbanevskaya agreed to state, orally but not in writing, that she had committed criminal acts while suffering from mental illness and that she would not repeat them after her release. That was sufficient for the KGB, which by then was keen to put an end to the negative publicity her case and several others had elicited in the West. "If you are let out," warned the psychiatrist Daniil Lunts, a KGB employee who led the diagnostic division of the Serbsky Institute, "your friends will quickly drag you back into your previous activity even though you promise now that you will not resume it. It would be better to change your circle of friends."[80]

Released in February 1972, Gorbanevskaya did not change her circle of friends. But she kept her dissident activities out of view, declining to sign protest letters or attend demonstrations. Instead, she wrote anonymous pieces for the *Chronicle of Current Events* and distributed funds collected for the families of political prisoners. Once again she was placed on probation, subject to constant surveillance and periodic psychiatric examinations, under threat of reconfinement at any moment. Finding this arrangement intolerable, in December 1975 Gorbanevskaya accepted the KGB's proposal that she leave the Soviet Union with her two young sons.[81]

Andropov calculated that it was more "expedient," as he put it in reports to the Central Committee, to arrange for certain dissidents to emigrate rather than trying to silence them via intimidation, imprisonment, or internal exile. That calculus was influenced by the intensity of public criticism from abroad, including by communist parties in Western Europe, against the Soviet practice of punitive psychiatry. Archival sources leave no doubt that Soviet leaders were acutely sensitive to such criticism, which was often triggered by dissidents' transmission to the West of information about specific cases. When it came to Bukovsky, both transmission and criticism took an especially spectacular form: an interview filmed in Moscow by CBS journalist William Cole and broadcast on *60 Minutes*.[82] In calm, lucid responses, Bukovsky recounted the horrors of his own multiple periods of confinement in psychiatric hospitals, among other things for helping Volpin organize the first transparency meeting in December 1965.

Within weeks of the broadcast of the interview, Soviet authorities ordered Cole to leave the Soviet Union—one of roughly twenty Western correspondents expelled in the early 1970s "for activities incompatible with the status of a foreign journalist," which in many cases

FIGURE 14.4. Natalya Gorbanevskaya with fellow dissident Venyamin Yofe, shortly before her emigration in 1975

meant excessive contacts with dissidents.[83] Bukovsky, who by age thirty-five had spent a dozen years in prisons, psychiatric hospitals, and labor camps, was deported in 1976 in exchange for the general secretary of the Communist Party of Chile, Luis Corvalán, who was imprisoned without trial under the dictatorship of Augusto Pinochet. Corvalán received a warm welcome in the Kremlin, Bukovsky in the White House. In response to the Soviet press describing Bukovsky as a "hooligan," Vadim Deloné, one of the participants in the August 1968 demonstration on Red Square, circulated the following ditty:

> They exchanged a hooligan
> For Luis Corvalán.
> Where can we find a whore,
> To trade Brezhnev for?![84]

Deloné too was now an expatriate, having accepted an offer to leave the Soviet Union together with his wife, Irina Belogorodskaya in 1975.

On May 29, 1970, the well-regarded geneticist Zhores Medvedev (named after the French socialist Jean Jaurès), twin brother of Roy (named after the Indian socialist Manabendra Roy), was seized from his apartment in Obninsk, outside Moscow, and incarcerated in a psychiatric hospital. No charges were filed against him; there was no arrest and no trial. The KGB was punishing Medvedev for his publication abroad of a critical history of Soviet genetics (*The Rise and Fall of T. D. Lysenko*, 1969) and for more recent samizdat works on the legal right to privacy of correspondence and international scientific cooperation.[85] By his own account, Medvedev accepted the legitimacy of the Soviet state while disagreeing with some of its policies. He had never written or signed statements addressed to the Soviet leadership.[86] "This can happen tomorrow to any of us," Alexander Solzhenitsyn warned in an open letter titled "This Is How We Live." He then touched the third rail, likening Brezhnev's Soviet Union to Hitler's Germany: "The incarceration of free-thinkers in insane asylums is *spiritual murder*, a variation on the *gas chamber*, and even more cruel: the torments of those put to death in this way are more evil and last longer. Like the gas chambers, these crimes will *never* be forgotten, and *all* those who take part in them will be condemned in perpetuity, during their lives and after their deaths. Even in lawlessness, even in evil deeds one must remember the boundary beyond which a man becomes a cannibal!"[87] For the intelligentsia, Solzhenitsyn later wrote, punitive psychiatry was "more dangerous and more a matter of principle" than the invasion of Czechoslovakia—it was "a noose around our own throats."[88]

Thanks to letters of protest from prominent Soviet scientists and cultural figures, as well as pressure from scientists abroad, Zhores Medvedev was released after three weeks in psychiatric confinement. In January 1973, the KGB allowed him to travel to London, together with his wife and the younger of two sons, to take up a one-year research fellowship in genetics. "Don't build up any illusions," Solzhenitsyn told him on the eve of his departure. "They're not going to let you back in."[89] Several months later the Presidium of the Supreme Soviet stripped him of his Soviet citizenship.

As for Solzhenitsyn, his one-man war against the Soviet system was approaching its climax. Unpublishable at home, his novels *Cancer Ward* and *The First Circle* had appeared abroad to critical acclaim, emboldening their author to openly call for an end to censorship of literary works in the USSR. The Union of Soviet Writers responded by expelling Solzhenitsyn from its ranks. In 1970, in what felt to many like one of the Cold War's more rarefied proxy wars, the Swedish Academy awarded Solzhenitsyn the Nobel Prize in Literature. He declined to attend the ceremony in Stockholm for fear that the Kremlin would bar his return.

By now, the KGB was aware of a monumental history of the Gulag that Solzhenitsyn had been composing—and hiding—since shortly after his return to Moscow from internal exile in the mid-1950s. Determined to prevent its publication and if necessary to silence Solzhenitsyn himself, the Soviet leadership, including the KGB, began to weigh its options. A smear campaign in the Soviet press following the Nobel announcement included comparisons of Solzhenitsyn to Valery Tarsis and Anatoly Kuznetsov, writers who had been allowed to leave the USSR and then stripped of their citizenship, effectively barring their return, as well as to Bukovsky and Amalrik, well-known dissidents sentenced to psychiatric prison and labor camp—implying that these were the options Solzhenitsyn himself now faced.[90]

At the latest by March 1972, Andropov together with the Soviet procurator-general, Roman Rudenko, had drawn up plans for Solzhenitsyn to be expelled from the USSR. Citing a litany of anti-Soviet works, they highlighted a passage from Solzhenitsyn's unpublished verse play *Feast of Victors* in which a character proclaims: "The USSR! It's a dense forest after all! A dense forest! No laws! Just power—to seize people and torture them according to the constitution—or without it." Their recommendation to seize and expel Solzhenitsyn via administrative fiat, rather than putting him on trial under formal charges, only confirmed the passage they quoted. To be sure, expulsion was not without its costs. In Andropov and Rudenko's view, it would inevitably "be used by our enemies to stimulate anti-Soviet activities." But more important was the advantage that "in Soviet peoples' eyes, Solzhenitsyn would thereby cease to be an ambiguous figure, because the Soviet state's attitude toward him will have been clearly established."[91] Apparently, one could not count on a courtroom trial to produce that effect.

The dissident problem—or, as Brezhnev put it after reading Andropov and Rudenko's recommendation, "the question why these kinds of sores keep forming on the generally healthy organism of our society and why they are eating away at this healthy organism"—was taking up more and more of the Party leadership's attention. Brezhnev seems to have been genuinely perplexed, perhaps because his sense of the sources of dissent was vague and cliché-ridden. "It's a result of our shortcomings," he announced at a Politburo meeting on March 30, 1972, "of the fact that we let down our guard and in certain arenas we stopped conducting a genuine Bolshevik struggle against this phenomenon." The problem, he said, had to be understood as "a form of class struggle, on an international scale as well as inside our country, thanks to the influence exercised on a certain backward portion of people."[92]

What would a genuine Bolshevik struggle against dissidents look like? There were still some Old Bolsheviks around, people who had joined the Party before the revolution and who were intimately familiar with the old ways of dealing with troublemakers. One of them was Vyacheslav Molotov, Stalin's long-serving foreign minister. Now in his eighties and comfortably retired, Molotov expressed his views in conversations with various acolytes. Letting disaffected Jews leave the Soviet Union, for example, was fine, according to Molotov: "To hell with them, if they don't want to stay. They have their own homeland. But to allow a Russian to leave—that in my view is a sign of weakness." The bottom line, for Molotov, was that "politics decides." "It would be easy to kill Solzhenitsyn," he added. "But I think it would be better to put him in prison." Sakharov, on the other hand, "must be reckoned with. It seems, however, that he will not come to his senses, neither he nor his friends, acquaintances, those in the circle around him. To a certain extent one has to work on people, on their milieu. It's not always so easy. Even if you crush that swarm all at once," he concluded, "your troubles will not end. One has to keep fighting it."[93]

In Brezhnev's Politburo, the debate about what to do with dissidents similarly oscillated between imprisonment and various forms of exile, whether within the Soviet Union or beyond its borders. Political—including reputational—consequences, or what Andropov called "expediency," remained the decisive factor. Only rarely did issues of legality enter the conversation. In contrast to Molotov, Brezhnev's colleagues gave relatively little weight to the nationality of individual dissidents, despite their sensitivity to the threat of nationalism within the multinational Soviet state. Available transcripts of Politburo discussions about dissidents do not mention potential assassinations, though that technique evidently remained in the KGB's repertoire.

Debates over what to do with dissidents were becoming more frequent. In the early 1970s, Sakharov alone was discussed by the Politburo, the apex of power in the Soviet system, on average once per month.[94] "We need to draw very serious conclusions," intoned Viktor Grishin, head of the Communist Party's Moscow City Committee, "and take practical measures." Sakharov, Petr Yakir, and Solzhenitsyn needed to be removed from Moscow, but how and where to? "Maybe there needs to be a conversation with Sakharov, I don't know, but we have to somehow put an end to this business, because he's gathering people around him. Although the group is not large, it's harmful." Mikhail Suslov, the Politburo's longest-serving member, agreed: "Sometimes in life, a thousand people can put up a building but all it takes is one person to destroy it."[95]

At a Politburo meeting on March 30, 1972, Suslov and Mikhail Solomentsev, chairman of the Council of Ministers of the Russian Soviet Republic, blamed Khrushchev. The Party's ideological work was supposed "to inculcate the entire people with the spirit of intolerance" toward dissidents, but Khrushchev had sown confusion in this regard. "He's the one who cleared a path for Yakir, associated with him, encouraged him. He cleared a path for Solzhenitsyn too, and elevated him, that bastard. And look what came of Khrushchev's loose comment about how there are no longer any political enemies here."[96] The latter charge referred, slightly inaccurately, to Khrushchev's declaration at the Twenty-First Party Congress in 1959, following the liquidation of Stalin's personality cult, that there were no longer political *prisoners* in the Soviet Union.[97] Khrushchev's claim, added Suslov, that "there are no political opponents or political enemies" in the Soviet Union had "harmed the people." It was dangerous to pretend that the USSR's internal enemies had been vanquished. The Soviet leadership needed to call things by their name, to declare Solzhenitsyn and Yakir enemies of the people, and expel them from the USSR.[98]

At this point Brezhnev felt the need to chime in. "Actually, as comrade Andropov mentioned, we don't have a law for punishing political jabbering." How could that be, wondered Boris Ponomarev, head of the international division of the Central Committee. How could the Soviet Union, on the eve of the fiftieth anniversary of its founding in 1922, its global authority steadily advancing, and having helped liberate over seventy countries from colonialism—how could such a country "be unable to deal with some Solzhenitsyn, some Yakir, and Sakharov"? There was emphatic agreement that something needed to be done, but no consensus on what, when, or how. Finally Alexei Kosygin proposed a solution:

> KOSYGIN: Let comrade Andropov resolve the question of what to do with these people in accordance with those laws that we have. We in turn will monitor how he resolves this question. If he resolves it incorrectly, then we'll correct him.
>
> ANDROPOV: That's precisely why I'm consulting with the Politburo.
>
> BREZHNEV: Judging by the previous comments of our comrades, one thing is clear—that we are unified in our principled assessment of the question. Therefore, perhaps we should end our discussion here, unless a comrade objects. Does any comrade insist on speaking further?
>
> VOICES: We are completely "for," and we can end the discussion here.
>
> BREZHNEV: As for Yakir and Solzhenitsyn, I agree with the comrades' opinions. We must absolutely remove them from Moscow.[99]

Everyone was unified and nothing was decided, except that the KGB should handle the problem.

The Politburo revisited the Solzhenitsyn case again and again. A turning point came in September 1973, when the KGB interrogated the elderly

FIGURE 14.5. Alexander Solzhenitsyn

Elizaveta Voronyanskaya, one of Solzhenitsyn's trusted typists, and coerced her into revealing the hiding place of one of the three extant typescripts of the still unpublished *Gulag Archipelago* (a copy that, ironically, Solzhenitsyn had implored her to burn). Shortly thereafter she hanged herself in her apartment in Leningrad. Having waited six years since completing the manuscript, Solzhenitsyn decided the time had come to unleash his grand indictment of the Soviet system. A Russian edition of *Gulag* was published in Paris that December. Soon the Voices began broadcasting excerpts to Soviet listeners.[100]

At a Politburo meeting on January 7, 1974, the familiar alternatives were debated yet again: put Solzhenitsyn on trial or expel him from the USSR? Once more, the decisive factor was politics, or, as Nikolai Podgorny, chairman of the Presidium of the Supreme Soviet, put it, "Let's look at what will be more advantageous for us: a trial or expulsion.

I consider our laws humane and at the same time merciless with regard to enemies. We should try Solzhenitsyn according to our Soviet law in our Soviet court and force him to serve his sentence in the Soviet Union." That would put the USSR in a favorable light, for as he noted, "In China they publicly execute people; in Chile the fascist regime tortures people and shoots them." Andropov reiterated his preference for expelling Solzhenitsyn from the Soviet Union with or without his cooperation. "Back in the day they expelled Trotsky," he reminded his colleagues, "without asking for his agreement." Ivan Kapitonov, another Politburo member, took Podgorny's side. "I would like to judge the situation as follows: if we expel Solzhenitsyn from the USSR, what are we demonstrating—our strength or our weakness? I don't think we would be demonstrating our strength." Kosygin took a similar approach: "What will cause the least damage?" Expelling Solzhenitsyn was likely to stir up a fuss in the West. By contrast, "We have nothing to fear from applying the severity of Soviet jurisprudence to Solzhenitsyn. Look at England: they execute hundreds of people. Or Chile—the same thing. We should put Solzhenitsyn on trial and tell everyone about him. We can send him to Verkhoyansk to serve his sentence. None of the foreign correspondents will go there. It's very cold."[101]

Above all, it was the anticipated Western response that shaped the Politburo's calculus. The USSR was deeply invested in the success of the Conference on Security and Cooperation in Europe, which was approaching final agreement on what would come to be known as the Helsinki Accords, ratifying the borders set by Stalin in Eastern Europe after World War II. As for the domestic Soviet audience, according to Kapitonov, "we need to begin work on unmasking Solzhenitsyn, and then whatever steps we take will be understood by our people." Podgorny concurred: "The people will support any action we take." Brezhnev summed up what was now the majority position:

> I think the best method is to act in accordance with our Soviet laws. The investigation must be conducted openly. We should demonstrate to the people Solzhenitsyn's hostile anti-Soviet activity, his desecration of our Soviet system, his blackening of the memory of our great

leader, the founder of our party and state V. I. Lenin, his desecration of the memory of the victims of the Great Patriotic War, his justification of counter-revolutionaries, his direct violation of our laws. He must be tried on the basis of our laws. There was a time when we were not afraid to act against the counter-revolution in Czechoslovakia. We were not afraid to let [Stalin's daughter Svetlana] Alliluyeva out of the country. We survived all that. I think we'll survive this too.[102]

The stenogram indicates that "everyone" responded with "Correct. Agreed." Brezhnev's assertion that the USSR had survived the very public defection of Stalin's daughter to the United States (via India) as well as the embarrassment of having invaded its own socialist ally was of course correct. The salient point, however, is that he assumed the trial of Solzhenitsyn would produce an analogous effect.

There was no trial. Instead, the KGB orchestrated a propaganda campaign in the Soviet mass media, branding Solzhenitsyn a renegade, traitor, Judas, blasphemer, and counter-revolutionary. He was pilloried as an ally of both Mao Zedong and the Pentagon.[103] There was no "open investigation" and the government did not "try him according to our Soviet law in our Soviet court and force him to serve his sentence in the Soviet Union." On February 7, 1974, Andropov wrote a personal note to Brezhnev (rather than to the Politburo as a whole) informing him that "the Solzhenitsyn question has now moved beyond being a criminal matter and is becoming a significant problem with a certain political character." That, as previous Politburo discussions made clear, is what it had always been. Was Andropov, who had consistently favored expulsion over a trial, manipulating Brezhnev into overruling the Politburo, or did he genuinely perceive a shift in the political calculus? Citing surveillance information from the Fifth Directorate, he informed Brezhnev that while the overwhelming majority of Soviet people "correctly evaluated the criticism of Solzhenitsyn" (that is, were persuaded by the official propaganda campaign), within that majority many wanted to know "why the authorities do not take measures" against him. Certain members of the "creative intelligentsia," moreover, along with some students and workers, were expressing sympathy for Solzhenitsyn's arguments.

The crucial arena, according to Andropov—Hungarian complex intact—was no longer foreign but domestic. "Despite all our desire to avoid damage in international affairs," he argued, "it is simply not possible to put off the resolution of the Solzhenitsyn question. Further delay could trigger consequences inside the country that would be extremely undesirable for us. Unfortunately, we have no other way out, insofar as Solzhenitsyn's unpunished behavior is already costing us far more inside the country than what we will be required to pay in the international arena."[104]

It is not clear what led Andropov to identify such a critical threat lurking within Soviet society scarcely four weeks after the Politburo had confidently agreed that "the people will support any action we take." Whether or not he was performing an end-run around his fellow Politburo members, Andropov appears to have been driven by one central concern: putting Solzhenitsyn on trial would court disaster. It wasn't just that the author of *The Gulag Archipelago* was sure to be a maximally defiant defendant or that his global fame guaranteed wall-to-wall coverage in the international press. The charges against Solzhenitsyn— "slander against the Soviet order, the Communist Party and its internal and external policies," "desecration of the sacred memory of V. I. Lenin and other Party leaders"—would inevitably draw more attention to his historical indictment of communism. Moreover, the only actual law Solzhenitsyn was directly accused of violating (in the Soviet procurator's preliminary list of charges), was the Universal Copyright Convention, ostensibly for the manner in which his works were published abroad.[105]

On February 11, Solzhenitsyn received a summons to appear at the procurator's office. Rather than signing to acknowledge receipt, he attached a typed statement:

> Under the conditions of unrelieved, pervasive lawlessness that has reigned for many years in our country . . . , I refuse to acknowledge the legality of your summons and will not appear for interrogation by any agency of the state. Before requiring citizens to answer to the law, learn to fulfill it yourselves. Liberate the innocent prisoners. Pun-

ish those guilty of mass exterminations and false denunciations. Punish the administrators and the special units that carried out genocide (the deportation of *entire peoples*). Today, deprive local and departmental satraps of their unlimited power over citizens, their arbitrary use of courts and psychiatrists. Satisfy the *millions* of lawful but suppressed complaints.[106]

On the following day, eight officers showed up and took Solzhenitsyn to Lefortovo Prison, where officials informed him that he was being charged with treason under Article 64 of the Criminal Code—a charge that had not appeared in the procurator's brief four weeks earlier. Was the KGB bluffing? Solzhenitsyn never found out. Twenty-four hours later, he was placed on an Aeroflot flight bound for West Germany.[107]

Whether the expulsion of Solzhenitsyn was a symptom of weakness (Soviet leaders feared the power of a lone writer, even one whose works were barred from publication in the USSR) or of strength (Soviet leaders could seize and immediately deport anyone, whenever they pleased), the fact that he was never put on trial betrayed a lack of confidence in the Soviet judicial system. Mikhail Malyarov, the first deputy procurator-general, had once asked Sakharov, "Who gave you the right to doubt our system of justice?"[108] Evidently, the KGB's doubts were even deeper than Sakharov's.

We should not be surprised, of course, that political factors shaped the Kremlin's approach to the phenomenon of dissent, including its legal dimensions. The Soviet government was hardly unique in this respect. What is noteworthy is the ease with which legal considerations were brushed aside as well as the absence of pushback by Politburo members who had favored putting Solzhenitsyn on trial, when faced with Andropov's sudden reversal of that plan. The decision not to bring Solzhenitsyn to justice was emblematic of a larger retreat from "socialist legality" as a tool of governance in favor of a repertoire of informal, extra-judicial techniques that sought either to preempt dissident activity (via prophylaxis) or to extrude dissidents from Soviet society (via forced emigration or the use of psychiatric confinement). Couched in the language of progressive social and medical science, prophylaxis and punitive psychiatry allegedly helped their subjects by stopping criminal activity before it happened or diagnosing and treating incipient mental illness. In reality, they opened the door for the KGB to operate almost entirely outside any legal constraints.

By creating the Fifth Directorate, Andropov signaled an awareness of a qualitatively new challenge to the USSR, one that rejected armed force and underground conspiracy in favor of deploying Soviet and international law. The sheer scope of the response to the dissident movement—the thousands of phones tapped, apartments searched, suspects tailed, prophylactic conversations held, prisoners interrogated—indicates the heightened level of concern. The goal was now to remove dissidents from Soviet society rather than attempting to transform them into upstanding "Soviet people." That, in fact, was the puzzle: by anyone's definition, they already *were* Soviet people. Brezhnev was by no means alone in his struggle to understand where dissidents came from. As one of the letters from ordinary citizens published by the Soviet mass media put it, responding to the official campaign against Sakharov and other dissenters, "We cannot understand how someone who has been raised under Soviet rule, whose education and professional status were provided for by the toil of common laborers, can so unscrupulously libel our way of life."[109] For the post-Stalinist security services, giving up mass terror as an instrument of social control seems to have gone hand in hand with diminishing interest in finding answers to such questions, a loss of desire to understand how and when certain Soviet citizens had gone wrong. It is as if the much-discussed apathy of postwar Soviet youth found a counterpart in the KGB itself, in the form of declining ideological élan—couched, to be sure, in the language of professionalism. Extra-judicial methods were regarded by their practitioners as "sophisticated" and "ingenious," if only by contrast with Stalin's indiscriminate terrorism.[110] Caught up in the global Cold War, the KGB fell back on a stock answer to the question of how to account for Soviet dissidents: knowingly or not, they were agents of hostile foreign powers, tools and instruments rather than home-grown, self-propelled entities.

That assumption helps account for a second noteworthy aspect of the Soviet government's handling of the dissident movement, namely, the extent to which it was tethered to the USSR's relations with the West. Internal Politburo discussions make clear that Brezhnev and other Soviet leaders were acutely sensitive to Western criticism of the USSR's internal affairs, including violations of international human rights norms, about which the dissident movement seemed to generate an endless supply of damaging evidence. The Fifth Directorate's shift toward stealthier, extra-judicial tactics against dissidents was driven to a considerable degree by the negative publicity that political trials stirred up in Western media. So were decisions about how to manage individual cases: there was less need to stand on ceremony with dissidents unknown to the outside world. To be sure, Soviet leaders often dismissed Western criticism as "noise," fake indignation designed to serve the purposes of psychological warfare. The frequency with which they cited such criticism, however, and the effort they expended in combating it suggest that it found a significant place in their political calculus. The rise of East-West détente only exacerbated these issues insofar as greater openness and interaction raised Western expectations that the Soviet regime would relax its rule at home as well. The Kremlin, by contrast, harbored no such expectations.

15

Fallen Idols

"You see how Yakir and Krasin are testifying?" [KGB lieutenant-colonel] Tarasov said to me. "They're thinking of themselves. But you're afraid to say a single word."

"I'm thinking of posterity," I replied, unaware that Krasin was calling on dissidents to denounce each other to the KGB for the sake of posterity. "I'm thinking that in the future, people will read all these interrogation transcripts and will judge us."

"But what people! What people!" demanded Tarasov, starting to get agitated. His entire body language was trying to prove that those future readers would be tried-and-true comrades just like him. Those bastards really do believe in their future.

—ANDREI AMALRIK

For most people today, whether in Russia or elsewhere, the names Petr Yakir and Viktor Krasin have been eclipsed by those of Andrei Sakharov and Alexander Solzhenitsyn. That displacement happened not because the renowned physicist and writer were vying for leadership of the dissident movement. For different reasons, neither one was interested in that job. Rather, Yakir and Krasin's leading role in the movement's history, including the founding of the first dissident organization, sank into oblivion because of their spectacular downfall, meticulously stage-managed by the KGB in what was undoubtedly its most, and perhaps only, successful dissident trial.

FIGURE 15.1. Petr Yakir and Viktor Krasin on the balcony of Yakir's apartment in Moscow

Yakir's personality and biography, according to Krasin, "drew people to him like a magnet." By the late 1960s, he had come to be known among certain dissidents as "the father of Russian democracy" and the "patriarch of the movement."[1] Young people in particular were drawn to his combination of revolutionary pedigree, Gulag survivor mystique, and Bohemian lifestyle. He was also the only dissident who could pick up the phone and call the involuntarily retired Nikita Khrushchev to talk about current events. According to the *Times* of London correspondent David Bonavia, Yakir "was constantly trying to recruit new people," and in contrast to strict rule-of-law advocates such as Alexander Volpin or Valery Chalidze, who often came across as punctilious, Yakir was "earthier, he had more immediate rapport." Yakir and Krasin seemed a perfect match, the one's charisma complemented by the other's formidable organizational drive. They saw each other or spoke by phone nearly every day.[2]

Nobody approached foreign journalists more boldly than Yakir and Krasin. They had "a frightful habit of telephoning suddenly to demand late-night meetings," recalled Bonavia.[3] Yakir considered it undignified to rendezvous surreptitiously with Western correspondents on subway platforms or street corners. "If we're going to do them the honor of meeting with them, they should receive us in their homes as guests," he told Krasin, despite the inevitable presence there of KGB listening devices.[4] Krasin, drawing on his enormous samizdat collection, assiduously supplied correspondents (or *kory*, as they were known among dissidents) with the texts he thought they should have. Yakir went even further, agreeing to be interviewed in June 1970 by CBS journalist William Cole, who introduced him to millions of American viewers as "the non-titular head of the democratic movement."[5] Bonavia wrote that Yakir "is regarded both by the KGB and by many Soviet dissidents as the *vozhd* (Great Leader) of what has come to be known as the Democratic Movement, or, more simply, the Movement."[6]

Like the KGB but with entirely different sympathies, Western media contributed to the fashioning of Soviet dissident reputations inside as well as outside the USSR. But Cole and Bonavia were not merely indulging in journalistic hyperbole. Yakir's apartment had become the preeminent gathering spot for dissidents, particularly after the arrest of Pavel Litvinov in August 1968 and Petr Grigorenko less than a year later eliminated the leading alternatives. "One quarrelled with him, laughed with him, drank with him, got angry with him, or admired and marvelled at him," recalled Bonavia. "He seemed to want drama around himself, and he certainly got it."[7] To be sure, many rights-defenders continued to oppose having leaders at all, but for others—and certainly for Western journalists and the KGB—it seemed only natural that certain individuals would exercise authority over the movement.

In his samizdat reply to Andrei Amalrik's gloomy prognosis of the dissident movement's (and the USSR's) future, Yakir insisted that brighter days awaited the movement, thanks to "an irreversible process of self-liberation."[8] Inside Yakir himself, however, that process turned out to be eminently reversible. Behind the high-spirited, fearless persona, according to Amalrik, Yakir was "lacking in depth, easily provoked, prone to going sour at the first setback. His irascibility required regular fixes in the form of prodigious quantities of vodka." The phrase "Petr is on a binge" (*Pyotr pyot*) made the rounds in the dissident movement. The man whom Amalrik's wife referred to as "Bacchus" was, at least according to Krasin, "not an alcoholic, but what they call a 'drunkard.'"[9]

FIGURE 15.2. Petr Yakir

The KGB arrested Yakir on June 12, 1972, on charges of anti-Soviet agitation and propaganda.[10] Sudden lack of access to alcohol inevitably influenced his mental and emotional state. So did KGB threats to arrest Irina Yakir, his only child, born while her father was in Stalin's Gulag and her mother in Siberian exile, now twenty-four and pregnant. Such threats evidently triggered Yakir's traumatic memories of his own arrest as the offspring of an "enemy of the people" and the nearly two decades of violence and deprivation in the camps that followed. Because Yakir's memoir, *A Childhood in Prison*—one of the earliest autobiographical texts by a Soviet dissident—was already on its way to publication in the West at the time of his arrest, we have no firsthand account from Yakir himself of what transpired behind the walls of the KGB's Lefortovo Prison.[11] Shortly before his arrest, however, he made a prophetic tape recording and gave it to Bonavia. "If they beat me," Yakir said, "I will say anything. I know that from my former experience in the camps. But you will know it will not be the real me speaking."[12]

The KGB did not beat Yakir. But memories of the camps, anxiety over his daughter's safety, and alcohol withdrawal did. The KGB officers in charge of handling him, senior investigators Pavel Alexandrovsky and Gennady Kislykh, patiently plied their trade. Although Yakir initially insisted that his dissident activities had not violated Soviet law, "gradually, from interrogation to interrogation," they wrote in their report on the case, "he began to concede certain facts, then confessed to crimes that he himself had committed, and finally revealed crimes committed by his accomplices."[13] Within weeks Yakir broke down completely, seemingly retreating into a state of learned helplessness. "We are the ones upholding your father's cause," the KGB drilled into Yakir—"not you."[14] He was ready to give his interrogators virtually any information they wanted in exchange for promises to spare Irina and to give Yakir himself a reasonable sentence. One can debate who the "real" Yakir was—the pathological optimist or the traumatized survivor—but it no longer mattered: he was now prepared to say and reveal anything.

Krasin, the craftier of the two, took longer to break. For two months following his arrest in September 1972, he refused to answer questions during daily interrogations, which often lasted twelve hours. "I do not consider myself guilty and I never will," he told Alexandrovsky. "I will not give any kind of evidence. You won't get a single word out of me."[15] To wear down his resistance, Alexandrovsky and Kislykh planted a stool-pigeon in his cell, shared Yakir's revelations, and exploited Krasin's concern about his young new wife, Nadezhda Yemelkina, who had been arrested for unfurling a banner in Pushkin Square reading "Freedom for political prisoners in the USSR!"[16] Krasin convinced himself that the KGB needed his cooperation for a planned show trial and therefore that he could successfully negotiate with his interrogators for favorable terms. "Outwardly I was not yet broken," he recalled, "but inside I was, though I would never have admitted it at the time. I was prepared to be broken, in fact I readied myself for it. I was just waiting for the right reason."[17] He began by giving what he considered harmless information about the location of this or that cache of samizdat, incriminating no one but himself. Pressured to name his accomplices, he decided to mention someone who wasn't active in the movement and therefore—so he reasoned—was not at great risk. "I gave him up, in cowardly and cruel fashion," Krasin later confessed to Yemelkina, "as if he were a pawn in a chess game, 'in order not to sacrifice others.' The others I gave up later."[18]

Alexandrovsky took every opportunity to exploit his subjects' vanity. He constantly referred to Krasin and Yakir as "leaders of the movement." People instinctively followed them, Krasin recalled him saying, and therefore they needed to recognize the enormous responsibility they bore for the future of their fellow activists: "We alone could stop them and save them—and this would be our justification."[19] It wasn't difficult to plant notions of power and influence in Krasin's mind; in certain respects they were already there. Alexandrovsky also skillfully leveraged Krasin's doubts vis-à-vis Volpin's rule-of-law strategy for reforming the Soviet state:

> Stop hiding behind the Soviet constitution and the UN declaration of rights. Unlike those who really believed they were fighting to democratize the Soviet public order, to observe the Soviet constitution and so forth (mistakenly believed, that is—they were deluding themselves) you had no such illusions. You were just taking cover behind those slogans, when in fact you were fighting against the Soviet order. . . . Who used to say that the Kremlin needed to be torn down and flattened with bulldozers? You. Who used to say that the October Revolution was a catastrophe for Russia and that the Bolsheviks had ruined the peoples of Russia, turning them into obedient herds? You. That in the USSR every kind of free thought had been strangled for the sake of communist ideology, and that the Communist Party had erected a dictatorship and a system of violence against human beings without precedent in history? That terror was the essence of Soviet power? You used to say all that.[20]

It was true: Krasin had been an ardent anti-communist ever since his first stint in the Gulag during Stalin's final years. Even his 1966 dissertation in economics subtly reflected his disgust with the Soviet system, compiling a wealth of statistical data to argue that, notwithstanding its cyclical booms and busts, capitalism had demonstrated robust long-term growth and that the Great Depression of 1929–39 was "a unique event that was never repeated." His adviser, troubled by these findings, demanded that Krasin at least mention Marx and Lenin. He refused: "I'm performing a purely statistical analysis—what do Marx and Lenin have to do with it?" He quit graduate school before defending his dissertation.[21]

It was as if Alexandrovsky had lifted a weight from Krasin's shoulders. The dissident movement's deliberate avoidance of "politics" in favor of legalism had never sat well with Krasin—neither in theory nor, as the founding of the Initiative Group for the Defense of Civil Rights showed, in practice. Finally his own politics were out in the open. He could admit not just his uncompromising hostility toward the Soviet state but his skepticism about the dissident movement's naive tactics and goals. The Fifth Directorate's strategy vis-à-vis Yakir and Krasin was now diabolically simple: let them do the KGB's work. Rather than merely go after the hub, use the hub to dismantle the entire wheel. Who better to persuade dissidents to cease and desist than their erstwhile leaders?

Using the time-tested technique of *ochnaya stavka*, or mutual confrontation of witnesses, Alexandrovsky arranged for Yakir and Krasin to meet in person with friends and allies. Before each encounter at Lefortovo, he would supply background information and tips on how to be "helpful"—ostensibly to the dissident who had been brought in for questioning, but in fact to the KGB. For example, Irina Belogorodskaya, a chemical engineer and signer of countless protest letters, initially refused to give information, scorning her interrogators as "Stalinist battering-rams." The KGB wanted her to talk, including about her cousin, Larisa Bogoraz. As she sat down with Alexandrovsky and Krasin inside the prison, Krasin presented his "new views": the dissident movement was in the process of being crushed, resistance was futile and could only produce more arrests and longer jail sentences. Having grasped that fact, Yakir and Krasin had decided to compromise with the KGB, and Belogorodskaya should do the same. The movement was a doomed fortress under siege, and the responsible thing at this point was to safely evacuate as many people as possible and then surrender. "Ira [diminutive for Irina] looked at me with surprise and dismay," Krasin recalled. "She hadn't expected such advice from me. 'Why? Why?' she asked, over and over, 'I don't understand—what changed?'"[22]

Yakir and Krasin delivered a dozen chess sets' worth of pawns to the KGB, some two hundred individuals who were brought in for questioning and witness confrontations. Among them was nearly every dissident mentioned in this book.[23] The Fifth Directorate compiled 151 volumes of interrogation transcripts and related documentation, hived off from Case 24 into a separate dossier for Yakir and Krasin known as Case 63.[24] Many of those who were interrogated or arrested (or both) refused to answer questions.[25] Others, including Belogorodskaya, eventually named names themselves. News from participants in various encounters at Lefortovo quickly made the rounds among Moscow intellectuals and eventually appeared in the *Chronicle of Current Events*, as the following announcements illustrate:

> Over the course of the spring, at interrogations of Irina YAKIR, Ludmilla ALEXEYEVA, and others, evidence given by I. BELOGORODSKAYA is starting to appear. It has become known that this evidence includes mention of the following names in a substantially "criminal" context: Irina YAKIR, Gabriel SUPERFIN, Yuri SHIKHANOVICH, Ivan RUDAKOV, Natalya KRAVCHENKO, Ludmilla ALEXEYEVA. There was a witness confrontation between ALEXEYEVA and KRAVCHENKO on the one hand, and BELOGORODSKAYA on the other.
>
> The apartment of Gyuzel MAKUDINOVA, wife of Andrei AMALRIK (convicted in 1970), was searched. During the search, personal correspondence and money were removed. There is reason to believe that the search was connected to testimony against her by YAKIR and KRASIN. On February 12, MAKUDINOVA was interrogated by the KGB following evidence given by KRASIN.
>
> Anatoly YAKOBSON was interrogated about the *Chronicle* and documents of the Initiative Group. He confirmed his participation in letters of the Initiative Group. He did not respond to other questions. Handwriting samples were taken for graphological analysis.

During his interrogation of Pavel LITVINOV, investigator Alexandrovsky reproached him for advising acquaintances not to testify, claiming that LITVINOV exerts a bad influence on them and puts them at risk, that he was behaving cruelly and "is on the side of Truth, but not of Christ." Litvinov did not provide testimony.

Tatyana VELIKANOVA, Tatyana KHODOROVICH, and Sergei KOVALEV refused to participate in the investigation, and specifically in the graphological analysis. As grounds for refusal they mentioned the systematic procedural violations associated with trials based on Articles 190 and 70 as well as the resulting unjust sentences. Not a single known trial [based on Article 190] had demonstrated that false, let alone deliberately false information was disseminated via samizdat. Not a single known trial involving Article 70 had demonstrated intent to undermine state power. While interrogating S. KOVALEV, investigator Alexandrovsky issued indirect threats against members of the Initiative Group.[26]

For many dissidents, such news was utterly demoralizing.[27] The movement seemed to be cannibalizing itself in front of their eyes, in a macabre play staged by Alexandrovsky and the Fifth Directorate. For discerning readers, however, the news also conveyed valuable information about what facts had already been revealed, who had already been compromised, and what techniques worked during interrogation. All this was useful knowledge for those facing possible interrogation themselves. It also constituted an evolving public report card on the moral fiber of individual detainees.

Alexandrovsky wanted Yakir and Krasin themselves to directly address the wider samizdat public. "Your new views deserve to be known out there," he urged. "You have nothing to be ashamed of. Your views are correct and were hard-won by means of a difficult struggle with yourself. If you save even one or two people from arrest, that will be a reward in itself."[28] Now samizdat would be deployed in reverse, as Yakir and Krasin dutifully circulated open letters (hand-typed) calling on their former comrades to repent. Yakir addressed one letter to Sakharov. Unlike most messages smuggled out of prison, this one didn't need to be swallowed (inside a plastic pellet) by someone's spouse during a conjugal visit and then excreted a day or two later. A KGB officer hand-delivered it to Sakharov's apartment.[29] While expressing deep respect for the eminent physicist, Yakir called on him to stop making public statements, since they were being used by bourgeois propagandists to harm the Soviet Union. As for Krasin, his open letter was addressed to "my friends at liberty" and similarly requested that they cease their harmful activity in recognition of the fact that the movement had collapsed (or, as Amalrik sarcastically paraphrased Krasin's letter, in recognition of the fact that Yakir and Krasin had been arrested). Dissidents needed to "overcome the psychological barriers" that kept them from cooperating with the KGB and instead testify not only about their own activities but about those of other dissidents. "The only thing one could do now," as Amalrik put it, "was to record the Movement's history for posterity. Since the KGB had nobly taken on the role of historian, everyone should honestly tell the KGB everything."[30]

At their trial in August 1973, Krasin and Yakir, having been coached by Alexandrovsky and Kislykh, described at length their supposed violations of Article 70 (anti-Soviet agitation and propaganda). During early interrogations, both had been threatened with prosecution under Article 64 (treason), for which the maximum penalty was death. Krasin, according to his wife, ought to have realized that the KGB was bluffing.[31] Whatever the charges, Alexandrovsky promised relatively light punishment in return for their cooperation, which they gave in spades. The defense attorneys requested leniency for their clients based on their previous "unlawful repression during the period of the personality cult." "I don't want to die behind barbed wire," Yakir pleaded in his final statement to the court.[32] As the two defendants awaited sentencing, a prison guard who was accompanying them smirked, "How about three years in the camps and three years exile? Would that be OK?" Back in their cell, Yakir began to scream hysterically: "We gave them everything, those bitches, and they're giving us six years!"[33]

The next day, the judge announced precisely the sentences mentioned by the guard. Within an hour, the same guard appeared at their cell and ordered Krasin to follow him. Krasin was handed over to the director of Lefortovo Prison, who led him to an office where a tall, heavy-set man introduced himself and extended his hand: "I am KGB chairman Andropov." Krasin shook his hand. "I recognize you from the portraits," he replied. Yuri Andropov had gotten wind that the two defendants were upset about their lengthy sentences and wanted to assure them that all would be well. "Submit an appeal," he advised, "and they'll reduce the sentences to match the time you've already served in detention. All that's left will be the exile. We're not planning to send you very far away. You can pick a city near Moscow. You'll spend about eight months there, during which you can apply for a pardon, and then you'll return to Moscow. We couldn't just let you walk out of the courtroom. As I'm sure you'll agree, you and Yakir did a lot of damage. Besides, we covered your trial extensively in the press. But we won't publicize the appeals."[34]

Andropov assured Krasin that there was no reason to fear a return to Stalinism. Soviet leaders remembered quite clearly how things were under Stalin and had no intention of reviving the old ways. He was well aware that both Yakir and Krasin had suffered unjustly during the Stalin era and that their fathers had perished. "Incidentally," he added, "after the war I too expected to be arrested at any time."[35] Then came the ask: Would Krasin and Yakir be willing to take part in a press conference for foreign journalists? Western media had been spreading so many lies about their case. It was important to set the record straight and show the West that the defendants were not pressured into pleading guilty and had freely repudiated their former actions. None of the future reductions of their sentences, Andropov hastened to add, was contingent upon agreeing to hold the press conference.

For several days, Alexandrovsky intensively coached the two convicts, practicing questions and answers, composing a crib sheet to calm their nerves. The KGB provided Krasin with a foreign-made suit; Alexandrovsky personally lent him one of his ties (Yakir, true to his Bohemian image, declined to wear one). On September 5, over two hundred reporters from foreign and Soviet newspapers, radio and television companies, and news agencies assembled inside the House of Journalists in the heart of Moscow. Backstage, a prison doctor performed a last-minute check of the convicted men's vital signs; it was important that they appear calm and in good health. During the hour-long press conference, questions were submitted in writing and then selected by Mikhail Malyarov, the first deputy procurator-general of the Soviet Union. Most of them were familiar from Alexandrovsky's practice sessions.[36] When asked to characterize the "democratic movement," Krasin insisted that that term was invented in the West:

> We knew that no such movement existed in our country. On some occasions I used this term for propaganda purposes. In actual fact the "movement" was nothing more than a small group of individuals who composed and circulated letters and statements of protest against the trials of persons who were charged with anti-Soviet activities. . . . By conveying these documents abroad we misled the global public and facilitated the establishment of an attitude hostile to the Soviet Union. We represented no one but ourselves: a small group, separated from the Soviet public and coming out against its interests, a group whose activities had been inordinately played up by reactionary propaganda for its own purposes. This is precisely why a situation developed where Western hostile circles lauded us to the skies, but the Soviet people looked upon our activities with anger and indignation.[37]

Yakir, for his part, declared that "much of what we had been doing were illegal acts of a concealed nature" and that "we knew that our acts were a violation of the country's laws." Claims that healthy individuals were being sent to mental hospitals as punishment, he added, were "deliberate slander."[38]

Andropov could only have been delighted with these spectacular confessions.[39] The two dissident "leaders" admitted to the world that they and their accomplices had knowingly broken the law, deliberately slandered the USSR, and had done so in cahoots with shadowy anti-Soviet organizations in the West. Stealing a page from the dissident playbook, the KGB had arranged its own press conference for foreign journalists—not in some cramped apartment, but openly, in the august auditorium of the House of Journalists, with television cameras rolling. Not even Stalin's show trials had achieved this level of transparency. As

FIGURE 15.3. Press conference with Viktor Krasin (left), an interrogator's tie around his neck, and Petr Yakir (right), Moscow, September 5, 1973

if in response to the unprecedented video interviews with Yakir and other dissidents smuggled out by Cole of CBS News three years earlier, the KGB was letting the West know that it too could press its case via television—including with the now repentant Yakir.

Within weeks of the televised confessions, Alexandrovsky and Kislykh contributed an article to the KGB's classified in-house journal, describing their masterful use of "logic and psychology" to break Yakir and Krasin.[40] The head of the department in the Soviet Procuracy responsible for overseeing investigations by the KGB agreed, praising their "brilliant work." The entire case "deserves to be used as a model for teaching all investigators in the state security services how to conduct an inquiry into similar types of criminal cases."[41] Not only had the press conference compromised the dissident movement's image at home and abroad, but behind the scenes, through dozens of witness confrontations with Yakir and Krasin, the KGB managed to poison relations among dissidents, sowing suspicion, doubt, and mistrust. The personal bonds of friendship that had made dissidents' civic engagement possible—and that to a great extent remained the glue that held the movement together—were now tested as never before.

Recalling the aftermath of the investigation, Ludmilla Alexeyeva covered her face with her hands and groaned, "It was horrible, horrible!"[42] As in 1968, numerous participants and observers assumed the movement was now dead. Shortly after Krasin and Yakir's press conference, the poet Ilya Gabai, whose witness confrontation with Krasin had focused on the discussions that led to the founding of the Initiative Group, leapt to his death from the balcony of his family's eleventh-floor apartment.[43] "Even people who were not close to them [Krasin and Yakir]," wrote Amalrik, exiled in Siberia at the time, "told me later that for half a year they were simply paralyzed."[44] The Fifth Directorate tracked the mounting despair in the dissident milieu by, among other things, using an undercover agent to pose as a courier between Vladimir Dremlyuga, one of the Red Square protesters in August 1968 currently serving a sentence in a corrective labor camp, and dissidents in Moscow. Dremlyuga dispatched a letter to his friends in which he denounced his "former idols" as traitors to the cause. "If even such authoritative figures as Yakir and Krasin agree to testify and recant," he lamented, "then that signifies 'total defeat, the collapse of illusions.'" Falling into a deep depression, Dremlyuga was persuaded by the KGB to publish a statement renouncing his former activities in exchange for early release.[45] For years, the KGB would urge dissidents to "follow Yakir's example" and cooperate with this or that investigation.[46]

The clearest sign of paralysis in the dissident movement was the sudden silence of the *Chronicle of Current Events*. Since its debut in April 1968, twenty-seven issues had appeared, roughly every other month, ranging in size from a dozen to a hundred sheets of onion-skin paper. The most recent issue (number 27) had been released in October 1972. The KGB regarded Yakir and Krasin as the *Chronicle*'s instigators, and like Lenin's underground newspaper *Iskra* (Spark), founded in 1900, the *Chronicle* was meant to be—in the Fifth Directorate's view—not only a propagandizer and agitator but an organizer of its readers. The *Chronicle* had enabled its two founders "to bring under their influence ... several hundred unstable people in different cities of the Soviet Union."[47] "After the arrest of Krasin and Yakir," Deputy Procurator-General Malyarov proudly announced at their confessional press conference, "the printing

FIGURE 15.4. Vladimir Dremlyuga

of the *Chronicle* ceased."[48] Chronologically that was true, but one event did not cause the other. Yakir and Krasin had contributed to and disseminated the *Chronicle* but were never in charge of editing it. Once again the authorities were operating with a hub-and-spokes model of centralized, top-down leadership. At one of her interrogations at Lefortovo, Alexeyeva was told by a KGB officer: "Everything would become much simpler if someone would call and give us the address where the *Chronicle* is produced. The caller need not identify himself. Just dial me directly, give the address, and hang up."[49] Alexeyeva declined the offer. She also declined to reveal that the *Chronicle* had not one address but many and that the job of editor had already migrated—informally, as usual—among half a dozen individuals, assisted by roughly fifty contributing editors (including Alexeyeva) and an unknown number of typists (also including Alexeyeva).[50]

The real reason for the apparent death of the *Chronicle* was that the KGB had issued a threat shortly after arresting Yakir and Krasin, conveyed by Yakir to his daughter Irina at one of their witness confrontations in November 1972. Every new issue of the *Chronicle*, the KGB let it be known, would lead to new arrests, in addition to adding years to Yakir's and Krasin's prison sentences. Those on the arrest list, moreover, need not be connected to the *Chronicle*; any dissident was fair game.⁵¹ Alexandrovsky and Kislykh may have considered this part of their strategy of "logic and psychology," but dissidents understood it for what it was: blackmail.

The activists around the *Chronicle* now faced a new crisis. One group insisted that the KGB's threat amounted to a policy of hostage-taking and that negotiating with terrorists was out of the question. The biologist Sergei Kovalev was emphatic on this point. Hostage-taking "was the foundation of the Soviet system of social control," he argued, as the regime deployed threats against individuals in order to induce conformity among those around them (family members, colleagues, fellow villagers, and others). "We had to reject the imposed system of collective liability," according to Kovalev, and act as "grown men and women. Each person ought to be responsible for his own behavior."⁵²

Others weren't so sure. "We [have] no moral right to risk the lives of innocent people" by disseminating new issues of the *Chronicle*, as Alexeyeva summed up the opposing position.⁵³ The KGB was likely to go after Anatoly Yakobson, whose family—including his gravely ill young son—was waiting for permission to emigrate. Who wanted that on their conscience? Information on arrests and interrogations, especially those connected with Cases 24 and 63, continued to be gathered, but for a year and a half there were no new issues of the *Chronicle*. After the Yakir-Krasin trial and press conference, however—and after Yakobson and his family left for Israel—Kovalev's position gained the upper hand.⁵⁴ On May 7, 1974, Kovalev, the mathematician Tatyana Velikanova, and the linguist Tatyana Khodorovich held a press conference for foreign journalists in Khodorovich's apartment. Insisting that the *Chronicle* was neither illegal nor guilty of slandering the Soviet state, they announced it as their "duty" to resume circulation: "We are convinced of the necessity of making available this . . . truthful information about infringements of basic human rights in the Soviet Union to everyone who is interested."[55] Journalists received copies of issues 28, 29, and 30 of the *Chronicle*, backdated to give the impression of uninterrupted publication. A brief editorial note informed readers of the KGB's threat and of the decision to ignore it. "There is no need to clarify the ethical quandary experienced by people who face the difficult necessity of making decisions not only for themselves," wrote Kovalev, Velikanova, and Khodorovich. "But continued silence would mean supporting—if only indirectly and passively—'hostage tactics' that are incompatible with law, morality, and human dignity. Therefore the *Chronicle* is resuming publication."[56] The ethical quandary struck at the heart of the dissident ethos. To "live like free people" meant making decisions for oneself but also letting others make decisions for themselves. It was bad enough when the state impeded people's freedom; now the KGB was giving dissidents little choice but to do the same, putting not only themselves but unknown others at risk of arrest.

We do not know whether the *Chronicle*'s resumption led the KGB to arrest people it otherwise would not have arrested. What we do know is that Case 24, which over the course of nearly two years ensnared Yakir and Krasin and, with their help, hundreds of other dissidents, failed to achieve its objective, namely, silencing the periodical that was the informational lifeline of the dissident movement and the chief source of knowledge about it abroad. For all the damage caused by the spectacular public recantations of Yakir and Krasin, moreover, their case failed to become a model for the Fifth Directorate's dealings with dissidents. Instead it was a one-off, the exception to a larger pattern of the KGB's retreat from using judicial trials to curb dissent. It is telling, in fact, that despite their guilty pleas, it was not the *trial* of Yakir and Krasin to which foreign journalists (and the Soviet television audience) were invited, but the ensuing press conference. Andropov evidently did not trust courtroom proceedings to make the desired impression. Neither did senior investigator Alexandrovsky. Once, during an interrogation, Krasin asked him why the KGB hadn't arrested him and put him on trial earlier, in 1968. The reply: "We didn't need yet another hero."⁵⁷ Experience showed that nothing turned dissidents into heroes more reliably than putting them on trial.

The case against the *Chronicle* posed a similar problem for the KGB. Despite years of labor-intensive investigation, the Fifth Directorate found no significant evidence to substantiate its claim that the paper had violated Article 70 by engaging in "slanderous fabrications" against the Soviet order. Its fact-checking report concluded that "the substance of the materials presented in the *Chronicle of Current Events* . . . essentially corresponds to reality."⁵⁸ That conclusion was carefully kept secret. But it confirmed Kovalev's testimony. Over and over he insisted to his interrogators on "the truthfulness and sincerity of the *Chronicle*," at the same time noting that "of course it cannot guarantee against mistakes." Kovalev had his own doubts concerning "the consistent truthfulness of official [Soviet] sources, and unlike you I have identified several instances of egregious departure from reality in those sources." He didn't stop there:

> I assert that the KGB's interest in the *Chronicle*, in samizdat, in the Initiative Group etc. is illegal, that for political reasons the authorities are playing an unworthy and hypocritical game of justice. I do not wish and cannot permit myself to take part in that game. I consider the activities [you accuse me of] socially useful, and in any case morally irreproachable and not contrary to the law. Prestige based on silencing, or on outright lying, is not something I would want to support, although it has not been my particular goal to damage it. I think it's people who behave arbitrarily who damage their prestige, not those who speak openly about that behavior. If the authorities believe that certain messages are causing it moral or political damage, then kindly have them stop violating their own laws. In that case, those messages will disappear by themselves. Or, if worse comes to worst, at least have them make public a list of questions which it is prohibited to discuss.⁵⁹

Few dissidents under interrogation were as defiantly clear-headed as Kovalev. But even fewer proved willing to cooperate in the manner of Yakir and Krasin. The KGB would never again stage a press conference

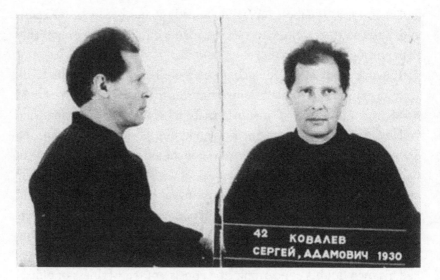

FIGURE 15.5. Sergei Kovalev, arrest photos, December 27, 1974

with convicted dissidents. When Kovalev was arrested in Moscow in 1974, his trial was moved to Vilnius, in Soviet Lithuania, to reduce attention from the foreign press.

Andropov kept his word. The relatively mild punishment meted out to Yakir and Krasin was exactly as he had promised during his meeting with the latter in Lefortovo. And no wonder: in trials of dissidents, it was the KGB that determined the verdict and sentence. After two years in prison (the year of pre-trial detention was allowed to count toward his three-year sentence), Krasin was allowed to emigrate to the United States. In the ensuing years, he would produce two memoirs— one penitent, the other defiant—as he struggled to explain what he had done to himself and to others. Yakir chose to remain in Moscow at his apartment near the Auto Factory metro station, where there were no more visiting days. He never wrote or signed another samizdat document.

Plenty of Soviet dissidents came from families in which the father or mother or other relatives had been sent to Stalin's Gulag. Yakir and Krasin were unusual within the dissident milieu in that they themselves had survived years of brutality in the camps. The aura of that experience heightened their authority in the movement and perhaps in their own minds. In Yakir's case, Western journalists magnified the effect by casting him as the dissident movement's de facto leader, an impression the Voices transmitted to Soviet listeners. In the hands of the KGB, however, the same traumatic experience that gave them their moral capital proved to be their undoing. Yakir and Krasin didn't merely fear a return to Stalinism. Inside them, it had never gone away.

16

How to Conduct Yourself

> It's not your job to think about the methods they use to squeeze you. Your job is to think about how to conduct yourself under pressure.
>
> —ARKADY AND BORIS STRUGATSKY

On May 30, 1972, a group of friends gathered at Moscow's Belorussky train station to bid farewell to Alexander Volpin. He was on his way to Vienna, Israeli visa in hand, but ultimately bound for the United States. The KGB had evidently had enough of him, along with other leading propagators of the legalist approach to dissent such as Boris Tsukerman (allowed to emigrate to Israel in 1971) and Valery Chalidze (allowed to travel to the United States in 1972 and then stripped of his citizenship). Having survived four episodes of psychiatric confinement, Volpin had had more than enough of the Soviet Union. During his most recent involuntary internment, he had asked the attending psychiatrist to explain on the basis of what law he had been seized from his apartment. "You're ill," she replied. "I'm a doctor, and legal questions don't interest me. I'm the one who asks questions here, and if you don't answer, then for me that's just a deep symptom of your illness."[1]

As he waved goodbye to his friends from the window of the departing train, Volpin called out, "Fight not only for the right to leave the USSR, but for the right to return!"[2] He must have sounded yet again like Don Quixote, armed with Article 13 of the Universal Declaration of Human Rights.³ Everyone knew that the Kremlin did not allow troublemakers to return, especially unrepentant ones. As the elderly Vyacheslav Molotov put it, when asked about the thousands of Soviet Jews being granted exit visas in the 1970s, "Don't let them return. What is our country—a hotel?"⁴

Volpin's parting gift was a samizdat pamphlet titled *Legal Memorandum for Those Facing Interrogations.*⁵ For years, he had been advising anyone who would listen on how to deploy the Soviet Code of Criminal Procedure during encounters with the KGB.⁶ There was no shortage of interested listeners: by the late 1960s, as chapter 14 showed, the Fifth Directorate was conducting roughly ten thousand conversations (prophylactic or otherwise) per year with individuals suspected of anti-Soviet activity. Such encounters were now a routine feature of dissident life.

Notwithstanding the fear they inspired, interrogations served for many dissidents as privileged moments of truth. Countless petitions and open letters addressed to Soviet authorities had produced not a single reply—apart from the reprisals against those who signed them. For all but a handful of well-connected figures such as Andrei Sakharov and Petr Yakir (prior to 1968), there seemed to be no hope of opening a dialogue with the state. Interrogation by a KGB officer was thus the closest most dissidents would ever come to a face-to-face encounter with a representative of the Soviet government. Here one finally had the chance to cut through the verbal fog of the lip-service state and speak truth to power. "For other-thinkers," recalled Gleb Pavlovsky, a student at Odesa State University who had been expelled from the Communist Youth League in 1969, "meeting with 'the organs' is a climactic moment. You've been waiting for this forever."⁷

Even hardened veterans of the Gulag were susceptible to the fantasy of direct communication with Soviet leaders. Sitting in a cell in Lefortovo Prison following his arrest in February 1974, Alexander Solzhenitsyn was suddenly handed a suit and tie and told, "Let's go":

Where to? I had no doubt: to the government, to that very Politburo of theirs. Now, finally, for the first and last time, we would have a conversation! I had at times expected this moment, when they would become enlightened, when they would take an interest in speaking with me—would it not be interesting for them? A serious conversation, perhaps the most important conversation of my life. There was no need to make a plan—it had long taken root in my soul and in my head. Arguments would come by themselves, I would speak freely, without constraints, not the way their subordinates talk to them. As I walked [away from the cell], I was conversing with the Politburo in my mind. I had the feeling that, over the course of two or three hours, I could move them in certain respects, shake them loose. It was impossible to get through to the fanatics of Lenin's Politburo or to the sheep of Stalin's. But these fellows? Absurd as it may sound, I thought it was possible.[8]

He never got the chance. Solzhenitsyn was whisked off not to the Politburo but to Sheremetyevo airport, stripped of his Soviet citizenship, and expelled to West Germany. He would never set foot in the Soviet Union again.

As Solzhenitsyn's account suggests, encounters with Soviet officials were also moments of truth about oneself. They tested one's character, one's loyalty to friends, mental agility, and, not least, physical and moral stamina in the midst of a verbal duel that could go on for hours at a time. Insofar as dissidents were engaged in partisan warfare with the Soviet state, interrogations represented the combat zone, scenes of intense psychological drama. They had long been an object of cultural fascination, from the encounter between Jesus and the Grand Inquisitor in Dostoevsky's *The Brothers Karamazov* (1880) to Arthur Koestler's haunting depiction of the Communist Nikolai Rubashov's confrontations with Stalin's NKVD in *Darkness at Noon* (written while the ex-Communist Koestler was himself undergoing repeated interrogations by the French police in 1940).[9]

The existential question was "how to conduct yourself under pressure," as the popular science-fiction writers (and brothers) Arkady and Boris Strugatsky put it in their 1976 novel *A Billion Years before the End of the World*.[10] For Boris Strugatsky, this was by no means a purely fictional scenario: in 1973, he was subjected to an eight-hour interrogation by the Leningrad KGB as part of a case against his friend Mikhail Kheifets, author of the introduction to a samizdat collection of poems by the exiled Joseph Brodsky. By his own account a "grotesquely inexperienced" witness, Strugatsky was caught lying and threatened with prosecution. The encounter "left an indelible impression and colored the entire atmosphere of *A Billion Years*."[11]

In the novel, the mathematician Vecherovsky offers the following advice: "It's not your job to think about the methods they use to squeeze you. Your job is think about how to conduct yourself under pressure.... No one is going to help you. No one is going to give you advice or make decisions for you. Not academicians, not the government, not even all of progressive humanity."[12]

For some dissidents, usually men, interrogations provided the functional equivalent of war stories told around kitchen tables, preferably highlighting the protagonist's heroism and cleverness, in contrast to the ham-fisted malevolence of the KGB. "How we laugh when we come home from interrogations," proclaimed the mathematician Vladimir Albrekht, who experienced more than a few himself.[13] Pavel Litvinov's transcript of his conversation in September 1967 with KGB agent Gostev became a samizdat sensation.[14] Alexander Podrabinek, an emergency medical technician who landed in the KGB's crosshairs after he began to publicize the USSR's practice of punitive psychiatry, took the documentary impulse even further, secretly tape-recording his interrogations in order to replay and dissect them afterward.[15]

Studying transcripts of interrogations, however, could only do so much to strengthen one's position during questioning by the Fifth Directorate. Interrogation transcripts are like chess move lists, the written records of matches that capture the two players' moves sequentially. What they fail to capture, of course, are the moves each player considered at any given moment, the larger strategies they sought to advance (including the use of bluff, intimidation, and distraction), and the strategies they believed their opponent was following—not to mention their blood pressure, hunger level, alertness, anxiety, exhilaration, and exhaustion, along with attendant body language and facial expressions. Like chess move lists, transcripts are two-dimensional representations of experiences involving far more than two dimensions.

For Volpin, interrogations had been an object of fascination ever since his first arrest in 1949 ("Under interrogation I shall invent no lies at all.... / I shall penetrate their protocols and their minds").[16] By the 1960s, he had concluded that the key to withstanding pressure during interrogations—pressure that could easily lead to unintended self-incrimination, or even worse, incrimination of others—was to be found in the Code of Criminal Procedure. Interrogators would typically reference procedural rules by number at the outset of an interrogation ("In accordance with Article such and such of the Code of Criminal Procedure...") but did not spell out or explain them. Volpin therefore composed a short samizdat guide to the legal rights of witnesses and defendants during criminal investigations. The resulting *Legal Memorandum for Those Facing Interrogations,* disseminated under the less formal title *How to Conduct Yourself at an Interrogation,* provides a host of practical suggestions, each accompanied by a citation to the relevant article of the Code.[17]

To begin with, Volpin advised, one should analyze in advance the events that presumably led to a summons by the KGB, in order to anticipate the likely criminal charges. To gain extra preparation time, one should respond to the summons as late as legally allowable. Pleading mental illness might help avoid criminal prosecution, but it came with the risk of indefinite incarceration in a psychiatric hospital. Pleading drunkenness—a common defense in cases of politically tinged misbehavior—did not offer protection from prosecution. In any interrogation not explicitly connected to a criminal investigation or trial, witnesses had the right, without penalty, to refuse to answer questions. Leading questions were prohibited. Both defendants and witnesses had the right to write down their answers in their own hand, rather than having the interrogator or an assistant do so. Writing one's answers was a good way to slow down the proceedings, creating more time to think and strategize. It also reduced the likelihood that, when one signed the completed transcript at the end of the interrogation (as required by law), unintended or inaccurate words and phrases would be included.

It was often helpful, Volpin wrote, to distinguish between juridical and everyday meanings of key terms. "Soviet power," for example, needed to be understood juridically as power derived from the Soviet Constitution (thus narrowing the range of activities that could qualify as being directed "against Soviet power"). In its strict legal meaning, the word "disorder" could not refer to demonstrations as such, since the right to hold demonstrations was inscribed in the constitution. The optimal time to draw these distinctions was when being asked whether one understood the charges in the indictment—a question typically posed at the beginning of an interrogation. It was important to answer not by saying "No, I don't understand the accusations"—which could lead to being ruled mentally incompetent to stand trial—but rather "Not entirely, because certain words in the accusation have a two-fold meaning."

If the interrogator asked about the activities of one's friends, it was best to respond with "I am testifying as a defendant, not as an informant." To avoid punishment when refusing to provide testimony, one should respond to questions not with silence, but with an explanation of the refusal. That way, one could plausibly claim to have responded to the question with something that qualified as "testimony."

Exasperated KGB investigators who found themselves entangled in such procedural fine points were known to throw up their hands and declare, "You've been reading Volpin!" *How to Conduct Yourself at an Interrogation* quickly became the samizdat equivalent of a best-seller, with a small-format tamizdat edition published in Paris in 1975, a *vade mecum* suitable for smuggling into the Soviet Union. As the memorandum itself noted, there would have been little need for such a guide had the KGB not restricted the pool of available defense attorneys by requiring special permission to work on "political" cases.[18] Just as samizdat circumvented Soviet censorship by turning readers into do-it-yourself publishers, Volpin's memorandum responded to restrictions on the legal profession by coaching defendants and witnesses to become do-it-yourself lawyers.

How to Conduct Yourself at an Interrogation inaugurated a distinct and widely consumed genre within the world of samizdat: the dissident conduct manual. By the end of the 1970s, half a dozen such guides had appeared, covering situations ranging from interrogations to apartment searches to interviews by psychiatrists to life in general. One samizdat activist reports having typed Volpin's manual "a hundred times . . . I learned it by heart." She wasn't the only one. The Fifth Directorate produced its own in-house copies of this and other dissident manuals, so that its employees could familiarize themselves in advance with their disruptive tactics.[19]

Unlike most conduct literature, dissident manuals were not part of a civilizing mission intended to transmit mores and manners to a benighted audience.[20] They were produced by people who themselves had struggled through the kinds of encounters they sought to analyze. Their readers had too—or anticipated doing so. The goal was to help readers manage the attendant fear and uncertainty and emerge with dignity intact. Apartment searches, for example, were a common KGB technique to attempt to disrupt the distribution of samizdat. They could last twelve hours or longer and often ended with one's possessions strewn on the floor, furniture damaged, books and manuscripts confiscated, and occasionally an arrest. "For a foreign journalist to work in Russia without witnessing a search," Andrei Amalrik observed, "is like visiting Spain without witnessing a bullfight."[21] According to Pavlovsky, who experienced nearly a dozen of them, an apartment search "changes your relationship to your home." "From now on you know that you no longer have a secure haven. It changes your sense of place; the sensation of physical safety disappears forever."[22]

An interrogation gone wrong could jeopardize many things, not the least of which was your ability to live with yourself. Giving in to pressure to name names or reveal the location of a cache of samizdat was known to induce deep depression or even suicide. "How should we live in order to respect ourselves?" asked Arseny Roginsky, a young bibliographer at the State Public Library in Leningrad who edited a samizdat journal devoted to the history of Stalinism (and who would eventually direct the Memorial Society in post-Soviet Russia). "This question was exceptionally important for us."[23]

Like all conduct literature, dissident manuals were prescriptive; they tell us more about norms and ideals than about actual behavior. And therein lies their particular value, as fragments of a conversation about

FIGURE 16.1. Arseny Roginsky

the kind of relationship an individual should have with the Soviet state. Most dissidents, like most Soviet citizens, took for granted that the USSR would be around for decades, if not centuries. The conversation therefore addressed not just the age-old, action-seeking question "What is to be done?" but the existential "How should you conduct yourself?" What were the requirements for self-respect? Was it ever permissible or advisable to lie to the KGB (as distinct from not revealing the truth)?

Did interrogation rooms and psychiatric wards demand a distinct code of behavior, a moral state of exception—or did they merely reproduce everyday Soviet reality in heightened form?

By the time the dissident movement's would-be leaders Petr Yakir and Viktor Krasin spilled hundreds of names of fellow dissidents to their KGB interrogators, it was clear that encounters with the security organs were fraught with risk to both the individual and the movement as a whole. In their case, the key vulnerability lay in what Krasin called the "mystical fear" instilled in Gulag prisoners. Soviet psychiatry dubbed it "camp fear." They had witnessed cell mates being taken away and shot; they had seen how the security forces could do anything they wanted to a prisoner. "This fear," Krasin confessed, "remains in the human soul forever." It was fine to comb the Code of Criminal Procedure for information about prisoners' rights and the rules of interrogation, but former prisoners were inclined to view "rights" and other abstractions as ephemeral. What endured were the denouncers and torturers, unpunished, some still occupying high offices two decades after Khrushchev's dethronement of Stalin.[24] "Notions like 'human dignity' or 'the inviolability of the person,'" wrote Andrei Sinyavsky after his return from the camps, "are to my mind a kind of gobbledygook, part of a generally accepted jargon or code serving the same sort of practical convenience as exclamatory phrases such as 'you don't say!' or 'goodness gracious!'— which are all very well in polite conversation, but make little real sense and cannot be taken very seriously."[25]

Most dissidents, however, did not carry "camp fear" inside them. They were more likely to be subject to what Valentin Turchin, drawing once again on his training in physics, called "the inertia of fear," the collective afterlife rather than the actual experience of Stalinist terror. For Amalrik, it was not so much fear, or even the residue of fear from the Stalin era, that induced widespread inaction, but the sense that "nothing will change in any case."[26] Forged in revolution, civil war, break-neck industrialization and collectivization, mass purges, terror, and the Nazi war of annihilation, battle-hardened Soviet socialism had reached its triumphant steady state. The USSR, according to the anthem adopted in 1944, was now "indestructible"; a verse added in 1977 reminded

FIGURE 16.2. Andrei Sinyavsky in 1971, the year he was released from the Gulag, with his wife, Maria Rozanova, and their son, Yegor

listeners that the idea of communism was "immortal." The Soviet Union was forever.

The apparent hopelessness of changing the Soviet system, however, did not preclude the possibility of changing oneself. Whatever its continuing world-historical ambitions, by the 1960s the Communist Party had largely lost its appetite for policing the inner lives of Soviet citizens. Emancipation from cognitive and emotional reflexes conditioned by the Soviet system—from what Vladimir Zelinsky called "ideocratic consciousness" and historians Jochen Hellbeck and Igal Halfin have termed "Soviet subjectivity"—was at least conceivable at the individual level, if not in public life.[27] In fact, it could proceed *only* at the level of the individual, interacting with like-minded individuals. Thanks to the miracle of samizdat, Vyacheslav Igrunov observed from Odesa, "spiritual and cultural life can exist with or without the state's permission."[28]

For many, the cultivation of inner freedom required abstinence from public life. This could mean a life of "inner emigration," as if having mentally departed the USSR despite remaining physically present. In extreme cases, it involved the willful non-recognition of Soviet reality, as if the latter simply did not exist. All these approaches had at least one thing in common: an element of make-believe, a dogged determination to live "as if" against the background of an immovable "as is." Volpin's legalist approach to dissent involved acting as if the Soviet state were subject to the laws and regulations inscribed in its constitution and Code of Criminal Procedure. He was known, for example, when encountering an acquaintance on a crowded bus or trolley car, to strike up a conversation about taboo political topics despite the supremely public setting.[29] As uncomfortable as that may have been, the alternative, as Zelinsky argued, was to inhabit another fictive "as if," this one captured in the Soviet state's implicit social contract with its citizens: "I will speak to you about your freedom and you will act as if there is no freedom."[30] Dissidents sought to act as if their freedom were real.

While it could serve as a powerful tool of psychic freedom, inner emigration was not much help during interrogations or apartment searches or other increasingly common encounters with agents of the Fifth Directorate. Nor, for that matter, was conscience, the moral intuition prized by the anti-organizational wing of the dissident movement. Under intense, methodical, hours-long interrogation, one's inner voice was not always the best counselor. "They're constantly sending the message 'we know everything, you just need to confirm,'" recounted Natalya Gorbanevskaya. "But they're far from knowing everything, and you shouldn't fear them. You should fear yourself."[31] Pavlovsky, who like the young Gorbanevskaya was maneuvered by interrogators into betraying a friend (in his case, Igrunov, the founder of Odesa's clandestine samizdat library), agreed: "The main threat is inside yourself."[32]

Consistent with his dream of emancipating reason from emotion, Volpin urged readers to use juridical techniques to prevent misjudgments driven by fear or panic—or faulty intuition. Such techniques had the advantage of being grounded in official Soviet documents that could be cited chapter and verse. But they also had notable drawbacks, especially if you were neither a lawyer nor a logician. Volpin's methods were highly technical and required extensive memorization. Relying on the Code of Criminal Procedure, moreover, significantly limited one's repertoire. More to the point, Volpin's approach sought to constrain intuition in a setting inescapably charged with emotion. "Everyone says you have to know certain principles of criminal law," complained Vladimir Albrekht in his manual on how to behave during an apartment search. "Aren't the laws of ordinary morality enough? If there is such a thing as truth, I think it must be naive. Not for nothing was it a child who first said 'The emperor has no clothes!' In some sense we too are children." There was nothing wrong with knowing and invoking the law; in some situations it was vital. But it was equally important to appeal to justice and common sense.[33]

Though trained in mathematics like Volpin, Albrekht was cut from a very different cloth. Whereas Volpin prized logic and reason, Albrekht was drawn to paradox and irony. His Polish father, the Bolshevik worker Jan Albrekht, was arrested in Moscow in 1937, swept up by the NKVD's "Polish operation," and secretly executed in March 1938, shortly after Vladimir turned five. His Jewish mother, Ida Tiagai, was fired from her job as a foreign language instructor and, as the spouse of an "enemy of the people," barred from teaching. After gaining a personal audience with Lenin's widow, Nadezhda Krupskaya, she was able to enroll in a nursing program. She had her son write a letter to Stalin asking the whereabouts of his father. The answer came nearly a decade later, after the war: Jan Albrekht had died "of heart failure." In 1955, he was formally rehabilitated "due to the absence of a crime." It would take another three decades for the family to learn the true cause of death.[34]

Vladimir Albrekht grew up in Moscow and taught high school there. His first brush with the KGB came in 1972 in connection with the investigation of the *Chronicle of Current Events*. He subsequently underwent multiple interrogations and became a dedicated collector of stories of run-ins with the Fifth Directorate. And for good reason: during the 1970s he produced four conduct manuals, at least two of which were published in Paris, designed for smuggling back into the USSR. The most famous, *How to Be a Witness*, went through multiple editions.

Albrekht's manuals pushed the boundaries of the genre in new directions. To begin with, unlike Volpin's, they are often funny. "We were born to make Kafka real," announces one, parodying the famous opening line

FIGURE 16.3. Vladimir Albrekht

of the hymn of the Soviet Air Force ("We were born to make a fable [*skazka*] real, / To overcome space and expanse. / Reason gave us winged arms of steel, / And in place of our heart, a fiery engine"). They also assume a certain intimacy with the reader, addressing him or her directly, in the manner of the nineteenth-century mock-conversational literary style known as *skaz*. "Did you know, reader, that the investigator obtains his results with the help of no fewer than eighteen techniques? ... You might enjoy asking which technique he's trying so unsuccessfully to use with you and advise him to use a different one."[35] Albrekht would rehearse the manuals in private readings among friends, leading one to comment that by "carnivalizing interrogations," Albrekht had created "a new genre."[36]

The most striking thing about Albrekht's manuals, apart from their playful authorial voice, is their attention to the inner state of the participants. What started out as a purely documentary genre—a move list in

FIGURE 16.4. Cover of the pocket-sized, second foreign edition (Paris, 1983) of Vladimir Albrekht's *How to Be a Witness* (Kak byt' svidetelem). Artwork by Alexander Kosolapov, "Matches (Bather)" (Spichki [Kupal'shchitsa]), 1974.

two dimensions—had acquired, with Volpin's incorporation of extra-textual information about Soviet law, a third dimension.[37] Albrekht added further dimensions to create an almost novelistic effect, as in this account of the prelude to his interrogation about the activities of Irina Belogorodskaya in 1972:

> Unfortunately, I am afraid. . . . Fear arises when it becomes clear that neither hearing nor sight nor smell nor touch are of any use. There's just no way to deploy them. How do you block fear? With a good joke? No! Only with a good thought! One will come to me momentarily.
> Now we've arrived and we're inside the interrogator's office. On the wall, there's Lenin. I feel better already.
> "Take a seat, please. My name is Anatoly Gavrilovich Zhuchkov. I'm handling Lyudmila Mikhailovna Belogorodskaya's case (he was mistaken: her first name is Irina). My job is to interrogate you as a witness. But first I will explain your rights and obligations."
> While he speaks, I try to lower my blood pressure.
>
> "Is that all clear?"
> "Yes. Everything. Absolutely. And I think it would be good to begin right away writing everything down."
> "Why right away? I'm required by law to start off by speaking with you about the background of the case. Why do you insist on immediately writing everything down?"
> "For purposes of expediency—nothing more. You ask the questions, I give the answers. What could be simpler?"
>
> Zhuchkov did not agree. It turns out he is supposed to create a relaxed atmosphere, thank you very much. Conversing with him is supposed to help me remember what I have forgotten. He will help me, prompt me.[38]

The two men get into an elaborate argument about whether Albrekht is allowed to enter the case number (24) and Belogorodskaya's name in the transcript. Zhuchkov, still trying to relax his witness, tries to draw Albrekht's attention to a specific day:

"Think about a certain event that happened on your 35th birthday."

This is probably a hint of some kind.

He paused in a way meant to suggest significance. I paused too. There was hardly anything for me to remember or forget. My 35th birthday didn't mean much to me. The last time I celebrated my birthday was on February 23, 1937, when I was four. My father always marked my birthday three days early so that it would coincide with Red Army Day. But on the night of August 5, 1937, he was arrested. He said it was a "mistake," that he'd return soon. But he didn't return. On March 14, 1938, he was executed "by mistake." Since then I have never marked my birthday. In contrast to March 14. If only [Zhuchkov] would ask about *that* day. Last year I was taken to a police station on March 14. They checked my documents, saying they were looking for a criminal. I had just come from the Yakirs' apartment and was standing at the Auto Factory metro station. . . . But none of this was of any interest to Zhuchkov.[39]

Alternating dialogue with stream of consciousness, Albrekht offers a subjective (and possibly fictionalized) but in certain ways more realistic version of the interrogation experience, including arguments, silences, and mental associations uncaptured by the official transcript.[40] Zhuchkov mentions his birthday, triggering a flashback to his father's arrest by Zhuchkov's predecessors in the NKVD, which reminds Albrekht of his recent detention on the anniversary of his father's execution. Readers were thus perhaps less likely to be caught off guard by the multiplicity of their own associative thoughts and emotions when under interrogation.

Albrekht's manuals also provided memorable advice on how to respond—or not respond—to potentially fraught questions. Never acknowledge that a given samizdat text is anti-Soviet or claim—if said text is discovered during an apartment search—that it doesn't belong to you. Both would signal guilt of one kind or another. "You never have to justify yourself or prove anything," he warned. "The burden of proof is on them and them alone."[41] Rather than invoking Volpin's procedural fine points during an interrogation, Albrekht favored an artful naïveté, as, for example, in this exchange:

"Where did you meet with Belogorodskaya?"

I panicked. I did not want to name any names, absolutely none! I pretended to be wracking my brain. Well, where could I actually have seen Belogorodskaya? In various people's homes. At birthday celebrations . . .

"At farewell gatherings," Zhuchkov hinted.
"Which ones?"
"Well, for example, the one for Alik Volpin. Remember? At the Belorussky train station. You wanted to pass on a photo album of Russian and Soviet writers to a certain someone."

Even though I lowered my very guilty eyes, I could feel the penetrating, painfully familiar stare.

"No, I don't remember." I tried very hard to say that calmly.

Zhuchkov telegraphed the message "We know everything." It practically dripped from his face. I could feel the pleasure with which he was now scrutinizing mine.

"And then again at the New Year's party."
"What New Year's party?"
"The children's party."
"Which children's party?"

I was genuinely surprised. He too was surprised as well as annoyed by my forgetfulness.

"The New Year's party for children of those convicted under Articles 190 or 70," the investigator slowly enunciated. His cunning look once again said: "We know everything. Everything."

Out of surprise, I laughed. "Of course, I remember. I confess to being at the New Year's party. But was Ira there?"
"Absolutely. She was there," Zhuchkov confirmed.
"How do you know? Were you there too?"[42]

If Volpin advised channeling one's inner lawyer, Albrekht illustrated how to channel one's inner Holy Fool. The best answers, he advised, were "honest but not concrete, bold but naive."[43] Sometimes, as in the above exchange, the best answer might be a question.

In place of Volpin's reliance on the dense Code of Criminal Procedure, Albrekht constructed his own simplified code, encapsulated by the easy-to-remember acronym P-L-O-D ("fruit"). "P" signified *protokol*, the default response when facing difficult questions: "Write the question in the protocol and I will answer it." This bought time and deterred interrogators from asking leading or other improper questions. "L" stood for *lichnoe*, or personal, matters, a reminder not to be afraid to use arguments based on personal relationships or personal ignorance. Refusing to characterize someone else's views, refusing to make statements beyond one's assigned role as witness (as opposed to defendant), citing one's ignorance of the law when asked to comment on alleged criminal activity—all these might feel uncooperative or cowardly but were perfectly legitimate. "O" signaled *otnoshenie k delu* (relationship to the case), or the absence thereof. During an interrogation regarding the physicist Andrei Tverdokhlebov (one of the founding members of the Human Rights Committee), a witness was asked to name the author of a certain letter. He refused:

> INVESTIGATOR: Do you refuse to say that you know who the author of the letter is?
> WITNESS: No, I do not refuse.
> INVESTIGATOR: What then prevents you from answering the question "who is the author of the letter"?
> WITNESS: What prevents me is the necessity of remaining within the bounds of the case under investigation.

Valery Chalidze had also been known to use this technique. When asked by an interrogator why he refused to answer a certain question, Chalidze allegedly replied, "Your question does not relate to the present case. It relates to a case that has not yet been opened, concerning my refusal to answer the preceding question."[44]

Finally, "D" stood for *dopustimost*, or permissibility from a moral standpoint. Aggressive or unacceptable questioning during an interrogation needed to be called out and, if possible, entered into the protocol. "I request that the investigator not circle around me," Albrekht offered by way of example, "not intimidate me, not blow smoke in my face, not raise his voice, not push me to answer quickly. In a word, not subject me to pressure." Conversely, witnesses should not shy away from invoking moral (not just legal) standards of behavior—if only because defendants were likely to read investigation protocols in the lead-up to their trial and would be heartened by expressions of moral solidarity from witnesses.[45]

Did any of this matter during actual interrogations with actual KGB investigators? In a conversation with Albrekht shortly after his release from prison in 1987, Pavlovsky maintained that most people who read his manuals, including Pavlovsky himself, did not apply the "PLOD" system or Albrekht's other ideas when undergoing interrogation or apartment searches. "For a thousand reasons," Pavlovsky claimed, "personal and otherwise—the way they were raised, fear, the legal setting—people don't conduct themselves that way." It was all a game, and as everyone learned sooner or later, the KGB did not play games. Pavlovsky was in a position to know: in the course of his own interrogations by the KGB he had incriminated multiple friends and was pressed into collaborating.[46] Albrekht begged to differ. Interrogations were a kind of theater, interrogators were aspiring playwrights, and witnesses could steer the script in favorable or at least less damaging directions.[47] He was proud that his court sentence included the charge that he had "written a textbook for criminals and ... conducted himself during the investigation according to his Jesuitical recommendations."[48]

It is no simple matter to assess the impact of conduct literature on actual conduct. Most of the examples of interrogations and searches featured in Albrekht's guides stem from accounts by people on the receiving end of such procedures, who presumably were inclined to favor scenarios in which they got the upper hand, and who were not above embellishing accounts of their own experience. There is, however, another means of assessment: official transcripts of thousands of interrogations preserved in KGB archives. To be sure, only a tiny fraction of those transcripts are accessible, making generalizations risky.[49] As the preceding discussion illustrated, moreover, two-dimensional move lists, with alternating utterances by interrogator and witness, hardly capture the full experience, or perhaps even the full extent, of what was said by the two sides.[50] Nonetheless, the limited corpus of available official transcripts gives the impression that, despite anecdotal evidence of KGB frustration with witnesses who applied the "Jesuitical" methods championed by Volpin and Albrekht, those methods found at best modest expression in actual encounters between dissidents and the KGB. To be sure, thanks to Volpin, witnesses knew they had the option of refusing to answer a question—a nearly unthinkable tactic in an era when most people assumed that such a refusal was itself a criminal act.[51] A handful of witnesses applied the tactic relentlessly, as a page from the transcript of biologist Sergei Kovalev's interrogation in Vilnius in February 1975 shows (figure 16.5). Few of the other techniques outlined in conduct manuals, however, have left traces in the available interrogation transcripts.

The apparently limited impact of dissident conduct manuals on the pragmatics of interrogations and house searches, however, hardly exhausts their significance for the dissident movement. Conduct manuals were widely reproduced via samizdat and went through multiple editions in tamizdat; they were the subject of intense discussion and debate. Their chief impact, it seems, lay in the dissemination of an ideal model of dissident comportment: self-reflective, methodical, bold, stubborn, and occasionally puckish. Conduct manuals can be understood as instruments designed to codify and disseminate the characteristics of "outstanding personalities," the dissident virtuosi whose example was supposed to act as a magnet for others. In this sense, they were guides not just for how to behave in particularly acute encounters with Soviet power, but for how to live in the Soviet setting generally. "I'm not sure you need to conduct yourself in some special way at an interrogation," Albrekht advised. "In my view, at an interrogation you should conduct yourself like everywhere else. Everywhere, you should conduct yourself deliberately, i.e., thinking about each step."[52]

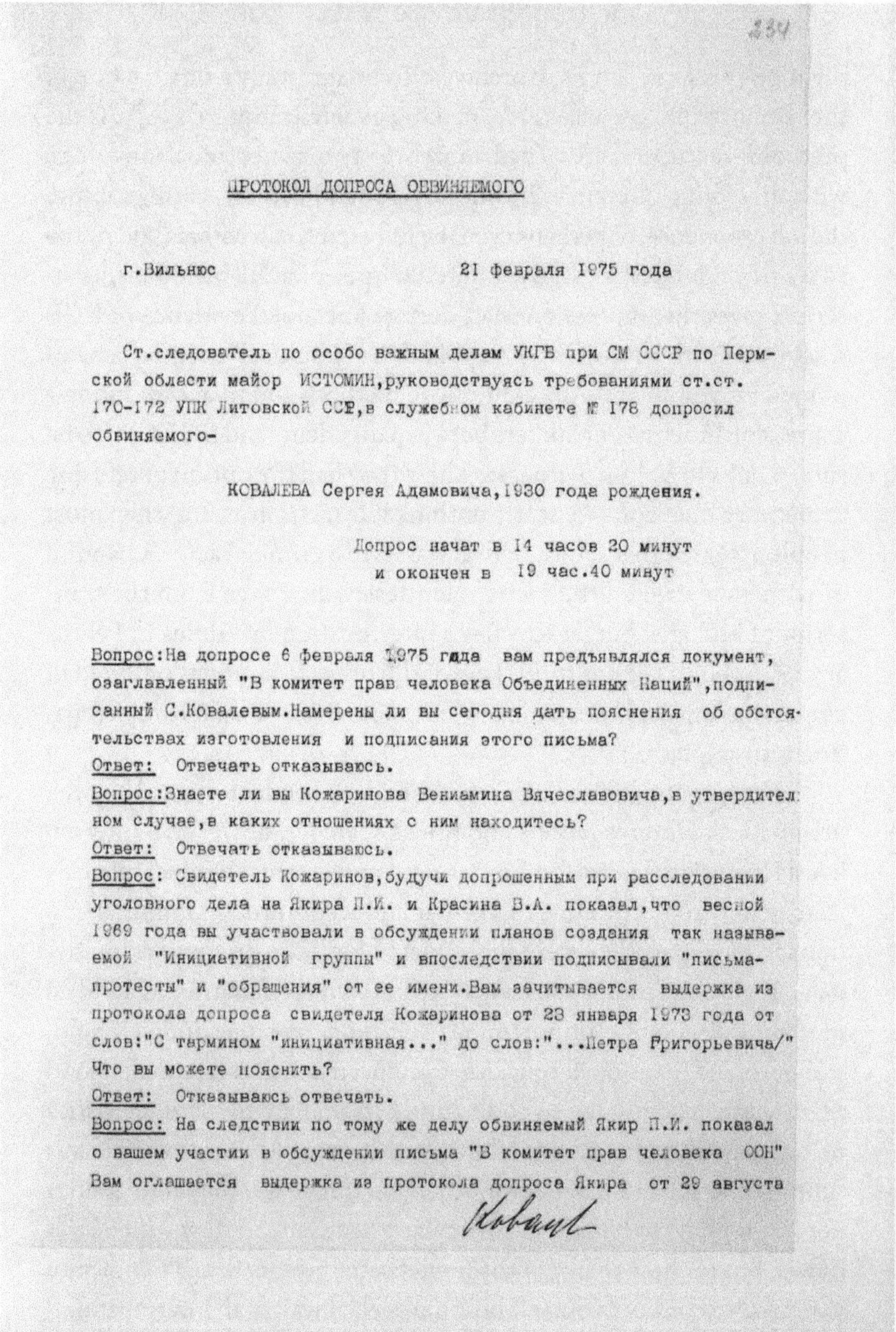

ПРОТОКОЛ ДОПРОСА ОБВИНЯЕМОГО

г.Вильнюс   21 февраля 1975 года

Ст.следователь по особо важным делам УКГБ при СМ СССР по Пермской области майор ИСТОМИН, руководствуясь требованиями ст.ст. 170-172 УПК Литовской ССР, в служебном кабинете № 178 допросил обвиняемого-

КОВАЛЕВА Сергея Адамовича, 1930 года рождения.

Допрос начат в 14 часов 20 минут
и окончен в 19 час. 40 минут

Вопрос: На допросе 6 февраля 1975 года вам предъявлялся документ, озаглавленный "В комитет прав человека Объединенных Наций", подписанный С.Ковалевым. Намерены ли вы сегодня дать пояснения об обстоятельствах изготовления и подписания этого письма?
Ответ: Отвечать отказываюсь.
Вопрос: Знаете ли вы Кожаринова Вениамина Вячеславовича, в утвердительном случае, в каких отношениях с ним находитесь?
Ответ: Отвечать отказываюсь.
Вопрос: Свидетель Кожаринов, будучи допрошенным при расследовании уголовного дела на Якира П.И. и Красина В.А. показал, что весной 1969 года вы участвовали в обсуждении планов создания так называемой "Инициативной группы" и впоследствии подписывали "письма-протесты" и "обращения" от ее имени. Вам зачитывается выдержка из протокола допроса свидетеля Кожаринова от 23 января 1973 года от слов: "С термином "инициативная..." до слов:"...Петра Григорьевича/" Что вы можете пояснить?
Ответ: Отказываюсь отвечать.
Вопрос: На следствии по тому же делу обвиняемый Якир П.И. показал о вашем участии в обсуждении письма "В комитет прав человека ООН" Вам оглашается выдержка из протокола допроса Якира от 29 августа

Ковалев

FIGURE 16.5. Page 1 of the transcript of Sergei Kovalev's interrogation on February 21, 1975. The interrogation, conducted by KGB major Istomin in Vilnius, capital of the Lithuanian Soviet Socialist Republic, lasted five hours and twenty minutes. To each of Istomin's questions, Kovalev responded, "I refuse to answer." Kovalev's signature appears at the bottom of every page of the protocol.

The Soviet Union was not going away, and there was no use pretending otherwise. A life of dignity required not just living with oneself, but doing so while coexisting with Soviet power. The question, as Albrekht formulated it, was "how a normal person should conduct himself in a non-normal environment, how a good actor should act in a bad, shameful play."[53] It was the old Christian conundrum: how to live virtuously in a fallen world. Or, in Amalrik's secular version, how free people should conduct themselves in an unfree country.

The most unfree places in the Soviet Union were the psychiatric institutions where the government indefinitely confined dissidents whom it preferred not to put on trial. Vladimir Bukovsky, Natalya Gorbanevskaya, Petr Grigorenko, Leonid Plyushch, Alexander Volpin, and Ivan Yakhimovich were among the hundreds of other-thinkers who experienced such punishment. However grim the conditions in Soviet prisons and corrective labor camps, few could rival psychiatric institutions, with their use of mind-altering drugs, for their capacity to suffocate even the sturdiest inner freedom. Loss of agency meant loss of dignity. For Solzhenitsyn, as we have seen, it was nothing less than "spiritual murder." What forms of conduct could shield one from such an experience or, in the worst-case scenario, preserve one's dignity while enduring it?

Shortly after Grigorenko was lured to Tashkent in the spring of 1969 and arrested by the KGB—the event that triggered the founding of the Initiative Group for the Defense of Civil Rights—he underwent a psychiatric evaluation and was found fit to stand trial. Evidently unsatisfied, the KGB transferred Grigorenko to the Serbsky Institute in Moscow, where a second evaluation overseen by Daniil Lunts, the KGB's psychiatrist on call, produced the desired finding: the retired major-general "suffers from a psychic illness in the form of pathological (paranoiac) development of the personality with the presence of ideas of reformism" and therefore could be detained without trial indefinitely.[54] While Bukovsky, who had befriended Grigorenko during their overlapping incarceration at the Serbsky Institute in 1965, was arranging to smuggle Grigorenko's medical evaluation to psychiatrists in the West, another psychiatrist was not waiting for those results. Semyon Gluzman had just completed his training in psychiatry at the Kyiv Medical Institute and was keen to launch his career. He was an avid reader of samizdat, including statements claiming that his chosen field of psychiatry was being used to punish dissidents. "My beloved," he wrote, "was being turned into a prostitute." After reading open letters by Solzhenitsyn, Sakharov, and other intellectuals protesting against punitive psychiatry, he wondered why not a single Soviet psychiatrist had spoken out.

Keen to defend the honor of his new profession, Gluzman decided to perform his own evaluation *in absentia* of Grigorenko's mental health. Eventually, he secured the cooperation of Grigorenko's family, who provided medical records secretly hand-copied by Grigorenko's lawyer, Sofia Kallistratova. The more earnestly Grigorenko had tried to explain his thinking to Lunts and other doctors at the Serbsky Institute, Gluzman noted, the more they twisted his ideas into symptoms of psychopathology. On the basis of Grigorenko's medical records and his samizdat essays on topics ranging from Stalin's role in the Second World War to the campaign by exiled Crimean Tatars to return to their ancestral homeland, Gluzman determined that he was fully sane and fit to stand trial. He sent his evaluation to Sakharov with the request that it be distributed in samizdat—unsigned, so as not to jeopardize his career.[55] It was too late: for daring to offer a second, or rather third, opinion (which confirmed the original assessment by psychiatrists in Tashkent), Gluzman was arrested, tried, and sentenced to seven years of hard labor followed by three years of exile. He was twenty-six.

Gluzman had not risked his career and his freedom for a friend; he and Grigorenko had never met and never would. Grigorenko's case attracted him because of how well documented it was and how forcefully it demonstrated the willingness of leading Soviet psychiatrists to bend their expertise for political purposes. The personality that loomed largest in Gluzman's mind, in fact, was not Grigorenko but Lunts, the "psychiatrist-executioner." It was bad enough that Lunts was turning psychiatry into a tool of state persecution. What made it unbearable was that, like Gluzman, he was Jewish. "I was raised by my father," Gluzman recalled, "in full awareness of the role Jews played in repressions by the Stalinist regime."[56] Fishel Gluzman, a physician-scientist, and Galina

FIGURE 16.6. Semyon Gluzman, ca. 1971, after receiving his medical degree

Gluzman, a physician, had both served as medics in the Soviet Army during World War II. One of Semyon Gluzman's earliest memories was of overhearing his parents in the kitchen, on the day Stalin's death was announced in 1953, repeatedly whispering, "If only it had happened sooner." Loyal Party members, they eagerly consumed every issue of *Novyi mir*, the favored journal of the progressive Soviet intelligentsia. Sometimes Fishel Gluzman would allow his son to join him late at night to listen to the Voices, "to get information about our own country."[57]

One of the first people Gluzman met in the Perm-35 corrective labor camp was Bukovsky. Both had embarked on a process of self-definition by differentiating themselves from their communist parents. Bukovsky, as we saw in chapter 6, took his parents' claim not to have known the full truth about Stalin's terror as grounds for insisting on full transparency regarding the current actions of the Soviet government, "so that no one could say later, 'I didn't know.'" Gluzman's situation was different: his parents knew. They had privately opposed Stalin while remaining loyal to the Party's ideals. And that, for Gluzman, was the problem: their criticism was private. They were too willing to compromise and wait. With Lunts now doing the KGB's dirty work, he thought, "here we go again."[58] He was not going to wait.

These and other tensions surfaced in an extraordinary letter Gluzman smuggled to his parents in August 1974, a year and a half into his sentence in Perm-35. And not only to his parents: the letter also circulated in samizdat, by all indications with Gluzman's approval, thereby transforming a filial *cri de coeur* into something more like a generational credo. Gluzman had just learned, in a letter from his parents, that they had tried to visit him in the labor camp, having traveled over a thousand miles from Kyiv, only to be turned away by the camp administration, which did not bother to inform their son of their arrival. Gluzman had earned a previous visit by agreeing to dig trenches around the border of the camp, part of the construction of a new security perimeter—labor that he described to his parents as "immoral from a prisoner's point of view." The local KGB officer had told Gluzman that he had one weak spot: his parents. He had compromised his principles only to satisfy their—not his—desperate desire for a visit. He would not make that mistake again. The KGB, he informed his parents, was wrong: "I have no weak spots. I am not permitted this luxury. Just as I am de facto deprived of the right of correspondence, of qualified medical care, of visitations with relatives, of basic human dignity, so am I deprived of the right to emotions. Such are my days in the camp—cold, hungry, unsentimental."[59]

Gluzman's parents had begged him to "reconsider his values." He was well aware that his arrest was "unspeakably difficult" for them:

> Your expectations [for me] have been shattered: prison instead of a career in medicine and science, hopes of starting a family. . . . all came to nothing. But what values should I reconsider? My critical position regarding the cult of personality, which I learned from my teachers in school and at the institute, from books, films, and not least, from official materials produced by your Party? Should I forget about the

dozens of people I know who have tasted the full horror of our contemporary *oprichnina*? Is all really well in the "state of Denmark"? You are communists. Why are worthy citizens of the "land of socialism and democracy" such as yourselves not permitted even a superficial look at the materials of my case? Why were you barred from the trial of your own son and why were you not shown the verdict? In your letters you lecture me, you believe the words of a professional punisher [the commander of Perm-35] who lacks all conviction, thereby making it easier for him to exert pressure on me. This guy once announced to a friend of mine, "I can stand you on your head if I feel like it." This kind of acrobatics perfectly captures the humanism of the socialist penitentiary system. I experience it on my own skin. Did you know that one misstep or the slightest lagging behind during a prisoner transport is treated as an attempt to escape (I know this from personal experience), that at night, when it was minus 50, they pushed me into the snow and let a guard dog on a leash have at me—"just in case"? Am I supposed to forget how the transport boss "educated" a convicted prostitute by bringing her to his train car for the night?[60]

To reconsider his values, Gluzman informed his parents, would mean "renouncing myself, renouncing the moral foundations I received as a child in your family." The values that had landed him in Perm-35 were in fact their values. Some found expression in works approved by the Soviet state; others were bound up with Gluzman's sense of his inheritance, passed down from his father's father. "I am a Jew," he reminded his parents, "and my Judaism does not consist solely of memory of the victims of genocide, of persecutions driven by a prejudice elevated into dogma. My Judaism includes knowledge of the contemporary life of a people that has its own state, its own history, future, and, thankfully, its own armed forces. My grandfather Abram, executed in Babi Yar, did not ask me to 'reconsider.'" The message was clear: unlike his parents, Gluzman had acted on values that his family and history itself had instilled in him. He had chosen to live with himself. "Whatever happens to me," he wrote, "I have no regrets. I am truly content with my fate. It's difficult for you to understand this. Your generation was shell-shocked by the year [19]37 and the years that followed. Fear, fear, fear. It's unbearable—to fear your own desires."[61]

Born in 1942 and 1946, respectively, Bukovsky and Gluzman belonged to the first Soviet generation to come of age after Stalin. They also shared, as Bukovsky observed shortly after they met in Perm-35, insider knowledge of Soviet psychiatry, one as involuntary subject, the other as professional practitioner. The two teamed up to give a series of lectures on Soviet psychiatry to fellow prisoners, with barracks as an auditorium. It was Bukovsky's idea to turn the lectures into a written guide that would supplement earlier advice literature by Volpin and Albrekht. The resulting *Manual on Psychiatry for Other-Thinkers* was hand-written in tiny letters on pieces of onion-skin paper, rolled into pellets, stuffed into a small plastic capsule, and swallowed by Valery Rumyantsev, a former KGB officer about to complete his fifteen-year sentence for treason. In 1974, Rumyantsev was released from Perm-35, made his way to Moscow, excreted the capsule, and delivered it as instructed to Yuli Daniel, who arranged to have it copied and released into samizdat.[62]

Like other dissident guides, the *Manual on Psychiatry* presented its goal as eminently practical: to identify "correct behavior" for minimizing the chances of being diagnosed as insane by Soviet psychiatrists. In *How to Conduct Yourself at an Interrogation*, Volpin had warned readers against assuming that a diagnosis of mental illness was preferable to a prison sentence, noting the unique lawlessness of psychiatric institutions.[63] His manual, however, which Bukovsky and Gluzman acknowledged as "widely known to the samizdat reader," was designed solely for encounters with interrogators rather than psychiatrists. The danger, they wrote, was that Volpin's legalistic argumentation could itself become fodder for a diagnosis of mental illness. Frustrated by a Volpinesque insistence on procedural fine points, an interrogator might decide instead to request a psychiatric examination. The exam needn't culminate in a finding of "sluggish schizophrenia"; it could always fall back on diagnoses of "litigation mania" or "reformist delusions." In the presence of a professional psychiatrist ordered to produce a diagnosis of mental illness, Bukovsky and Gluzman warned, dissidents armed only with Volpin's techniques would find themselves "absolutely powerless."[64]

The first step was to puncture the mystique surrounding psychiatry itself. "Remember, a psychiatrist is an ordinary person," Bukovsky and Gluzman assured their readers, "with no supernatural abilities. The notion, common in certain circles, that a psychiatrist can 'look into your soul, read your thoughts, or force you to tell the truth' is absurd."[65] As the literary scholar Rebecca Reich has shown, the *Manual on Psychiatry* in effect reverses the diagnostic gaze, categorizing the distinct personality types commonly found among psychiatrists.[66] The "Beginner" is benign but suggestible (especially by superiors). The "Academic" and the "Voltarian" regard the use of punitive psychiatry against dissidents as a dirty business in which they prefer not to take part. "Help [them] by using the right tactics," the manual advises. Most dangerous are the "Philistine" and the "Professional Hangman." The former "does not understand phenomena such as surrealist art ('horses don't really fly') or modern poetry ('where are the rhymes?')." It was best to avoid symbolic language when dealing with Philistines. When Grigorenko, for example, was asked to explain why he engaged in "anti-social activity," he responded, "I couldn't breathe"—which the attending psychiatrist took as a symptom of paranoia. The paradigmatic "Hangman" was Lunts, who "deliberately pronounces criminal diagnoses of insanity upon healthy people."[67]

Given the unmistakable element of caricature in this taxonomy, it seems unlikely that Bukovsky and Gluzman actually expected those undergoing evaluation to diagnose which type of psychiatrist they were dealing with. Rather, their aim was to diminish the sense of intimidation, to remind subjects to think tactically as they interacted with medical personnel. Dissidents undergoing psychiatric evaluation needed to keep in mind that their interlocutors operated according to a pseudo-scientific "class morality," which meant that whatever served the interests of the ruling class (in the Soviet case, "workers") qualified as ethical. Against this "ends justify the means" approach, the dissident "non-class" (that is, universal) ethics of "truth, sincerity, and compassion," when practiced in the lawless setting of Soviet forensic psychiatry, was simply untenable. To give honest answers about one's life—one's childhood, relations with parents, emotional struggles, dreams—was to invite ruin. "Lying is vile," Bukovsky and Gluzman conceded. "But bear in mind that your fate depends on your desire and ability to be amoral vis-à-vis individuals and organizations that profess the morality of Hottentots." In this setting, dissidents' "abstract morality" had to yield to a "worldly" or "grounded" morality.[68]

While lying under interrogation was a punishable offense, lying to a psychiatrist, according to the *Manual on Psychiatry*, was not. The question was *how* to lie. What fictitious version of the self would best shield one from a diagnosis of mental illness? Soviet psychiatry, like its counterparts elsewhere in the 1970s, was overwhelmingly focused on mental disorders and pathologies. Beyond the absence of mental illness, it lacked explicit criteria for psychic health and well-being. Implicitly, however, it relied on a model of normalcy that Bukovsky and Gluzman characterized as petit bourgeois. The mentally healthy "Soviet person," in fact, could be summed up with a phrase plucked from Lenin's magnum opus, *Imperialism, the Highest Stage of Capitalism*. According to Lenin, imperialism generated an enormous accumulation of capital in a handful of countries. This led to the "extraordinary growth of a class or rather a stratum of *rentiers*, i.e., individuals who live off of 'dividend vouchers' [dividends from capital investments]."[69] The "conventional standard of psychic health," according to Bukovsky and Gluzman (paraphrasing Lenin but without attribution), could be found precisely in the personality of this "*rentier* with his 'dividend vouchers'":

> A *rentier* is a person of mediocre intellect, bourgeois in taste; more civilized than cultured, risk-averse. He is satisfied with his modest but stable social position ("the higher you fly, the harder you fall") and doesn't let himself get carried away. He lacks the capacity for any sort of creativity and is a bulwark of whoever is in power. The guiding force in his life is the instinct for self-preservation. His life is monotonous but peaceful. He considers his lifestyle the only correct one, the

wisest and safest in the sea of our existence, full of adversity, pitfalls, and cataclysms.⁷⁰

To be considered psychologically healthy in Brezhnev's Soviet Union, therefore, it was best to mimic the shallow, conservative *amour-propre* of the petite bourgeoisie. The half-century spent forging an improved, post-capitalist edition of humanity had apparently arrived where it started, at the very character traits against which "new Soviet people" were supposed to define themselves. Or as a late Soviet joke put it, in response to the question "What is socialism?": the longest path from capitalism to capitalism.

For Bukovsky and Gluzman, the true antipode of the Soviet *rentier* was not the mythical *homo sovieticus* but the Soviet dissident. Those eager to avoid psychiatric internment therefore needed to prepare anodyne accounts of their childhood and adolescence, free of excessive reading, independent-mindedness, and non-conformist behavior. Had Don Quixote not become delusional from reading too many books of chivalry? The *Manual on Psychiatry* recommended tailoring one's biography to the narrow, self-interested norm (quite the opposite of the heroic biography that Anna Akhmatova saw the KGB inadvertently fashioning for dissidents), as per the following guidelines:

- You show no interest (and never have) in philosophical problems, psychiatry, parapsychology, mathematics.
- You have no interest in contemporary "modern" art or any understanding of it. You do not devote your free time exclusively to solitary pursuits such as reading, gardening, or the contemplation of nature and works of art.
- You have hobbies and are interested in sports (if only as a spectator or fan).
- If you're single, do not explain that fact by citing a lack of attraction to the opposite sex or an antipathy toward family life. Find some other reason (lack of an apartment, low salary, you were planning to marry but were prevented by your arrest...).

- Try to ground your opinions not on the basis of personal experience or by analyzing reality, but by referring to literary sources, statements by authorities, etc. (otherwise the phrase "overestimates his capabilities" will appear in the experts' findings).
- However unpleasant, the best motive you can offer for the actions being imputed to you is: "I wanted to become famous; I didn't understand the seriousness of the consequences, I didn't stand back and look at myself; I didn't understand that I had gone too far," and so on. Unfortunately it is precisely (ugly) motives like these which will be regarded in a positive light at your evaluation.[71]

The parodical undertone in these instructions in no way diminishes the seriousness of their intent. Like Albrekht's guides, which also include satirical elements, the *Manual on Psychiatry* charts the limits of Volpin's legalist approach to relations with the Soviet state. Without repudiating that approach, Bukovsky and Gluzman positioned Soviet psychiatry as the sole remaining arm of state power that sought to probe the inner lives of Soviet subjects (the KGB having lost interest). It was therefore an arena where strictly logical thinking risked being cast as pathological and lying, notwithstanding Volpin's view, was not merely permissible but imperative.[72] Their advice to strategically mimic the alleged cynicism, egotism, and mediocrity of the *homo sovieticus*—or Soviet psychiatry's normative version thereof—only reinforced the notion of dissidents as virtuosi. To live like free people, it seemed, meant to shun vulgar, bourgeois instincts, and thus to be part of the elect.

The *Manual on Psychiatry* described itself as aiming at "the widest possible readership." According to Gluzman, a KGB officer once told him that copies were found wherever they did apartment searches of dissidents.[73] With samizdat, of course, the absence of data on print-runs or book sales makes it impossible to know how many copies of any given text were produced, much less read. But the contents of the conduct manuals explored so far in this chapter suggest that their target audience was limited to the dissident milieu, that is, people for whom interrogations and apartment searches were already a lived experience or at least a real possibility.

FIGURE 16.7. Semyon Gluzman in Siberian exile in 1979, after completing his seven-year sentence in a corrective labor camp

By contrast, Solzhenitsyn's *Live Not by the Lie*, released on the eve of his expulsion in February 1974 (and published in a Russian-language tamizdat edition in Paris soon thereafter), sought a much wider audience—"the people," as he put it in his memoir.[74] In characteristically emphatic prose, Solzhenitsyn spelled out a straightforward code of personal conduct applicable, in theory, to all Soviet citizens: "Let us refuse to say what we do not think!"[75] The Soviet system used to rely on brute force but had grown old and less confident; to preserve itself, it now relied on mass participation in lies. Like a virus, lies required a large number of host organisms in order to survive and spread. "The simplest and most accessible key to our liberation" was therefore "PERSONAL NON-PARTICIPATION IN LYING! Let the lie conceal everything, let it control everything, but we will insist on one small point: IT WILL NOT EXERCISE CONTROL THROUGH ME!"[76]

Concretely, this meant refraining from speaking, writing, or signing even "a single phrase" that one regarded as a distortion of the truth, whether in public or private. It also meant refusing to deploy "authoritative" quotations (a reference to the ubiquitous practice of padding introductions and conclusions with passages—often irrelevant—from Marx, Lenin, and/or the current general secretary of the Communist Party) as a form of ideological insurance, unless one "completely shares the quoted idea" and it is precisely relevant to the issue at hand. One should neither participate in meetings and demonstrations in whose purpose one did not believe, nor raise one's hand in favor of a proposition or candidate whose honesty one doubted. If a speaker in a meeting, assembly, lecture, performance, or film uttered "a lie, ideological nonsense, or shameless propaganda," one should leave immediately. According to Solzhenitsyn, there were already "dozens of people" who had been following this code for years. "If we become thousands," he urged, echoing Volpin's arithmetic of resistance, "they won't be able to take measures against anyone. If we become tens of thousands, we will not recognize our own country!"[77]

Live Not by the Lie did not call on readers to speak truth to power. It urged them to stop "speaking Bolshevik," even in the attenuated form of lip-service.[78] "We have not matured enough to go to the squares and shout the truth," Solzhenitsyn announced, "or to say out loud what we think." "We," of course, did not include Solzhenitsyn, whose no-holds-barred indictment of the Soviet project, *The Gulag Archipelago*, triggered shock waves on both sides of the Iron Curtain. Nor did it include other outspoken dissidents, who were already doing precisely what Solzhenitsyn informed his readers they were not yet mature enough to do (in language strikingly similar to the stock Bolshevik charge of "political immaturity"). The Soviet population, according to Solzhenitsyn, could in effect be sorted into three groups: a tiny number of virtuosi, who said out loud what they thought; those who were at least honest enough not to say what they didn't think; and the "hopeless and worthless" rest, whose "submissiveness," "cowardice," and "timidity" kept them in "spiritual servitude."[79] In place of the standard Soviet trinity of labor (workers, peasants, and intelligentsia), Solzhenitsyn offered his own, grounded in ethics: truth-tellers, non-liars, and liars.

Despite its appeal to ordinary citizens (not just dissidents) and its code of conduct applicable to a host of everyday situations (not just interrogations and apartment searches), *Live Not by the Lie* elicited a muted response in the Soviet Union, at least in the short run.[80] Well before its release, its author had been subjected to a sustained campaign of vitriol in the Soviet press, his novels and short stories blocked from publication in the USSR for over a decade. After receiving the Nobel Prize in Literature in 1970, Solzhenitsyn found himself lionized in the West and, apart from significant portions of the intelligentsia, publicly anathematized in his own country.[81] Even among the truth-tellers, however, the essay seems to have encountered a rather cool reception. Leonid Plyushch admired the idea of non-participation in lying, but considered it "almost impossible to carry out" in a country like the Soviet Union. In an extended analysis of the Nobel laureate's oeuvre, Amalrik, whose pungent criticism of the Soviet system rivaled Solzhenitsyn's, did not even mention *Live Not by the Lie*. Perhaps it was the essay's hectoring tone, so at odds with the dissident movement's ethos of non-coercion, or its peremptory dismissal of "cowards" who made even minor accommodations to Soviet reality. In contrast to Solzhenitsyn's fictional works, noted Pavel Litvinov, *Live Not by the Lie* demanded "a closed system of normative ethics," with no possibility of a "third way," no room for moral ambiguity or skepticism, and little if any compassion for those unable to meet its rigorous requirements.[82]

The dissident movement's hopes of initiating a dialogue with the Soviet government did not succeed. There is nonetheless a kind of dialogic quality—highly asymmetrical, to be sure—in the relationship between the two. "To a considerable extent," Amalrik noted, "the authorities themselves determine the style of the opposition."[83] As this and the preceding chapters have shown, the repertoire of contention developed by dissidents in turn helped shape the evolving policies of the Soviet government. It was the Kremlin's retreat from mass terror after 1956 that unintentionally opened up space for dissent (permitted or not). The government's initial reliance on the legal system and "socialist legality" as an antidote to the "cult of personality" (and to maintain limits on non-conformism) helped inspire Volpin's law-based tactics of resistance. Those tactics—transparency, civil obedience, and samizdat—in turn led Yuri Andropov and Filipp Bobkov, architect and director, respectively, of the KGB's new Fifth Directorate, to seek alternatives to Soviet courts, which time and again delivered the desired legal outcome but the wrong political effects. Extra-judicial punishment and threats thereof were never absent from the Soviet system, but in the form of the Fifth Directorate's "prophylactic conversations," they mushroomed by the early 1970s into a mass phenomenon. Dissident conduct manuals are best understood as a response to this latest development in the mutual evolution of the movement and its official antagonists, a new samizdat genre reflecting and feeding a preoccupation with unmediated encounters between rights-defenders and Soviet authorities. (It is telling that no one produced a manual on 'how to conduct yourself at a trial'). This ongoing series of mutual adaptations constitutes the "meta-move" list in the contest between dissidents and the Soviet state.

17

Allies, Bystanders, Adversaries

I'm no Solzhenitsyn, of course. Does that deprive me of the right to exist?
—SERGEI DOVLATOV

The annual transparency meetings on Moscow's Pushkin Square on Soviet Constitution Day (December 5) offer a suggestive visual metaphor for the place of dissidents in Soviet society. At their core was a cluster of men and women, typically ranging from a few dozen to 150 or so, gathered at the monument to Russia's revered national poet, from whose pen had come some of the most deeply resonant expressions of Russian humanism—along with a profound skepticism of democracy, formal rights, and the rule of law.[1] Bound by ties of friendship and trust, the individuals assembled around Pushkin communicated with onlookers textually, silently holding handmade signs and banners, or, when that proved too risky, removing their hats in wordless homage to the Soviet Constitution and victims of state lawlessness.[2] A circle of spectators, standing at some remove from the gathered activists, usually included an unknown number of the demonstrators' friends and colleagues who shared their critical view of the Soviet order and admired their courage but were reluctant to jeopardize careers and families by entering the inner core of the gathering. Among the bystanders were also militia and plainclothes officers from the Fifth Directorate who occasionally filmed or harassed participants. Passersby constituted by far the largest element, waves of Muscovites flowing into and out of the adjacent Pushkinskaya and Tverskaya metro stations, some briefly lingering to take in the unusual spectacle, others hurrying past without seeming to notice.

Such were the circumstances in which dissidents found themselves in Soviet society: a small core of activists with an unknown and unknowable number of silent allies among those paying attention (referred to informally as "semi-dissident" or "dissident-proximate"), against the backdrop of the 250 million Soviet citizens whose sympathies were uncertain but who appeared either indifferent or hostile to the dissident cause.[3] To be sure, the ability to convey dissenting views to sympathetic bystanders textually (via samizdat, amplified by the Voices) constituted a critical breach of the Soviet government's information monopoly. But dissidents had few if any means to gauge how those views were received beyond the clandestine world of samizdat debates. The absence of public opinion polls and multi-candidate elections (which would have enabled at least a crude assessment of the distribution of popular opinion) left dissidents along with everyone else to rely on anecdote and speculation.

The Soviet state, meanwhile, used its control of mass media—print, radio, and television—to press home the idea that, as a ubiquitous slogan put it, "The People and the Party are One." Dissidents were hardly a frequent topic in Soviet newspapers, but when they did appear, they were invariably cast as figures alien to Soviet life. A speech by Leonid Brezhnev published in *Pravda* in 1977, for example, described dissidents as tools of foreign "subversive centers." "Our citizens," Brezhnev said, "demand that these allegedly prominent figures be treated as opponents of socialism, as traitors, and as accomplices, if not agents, of imperialism."[4] "What is behind the furor over 'human rights'?" asked a *Pravda* editorial. The answer: "a pitiful handful of characters with anti-Soviet sentiments, an insignificant and tiny group of people who do not represent anyone or anything, are remote from the Soviet people and exist merely because they are supported, paid, and praised by the West."[5]

It may have struck some readers as strange that a pitiful and insignificant group of individuals should periodically surface in speeches by Brezhnev, Yuri Andropov, and other Soviet leaders, but that could perhaps be explained by the inordinate attention certain dissidents received from Western media (which, it was assumed, were carrying out orders from ruling elites, as Soviet media did). Indeed, attacks on dissidents in Soviet newspapers were typically precipitated by praise and attention from the West, rather than by their own writings and pronouncements, which were rarely mentioned, much less debated. *Literaturnaya gazeta* assured its readers that dissidents were far from the fearless fighters for human rights so frequently lauded by Westerners as "models of rectitude and purity." As Andrei Amalrik overheard a KGB investigator tell a subordinate, "Don't go around thinking they're saints! They drink vodka, they have broads!"[6]

Saints or renegades: Cold War images of dissidents tended to the extremes. For many inhabitants of the USSR, praise by the West confirmed the dissidents' status as traitors; for others, condemnation by the Soviet press made them heroes. For Soviet citizens whose social or professional networks—or samizdat reading and listening habits—brought them into proximity with rights-defenders, the reality was considerably more complex, morally as well as psychologically. The present chapter probes the relationships between dissidents and their onlookers in Soviet society, between rights-defenders and those one might call allies, bystanders, or adversaries. The result will be not a tallying of sympathy, indifference, and opposition to the dissident cause—something the available evidence hardly allows—but rather a study of modes of relating to the dissident phenomenon by those adjacent to it, and the impact of those relationships on the phenomenon of dissent itself. Were other-thinkers succeeding in Alexander Volpin's mission of spreading legal consciousness to their fellow citizens? Were they revolutionizing the way revolutions are accomplished?

"Movement people need to be present in all spheres of life. If only by means of our irreproachable conduct, we need to set an ethical example for those around us."[7] Vyacheslav Igrunov's vision of dissidents as charismatic virtuosi, capable of exerting a magnetic pull on bystanders, was not entirely far-fetched, as his own experience suggested. On August 25, 1968, the nineteen-year-old Igrunov was one of many Soviet citizens who learned via the BBC of the demonstration on Red Square that afternoon. Listening from his communal apartment in Odesa, he took special note when it was reported that one participant, Larisa Bogoraz, was the wife of the convicted writer Yuli Daniel, then midway through his five-year sentence in a strict-regime labor camp. "For me, family tradition carried enormous significance," Igrunov noted, describing himself as "an idealist." "If the husband was arrested for the cause, if the wife was arrested for the cause, that meant things were good in the family." He began to read samizdat essays by Bogoraz, captivated by their lively and humane sensibility. When he obtained a copy of her final statement at the trial of the demonstrators later that year, with its declaration of personal responsibility for the actions of one's government and its equation of silence with complicity, he "became her fan."[8]

The human networks that connected dissidents in Odesa with those in Moscow were fragile. The arrest or relocation of a single individual could sever the tie. This is what happened in 1972, when the Fifth Directorate's operation against the *Chronicle of Current Events* paralyzed the transmission of samizdat across much of the Soviet Union, leading many to assume that the movement was dead. The "very painful" rupture in communication with Moscow ended two years later, according to Igrunov, when he himself became a courier between the Soviet capital and Odesa. The first person he requested to meet in Moscow was Bogoraz, recently returned from four years of Siberian exile. A friend of a friend provided her address and phone number, and in January 1974 Igrunov stood at the threshold to her apartment, tingling with anticipation. The person who opened the door, "about whom so many legends had formed," looked shockingly old (Bogoraz was forty-four at the time). Even though the norm among most dissidents was to quickly switch to *ty* (the informal "you"), Igrunov could not bring himself to address Bogoraz with anything less than *vy* (the formal "you") or the ultra-respectful "Larisa Yosifovna." During their conversation, Igrunov, then twenty-six, had the sensation of being "a little boy" conversing with "a person full of life's wisdom." The feeling of infantilization seems to have derived from the superhuman moral authority he had invested in her from afar, now brought to the surface by sitting at a kitchen table with the actual Larisa Yosifovna Bogoraz.[9] Suddenly, he was aware of his own child-like wish to be like her.

Myroslav Marynovych, a young Ukrainian "hungry for a monumental accomplishment," experienced a similar charismatic aura in the dissident circle around the journalist Viacheslav Chornovil and the poet Atena Pashko. Chornovil had repeatedly gotten into trouble with the KGB for publicizing the imprisonment of Ukrainian intellectuals. For Marynovych and other restless youth, the tormented love affair between Chornovil and Pashko had become "a symbol described variously as poetic, heroic, anguished, haunted, and confined," as Chornovil was repeatedly threatened or imprisoned and the KGB pressured Pashko to renounce her partner. Visiting their apartment in Lviv for the first time in 1973 (Chornovil had just been sentenced to six years of hard labor and three years of internal exile), the twenty-four-year-old Marynovych came "not as a guest, but as a pilgrim." Like Igrunov, he felt overwhelmed in the presence of dissidents who were "taking part in something monumental." "The suffering of these people was sacred to me, they themselves were beyond suspicion, and the path they traveled on was my path."[10]

These are accounts by young men from the provinces who were drawn to dissidents as much by their aura of personal heroism and suffering at the hands of the state as by the substance of their dissent.[11] Ideas of civil or human rights or holding the Soviet state to its own laws barely figure in such narratives. Many of the letters and telegrams—and occasionally unannounced visitors—that reached Pavel Litvinov, Petr Yakir, and Andrei Sakharov similarly gave the impression that the dissidents' "ethical example," as Igrunov put it, was indeed capable of attracting would-be allies. In a handful of cases, such figures could be found inside the Communist Party apparatus or even the KGB. Anatoly Chernyaev, a member of the Central Committee and an assiduous diarist, could not help but admire the "special community" of dissidents who were willing "to engage in great self-sacrifice, to demonstrate a devotion to each other that is unusual for today's social norms."[12]

Inside the Fifth Directorate, Viktor Orekhov's job analyzing samizdat and surveilling Moscow dissidents similarly elicited unexpected sympathies. As a plainclothes officer in the early 1970s, he was assigned to monitor a December 5 transparency meeting on Pushkin Square, watching in disgust as fellow undercover agents dumped snow on Sakharov's head. At a gathering at Ludmilla Alexeyeva's apartment, Orekhov—still a loyal Party member—encountered "a different world." With their "carefree style, openness, and freedom of thought," dissidents appeared to him like "potentially dangerous extra-terrestrial beings."[13] A trip with the Bolshoi Ballet to Japan, where Orekhov served as the KGB's eyes and ears vis-à-vis the USSR's star dancers, exposed him to another unknown world, the world of capitalism, which looked nothing like what he had been led to expect. The turning point came when, back in Moscow, Filipp Bobkov, head of the Fifth Directorate, placed Orekhov on a team of agents responsible for vetting the thousand-page typescript of Alexander Solzhenitsyn's *Gulag Archipelago*, which the KGB had just tracked down and confiscated after the successful five-day interrogation of Elizaveta Voronyanskaya, one of Solzhenitsyn's typists. If Solzhenitsyn's magnum opus stunned Orekhov with its portrait of "an entire country under the yoke of totalitarianism," then Voronyanskaya's suicide just days after being broken by the KGB showed that the yoke was still in place—and that Orekhov was indirectly helping keep it there.[14]

Over the next several years, Orekhov went rogue. It began with clandestine warnings to dissidents that the KGB was planning to search their apartments, so that samizdat and other incriminating materials could be removed in advance. Yuri Orlov and Alexander Podrabinek were tipped off about their impending arrests. Orekhov passed on information about prison sentences decided by the KGB before trials began, so that dissidents could publicize the sentences in advance and thereby discredit the judges. He surreptitiously arranged to provide courtroom passes to friends and relatives of dissident defendants. In most cases, his supremely risky overtures failed: their intended recipients refused to believe him, refused to resort to clandestine methods, let alone collaborate with a KGB officer. Convinced that he was a provocateur, some dissidents, according to Orlov, "spoke rather openly about the mysterious KGB captain in the presence of listening devices planted in every dissident apartment." The Fifth Directorate caught Orekhov in 1978; he was sentenced by a military tribunal to eight years in a corrective labor camp.[15]

Unlike Orekhov, the vast majority of sympathetic bystanders preferred to remain bystanders. The biologist Sergei Kovalev, who helped compile the *Chronicle of Current Events,* served as a member of the Initiative Group for the Defense of Civil Rights, and signed countless open letters and petitions until his arrest in 1974, divided the non-dissidents in his orbit into four categories.[16] One comprised quite a few of his professional acquaintances—biologists, physicists, and mathematicians—who sympathized with and indirectly supported rights-defenders. When Kovalev was forced to resign his position at Moscow State University in 1969, colleagues arranged for him to work as a research assistant or technician in their labs so that he could continue to support his family. Others offered their apartments as workspaces or gathered monthly donations to supplement Kovalev's meager income. Friends volunteered to type and reproduce his samizdat statements as well as the *Chronicle.*

The second category consisted of colleagues, especially those of an older generation, who disapproved of Kovalev's dissident activities. Soviet science, they argued, depended on the state, and the state took good care of its scientists with generous funding and access to coveted dachas. In return, a certain gratitude was to be expected. Whatever problems arose from faulty policies or directives were better managed piecemeal and off-stage rather than through open confrontation. As for certain regrettable instances of state lawlessness, was it really the job of scientists to fix such things? Hadn't the situation improved dramatically since Stalin's time? Dissidents dirtied the collective nest by discrediting the state and compromising national security. They were rightly punished. The entire discourse of human rights was just a propaganda weapon in the hands of the West against the USSR.

A third cohort argued that social and political issues should be managed by professionals—not dilettantes. "Imagine, Sergei," explained a colleague from a neighboring lab, "that someone started lecturing you about cell membranes" (one of Kovalev's areas of expertise). "You'd tell him to get lost, and you'd be right. So why do you consider yourself an expert when it comes to politics?" Kovalev's objection that his activism was in the realm not of politics but of human rights, which concerned everyone, did not convince his interlocutor (in post-Soviet Russia, Kovalev himself would abandon the idea that the defense of rights was apolitical). This colleague, according to Kovalev, had been shaped by the conviction "implanted in all of us over the course of decades" that "certain questions were best decided for us."[17]

The fourth and final group, the largest within Kovalev's taxonomy of onlookers, shared with dissidents the sense that the Soviet system faced serious problems, yet considered open resistance not only futile but likely to intensify the state's repressive tendencies. Better to work quietly, taking "small steps." For some, those steps involved liberalizing the Soviet economy, from which political liberalization would then naturally flow. For others, "small steps" meant joining the Communist Party and working from inside the machinery of power. Neither of these options struck Kovalev and other dissidents as remotely realistic.[18] Otherthinkers did not think very much about political economy. Enduring contradictions between Marxist theory and Soviet reality had cured them of the assumption that the socio-economic "base" determined the political-legal "superstructure." As for joining the Communist Party— that signified giving in to cynical careerism, not least because the Party's central function following Khrushchev's ouster in 1964 seemed to be to stymie structural reform of any kind.

Since the break-up of the USSR, all sorts of former Soviet citizens have recorded their experiences in memoirs, including their encounters with and attitudes toward dissidents.[19] For all their diversity, one common feature of most such accounts is their relative silence when it comes to dissident ideas. What turns out to have mattered most to those Soviet citizens who crossed paths in one way or another with people like Kovalev was the disruptive impact dissent had on their way of life. Dissidents, as the ex-wife of one of them put it, were "disturbers of the peace."[20]

Peace was something most people in the Soviet Union did not take for granted. The first half of Soviet history was an almost unending saga of mass violence: global wars, civil wars, class war. The crusade to build socialism proceeded along a series of war-like "fronts," whether in agriculture, housing, education, or culture. Stalin may have rejected Trotsky's theory of "permanent revolution," but for a quarter-century he kept Soviet society on what amounted to a permanent war footing against perceived enemies inside and outside the country. Even Khrushchev, who presided over an extraordinary retreat from state-sponsored violence, kept Soviet citizens on edge with his frequent zig-zags in both domestic and foreign policy. Brezhnev, by contrast, delivered not just "stability of cadres," that is, an end to Khrushchev's practice of periodically reorganizing the institutions of governance, but social and economic stability for the Soviet population as a whole. Not for nothing are Brezhnev's nearly two decades as Soviet leader (1964–82) remembered as a golden era, a "respite from history."[21]

The Brezhnev era's unprecedented social peace rested in part on the blossoming of informal ways of managing the everyday challenges and contradictions of Soviet socialism. Revolutionary blood-letting yielded to mutual back-scratching. Soviet life was characterized not just by pervasive informality (as described in chapter 12), but by a vast repertoire of illegal or semi-legal practices at virtually every level of society. The most widespread of these belonged to the so-called second economy or shadow economy, in which goods and services were reallocated by private means according to private preferences.[22] The USSR's shadow economy was not a fringe phenomenon; during the Brezhnev years, it is estimated to have constituted nearly a quarter of average household expenditures and somewhere between 12 percent and 22 percent of the USSR's gross domestic product.[23] It encompassed the barter, theft, and repurposing of all manner of state property from food to apartments to construction materials. It included the exchange of favors, the giving and receiving of bribes, and more. Just as samizdat was a response to the deficits of official Soviet print-culture caused by censorship, an attempt to satisfy unmet demand for information, ideas, and culture, so the shadow economy was largely a response to the myriad deficits created by the USSR's centralized command economy, a way of satisfying unmet demand for consumer goods and services. Both operated in what one economist has aptly called "the penumbra of the legal."[24] Whereas samizdat was largely the province of the intelligentsia, the shadow economy counted among its participants peasants on collective farms, factory workers, intellectuals, members of the Communist Party—the entire spectrum of Soviet society.

Informal practices in the USSR were not just techniques for obtaining goods and services. They expressed a broader cultural norm that validated and in certain respects celebrated the bending or circumvention of rules in order to make everyday life more bearable in the presence of an overweening state and an inefficient economy. It would not be an exaggeration to call such practices a national pastime, especially during the Brezhnev era. "Law is like a telephone pole," a Soviet saying went. "You can't jump over it, but you can get around it." Getting around the law was often a group endeavor, in many cases predicated on and reinforcing bonds of trust and loyalty—the same small-scale, highly personalized form of sociability from which the dissident movement had emerged in the mid-1960s and to which it kept returning, as if to a default setting. No wonder, then, that in a society consisting of small, internally loyal and externally mistrustful coteries, the Soviet government found itself calling over and over again for the "strengthening of legality," suggesting how elusive that goal remained—not least for the government itself, which similarly consisted of patronage networks defined by internal loyalty and external mistrust.

For millions of Soviet citizens whose well-being depended on a flexible approach to rules, the dissident call for transparency and the rule of law had unwelcome implications. "When in Russia they say 'It is necessary to strengthen legality,'" observed the sculptor Ernst Neizvestny, "all Russia begins to tremble."[25] When that call came not from the Kremlin but from dissidents, the reaction was more likely to involve irritation than trembling. The dissident agenda, even if it stopped well short of Volpin's dictatorship of reason, threatened to disrupt the entire post-Stalinist dispensation, according to which the lip-service state demanded outward conformity in exchange for diminished scrutiny of citizens' private lives. In comparison to the typical Western pattern of conformism, which involves both cognitive and behavioral compliance with dominant values, Soviet conformism was more flexible, often combining compliance in behavior with non-compliance in viewpoints.[26] Doublespeak—or, as the teacher Valeria Gerlin was instructed by her boss, "taking our grievance no further than the kitchen"—seemed a small price to pay for the peace and predictability of the Brezhnev era.[27] Why insist on a single set of ethical principles for the private kitchen and the public square? Are you really lying if your listeners know you're lying, and they do the same? To lie, after all, is "to intentionally mislead others when they expect honest communication."[28] If they don't expect honest communication—by actors in a play, for example, or by co-workers at a mandatory Party meeting (also a kind of theater)—then are you in fact lying? Why make a fuss about the government not obeying its own laws when everybody understood that those laws were never meant to be taken literally? Why antagonize the Soviet leadership with protests about rights violations when protests didn't seem to accomplish anything?

Individuals sympathetic to the dissident movement may have gotten over the fear of encountering former acquaintances who returned from the camps, unlike an earlier generation that routinely shunned survivors of Stalin's Gulag. But they were wary of bumping into Volpin on the bus. An art historian recalled a fellow university student with "dissident tendencies" who "was constantly saying something skeptical about the Party, about the Soviet system." These pronouncements "made a strong impression—they elicited not so much fear as unpleasant sensations. It's one thing to read Dostoevsky and quite another to interact with his protagonists.... When a real person is standing in front of you constantly saying something skeptical, it becomes unpleasant. He expects some sort of reaction from you, but you have nothing to say. Not because you aren't capable of analyzing the way he does, but because you don't want to."[29] Dissidents didn't deny this effect; if anything, it confirmed their cherished belief that they were saying out loud what many others were thinking. In the 1970s, covert emissaries to the USSR from the anti-Soviet émigré organization People's Labor Alliance (NTS) periodically reported their impressions of the public mood in Moscow and other cities. "One often hears," noted one report, "that the dissidents are trying to create unnecessary noise and confusion, unnecessary commotion." There were complaints about "the small group of noisy, half-crazy disturbers of the peace."[30]

Soviet socialism offered fertile ground for the cultivation of ironic detachment from official norms. The post-Stalin retreat from policing Soviet souls made such detachment easier, provided, of course, that it remained private. Dissidents were thus not the only ones pursuing inner freedom. Among its other effects, an ironic stance vis-à-vis official ideological discourse expanded the space for humor, of which there was no shortage in the USSR. As the anthropologist Alexei Yurchak has persuasively shown, late Soviet humor rarely aimed to undermine or resist the existing order. Its purpose was rather to signal a knowing awareness of the yawning gulf between Soviet ideology and Soviet reality. One could thereby confirm—to oneself, but also to one's fellow joke-tellers, since by the Brezhnev era joke-telling had become a common group ritual—that one's lip-service was cynical rather than naive. Lip-service in this sense was the public façade that enabled a zone of private freedom (including humor) to operate with minimal disturbance. Freedom in secret became the secret of freedom.[31]

This was very different from the dissident view that genuine inner freedom required the abandonment of double-speak in the name of transparency. As we saw in chapter 11, Vladimir Zelinsky's influential samizdat essay "Ideocratic Consciousness and the Individual Person" sharply distinguished between two forms of self-emancipation from Soviet ideology: the "spiritual" and the "cynical." In the first, having realized that their public face was in fact a mask (in the form of ideologically conditioned reflexes) rather than their true self, certain individuals chose to remove the mask in the name of authenticity and wholeness. The second and much larger group, having come to the same realization, chose to keep their masks on.

The cynics could cite eminently pragmatic reasons for continuing to wear their masks. Masks served as shields, protecting their bearers from excessive scrutiny; masks made a certain private freedom and a rich inner life possible, which for many people was well worth the price of hypocrisy. Mask-wearing, after all, was by now so ubiquitous that most people, or at least most grown-ups, understood perfectly well that public utterances were merely performative. As one of Yurchak's subjects put it, when asked about the dissident movement, "Everything was already clear to everyone, so why speak about that? It wasn't interesting."[32] The ritual of unmasking had once been an integral part of Bolshevik campaigns against hidden enemies, but now masks were an accepted accoutrement of Soviet life. Why return to the earlier practice?

Leonid Plyushch argued about this with his cynical friends, who eventually became former friends. Plyushch was not given to compromise. Members of the Initiative Group described him, during his nearly three-year imprisonment in a psychiatric hospital in Dnipropetrovsk, as having "acted according to the well-known logic of a pre-revolutionary intellectual. Whether a Tolstoyan, a Populist, or a Marxist, [such an intellectual] always pushed things to the limit, sacrificing himself and the peace of those close to him."[33] His cynical former friends, according to Plyushch, saw things differently:

> "There is always shit, there always has been and always will be. There is shit in people, in the state, in struggle." They savored the absurdity, the despair, and the pathology of society and hurled accusations against participants in the resistance: "They will bring new blood, a new Gulag, shit that stinks. Whereas Soviet shit has already dried up, and if you don't touch it, you won't smell it."[34]

On this topic, too, there was an illuminating *anekdot*:

> A crowd of people is silently standing in a pond full of shit that comes up to their chins. Suddenly a dissident falls in and starts thrashing his arms and shouting in disgust: "How can you put up with this? How can one live in such conditions?" To which the crowd quietly responds, "Stop making waves."[35]

In this allegory of the dissident as disturber of the peace, the Soviet population lives in miserable conditions, but the dissident's response to those conditions only makes things worse. It is telling that the crowd does not dispute the dissident's assessment of the situation, but merely his (or her: in Russian the dissident's gender is not indicated) reaction to it. It is also telling that the disturbance—ripples of sewage washing up against people's faces—is caused directly by the dissident, rather than by the state's punitive response to the dissident's public criticism. The state is entirely absent, leaving the focus exclusively on relations between dissidents and the rest of Soviet society. The dissident appears in this allegory, moreover, as new to the pond, an outsider suddenly plunged into the conditions of Soviet life, rather than someone who grew up in those conditions and has come from within to find them intolerable. You didn't need to be a stranger to experience the estrangement that lay bare the reality of Soviet life. But to publicly criticize that reality, apparently, you did. Functionally, the dissident in this allegory has been transformed into a foreigner.

In his influential essay "How Would I Have Conducted Myself?," the scholar Jan Philipp Reemtsma notes the tendency, when reading or thinking about Nazi Germany, to wonder how one would have acted in that setting. Would I have participated in mass murder? Would I have hidden Jews and other hunted individuals from their would-be murderers? Or perhaps joined the resistance? Or would I have sought to remain as safely neutral as possible?[36] All reflection about the past involves a certain degree of conscious or unconscious identification with its protagonists, of course; without such identification, it would be impossible to understand how they viewed themselves and their world. The stark moral antinomies that characterize historical phenomena such as Nazism or the American abolitionist movement (to name only two prominent examples) tend not just to intensify the process of identification but to turn it into an occasion for self-scrutiny. Imagining oneself as an historical actor in such a setting becomes a vehicle for asking: What kind of person am I? What kind of person do I hope I would be under such acute circumstances? How would I conduct myself if what I consider to be my deepest values were put to the test?

Traces of such self-scrutiny surface in contemporary Western writing about Soviet dissidents, not least because dissidents struck many observers as fearless champions of their own deepest values.[37] Older generations of émigrés, too, wondered how they would have conducted themselves. "What would you have done, if you had remained in the motherland?" the ninety-two-year-old poet Sophia Dubnov-Erlikh asked herself, after reading Efim Etkind's account of his forced emigration from the USSR in 1974 for supporting Joseph Brodsky and Solzhenitsyn. "I have been interrogating myself in the style of the Inquisition," she wrote to Etkind."[38] Dissidents, too, debated "how to conduct oneself," not as a hypothetical proposition, as in Reemtsma's essay and Dubnov-Erlikh's letter, but in the here and now. Many Soviet bystanders pondered that question too, with dissidents as an inescapable reference point. A decade and a half after the Soviet collapse, for example, the Soviet historian Aron Gurevich praised the dissident movement as "absolutely essential for our spiritual life and the life of our society." Then came the confession:

> The guiding principle of my public conduct was non-participation. . . . I was never a dissident and didn't join the rights-defense movement. The fact that, over time, my trustworthiness began to come under suspicion in certain [official] circles says less about me than about those Party and academic functionaries who kept me "under the lid," refusing to let me travel abroad and firing me from this or that position. Therefore, in response to the question "Was I a citizen?" I must repent: I was a passive subject. Non-belonging and non-participation hardly constitute a worthy civic stance. When dissidents were persecuted, when Solzhenitsyn and Sakharov were publicly denounced, where were we? We sympathized with them, but in a cowardly manner, in silence, thereby facilitating the arbitrariness of the authorities. "Almost everybody conducted themselves like this," one might object. Yes, but firstly, not everybody. And secondly—and this is the main thing—everyone has to answer to his own conscience. References to the herd instinct cannot serve as an excuse.[39]

The phrase "I wasn't a dissident, but . . ." is a commonplace in post-Soviet recollections by individuals who sympathized with the movement but were unwilling to expose themselves and their families to the enormous costs of active participation.[40] For Gurevich and others, the phrase served as a kind of apology, as if being a dissident would have been more honorable—if only the author had had the courage. More than a few Soviet émigrés retroactively recast themselves as dissidents to raise their cachet abroad, leading the satirist Alexander Zinoviev to remark that "many more dissidents arrived in the West than left the Soviet Union."

Other bystanders were less apologetic. The writer Lev Navrozov, for example, was acquainted with a number of dissidents and admired "their personal altruism, kindness, and integrity, their devotion, pride, and fortitude." But he worried that "those who dared to express their dissidence openly" would draw him into an action that could get him arrested. Refusing to sign protest letters, however, would cause him to be regarded "by a certain public opinion" as a coward, an obsequious careerist. Navrozov concluded that he belonged to a different "spiritual vintage."[41] The fate of Winston Smith, the doomed dissident protagonist of George Orwell's novel *1984*, loomed large. "The goal of my life," Navrozov declared, "was to prove Orwell wrong, to prove that the individual can develop a defense mechanism matching the omnipotence of the state. I developed this defense mechanism. Why should I then disarm, let the state-security office get at my real self and thus perhaps triumph over me, since my actions would be sufficient proof of a political trespass? Anyway, all my writings would have been confiscated in the very first house search."[42] Like many intellectuals, Navrozov was more intent on insulating himself from the Soviet state than on reforming it, which in any case appeared all but impossible. A similar approach guided the editors of the Leningrad samizdat journal *Thirty-Seven*: "The flat spirit of opposition is incapable of inspiring any of us." True freedom required that one *not* be in dialogue with the state, including the state's laws.[43]

Self-scrutiny triggered by the presence of dissidents could elicit not just repentance, as with Gurevich, or an alternative self-definition, as with Navrozov, but disdain. Even sympathetic bystanders were known to use the term "professional dissident" to mark their distance from someone who seemed to turn protest into a career, with all the attendant implications of self-aggrandizement and profit-seeking. "I don't call myself a dissident," recalled one Soviet observer, "because for me it was not a profession."⁴⁴ "Professional" dissidents allegedly had no other source of identity and social status. They protested in order to acquire renown, or to raise their chances of being allowed to leave the Soviet Union, or to tap into the network of mutual aid, including financial aid, established by dissidents and their supporters in the mid-1970s.⁴⁵ Even bona fide dissidents used the term "professional dissident" along with references to Dostoevsky's "devils" to refer to activists whose motives they considered less than admirable. As Sakharov pointed out, it was the Soviet government that created "professional dissidents" by derailing the careers of those who signed protest letters or produced critical essays in samizdat.⁴⁶

Bystanders were sometimes presented with a difficult choice: publicly repudiate their dissident acquaintances or face extra-judicial punishment. The most acute example of this dilemma arose in the early 1970s in connection with the Kremlin-orchestrated media campaigns against Sakharov and Solzhenitsyn. Hundreds of scientists, writers, and other intellectuals were pressured to sign letters condemning the allegedly treasonous behavior of the two men, under threat of losing high-status appointments, access to coveted dachas, permission to travel abroad, and other state-controlled privileges. For dissidents it was a no-win situation. Colleagues who signed such letters—among them eminent figures such as the composers Dmitry Shostakovich and Aram Khachaturian, the writers Konstantin Simonov and Konstantin Fedin, the violinist David Oistrakh, and Sakharov's fellow physicists Mstislav Keldysh and Yuli Khariton—resented dissidents for turning them into public accomplices of state repression (signatories' names appeared under the denunciatory letters in leading Soviet newspapers). Some of those who refused to sign—among whom were the nuclear physicists Petr Kapitsa and Anatoly Alexandrov—resented dissidents for precipitating their loss of privileges. Signers and non-signers resented each other. "It poisoned the atmosphere," recalled Central Committee member and "in-system reformer" Georgi Arbatov.⁴⁷

More than a few friendships came to an end. "The liberals' departure from the Movement after 1968," recalled Gleb Pavlovsky, "was experienced—and not only by me—as a betrayal. It was unbelievably painful."[48] "A Soviet dissident," according to Ludmilla Alexeyeva, "quickly became a pariah even among people who privately shared his views. He served as a silent, or not so silent, reminder that some people in Soviet society had chosen to act as citizens. By just being there, a dissident could induce guilt."[49] Dissidents' pronouncements could leave the impression that they considered themselves the only true citizens of their country; the rest of the population were "subjects" (Gurevich), "cretins" (Bogoraz), "the masses" (Alexeyeva), or, at best, "fellow citizens, cows and oxen!" (Volpin). Subjects, cretins, and masses were nonetheless given to understand that they shared responsibility for all the Soviet government's repressive measures. This cannot have been a welcome message for most Soviet people, especially when their government was constantly reminding them that what they were in fact collectively responsible for was defeating Nazi Germany and leading the struggle to liberate humankind from capitalism. As Zinoviev put it, "The dissidents were the bad conscience of Soviet society."[50]

The fraught relations explored in the preceding pages, it should be emphasized, pertain to bystanders who to varying degrees shared the dissidents' position. But it wasn't only "liberals," to use Pavlovsky's term, who distanced themselves from the movement after 1968. Rights-defenders, as we have seen, prided themselves on their ecumenicism, rejecting ideological or political litmus tests in favor of a minimum program of inner freedom and strict rule of law. For some, however, that minimum program itself lost its allure after 1968. Tatyana Goricheva, a philosophy student and feminist who was baptized in the Russian Orthodox Church in 1974, had admired dissidents for opposing "the decline of the people into hopelessness," giving them "faith in their own possibilities," and "freeing [the human] personality" from all-encompassing state control. For Christians, however, the intelligentsia "cult" of individualism, resistance, and overcoming fear harbored the "mother of all sins—pride." Dissident heroism concealed a dangerous "self-love, theatricalism, egocentric self-confidence." Rather than fearlessness, which risked "being diverted into a self-satisfied and superficial Soviet humanism," Goricheva recommended the fear of God. "Merely heroic people," she concluded, "need an enemy, a borderline situation, risk, standing at the edge. I also loved all that but at the same time I am ashamed of it because God descends into the heart which is purified of passion," of the desire for individual agency. Whereas dissidents, in order to combat civic passivity, called on Soviet citizens to accept responsibility for all the actions of their government, Goricheva found herself drawn instead to "the beautiful and liberating feeling of guilt, about everyone and everything," as an antidote against "human self-regard."[51]

For Gennady Shimanov, the rupture was even sharper. The largely self-taught son of Russian peasants, Shimanov came into the orbit of the dissident Orthodox priest Dmitry Dudko in the early 1960s. As Khrushchev's anti-religious campaign was nearing its zenith, Shimanov spent several months at a psychiatric hospital being treated with insulin injections to cure him of his belief in God. His memoirs, published two decades after the Soviet collapse, describe his thinking at the time as follows: "I'm for God; Soviet power is against God; therefore, I have to be against Soviet power." He met Vladimir Bukovsky at the Mayakovsky Square poetry readings and soon thereafter joined Moscow dissidents in their efforts to contain the Soviet state within the limits of the Soviet Constitution. Volpin's ideas about rights were "true," Shimanov later recalled, "but they weren't the only truth, they weren't the full truth or the main truth in the lives of Soviet people." They were dangerously value-free. By drawing attention to the enormous contradictions between theory and practice in the Soviet system, Volpin "made people blind both to the vices of the West and the merits of the USSR."[52] By the early 1970s, Shimanov had come to regard civil rights in the West as no less a sham than those contained in the Soviet Constitution, and Soviet rights-defenders as seeking to "hypnotize" Soviet citizens with the idea of human rights. From here it was only a short step to the claim that an emphasis on individual rights would inevitably lead to a "decline of morality, of the family and national culture, without which the people would be transformed into a mass of solitary individuals, egotistically cut off from one another." The solution, Shimanov decided, was to make Russian Orthodoxy the state religion of the Soviet Union. This was the only way to ensure that the USSR remained an "ideocratic state," which

Shimanov, in stark contrast to Zelinsky, considered necessary for the salvation of the Russian people.⁵³

The turning point for Shimanov came in 1971. At the time, he was a "respected dissident," welcomed in the rights-defending milieu. He didn't like what he found:

> I discovered a lot of Russophobia there. Before, I hadn't considered it significant, since my conception of the Jews came from Soviet sources, according to which they were a people that had been persecuted before the revolution, an unhappy people. They feel offended, and they express this feeling in a way that leads to Russophobia. But when I saw them in real life, I understood that this wasn't the case at all. It was their inborn chauvinism. And then my inexhaustible patience snapped. True, the conflict occurred not with Jews, but in their presence—at the apartment of the Chuvash poet Gennady Aygi. He too was anti-Soviet, but until then I hadn't noticed his Russophobic attitudes, and we were actually friends. But suddenly he got carried away. I banged on the table and said, "I'm done with you guys" (or something like that) and broke off relations with him and the Jews who were present.

By the end of his life, Shimanov, a self-proclaimed anti-Semite, had fully repudiated his former dissident colleagues along with the ideas derived from "their Masonic Talmuds."⁵⁴

From his exile in Vermont, Solzhenitsyn expressed a similar skepticism of the dissident movement's emphasis on the rule of law while distancing himself from Shimanov's anti-Semitism. Like other émigré groups, exiled Soviet dissidents often suffered from a toxic impotence, compounded by a lifting of the unspoken rule against attacking each other. In Solzhenitsyn's case, this meant condemning émigré dissidents who continued to endorse democracy, human rights, and what he called the "cult of pluralism," none of which, he argued, offered actual solutions to actual problems:

> [Rights-defenders] have diverted the world's attention from the basic conditions of the common people's life in our country. All that

matters [to them] is whether that murderous regime is obeying its own lying laws. After their timely exit what now concerns them is whether unlimited freedom of speech will triumph the day after somebody (who?) overthrows the present regime. And how extensive are the territories over which their free thought will flutter tomorrow? They do not think prudently about how to build a *house* for this thought, whether there will be a roof over their heads or real butter in the stores.[55]

Rights-defenders, of course, were not aiming to overthrow the Soviet state, much less change its borders. But for Solzhenitsyn, their obsession with "infringements of Soviet legality" missed the fundamental problem: communism. "Pluralists" insisted that "nothing remains of communism but words" (that is, the lip-service state) when in fact it was "marching over mountain ranges and oceans, crushing yet more nations with every step, and soon will smother all mankind." Worst of all, rights-defenders blamed the Soviet order on Russia and Russians instead of Marxism-Leninism:

> The mightier a people, the more freely it laughs at itself. We Russians ... have always made fun of our country and scolded it mercilessly, always considered everything in Russia worse than anywhere in the world—but we do this with love. As with our literary classics, we suffer with Russia. Now we are being shown how it is done with hatred.... If the Party bureaucracy were to collapse tomorrow, these "cultural forces" would rise to the surface, and we would hear their thousand-strong roar, not about the people's needs, not about the land, not about the dying population, not about the responsibilities and obligations of each of us, but about rights, rights.... They would smash whatever remains of us in another February [1917], another collapse.[56]

If prior to his exile Solzhenitsyn regarded civil or human rights as a necessary but insufficient ingredient of Russia's future, he was now inclined to regard them as hollow and tainted by association with an equally hollow West.

Solzhenitsyn never considered himself part of the dissident movement in any of its incarnations, and he was not inclined to move closer to it from his isolation in Vermont. He was above all a writer, not a joiner. Nonetheless, his claim that Soviet dissidents lacked an agenda beyond the rule of law and the protection of civil and human rights, and that this absence was in part traceable precisely to the pluralism they valued—which made consensus on anything beyond human rights and the rule of law difficult—was on target. Nor was he wrong to note the rising estrangement of many rights-defenders from the Soviet population. Much the same could be said of the Soviet intelligentsia as a whole. Decades of exhortations by Soviet leaders for intellectuals to "merge" with the masses could not overcome the long-term trend at work in modern industrial societies the world over: education, especially higher education, was now the great stratifier.[57]

There is a well-known story that dissidents liked to cite as a way of making sense of their strange place in Soviet society, simultaneously shunned and talked about because they said out loud what others were (allegedly) thinking.[58] In Hans Christian Andersen's literary folktale "The Emperor's New Clothes," a pair of swindlers persuade a vain emperor that they can weave him a new outfit that not only will look magnificent but will be invisible to anyone who is unfit for office or just plain stupid. The emperor, captivated by the idea of being able to expose incompetence in this manner, readily agrees. One after another, the emperor's officials and then the emperor himself observe the weavers at work. Although the looms appear to be empty, each observer, not wanting to be thought stupid or incompetent, pretends to admire the nonexistent clothing. When the emperor leads a celebratory procession through the city, the residents go along with the pretense, praising the emperor's new clothes as superior to all others, until a small child blurts out, "But he hasn't got anything on!" The townspeople begin to whisper the child's words to each other and then, at last, cry out together: "But he hasn't got anything on!"

"I was that child," Andrei Amalrik declared, a decade after the sensation triggered by his *Will the Soviet Union Survive until 1984?* Semyon

Gluzman, insisting that he "was never a revolutionary," offered a similar summation of his dissident activity: "I cried out in a weak voice, 'The emperor is naked!'" In his advice manual on how to conduct oneself during an apartment search, Vladimir Albrekht encouraged readers to invoke not just statutes from the Soviet Code of Criminal Procedure but intuitive moral truths: "Not for nothing was it a child who first said 'The emperor has no clothes!' In some sense we too are children." If every Soviet citizen would cease to lie, Solzhenitsyn insisted, "we would be amazed how quickly ... that which ought to be naked will appear naked to the world."[59] Andersen's folktale resonated with dissidents because it captured their conviction that the truth they spoke to power was already evident to everyone, in fact bordered on self-evident, and that what distinguished dissidents as truth-tellers was less insight than the absence—as in young children—of accumulated habits of dissimulation, self-censorship, and lying.

A Soviet version of Andersen's folktale would need a different ending. The child's declaration that the emperor hasn't got any clothes on would elicit a scolding from the townspeople: "Everything is already clear to everyone, so why speak about that? It isn't interesting." There would be complaints about disturbing the peace. Too young to stand trial, the child would be found to suffer from "political immaturity" and held indefinitely in a psychiatric hospital. Perhaps the Soviet emperor would secretly suspect that the child was right about the non-existent clothes, but like the emperor in Andersen's story would think to himself, "This procession must go on," and would walk more proudly than ever as his comrades held high the magnificent royal mantle that wasn't there at all.[60]

18

Rights-Defenders among the Nations

> Men, it has been well said, think in herds; it will be seen that they go mad in herds, while they only recover their senses slowly, and one by one.
>
> —CHARLES MACKAY

Soviet dissidents invariably understood their path to the defense of constitutional rights as a solitary one, set off by a stumbling block involving a morally unsettling encounter with Soviet power. For Natalya Gorbanevskaya, it was being pressed to testify against fellow students; for Andrei Sinyavsky, it was serving as an informant for the KGB. For Gleb Pavlovsky, an order by officials at Odesa State University to shave off his beard triggered an initial "personal rebellion" that was "existential rather than political."[1] Ludmilla Alexeyeva encountered her stumbling block when reading Lenin by herself, without the ideological guardrails erected by schoolteachers. Andrei Sakharov met his when he realized that people far less refined than he would control the weapons of mass destruction he had designed.

Chapter 17 explored how individual bystanders responded to the dissident movement with a mixture of admiration, guilt, and annoyance, the net effect of which was a rising resentment of rights-defenders as troublers of both social peace and individual conscience. Dissidents, simply put, made quite a few people feel ashamed of being afraid and ashamed of lying. Not everyone responded this way, of course. And just as significant, not everyone responded one by one. Soviet society included numerous groups whose collective grievances against the state brought them into the dissident orbit. Those grievances typically predated the chain reaction, and in certain cases—Baptists resisting state policies that limited their ability to raise their children as Christians, Crimean Tatars seeking to return to their homes following mass deportations by Stalin—the aggrieved groups had organized themselves prior to the emergence of the dissident movement. Some of their techniques, moreover, anticipated those deployed by rights-defenders, including the circulation of homemade news bulletins and the gathering of signatures on public letters of protest.

Alexander Volpin's "meta-revolution" aimed at a transformation of consciousness in which Soviet citizens would learn, one by one, that "without universal respect for rights there cannot be freedom for anyone." Individualism went hand in hand with universal applicability: as Volpin put it, "We must teach the broad public to see in the laws the guarantee of their [own] rights," regardless of whether they were writers, veterinarians, or tailors.[2] Soviet law itself had already begun to move in this direction, abandoning nearly all legal distinctions based on class, occupation, and social origin in favor of a society envisioned as a single collective of citizen-workers. The defense of constitutional rights was therefore relevant to the entire Soviet population. Volpin's algorithm tracked the meta-revolution's progress according to its ability to expand beyond face-to-face communities bound by ties of personal friendship, into a movement capable of attracting strangers bound by nothing more—or less—than shared citizenship, as two engaged citizens became four, four became eight, and so on.

Invisible to this algorithm were not just distinctions (and solidarities) based on education and occupation (writers, veterinarians, or tailors) but those grounded in ethnicity and religion. In his pioneering analysis of the dissident movement in *Will the Soviet Union Survive until 1984?*, Andrei Amalrik paid close attention to the class and occupation of the roughly one thousand signers of open letters of protest. Tellingly, however, he paid no attention to their ethnicity or religion—or gender.[3] And yet the collective grievances of certain groups defined precisely on these lines gave them the potential to become the dissident movement's numerically most significant allies. Lithuanian Catholics seeking religious freedom, Ukrainians seeking cultural autonomy from Moscow, or Jews seeking to emigrate from the USSR (to name some of the more prominent examples) proved far more receptive to the dissidents' message and tactics than did most individual bystanders. Rights-defenders offered such groups a repertoire of contention attuned to the post-Stalin order, shunning violence and strident anti-communist rhetoric in favor of law and international human rights norms. Most important, the dissident movement possessed a prized asset: access to foreign journalists in Moscow and, through them, to Western publics as well as millions of Soviet listeners to the Voices.

Members of repressed minority groups, whether ethnically or religiously defined, had one thing in common: their grievances—their stumbling blocks—were collective, shaped by group experience and group interest. The metropolitan dissidents, by contrast, had followed solitary paths to the doctrine of individual rights. Could they find common cause, or even a common language?

The roughly three million Lithuanians in the USSR during the Brezhnev era constituted just over 1 percent of the Soviet population. The overwhelming majority were Catholic. Along with the other Baltic republics of Estonia and Latvia, Lithuania was one of the least Sovietized territories of the USSR, having been annexed by Moscow initially in 1940 and then definitively, following the Nazi interlude, in 1945. For nearly two decades, underground Lithuanian partisans known as "forest brothers" fought against and refused to recognize Soviet rule. The decline of armed guerrilla units as a mode of opposition in Lithuania paralleled the eclipse of revolutionary Marxism-Leninism as the dominant idiom of resistance in the Russian republic. By the early 1960s, both were largely spent.

As the seeming permanence of the Soviet order sank in, techniques of transparency and appeals to Soviet law and international human rights—

the dissident repertoire, in other words—found fertile soil in Lithuania.⁴ When three priests were arrested in 1970-71 for teaching the catechism to children (at the request of their parents), Catholic activists composed an open letter to Brezhnev and UN secretary-general Kurt Waldheim demanding that the freedom of conscience guaranteed by the Soviet Constitution be afforded to religious believers. In two months in 1972 they gathered 17,054 signatures from across Lithuania. Subsequent petitions to the Lithuanian Ministry of Education and the Commissioner of Religious Affairs gathered 14,604 and 16,800 signatures, respectively. Of fifty-one documented public demonstrations in Lithuania between 1965 and 1978, a dozen involved a thousand or more participants.⁵

These numbers stunned dissidents in Moscow, where, according to Sergei Kovalev, "we never managed to gather more than three or four hundred" signatures (out of a metropolitan population of roughly seven million). Part of the difference could be chalked up to what Kovalev called the Muscovites' "sloppy ways" in comparison with the more orderly Lithuanians. Alexeyeva agreed: the juxtaposition highlighted "our genuinely Russian lack of organizational skills."⁶ The Lithuanians, however, were working with a far more receptive population. True, none of their open letters and petitions received a response, nor did they prevent the three priests—Antanas Šeškevičius, Juozas Zdebskis, and Prosperas Bubnys—from being convicted of engaging in "religious propaganda." But they generated considerable publicity abroad, thanks to coverage by the *Chronicle of Current Events*. Soviet authorities, moreover, declined to punish the signers, apparently because there were simply too many.⁷ It was a striking lesson in the strength of numbers.

The petition campaigns marked the beginning of regular coverage by the *Chronicle* of repressions in Lithuania. In 1972, Catholic activists founded their own periodical, *Lietuvos katalikų bažnyčios kronika* (Chronicle of the Lithuanian Catholic Church), closely modeled on the *Chronicle of Current Events*. Arimantas Raškinis, a biologist and practicing Catholic who befriended Kovalev when they worked in the same Moscow laboratory, served as conduit for the exchange of copies of the two *Chronicles* between Moscow and Vilnius, the Lithuanian capital.⁸ To reach supporters not just in the West but in rural Lithuania, Raškinis and other couriers would deliver copies of the *Chronicle of the Lithuanian Catholic Church* to dissidents in Moscow, who would help smuggle them out of the country via foreign journalists and tourists, who in turn would forward copies to the various Western shortwave radio broadcast services, including Vatican Radio, whose programs reached listeners across Lithuania.[9]

For Lithuanian activists, Kovalev, Sakharov, and other Moscow-based dissidents offered a rhetoric and an accompanying set of practices that made it possible to articulate a religious and national cause in the idiom of universal human rights, as well as to dramatically expand their audience in Lithuania and the West. "The sacrifices of Russian dissidents," noted the *Chronicle of the Lithuanian Catholic Church*, "have helped Lithuanians see the Russian nation in a new light."[10] Kovalev and Sakharov had never regarded themselves as representatives of the Russian nation. They envied their Lithuanian counterparts, situated in a tiny republic at the USSR's European border, for their ability to mobilize far larger numbers of overt supporters, so conspicuously lacking among bystanders closer to home.

Not all Lithuanian dissidents were ready to see the Russian nation in a new light, however. The more uncompromising nationalists among them insisted that Soviet dissidents endorse the goal of full Lithuanian independence as a precondition for mutual cooperation. Anything short of that would signify a tacit endorsement of Russian imperialism, which in their view was the core problem—not communism or the absence of individual rights. As one group put it,

> One cannot liberate man and leave nations in slavery. Russian imperialism is incompatible with democracy. To some Russian dissidents it seems that it would be sufficient to implement human rights in the empire and there would be no need for the enslaved nations to create their own states. But we do not want voluntary slavery. And our goals will not coincide until the Russian dissidents will come out for complete freedom for the colonies. The right of national self-determination is now recognized by separate individuals, but this does not change the general outlook of the dissidents.[11]

Dissidents were not in the habit of issuing programmatic statements beyond the demand that the Soviet government observe its own laws along with those contained in international human rights covenants. The absence of a statement endorsing Lithuanian independence or condemning Russian imperialism, therefore, whether from the Initiative Group for the Defense of Civil Rights or the Human Rights Committee or any other dissident entity, tells us little about "the general outlook" of the movement. The Soviet Constitution certified the right of every Union republic, including Lithuania, to "free secession from the USSR."[12] Advocating such a move, however, was invariably treated as treason. Dissidents' silence on the matter highlights their selectivity regarding which constitutional rights the state should be pushed to observe: those of individuals took precedence over those of nations or other collective entities.

A similar tension between collaboration with rights-defenders and wariness of their Moscow-centrism characterized other national minority dissident communities. As in Lithuania, resistance to Soviet policies in Ukraine underwent a sea-change during the decade following Stalin's death, as armed partisan battles gave way to non-violent, text-centric tactics characteristic of the late Soviet intelligentsia. An autonomous chain reaction of arrests, trials, and protests began in Ukraine nearly simultaneously with that triggered by the arrests of Andrei Sinyavsky and Yuli Daniel in Moscow. Like its Lithuanian counterpart, Ukrainian dissent was permeated by national sentiment, ranging from calls for greater cultural autonomy to demands for outright secession from the USSR.[13] Ukrainians' experience of political independence from Moscow was briefer than Lithuanians': three years (1918–21) versus nearly a quarter-century (1918–40). Ukraine's cultural proximity to Russia, together with the enduring Russian conviction that Ukrainians were not a separate nation (and that Ukrainian was no more than a dialect of Russian), heightened the existential threat of Russification for Ukraine. The single most widely read Ukrainian dissident text, the literary critic Ivan Dziuba's *Internationalism or Russification?* (originally circulated via *samvydav* [samizdat] in 1965), took up precisely this issue, arguing that

Soviet nationality policy had deviated from Leninist internationalism in favor of Russian chauvinism.[14]

The specter of cultural death by assimilation helps account for the leading role of writers and poets among Ukrainian rights-defenders (and the appeal of what they called "philological nationalism"), in contrast to the predominance of scientists and mathematicians among their Russian counterparts.[15] It also helps explain Ukrainian dissidents' abiding skepticism of claims that individual rights, the rule of law, or democracy were enough to stem the tide of Russification in a country where Russians outnumbered Ukrainians by more than three to one (137 million as compared to 42 million). As the Ukrainian writer Viacheslav Chornovil noted in the samizdat journal *Ukrainsky visnyk* (the *Ukrainian Herald*, founded in 1970), the Human Rights Committee in Moscow produced studies of many different kinds of rights but not those of nations. On the contrary, Sakharov and others advocated individual human rights in part as an antidote to "the threat of nationalism," proposing among other things that Soviet passports should no longer indicate the bearer's nationality (Ukrainian, Armenian, Russian, and so forth), but simply list everyone as "citizen of the USSR"—a move that Chornovil suggested could undermine the status of Ukraine and other republics. The *Chronicle of Current Events*, he argued, while giving welcome attention to the persecution of Ukrainian activists, "engages in no programmatic advocacy beyond freedom of speech and information": "It willfully claims some sort of supra-national or all-Union character, but in fact is produced by Russian (or possibly partly Jewish) circles. One notes that the bare-bones information from the [non-Russian] republics is positioned as an addendum to the detailed description of events in Russia, above all in Moscow, which in and of itself gives an inaccurate picture of the situation in the USSR."[16] Accurate or not, the *Chronicle* faced steep challenges in attempting to gather information across the enormous expanse of the USSR, given the repressive conditions under which it operated. For her part, Alexeyeva charged Ukrainian activists with taking up "the defense of only one right—the right of equality on the basis of nationality," recording "only violations of that right, and only if Ukrainians were involved." The *Chronicle* alone attempted to publicize repressions against a wide variety of individuals and groups.[17]

Ties between metropolitan rights-defenders and activists in various movements for national or religious autonomy often emerged from personal relationships in ways that were unpredictable. Petr Grigorenko, for example, a Russified Ukrainian, became the dissident contact for Tatars seeking to return to their Crimean homes. Kovalev, a secular Russian, served as a link to Lithuanian Catholics via Raškinis. Ukrainian activists often relied on Leonid Plyushch, the Ukrainian mathematician, for communication with dissidents in Moscow, despite the fact that Plyushch, a Marxist, was skeptical of national movements. It was yet another illustration of the dissidents' remarkable pluralism in action.

Relations with the Jewish emigration movement were different.[18] True, the applied mathematician Anatoly (Natan) Sharansky and the sinologist Vitaly Rubin performed a bridge function similar to that of Grigorenko, Kovalev, and Plyushch. The difference lay in the striking proportion of individuals of Jewish or partly Jewish descent within the dissident movement itself, a proportion far above that of Jews in the Soviet population. They include many of the people whose stories weave their way through this book: Vladimir Albrekht, Raissa Berg, Larisa Bogoraz, Yelena Bonner, Yuli Daniel, Ilya Gabai, Alexander Ginzburg, Yuri Glazov, Semyon Gluzman, Natalya Gorbanevskaya, Dina Kaminskaya, Lev Kopelev, Viktor Krasin, Pavel Litvinov, Roy and Zhores Medvedev, Raisa Orlova, Grigory Pomerants, Boris Shragin, Boris Tsukerman, Vladimir Voinovich, Alexander Volpin, and Petr Yakir.[19] Inside the movement, this circumstance tended to inspire nervous humor, especially in Jewish dissidents. When Grigorenko put together a list of potential members of the planned Committee in Defense of Ivan Yakhimovich, Krasin responded, tongue in cheek, "This isn't a committee—this is a cabal of yids headed by a Russian general!"[20] When bystanders at the demonstration on Red Square in August 1968 yelled, tongues definitely not in cheeks, "They're all yids!," they weren't far off: five of the eight demonstrators could claim one or more Jewish parents. A third of the members of the Initiative Group for the Defense of Civil Rights were of

Jewish descent. Even among Russian Orthodox dissidents, three of the most prominent were of Jewish ancestry: Alexander Men, Anatoly Levitin-Krasnov, and Mikhail Meerson-Aksenov. "I'm outraged that 85 percent of Democratic Movement members are Jews!," the poet Naum Korzhavin (né Mandel) joked. "The percentage ought to be fair: fifty-fifty!"[21]

Among dissidents, ethnic differences, like political differences, were either celebrated as a welcome sign of diversity or left off-stage so as not to distract from the shared goal of defending individual constitutional rights against an overbearing state. Outside dissident circles, not surprisingly, the over-representation of Jews nourished the view that the dissident movement was essentially a Jewish operation—much as the revolutionary movement in the tsarist empire, including its Bolshevik faction, had been attacked as a Jewish conspiracy against Holy Russia. In his satirical novel *Homo Sovieticus*, Alexander Zinoviev has an interrogator ask whether any of a defendant's grandparents were Jewish, to which the defendant responds, "In Russia, peasants couldn't be Jews. . . . But under Soviet power, all dissidents are required to be Jews."[22] Soviet authorities were only too happy to nourish the conflation of rights-defenders and Jews by publicly ascribing the dissident movement to "Zionist" forces and requiring that dissidents allowed to leave the USSR do so with Israeli visas in hand, regardless whether they were Jewish or wished to live in Israel. False rumors circulated, possibly with help from the KGB, that even Alexander Solzhenitsyn and Sakharov were Jewish, having allegedly Russified their last names from the original Solzhenitsker and Tsukerman (the latter a Yiddish variant on *sakhar*, the Russian word for sugar).[23] It did not help that the second wives of both men—the mathematician Natalya Svetlova and the physician Yelena Bonner—had Jewish mothers. Amalrik, neither a Jew nor married to one, was also suspected by Russian nationalists of being Jewish in light of his scathing comments about Russians ("this people with no religion or morality") in *Will the Soviet Union Survive until 1984?*[24] No dissident, it seems, was entirely insulated from rumors of being Jewish—not even the self-professed anti-Semite and Russian nationalist Gennady Shimanov.[25]

In 1979, the Soviet census recorded 1.7 million citizens of Jewish nationality, less than 1 percent of the USSR's population. Dispersed across multiple ethnic republics, Jews were concentrated in major urban centers such as Moscow, Leningrad, and Kyiv. They were also concentrated in the intelligentsia, with a higher proportion of advanced degrees among them than among any other Soviet ethnic group.[26] In many ways, Jews were one of the USSR's success stories, having achieved remarkable upward mobility and entrance into a wide range of elite professions, a process that had begun in the late tsarist era but accelerated dramatically during the early decades of Soviet rule.[27] Under Stalin, that success had begun to generate a backlash aimed at diminishing the Jewish presence in the Communist Party and elite institutions of higher education. By the 1950s, the USSR's ethnic affirmative action policies caught up with the Jews, effectively blocking and in some cases reversing their meteoric rise. Once known as the world's first anti-anti-Semitic state, the Soviet Union now engaged in its own suppression of those it strategically branded "cosmopolitans" and "Zionists."

Unlike Lithuanians or Ukrainians, Jews had never formed armed units opposed to Soviet power. On the contrary: most Soviet Jews regarded the USSR—correctly—as the country most responsible for crushing their mortal enemy, Hitler's Germany. In the years following that historic victory, however, the Soviet government appeared to perform a volte-face, presiding over widespread discrimination against Jews—and worse. Stalin's final years witnessed the execution of leading Yiddish writers and poets who had been members of the Jewish Anti-Fascist Committee, as well as official accusations of a "doctors' plot" involving a group of physicians—many of them with recognizably Jewish surnames—who allegedly sought to poison top Soviet leaders, sparking rumors of an impending mass deportation of Jews to the Soviet far east.

It was one thing to contend with popular anti-Jewish prejudice, which the Bolsheviks had made considerable efforts to eradicate along with other symptoms of "backwardness." It was quite another to confront a Soviet government that now appeared to harbor such prejudices itself—a government that, notwithstanding its post-Stalin retreat from policing its citizens' souls, still exercised enormous power over their housing, education, and careers. Compounding this unexpected reversal was the USSR's retreat from its initial support for the new state of Israel (Stalin was the first leader to grant Israel diplomatic recognition) and its military assistance to Arab countries before and during the Six-Day War in June 1967, which concluded with a decisive Israeli victory. That victory, along with the shock of thinly veiled restrictions on Jewish upward mobility, helped jump-start the idea of Jewish emigration.

Jews seeking to exit the USSR borrowed freely from the dissident repertoire, couching their requests in terms of law and rights, staging public demonstrations, and publicizing their predicament via samizdat, usually in Russian. An open letter signed by thirty-nine Moscow Jews in March 1970, for example, included the following statement: "The present state of our citizenship includes the right of the state to demand from us no more than obedience to the law, and our claims to freedom of repatriation are based on Soviet laws and guarantees of international law."[28]

That same year, the *Chronicle of Current Events* introduced a regular feature called "News from the Jewish Community," curated by Sharansky and Rubin. Readers learned about individuals whose applications to leave the USSR had been rejected ("refuseniks"), arrests and trials of Jewish activists, and the fate of those who had been sent to the camps. Sakharov and other members of the Human Rights Committee added their signatures to some of the earliest petitions by refuseniks and attended public demonstrations calling attention to their plight. The dissident movement provided much-needed publicity, although with Jews in the United States, Canada, Israel, Great Britain, and Australia avidly lobbying their governments on behalf of Soviet Jewry, refuseniks were perhaps less reliant than other minorities on Moscow dissidents to transmit their story to the world.[29]

Like their Lithuanian and Ukrainian counterparts, activists in the Jewish emigration movement were nonetheless ambivalent when it came to cooperating with dissidents. Association with them risked giving the impression that refuseniks shared the goal of reforming the Soviet system, when in fact their desire was to leave it, one family at a time. Why become entangled in domestic Soviet disputes and thereby incur additional harassment? Jewish nationalists invoked the Zionist idea of a return to the Jewish homeland—an idea at least nominally consistent with the Soviet principle of repatriation, according to which the USSR's ethnic and national minorities should be gathered in their putative republican or regional homelands.

Rights-defenders, by contrast, sidestepped both the Zionist and the Soviet rationales for Jewish emigration in favor of a much broader principle: freedom of movement as a fundamental component of human dignity and therefore of human rights. "It should be remembered," Valery Chalidze wrote, "that this issue concerns emigration by people in general. It is a matter of forcibly keeping people under the jurisdiction of a state whose territory they want to leave."[30] Not just Jews but every Soviet citizen, he argued, should have the right to leave the country, whether for repatriation or any other collective or individual purpose. Furthermore, as Volpin had urged his friends from the window of his departing train in 1972, they should have the right to return if they wished.

This was not a cause for which most Soviet Jews seeking to emigrate were prepared to fight. It would make Jewish emigration captive to a far more ambitious agenda in the hands of a dissident movement whose record of practical victories appeared modest at best. What if Volpin's "meta-revolution" ended up resembling previous Russian revolutions? The physicist Alexander Voronel, who had taken part in protests at the trial of his wife's friends Sinyavsky and Daniel, gave voice to such fears: "We must now remain aloof, while it is not yet too late to separate our problems from [the Russian people's] problems. Otherwise they will solve our fate together with theirs, and this solution will be radical." Voronel took his own advice and left for the ancestral homeland in 1975.[31]

Technically, there was already a Jewish homeland inside the Soviet Union: the Jewish Autonomous Region of Birobidzhan, established in 1928 near the USSR's remote border with China. But very few Jews moved there, and of those who did, most did not stay long.[32] If there had been a viable Jewish republic in the Soviet Union, akin to the Lithuanian,

Ukrainian, and other ethnic republics, it is quite possible that the proportion of Jews among those fighting for Soviet constitutional rights and international human rights would have been substantially lower. Their energies would instead have been directed toward achieving greater Jewish collective autonomy within the Soviet order. Instead, Soviet Jews were effectively a diaspora nation *within* the USSR, a country conceived as a federation of ethnic republics, with a specific national identity inscribed in every citizen's passport. Like Jews in the global diaspora, their status as internal strangers predisposed them to favor the rule of law and universal rights.[33]

What should we make, then, of the striking over-representation of Jews in the dissident movement (somewhere between a third and a half) as compared to their proportion of the Soviet population (less than 1 percent)? The degree of over-representation looks somewhat lower if the baseline is not the Soviet population as a whole but the intelligentsia, where Jews were disproportionately concentrated. It looks lower still if the baseline is restricted to major cities such as Moscow, Leningrad, and Kyiv, where the majority of postwar Soviet Jews lived.

Moreover, just as the prominence of scientists in the dissident movement does not change the fact that the overwhelming majority of Soviet scientists were never dissidents, so the prominence of Jews in the movement does not change the fact that the overwhelming majority of Soviet Jews were never dissidents. If "science" as a set of mental habits or practices offers at best a weak explanation for who became a dissident, then "Judaism" is even weaker, for the simple reason that Jewish dissidents had little to no familiarity with Jewish religion, culture, and language, notwithstanding Shimanov's claim that rights-defenders' law-centered approach could be traced back to "their Masonic Talmuds."[34] In contrast to Jews in another civil rights movement (in the United States) in which they were also over-represented, Soviet Jewish rights-defenders did not cite Judaism or Jewish history as motivating their participation.[35] Almost without exception, they were raised in secular, Russified, often devoutly communist families, in which a few phrases in Yiddish spoken by a grandparent constituted the sum total of their Jewish education. That was certainly the case with Bogoraz, who described herself as having "an unquestionable genetic tie with Jewry" even as her entire sensory apparatus—"what the eye sees, the ear hears, the skin feels"—led her to self-identify as Russian.[36]

To paraphrase Karl Marx, people may fashion their own identities, but they do not fashion them however they please.[37] Bogoraz's sensory apparatus may have led her to self-identify as Russian, but her "genetic tie with Jewry"—like that of Volpin, Litvinov, Orlova, and others—meant that a Jewish identity was also imposed upon her by Soviet society and the Soviet state, an identity that predated the USSR and was hardly unique to it or to Russia. Sally Belfrage's account of conversations with Volpin in 1957 captured a similar dichotomy. As a self-declared "enemy of any religion or nationalism," Volpin had imbibed the Soviet ethos, yet he was constantly required to identify as Jewish. He could not even fill out an application for a library card "without proclaiming to the world: nationality—Jew," with all the liabilities that went with it. Why, he wanted to know, should he "be penalized for something he was so indifferent to"?[38]

The sharp postwar reversal of the Kremlin's policy toward Jews, all the more jarring given their manifold participation in building Soviet socialism, placed additional stumbling blocks in their paths.[39] The parents of quite a few Jewish dissidents had been loyal Party members or employees of the security services—part of the Soviet ruling class, in other words, a status that likely made their fall from favor even more acute. Prior proximity to power may also have emboldened some of their children to challenge that power. Whatever the case, more than any other single factor, the volte-face in state policies toward Soviet Jews helps account for both the Jewish emigration movement and the high proportion of Jews in the movement to induce the Soviet state to abide by its own laws. Being Jewish, of course, was hardly a prerequisite to becoming a dissident. There were many routes by which Soviet citizens, Jewish or not, could come to perceive themselves as "other-thinkers" or, to use Bonner's description of her fellow rights-defenders, as "foreigners at home."[40] Jews' inherited role as internal strangers and disturbers of the peace aligned well with widely held perceptions of dissidents.

"I always felt like a stranger," wrote Gluzman, "first as a Jew and then as an other-thinker."[41]

Whether they took the form of a chain reaction, a social movement, or an NGO, Soviet dissidents had always been a loose ensemble of individuals dedicated to the freedom of the human personality and convinced that defending the legally enshrined rights of individual citizens was the surest way to foster that freedom. While they developed their own collective rituals—weekly visiting days, the production and consumption of samizdat, transparency meetings on December 5—they resisted thinking of themselves as a collective, preferring to celebrate their diversity as a collection of individual "other-thinkers." Their approach to Soviet society mirrored these libertarian instincts: people would be drawn to the movement one by one, driven by conscience, a quest for dignity, and civic courage, rather than herd instinct or groupthink.[42] Unlike the "going to the people" movement of the 1870s, in which several thousand university students fanned out to the Russian provinces to live among peasants and prepare them for revolution, dissidents mostly expected people to come to them.

Only in retrospect did Kovalev begin to question this approach. Looking back on his relations with individual friends, colleagues, and other bystanders, as well as with Lithuanian Catholics, Kovalev came to the conclusion that "a movement for the defense of human rights seems to have a chance at success only when the infringement of those rights simultaneously affects the interests of an entire group, be those interests national, religious, or socio-economic. The struggle for the rights of the individual appears at best to be a cause for loners or small groups with a pronounced legal consciousness—at least in our country."[43] When Alexander Malinovsky, in his samizdat essay "Considerations concerning the Liberal Campaign of 1968," called on the dissident movement to build bridges to Soviet society, he meant building alliances in the classic Leninist sense: with workers and peasants. But aside from the handful of neo-Leninists in their ranks, dissidents dismissed this approach as having been exhausted by the Soviet practice of "popular mobilization," a practice they sought to transcend via the idea of citizenship as the basis for civic engagement. Their strategy required, as Sakharov put it, "blunting the class-based approach."[44]

To the extent that the dissident movement found allied groups in Soviet society, they defined themselves not as workers or peasants (even if they were) but as members of aggrieved national or religious minorities. By the mid-1970s, despite their all-embracing rhetoric of civil and human rights, dissidents had developed a reputation of standing up for three kinds of people: rights-defenders (that is, themselves), Soviet national and religious minorities, and foreigners (reform-minded Czechs and Slovaks).[45] To many ethnic Russians—the majority of the Soviet population—these groups had one thing in common: they were outsiders hostile to the Soviet state. Their dissident defenders therefore also had something in common: a whiff of treason.

19

Dissident Fictions

"I don't care," Yefim snapped. "I say what I think. I have nothing to hide." This annoyed me. He had always been so careful, with his code words and hints, that you couldn't understand him half the time. But now, you see, he had nothing to hide. If others maybe had something to hide, that was not his concern.

—VLADIMIR VOINOVICH

For bystanders and passersby too distant from the dissident world to be acquainted with its inner workings, information about rights-defenders came largely from Western radio broadcasts and, to a much lesser extent, the Soviet press. But there was another window into that world: literary fiction. This was to be expected in a country where literature had long constituted a critical, at times preeminent forum for debates about controversial social and moral questions. Novels and short stories by Andrei Sinyavsky and Yuli Daniel, after all, had triggered the events that launched the dissident movement. Alexander Solzhenitsyn, Lydia Chukovskaya, Venedikt Erofeev, and dozens of other writers produced literary works on taboo themes that circulated exclusively via samizdat and tamizdat. But these were works *by* dissident writers; they were not *about* dissidents. To be sure, historical novels set in the tsarist era, such as Vladimir Voinovich's *A Degree of Trust* (about the socialist revolutionary Vera Figner) or Natan Eidelman's *Lunin* (about Mikhail Lunin and other participants in the Decembrist uprising) were published in the USSR and read, at least by some, as sympathetic allegories about Soviet dissidents. The first literary work to take up contemporary dissent, however, was anything but sympathetic. This was Vsevolod Kochetov's 1969 novel *What Is It You Want?* (*Chego zhe ty khochesh?*).

What Kochetov wanted was similar to what the Fifth Directorate wanted: to expose dissidents and liberals as tools of Western imperialism in its relentless campaign to destroy the Soviet Union. As editor (from 1961 until his death in 1973) of the arch-conservative journal *October*, Kochetov waged a relentless campaign against critics of Stalinism. *What Is It You Want?*, which was serialized in *October*, featured a rogues' gallery of Russian émigrés, their spouses, and hangers-on, all employed by the CIA as puppet-masters of dissolute, unpatriotic Soviet intellectuals.[1] The novel's main villains, Benito Spadi and Portia Brown, are thinly disguised makeovers of the Italian communist literary critic Vittorio Strada and the American journalist and translator Patricia Blake, both of whom wrote extensively about Soviet culture in the post-Stalin era. The CIA's plan, according to the unfailingly seductive "Miss Brown," called for the spiritual dismantling of communism via a three-pronged assault. The older generation would be targeted with religion: "Toward the end of life," she explains, "people can't help thinking about what awaits them there, on the other side." For middle-aged people, "so-called adults," there was a different strategy. Soviet citizens' standard of living had risen in recent years "thanks to the efforts of their government," with the result that more and more people had money to spend. Using all available channels—shortwave radio broadcasts, exchanges of illustrated books and magazines, and especially movies with images of the high life—"we are awakening in them a craving for comfort, for acquisitiveness. In every way we are instilling a cult of things, purchases, accumulation." This would inevitably weaken "the spirit of collectivism, which makes them strong and invulnerable."[2]

The key to the CIA's operation was Soviet youth and, more specifically, youthful sexuality. "Here is the richest soil for us to sow," Miss Brown explained. "The young mind is arranged in such a way that it protests against everything that limits its impulses. And if you beckon it with the possibility of complete liberation from any restrictions, from any duties, let's say—to society, to adults, to parents—from any morality, then it's yours." It was not an easy strategy; so far it had taken root only in Moscow, Leningrad, and a few other cities. "But we're working, working, working," Miss Brown assures her Soviet contact ("placing her warm palm on his arm"). Working to stir up young minds in the universities, to support "underground journals and leaflets," to discredit former idols and authoritative figures. Working to get young people to regard valor as requiring audacity. Working with mini-skirted young women, "one of our weapons":

> They are sexualizing the Russian atmosphere, leading young people away from civic issues into a purely personal world, a world of niches. This is what we need. This is how to weaken the Communist Youth League, to turn their meetings and their political education into a formality. Everything will be just for show, for propriety, beyond which there will be private life, sexual life, life liberated from duties. And then, in an atmosphere of indifference toward social issues . . . , it will become possible to gradually advance people who prefer the Western system, not the Soviet, not the communist system, toward positions of power in various leading organizations.[3]

The CIA was thus the hidden architect of the false promise of inner freedom. *What Is It You Want?* endowed the West (as embodied by Portia Brown) with nearly limitless powers to corrupt Soviet society. The actual work of women such as Larisa Bogoraz, Natalya Gorbanevskaya, and Ludmilla Alexeyeva was apparently beyond Kochetov's imagination. For him, the dissident movement was designed to accomplish what Western military intervention against the Bolsheviks during the Russian civil war (1918–21) had failed to achieve, now using stealth rather than force, and over a much longer period of time.

There would be patriotic resisters at all levels of Soviet society, of course, but such people could gradually be neutralized by branding them as Stalinists—"a term cleverly invented," Miss Brown notes, "by Mr. Trotsky." Were they actually Stalinists? Miss Brown's interlocutor, the earnest, soul-searching Petr Saburov wanted to know. What was wrong with being a Stalinist, anyway? Was Stalin not a Marxist, a revolutionary, a Bolshevik? Do Stalinists have their own special position that somehow contradicts the general position of the Bolsheviks? Miss Brown had to spell it out for him:

> What difference does it make, Stalinist or not! What matters is that you can beat people with this word. You can use it against popular writers, artists, composers, scholars, movie and theater directors, all those who despite having been labeled Stalinists continue to act on the principle of Party-mindedness in art, who continue to work according to the renowned method of socialist realism. For the time being, they are still masters of thought among the people, they shape people's spiritual world. It's fine to use all means against them.[4]

Kochetov could not have put it better. In the meantime, having introduced the art of striptease to various anti-Stalinists from the Moscow intelligentsia (among them a lightly fictionalized version of the poet Andrei Voznesensky), Miss Brown proceeds to help a young writer smuggle his works abroad and have them published in London, including a novel in which, following her advice, he tweaks the names of two well-known Party-minded writers and casts them as bumbling KGB agents. "Can you imagine," she exults, "how Russian listeners will laugh when they hear this on the BBC or Voice of America! A pair of popular Party writers, orthodox types, making a vulgar cameo appearance!"[5]

Kochetov's scabrous novel unleashed a furor. The three fall issues of *October* where it was serialized immediately sold out. Before long, the samizdat sphere was thrumming with outrage—but also with laughter. "People are reading it and re-reading it," noted one observer, "they resent it, make fun of it, are horrified by it."[6] "Kochetiana," a satirical poem by the physicist Grigory Podyapolsky (the only person to be a member of both the Initiative Group for the Defense of Civil Rights and the Human Rights Committee), proclaimed: "Skinny daddy Samizdat / From Kochetov is getting fat."[7] The *Chronicle of Current Events* did its best to keep readers informed of the novel's reception.[8] Before long, two samizdat parodies were making the rounds, one titled *What Is It You're Laughing At?* (*Chego zhe ty khokhochesh?*), the other *What Is It, Rooster?* (*Chego zhe ty, kochet?*, a play on the author's last name).[9]

That *What Is It You Want?* was a *roman à clef* was lost on no one, even if Kochetov, overeager for verisimilitude, confusingly referred to certain figures by their actual names, including the Gulag memoirist Yevgenia Ginzburg and the Berkeley professor Gleb Struve. For those in the know, the novel seethed with personal score-settling. The vulgar cameo appearances engineered by Portia Brown to demean a pair of popular Party writers, for example, was Kochetov's revenge against Sinyavsky, whose tamizdat novel *Lyubimov* featured precisely such a bungling pair bearing the names Kochetov and Sofronov.[10] Portia Brown herself, or rather Patricia Blake, had published an unflattering interview with Kochetov in 1963, in which she referred to his novels as "the essence of Soviet philistinism."[11] It was payback time.

Leonid Plyushch was one of the first to note that Kochetov's payback against Blake took the form of a spanking. The author of *What Is It You Want?* seemed inordinately fond of the word "behind" (*zad*), in keeping with his apparent sexual and scatological fetishes. Plyushch, an avid reader of Freud, diagnosed Kochetov as suffering from narcissism and persecution mania.[12] In the novel's climactic scene, the character Bulatov—a prominent Soviet writer who serves as Kochetov's alter ego—decides to give Miss Brown her just reward for producing so many poisonous articles about the USSR's dedicated practitioners of socialist realism:

> He didn't argue with her, he didn't accuse her of writing inaccurately and tendentiously about this person and that person, he didn't get drawn into any sort of dialogue. Rather, with a grin and with condescending regret designed to offend her, he simply slapped her backside.... Rumors will spread about this, of course; she will be ridiculed and compromised. Her competitors and those who envy her will say, "Ah, Portia! She's the one who got slapped on her behind in Moscow!" Then the gig will be up, she won't be able to count on anything: no more weighty assignments, no more visas, no more excessive honoraria. The slap was light and symbolic. It did not leave the imprint of Bulatov's hand, and obviously no medical investigation was going to bother to write up a report about black and blue marks.[13]

Viktor Sokirko, recently expelled from graduate school in engineering as punishment for signing multiple open letters in 1968 on behalf of arrested dissidents, interpreted this scene as signaling something more than Kochetov's fantasy of how he wished he had concluded his interview with Blake in 1963. Bulatov's hand, according to Sokirko, stood for "the hand of Moscow," a phrase that appears repeatedly in *What Is It You Want?* "Don't argue, don't get involved in dialogues," Sokirko paraphrased, "just give all the foreigners and all our other-thinkers a good spanking, and all our troubles will be liquidated. This is the thinking of Bulatov-Kochetov and his high-ranking friends. . . . He doesn't conceal his dream: to destroy all other-thinkers, if not with his pen, then with the help of 'people from the Lubyanka.'" Spanking instead of dialogue: this, according to Sokirko, was the model guiding Moscow's use of "draconian measures . . . unconnected to any legal framework," the kind of punishment with which the Fifth Directorate had produced a significant retreat from the dissident movement in 1968.[14]

As the preceding chapters have shown, things were not quite so simple. The Fifth Directorate deployed what amounted to a combination of dialogue (in the form of prophylactic conversations) and spanking (extrajudicial punishments such as termination of employment, expulsion from university, and harassment and intimidation of family members). More significant, perhaps, than Sokirko's claims about relations between dissidents and the security organs was his view of relations between dissidents and the rest of the Soviet population:

> [Kochetov's] message frequently falls on fertile soil, nourished already in Khrushchev's time, of yearning for a strong personality, for Stalin. This yearning comes not so much from above, but from the middle and lower layers of society, precisely from those who lived in ignorance under Stalin, who relied not on their own mind but on the "higher" one. These are people who, while suffering themselves, gave thanks for the blessing of the whip. We mustn't delude ourselves: there are many such people, a great many. They are potential fanatics, Red Hundreds, the spearhead of a future "cultural revolution." And not just among workers, but among all our classes and groups.[15]

Sokirko's appraisal of *What Is It You Want?* gave voice to deep-seated fear and mistrust of the Soviet population, a sense of almost unbridgeable moral and cultural distance between them and an unidentified "we" ("We mustn't delude ourselves"). Just as Kochetov's literary spanking of Portia Brown echoed the crack of the whip wielded by mounted Cossacks against protesting students in pre-revolutionary Russia—iconic instances of humiliating, extra-judicial, corporal punishment—so Sokirko's wariness of the Soviet "masses" recalled the chastened response by Russian intellectuals to the savage violence of the revolution of 1905, captured in the renowned *Landmarks* collection of essays. Deep down, the "democratic movement" did not trust the *demos*.[16]

Despite its scandalous vilification of signers of open letters and purveyors of samizdat, *What Is It You Want?* was so eager to depict them as pawns of the CIA that it ended up putting foreigners rather than dissidents at the center of the plot. As Sokirko noted, "the exclusive source of the entire oppositional genre" appeared to be Portia Brown's bed.[17] So caricatured were Kochetov's "other-thinkers" that even his ideological allies were embarrassed. Mikhail Sholokhov, the Nobel Prize–winning writer who shared Kochetov's nostalgia for the days of revolutionary justice, sent a private note to Leonid Brezhnev urging him to quash attacks against his fellow author. Kochetov, he wrote, "tried to do something important and necessary, to expose the penetration of our society by ideological saboteurs." His writing was "not always at the required level," Sholokhov conceded, but to allow Kochetov to be ridiculed "is hardly useful for our cause."[18]

A few years later there appeared a very different fictional rendering of the dissident world, this time by an author who, unlike Kochetov, knew that world intimately. Anna Gerts's 1972 samizdat novel *Sacred Paths to Willful Freedom* used a simple but ingenious device to convey the perspective of an extremely close bystander to the life of Soviet dissidents: its narrator, Nadezhda Fedorovna Aksanova, is married to one. Through Aksanova, or Nadia as she is known (apart from KGB agents, everyone in the novel uses diminutive first names), readers vicariously experience the sharing of samizdat, apartment searches, interrogations, arrests, vodka-inspired toasts "to the success of our hopeless cause," interminable discussions of inner freedom and the meaning of life, and other rituals of dissident life. Most of the men are unemployed, having been fired from their jobs for signing open letters or distributing samizdat; most of the women work, including Nadia, a hospital physician. Almost everyone is sleeping with someone other than their spouse.

Anna Gerts was the pseudonym of Maya Lazarevna Zlobina, a literary critic and translator who freelanced at the journal *Novyi mir*. She had known Sinyavsky and Daniel since before their arrest and participated in the chain reaction, signing her name on at least one open letter in 1967, demanding the right to attend the "Trial of the Four." She does not appear to have authored any samizdat texts before or after her novel.[19] Zlobina composed *Sacred Paths to Willful Freedom* in 1972, as the dissident movement appeared to be in one of its death spirals. Given the subject matter, there was never any question of publishing her text in the USSR; the dedication to Andrei Amalrik on the title page alone put the nail in that coffin.

More acutely than any other writer inside or outside the movement, Zlobina explored how dissidents disturbed the peace of late Soviet society, not merely by irritating or endangering those around them, but by unsettling them. When we meet Nadia, she is struggling with the dilemma that will define the entire novel: whether to remain with her dissident husband, Alexander Moiseevich ("Alik"; we do not learn his last name, but we don't need to, since his patronymic, "son of Moses," flags him as Jewish), or dump him in favor of the man with whom she has been having an affair for several years, the gainfully employed, eminently practical, utterly reliable Gleb Pavlovich Loginov. Alik is charismatic, self-sacrificing, and fragile; he used to teach at a high school but was fired after giving a lecture about the poet Osip Mandelstam, who perished in Stalin's Gulag. Now he occasionally gives private lessons, the income from which does not cover his drinking habit (getting drunk, he insists, is a specifically Russian form of resistance to state power). He is full of ideas but can never seem to finish anything, not even short entries for the *Chronicle of Current Events*. He is also a relentless womanizer—a major trigger of Nadia's affair with Gleb—as if to compensate for his failed career and his utter dependence on his physician-wife to keep the family afloat.

Everyone, it seems, is urging Nadia to divorce Alik. Her much-loved nanny puts it succinctly: "He doesn't look after his own. He's always cooking up something for strangers. And mostly with words—he sure knows how to talk. A dumpling with no filling."[20] Gleb regards Alik as a childish ne'er-do-well "incapable of weighing the consequences of his actions." Even if Alik spoke the truth and was trying to teach his students to think for themselves, why did he put them and his own family at risk by talking about Mandelstam? "He wanted to show off!" Gleb insists. "He's irresponsible, infantile, incapable of adapting to life." When Nadia responds that Alik doesn't want to adapt to Soviet life, Gleb counters, "That's a very convenient way to justify being a loafer."[21]

It's not that Nadia's lover opposes everything her husband stands for. Like Soviet citizens up to their chins in a pond of sewage, Gleb shares many of Alik's criticisms of the Soviet system. He acknowledges that Soviet schools "fill students' brains with lies." Indeed, following the trial of Sinyavsky and Daniel, Gleb had shown a keen interest in the emerging dissident movement and wanted Nadia to introduce him to Alik. He was looking for "a constructive idea, a theory, a program, . . . a concrete alternative" to the current system, but the movement didn't seem to have one. "Nobody even thought about that at the time," Nadia recalls. "They were simply happy that people had stopped being afraid and were doing what they could." That strikes Gleb as unserious, and he soon loses interest in the dissidents' "amateurism." Nadia understands: her lover belongs to that category of people who are incapable of taking part in a cause they consider hopeless. Gleb comes to believe that "the Soviet system is fully consistent with Russia's historical traditions" and that, in any case, nothing can change it. "Every people gets the government it deserves," he likes to tell Nadia. Having concluded that nothing will change, he stops troubling himself, adopting a tone of condescending irony toward those who continue to bang their heads against the wall. "Nothing is more hopeless than trying to humanize Soviet power," he announces. "The only thing we can do is block that power from entering our soul, try not to notice it and keep our contact with it to a minimum."

Gleb remains an eager consumer of samizdat—Orwell, Solzhenitsyn, the philosophers Nikolai Berdyaev and Herbert Marcuse—but he considers them "leisure" reading, for pleasure, along with playing tennis three days a week.[22]

Nadia's close friend Marishka (Marina) has her own theory about why Nadia foolishly continues to support Alik instead of accepting Gleb's offer to share his apartment. Objectively, of course, there's no comparison between the two: Gleb is financially secure, is ambitious, has a beautiful apartment and a car; Alik can offer only high-minded poverty and endless troubles. Alik is sensitive, wounded, agitated, prone to dark moods and outbursts. He may be capable of great feats, but the smallest setbacks can throw him into despair. "He needs support and comfort," says Marishka. "He needs to be *saved*, darling, and that is something no Russian woman can resist."[23] Nadia has been saving him for most of the ten years they've been married, serving as the breadwinner, raising their son (Sasha), cooking and cleaning, and picking up Alik's dirty laundry, hiding his samizdat in anticipation of yet another apartment search, putting up with his affairs.

Marishka, it later turns out, has her own motive for advising Nadia to leave her husband: she has begun an affair with him. In the meantime, however, her critical appraisal of Alik, like Gleb's, pushes Nadia to clarify her own feelings. At night, lying in bed next to Alik, she despises him, despises his cheating, his lying about cheating, his constant unreliability. During the day, "pity rises to the surface" as she opens herself to Alik's tenderness and suffering. The eminently sensible reasons for leaving him, the long list of grievances, the self-justifications—none of this can suppress what she calls "the raw feeling of guilt." It's not so much that Nadia wants to save Alik, as Marishka would have it; rather, she feels guilty watching his doomed struggle against the Soviet state. Dissidents are hounded out of their jobs, chased into a corner, under constant threat of searches, interrogations, and arrests. "They should be pitied, not judged," Nadia tells herself. "And I pity them, don't I? But all the same they irritate me, and there's nothing to be done about it." Whatever one might say about the dissidents, "as long as such people exist, there is still hope that all is not lost. That helps me live." And yet, if the end result is "self-destruction, the breakdown of the individual, what is the point of all these sacrifices?"[24]

Zlobina's novel leaves this conflict unresolved, just as it leaves Nadia suspended between Alik and Gleb. Her feelings of admiration, irritation, and guilt vis-à-vis Alik express in heightened form responses to dissidents that we encountered in chapter 17. Those emotions are intensified, of course, by Nadia's extreme proximity as a bystander-spouse. Whatever the degree of social distance, it was impossible to know how such feelings would play out in any given person. Vyacheslav Igrunov and Myroslav Marynovych were drawn into the movement; Aron Gurevich and Lev Navrozov admired it from the sidelines; Tatyana Goricheva and Gennady Shimanov withdrew from it. Referring to Alik and his dissident circle, one of Nadia's friends tells her, "I used to consider it a stroke of good fortune that I knew them. The best people, the conscience of Russia. Now I sometimes think they're no better than those who persecute them. They're even worse."[25] Was the original devil not a former angel?

As with Kochetov's *What Is It You Want?*, there was no shortage of reader responses to Zlobina's novel. According to the *Chronicle of Current Events*, it circulated widely in samizdat following its release in 1972. Yet it took four years for the *Chronicle* to mention *Sacred Paths to Willful Freedom* in its "News of Samizdat" section, by which time it was hardly news. The precipitating event was the decision in 1976 by the émigré editor Roman Gul—whom Kochetov had dismissed (by name) as "pathetic" in *What Is It You Want?*—to publish Zlobina's text serially in the New York–based *Novyi zhurnal* (New Journal). Gul, the *Chronicle* informed its readers, was predicting a best-seller.[26]

Larisa Bogoraz was not happy. Having returned from Siberian exile in 1972, she had kept a low profile, refraining from public statements and dissident activism in general. Zlobina's novel, with its darkly intimate portrait of the dissident movement's "home front," evidently tested that restraint.[27] Bogoraz's initial silence may have reflected her desire not to provoke the Fifth Directorate so soon after her return from exile or the sense that open criticism of Zlobina's work would only draw additional attention to it. Whatever the case, the novel's publication in New York seems to have tipped the scale. In a caustic review titled "Petty Devils"—an excerpt of which the *Chronicle* made sure to include along with its four-year-late acknowledgment of the novel's appearance—Bogoraz dismissed *Sacred Paths to Willful Freedom* as a "women's novel" whose central moral dilemma was whether "to sleep with the guy one wants or the guy with whom duty requires one to sleep." Zlobina, Bogoraz charged, had sprinkled the unpublishable names of actual dissidents across her text—Sinyavsky, Daniel, Amalrik, Petr Grigorenko, Andrei Sakharov (and Bogoraz too, although she didn't mention this)—in order to fashion a patina of authenticity and to flag the text as belonging to the exalted realm of the "free Russian word." "God forbid," Bogoraz added, "that we should forget we're reading an uncensored work." Mingling names of real-world dissidents with those of Alik and his friends allowed Zlobina to engage in a cheap literary sleight of hand, conveying the impression of intimate knowledge of the dissident movement: "You know these people from the stage, from the heroic spectacle. Read Anna Gerts's novel and you will see them behind the curtains, with traces of make-up, half undressed, a glass of vodka in hand. You will find out who is sleeping with whom and who is living at whose expense. Absolutely fascinating—'Pushkin in real life,' so to say."[28] If anyone was familiar with the heroic mythologizing of Soviet dissident life, it was Bogoraz. She had watched her first husband, Yuli Daniel, be transformed into an icon following his 1966 trial, and she herself was vaulted into the dissident pantheon two years later in Red Square. The problem wasn't that Zlobina's novel brought the sainted dissidents down to sinful earth. Bogoraz had no illusions regarding some of their personal moral failings, not least her former husband's. It was that, in her reading, *Sacred Paths to Willful Freedom* sent a more damaging message: "The heroic spectacle failed and wasn't worth it."[29]

Zlobina, who was not allergic to clichés, had one of her characters invoke *The Devils* when speaking of Alik's circle of dissidents. It had become something of a tradition in Russia, Bogoraz noted, ever since Dostoevsky's novel appeared in 1872, for this epithet to be used against virtually every social movement. A century later it had become downright fashionable:

Alas, we have grown accustomed to passing judgment via analogy and to bowing before authorities: a sign of pettiness of feeling and laziness of thought. Armchair critics don't even attempt to consider the essential properties—good or bad—of a new social phenomenon, with all its multiplicities of meaning and form, with its unexplored genesis and unfamiliar structure. Instead they seek out its real and imagined vices, gloat when they discover foam and debris on its surface, crave new revelations and evidence against it. For these critics, Anna Gerts's novel, which recycles salon chatter about the Resistance, will be a pleasing gift.[30]

The Devils drew on a real incident involving the underground revolutionary cell led by Sergei Nechaev, a serial manipulator and liar who in 1869 murdered a former co-conspirator. Dostoevsky's novel quickly became the canonical account of the Nechaev affair and, for many readers, of Russia's entire revolutionary underground. What if Zlobina's novel—which, to be sure, no one would have placed in the same literary league as Dostoevsky's—ended up exercising an analogous pull on popular understanding of the dissident movement, shaping the narrative and benefiting from the undeserved halo of having appeared in samizdat?[31] Compared to this, the dissident biographies unintentionally fashioned by the KGB looked better than ever.

Much of Bogoraz's antipathy for *Sacred Paths to Willful Freedom* seems to have drawn on precisely this fear (though it can't have helped that the novel's title was lifted verbatim from a poem by her ex-husband, Daniel).[32] Like many rights-defenders—women as well as men—she considered topics such as childcare, family life, and sexuality to be secondary to civil liberties, or as the literary scholar Ann Komaromi put it, "banal and unworthy of debate."[33] It may therefore seem strange that Bogoraz went out of her way to emphasize how factually accurate the novel's details were. Zlobina, she noted, outdid Dostoevsky in this regard: "In her novel there is not a single invented detail." The way dissidents meet in the bathroom with the tap water running so as not to be overheard, the way they move samizdat from apartment to apartment when a KGB search seems imminent, the way they collect funds for families of political prisoners: "Everything, everything is reproduced with photographic precision." Bogoraz even recognized specific topics of conversation, toasts, anecdotes, familiar phrases. "I know by whom, when, and under what circumstances they were spoken." Some of the novel's episodes, too, were taken "from life, without the slightest alteration."[34]

All of this was ammunition for Bogoraz's coup de grâce: *Sacred Paths to Willful Freedom* slavishly traded in superficial facts and failed to do the imaginative work that raises literature above mere empirical knowledge. Zlobina had committed the blunder of getting the facts right and the story wrong, or, as Dostoevsky once put it, she practiced a realism that sees no further than its nose. And she did so in a spirit of malice, focusing exclusively on "the vacuous, the superficial, the secondary, the extraneous." *Sacred Paths to Willful Freedom* was in fact no better than Kochetov's *What Is It You Want?* If Zlobina had not cleverly altered the names of certain characters (insiders instantly recognized Anatoly Yakobson in Alexander Moiseevich) or created crude composite figures ("Vasya's nose + Kalina's ears + Fedin's hat"), she could have been charged with slander, according to Bogoraz. In any event, the damage was done: "Plenty of idle hunters will learn from this novel previously unknown details from the biographies of real people. For them this will be an entertaining pastime."[35]

Natalya Gorbanevskaya's name also figured in Zlobina's novel, as one of the half-dozen off-stage "good dissidents." What Bogoraz condemned as a crude attempt at verisimilitude, however, Gorbanevskaya was inclined to see as something more sinister. Even before she obtained a samizdat copy of Zlobina's novel, Gorbanevskaya had been told by an acquaintance that "it's going to deliver a good dressing down to your whole Democratic Movement!" The smell of *Schadenfreude* was unmistakable. Gorbanevskaya wrote her own critical review, arguing that the casual dropping of names such as hers and Bogoraz's and Amalrik's was meant to serve as an alibi against charges that the author was in the same camp as Kochetov and other neo-Stalinists. Accusations that the novel was "anti-dissident" could then be rebuffed: Did it not preserve the halo around Amalrik and Gorbanevskaya and others? Meanwhile, all the fictional dissidents who actually populate the novel are motivated exclusively by a desire for self-promotion, an inability to find a place in life, and pathological despair. Readers didn't just find this entertaining, as Bogoraz suggested; they were using it, Gorbanevskaya argued, "as self-justification, justification for their own passivity—by exposing the emptiness, the pointlessness, the sordid motives behind any kind of activism."[36]

Gorbanevskaya leveled these charges, it should be remembered, against readers of samizdat, the only medium through which Zlobina's novel was available at the time. Her review addressed not the adversaries of the dissident movement, not the millions of passersby who either didn't notice or pretended not to notice it, but bystanders close enough to be embedded in networks of samizdat. *Sacred Paths to Willful Freedom* drew to the surface Gorbanevskaya's anguish and rage—and not only hers—at having been abandoned by so many of those bystanders following the public disavowal of the movement by Petr Yakir and Viktor Krasin:

> Anna Gerts's novel was written after the severe crisis experienced by our movement in 1972–3. The hardest thing of all is to emerge from a crisis in a genuine way, by reflecting on what was in the very soil of the movement that allowed typical human weaknesses to swell into tragic results. By searching for what each of us is guilty of, not just the heroes and victims of the show trials. By turning inward, to one's personal responsibility, forming and reforming it for the future to make it more severe and uncompromising, rather than turning to any sort of external forces.
>
> The imaginary way out of the crisis is much easier. To become estranged: we're the pure ones, we're bystanders, we already understood everything long ago and have stepped away from all this commotion and nonsense. You can't lure us or drag us in. We are bearers of moral values and messengers of moral judgments, which, of course, we will convey in exclusively artistic form (another kind of alibi—you can't fault it). "We" didn't step away from "all that" because we feared repressions (listen: is being afraid really something to be ashamed of? Simply being afraid, stepping back and living one's life, without praising anybody's heroism or denouncing those whom you

abandoned?). No, we distanced ourselves in a supremely ethical manner from a fundamentally rotten cause. We will continue to open people's eyes to this devil's lair.[37]

These were the two available outcomes, according to Gorbanevskaya, following the movement's most recent near-death experience. One demanded self-scrutiny. How had the movement become so vulnerable to the authoritative but unscrupulous Yakir and Krasin, whose undoing by the KGB managed to pull so many others down with them? Only by returning to the primacy of individual conscience could the movement rededicate itself to the cause of defending civil and human rights. The other outcome favored retreat. Because public activism was inherently corrupting and tended to attract people with personal pathologies (and dubious ethical standards), and because, in any case, years of dissident activity had achieved almost nothing, it was better to devote oneself to the cultivation and preservation of moral and spiritual values, which after all were more important and more enduring than any particular set of legal or political arrangements. True freedom required keeping one's mask on.

In spending time with novels by Kochetov and Zlobina, I have sought not so much to mine them for insights into the dissident world, but rather to treat them as occasions for critical reflection on that world by contemporaries. Like memoirs and other non-fictional sources, they illuminate the fraught relationship between Soviet dissidents and Soviet society. I do not share Bogoraz's and Gorbanevskaya's conviction that *Sacred Paths to Willful Freedom* delivers an unremittingly negative verdict on the dissident movement along with a seal of approval for those who distance themselves from it (while continuing to read samizdat for intellectual stimulation). In my reading, Nadia genuinely struggles between her dissident husband and her pragmatic lover precisely because both are flawed. In a telling scene near the end of the novel, Nadia and Gleb pass by an apartment window from which they can hear a young boy begging his father not to beat him. Gleb is unsure how to react and proposes looking for a policeman. Nadia realizes that Alik would have intervened himself, on the spot. She admires his "Don Quixotism," telling Gleb, "that's the only way you can remain an ethical person in this vile world." Gleb mumbles something about it being better not to interfere lest one make matters worse.[38]

What is most striking about the readings of *What Is It You Want?* by Plyushch and Sokirko, and of *Sacred Paths to Willful Freedom* by Bogoraz and Gorbanevskaya, is the way they evoke a shrunken dissident world under siege not just by the KGB and the Soviet government, but by Soviet society at large. All exude a bitter sense of abandonment by allegedly sympathetic bystanders. "I have no illusions," wrote Valentin Turchin. "My conflict is not merely with the authorities, or even primarily with the authorities, but with society."[39] Whereas Kochetov presents "other-thinkers" as little more than puppets of foreign intelligence services, Zlobina dwells on their drinking, extramarital affairs, and other personal failings, along with the disruption they cause those around them. Neither engages at any level with the substance of dissent—the defense of civil and human rights, the rule of law, assistance to political prisoners and their families. Zlobina's portrayal of the dissidents' personal failings, which Bogoraz essentially confirmed but insisted was beside the point, might not be beside the point when juxtaposed with dissidents' self-declared role as "ethical examples" to the rest of the Soviet population. While there is little reason to believe that dissidents drank or betrayed their spouses more frequently than the rest of the intelligentsia, what stood out to bystanders was the contrast between that lifestyle and dissidents' self-cultivated reputation as "outstanding personalities," the conscience of their country. What did not stand out or even get noticed, it seems—not by Kochetov, Zlobina, their critics, or even the Fifth Directorate—was the contrast between dissidents' demand for the rule of law and their resistance to formal rules and procedures among themselves.

Rights-defenders, the *Times* of London correspondent David Bonavia noted in 1973, "have mostly already been extruded from Soviet society—sacked from their jobs or dismissed from their institutes." Now they found themselves shunned by many bystanders as well. In the wake of the Yakir-Krasin debacle, "our movement lost much of its prestige among the intelligentsia," recalled Ludmilla Alexeyeva. At a gathering of intellectuals she overheard someone say, "Some heroes. [The authorities] step on their tails, and what do they do? They crack, they plead, they name names." Thousands of bystanders "suddenly began to feel good about their decision to keep their distance from the democratic movement. Prisons would have provided a less oppressive environment than the Moscow of that time."[40] A process of mutual estrangement seemed to be setting in, as dissidents reformatted themselves into small-scale, Western-oriented associations, and bystanders—even sympathetic ones—found reasons to turn away.

Dissidents now lived in a "ghetto," according to Alexeyeva, with "its own traditions, literature, celebrations, etiquette, even institutions." While the rest of the country celebrated the "day of solidarity of the working masses" and the anniversary of the Bolshevik Revolution on May 1 and November 7, dissidents gathered on March 5 and December 5, the anniversaries of Stalin's death and Stalin's constitution. They were, as Yelena Bonner put it, "alone together."[41]

PART V
From the Other Shore

20

The Kindness of Strangers

> It was one thing to tell the truth at home. It was something else to tell it to outsiders, many of whom were genuine enemies of our country.
>
> —LUDMILLA ALEXEYEVA

Do Soviet dissidents belong to the broader history of protest that enveloped much of the industrialized world in the 1960s? This notion has tempted historians of social movements and international relations alike. From Paris and Berlin to Prague and Moscow (and, in some versions, in Berkeley, Beijing, and Mexico City as well), it has been argued, a new generation of "sophisticated rebels" developed a new language of dissent, a "global lingua franca of protest," an "international dissident culture," culminating in 1968 with the "first global revolution of the twentieth century."[1]

These ambitious claims for a transnational history of dissent deserve close scrutiny.[2] The Soviet dissident repertoire of contention, with its emphasis on civil obedience and legalism and communication via samizdat, diverged sharply from those of protesters in the West (not to mention Mao Zedong's Red Guards). Soviet dissidents had neither the numbers nor the freedom (nor, generally speaking, the desire) to stage mass demonstrations. Most were older and more professionally established than protesters elsewhere. In contrast to Berkeley, Mexico City, Paris, and Berlin, Soviet universities did not serve as hubs of recruitment and activism. To be sure, mass media's insatiable hunger for "leaders"—the more oppositional the better—significantly shaped the evolution of protest in both the West and the USSR.[3] The difference lay in the fact that, whereas the media spotlight on protesters in the West was wielded by Westerners and operated largely via television, in the Soviet Union it was wielded by Westerners and operated largely via shortwave radio. "Radio mentioned me!" exclaims a character in Sergei Dovlatov's 1981 novella *The Compromise*. "I'm sort of a dissident!"[4] Apart from occasional diatribes against dissidents as foreign agents seeking the USSR's destruction, Soviet mass media mostly passed over them in silence.

In certain respects, however, claims for a global movement of dissent do not go far enough. For Soviet dissidents, something more than a lingua franca with activists in the West was at work. There were actual conversations and mutual influencing, carried out via texts smuggled into and out of the USSR, arduous long-distance phone calls, and face-to-face meetings in Moscow apartments. These conversations, however, were not conducted with the Mario Savios and Daniel Cohn-Bendits and Rudi Dutschkes of the world, or with other protesters in Berkeley, Paris, and West Berlin. Nor were they conducted with representatives of communist parties in Western Europe, who ceased to be the default international interlocutors as neo-Leninism lost its status as the preeminent idiom of internal opposition to the Soviet state. Instead, Soviet dissidents eventually found interlocutors—and models of organization—among what were then relatively obscure Western NGOs.[5] Even then, a common language across East and West was by no means a given; it was worked out in fits and starts not in the 1960s but in the 1970s, with imperfect results. Insofar as a shared transnational language emerged, moreover, it was not one of youthful rebellion, anti-militarism, or civil disobedience, but the "thin" lingua franca of human rights.[6]

Initial encounters across the Iron Curtain did not go well. When approached in 1969 and 1970 by representatives of citizens' groups from Italy, France, and Belgium, as described in chapter 12, wary members of the Initiative Group for the Defense of Civil Rights declined offers of collaboration. Valery Chalidze's Human Rights Committee actively sought out partners in the West but did not last long enough to bring those relationships to fruition.

Among the various Western NGOs that took an interest in Soviet dissidents, none would come to play as direct a role in their movement as Amnesty International. Founded in London in 1961 by the attorney Peter Benenson, Amnesty introduced itself to the world as a "movement composed of peoples of all nationalities, politics, religions and social views who are determined to work together in defense of freedom of the mind."[7] Amnesty focused its efforts on securing the release of specific individuals jailed solely for their opinions or beliefs, people it called "prisoners of conscience," provided that they had not advocated or used violence. "Pressure of opinion a hundred years ago," Benenson announced on the pages of the London *Observer*, "brought about the emancipation of the slaves. It is now for man to insist upon the same freedom for his mind as he has won for his body."[8]

Notwithstanding its self-description as a movement composed of peoples of all nationalities and religions, for much of its early history Amnesty's membership consisted almost exclusively of individuals from English-speaking countries and the Western half of Europe. Its annual dispatch of thousands of Christmas cards to prisoners of conscience around the world unintentionally exposed the gap between universal aspirations and an assumed Christian sensibility.[9] And yet, in its adoption of prisoners whose cases it sought to publicize, Amnesty went to extraordinary lengths to maintain not just the appearance but the practice of universalism and non-partisanship. Guided by the geography of the Cold War, Benenson and his fellow activists scrupulously sought out prisoners of conscience in equal proportion from capitalist, communist, and developing countries—a goal more easily articulated than met. At the grass-roots level, each local group of Amnesty volunteers was required to work simultaneously on behalf of three prisoners, one from each category, and never on behalf of a prisoner from their own country. "There are other organizations," Amnesty noted, "that are working within some more limited ideological framework for aid to those of their own particular persuasion. Our strength is that in such matters we take no sides at all."[10] With its peculiar blend of engagement and detachment, Amnesty practiced, indeed institutionalized, a deliberately distanced empathy, a novel brand of kindness that could come *only* from strangers. In the eighteenth century, the historian Lynn Hunt has argued, sentimental novels expanded the range of their readers' empathy, thereby laying the groundwork for the language of universal rights that accompanied the American and French Revolutions at the end of that century.[11] In the second half of the twentieth century, Amnesty taught its members to cultivate multiple, transnational circles of empathy and to keep them in check against each other so as to transcend the ideological chasm of the Cold War. "We are beginning to become what we wish to be," the organization's chairman, Lionel Elvin, reported to members in 1963: "the Ombudsman of the imprisoned conscience everywhere."[12]

Distanced empathy posed special challenges, beginning with how to obtain reliable and timely information about individuals held in prisons thousands of miles away, often guarded by secretive governments. Lawyers and writers from the Soviet bloc—the presumed counterparts of Amnesty's members in Britain and elsewhere—had declined invitations to open a dialogue. Communist countries, observed the scholar and attorney Leonard Schapiro, "insist on their right to criticize the administration of justice in the West, and so they should," yet they resented even the slightest critique of their own legal system. "Since the Russian public is denied that right," Schapiro continued, the West had not just a right but "a duty to criticize" Soviet miscarriages of justice.[13] Criticism, however, needed to be informed. As Benenson—like Schapiro the offspring of Jewish immigrants from tsarist Russia[14]—noted in his closing remarks at an Amnesty conference on prisoners of conscience in communist countries, the first challenge was precisely the dearth of information. It was difficult even to obtain names of individual prisoners of conscience in the USSR, let alone to reliably estimate their total number. Internal Amnesty estimates from the early 1960s ranged from five thousand to several hundred thousand.[15]

The conspicuous rise of human rights advocacy in the 1970s, in which public "naming and shaming" became an indispensable tactic, makes it easy to forget how uncertain the role of publicity was during Amnesty's first decade. Today most human rights organizations focus almost exclusively on how, rather than whether, to maximize public attention to rights violations, while governments—or at least those governments that make it their business to monitor the state of human rights in various countries—weigh the risks and benefits of behind-the-scenes diplomacy versus open criticism. The evolving relationship between Soviet dissidents and Amnesty International sheds considerable light on the origins of this informal division of labor. To understand that relationship, one needs to revisit an era when NGOs themselves had to assess the risks and benefits of publicity, if only because no other entities—neither sovereign states nor the United Nations—had yet gotten into the business of systematically and impartially monitoring compliance with international human rights norms, let alone speaking out against their violation.[16] As late as 1968, Amnesty's International Executive Committee, the body responsible for setting the organization's global policies, registered "strong doubts" about whether extensive publicity regarding prisoners of conscience "can do any good in Soviet cases." On the contrary, it might trigger retribution against prisoners' families and friends or against prisoners themselves.[17]

Whatever the Committee's doubts, for much of the 1960s the question whether to publicize the plight of prisoners in the USSR was mostly moot: apart from a handful of cases, there was a near total information blackout regarding convicted individuals. Requests for permission to send observers to "political" trials, one of Amnesty's standard fact-finding techniques in other parts of the world, were met with silence or a terse message from Soviet embassies that the requests "were without foundation."[18] When a Norwegian lawyer and Amnesty member, Ingjald Orbeck Sörheim, arrived in Moscow in January 1968 on a tourist visa and announced his intention to attend the "Trial of the Four" (Alexander Ginzburg, Yuri Galanskov, Vera Lashkova, and Aleksei Dobrovolsky), he was barred from entering the courtroom.[19] No Amnesty volunteers had ever received a reply to cards addressed to Soviet prisoners; these were usually stamped *inconnu* or *adresse inexacte* and returned to the senders. Nor had Soviet authorities demonstrated any willingness to provide information about prisoners in response to hundreds of letters from local Amnesty groups or from the organization's London leadership.[20]

Moscow's silence in response to inquiries from Amnesty should not be taken as a sign of special hostility toward the organization itself. True, the Russian-language letterhead used by the Amnesty leadership for letters to Soviet officials, with its prominently displayed epigraph (erroneously attributed to Voltaire)—"I despise your ideas but am prepared to die for the sake of your right to express them"—displayed what can only be called an egregious tone-deafness vis-à-vis its audience.[21] But Amnesty's work during the 1960s on behalf of imprisoned leftists in the West did not go unnoticed. A 1968 article in *Pravda* praised Amnesty as an "authoritative international organization" working to stop the arrest and torture of communists by the "fascist" junta in Greece.[22] The first Soviet handbook on international NGOs, while classifying Amnesty as "an international bourgeois organization," described its goals as "promoting guarantees of free speech and religion in accordance with the Universal Declaration of Human Rights" and "assisting in the liberation of individuals imprisoned for their political or religious convictions."[23]

But along with these admirable activities, the handbook warned, Amnesty "often adopted reactionary positions under the flag of objectivity." In 1966, it was revealed that Benenson, Amnesty's founder, had knowingly received covert British government funds to support Amnesty's work on behalf of political detainees in the former colony of Rhodesia (today's Zimbabwe). The International Commission of Jurists, another NGO that Benenson had helped found in the 1950s—and which for several years shared office space in London with Amnesty—was found to have unknowingly received covert CIA funding via its American affiliate. Even worse, the head of the International Commission of Jurists at the time, the Irish statesman Seán MacBride, was also chairman of Amnesty's International Executive Committee.[24] The ombudsman of the world now appeared entangled in the politics of neo-colonialism and the Cold War.

Amnesty emerged from this affair with a heightened sensitivity to issues of non-partisanship and a good deal of confusion as to what kind of working relationship it ought to have with the world's sovereign states. Financial ties were clearly out of the question. But what about other kinds of relationships? Prisoners of conscience, after all, were almost always prisoners of states, and only states could make the decision to release them. Amnesty had been founded on the proposition that not just states but ordinary people could help realize the rights enshrined in the Universal Declaration of Human Rights. What, if any, forms of dialogue or perhaps even cooperation could take place between an NGO such as Amnesty and an authoritarian, secretive state such as the USSR? Despite Amnesty's professed ideological neutrality and its record of engagement on behalf of persecuted communists, Soviet authorities now had good reason to doubt its impartiality. To the Soviet mindset, professions of non-partisanship only fueled suspicion of masked purposes, especially given that the map of Amnesty's membership bore a striking resemblance to the map of the NATO alliance. Small wonder, then, that Amnesty's letters to the Soviet government went unanswered—for now.[25]

In the midst of Amnesty's funding scandal, an unexpected document from behind the Iron Curtain arrived at the organization's London headquarters. Compiled by a group calling itself the Council of Prisoners' Relatives, it contained the names of two hundred imprisoned Soviet Baptists.[26] In one of the earliest non-literary uses of samizdat, an "initiative group" of Baptist resisters, formed in August 1961, circulated a series of petitions addressed to Khrushchev and other Soviet leaders, criticizing the All-Union Council of Evangelical Christian Baptists for abetting in the anti-religious campaign and requesting permission to convene a congress of independent Baptists. When the "initiative group" announced soon thereafter its intention to form a separate evangelical organization, hundreds of participants were arrested.[27]

Five years had elapsed before the list of imprisoned Baptists arrived at Amnesty's headquarters. Most of the information, according to Amnesty's internal assessment, was "too old to be used" by the time it was received in 1966.[28] In the same year, copies of a letter sent by imprisoned Ukrainians to Soviet procurator-general Roman Rudenko, protesting policies of Russification and demanding that Ukraine be allowed to exercise its constitutional right to peacefully secede from the Soviet Union, reached the West (including Amnesty) with a similar time-lag of roughly five years following the protesters' trial in Lviv.[29]

Thanks to the dramatic growth of samizdat in the 1960s and the Soviet Union's increasingly porous borders, the number of such documents was growing even as the time it took them to reach the West was beginning to shrink. A second petition concerning a new round of arrests of dissenting Baptists in 1964, for example, took just over two years to arrive in London. The arrival of information about imprisoned Baptists made it possible for the first time for Amnesty to adopt, on the basis of reasonably systematic data, nearly a hundred prisoners of conscience inside the Soviet Union. For years, Amnesty had been stymied by its inability to deploy in the USSR its usual methods for gathering data on prisoners of conscience. The Soviet press had proved all but useless. Repeated appeals to Soviet officials, like those from Soviet dissidents, had gone unanswered. Attempts to send observers to "political" trials, let alone fact-finding missions to prisons, labor camps, and psychiatric hospitals, had gotten nowhere.[30] By the late 1960s, however, facts began finding their way to Amnesty. Amnesty, in turn, began to adopt Soviet cases on the basis of information pre-selected and arranged by Soviet prisoners themselves, or by local activists working on their behalf.

Those who became known as rights-defenders were not the first to publicly invoke rights enshrined in the Soviet Constitution, to make use of samizdat, or to organize petition campaigns and public demonstrations on behalf of victims of state repression. But they were the first to apply these techniques systematically and to people other than themselves. It was the metropolitan dissident intellectuals who, beginning in the mid-1960s, captured Western attention and from whose comparatively thinner ranks iconic individuals emerged. When it came to crafting and exporting the narrative of internal resistance to the Soviet state, it was they, not their predecessors, who prevailed. The earliest documents from arrested Baptists and Ukrainians were eclipsed by texts fashioned by rights activists, whose breadth of coverage, frequency of reporting, and geographical location (Moscow) helped elevate them to positions of unprecedented visibility from the West. It was not just that more and more samizdat texts—hundreds of documents annually—were seeping out of the USSR. Increasingly, rights-defenders were selecting and organizing the many trickles of data on the repressive activities of the Soviet state into a coherent stream before directing that stream westward.

What had begun as trickles of data into Amnesty's offices swelled, over the course of just a few years, into a more or less steady current. Prior to 1968, as internal Amnesty documents noted, "the central difficulty of work in this area [was] lack of information": "[We] had nothing substantive on the Soviet Union." By 1970, Amnesty officials were noting "the tremendous improvement in the flow of information" from the USSR.[31] So great was Amnesty's regard for (and reliance on) the *Chronicle of Current Events* that in 1970 it took the controversial decision to translate and publish over a thousand copies of each successive issue as it arrived in the London office—the only known instance in which Amnesty published work produced by individuals outside the organization.[32] This step significantly strengthened the *Chronicle*'s credibility in the eyes of Western journalists.[33] By 1975, Amnesty was confronting "an increasing problem of too much information (samizdat) reaching the International Secretariat from the Soviet Union," straining its capacity to translate and distribute incoming texts to the hundreds of local Amnesty groups that had adopted individual Soviet prisoners of conscience.[34] Indeed, the flood of communication from the Soviet dissident movement helped catalyze the professionalization of Amnesty's growing research department, which in 1971 established a working group in London to collect and translate documents arriving from the USSR.[35] In 1974, representatives of local Amnesty groups from various countries held the first annual meeting to coordinate efforts specifically on behalf of prisoners in the Soviet Union.[36] Within a few years, 630 local Amnesty groups in eleven countries were working on behalf of some five hundred Soviet prisoners of conscience. According to Amnesty officials, this represented "an upper limit for the balance of [our] work as a whole and for the capacity of the research team."[37] For most of the 1970s, the Soviet Union was at or near the top of the list of countries with the most adopted prisoners of conscience.

The *Chronicle* regularly reported on rights violations involving Baptists, Ukrainians, Tatars, Jews, and ethnic Germans, providing them a degree of visibility they were unable to achieve on their own.[38] As Amnesty's list of adopted Soviet prisoners of conscience expanded, those jailed or exiled for religious activities (as opposed to the "politicals") continued to constitute roughly half the total. When it came to highlighting individual cases of persecution, however, Amnesty's leaders, like Western journalists, displayed a marked preference for urbane intellectuals who spoke in a universal idiom of secular rights—which is to say, for people more or less like themselves. When Amnesty launched its "Postcards for Prisoners" program in 1965—one of a series of increasingly ambitious international publicity campaigns—the USSR was represented by Alexander Volpin. Over the course of the next five years, the postcard campaign took up the cause of Vladimir Bukovsky and Yuri Galanskov as well as Petr Grigorenko—all activists for civil and human rights. By comparison, two Baptists were the object of postcard campaigns: Aida Skripnikova and the preacher Georgi Vins, who described the encounter of evangelical Christians with the Soviet state as a contest "between God's servants and Satan."[39]

For its 1968 "Prisoner of the Year" campaign, Amnesty set itself the task of selecting three iconic prisoners of conscience, one each from the capitalist, socialist, and developing worlds. The pragmatists on Amnesty's International Executive Committee argued for choosing individuals based on the likelihood of their being released; others asserted that longevity of imprisonment should be the deciding factor. In the USSR, the Committee noted, "there are very many candidates one could have presented from various categories of prisoner of conscience, e.g., imprisoned writers, Ukrainian nationalist intellectuals, dissident Baptists and so on. The [Alexander] Ginzburg case is a good one, as it provides a link with Sinyavsky and Daniel; it is also representative of many other similar ones."[40] Ginzburg, of course, had distinguished himself as compiler (not writer) of samizdat poetry anthologies and the *White Book* of documents concerning the trial of Andrei Sinyavsky and Yuli Daniel. If the ability to stand for a large number of analogous cases had been the deciding factor, however, an imprisoned Baptist would surely have been chosen, given that more than half of Amnesty's adopted Soviet prisoners that year were evangelical Christians.

When considering the possibility of a subtle selection bias in an organization dedicated to a set of ostensibly universal rights, one should recall that Amnesty faced daunting hurdles in gathering and evaluating evidence of abuses on an increasingly global scale. Evidence of preferences in its relationship with various targeted communities in the Soviet Union should be seen against the background of the organization's strenuous efforts to maintain balance across the Cold War's fault lines. In fact, a moral as well as a utilitarian case could be made for heightened attention to those who were not only victims of rights abuses, but who themselves actively defended the rights of others, or what Hannah Arendt called "the right to have rights," just as many legal systems impose more severe punishment for assaulting a police officer than a civilian.[41]

But neither in public nor behind closed doors did Amnesty make that case. Rather, it was beleaguered Soviet dissidents who, by articulating a cosmopolitan, secular idiom of struggle for civil liberties, exerted an unspoken pull on the perceptions and priorities of what was becoming the most influential human rights NGO of the Cold War era. Dissident texts invoked the Soviet Constitution and the Criminal Code as well as the Universal Declaration of Human Rights and other international agreements—precisely the genre of documents Amnesty cited in its appeals. Baptist documents, by contrast, deployed biblical quotations and references to divine commandments: "It is time to act for the Lord, for they have violated Your law" (Psalm 119:126) and "Render unto Caesar the things that are Caesar's, and unto God the things that are God's" (Matthew 22:21).[42] Ukrainian nationalists, for their part, even when armed with impeccable legal arguments (Soviet as well as international) for linguistic and cultural autonomy, could not entirely free themselves, in the eyes of many Western observers, from the tangled legacies of ethno-nationalism and the undeniably political nature of the demand for secession.

Dissident narratives also reinforced the individualizing instinct that governed virtually all of Amnesty's public campaigns, with their emphasis on "putting a face" on human rights issues. That instinct drew on the Enlightenment conviction that individual agency, or sovereignty over the self, was the sine qua non of human dignity. Although Baptists and

Ukrainian nationalists were a substantial presence among Amnesty's adopted prisoners of conscience—thanks in no small part to coverage of their plight by the *Chronicle of Current Events*—they stood for discrete collectives based largely on ties of kinship and ethnicity, while the dissidents appeared as full-fledged individuals drawn together by purely elective affinities—again subtly mirroring Amnesty's own members.[43] The fact that Baptists and Ukrainian nationalists defined themselves as groups did not sit comfortably with Amnesty's assumption that individual figures—such as those selected as "Prisoner of the Year"—generated the most potent publicity. Imprisoned Baptists, as an internal Amnesty report noted, "nearly all have a similar background," making them easier to categorize for adoption but more difficult to deploy as fully fleshed-out icons of persecution. In a kind of self-reinforcing logic, it was concluded that letter-writing campaigns were "more or less meaningless in the case of less publicly known prisoners such as the majority of the dissident Baptists."[44] Distanced empathy, not surprisingly, had preferences.

Alongside its moral solidarity with prisoners of conscience around the world, Amnesty kept a pragmatic eye on power, that is, on the governments that controlled the fates of those prisoners. For most of the 1960s, efforts to communicate directly with Soviet authorities had come to naught: invitations to Soviet jurists to attend conferences sponsored by Amnesty had failed to elicit a response, as had the overwhelming majority of letters from Amnesty's many adoption groups inquiring about this or that prisoner. The first to break through the wall of silence was Seán MacBride, since 1965 the chairman of Amnesty's International Executive Committee. Born in Paris to exiled Irish republicans, MacBride had, among Amnesty's leaders, unique credentials with which to approach the Kremlin. The British government had executed his father for participating in the 1916 Easter Uprising in Dublin and imprisoned MacBride himself three times for revolutionary activities during the 1920s. After serving as chief of staff for the Irish Republican Army (and defense attorney for several of its arrested leaders), MacBride rose to become independent Ireland's minister of external affairs in 1948. A leading figure in the postwar international peace and disarmament movements, he was a frequent visitor to the foreign ministries of communist countries, where he also worked to build bridges between associations of lawyers across the Iron Curtain.[45]

In the early 1970s, MacBride visited the USSR several times in his capacity as chairman of the International Peace Bureau (an NGO) in order to help plan the World Congress of Peace Forces, to be held in Moscow in October 1973. This connection allowed him to initiate a conversation about Amnesty's activities with Soviet officials, who arranged for MacBride to deal directly with what they considered to be Amnesty's closest equivalent in the USSR, the Association of Soviet Lawyers. Like all "public" organizations in the Soviet Union, the Association was controlled by the Communist Party. This was of course not all that different from how Soviet officials understood Amnesty and many other Western NGOs, whether because of the history of covert funding by Western intelligence services or because Soviet officials instinctively projected certain of their own local arrangements onto groups operating in other countries.

For Amnesty, talks in Moscow presented an opportunity to inform Soviet authorities about Amnesty's work around the world, to pursue cooperation in the defense of jailed communists in countries such as South Vietnam and Indonesia, and to discuss political imprisonment in the USSR.[46] For the Soviets, Amnesty represented a new player in the international contest for "global public opinion." Soviet media had on occasion praised Amnesty's activities when they cast a critical light on Moscow's opponents—especially when those opponents were clients of the United States such as South Vietnam and (until 1979) Iran. But most Soviet press coverage of Amnesty in the years leading up to MacBride's overture had been unmistakably negative, even as it conveyed the impression that Amnesty exercised significant influence on Western publics. The Ukrainian edition of *Pravda* accused Amnesty of occupying a "leading position among organizations that conduct anti-Soviet propaganda," "disseminating falsified materials [a veiled reference to the *Chronicle of Current Events*] in capitalist countries," and being "closely allied with the intelligence services of the USA and England."[47]

In response to Amnesty's claims regarding the forced confinement of Soviet dissidents in psychiatric hospitals, *Izvestiia* attacked the "notorious" and "unscrupulous" organization as a "malignant slanderer concerned with one thing only: to depict the mentally sick person as a great 'fighter for an idea.' It is not given to every film star in the West to receive such 'publicity.'"[48]

The Kremlin apparently regarded MacBride's initiative as a chance to steer Amnesty's powers of publicity in a direction more favorable, or at least less damaging, to Soviet interests. The recent increase in Amnesty's knowledge of and attention to the plight of prisoners of conscience in the USSR may well have added a sense of urgency. In any event, Soviet authorities recognized that MacBride, the only former revolutionary and former prisoner of a capitalist country within Amnesty's leadership, was their best hope. In addition to asking MacBride to serve as deputy chairman of the World Congress of Peace Forces, they invited representatives from Amnesty to take part in sessions at the Congress dealing with human rights—the first time Amnesty had been permitted to send staff members to the Soviet Union in their official capacity, as opposed to visiting as tourists. Furthermore, meetings were arranged with prominent Soviet jurists such as Arkady Poltorak (former secretary of the Soviet delegation to the Nuremberg Trials), Igor Blishchenko (a leading specialist on international law at the Ministry of Foreign Affairs), and Samuil Zivs (an expert on law in capitalist countries and vice-president of the Association of Soviet lawyers). A confidential report by Nigel Rodley, Amnesty's legal officer, registered the historic new phase in Amnesty's relationship with the USSR: "Bilateral negotiations took place between the Soviets and Amnesty. These negotiations were conducted by lawyers."[49]

Whether they qualified as negotiations was in the eye of the beholder, given the vast asymmetries of power and resources. This was not just David meeting Goliath; this was an unarmed David meeting Goliath. Nonetheless, a face-to-face conversation with Soviet officials was precisely what a lawyer and statesman—and man of the Left—such as MacBride had long sought in dealing with Moscow: diplomacy rather than denunciation, sober calculation of complementary interests rather than

Cold War confrontation. In an interview with the *New York Times*, Rodley struck a conciliatory note: "Because the human rights problems here have been exploited for Cold War purposes in the West, it became necessary to try to establish an atmosphere of good faith and objectivity." MacBride went further. While acknowledging that "a number of things are done here which do not conform with the norms we would expect in the West," he emphasized the need for "correct and accurate assessment of all the relevant facts." "There has been a good deal of exaggeration in the foreign press," he added, "in regard to the extent, if any, to which psychiatric hospitals are being used in dealing with political prisoners."[50]

Such reports were to be found not just in the press, however. Amnesty itself, using information compiled by Vladimir Bukovsky, Semyon Gluzman, and other dissidents, had repeatedly drawn attention to the political use of psychiatry in the USSR. As the *Times* noted, MacBride's skepticism about accusations of punitive psychiatry was "not entirely shared by others close to Amnesty International." That was an understatement. MacBride's public questioning of Amnesty's own claims sprayed fuel on long-smoldering tensions within the organization between those strategically sympathetic and those implacably hostile to the Soviet Union, and between proponents of direct and discreet negotiations with states (even, or especially, with unsavory ones) and advocates of public diplomacy, including public shaming. Amnesty's internal divisions were about to explode into the open—with help from Soviet dissidents.

In the early 1970s, Andrei Tverdokhlebov, one of the co-founders of the Human Rights Committee along with fellow physicists Valery Chalidze and Andrei Sakharov, had established contact with Amnesty—not with the organization's headquarters in London, but with the Dutch national section and local Group 11 in New York City, both of which had a particular interest in Soviet prisoners of conscience. At Tverdokhlebov's request, members of the "Madison Avenue Group" (as Group 11 was informally known) had supplied him with a wide range of Amnesty publications, including annual newsletters and the *Handbook for Groups*. Tverdokhlebov and the engineer Vladimir Arkhangelsky, having become voracious consumers of Amnesty materials, began to circulate

FIGURE 20.1. Andrei Tverdokhlebov, 1976

translated excerpts, along with other human rights–related texts, in a samizdat periodical called *Amnesty International*.[51] The translation and flow of data were now two-way: Amnesty published the *Chronicle of Current Events* in English, while Tverdokhlebov circulated typewritten collections of Amnesty documents in Russian.[52] This was evidently too much for the Fifth Directorate. Already under investigation for his ties to both the *Chronicle* and the Human Rights Committee, Tverdokhlebov was confronted on the night of August 27, 1973, by security police who searched his apartment, seizing thousands of documents, including the archive of the Human Rights Committee and voluminous materials from Amnesty, the United Nations, and various organizations devoted to international law.[53]

Two weeks later, Tverdokhlebov was on the phone with Leonid Rigerman, a recent Soviet émigré affiliated with Amnesty's International Secretariat in London. Despite, or perhaps because of, the KGB's search and seizure, he and three colleagues—Arkhangelsky, Ilya Korneyev, and Vladimir Albrekht (author of multiple dissident conduct manuals)—had decided to form a new organization called Group 73, modeled on

Amnesty's local chapters of volunteers. Or rather, modeled on what they imagined Amnesty International to be: a network of individuals working globally to provide material and legal assistance to political prisoners and their families.[54] Arkhangelsky submitted an application to the Moscow City Council, requesting permission to set up a bank account to accept voluntary donations. The KGB, aware of Tverdokhlebov's broader intentions, had instructed City Council bureaucrats to deny official recognition to Group 73 and to demand information about its relationship to Amnesty.[55] After relating this news by telephone to Rigerman in London on September 10, Tverdokhlebov dictated a message to be conveyed to Amnesty's Sixth International Council, due to convene shortly in Vienna. On September 15, hundreds of Amnesty delegates from twenty-four countries who attended the meeting at Albert Schweitzer Haus listened to the following words from Moscow:

> Greetings to the International Council of Amnesty International. Since childhood we have been accustomed to hearing such phrases as "political action of the masses," "the active foreign policy of the government and party," "the struggle for social rights and the social reconstruction of society," "the scientific and technological revolution"—and these are things we imagined the world was preoccupied with. As for words like "conscience," "dignity," "conviction"—we are accustomed to apply them exclusively to the exertions and strivings of individual human beings. For who can help one to value such words, and to preserve their value, other than oneself and those to whom one is closest? At first we were astonished, and could not grasp, that in fact total strangers can help, people who live in the most distant countries, in conditions utterly different from one's own, in other cultures. It is this above all that we value in your example and your activity, insofar as we are in a position to judge them. Please accept our best wishes.[56]

The statement begins with a litany of phrases drawn from the lexicon of official Soviet discourse. By the time they reached adulthood, Soviet citizens had heard and read such phrases thousands of times—in radio and television broadcasts, in newspapers, in wall-posters at their schools, neighborhoods, and places of work. Indeed, many of them had repeated these phrases themselves, in student exercises or obligatory political education seminars in factories and offices—all part of their contribution to the lip-service state. As the statement notes, the "political action of the masses" and the "struggle for social rights" are "things we imagined the world was preoccupied with"—not things with which the authors themselves were preoccupied.

The statement then shifts to a different register: single words denoting personal ethical norms. Could concepts such as "conscience," "dignity," and "conviction," which applied to relationships within face-to-face communities, avoid the fate of official phrases, which had long since been drained of meaning and lost their moral force? Could such concepts exercise any real influence outside circles of kinship and friendship? The distant strangers of Amnesty International seemed to suggest that they could. For Tverdokhlebov, Arkhangelsky, and others, Amnesty's language—the language of human rights in action—had the effect of reinvigorating certain ethical norms by transposing them from the private sphere into realms of public, indeed global significance. Like Alexander Volpin's repurposing of the term *glasnost* (transparency), Amnesty's language produced a defamiliarizing effect by pulling certain ethical categories out of ordinary usage.[57]

Terms such as "conscience," "dignity," and "conviction" were of course not new. On the contrary, they were well-established but by now largely hollow elements of official phraseology. By the second half of Soviet history, such values were widely understood to operate primarily in the private sphere. It was a Soviet variant of the process Max Weber called disenchantment, in which "the ultimate and most sublime values retreat from the public sphere into either the transcendental realm of mystical life or the brotherliness of direct and personal human relations."[58] Amnesty's novelty—and its appeal to Soviet dissidents—lay precisely in its apparent reversal of that trend, extending the private, apolitical ethics of family and friends to complete strangers. For Volpin, seeing unfamiliar faces at transparency meetings on Pushkin Square was a sign that the dissident movement had moved beyond circles of friendship. Amnesty was now enacting that transformation on a scale he could hardly have imagined.

For members of Group 73, not just Amnesty's re-enchanted language but its organizational form exercised a magnetic pull. Valentin Turchin, inventor of REFAL (Recursive Functions Algorithmic Language), the USSR's leading computer programming language, and compiler of the book *Jokes Physicists Tell*, regarded Amnesty International as a harbinger of a possible future global order:

> Marxism-Leninism preaches in theory, and puts into practice, an extreme politicization of all aspects of social life. It sees everywhere (and where it does not see, it plants!) the struggle of economic interests and the struggle for political power. What the world needs now is the opposite approach—the *depoliticization* of the most important aspects of life. The success of Amnesty International, its growing membership and influence, show that more and more people on planet earth are beginning to understand this. The fundamental idea of Amnesty International is to depoliticize our understanding of individual civil and political rights, of the impermissibility of torture and other forms of inhuman treatment of human beings. From time immemorial these issues have been classified as political. Amnesty International is shifting them to the sphere of universal morality.[59]

Turchin found the ideal model of social organization not in the West's unruly capitalist democracies, but in Amnesty. Human communities, he argued, develop ever more refined instruments of control, from physical violence to economic necessity to spiritual culture. The Soviet Union, stuck in its post-Stalin inertia of fear (that is, memory of mass violence), would be best served not by developing a multi-party democracy, but by expanding the Communist Party into a pluralist, inclusive network. The highest form of ties among human beings, which Turchin called valence bonds (by analogy with the bonds between atoms), are personal and individualized, fostered in small groups based on mutual understanding, trust, friendship, and creativity. Such bonds had been the animating force within the circles of friends that constituted the dissident movement in its earliest incarnation, and they continued to work as the movement reformatted itself into various associations not

FIGURE 20.2. Valentin Turchin

unlike NGOs, starting with the Initiative Group for Civil Rights in the USSR.

Turchin saw valence bonds at work in the small local chapters that constituted Amnesty's transnational network of activists. So did Yuri Orlov, for whom the formation of an Amnesty chapter in the Soviet Union was "part of a plan—to enable the formation of as many unofficial rights groups as possible, drawing people into diverse, peaceful activity independent of the government." "[We] wanted to give Soviet citizens an example of devotion to pluralism and to tolerance of diverse ideas unconnected to violence."[60] Such groups, internally bound by personal friendships, according to Turchin, were "the most human mode of social intercourse," the one in which "what is properly human unfolds to the full." According to the Ukrainian poet, novelist, and philosopher Mykola Rudenko, Amnesty was "the most humanist of all organizations."[61] By contrast, the West's adversarial political system, in which all manner of decisions were made by majority rule, was little more than a refined version of civil war, replacing one form of coercion (fighting) with another (voting). "Voting does not solve problems," Turchin insisted, because voting was merely "a kind of weapon," whereas valence communities were a "tool." "The kind of work that must be done to solve complex social problems can only be done by small groups of people" internally linked by valence bonds.[62]

It was precisely this kind of forward-looking association, both practical and ethical in its approach to the world, that Amnesty embodied for Orlov, Rudenko, Turchin, Tverdokhlebov, and others. By "refraining from the *struggle for power* and everything associated with it," Amnesty was detaching the problem of individual rights from any particular type of government or ideology, and therefore transforming rights into a purely moral issue. "This," Turchin concluded, "is one of the most promising developments of our time":

> Amnesty International ... scrupulously refrains from expressing any political preferences. Of course this does not mean that a member of Amnesty should not *have* political preferences. It is clear to everyone that the people working for Amnesty sympathize with political regimes which more or less respect human rights, and are indignant toward regimes which crudely trample on those rights. But in their work within that organization, they ignore those sympathies and antipathies. Such, it seems, is the only road to the integration of humanity.[63]

The integration of humanity required individuals to suspend their political passions even as they cultivated fellow feeling toward strangers. Such was the path to distanced empathy.

Less than a month after sending their greetings to Amnesty delegates assembled in Vienna, the four members of Group 73—Tverdokhlebov, Albrekht, Arkhangelsky, and Korneyev—along with Turchin, Orlov, and five others, applied to London for registration as a national section of Amnesty International.[64] The timing could hardly have been worse for MacBride, whose carefully cultivated back-channel to Soviet authorities seemed to augur a future dialogue with leading figures in the socialist world. There had already been one close call: the search of Tverdokhlebov's apartment on August 27, during which large quantities of Amnesty materials were seized, had taken place precisely (and perhaps not accidentally) in the midst of, and just a few miles from, MacBride and Rodley's "bilateral negotiations" with Soviet jurists. How would it look now to suddenly announce the formation of a Moscow chapter of Amnesty, especially one composed of dissidents? The London leadership had already privately expressed concerns that any unauthorized Soviet Amnesty group would be subject to severe persecution by the state, as had happened with a short-lived Amnesty chapter in Park Chung-hee's authoritarian South Korea. Amnesty's goal, after all, was to reduce, not add to, the number of prisoners of conscience in the world. If, conversely, Soviet authorities were given a say in the establishment of an Amnesty group, would they not seek to fashion another Potemkin NGO, tightly controlling its activities to suit their own purposes and thereby subverting Amnesty's mission?[65]

The Madison Avenue Group did not share these concerns. On the contrary, its members encouraged Tverdokhlebov to meet with MacBride while the latter was in Moscow for the World Congress of Peace Forces in late October, regarding such a meeting as "an opportunity for direct, personal contact" between the London leadership and Soviet citizens who were interested in forming an Amnesty chapter.[66] This was precisely the kind of contact that the KGB feared would be fostered by the Congress and which it was determined to prevent.[67] In Amnesty's case, there was no need for preventive measures: reluctant to jeopardize his ongoing conversations with Soviet officials, MacBride refused to receive Tverdokhlebov when the latter appeared at his hotel.[68] Coming on the heels of his public skepticism regarding claims of Soviet psychiatric abuse, this rebuff unleashed a firestorm inside Amnesty, including calls for MacBride's resignation. His supporters, in turn, suggested that such calls came from CIA moles within the organization.[69] As if that were not enough, when Amnesty's secretary-general, Martin Ennals, arrived at work in Amnesty's London headquarters on October 31, he found on his desk a press release from Valery Chalidze—now head of Khronika Press in New York—featuring a copy of Tverdokhlebov's application under the headline "11 Moscow Intellectuals Found USSR National Section of Amnesty International."[70]

Ennals and the London leadership were "very disturbed."[71] Only the International Executive Committee, led by MacBride, could authorize the creation of a national section, and it was therefore difficult to view Chalidze's gesture as anything but a crude attempt to force Amnesty's hand. Another way to see it, of course, was as an example of the dissidents' "do it yourself" ethos—the same ethos that inspired samizdat's circumvention of Soviet censorship. A scathing letter to Chalidze from Rigerman—the émigré physicist who coordinated telephone communications between Tverdokhlebov in Moscow and Amnesty's London headquarters—captures some of the fallout:

> Amnesty's leaders are now seriously asking themselves whether the Soviet group wishes to function as a national section of Amnesty or is merely trying to use Amnesty's name for cheap publicity. For my part, I am asking myself, do you understand that the ease of publication in the West does not in any way mean that everything should absolutely and immediately be published? Do you understand that your friends in Russia write letters and statements to various western organizations not merely to boost your authority as the preeminent [émigré] source of information, but primarily in order to establish genuine, fruitful connections with those organizations? ... The fact that you are in Russia or are an émigré from Russia does not in any way give you the right to act with that characteristically Russian presumptuousness.[72]

Rigerman's fury notwithstanding, Amnesty's predicament hardly stemmed from "Russian presumptuousness," since it was in fact Edward Kline, a businessman and member of Amnesty's Madison Avenue Group, who had initiated the press release—not Chalidze.[73] Moreover, news of the application from Moscow for affiliation with Amnesty was spreading via the same samizdat itineraries that had already opened countless unauthorized windows onto the USSR—and that had so dramatically transformed Amnesty's advocacy on behalf of Soviet prisoners of conscience. In this sense, the application to become an Amnesty chapter differed little from the thousands of "open letters" that, notwithstanding their particular addressee (general secretary of the Communist Party Leonid Brezhnev, procurator-general Roman Rudenko, secretary-general of the United Nations Kurt Waldheim, and others) had circulated in unregulated fashion within the Soviet Union and seeped across its borders. Sooner or later, even without Chalidze's or Kline's involvement, news of the Moscow group's initiative would have become public, thereby requiring a response from the Amnesty leadership.

Soviet dissidents were engaged in a learning process. The Initiative Group for Civil Rights in the USSR had shunned partnerships with foreign NGOs, fearful of feeding the accusation that it was a pawn of the Soviet Union's enemies. The Human Rights Committee sought out such partnerships, convinced that it could replicate in the USSR the kind of impartial expertise cultivated by the International League for the Rights of Man and similar organizations in the West. Tverdokhlebov, Turchin, and other dissidents were attempting to go even further, seeking to open a branch of Amnesty International in Moscow, in effect importing the practices of a Western NGO into the capital of the socialist world.

Amnesty International was learning too. Its leaders were confronting the unexpected consequences of their own universalism, as they contemplated what it would mean to have a branch of their organization behind the Iron Curtain. Rather than a shared language of protest, what connected Amnesty and the dissidents was a common set of reference points in the form of selected articles from the Universal Declaration of Human Rights, along with a shared conviction that such rights transcended the realm of politics and ideology. Their repertoires of contention also overlapped, in the sense that both relied on letter writing by engaged citizens. But Amnesty was still feeling its way as regards the use of publicity, especially vis-à-vis the Soviet Union, whereas Soviet dissidents had already learned to use publicity, or transparency, as their primary mode of struggle. The conditions under which the two sides worked, of course, could hardly have been more different: Amnesty's members, at least in the West, did not fear losing their jobs, or seeing their children denied entrance to a university, much less arrest or exile.

They were instructed to labor only on behalf of other people's human rights, never their own. Only under such conditions did Amnesty's highly formalized procedures of Cold War impartiality make sense.

Amnesty's London leadership now faced an awkward set of options. To recognize the Moscow chapter might give the appearance of succumbing to manipulation and could disrupt MacBride's delicate diplomatic overture to the Kremlin. To refuse recognition would compromise Amnesty's carefully tended reputation for universalism and non-partisanship, as well as risk alienating the very Soviet dissidents Amnesty was trying to protect. On a deeper level, the Moscow initiative forced Amnesty to confront a fundamental question: Was it willing not just to stand up for Soviet dissidents but to embrace them as partners in the enterprise of defending human rights?

21

Adoptees at the Gate

The principle of non-intervention in matters essentially within the domestic jurisdiction of states is a rule of law binding on *states*, who are the creators and addressees of international law. It is a principle regulating the conduct of governments in their mutual relations, not the conduct of individuals lawfully expressing concern for their fellow human beings. The mere accumulation and publication of information can in no way be categorized as *intervention*.

—AMNESTY INTERNATIONAL

The Moscow group's application arrived as Amnesty confronted mounting accusations that it was not a genuinely international entity defending universal rights but a Western organization serving Western (that is, Cold War and/or neo-colonial) interests. As early as 1969, Amnesty's annual report had expressed the need to "strengthen [our] links in countries where so far we have failed to gain a footing."[1] In the early 1970s, Amnesty remained an almost exclusively First World organization. The oft-invoked equilibrium among the three global zones (capitalist, socialist, and developing) had been more or less achieved, at least symbolically, when it came to adopted prisoners. But the same could hardly be said of the 1,817 local groups that undertook the day-to-day work of advocacy on their behalf, of which 1,801 (99 percent) were in First World countries.[2] There were no Amnesty groups anywhere in the socialist world.³ Outside the West, those who became involved in Amnesty's work were overwhelmingly on the receiving end, "adopted"—to use Amnesty's parental metaphor—by Westerners for whom human rights were less a means of self-defense than an instrument of global moral improvement.⁴ In 1972, Amnesty's Long-Range Planning Committee had concluded that "to become a truly international organization we will have to pursue a true international objective in terms of future National Section development." In the past, the Committee noted, "the growth of National Sections took place in a more or less haphazard way." The formation of chapters in countries with authoritarian governments would require especially close attention: "No premature steps should be taken in this direction without a carefully considered development plan."⁵

The Moscow initiative, and the fact that it immediately became public, rendered such planning moot. Amnesty was now forced to navigate what everyone regarded as a potential minefield involving one of the world's two superpowers, a colossal state that regularly sat near the top of Amnesty's list of countries with the most prisoners of conscience. The ensuing internal debate was driven by diverging assessments of the Soviet government's likely response to the establishment of an Amnesty group in the USSR, by contrasting views on whether Amnesty's model of engagement by ordinary citizens in human rights advocacy was in fact exportable to non-democratic countries, and, more broadly, by disagreements over where and by what means Amnesty ought to invest its moral capital.

Three basic positions emerged. The first, represented by secretary-general Martin Ennals and Dirk Börner of Amnesty's Hamburg group, rested on the hope that Amnesty might be able to have its cake and eat it too, recognizing the Moscow group and carefully guiding its human rights activities while continuing to seek common ground with the Kremlin. The second, represented by Seán MacBride, chairman of the International Executive Committee, and Lothar Belck, the Committee's Swiss treasurer, insisted that the cake could either be had or eaten, but not both, and advocated cultivating ties with the Soviet government as the best route by which to advance Amnesty's agenda in the USSR and globally. The third, articulated most forcefully by the British Sovietologist Peter Reddaway together with the Madison Avenue Group, agreed with the second as regards the cake, but dismissed the possibility of cooperation with Soviet authorities, preferring instead to throw Amnesty's full support to Andrei Tverdokhlebov and his fellow applicants.

In a June 1974 meeting in Moscow with the jurist Igor Blishchenko, Amnesty officials discussed the possibility of designating a contact organization (presumably the Association of Soviet Lawyers) that would henceforth field inquiries from Amnesty groups abroad, explain the legal background of individual cases, and, where appropriate, conduct inquiries. Blishchenko promised to provide information on Soviet prison regulations, including guidelines for contacting prisoners and sending care-packages. In return, he asked that Amnesty work with Soviet organizations in its campaigns against human rights abuses in Latin America and South Africa and that a Soviet observer be invited to the next international meeting of Amnesty representatives—another unprecedented move from the Soviet side.[6]

At that same meeting, however, Blishchenko made it clear that the Association of Soviet Lawyers would be "in an impossible position" if Amnesty were "to maintain formal ties with dissidents." To both MacBride and Reddaway, it was clear that Amnesty faced a stark choice: the possible forms of cooperation dangled by Blishchenko and other Soviet officials were intended as a quid pro quo in return for Amnesty's rejection of the Moscow group's request for affiliation. For MacBride and his allies on the International Executive Committee, it was a deal worth making, since they were convinced that accepting Tverdokhlebov's application would jeopardize not only Amnesty's emerging dialogue with the Soviet government but its reputation as an apolitical organization. There was more than a little irony in the latter concern. Soviet rights-defenders, after all, had long insisted with equal vehemence on the apolitical nature of *their* activism and often expressed frustration at being perceived by their supporters in the West—not to mention the KGB—as an "anti-Soviet" movement.

MacBride was convinced that Amnesty groups could not function in countries "where democratic institutions and civil liberties are either weak or non-existent."⁷ Vulnerable to manipulation by the state or by forces hostile to the state, such groups would be unable to maintain the ideological neutrality central to Amnesty's work. The Kremlin might tolerate Soviet citizens writing letters on behalf of prisoners of conscience in capitalist countries, but it could hardly be expected to do so when those prisoners were in communist countries, especially those allied with the USSR. Human rights might be universal, but Amnesty's model of human rights activism via distanced empathy was apparently a luxury only citizens of liberal democracies could afford.⁸

The Madison Avenue Group took issue with nearly every facet of that argument. "If Amnesty is not to be solely a West European [and] American organization," the group's chair, Yadja Zeltman, wrote, "it must find ways to operate within the framework of different social and political systems and in different environments." Gaining an affiliate group in the USSR would offer "an opportunity for the growth of Amnesty not only in the number of countries represented, but also in effectiveness and reputation."⁹ As Andrew Blane, professor of Russian history at the City University of New York and a member of Amnesty's International Executive Committee, put it, "Even the appearance that grass-roots association is being sacrificed to high-level dialogue strikes me as not the stuff of which Amnesty is made."¹⁰ Reddaway, Amnesty's leading authority on Soviet affairs, similarly took aim at the anti-recognition argument. For all its apparent *realpolitik*, he wrote in a blunt letter to the Amnesty leadership, that argument was "naive" and "barren." True, in contrast to the Stalin era, when even the rumor of affiliation with a foreign organization could lead to a person's summary execution or decades in the camps, the current Soviet leadership had so far taken no decisive action—possibly waiting to see how Amnesty itself would react to Tverdokhlebov's request for affiliation. But this, in Reddaway's view, by no means signaled the Kremlin's readiness to negotiate. Even if it did, Amnesty could not afford to enter into a "bargaining relationship" with Moscow. "In such a game," he wrote, "Soviet authorities would hold much the stronger cards, and they have the experience and skill to outwit [Amnesty] decisively." They would pretend to negotiate in good faith while extracting concessions from Amnesty, would "fail to deliver their side of the bargain, and then would skillfully maneuver for a long time to avoid paying the price."[11] Any concessions on behalf of Soviet prisoners of conscience would come strictly as a result of "unpleasant pressures (publicity, etc.)." Recognition of an Amnesty group in Moscow would be an effective form of such pressure and would, according to Reddaway—and contrary to fears expressed by Ennals and others—only strengthen the personal safety of its members: "If they are recognized, it is unlikely that *any* of them will be arrested in the foreseeable future (or at most only one or two of them)."[12] Irmgard Hutter, head of the Austrian national section of Amnesty, argued that *not* recognizing Tverdokhlebov and the other applicants might endanger them.[13]

Moral authority, or, to put it somewhat differently, the accumulation and investment of moral capital, was essential to Amnesty's work, and it was here, according to supporters of recognition, that a Soviet affiliate could offer the most valuable dividends. MacBride might be correct in assuming that an Amnesty group in Moscow would have no access to the Soviet press and therefore to Soviet public opinion. But it would, in the view of its supporters, have "tremendous access to the international press for which it would be envied by many other [Amnesty] groups." Börner, the head of Amnesty's Hamburg group, predicted that "the British government will certainly not be impressed by any protest against Pat Arrowsmith's imprisonment printed in *Pravda*"—Arrowsmith, a well-known British peace activist and co-founder of the Campaign for Nuclear Disarmament, had been arrested multiple times—"but it would be truly embarrassed if this [were to come] from a [Soviet] dissident group and [be] printed in the *Times* [of London]." In this case, Soviet dissidents could "do more than ten groups in Germany."[14] Other Amnesty leaders, adopting a strikingly instrumental approach, agreed: "Arrest of a member of the Soviet section for defending prisoners in Spain or in Greece will be very good publicity for Amnesty"; "A USSR section would be very good for the moral position of Amnesty."[15] An affiliate in Moscow represented not just a presence in the epicenter of the socialist world, but a potentially powerful form of moral leverage outside that world.

Exploratory talks between Amnesty (represented by MacBride) and the Association of Soviet Lawyers (represented by Samuil Zivs) began in the summer of 1973. Lists of adopted prisoners in the USSR were handed to Soviet representatives for comment; those representatives in turn provided Amnesty with a copy of Soviet regulations for prison and labor camp conditions. Amnesty informed Zivs and others that it was preparing a report on the treatment of Soviet prisoners of conscience and that Soviet officials would be given an opportunity to comment before it was published. The Soviet side warned Amnesty that Valentin Turchin, Tverdokhlebov, and their associates were trying to "start an opposition movement under the label of Amnesty."[16]

Following a July 1974 meeting at Zivs's office in Moscow, Amnesty leaders Ennals, Thomas Hammarberg, and Börner gathered in Turchin's apartment on Butlerov Street, just a few metro stops away, to hash out the still unresolved issue of recognition. Present at that gathering, in addition to Turchin and the Amnesty officials, were Tverdokhlebov and Yuri Orlov, with Tatyana Litvinova (aunt of Pavel Litvinov, mother-in-law of Valery Chalidze) acting as translator. Orlov recalled the conversation as follows:

> Our guests delivered arguments against giving Soviet members of Amnesty the status of a "[national] section." It was difficult, they said, to deal with a totalitarian country; we might run up against provocations by the KGB. It would make more sense for Tverdokhlebov, they advised, to devote himself to some activity more effective than Amnesty, "if you wish to overthrow this system." "We are not setting ourselves any such goal," I remarked to the ceiling, just in case. After many hours of negotiating they agreed to a compromise: Amnesty would register us as a "group"—the lowest status, which did not give us the right to send delegates to international congresses of Amnesty International. I could see that they didn't want trouble from us. It's also possible that, bewitched by the Soviet political game, Amnesty's leadership had decided not to complicate its relations with the Soviets by establishing excessively close ties with dissidents.[17]

For Orlov and other dissidents on the receiving end of Amnesty's distanced empathy, it was hard to know which was worse: the ignorance of Amnesty's representatives, unaware that Turchin's apartment was bugged and that, in any event, Soviet dissidents adhered to strictly legal forms of activity, or their condescension in explaining to dissidents what it would be like to deal with a totalitarian state. Relegated to second-class status as a "group," Orlov and others were left to wonder just how "international" Amnesty really was. The encounter felt a long way from the new, creative global arrangements Turchin had imagined.

As Orlov correctly intuited, Amnesty—with Ennals in charge—was still attempting to have it both ways with the Kremlin. Inside the Kremlin, however, interest in cooperation was waning. Whatever feelers had been extended by Zivs and others—acting on instructions from above—were almost certainly, as Reddaway had argued, designed to dissuade (or at least delay) Amnesty from formally recognizing the Moscow applicants as an affiliate. It was not just that Amnesty—which KGB chairman Yuri Andropov referred to as a "'non-governmental' organization," using scare quotes—was regarded by the Kremlin as an agent of Western imperialism.[18] Rather, as Andropov and Soviet procurator-general Roman Rudenko stated in an April 1975 memorandum to the Central Committee, by applying for affiliation with Amnesty International, Tverdokhlebov and others "have set out to legalize, and as far as possible to establish immunity for, the anti-Soviet activity of a cohort of renegades in our country."[19] In a striking (mis)interpretation of its mission, or perhaps its name, Andropov and Rudenko cast Amnesty as seeking a kind of extraterritorial status for its members around the world, not unlike the legal immunities established by nineteenth-century European colonial powers on behalf of their citizens in China and Japan. This was not an isolated misreading: when Hammarberg, head of Amnesty's Swedish section, met that same month with an officer from the Soviet embassy in Stockholm, he was similarly told that dissidents "only wanted Amnesty membership as a guarantee against arrest." To Soviet authorities, the kindness of strangers looked like a stealthy form of neo-colonial interference in the domestic affairs of a sovereign state.[20]

This was a rather fanciful view of Amnesty's mission. The organization had never offered, or even claimed to be able to offer, anything like immunity, much less transnational immunity, for its members or adopted prisoners. But its name could give the impression—as some of Amnesty's early supporters had worried—that adopted prisoners were guilty and therefore seeking amnesty or that Amnesty International was asking states to overturn convictions.[21] Amnesty's would-be members in Moscow, moreover, as well as their supporters abroad, did harbor hopes that official recognition by the organization's headquarters in London would help shield them from arrest, on the theory that the Soviet regime was keen to avoid the negative publicity that arrest or persecution would stir up. As for charges of violating state sovereignty, Amnesty officials insisted that the principle of non-intervention in domestic affairs applied to states—the creators and addressees of international law—but "not to individuals lawfully expressing concern for their fellow human beings." In any event, gathering and publishing information, Amnesty insisted, did not even qualify as intervention, regardless of whether it was undertaken by a state or by an NGO like itself.[22] Amnesty and the Kremlin now appeared headed for a collision.

MacBride's resignation from the chairmanship of Amnesty's International Executive Committee in September 1974—one month before he was awarded the Nobel Peace Prize for decades of work on behalf of global disarmament and human rights—triggered a more confrontational turn in Amnesty's relations with the Soviet government.[23] Later that month, almost exactly a year after Tverdokhlebov and other dissidents applied to become affiliated with Amnesty, the International Executive Committee formally granted recognition—not as a national section, but as a local group, in accordance with the agreement reached in Turchin's apartment two months earlier. The Moscow group elected Turchin as chair and Tverdokhlebov as secretary and was promptly assigned adoption cases from Spain, Yugoslavia, and Sri Lanka. The Soviet government immediately began blocking the group's incoming mail (mostly from London) but did little to stop members from advocating on behalf of their adopted prisoners. Predictions of the dissidents' outsized moral leverage were quickly confirmed—in spades. The group's Spanish adoptee, the imprisoned playwright and critic Alfonso Sastre, was released shortly after the Moscow chapter sent letters to Francisco Franco and presented a petition to a Spanish representative in Moscow.

In response to the group's inquiry about its Sri Lankan prisoner (an ethnic Tamil calling for greater cultural autonomy), that country's ambassador personally invited Turchin for tea at the embassy—on the mistaken assumption that the Moscow Amnesty group was a typical Soviet "public" organization operating de facto on orders from the Kremlin. Even after Turchin explained that that was not the case, the ambassador, not wishing to take chances, arranged for the prisoner's release.[24]

Spain was soon replaced on the Moscow group's adoption list by Uruguay, where communists and other leftists had been subject to mass arrests and torture following a right-wing coup d'état in 1973. By now, Turchin and others had learned to assimilate Amnesty's language as well as to attract a who's who of Soviet dissidents to Amnesty's cause. In a petition to Juan María Bordaberry, Uruguay's president-turned-dictator, forty signatories—including Petr Grigorenko, Yuri Orlov, Andrei Amalrik, Larisa Bogoraz, Alexander Ginzburg, and Vitaly Rubin (all former Soviet prisoners of conscience) as well as Andrei Sakharov, Yelena Bonner, and the refuseniks Vladimir Slepak and Aleksandr Lunts—expressed their dismay:

> No one should be subjected to degrading treatment and torture. Every human being should have the right to express his opinions and to move about freely. Only if people of all the world's countries, all nationalities and political convictions will support each other in the struggle for the observance of these great principles can humanity count on a better future. We ask you, Mr. President, to consider this petition as evidence of our deep respect and friendly feelings for the people of your country.[25]

There was no denying the gravitas of a letter protesting human rights violations against communists in one country (although the letter studiously avoided any mention of politics) by victims of analogous violations on the part of their own communist government.[26] A press release from Amnesty's London headquarters brought the point home:

> The Uruguayan government and media have repeatedly denounced Amnesty ... as "communist inspired"—a charge which Amnesty has

rejected, pointing to its record of unceasing work for prisoners of conscience in all parts of the world, including the Soviet Union and other socialist countries. The fact that so many prominent Soviet citizens who have fought for human rights in the USSR, often at considerable risk to themselves, have signed this petition to President Bordaberry underlines the universal humanitarian concern about the torture of detainees in Uruguay.

The intervention by Soviet dissidents, Amnesty proclaimed, "makes nonsense" of charges that the organization's campaign in Uruguay was "communist motivated."[27]

As Amnesty's leaders learned, the Moscow group was capable of spending its moral capital in other ways as well. While the existence of an Amnesty affiliate in the USSR was a potent symbol of the organization's newfound reach beyond the capitalist West, criticism of Amnesty's lack of presence in the Second and Third Worlds had in the meantime expanded beyond the issue of membership. In 1973, the Dutch national section submitted a memorandum to the International Executive Committee titled "Is Amnesty Impartial Enough?," arguing that the dearth of members outside the West was due in part to the fact that Amnesty worked exclusively for civil and political rights, neglecting the social and economic rights that were foregrounded by socialist and developing countries.[28] It was a position captured by the oft-quoted phrase "Human rights begin with breakfast."[29] By working exclusively on behalf of the kinds of rights celebrated by capitalist countries, the organization was de facto forfeiting its claim to non-partisanship.

The Dutch memorandum urged that Amnesty members from Second and Third World countries be asked to give their opinions on the subject of impartiality. Turchin was happy to oblige. In a blistering samizdat essay released in 1975 (and published in tamizdat) under the title "What Is Impartiality?," he informed his fellow Amnesty members in the West that Soviet-bloc countries cynically offered limited social and economic rights "to the degree that is necessary for the functioning of the state machinery" and in order to distract from the absence of genuine civil and political rights, to which they paid lip-service only because such rights were "currently in fashion." In the West, Turchin argued, well-anchored individual liberties had made possible the advances in science and technology that enabled the material well-being of large populations. Later, those advances had been imported to non-Western countries, in particular the Soviet Union, allowing rulers to satisfy the basic material needs of their populations without granting civil and political freedoms. This "totalitarian approach" to modernization dehumanized society and led to an "evolutionary dead-end." There were indeed, as the Dutch Amnesty members claimed, contrasting Eastern and Western approaches to human rights, two "intellectual platforms." What would it mean, Turchin asked, to be "impartial" in this context? "Should Amnesty International refrain from choosing between these two conceptions and declare them equal? That would indicate an absence not of partiality but of thought and principle. Impartiality in an international organization does not mean that every human being on the planet shares its ideas and principles. No such ideas and principles exist. It means merely the equal application of principles to all countries, communities, and people."[30] Soviet ideology, Turchin noted, denied the value and even the possibility of impartiality in matters of social life, regarding it as a form of deception or self-deception. Western liberal thought, by contrast, celebrated the capacity of the human mind "to observe itself as if from an external position, to scrutinize its own system of thought, value, and aspirations from the perspective of a broader metasystem."[31] For Turchin, Amnesty itself represented, in embryo but on a global scale, precisely such a metasystem.

One suspects that this was not quite what the Dutch authors of the memorandum had in mind when they asked non-Westerners to weigh in on whether Amnesty was sufficiently impartial. Even in their private correspondence—let alone in public pronouncements—Amnesty members rarely displayed such uninhibited criticism of a government or political system as Turchin had done. Yet here again an aura of moral authority hovered around the words of Soviet dissidents, words that were smuggled out of the USSR at considerable risk to their authors. It was one thing for Amnesty to defend people like Turchin—that of course was the organization's *raison d'être*. It was quite another to have them as members. And still another to do both at once.

Within a year of Turchin's circulating his essay, Amnesty's London-based Research Department, the nerve center of the organization, produced a trenchant report that, in somewhat cooler language, directly echoed Turchin's central argument. Amnesty was committed to both impartiality (working "irrespective of political consideration") and balance (between "different world political ideologies and groupings"). It had traditionally placed greater emphasis on balance, as reflected in the stipulation that every local group should work on behalf of three adopted prisoners of conscience, one from each of the world's main ideological zones. Turchin's essay highlighted the political limits of impartiality itself. "Our attempts to adhere to the concept of balance," Amnesty's report concluded, "may, ironically, have damaged our impartiality."[32]

Predictions regarding the power of appeals by Amnesty's new Moscow chapter proved correct. Those concerning the added protection for dissidents affiliated with Amnesty did not. Within months of Amnesty's admitting the Moscow group, the KGB arrested members Sergei Kovalev and Tverdokhlebov and briefly detained Mykola Rudenko. Turchin was fired from his job and blacklisted, putting him at risk of arrest on charges of parasitism (unemployment). For Andropov and Roman Rudenko, the decision to arrest members of Amnesty's Moscow group was based on the same calculation of cost and benefit that had governed the decision to expel Alexander Solzhenitsyn. The crackdown "will provoke a loud anti-Soviet campaign abroad," they reported to the Central Committee. "However, the costs of curtailing [Tverdokhlebov's] hostile activities will undoubtedly be lower than [allowing] the continuation of his criminal behavior."[33]

With MacBride's resignation and the arrest of three of its Soviet members, Amnesty's back-channel to the Kremlin evaporated. When Amnesty, as promised, sent a draft copy of its report on prisoners of conscience in the USSR to the Association of Soviet Lawyers, seeking "comments on the accuracy or interpretation of the facts in the report" prior to publication, Lev Smirnov, president of the Association and

FIGURE 21.1. Vyacheslav Sysoyev, caricature of a Soviet prison camp guard with "human rights" written on his knuckles. "Human" is misspelled.

chairman of the Soviet Supreme Court, refused to discuss what he called a "vulgar falsification and defamation of Soviet reality and socialist legitimacy."[34] Amnesty published *Prisoners of Conscience in the USSR: Their Treatment and Conditions* in November 1975 in eight languages, accompanied by an enormous publicity campaign. Laying out statute by statute how Soviet law enabled the prosecution of non-violent expression, the report analyzed the inconsistencies between Soviet law and international norms and covenants, including those ratified by the USSR, and, above all, the gruesome conditions in which prisoners of conscience were forced to live along with other prisoners. Much of the exposé's evidence—including photographs of prisoners and camps—derived from samizdat and other dissident sources.

Smirnov had good reasons for refusing to comment on the manuscript of Amnesty's report, beyond the fact that the USSR officially denied the existence of political prisoners or prisoners of conscience in the Soviet Union. Having served as judge at the 1966 trial of Andrei Sinyavsky and Yuli Daniel, he could hardly have forgotten the experience of presiding over an international public relations debacle. But his reluctance to publicly contest Amnesty's claims captured a deeper conundrum in which the Soviet government found itself—a conundrum that would lead to uncharacteristic flip-flopping over the next decade. No organization and no government had ever produced, much less released with such fanfare in multiple languages, this kind of detailed exposé on the fate of arrested dissidents in the USSR. Two years earlier, Solzhenitsyn's *Gulag Archipelago* had shocked readers with its revelations about the vast network of forced labor camps inside Lenin's and Stalin's Soviet Union. To be sure, *Prisoners of Conscience in the USSR* was no match for the incandescent fury of Solzhenitsyn's eye-witness account. But Amnesty's report was about conditions *now*, in the era of détente between the USSR and the West.

Publicly responding to *Prisoners of Conscience in the USSR*—even to statistics that they knew were inaccurate—put Soviet officials at risk of entering into an unwelcome debate. Countries such as China or Singapore might deflect charges of violating international human rights norms by denying their universality and invoking cultural differences. In contrast to the "Asian values" critique of human rights as an ideology of Western individualism, the Kremlin had always insisted that human rights found their most genuine fulfillment precisely under socialism, in the USSR.[35] That same logic helps explain the extraordinary sensitivity of Soviet officials to Amnesty's (and not only Amnesty's) criticism of the USSR's human rights record.

While Smirnov declined to comment on Amnesty's report, first deputy minister of justice Alexander Sukharev gave a lengthy interview to the Soviet journal *New Times*, under the headline (in the English-language edition) "Human Rights in the Soviet Union: Putting the Record Straight." "Our country," he stated, "long ago reached a level in the matter of tangible guarantees and defense of human rights that the average citizen in the so-called 'free world' can only dream of." This was a reference to the Soviet Constitution's distinctive formulation of the rights of free speech and assembly, according to which the state was obliged to provide the material means—printing presses, meeting halls, and so forth—to realize those rights. "The very concept of 'political prisoner,'" he assured readers, echoing Khrushchev, "does not exist in our country."[36]

One of those readers was Amalrik, who was not assured. Amalrik immediately responded in a samizdat essay that combined, as he put it, "delight and disgust": "Delight, inasmuch as the very fact of this longwinded attempt to justify Soviet penal policy indicates just how sensitive the Soviet government is to the reaction of global public opinion to violations of human rights in the USSR and, consequently, just how much [Soviet] public opinion could achieve here. Disgust, because almost all of Mr. Sukharev's assertions are lies, and lies are always unpleasant to read."[37] According to Sukharev, Soviet citizens could not be prosecuted for their political or religious beliefs, only for their actions. "Views," he noted, "are not within the realm of jurisprudence." To this Amalrik countered, "Once an unorthodox thought has been expressed, whether printed, written down or spoken, it is regarded as an indictable act." Thousands of Soviet citizens, furthermore, had been prosecuted not for their own verbal acts, but for reading or possessing those of others, despite the fact that "there is no published list of books which it is forbidden to possess."[38]

Smirnov's colleague Zivs offered his own response to *Prisoners of Conscience in the USSR*. In a letter published in the West German *Frankfurter Rundschau*, Zivs, who had met multiple times with Amnesty representatives over the previous years, ridiculed the report for relying on "invention and speculation" to cover up its "lack of information concerning the actual state of affairs" in Soviet prisons. Particularly galling to Zivs was the report's estimate that there were "at least 10,000 political and religious prisoners in the USSR today."[39] He mocked the report's claim that there was no known case in which a political trial in the Soviet Union had ended in an acquittal. "The fact that the authors of the 'report' lack information," Zivs noted, "in no way shows that in this very narrow category of trials there have been no acquittals, no withdrawal of charges before trial, or no reversal of the verdict after appeal."[40] As Amnesty noted in its response to Zivs's letter, however, "Dr. Zivs offered no substantive information to challenge our findings on these matters."[41]

Both Sukharev and Zivs showed themselves to be exceptionally hamfisted practitioners of public relations, mocking their opponents' lack of evidence but failing to provide even an iota of their own. They were, in effect, trapped by the Soviet Union's own propaganda. Having denied for years that there was even a single person imprisoned in the USSR for his or her political or religious beliefs, they were in a poor position to challenge Amnesty's estimate of "at least 10,000" such prisoners (an estimate that was almost certainly much higher than the actual number) with a more realistic, let alone evidence-based, figure.[42] The Kremlin's fortress mentality, which kept the total number not just of political but of *all* prisoners shrouded in secrecy, poorly prepared its officials for the art of public diplomacy. Even in the United Nations General Assembly, where the USSR generally found a sympathetic audience, a Soviet delegate could only rail against Amnesty for "poisoning détente" and engaging in "slanderous accusations" against the socialist countries, without addressing the specific allegations of *Prisoners of Conscience in the USSR*.[43] In the era of "peaceful coexistence," when vast nuclear arsenals on both sides raised the incentive to seek non-lethal forms of competition, public relations had become a crucial arena of global struggle. On the topic of human rights, the Kremlin was losing.

Andropov and Soviet foreign minister Andrei Gromyko sensed the damage being done by Amnesty's work. In June 1977—halfway into Amnesty's "Year of the Prisoner of Conscience"—their memorandum to the Central Committee described Amnesty's report as "slanderous" and announced a more aggressive policy by the KGB to "monitor [Amnesty's] activities," to "interrupt its anti-Soviet actions," and to prevent its members from entering the territory of the USSR. They also recommended that Soviet institutions and public organizations adopt "the tactic of completely ignoring [Amnesty] in the course of their contacts with foreign non-governmental circles." The memorandum applied to Soviet media as well: "It seems expedient to refrain from publishing in the Soviet press any material about the activities of Amnesty International, in order not to give the opponent an opportunity to draw us into an undesirable discussion."[44]

Andropov and Gromyko's gag-order did not last long. Within months of their 1977 memorandum, Amnesty International was awarded the Nobel Peace Prize, catapulting the organization onto the world stage and cementing its reputation as the largest and most influential human rights organization of the postwar era. In its award presentation, the Norwegian Nobel Committee not only quoted Solzhenitsyn but leveled a thinly veiled rebuke against the USSR for claiming that the act of calling attention to human rights violations in another country constituted interference in that country's domestic affairs. "I cannot believe," Committee chair Aase Lionæs said, "that a ruse of this nature aimed at glossing over injustices perpetrated in one's own country will be countenanced by international opinion today."[45] While it is impossible to measure the moral capital that flowed to Amnesty from the Nobel award, the effect on its finances was unmistakable: having gradually expanded since 1970, the organization's budget doubled in size within a year of the award, reflecting a surge in membership, especially in the United States, which previously had lagged behind Europe. The following year's budget witnessed another jump, of nearly 50 percent.[46]

Heightened visibility and bigger budgets enabled a significant expansion of Amnesty's work. Between 1973 and 1979, the full-time staff of the International Secretariat nearly tripled, from 53 to 146.[47] New tactics emerged. Dozens of professional organizations in Western countries, including psychiatric associations and labor unions, were mobilized to send protests to their Soviet counterparts.[48] Inspired by the divestment campaign against the apartheid regime in South Africa, American Amnesty groups lobbied major U.S. corporations that had signed trade agreements with the Soviet Union to "make inquiries to their Soviet contacts" regarding adopted prisoners.[49]

The crescendo of protests over Soviet human rights violations was enough to cause Andropov to change his mind about the silence imposed in 1977 on Soviet media regarding Amnesty's activities. In July 1980, together with deputy minister of foreign affairs Georgy Kornienko, Andropov sent a memorandum to the Central Committee outlining a new—or, rather, a return to an old—approach. Even as it described the policy of ignoring Amnesty as "fully justified," the memorandum urged that, in light of Amnesty's "hostile actions," the editors of *Izvestiia*, with assistance from the KGB and the Ministry of Foreign Affairs, should "publish a series of exposés with the goal of compromising [Amnesty] vis-à-vis the publics of socialist and capitalist countries."[50] *Izvestiia* promptly obliged with a string of articles attacking Amnesty as a "troubadour of 'psychological warfare.'"[51] Under headlines such as "Tearing off the Masks," "Impartiality with a Secret Compartment," and "Handmaiden of Secret Services," its journalists rehashed the charges of Amnesty's covert financing by the CIA and its adoption of various criminals (including war criminals) under the guise of "prisoners of conscience." Rather than directly rebut Amnesty's inaccurate claims regarding the numbers of political and religious prisoners in the USSR, Moscow's state-run media focused on impugning the organization's motives or arguing that in the West there were prisoners of conscience as well as mentally healthy individuals confined in psychiatric hospitals, about whom Amnesty remained studiously silent.[52] Readers could be forgiven for wondering whether such statements amounted to a tacit admission of analogous practices in the Soviet Union.

It was left to none other than Zivs, formerly the point man on the Soviet end of MacBride's back-channel, to produce the most sustained and widely disseminated Soviet counter-attack on Amnesty and what Zivs called "the little band of dissidents."[53] The rebuttal began with a slender book titled *Human Rights: Continuing the Discussion*, in English, French, German, Russian, and Spanish editions, all published in Moscow. Soon thereafter came *The Anatomy of Lies*, in English and Russian editions also published in Moscow. Taken together, they form a kind of *summa* of the Soviet government's lumbering efforts to sway international public opinion on human rights and dissent in the USSR.[54] As with *Izvestiia*'s orchestrated campaigns and Zivs's letter to the *Frankfurter Rundschau*, the two book-length attacks ridiculed Amnesty's estimate of ten thousand Soviet prisoners of conscience without offering direct counter-evidence.[55] The dramatic rise in Amnesty's budget served as proof of covert funding. Smear tactics against individuals from Amnesty—above all, Reddaway and Hammarberg—stood in for evidence-based responses to their criticisms of Soviet practices. Amnesty and other Western organizations were portrayed as the masterminds behind the production of samizdat, encouraging dissidents "to claim immunity before the law" (in contrast to their actual argument that samizdat as a mode of disseminating information was not forbidden by

Soviet law). Once again, Amnesty's critique of Soviet courts and prisons was dismissed as "psychological warfare."[56]

"Is there any object [sic] for discussion with dissidents?" Zivs asked in *Human Rights: Continuing the Discussion*, seemingly ignoring the fact that a negative response would sit poorly with the book's title. "We shall allow ourselves to state that there are some points of view on which history itself has closed debate. It has issued its final verdict on them.... We do not intend to argue with dissidents."[57] But dissidents were intent on arguing with him. Just as Amalrik publicly rebuffed Sukharev's attempt to undermine Amnesty's claims, so in Zivs's case Sakharov delivered a sharp rebuttal—in samizdat as well as on the pages of the *New York Review of Books*. For all of Zivs's efforts to discredit Amnesty's claims about the Soviet Union, Sakharov observed, Moscow had never permitted an international body such as the Red Cross or the World Health Organization to visit Soviet prisons or camps. "That," he concluded, "better than any words, demonstrates how groundless are Zivs's efforts at refutation."[58]

Valery Chalidze once joked that for the Soviet Union there existed not one but two global human rights organizations: the good "Mezhdunarodnaya amnistiya" (Amnesty International) and the evil "Emnisti Interneshunel."[59] The 1967 Soviet handbook of international NGOs reflected that stance, simultaneously lauding Amnesty's defense of the rights of free speech of imprisoned leftists in the West while warning of its sinister manipulation by the capitalist ruling classes. By the time an updated version of the handbook appeared in 1982, the good Amnesty had disappeared. What remained was "a reactionary organization which has de facto turned into a mouthpiece for 'psychological warfare' waged by imperialist secret services against the countries of socialism."[60] Noting the existence of Amnesty chapters in Great Britain, the United States, West Germany, Sweden, and elsewhere in the West, the handbook declined to inform its readers that there had once been a chapter in their own country.

The courtship of Soviet dissidents and Amnesty International, distant strangers communicating across the Iron Curtain via the thin language of human rights, was entirely unplanned. Raised by the Soviet system in the spirit of internationalism, dissidents initially addressed their appeals to fraternal communist parties across Europe, to the United Nations, to the Red Cross, or to the hazy entity known as "the global public." Amnesty was not among the Western NGOs that sought cooperation with the newly founded Initiative Group for Civil Rights in the USSR. When Chalidze went searching for potential partners for his newly founded Human Rights Committee, Amnesty was not among them either. Nonetheless, the swelling current of information on state persecution compiled and curated by Soviet dissidents transformed Amnesty's work. To be sure, the USSR was one of many dozens of countries around the world whose violation of human rights norms Amnesty sought to monitor. But the quantity and quality of data on the USSR, coupled with its status as Cold War superpower, ensured that Soviet prisoners of conscience reached and remained at the forefront of Amnesty's and eventually the Western public's attention.

Amnesty in turn captured the imagination of Soviet dissidents. In a message sent to an Amnesty gathering in Strasbourg in 1976, Turchin and Vladimir Albrekht described the organization as "an important new phenomenon in international life." Amnesty "stands above political struggle and at the same time works for the resolution of the most important political problems, which are connected with the rights of the person ..., a great undertaking destined for a great future." "Only if people base their activities on those principles and methods which are adopted by Amnesty," the message continued, "can they achieve genuine integration of free people, whether on the scale of the national state or on the scale of the whole planet."[61] Amnesty's visionary founder Peter Benenson had harbored similarly ambitious dreams, in which Amnesty would foster an "awakened and vigilant world consciousness" that would tackle not just the symptoms of political imprisonment, but its causes.[62]

In addition to disseminating human rights norms, Amnesty was implicitly in the business of spreading norms of advocacy, a cluster of techniques that I have called distanced empathy. In certain ways, the ethos behind those techniques mirrored rights-defenders' emphasis on legal formalism and their effort to move beyond face-to-face communities toward more impersonal networks of trust. Nonetheless, when dissidents sought to participate in Amnesty's work as partners rather than merely as adoptees, alarm bells went off in London. Could Soviet dissidents practice the political neutrality so prized by Amnesty? Under siege by their government and largely shunned by their fellow citizens, could they exercise the emotional discipline necessary for distanced empathy? London's hesitant response to the proposed affiliation led dissidents to develop their own doubts: How international *was* Amnesty? Was its vaunted neutrality merely a euphemism for timidity?

Despite these doubts and questions, Soviet dissidents and Amnesty International engaged in a process of mutual learning that proved highly effective in focusing global attention on human rights violations by the Soviet government. If we understand human rights as a politics of the Information Age (using information to name and shame rights violators), then human rights activism, as pioneered in the 1960s and 1970s, was the art of selecting, shaping, and making public that information. It was an art at which the Kremlin proved remarkably unskilled, its decades of experience producing propaganda seemingly overwhelmed by centuries-old habits of secrecy as well as the sheer novelty of Alexander Volpin's strategy of civil obedience.[63]

Did the contest over that elusive entity known as "public opinion" actually matter? And if it did, in what way and with what consequences? "Pressure of opinion a hundred years ago," Benenson had declared in Amnesty's founding text, "brought about the emancipation of the slaves."[64] In a gesture of Cold War ecumenism, Benenson noted that Amnesty's founding in 1961 coincided with the centenary of both Abraham Lincoln's inauguration and Russia's emancipation of its serfs. To be sure, it took not only the "pressure of opinion" but a civil war (in the American case) and a humiliating military defeat (in tsarist Russia's Crimean War) to undo entrenched systems of forced labor. But there is no denying that a changed climate of opinion was a necessary if hardly sufficient cause of the demise of those systems—as it would be of the Soviet Union in 1991.

22

Final Act

We all must choose: either Don Quixote or Sancho Panza. You have to be true to yourself. Don Quixote appears to be morally irreproachable in all respects, but we simply don't take seriously the grief he causes his niece. And what if Don Quixote has children? One can live without glory, but not without honor. This is the model of Sancho Panza: to sympathize with Don Quixote, but not to tilt at windmills.

—GRIGORY POMERANTS

August 1, 1975, was a day of triumph for Leonid Brezhnev. Heads of state from thirty-two European countries on both sides of the Iron Curtain, along with their counterparts from Canada, the United States, and the Soviet Union, gathered in the Finnish capital to sign what would become known as the Helsinki Final Act or Helsinki Accords. After years of grinding negotiations initiated by the USSR under the auspices of the Conference on Security and Cooperation in Europe, Brezhnev achieved a cherished goal: international recognition of the Soviet Union's expanded western borders as well as the borders of its East European satellite states, as drawn by Stalin following the Soviet Army's destruction of Hitler's European empire in 1945. In contrast to World War I, the new territorial arrangements forged at the end of World War II had not been ratified by a peace treaty, prospects for which were swept away by the Cold War that split the victorious allies with stunning speed.[1]

Thirty years later, the Helsinki Accords, while not a legally binding treaty, legitimized the Soviet order in the eastern half of Europe. In return, the USSR and all other signatory states pledged to honor human rights including freedom of speech, assembly, exchange of information, and emigration. These were rights that Soviet leaders had already pledged to honor in previous accords, including the legally binding International Covenant on Civil and Political Rights (ratified by Moscow in 1973), and that, as the preceding chapters have illustrated, they freely ignored when it suited them. The Final Act thus gave the USSR something of significant value—formal recognition of postwar borders—at virtually no cost.

This is how Helsinki's quid pro quo appeared to Brezhnev, its most enthusiastic sponsor, who made sure the full text of the Accords was published in Soviet newspapers. This is how it appeared to its many opponents in the West, too, who regarded the Final Act as validating the Kremlin's sphere of influence in exchange for worthless and unenforceable pledges of peace and human rights. And this is how it appeared to many Soviet dissidents: naive liberal democracies once again appeasing tyrants who, like Hitler in 1938, had no intention of keeping their promises.[2] By pretending that Moscow's assurances on human rights were something worth bargaining for, Western governments signaled that they were willing to sacrifice the dissident cause on the altar of détente with the Soviet Union.[3]

Anatoly (Natan) Sharansky saw things differently. A chess prodigy who grew up in the Ukrainian city of Donetsk (or Stalino, as it was called at the time of his birth in 1948), Sharansky worked at the All-Union Institute for Research on Oil Processing until 1973, when he applied to emigrate to Israel. He was promptly fired and denied an exit visa. To support himself, he gave private lessons in English and chess; among his English students were Yuri Orlov and Andrei Amalrik. Orlov, one of the founders of the Moscow chapter of Amnesty International, was in favor of forming more and more unofficial rights groups, to break down the taboo on independent voluntary associations.[4] After the signing of the Helsinki Accords, he and Amalrik drafted a declaration of principles for what they called the Human Rights Movement in the

USSR. They also composed an appeal to the Politburo, offering to "begin a dialogue about the future of the country" in which representatives of the Communist Party would meet with representatives of the Human Rights Movement. Both initiatives went nowhere.[5]

Sharansky had another idea: an appeal to the publics of Western states that had signed the Helsinki Accords, urging them to form committees to study and promote observance of the Final Act's human rights articles.[6] If the governments of Western Europe and North America had been naive enough to sign the Accords and were now too feckless to demand that the Kremlin honor its part of the quid pro quo, then perhaps public opinion could be mobilized to induce them to do so. Like Pavel Litvinov's August 1968 letter to Stephen Spender, Sharansky's proposal was an audacious attempt to leverage support in the West in order to advance the dissidents' grand strategy of containing Soviet power at home. Only now, the chosen instrument was not Soviet law but international human rights, linked via the Final Act to the territorial arrangements imposed by Moscow in the eastern half of Europe. The executor of Sharansky's proposal was not going to be Spender and other Western intellectuals and artists, but public associations whose composition and legal standing were left unclear.

Litvinov's proposal quickly bore fruit; Sharansky's did not. Western publics were not in a rush to constitute the Helsinki-monitoring committees he suggested. During the seven years (1968–75) that separated the two proposals, however, Soviet dissidents had broken the taboo against founding their own independent organizations. The fizzling out of the chain reaction had turned a spontaneous social movement into a series of NGO-like associations. If Western publics failed to embrace Sharansky's idea, then in the view of Amalrik and Orlov, Soviet dissidents would have to create their own committee first. Since the Soviet Union had been the driving force behind the Helsinki Accords, Amalrik reasoned, it made sense that Soviet citizens should take the lead in forming a national committee to monitor their fulfillment.[7] Built into the Accords themselves, moreover, was a mechanism for monitoring the implementation of the various articles, in the form of follow-up conferences, the first of which was scheduled for 1977 in Belgrade. Despite its name, the Final Act was conceived not as a conclusive one-time event, but as part of a continuing process.[8]

The anodyne name that Orlov came up with in May 1976 for the proposed citizens' committee—the Public Group to Assist in Implementing the Helsinki Accords in the USSR—was too much for some of his fellow dissidents, who balked at the idea of even appearing to assist the Soviet government in any way.[9] "All of this scheming with the Helsinki Act—it's all too Soviet and will only benefit this regime!" declared the linguist and Orthodox Christian Tatyana Khodorovich.[10] But Orlov, following Sharansky's lead, had in mind assistance to *Western* governments as stakeholders in the Final Act. The "Helsinki Group," as it came to be known, would continue the dissident tradition of exporting information on violations of human rights in the Soviet Union, but rather than addressing that information to the United Nations, Amnesty International, the International Red Cross, or the World Psychiatric Association—none of which had any direct leverage on the Kremlin—it would go straight to the USSR's Cold War antagonists, who were required to monitor the implementation of the Final Act by the terms of the Act itself.

Orlov had always refused to join in the traditional dissident toast "To the success of our hopeless cause." "If I considered it hopeless," he would say, "I would devote my time to something else."[11] He also disliked the widespread dissident allergy to politics. In his view, "political levers can be used to defend human rights," and therefore "the element of political game-playing" was inescapable.[12] On the morning of May 12, Orlov found a summons in the mailbox at his apartment building on Trade Union Street, instructing him to appear at the KGB district office for a "conversation" at 11 A.M. He decided to ignore it. Evidently, the Fifth Directorate had gotten wind of his plans. Together with the historian Mikhail Bernshtam, whose father worked for the KGB and who had recently been released from a mental hospital after being incarcerated for his dissident activities, Orlov hastily organized a preemptive press conference that evening at Andrei Sakharov and Yelena Bonner's apartment.[13] He was hoping that Sakharov might agree to serve as chair, but the request was declined. Sakharov cited his "rather negative experience with organizations" and his preference for "the freedom of speaking out as an individual."[14] At 11 P.M., with a single foreign journalist present, Orlov announced the formation of the Public Group to Assist in Implementing the Helsinki Accords in the USSR. The group consisted of eleven members, a who's who of dissidents, or rather of those dissidents who were not then in prison, in corrective labor camps, or exiled from the Soviet Union. Besides Orlov and Bernshtam, they included Ludmilla Alexeyeva, Bonner, Alexander Ginzburg, Petr Grigorenko, the astrophysicist Alexander Korchak, the geologist Malva Landa, Anatoly Marchenko, Vitaly Rubin, and Sharansky. Consistent with previous dissident patterns, three of the eleven were women, six had one or two Jewish parents, and none seemed interested in the group's gender or ethnic composition.

Other familiar dissident traits were also on display. There was no plan to expand the group's membership, hence no application procedure. The new group preferred the intimacy of small numbers, bound by friendship, mutual trust, and a spirit of creativity—Valentin Turchin's valence bonds. Like the Initiative Group for the Defense of Civil Rights in the USSR, it had no formal structure and no voting procedures. Only three signatures were required to issue a document or statement in the group's name. Marchenko was not asked in advance whether he would join. From his Siberian exile, he gave his assent post facto.[15]

Around midnight, after the press conference ended, Orlov went into hiding, fearful that the KGB might try to force him to retract the announcement of the Helsinki Group's founding. The Voices began broadcasting the news the next morning. Soviet listeners were informed that they could submit written statements regarding violations of rights spelled out in the Final Act to Professor Orlov, whose home address was also announced.[16] Satisfied with the public launching of the new association, Orlov returned to his apartment on May 14. The next day, while crossing the street, he was seized from behind, pushed into a car, and driven to a nearby KGB building, where an official warned Orlov that if the new group took any actions, he would be arrested. That same day, the Soviet news agency TASS released an unusual statement in English, meant for readers outside the USSR, declaring that Orlov had been warned about the "unconstitutional" nature of his activity, which he had undertaken "to raise his popularity among opponents of détente and enemies of the Soviet Union."[17]

By the end of 1976, the Moscow Helsinki Group managed to produce fifteen dossiers from the mound of materials received from Soviet citizens.[18] Forty-five copies of each dossier were typed following the time-tested samizdat method of onion-skin and carbon paper—one for each of the thirty-five countries that had signed the Helsinki Accords plus nine for foreign correspondents in Moscow and one for Alexeyeva's private archive—thousands of pages of documentation in all. The dossiers urged citizens in other countries to "follow our example" by forming, as Sharansky had originally proposed, their own committees to assist in the Final Act's implementation. "We do not presume," noted one of the documents, "that violations of the Helsinki Accords are possible only in the Soviet Union."[19] The idea was to foster contacts among independent citizens' groups in different countries, and eventually to coordinate international pressure on the participating governments—a model not unlike Amnesty International.

That is not quite what happened. The citizens who initially answered the call of the Moscow Helsinki Group were not in other countries but in other Soviet republics. In a reprise of the chain reaction, but now at a higher level, Helsinki monitoring groups sprang up in 1976 in Ukraine and Lithuania and in 1977 in Georgia and Armenia, all modeled on but independent of the Moscow group. The republic-level Helsinki associations each consisted of fewer than a dozen members, almost all of them previously active in movements for national autonomy, which remained their chief concern.[20] The structural similarity across Soviet republics suggests that the shift from nascent social movement to NGOs was not peculiar to metropolitan dissidents, but rather reflected the pervasive pull of Western models of citizen engagement for rights-defenders generally. The Helsinki Group in Moscow, which never self-identified as Russian, retained its position, like that of the *Chronicle of Current Events*, as a Soviet Union–wide focal point connecting victims of all manner of human rights violations with interlocutors in the West.

In the West, too, things took an unexpected turn. West European diplomats, not Americans, had taken the lead during years of negotiations with Soviet counterparts at the Conference on Security and Cooperation in Europe. American secretary of state Henry Kissinger dismissed the forum as "a meaningless psychological exercise" that would only distract from the imperative of arms control and superpower détente.[21] George F. Kennan was similarly skeptical, characterizing the Helsinki Accords as "high-minded but innocuous."[22] In a private phone call, the architect of détente commiserated with the architect of containment over the obstacles placed by Soviet dissidents in the way of rational superpower relations. Western media were exhibiting "hypocrisy" (Kissinger) and "hysteria" (Kennan) in their coverage of Soviet dissent. Having recently returned from a trip to Leningrad, Kennan reported that dissidents were "driving it very unwisely hard," making it possible for the regime "to split the whole Russian intellectual and aesthetic community so that a lot of the most important other Russian intellectuals have turned against them." "In any case," Kennan continued, "I don't think it's right for a great government such as ours to try to adjust its foreign policy in order to work internal changes in another country." "I'm so glad you are of that view," Kissinger replied.[23]

First-term congresswoman Millicent Fenwick, Republican from New Jersey, was not of that view. She visited Moscow with a congressional delegation in August 1975, shortly after the signing of the Final Act, and met with Brezhnev in the Kremlin as well as with Orlov and Turchin in the latter's apartment.[24] Brezhnev seemed not to regard all elements of the Helsinki Accords as equally subject to enforcement. Orlov impressed Fenwick with his belief that the Accords could function as a "lifeline" for Soviet citizens—provided that Western governments would monitor Soviet compliance. In a follow-up letter, Orlov informed Fenwick that "we dissidents are throwing ourselves on barbed wire in the hope that others can walk safely over our bodies."[25] That was enough to inspire Fenwick, back in Washington, to propose the creation of a commission within the U.S. government to oversee compliance with the Helsinki Accords. With bipartisan support, and over the objections of Kissinger and President Gerald Ford, the Commission on Security and Cooperation in Europe was established in June 1976. Its chief of staff indicated that it would serve, among other things, as "the voice of the dissidents."[26]

This was beyond what Sharansky, Orlov, or Amalrik could have imagined just one year earlier. The U.S. government had now set itself the task of directly monitoring Soviet (and East European) compliance with the Final Act. Shortly thereafter, Robert Bernstein, president of Random House publishing company and an influential free-speech advocate, founded the NGO Helsinki Watch as an adjunct to this effort.[27] In the wake of the American debacle in Vietnam, the U.S. Congress was keen to exert greater control over foreign policy, and human rights advocacy appeared to hold out the hope, for politicians as well as concerned citizens, of national redemption. With the election of Jimmy Carter in November 1976, that hope found a home in the White House.

In form, content, and strategy, the Moscow Helsinki Group closely resembled previous dissident organizations, whether the Initiative Group for the Defense of Civil Rights in the USSR, the Human Rights Committee, or the Moscow chapter of Amnesty International. As Orlov noted in one of the Helsinki Group's early public statements, however, the Final Act itself distinguished the group's work: "Unlike previous declarations containing human rights obligations, in the Final Act those obligations were agreed to by the Soviet government 'in exchange' for important political concessions by Western governments. This has made possible what are still very tentative but nonetheless unprecedented attempts by Western leaders to insist on the fulfillment of those obligations."[28] In contrast to the United Nations or Amnesty International, the United States and West European governments could apply to the USSR a wide array of carrots and sticks—economic, diplomatic, technology-related—within the framework established by the Final Act. Exposure of Soviet human rights violations therefore carried greater potential weight than at any time in the past, far beyond naming and shaming. For the same reason, those who called attention to such violations faced greater risks than ever before.

Most of the cases transmitted by the Moscow Helsinki Group fell into familiar categories, reflecting existing networks of communication with persecuted parties. These included Baptists and other Christians; Jews, ethnic Germans, and other Soviet citizens seeking to emigrate; political prisoners; victims of punitive psychiatry; and people arrested for exchanging lawful information via samizdat.[29] New groups were also finding their way into the Helsinki orbit: workers seeking to create independent trade unions; peasants seeking to leave collective farms; the disabled; retirees unable to survive on their state pensions. Grievances related to social and economic rights—the pride of the Soviet rights regime—now found a place in the petitions taken up by metropolitan dissidents. "Individual rights are closely and directly related to socio-economic rights," Bonner and other members of the Moscow Helsinki Group reported. "Without the guarantee of individual rights, it is impossible to successfully defend socio-economic rights. This is demonstrated by the actual state of affairs in the USSR."[30]

The Public Group to Assist in Implementing the Helsinki Accords in the USSR, and its unexpected resonance in the U.S. Congress and the American public, punctured the celebratory post-Helsinki mood in the Kremlin. "These people are pursuing the provocative goal of questioning the sincerity of efforts by the USSR to implement the Final Act," Andropov informed the Central Committee. "The 'group' is counting on Western public opinion to exert pressure on the Soviet government and will not, in Orlov's words, 'seek support among the [Soviet] people.'"[31] In the KGB's view, Orlov was using the Moscow Helsinki Group to "legalize his anti-Soviet activity," much as the founders of the Moscow chapter of Amnesty International had attempted to "legalize" theirs. "Insolence born of the attempt to ignore Soviet laws" had led Orlov, Ginzburg, and others to "believe in their own impunity." The Fifth Directorate's prophylactic measures were not producing the desired effect. It was all part of "a plan by the CIA to create organizational links on the territory of the USSR to carry out hostile activity aimed at undermining the Soviet order."[32]

Andropov believed—correctly—that the various Soviet Helsinki Groups sought to influence Western rather than Soviet public opinion. Together with the East German Stasi and other intelligence services of Soviet-bloc countries, the KGB therefore launched a campaign of "active measures" vis-à-vis Western publics, highlighting socialist countries' fulfillment of the Final Act and publicizing "gross violations" by capitalist countries, including the United States' suppression of Puerto Rican independence, "political genocide" against Native Americans, and illegal wire-tapping by the Federal Bureau of Investigation.[33] The KGB and the Stasi developed a plan to create an international "Foundation for Victims of Imperialism and Colonialism" masquerading as an NGO, based in a neutral country and run by "progressive activists." Aggrieved parties could submit petitions and requests for help to the Foundation, which would make it possible to "incorporate the incoming information into an active plan" of Soviet propaganda.[34]

Andropov's transnational public relations campaign failed, just as the campaign to discredit Amnesty International failed.[35] The awarding of the Nobel Peace Prize to Sakharov in December 1975 renewed global attention to the Soviet dissident movement. His Nobel lecture—delivered by Bonner, since Sakharov was denied permission to travel outside the USSR—in turn drew attention to the Helsinki Accords, noting that "there have been no essential improvements" in the Soviet Union's observance of the human rights spelled out in the Final Act.[36] The following fall, Sakharov conveyed a private message to Ford and Carter, the Republican and Democratic nominees for the U.S. presidency, urging them to support "the principle of active struggle for human rights around the world."[37] After winning the election, Carter took the unprecedented step of exchanging open letters with Sakharov, followed by an invitation to the exiled Vladimir Bukovsky to the White House.[38]

There were multiple reasons for the failure of the KGB's campaign to influence Western opinion regarding Soviet compliance with the Final Act. Perhaps the most obvious was that, rather than merely deny or dismiss the evidence delivered by Helsinki Groups in Kyiv, Moscow, Tbilisi, Vilnius, and Yerevan, Andropov decided to imprison the messengers. During the months leading up to the first international follow-up conference in Belgrade in October 1977, ten Helsinki Group members were arrested: Ginzburg (February 3, Moscow; sentenced to eight years in a labor camp); Oleksa Tykhy (February 4, Ukraine; ten years labor camp plus five years exile); Mykola Rudenko (February 5, Ukraine; seven years labor camp plus five years exile); Orlov (February 10, Moscow; seven years labor camp plus five years exile); Sharansky (March 15, Moscow; three years prison plus ten years labor camp); Merab Kostava (April 7, Georgia; four years exile, then re-sentenced to an additional six years); Zviad Gamsakhurdia (April 7, Georgia; recanted and was pardoned); Mykola Matusevych (April 13, Ukraine; seven years labor camp plus five years exile); Myroslav Marynovych (April 26, Ukraine; seven years labor camp plus five years exile); and Viktoras Petkus (August 21, Lithuania; three years prison, seven years labor camp, five years exile).[39]

Over the following two years, the number of arrests more than doubled. Nearly all Helsinki Group activists in the non-Russian republics, where the crackdown was especially severe, were now in prison, camps, or exile. The Moscow Helsinki Group, whose ranks were also decimated, created a dossier for each arrest, with the result that—as happened with the Initiative Group for the Defense of Civil Rights in the USSR—imprisoned rights-defenders soon constituted the single largest category of advocacy, outnumbering religious believers and those seeking permission to emigrate.[40] The number of arrests rose more quickly than replacements could be found. Ivan Kovalev, the son of dissident biologist Sergei Kovalev, joined the Moscow Helsinki Group in 1979 and helped assemble dossiers on two dozen cases, including that of his father (serving a seven-year sentence) and his wife, Tatyana Osipova (a fellow Moscow Helsinki Group member arrested in 1980), until he himself was arrested in 1981. By then, only three members were left to sign his dossier and forward it to the West: Bonner, the lawyer Sofia Kallistratova, and the mathematician Naum Meiman.[41]

Nothing confirmed the truth of claims about Soviet non-compliance with the Final Act's human rights clauses more eloquently than Moscow's crackdown on those who dared to make such claims. No "active measures" or public relations campaign by the KGB, no matter how sophisticated, could counteract the impression made by the extraordinarily harsh punishments meted out to arrested Helsinki Group members. Once again the KGB was turning dissidents into martyrs, infusing

FIGURE 22.1. Tatyana Osipova and Ivan Kovalev

them with moral authority before a Western (but not Soviet) audience. The effect was not lost on Orlov: "When you are trying to influence a community as enormous as the Western world, a community with an enormous number of internal interests, internal passions, internal controversies, how can you affect it except by ascending a cross? There is no other way."[42]

When the KGB learned that the imprisoned Orlov had been nominated for the Nobel Peace Prize, Andropov personally oversaw a covert campaign to compromise Orlov's reputation with members of the Nobel committee and various Norwegian political figures.[43] Meeting in September 1977 with President Carter in the White House, Soviet foreign minister Andrei Gromyko expressed the hope that the upcoming Helsinki follow-up meeting in Belgrade "would become a constructive forum instead of a place of mutual accusations."[44] He was soon to be disappointed. The Soviet delegation in Belgrade insisted on denying journalists and NGOs access to the deliberations. The American delegation, led by the combative former Supreme Court justice Arthur Goldberg, countered with regular press conferences detailing the meeting's progress and drawing on information provided by the Moscow Helsinki Group. Soviet delegates insisted that human rights were an internal matter not suitable for discussion at Helsinki follow-up conferences. The Americans, now much more involved in the "Helsinki process" than they had been during negotiations over the text of the Final Act, broke with diplomatic custom by singling out the USSR for criticism and mentioning Orlov and other arrested Helsinki activists by name. Soviet delegates responded with their own accusations of human rights violations by the U.S. government, thereby tacitly endorsing the notion that human rights were indeed suitable for discussion.[45]

The next follow-up meeting, in Madrid in 1980, was even more contentious. Other Western countries joined the United States in criticizing the Soviet crackdown on dissidents, and the Spanish authorities, unlike those in Yugoslavia three years earlier, refused to bar human rights activists from protesting at the conference.[46] Among the activists scheduled to attend was Amalrik, who together with his wife had left the USSR in 1976 and settled in France. Fifty miles outside Madrid, Amalrik's car swerved and struck an oncoming truck, killing him instantly. He was forty-two, four years shy of the year he predicted the Soviet Union would cease to exist, and eleven years shy of the actual event.

The collapse of communism starting in 1989 has tempted some historians to posit a "Helsinki effect" whereby international human rights norms spread to the East bloc, undermined the Soviet order from within, and eventually caused it to crumble.[47] For over a decade, however, no such effect was visible. On the contrary, Moscow and especially Washington slid further away from the Helsinki Accords' stated goal, namely, to reduce East-West tensions and eventually to move beyond the Cold War. The follow-up conferences in Belgrade and Madrid witnessed a pronounced sharpening of tensions between the two superpowers, as new arguments broke out over how to interpret and verify various commitments made in the Accords, especially those dealing with human rights. Other factors, too, helped revive Cold War hostilities, notably the exit from the political scene of détente's principal American sponsors, Richard Nixon and Kissinger, as well as increased Soviet military support for revolutionary movements in the developing world. The Soviet invasion of Afghanistan in December 1979 put the nail in the coffin of détente. That détente was already in the coffin, however, was due in no small measure to the toxic atmosphere generated by the cycle of Soviet rights-defenders producing evidence of the Kremlin's non-compliance with the Final Act's human rights provisions, the American leveraging of that evidence, and the KGB's brutal repression of those who delivered it. Never before had dissidents had such a tangible impact on the USSR's relations with the West. They became "a constant source of irritation" between the two superpowers, according to Anatoly Dobrynin, the long-serving Soviet ambassador to Washington, and "did great damage to détente."[48]

The damage they did to détente goes a long way toward explaining the ferocity of the KGB's final crackdown on dissidents in the late 1970s.[49] Their links to Western audiences—not just Amnesty International, but the immeasurably more powerful U.S. Congress—turned dissidents into "disturbers of the peace" on a new level. Now they were disturbing not just the peace of mind of (some of) their fellow Soviet citizens, not just the wary social peace between state and society that characterized "mature socialism," but the principle of "peaceful coexistence" between the superpowers, the mantra of Soviet foreign policy in the post-Stalin era. "Interfere as much as you can," Alexander Solzhenitsyn prodded Western leaders. "We beg you to come and interfere."[50] In the short run, therefore, the "Helsinki effect" intensified rather than inhibited the hounding of human rights advocates in the USSR. Contrary to expectations, it became the Soviet dissident movement's final act.

Writing shortly after the Soviet collapse, Ambassador Dobrynin conceded that "we managed the dissident issue in a thoroughly clumsy way: first we would bring things to the point of open conflict, scandal, and a sharpening of relations [with the United States], and then, in the end, we would give in." The main problem, according to Dobrynin, "was the policy of repression itself, which the Soviet leadership stubbornly pursued without thinking very much about the consequences for our domestic and foreign policies."[51] Leaving aside whether the devastating punishments imposed on Orlov and other dissidents amounted to "giving in," transcripts of Politburo discussions indicate that Soviet leaders in fact repeatedly weighed the consequences of various methods for handling dissidents.[52] For over a decade, Andropov, Brezhnev, and other senior figures debated what to do about Sakharov's increasingly vocal criticisms of Soviet policies. Everyone agreed that, as Andropov put it in 1970, "the time has come to conduct a detailed conversation with Sakharov... to persuade him to cease his politically harmful activity." A year later, Andropov again requested that Brezhnev "arrange a conversation with [Sakharov] at the Central Committee as soon as possible." Politburo members Alexei Kosygin and Nikolai Podgorny endorsed the idea, and Andropov repeated his request yet again several months later.[53] The problem was that no one wanted to be the one to talk to Sakharov. By 1975, Andropov was recommending punitive action. Unlike Solzhenitsyn, Sakharov could not be expelled from the USSR; he knew far too much about Soviet nuclear weapons. But as Andropov and procurator-general Roman Rudenko noted, "Punishment of Sakharov through court proceedings may evoke a serious negative reaction." Filipp Bobkov, the head of the Fifth Directorate, cautioned against making a "martyr" of Sakharov. Instead, Andropov proposed banishing Sakharov and Bonner via administrative decree to a Soviet city closed to foreigners. That step, he acknowledged, "will evoke an anti-Soviet campaign in the West. However, this will entail smaller political costs than permitting Sakharov to act with impunity in the future or putting him on trial for criminal activities." A corresponding decree was drafted along with statements for the Soviet mass media.[54]

Five years later, in January 1980, following Sakharov's condemnation of the invasion of Afghanistan and his call for an international boycott of the upcoming summer Olympics in Moscow, the decree was finally put into effect with unanimous support by the Politburo.[55] In the early morning hours of January 22, 1980, the writer Georgi Vladimov, who led what was left of the Moscow chapter of Amnesty International, reported to Sakharov that he was going to be expelled from Moscow. Later that day, as his driver was taking him to a special grocery store for members of the Soviet Academy of Sciences, the car was suddenly surrounded by police, two of whom got in the backseat and ordered the driver to take Sakharov to the Procurator's office. There he was informed that he was being exiled without trial to the city of Gorky, 260 miles east of Moscow and closed to foreigners. The piece of paper with the text of the decree had Brezhnev's name on it, but no signature.[56]

Sakharov's banishment presaged the end of rights-defense as the governing strategy of Soviet dissent. What was left of the Moscow Helsinki Group somehow managed to smuggle reports on his case to the Final Act signatories, noting that according to Article 160 of the Soviet Constitution, "No one may be deemed guilty of committing a crime and subjected to criminal punishment other than by the judgment of a court and in accordance with the law." Sakharov's exile, the Moscow Helsinki Group concluded, set a "horrifying precedent for every free-thinking person in our country. Anyone can be seized on the street, at work, at home."[57] Sakharov's fellow scientists remained silent, apart from a handful who publicly endorsed his forced removal from the Soviet capital.[58] On January 25, a lone mathematician named Vazif Meilanov appeared on Lenin Square in the city of Makhachkala, on the Caspian Sea, with a homemade sign reading "I protest against the persecution of Academician Sakharov. One should oppose ideas with ideas, not the police. Our society needs people like Sakharov, because they carry out genuine unofficial monitoring of state organs." His silent one-man demonstration lasted seventeen minutes. Following his arrest, Meilanov was sentenced to seven years in a labor camp and two years of exile.[59]

On September 6, 1982, the Public Group to Assist in Implementing the Helsinki Accords in the USSR announced the cessation of its work. Seventy-four-year-old Sofia Kallistratova, one of three remaining members, had just been indicted under Article 190 for disseminating "deliberately false statements slandering the Soviet state." Having brought nearly two hundred cases of Soviet human rights violations to the attention of the countries that had signed the Helsinki Accords, the group could no longer muster the minimum number of signatures for outgoing dossiers.[60]

Other Helsinki Groups in the Armenian, Georgian, Lithuanian, and Ukrainian republics had already stopped functioning due to the arrest and exile of their members. The Moscow chapter of Amnesty International closed for similar reasons. The Initiative Group for the Defense of Civil Rights in the USSR issued its final appeal in November 1981, signed by Tatyana Osipova and Tatyana Velikanova, both already inmates in the women's zone of the Mordovian labor camp complex. Valery Chalidze's Human Rights Committee had long since ceased to operate due to his and Alexander Volpin's departure from the Soviet Union. In his summary of KGB activities for the year 1982, Viktor Chebrikov, the new KGB chairman, triumphantly reported to Andropov—who had recently succeeded the deceased Brezhnev as general secretary of the Communist Party— that "the so-called 'Russian section of Amnesty International,' 'Helsinki groups,' and other similar groupings, having attempted to operate in the USSR, have been dismantled."[61]

On the pages of the *Chronicle of Current Events*, the dissident movement covered its own demise.[62] Along with a litany of apartment searches, arrests, "prophylactic conversations," trials, hunger strikes, and news from camps and psychiatric hospitals, the sixty-fifth and final issue (December 31, 1982) announced the suspension of the Moscow Helsinki Group and the death of the person once regarded by many as the movement's leader. "The most tragic events in the long-suffering life of Petr Yakir," Yuli Kim stated in a eulogy for his father-in-law delivered at a Moscow crematorium, "are marked by the years 1937 and 1972. He waited to be rehabilitated for 1937. I believe that rehabilitation will come for 1972 as well."[63]

The dissident movement had resurrected itself after previous near-death experiences, including its betrayal by Yakir and Viktor Krasin in 1972, but it would not do so this time.[64] Too many participants had been driven out of the Soviet Union, imprisoned, or silenced. The post-1968 reversion to small, face-to-face groups, now modeled on Western NGOs, had worked against the recruitment of new members to replace those who were lost. So had the cascade of extra-judicial reprisals by the Fifth Directorate, as well as the paucity of tangible successes in the campaign to strengthen the rule of law in the USSR.

What counted as "success" increasingly took the form of attention from the West—first from Western journalists and the Voices, then from NGOs such as Amnesty International, and finally from Western governments. Engagement with the West compensated—but only partly—for the Soviet government's refusal to enter into any form of dialogue with dissidents, even one designed merely to co-opt or deflect, as well as for the estrangement between dissidents and the rest of Soviet society. Prior to his departure from the USSR in 1972, Volpin had already concluded that "protests abroad have had a greater impact on the course of events in the USSR than actions by Soviet citizens in defense of their fellow countrymen. These protests ... served as clear evidence of the international significance of the repression that elicited them."[65] Support by the West, however, did nothing to advance Volpin's meta-revolution—the revolution in the legal consciousness of Soviet citizens.

By the time rights-defenders were vanquished in 1982, law-based dissent had ceased to function as the lingua franca of various aggrieved groups in the Soviet population. The unifying imperative of preventing a return to Stalinist lawlessness receded as it became clear that Khrushchev's successors, even if they aimed to put an end to the "Thaw," were not steering the USSR back toward mass terror. In the 1970s, the full range of dissident ideas about alternatives to Brezhnev's order came into view. In contrast to protest letters during the chain reaction, which typically bore anywhere from a dozen to several hundred signatures, the political programs of the 1970s tended to be single-author texts—Solzhenitsyn endorsing a return to rural values buttressed by religious authoritarianism, Sakharov (partly in reaction to Solzhenitsyn) calling for democracy informed by science, Roy Medvedev advocating a reformed socialism. Once various political preferences emerged, there was little evidence of coalition building. These were one-man pronouncements, their authors exclusively men.

The dissident repertoire—civil obedience, civic virtue, transparency, human rights—lost its magnetism. Symptomatic in this regard was the belated samizdat reception of Solzhenitsyn's impassioned call to renounce all forms of lying. Issued on the eve of his expulsion in February

1974, *Live Not by the Lie* had been overshadowed by the even more controversial *Letter to the Soviet Leaders*, in which Solzhenitsyn's antimodernism and mistrust of democracy became explicit.[66] Three years later, Viktor Sokirko, whose pungent replies to Amalrik's *Will the Soviet Union Survive until 1984?* and Kochetov's *What Is It You Want?* we encountered in previous chapters, launched a new debate over the virtue of private enterprise—even if it involved occasional lying and violations of laws governing socialist property. Writing under the pseudonym Burzhuademov (Bourgeois Democrat), Sokirko dismissed Solzhenitsyn's slogan as suitable only "for the righteous few, the holy among us"— dissident virtuosi, in other words. It would not lead ordinary people to a new way of life. The key to living like a free person in an unfree country was, first of all, "to stop hoping for some sort of cardinal, radical change, some 'liberation,' whether from above or below," and instead to follow one's own will, to think and work for oneself, rejecting the ethos of service to "the people," "the revolution," or "the Party." Citing Adam Smith and Max Weber, Sokirko called for a Protestant ethic of conscientious work and self-enrichment as the surest path to a freer country—even though that meant ignoring Soviet laws restricting private entrepreneurship and the private use of state property (the latter being the most common economic crime in the Soviet Union).[67] Gleb Loginov, the entrepreneurial, samizdat-reading, risk-averse lover of the heroine of Maya Zlobina's novel about the dissident world, *Sacred Paths to Willful Freedom*, would have been pleased.

Several dozen authors joined the conversation, eventually filling three samizdat volumes. From various standpoints, the majority of the contributors echoed Sokirko's argument that neither obedience to Soviet law nor the renunciation of lying were the highest imperatives. Valery Nikolsky, who had helped Volpin draft the Civic Appeal in 1965, praised Solzhenitsyn's prophetic voice and his historic role as an "uncompromising hero," but wondered whether "the earth would be a habitable place if everyone were uncompromisingly principled." By liberating Soviet society from fear, the dissident movement played an important role in the transition to a freer socialism with room for individual initiative, as in Hungary and Yugoslavia. Nonetheless, "the driving force in the process of improving the economy and society," according to Nikolsky, "will be self-interest."[68] Other authors agreed that individual initiative in the "gray economy" (also known as "left-handed business") was a necessary step toward greater freedom, even if it required breaking the law. "Can one really vouch for the perfection of laws, much less urge people to obey them," asked Sonya Sorokina, when they were made by "uneducated dimwits"? Sorokina, an economist, resigned from the Communist Party after her husband Viktor Sorokin, also an economist, was beaten by the police at the annual transparency meeting in December 1978.[69]

Grigory Pomerants, a veteran of Stalin's Gulag, prolific samizdat author, and critic of Solzhenitsyn's conservative nationalism, offered a philosophical reply to *Live Not by the Lie*—and to the fading ethos of the dissident movement:

> In our country, laws are not applied as they are in the West, with the intention of carrying them out in a dry and precise manner, in the spirit of Shylock. We apply them according to their utility in any given situation, following secret and unwritten instructions. There are laws that are meant to be observed one percent of the time, or five percent of the time. This looks unscrupulous. But only dissidents could seek strict fulfillment of Soviet laws, failing to understand that this is an anti-Soviet demand. Imagine for a moment that all laws on state crimes and embezzlement of socialist property would be applied to the fullest extent. There wouldn't be enough courtrooms or prisoner convoys. Who would be left to man the factories? To work the fields?

Skeptical of universal injunctions, whether to renounce lying or to abide by the law, Pomerants embraced a radical personalism, insisting that people should be true above all to their own selves. Soviet life was full of social conventions, from purely performative elections to various forms of public lip-service, with which the majority of the population made their peace. Dissidents, finding such conventions unbearable, branded them "lies" and "hypocrisy." Unable to live without the pursuit of glory, according to Pomerants, they chose the path of Don Quixote despite the severe consequences for themselves, their family members, and friends. "Even if Don Quixotism is necessary and justified," he wrote—and he was inclined to believe that for certain people it was—"it's impossible to be a good Don Quixote without being a poor family member." It was just as fine to be Sancho Panza, seeking not a glorious but an honorable life: "to sympathize with Don Quixote, but not to tilt at windmills."[70]

There were also signs from the other side of the Iron Curtain that dissidents were losing their aura. After three years as Moscow correspondent for the *Washington Post* (1974–77), Peter Osnos published an essay that dared to suggest that the dissident story was soaking up too much attention from American journalists who, to make matters worse, were insufficiently critical of its protagonists. "Are we encouraging dissent," Osnos wondered, "merely by writing about it?" Soviet dissidents expressed "what most Americans want—and expect—to hear about the evils of communism," but depending on them as sources of information about the USSR "creates a picture of that complex country as oversimplified in a way as Soviet reports about the United States being a land of little more than poverty, violence, corruption, and racism." Disproportionate attention to a handful of brave individuals came at the expense of covering broader trends in Soviet society.[71] Another American journalist, Fergus Bordewich, quantified that alleged disproportion, noting that during the Kremlin's crackdown on Soviet Helsinki monitors, fully half of the several dozen articles on the USSR published by the *New York Times* concerned dissidents. The result was "an image of a Soviet Union pulsating with internal discontent."[72]

Rights-defenders were being forced from the stage, but internal discontent itself could not be arrested or exiled. Even as the dissident movement faded, and with it the dream of taming the Soviet state via the rule of law, samizdat—the principal vehicle of "other-thinking"—continued to expand.[73] Initially, as the graph in figure 6.6 showed, the number of newly founded samizdat periodicals dropped precipitously during the post-Helsinki crackdown on dissent, from twenty-seven new titles in 1979 to just six in 1981. Journals focusing on the defense of civil and human rights never recovered from that plunge, as exemplified by the cessation in 1982 of the *Chronicle of Current Events*, the longest-running of them all. Other types of samizdat periodicals, however, not only recovered but became more numerous than ever, with a record thirty-one new titles founded in 1986. Many of them addressed specific communities: Catholics in Ukraine, rock music fans, feminists, workers, Jews. Others helped form interpretive communities focused on uncensored poetry and art, religious philosophy, or the future of socialism.[74] Samizdat was developing an entire ecosystem of secondary literary institutions (beyond the primary institutions of writer and reader, each of whom could double as typist/publisher) mirroring those of official print-culture: samizdat bibliographies of samizdat, samizdat journals devoted to reviewing new works of samizdat, letters to the editor of samizdat journals, even samizdat crossword puzzles.[75] By 1980, Vyacheslav Igrunov's samizdat lending library in Odesa was serving readers in Kyiv, Leningrad, Moscow, and Riga, with branches in Novosibirsk and Zaporizhzhia.[76] The dissident movement was broken, but so was the Kremlin's monopoly on information.

The cumulative effect of the dissident epoch was to implant the idea that the Soviet system, even while boasting of the "real rights" fostered by "socialist legality," was congenitally unable or unwilling to uphold its own laws. In the short run, states can achieve legitimacy in the eyes of their population through charismatic leaders, or successful economic policies, or victories on the battlefield, without necessarily operating within a system of legal constraints. In the long run, however, legitimacy by definition requires conforming to laws. Not just to Marxism's laws of history, but to positive law fashioned by human beings.[77] Volpin's quixotic mission to contain the power of the Soviet state by spreading the ideal language of law among his fellow citizens did not achieve its intended result. But even as the Helsinki Final Act helped legitimize the USSR's external borders, the dissident movement sparked by Volpin helped drain the Soviet system of legitimacy inside them.

Epilogue

BREAKING THE FOURTH WALL

SON: Doesn't it seem to you that at the end of the play the author is compromising with what he had initially attacked, and that he's trying to curry favor with the authorities by tacking on a happy ending—the move to a new apartment—so that he can smuggle in all the other stuff?

CRITIC: I totally disagree with you. You've completely missed the irony here. That's the most powerful moment in the play.

DAUGHTER: What's so powerful about it? He's just being obsequious and trying to show that people are well-off in our country, since they get new apartments.

LADY: But you've got it all wrong! On the contrary, what he's trying to say is that the new apartments are in no way better than the old ones—they're even worse.

SON: Still, the author shows a tendency to conformism.

CRITIC: Not at all! He's a genuine nonconformist.

SON (*confronting the* CRITIC): I assure you he's a conformist.

CRITIC (*confronting the* SON): And I'm telling you he's a nonconformist.

SON: A conformist! A conformist!

CRITIC: A nonconformist! A nonconformist!

SON (*enraged*): A conformist!

CRITIC (*enraged*): A nonconformist!

HOSTESS (*politely to the* LIEUTENANT COLONEL): And you? Did you enjoy the play?

—ANDREI AMALRIK, *IS UNCLE JACK A CONFORMIST?*

None of the characters in *Is Uncle Jack a Conformist?* thinks the play-within-the-play has, or should have, a happy ending. There were plenty of happy families in the Soviet Union (and Tolstoy notwithstanding, they were not all alike), but in general one doesn't look to that country for happy endings—especially if one is Andrei Amalrik. Composed in 1964 and confiscated by the KGB during an apartment search the following year, Amalrik's one-act drama probed the consequences of non-conformism, as if anticipating the movement that was about to erupt from the trial of two non-conformist writers, Andrei Sinyavsky and Yuli Daniel.[1]

Soviet dissidents understood themselves as actors in a drama unfolding on the world stage. Written and directed by the Communist Party of the Soviet Union, the play told the triumphant story of a revolutionary struggle to build the first truly enlightened, just, and equal society. By the 1960s, the cast of second-generation Soviet citizens was content to utter their assigned lines, carry out their roles and maintain the conventions that make theater's sovereign illusion of reality possible. Many of the actors were devoted to the story, and even if they weren't, their performances provided a stable income and left them plenty of time for their own pursuits. Among them, however, a handful grew to dislike the performances, finding them unconvincing and oppressive. The question, as Vladimir Albrekht put it in one of his conduct manuals for dissidents, was "how a good actor should act in a bad, shameful play."[2]

Rather than deliver their lines without conviction, or attempt to rewrite the play, or picket the theater, dissidents chose a spectacular form of disruption: they broke the fourth wall, speaking from the stage directly to the audience. "Our compatriots convene as if they were actors in a unified spectacle," wrote Boris Shragin. "It is considered a strict rule of the game not to address the auditorium in a direct and artless manner and thereby spoil the entire performance." Soviet ideology, Alexander Solzhenitsyn declared, was "a fake plywood veneer used to build columns in a theater set." Vladimir Zelinsky, describing the effects of overcoming ideologically conditioned reflexes and removing one's mask, found that "all our triumphal words and rituals have started to look fake, like props."[3]

Most of the dissidents' fellow cast members, not to mention the play's director and producers, did not appreciate having their performance spoiled by unscripted remarks to the audience, just as they disliked little boys (or girls) telling them the emperor had no clothes, even if they had privately come to the same conclusion. Theater, after all, requires the suspension of disbelief. Nor did they like it when an invisible Greek chorus known as "the Voices" served as uninvited witness and commentator, amplifying dissident actors' remarks. Breaking the fourth wall didn't merely interrupt the performance of the play. It lay bare for all—long before the wall of concrete and barbed wire at the edge of the Soviet empire, in Berlin, was torn down—the make-believe quality of the performance itself.

As this book has shown, the strategy of law-based dissent and conducting oneself like a free person was hardly "artless," despite Shragin's claim. The dissident strategy too required an element of make-believe. One had to act as if Soviet law were what it purported to be, a binding, unambiguous code of behavior; as if all Soviet citizens were responsible for the policies of their government; as if one could place significant constraints on the Soviet state by legal means alone, without altering existing institutions; as if it were possible to live freely through sheer force of will and independence of mind. One had to imagine that a few hundred "movement people," outstanding dissident personalities, could "change the country's moral atmosphere and its governing tradition."[4]

What drew people to these seemingly hopeless fictions? As we have seen in the preceding chapters, many dissidents came from families in which a parent or other relative had been imprisoned or executed during the Stalin years, but that was hardly a distinguishing feature in a society where tens of millions of people had such relatives. Some dissidents had themselves spent time in jail or labor camps prior to joining protests or signing open letters, but most had not. Instead, paths to the dissident movement, as Sinyavsky wrote in "Dissent as a Personal Experience," typically began with an inner struggle, a moral "stumbling block" to which "high-minded Soviet people" imbued with Soviet ideals were especially prone.[5] Like all orthodoxies, the Soviet creed contained labile elements capable of mutating in unforeseen ways that were reflexively condemned as heretical—as orthodoxies are hardwired to do. Stumbling blocks came in many shapes and sizes but frequently involved unsupervised encounters with Lenin or pressure from the KGB to inform on one's friends.

The title of Sinyavsky's essay could just as accurately be translated as "Dissent as a Personal Experiment." For many participants in the movement, dissent was as much about trying out new ways of living enabled by Khrushchev's Thaw as about conflict with the Soviet state. Western social movements in the 1960s and 1970s also emphasized new ways of living, new states of consciousness, and new values. What distinguishes the Soviet case is that these innovations emerged from the specific desire to break with the legacy of Stalinism. For most Soviet citizens, the USSR's retreat from Stalin's participatory dictatorship to Brezhnev's lip-service state brought the freedom to withdraw from all but the most routinized kinds of civic engagement. For those who joined the dissident cause, it signaled the possibility of experimenting with novel forms of engagement that, counter-intuitively, harnessed the expansion of freedom to the public defense of Soviet legal norms.

Linking freedom to strict observance of the law set up a certain tension with Soviet tradition, insofar as the latter never completely abandoned its vision of genuine freedom as the withering away of state and law, made possible by the internalization of all necessary ethical norms by socialism's new and improved citizens. That tension also found expression inside the dissident movement, which never fully confronted, let alone resolved, the contradiction between following the dictates of conscience (or, in Andrei Sakharov's version, the demands of Reason), on the one hand, and strict obedience to Soviet law, on the other. Hence what I have suggested was the hidden paradox of the movement: even as dissidents demanded strict fidelity to the Soviet Constitution and the Code of Criminal Procedure (what I call "'civil obedience"), they proved highly allergic to formal rules and procedures within their own milieu. Even after shattering the taboo against establishing independent organizations (beginning in 1969 with the founding of the Initiative Group for the Defense of Civil Rights in the USSR), most Soviet dissidents continued to resist, in their own work, formal rules or hierarchies of any kind. The point of drawing attention to this paradox is not to expose inconsistency or hypocrisy, but to suggest the persistence of informality and person-centered authority, and the relative weakness of institutions, even in a movement dedicated to combating those qualities in the Soviet state.

The dissident movement began as a collection of small face-to-face communities bound by strong ties of intimate friendship and trust. In certain ways it never lost those qualities. The movement itself had many lives, punctuated by several near deaths. It expanded with remarkable speed, via chain reaction, testing its capacity to function as a coherent social movement grounded in shared commitment to transparency and the rule of law. When faced with massive extra-judicial persecution, it contracted as quickly as it had grown, reinventing itself as a series of small NGO-like associations buttressed by samizdat networks of communication, all of which operated under a state of siege by the KGB's Fifth Directorate. Most of those NGO-like groups were outer-directed in the sense of speaking to sympathizers abroad rather than actively searching for allies in Soviet society. They remained an overwhelmingly intelligentsia phenomenon in which the intimate circle of friends was the default format. Those friendships helped the movement survive even its betrayal by two would-be leaders, Petr Yakir and Viktor Krasin.

When dissidents raised their glasses "to the success of our hopeless cause," what did they imagine success would look like? Alexander Volpin's meta-revolution sought to disseminate ideals of transparency, rights, and the rule of law—ideals already embedded in the Soviet Constitution—thereby making it more difficult for the government to engage in arbitrary or lawless behavior. While the dissident message seems to have resonated with a significant portion of the urban intelligentsia, many sympathetic bystanders came to regard dissidents as troublemakers who disrupted not just the performance of the Soviet play but a whole array of everyday mechanisms, many extra-legal, for coping with the Soviet system. The logician Alexander Zinoviev offered the following assessment in 1979, shortly after his forced emigration from the USSR: "Because of their way of life, the overwhelming majority of citizens in a communist society feel no need for civil liberties. And precisely for this reason they don't have those liberties. To put it crudely, they need civil liberties like a fish needs an umbrella."[6]

As Volpin himself recognized, when it came to applying pressure on the Kremlin, the dissidents' allies in the West proved more effective than their fellow Soviet citizens. The Communist Party, having exercised a de facto monopoly on political activity for more than half a century, had rendered the Soviet population politically anemic. Soviet citizens, including dissidents, lacked institutional mechanisms, whether an independent legislature, judiciary, or press (not to mention an autonomous church or private wealth and property that could translate into political influence), through which to channel interests distinct from those of the political leadership. As a result, Soviet leaders by and large operated with a sense of impunity vis-à-vis the Soviet population. "The people will support any action we take," Nikolai Podgorny remarked at a meeting of the Politburo to discuss what to do with Solzhenitsyn. No one disagreed.[7]

For all the stability it provided, the lip-service state, with its carefully scripted public performances, had the unsettling effect of making it nearly impossible—for members of the Politburo, dissidents, or anyone else—to know what the Soviet population actually thought, a common problem in authoritarian societies. Soviet leaders were perennially sensitive to even the slightest challenge to the Party's monopoly. They were especially sensitive to criticism from abroad, which helps explain why dissident contacts with the West, whether foreign correspondents in Moscow, NGOs such as Amnesty International, or members of the U.S. Congress, came to loom so large in the dissident world, eventually eclipsing contacts with Soviet society. "Without support from abroad," Yuri Orlov noted, "purely internal protests were of no use."[8] The accusation that dissidents were "hypnotized" by the West, however, misses the mark, if only because it ignores their misgivings about multi-party democracy, rampant consumerism, and perceived lack of firmness in confronting the Kremlin.[9] What enthralled dissidents was how Western attention to their cause invariably elicited a response from Soviet leaders, in stark contrast to the silence that met dissident letters and petitions directed to those same leaders. The movement's Western orientation was in fact overdetermined. The adversarial logic of the Cold War; the

Kremlin's refusal on principle to engage in any form of dialogue (even one designed to divert dissidents, as the Kremlin did with Seán MacBride and other representatives of Amnesty International); dissidents' increasing alienation from the Soviet population; and, not least, the Western example of real existing civil liberties—all these factors steered them to a sympathetic West.

Dissidents insisted that attention from the West—what Brezhnev and other Soviet officials dismissed as "noise"—did not put them in jeopardy, as Westerners sometimes feared. On the contrary, it seemed to limit repression by the KGB, prefiguring the kind of restraint that dissidents hoped to induce via transparency and respect for the law. That was the effect, at any rate, during periods when the Kremlin sought cooperation with the capitalist world. During times of retrenchment, however, or when it appeared that dissidents were disrupting the Kremlin's foreign policy objectives, Western attention provided little if any shield. Inside the movement, as we have seen, the Western gaze exerted a subtle gravitational pull, constantly seeking to identify leaders even as dissidents debated whether they should have leaders at all. Foreign correspondents in Moscow gave the movement a lifeline to Western publics by highlighting dissidents whom they could cast as heroes—preferably anti-communist and male.

Numerous moments in the history of the dissident movement highlight the way the West functioned as a ubiquitous point of reference, for dissidents as well as for the Soviet government. Sakharov was expelled from the Soviet nuclear weapons research program not when his *Reflections on Progress, Peaceful Coexistence, and Intellectual Freedom* was released in samizdat, but when it was published in the West. The leading samizdat journal, the *Chronicle of Current Events*, took its name from a BBC news program. Participants in the Red Square protest on August 25, 1968, choreographed their "sit-down demonstration" (a variant on the "sit-ins" of the American civil rights movement) with Western audiences partly in mind—even if miscommunication with Western correspondents derailed plans to capture the event in photographs. When Yakir and Krasin attempted to justify their shocking announcement of the formation of the Initiative Group for the Defense of Civil Rights prior to securing approval by the group's members, the reason they invented was a deadline allegedly imposed by Western journalists. When Amalrik, in the unforgettable opening lines of *Will the Soviet Union Survive until 1984?*, predicted (correctly) that readers would react to his book the way an ichthyologist would react to talking fish, the readers he was referring to were Western Sovietologists. When Amalrik half-jokingly appointed himself "communications officer" of the dissident movement, the communication he had in mind was not with the Soviet public but with foreigners in Moscow. Dissidents were Soviet people, and to be Soviet during the post-Stalin era meant having a version of the West forever in mind.

To what can the Soviet dissident movement be compared? The strongest affinity felt by dissidents themselves was to Russia's storied prerevolutionary intelligentsia, whose heirs they understood themselves to be. I have questioned that pedigree, even in the case of Sakharov, the usual star witness for those seeking dissident successors to figures such as the Decembrists or Alexander Herzen. Whether one judges by repertoire of contention, relationship to the existing regime (and its legal system), or dependence on the West, Soviet dissidents bear only a thin resemblance to nineteenth-century Russian intellectuals who indicted serfdom and autocracy. If there is a meaningful link between the two, it is more deeply submerged, in the realm of the psyche: the commitment to bold action even while harboring a sense of hopelessness. "I foresee that there will be no success," wrote Kondraty Ryleev, poet, Freemason, and veteran of the Napoleonic Wars, on the eve of the ill-fated Decembrist uprising in 1825, which sought to introduce constitutional rule in Russia. "But an upheaval is necessary, for it will awaken Russia, and we with our failure will teach others."[10] In a letter to the French historian Jules Michelet published in 1852, Herzen defended his fellow Russian revolutionaries: "We sacrifice ourselves without any hope, out of loathing and discontent. There is indeed a certain recklessness in our lives, but there is nothing banal, nothing stagnant or bourgeois."[11] Two revolutions later, the source of despair had shifted, as had the method of struggle against it, but the sentiment was uncannily familiar, somehow blending moral passion and fatalism, now tinged with the ubiquitous late Soviet irony: "to the success of our hopeless cause."

Comparisons to movements outside Russia and the Soviet Union but closer in time to the Soviet dissident phenomenon are instructive in a different way. "As many people here [in the USSR] listen to foreign radio stations as we did during the [German] occupation," reported Karel van het Reve in 1968 from the Soviet capital. He was addressing readers of the daily *Het Parool*, founded in 1941 as the underground newspaper of the Dutch resistance.[12] Amnesty International's Research Department regarded Soviet samizdat as comparable to "the literature produced by the resistance movements in occupied Europe during the Second World War."[13] Indeed, in the 1940s, across Nazi-occupied Europe, small bands of activists produced underground documents and newspapers, smuggling copies to the West where they were beamed back by the BBC and other shortwave radio broadcasters, foreshadowing the Cold War's media landscape. A recent study of European resistance movements during World War II found that, as with Soviet dissidents, "the most prevalent feeling among those who resisted was the discovery of the sheer variety of people who worked together during the war yet who in peacetime would have had no reason to be in contact with each other."[14] Resistance acted as a solvent of social distinctions, making it possible for participants to imagine a movement, and perhaps a future, without inherited hierarchies.

The fact that European resistance movements operated during wartime, and against a foreign occupier, is of course a crucial distinction: their goal was to overthrow Nazi rulers or local surrogates by any means available, including armed attacks. In the USSR, violent resistance to Soviet rule (also understood as a foreign occupation) had exhausted itself in western Ukraine and the Baltic states by the time the dissident movement emerged in the 1960s. European resistance movements, moreover, often harbored revolutionary ambitions for remaking their countries after liberation from German rule—not surprisingly, given the outsized proportion of communists in their ranks. Soviet dissidents considered talk of revolution, as Valery Chalidze put it, "silly."

The closest analogy to the Soviet dissident movement from the interwar era would seem to be found in Germany, where a home-grown participatory dictatorship found widespread support in the population, as it did in the USSR after 1945. Both settings (Germany under Hitler and the Soviet Union under Brezhnev) generated what the historian Hans Mommsen called "resistance without the people," small groups lacking an anchor in their respective societies.[15] That, however, is about as far as the similarity goes. The various pockets of active resistance in Germany—among communists, high-ranking military officers, students, and workers—were not only disconnected but virtually unknown to each other. An ongoing source of uncensored information such as the *Chronicle of Current Events* would have been unthinkable under the Nazi regime, as would the sustained public dissent by figures such as Volpin, Amalrik, Sakharov, or Solzhenitsyn. This has less to do with strategies of resistance than with the sheer bloodthirstiness of Hitler's police state, which paralleled the Stalinist order but diverged significantly from the "vegetarian," second-generation Soviet socialism of Stalin's successors.[16]

More than anything, what distinguished the Soviet dissident movement from the various forms of resistance under Nazi rule—whether in Germany or in the territories it occupied—was the strategy of transparency, operating within the bounds of legality and deploying Soviet law to limit the power of the Soviet state. No resistance movements in Europe appealed to Nazi law or set their sights on using Nazi law to tame the Nazi regime. Their goal was to topple or expel that regime and they were therefore obliged to work secretly. Opponents of Hitler within the German elite agonized, in fact, over how to justify extra-legal means (attempting to assassinate Hitler, for example) in their effort to return Germany to the rule of law.[17]

To find analogies to the Soviet dissident strategy of openly appealing to existing rights and laws, one has to look closer to home, namely, to other Soviet-style states, and to the period *after* that strategy first emerged in the USSR. Despite their many local and national variations, Soviet-bloc countries such as Poland, Czechoslovakia, and Hungary all witnessed the emergence in the mid-1970s of dissident groups deploying the language of rights and legalism.[18] In many cases, that language could draw on indigenous traditions dating from the interwar period, that is, prior to the Sovietization of East-Central Europe. The Helsinki Final Act, to which all the Soviet-bloc countries were signatories, gave new impetus to the defense of rights, as illustrated, for example, by Charter 77, the association founded in 1977 by Jan Patočka, Václav Havel, and other intellectuals. Charter 77 took Czechoslovakia's communist government to task for failing to observe its own constitution, the United Nations human rights covenants, and the Helsinki Accords. The several hundred signatories of Charter 77 insisted that they were acting within the bounds of Czechoslovak law and were neither "political" nor an "organization." As with Soviet dissidents, a dearth of lawyers among East-Central European activists fostered a loose approach to the distinction between civil and human rights. And like their Soviet counterparts, they resisted formal rules within their own ranks, preferring a flexible, anti-hierarchical working style.[19]

Samizdat flourished in Poland, Czechoslovakia, and Hungary in the 1970s, eventually outstripping its Soviet counterpart in volume, dispersion, and technological sophistication.[20] In East-Central Europe too, dissident movements were composed overwhelmingly of intellectuals (the great exception was Poland, where intellectuals, workers, and Catholics joined forces). There, too, the strategy of acting "as if" existing legal norms were real, of conducting oneself like a free person ("to play at being citizens," as the Czech writer Ludvík Vaculík put it) became the signature of the dissident persona.[21] And there, too, the short-term failure of this approach to constrain the behavior of the state was at least partially offset by the effect of breaking the communist theater's fourth wall. "I talk about rights," remarked the Polish ex-communist writer Kazimierz Brandys, "because they alone will allow us to leave this magic-lantern show."[22]

At greater remove from the Soviet dissident movement, chronologically and culturally, dissident activists in the People's Republic of China have exhibited a similar attraction to the language of human rights and the rule of law. The Weiquan (Rights-Defense) movement's strategy of "living as if" and "taking the play [Chinese law] for the real" echoes the stance pioneered by Volpin, Amalrik, and other Soviet dissidents.[23] The distinctive feature in China, at least since the early 2000s, is that Weiquan is almost entirely the work of professional lawyers, dozens of whom have been arrested for representing victims of rights violations by Chinese officials.[24]

Even after allowing for significant variations in dissent in Soviet-style countries (which are barely touched on here), one can't help but be struck by the shared rhetoric and tactics. Is this a case of dissident ideas and practices spreading outward from the epicenter of the socialist world (the USSR) to its various peripheries? Or are the similarities the result not so much of transnational influence as of a shared set of attributes in Soviet-style polities? The peoples' republics of Czechoslovakia, Hungary, Poland, and China were all built in the Soviet mold—the first three as Soviet satellite states following Stalin's victory in World War II, the last through Mao Zedong's emulation (with assistance from Moscow) of the Stalinist model following his victory in China's civil war. In all these countries, as in the USSR itself, dissident legalism did not emerge until after the "Stalinist" phase of socialist development—the carnivorous phase—had played itself out, to be replaced by a less belligerent "mature socialism." The new dispensation sought stability through at least the appearance of the rule of law and was increasingly content to exact tribute from its citizens in the performative mode of lip-service. Seen from a distance, the rights-based dissent introduced by Volpin looks very much like a symptom of second-generation socialism.

One final historical comparison helps clarify what was shared and what was distinctive about the Soviet dissident movement. It began, as we have seen, as a movement for civil rights, overlapping chronologically with the African-American civil rights movement. The Black struggle for equal rights, of course, dates back to the nineteenth century, but only in the post–World War II era did it begin to achieve notable success. Initially, as the historian Carol Anderson has shown, the movement understood itself as being about *human* rights, part of a global postwar (and post-Holocaust) campaign against racially motivated violence and genocide under the auspices of the newly founded United Nations and its Universal Declaration of Human Rights, which included—at Soviet insistence—social and economic rights. With the onset of the Cold War and McCarthyism, the African-American human rights movement, under attack for its international and alleged communist orientation, recast itself as a civil rights movement embedded in a purely American story. Turning to foundational American texts and the civil religion of equal opportunity, it preserved the vision of American exceptionalism and largely withdrew from entanglements in international human rights, shielding itself from accusations of compromising American sovereignty and American values.[25]

The Soviet dissident movement followed precisely the opposite trajectory. Initially invoking civil rights enshrined in the Soviet Constitution, it subsequently reinvented itself (in response to the failure of appeals to Soviet law) as a human rights movement, invoking the Universal Declaration and international human rights agreements endorsed by the Kremlin, while shifting its appeals to a global public. Dissidents became increasingly vulnerable to the kinds of charges of disloyalty and seeking interference by Cold War enemies that had dogged the early African-American rights movement (although both movements faced such charges even when appealing to purely domestic norms).

Even though it focused on combating discrimination against a minority within the U.S. population, the African-American civil rights movement was a mass phenomenon, attracting tens of thousands of active participants and millions of supporters. Soviet dissidents defended rights of speech and assembly and judicial transparency that, in theory, applied to the entire Soviet population, but never mustered more than a small fraction of that population as supporters. Whereas American civil rights activists engaged in civil disobedience in order to highlight the injustice of Jim Crow's legal segregation, Soviet dissidents engaged in civil obedience in order to highlight the Soviet government's violation of its own laws. Civil rights activists in the American South conducted boycotts, strikes, marches, and mass rallies; Soviet dissidents worked principally through the dissemination of texts via samizdat.

These contrasts notwithstanding, the grand strategies of the two movements, what one might call their legal choreography, displayed a certain resemblance. The central strategy in the American case was to trump Jim

Crow segregation laws in the South by invoking federal law, especially the Fourteenth Amendment to the United States Constitution, guaranteeing due process and equal protection under the laws to all citizens. Civil rights activists appealed to a national legal framework in order to preempt local frameworks. Soviet dissidents did something roughly similar, responding to the Soviet government's violation of its own laws by invoking international human rights norms, including the Universal Declaration of Human Rights (1948), the International Covenant on Civil and Political Rights (1966), and the Helsinki Accords (1975). The USSR had signed the latter two, and despite not having signed the Universal Declaration, the Kremlin often cited it when criticizing American racism and inequality.

The diverging outcomes of these two formally similar procedures are telling. In the American case, the principle of judicial review allowed federal courts, including the Supreme Court, to overrule local legislation deemed inconsistent with federal law. Washington also had military force at its disposal in the form of the National Guard, which it deployed in Alabama, Arkansas, and Mississippi to compel compliance with federal court decisions. The Soviet legal system, like that of most countries at the time, did not include the mechanism of judicial review.[26] Nor was there an equivalent mechanism at the international level. Most of the international human rights agreements invoked by Soviet dissidents did not have the status of international law, and, at the time, there were no international courts authorized to adjudicate allegations of Soviet (or American) human rights violations and no military force authorized to compel compliance with international human rights norms.

Civil and political rights, the kind that are capable of placing constraints on what states can do to their citizens, do not implement themselves. Regardless of setting, for rights on paper to become rights in practice has required both political pressure and relevant institutional mechanisms inside or outside the apparatus of the state. Soviet dissidents had none of these at their disposal, relying instead on samizdat, their homemade free press. They had words but not institutions, or, rather, the relevant institutions were all outside their country, on the other side of the Iron Curtain—Radio Liberty, the BBC, the Voice of America,

Amnesty International, eventually members of the U.S. Congress. The wonder of the dissident movement is how so few people with so few resources managed to discredit so profoundly the USSR's claim to be a modern, lawfully governed country.

Less than a decade after the dissident movement ceased to exist, the USSR imploded, an unprecedented end for a world power during peacetime. In the flush of communism's demise, certain Western commentators claimed that freedom-loving dissidents had brought down the Soviet behemoth, ignoring the fact that the movement's aim was to expand spaces of freedom *within* the Soviet order—to put limits on the Soviet state, not to topple it. History, to be sure, is teeming with unintended consequences. A subtler version of the argument casts Mikhail Gorbachev as the vessel that carried dissident ideas into the Kremlin, where for the first time they could shape policy-making at the highest level. Although Gorbachev presented himself to the Soviet people as a devoted disciple of Lenin (in his case, this was no mere lip-service), certain keywords in his reform program, including "transparency" and "democratization," sounded as if plucked from the dissident lexicon. That, at any rate, is how Ludmilla Alexeyeva heard them. "He borrows our slogans and draws on our ideas," she wrote during the height of Gorbachev's restructuring (*perestroika*) campaign. "We take no offense at Gorbachev and his associates for not citing us as sources. We are happy that our ideas have acquired a new life."[27]

At the time, citing the dissident movement as a source of ideas would have been unthinkable for a Soviet leader. The stigma against dissidents as renegades manipulated by the CIA was too deeply entrenched. In his memoirs, Gorbachev insisted that successful reform of the Soviet system was possible "only if it originated in the Party." Any attempt to initiate reforms from outside the Party "would have been doomed to failure, repulsed by the 'political core' of society, taken for a dissident attack on the existing order." Only after the USSR's collapse in 1991 could Gorbachev afford to pull back the curtain ever so slightly, noting that the dissident movement had "left traces, if not in the structures [of Soviet life], then in people's minds"—including, presumably, his own.[28]

Like the dissidents, Gorbachev did not seek to dissolve the USSR. It was his attempts to reform the Soviet system that, more than any other factor, brought about that unintended result.[29] The most consequential of Gorbachev's reforms, however, dealt with the economy, an arena, as we have seen, that dissidents had almost completely ignored. Insofar as dissident ideas entered Gorbachev's reform agenda, they were to be found in his abolition of state censorship, tolerance of political pluralism, and transparency about Soviet history. Like his dismantling of the USSR's centralized command economy, these reforms had the effect of unraveling the Soviet order, as long-suppressed grievances along with damning revelations of communist misrule entered a public sphere increasingly untethered to the Communist Party. The Gorbachev years accelerated the hollowing out of official Soviet ideology, to which the dissident movement had contributed so significantly during the preceding two decades.

The accumulated loss of legitimacy, largely obscured by constant lip-service to Soviet ideals, helps explain the suddenness of the Soviet system's collapse in 1991 as well as the notable weakness of resistance to that collapse. Seen in this light, 1991 belongs to a distinctive pattern in Russia's history, in which seemingly entrenched regimes of various kinds have been hollowed out and made brittle prior to suddenly imploding, with surprisingly little opposition by their erstwhile elites and beneficiaries. When serfdom, the bedrock of Russia's social order since 1649, was abolished in 1861, there was hardly any resistance, let alone armed resistance—in contrast to the violent opposition to slavery's abolition in the United States at roughly the same time. When the three-hundred-year-old Romanov dynasty collapsed in February 1917, hardly anyone was prepared to defend either it or the institution of monarchy. Multiple factors, of course, contributed to the hollowing out of these institutions: Russia's defeat by European powers in the Crimean War in 1856, the staggering burdens placed on imperial Russian society by the First World War, and in the case of the Soviet collapse, the lackluster performance of the Soviet economy in the 1970s and 1980s. In all three instances, however, the sapping of moral legitimacy (vis-à-vis serfdom, the monarchy, and the Communist

Party) was a prerequisite for the rapid change that followed. In the late Soviet era, the dissident movement advanced that process by exposing, over and over again, the regime's inability to abide by its own laws or by fundamental human rights norms to which it had repeatedly consented.

Any appraisal of the Soviet dissident movement's afterlife, including its influence in the countries that emerged from the break-up of the USSR, should begin with a simple fact: by 1991, many former dissidents were gone. Yuri Galanskov perished in a Mordovian camp in 1972 at age thirty-three; Vasyl Stus died in the Perm-36 camp in 1985 at age forty-seven; Anatoly Marchenko succumbed to a botched medical procedure in the Chistopol Prison in 1986 at age forty-eight. By the time the USSR imploded, many—perhaps half the movement at its maximum extent—had already left or been driven out of the country, scattered across Western Europe, the United States, and Israel. Raisa Orlova and Lev Kopelev settled in Cologne, West Germany, where Kopelev, a Germanist by training, became a public intellectual; Alexander Zinoviev did the same in Munich. Near the end of her life, the second half of which she spent in Paris, Natalya Gorbanevskaya retrospectively transferred the mantle of Don Quixote from dissidents to the Soviet Union, from those who had dreamed of the rule of law to those who had dreamed of a communist paradise:

> Донкихотская страна
> не родная сторона,
> хоть и не чужая.
>
> . . . . . .
>
> Кто на ослике верхом?
> Кто спевает петухом
> в свете дня жестоком?
>
> Quixotic country
> not my native side
> but then not foreign either.
>
> . . . .

> Who is perched on the donkey?
> Who crows like a rooster
> at the harsh light of day?[30]

Many other dissidents emigrated to France as well, including Andrei Sinyavsky (who became a professor at the Sorbonne), Alexander Ginzburg, Leonid Plyushch, Viktor Fainberg, Alexander Galich, Vadim Deloné, Irina Belogorodskaya, and, until his untimely death, Andrei Amalrik. Vladimir Bukovsky took up residence in Cambridge, England; when he returned to Russia in 2007 to run for president against prime minister Dmitry Medvedev, he resurrected the dissident toast "To the success of our hopeless cause."[31] He was barred from running.

Ludmilla Alexeyeva settled in Washington, D.C., and produced the first comprehensive history of the movement she had helped create. Valery Chalidze founded a Russian-language publishing house in New York with assistance from Edward Kline and Pavel Litvinov, who taught high school physics in nearby Tarrytown. Valentin Turchin taught computer science at the City University of New York; Yuri Orlov joined the physics department at Cornell. Alexander Solzhenitsyn spent two decades in seclusion in Vermont. Tomas Venclova became a professor of Russian, Polish, and Lithuanian literature at Yale. Alexander Volpin, Petr Grigorenko, Boris Shragin, Vladimir Albrekht, Andrei Tverdokhlebov, Mykola Rudenko, Viktor Krasin, and Nadezhda Yemelkina all settled in the United States. Vitaly Rubin emigrated to Israel and became a professor of Chinese philosophy at the Hebrew University of Jerusalem. Anatoly Sharansky took the Hebrew name Natan and became a prominent Israeli politician. Maya Ulanovskaya, Anatoly Yakobson, and Boris Tsukerman also emigrated to Israel along with, by 1991, several hundred thousand other Soviet Jews.

The vast majority of former dissidents did not join the expanding political arena during the Gorbachev era or thereafter. Many hoped to recover what was left of their professional careers and personal lives after having sacrificed both to their dissenting activities and the harsh repressions by the KGB that followed. In 1989, after more than a decade of separation, Larisa Bogoraz visited Alexeyeva in the United States—the two were "like sisters"—and the conversation turned to their unexpected life paths and the prospect of growing old. Bogoraz had just appeared at a convention of the staunchly anti-communist American Federation of Labor and Congress of Industrial Organizations (AFL-CIO), where an audience of thousands had applauded her. "After such things I am sick for days," she confided to Alexeyeva. "Too many people." "That is the kind of movement we were," Alexeyeva concluded. "We came from loud Moscow *kompanii* who met in small, crowded rooms, and that's where we would like to stay. It is up to the next generation to face the roaring crowds."[32]

Among the dissidents who remained in or returned to Russia and other countries of the former Soviet Union, only a few became politically active; the majority continued to regard politics as inherently corrupt and kept their distance. After being personally invited by Gorbachev to return to Moscow from his exile in Gorky, Andrei Sakharov successfully ran for a seat in the new Congress of People's Deputies in March 1989, only to succumb to a heart arrhythmia later that year. His widow, Yelena Bonner, remained an outspoken human rights advocate in Russia and abroad. Ivan Dziuba co-founded the political party People's Movement of Ukraine in 1989. Viacheslav Chornovil ran unsuccessfully for president of Ukraine in 1991, later founding an independent newspaper and gaining a seat in the Ukrainian parliament. Sergei Kovalev served briefly as human rights adviser to Russian president Boris Yeltsin and was a deputy in the Russian parliament for a decade. Gleb Pavlovsky became a "political technologist" in the administrations of Yeltsin and his successor as Russian president, Vladimir Putin, helping to establish the latter's system of "managed democracy" (managed, that is, by the Kremlin). Alexeyeva returned to Russia in 1993 and, notwithstanding her preference for the intimacy of dissident gatherings, helped reestablish the Moscow Helsinki Group as a public human rights organization, later serving on Putin's Human Rights Council (until 2012). Solzhenitsyn returned to Russia with great fanfare in 1994 and launched a television program called *Encounters with*

Solzhenitsyn that lasted six months before succumbing to poor ratings.[33] His writings, however, retained broad influence, thanks not least to Putin's decision to include portions of *The Gulag Archipelago* in Russia's mandatory high school curriculum.

In addition to the resurrected Moscow Helsinki Group, two new organizations helped promote the dissident legacy in post-Soviet Russia. The Memorial Society was founded in Moscow in 1989 with a dual mission: to document and increase public awareness of mass repressions during the Stalin era and to promote human rights and civil society in contemporary Russia. Both agendas were designed to "prevent a return to totalitarianism," as the society's mission statement put it.[34] By the early 2000s, there were independent Memorial organizations in over fifty Russian cities. The Andrei Sakharov Center, established in 1996, created a permanent exhibition on the history of the Soviet dissident movement and became an important forum for public events concerning the past, present, and future of human rights in the Russian Federation. In 2011, it sponsored a series of posters in Moscow subway cars, each of which began with the phrase "I'm not Sakharov, but I too ... ," after which came a variety of phrases: "am ready to take responsibility for my actions," "am ready to defend my position," "think that one must point out the truth." The campaign perpetuated the image of Soviet dissidents as virtuosi capable of extraordinary feats of moral courage, while suggesting that ordinary citizens could nonetheless aspire to analogous behavior in their everyday lives.

Volpin's technique of invoking existing laws and constitutional norms to contain the power of the state found expression in numerous protests during the first two decades of the Putin era, including those organized by Strategy-31, a coalition of opposition groups whose name refers to the article in the Russian Constitution that guarantees the right of peaceful assembly. An image of seventeen-year-old Olga Misik confronting riot police with a copy of the Russian Constitution quickly went viral.[35] New advice manuals appeared—now online—for those arrested at public protests, with practical suggestions on how to conduct oneself at a police interrogation.[36]

FIGURE E.1. A woman holding a Strategy-31 sign in Moscow in 2009. The sign reads "Authorities, observe the constitution."

FIGURE E.2. Constitution of the Russian Federation being held up to riot police in Moscow in 2021

For a time, the Putin government tolerated such messages—if not the protests themselves. No less a figure than Prime Minister Medvedev bemoaned what he called Russia's "legal nihilism," the tendency at all levels of society to bend or circumvent the law.[37] One indicator of the Kremlin's balancing act vis-à-vis the dissident legacy was the handbook for high school history teachers approved in 2007 (including by Putin himself, who was already showing a keen interest in deploying history for political purposes) for schools across the Russian Federation. "The influence of the rights-defense movement on the growing political crisis of the Soviet system is still the subject of intense disputes," the handbook acknowledged, while noting that what united dissidents was their "affirmation of human freedom as the highest value and their rejection of a system that suppresses such freedom. . . . The watchword of the rights-defense movement was the observance of Soviet laws and international agreements." In the end, the handbook concluded, "the impact of the phenomenon of dissidence itself on the spiritual climate of the last Soviet decades greatly exceeded both the number of dissidents and the scale of the actions they organized."[38]

The mass protests that broke out across Russia in response to widespread fraud in the parliamentary elections of December 2011 punctured what was left of Putin's tolerance for public opposition, even within the framework of "managed democracy." Kremlin television propagandist Vladimir Solovyev, known for his skill at turning the truth on its head, warned viewers: "In Russia, there is a culture of revolt, and this culture of revolt ends in bloodshed. In Russia, there is no culture of fighting for your rights within the framework of the law."[39] Anxious to avoid a Russian version of the "color revolutions" that had erupted in Georgia (2003), Ukraine (2004), Kyrgyzstan (2005), and other former Soviet republics, the Kremlin engineered a "foreign agent" law in 2012 designed to stigmatize and eventually strangle NGOs that received funding from abroad, as was the case with Memorial, the Moscow Helsinki Group, and the Sakharov Center. For the next decade, all three faced both persistent harassment by the Russian state and—like the Soviet dissidents whose legacy they carried—a largely unsympathetic or at best indifferent Russian population.[40] Russia's annexation of the Crimean peninsula from Ukraine in 2014, covert sponsorship of separatist movements in eastern Ukraine in the years thereafter, and full-scale invasion in February 2022 were accompanied by an intensified assault on independent civil society institutions that dared to criticize those moves. The awarding of the 2022 Nobel Peace Prize to the Memorial Society (together with individuals and organizations in Belarus and Ukraine) did nothing to protect it. By 2023, Memorial, the Moscow Helsinki Group, and the Sakharov Center had all been shut down, with scarcely a whimper of popular protest. The political apathy that KGB chairman Vladimir Semichastnyi feared was spreading across Soviet society in the 1960s was now treated as an asset by his successor, Vladimir Putin.

Under Russia's wartime regime, the ground for which was carefully prepared during the preceding decade, space for public dissent has shrunk to the vanishing point.[41] With the attack on Ukraine officially labeled a "special military operation," it became a crime to protest against "war." Russians have been arrested for holding a sign saying "No to war," for holding a sign saying "*** *****" (a stand-in for *Net voine*, or "No to war"), for holding a sign with nothing written on it, and for simply pretending to hold a sign.[42] Not even "criticism via silence," the Stalin-era practice repudiated by Soviet dissidents, is protected. Russian police have arrested or detained some twenty thousand anti-war protesters, vastly more than demonstrated against the Soviet invasion of Czechoslovakia in 1968, but still a tiny fraction of Russia's population of 143 million (of whom nearly a million have fled the country since February 2022).

In 2023, a new high school history textbook appeared under the imprimatur of Vladimir Medinsky, Putin's ghostwriter and former minister of culture. Russian students now encounter a version of the dissident movement seemingly lifted from Soviet-era newspapers. Without a word about their demand that the Soviet state observe its own laws, the textbook identifies "so-called dissidents" as people who "announced their disagreement with official ideology." "The West 'took good care' of them," it winks, "and therefore their activities were closely watched by the organs of state security"—as if the Fifth Directorate would have been happy to leave in peace Soviet citizens who voiced their disagreement without Western support. "This was important," the textbook adds, "given the growing threat of terrorism"—as if dissidents or Western countries had even a remote connection to terrorism inside the USSR. According to Medinsky, what brought down the dissident movement was not state repression and forced emigration, but "a severe crisis of ideas." "Everyone understood that in the realities of the Soviet system, transformations of society could begin only when initiated from above."[43]

The notion that successful reform can emerge exclusively from within the ruling elite belongs among Russia's most enduring and self-perpetuating myths, shared by figures as diverse as Gorbachev and Putin. Reforms initiated from outside the elite, so the argument goes, are doomed to failure, mass violence, or both. Never mind that such reforms have often been stymied by the elite itself, fearful of losing its monopoly of power, or that numerous reforms initiated from above have failed miserably. To be sure, unease with popular sovereignty is by no means unique to Russia. Most modern democracies include mechanisms that reflect such unease. But in Russia it has run deeper and lasted longer, encompassing both elites and the population at large, including more than a few Soviet dissidents, despite their willingness to challenge state power.

One of the consequences of the myth that effective reform can come only from above is that the ruling apparatus of the Russian state—the "above"—has itself remained largely unreformed, untamed, and undomesticated. The rule of law as advocated by Soviet dissidents may not be sufficient to contain the enormous power of modern states over their populations. Autocratic rulers, after all, can change laws at will, manipulating courts, legislatures, and the media. But the rule of law is surely an indispensable component of any such containment. Putin currently presides over what is essentially a feral state, where political opponents and those branded as traitors are as likely to be poisoned or assassinated as tried in a court of law, and where a single figure stands alone, unconstrained by a multi-party system or indeed by his own notional party, by a parliament made of something other than rubber, by a collective leadership like Brezhnev's Politburo, by an independent judiciary or press, or even, symbolically, by a visible family. Political systems governed by laws and subject to institutional checks and balances are of course capable of disastrous missteps. And feral power can occasionally project a certain sublime majesty. But the price in blood and treasure, as the latest chapter of Russia's history suggests, is staggering.

It took West Germans three decades, and the cultural watershed of the 1960s, to embrace those who had actively opposed Nazi rule. Should Russians find their way to political pluralism and the rule of law, they too will need a usable past. The story of the Soviet dissident movement will be waiting.

CHRONOLOGY OF THE SOVIET DISSIDENT MOVEMENT

1936

December — Promulgation of the third constitution of the USSR, known as the "Stalin Constitution," which remains in effect until 1977

1947

Hélène Peltier enrolls at Moscow State University and meets Andrei Sinyavsky

1948

December — The General Assembly of the United Nations votes to approve the Universal Declaration of Human Rights; the Soviet Union actively shaped the text but abstains from the final vote

1949

Alexander Volpin is arrested by the KGB and is sent to the Lubyanka prison and then to various psychiatric hospitals, followed by exile in Karaganda

1953

March — Joseph Stalin dies; Alexander Volpin is amnestied and returns to Moscow

1956

February — General secretary Nikita Khrushchev delivers his "Secret Speech" denouncing Stalin's crimes to the Central Committee of the Communist Party

October–November — Popular uprising in Hungary suppressed by the Soviet Army

November — The KGB pressures Natalya Gorbanevskaya into informing on two friends who had criticized the Soviet suppression of the Hungarian uprising

1957

July–August — World Festival of Youth and Students in Moscow; Alexander Volpin is arrested and involuntarily confined in a psychiatric hospital

1958

July	First unsanctioned youth gatherings at Mayakovsky Square, Moscow
October	Beginning of the public campaign against Boris Pasternak, following the publication abroad of his novel *Doctor Zhivago* and the decision to award him the Nobel Prize for Literature

1959

	Alexander Ginzburg launches the samizdat poetry journal *Sintaksis*
September	Alexander Volpin arrested and involuntarily confined in a psychiatric hospital

1961

	Alexander Volpin's *A Leaf of Spring* (including "A Free Philosophical Tractate") is published in New York
	Yuri Galanskov launches the samizdat journal *Phoenix*
	Andrei Amalrik and Vladimir Bukovsky are expelled from Moscow State University
May	Founding of Amnesty International in London

1962

November	Alexander Solzhenitsyn's Gulag novella *One Day in the Life of Ivan Denisovich* is published in the journal *New World*

1963

January	Alexander Volpin is involuntarily confined in a psychiatric hospital
June	Vladimir Bukovsky is confined in a psychiatric hospital
September	Alexander Volpin sues Ilya Shatunovsky for libel based on Shatunovsky's description of *A Leaf of Spring* as "anti-Soviet slander"
November	Major-General Petr Grigorenko founds the clandestine Union for the Struggle for the Rebirth of Leninism

1964

January	Roy Medvedev launches the limited-circulation samizdat journal *Months* (later known as *Political Diary*)
February	Joseph Brodsky is arrested in Leningrad for "parasitism"
February	Petr Grigorenko is arrested and confined in a psychiatric hospital, where he meets Vladimir Bukovsky
March	A transcript of Joseph Brodsky's trial by Frida Vigdorova circulates via samizdat
October	Nikita Khrushchev is removed from power "due to actions divorced from reality"

1965

January	Baptists in the USSR begin circulating the samizdat newsletter *Fraternal Leaflet*

January	Formation of the student literary society SMOG in Moscow
May	Andrei Amalrik is arrested for "parasitism"
September	Arrest of Andrei Sinyavsky (aka Abram Tertz) and Yuli Daniel (aka Nikolai Arzhak)
December	Vladimir Bukovsky is arrested and confined in a mental hospital
December	Alexander Volpin and others issue the Civic Appeal and organize the first "transparency meeting" at Pushkin Square in Moscow, calling for an open trial of Sinyavsky and Daniel
December	Ivan Dziuba circulates in samizdat the essay *Internationalism or Russification?*

1966

January–April	Prolonged campaign against Sinyavsky and Daniel in the Soviet mass media
February–March	Trial and conviction of Sinyavsky and Daniel
April	Over 120,000 exiled Crimean Tatars sign a petition demanding permission to return to Crimea
April–July	Petition campaign in defense of Sinyavsky and Daniel; beginning of the chain reaction
September	Alexander Ginzburg assembles the samizdat *White Book* documenting the trial of Sinyavsky and Daniel
September	Article 190 added to the Russian Criminal Code, expanding definitions of "slander of the Soviet system" and "violations of public order"
December	The General Assembly of the United Nations (including the USSR) votes to approve the International Covenant on Civil and Political Rights and the International Covenant on Economic, Social and Cultural Rights

1967

January	Arrest of Yuri Galanskov, Vera Lashkova, Aleksei Dobrovolsky, and Alexander Ginzburg ("the Four")
January	Vladimir Bukovsky organizes a demonstration on Pushkin Square in defense of "the Four"; he and others are arrested
May	Alexander Solzhenitsyn, in an open letter to the Union of Soviet Writers, denounces censorship of literary works as unconstitutional
May	Yuri Andropov becomes chairman of the KGB
July	Andropov establishes the Fifth Directorate within the KGB to manage domestic dissent
September	Beginning of the trial of Vladimir Bukovsky and other participants in the January 1967 demonstration on Pushkin Square
October	Pavel Litvinov disseminates a samizdat transcript of his conversation with KGB officer Gostev about Bukovsky's trial and conviction
November	Anatoly Marchenko completes his samizdat memoir *My Testimony*, about life in contemporary forced-labor camps

November	Pavel Litvinov circulates in samizdat the transcript of the January 1967 trial of Bukovsky and others, which is published abroad the following year

1968

January	The United Nations designates 1968 as "International Year of Human Rights" to mark the twentieth anniversary of the Universal Declaration of Human Rights
January	Beginning of the Prague Spring reform movement to create "socialism with a human face" in Czechoslovakia
January	Larisa Bogoraz and Pavel Litvinov release their letter "To the Global Public" in protest against the upcoming trial of "the Four"
January	British poet Stephen Spender and other prominent intellectuals publicly announce their support for Litvinov
February	Natalya Gorbanevskaya is involuntarily confined in a psychiatric hospital
February	Alexander Volpin is involuntarily confined in a psychiatric hospital
January–March	Trial of "the Four"; numerous collective open letters protesting the trial circulate via samizdat
March	Amnesty International selects Alexander Ginzburg as its "Prisoner of the Year" from the socialist world
March	Pavel Litvinov begins to hold "visiting days" every Tuesday evening for dissidents to gather at his apartment; Viktor Krasin and Petr Grigorenko do the same on different days of the week
April	Beginning of extra-judicial repression of hundreds of signers of open protest letters
April	Andrei Sakharov's *Reflections on Progress, Peaceful Coexistence, and Intellectual Freedom* is released in samizdat and published abroad
April	Natalya Gorbanevskaya and others found the samizdat journal *Chronicle of Current Events*, covering human rights violations in the USSR and the activities of the emerging dissident movement
May	Pavel Litvinov circulates via samizdat the transcript of the trial of "the Four" along with many of the collective protest letters that accompanied the trial; foreign editions are published beginning in 1971
July	Anatoly Marchenko is arrested
August	Pavel Litvinov responds to Stephen Spender's January 1968 letter, suggesting the creation of an international committee to support the Soviet dissident movement
August	Warsaw Pact troops occupy Czechoslovakia, ending the Prague Spring
August	Konstantin Babitsky, Tatyana Bayeva, Larisa Bogoraz, Vadim Deloné, Vladimir Dremlyuga, Viktor Fainberg, Natalya Gorbanevskaya, and Pavel Litvinov protest on Red Square against the occupation of Czechoslovakia
October	Trial of the Red Square protesters

1969

Natalya Gorbanevskaya circulates via samizdat the transcript of the trial of her fellow Red Square demonstrators; foreign editions are published beginning in 1972

Alexander Malinovsky (pseudonym A. Mikhailov) circulates his samizdat essay "Considerations concerning the Liberal Campaign of 1968"

January	Self-immolation of Jan Palach in Prague's Wenceslas Square in protest against the Soviet occupation
February	Alexander Volpin composes the manual *How to Conduct Yourself at an Interrogation*
March–May	Arrest of Ivan Yakhimovich, Ilya Gabai, and Petr Grigorenko
April–June	Andrei Amalrik writes *Will the Soviet Union Survive until 1984?*
May	Founding of the Initiative Group for the Defense of Civil Rights in the USSR; by the end of 1969, half its members are either imprisoned or confined in psychiatric hospitals
May	Founding of the Alexander Herzen Foundation (Amsterdam) by the Dutch historian Jan Willem Bezemer, the British Sovietologist Peter Reddaway, and the Dutch Slavicist Karel van het Reve, to publish select samizdat texts in Russian with author's copyright and payment of royalties
July–December	Various European human rights NGOs attempt to establish relations with the Initiative Group
July	Valery Chalidze launches the samizdat journal *Problems of Society*
September–November	Serialized publication of Vsevolod Kochetov's anti-dissident novel *What Is It You Want?*
November	Alexander Solzhenitsyn is expelled from the Union of Soviet Writers
December	Natalya Gorbanevskaya is arrested
December	The "transparency meeting" (December 5) becomes an annual event

1970

January	The samizdat periodical *Ukrainian Herald* begins to appear
March–June	CBS reporter William Cole records video interviews with Andrei Amalrik, Petr Yakir, and Vladimir Bukovsky; the interviews are televised in the United States in July
May	Andrei Amalrik is arrested; Zhores Medvedev is involuntarily confined in a psychiatric hospital
October	Solzhenitsyn is awarded the Nobel Prize for Literature
November	Founding of the Human Rights Committee by Andrei Tverdokhlebov, Valery Chalidze, and Andrei Sakharov
December	The Jewish samizdat periodical *Exodus* begins to appear

1971

Roy Medvedev's *Let History Judge: The Origins and Consequences of Stalinism* is published in the West; a Russian edition circulates in the USSR via tamizdat

Amnesty International begins publishing an English translation of the *Chronicle of Current Events*

March	Vladimir Bukovsky is arrested
June–July	The Human Rights Committee affiliates with the International League for the Rights of Man in New York and the International Institute of Human Rights in Strasbourg

1972

17,054 Lithuanians sign a petition protesting the imprisonment of Catholic priests Antanas Šeškevičius, Juozas Zdebskis, and Prosperas Bubnys and demanding that the freedom of conscience guaranteed by the Soviet Constitution be recognized for religious believers

Maya Zlobina writes the novel *Sacred Paths to Willful Freedom*, about the lives of Soviet dissidents; it circulates in samizdat and appears in tamizdat in New York in 1976

March	The samizdat periodical *Chronicle of the Lithuanian Catholic Church* begins to appear
March	Pavel Litvinov emigrates to the United States
May	Alexander Volpin emigrates to the United States
May	Stephen Spender, Michael Scammell, and other British intellectuals found the journal *Index on Censorship*, prompted by Pavel Litvinov's August 1968 letter to Spender
June–September	Arrest of Petr Yakir and Viktor Krasin; they implicate more than two hundred fellow dissidents in "anti-Soviet" activity
November	Yuri Galanskov dies at age thirty-three in a labor camp in Mordovia
November	Valery Chalidze is allowed to travel to the United States and stripped of his Soviet citizenship in December

1973

Vladimir Albrekht circulates in samizdat his manual *How to Be a Witness*

January	Zhores Medvedev is allowed to travel to Great Britain and stripped of his Soviet citizenship in August
August	Prolonged campaign against Andrei Sakharov in the Soviet mass media
August–September	Trial and conviction of Petr Yakir and Viktor Krasin; press conference featuring their public confession and condemnation of the dissident movement
September–October	Andrei Tverdokhlebov and a dozen other dissidents form Group 73, applying to join Amnesty International as the first affiliate in the socialist world

CHRONOLOGY OF THE SOVIET DISSIDENT MOVEMENT 647

December	Publication of Alexander Solzhenitsyn's *Gulag Archipelago* in the West
December	Andrei Sinyavsky and Maria Rozanova emigrate to France

1974

	Vladimir Albrekht circulates in samizdat his manual *How to Conduct Yourself during a Search*
January	Prolonged campaign against Alexander Solzhenitsyn in the Soviet mass media
February	Vladimir Bukovsky and Semyon Gluzman circulate in samizdat their *Manual on Psychiatry for Other-Thinkers*
February	Alexander Solzhenitsyn circulates in samizdat his manual *Live Not by the Lie*
February	Solzhenitsyn is expelled from the USSR
April	Using royalties from his *Gulag Archipelago*, Solzhenitsyn establishes the Russian Social Fund to Aid Political Prisoners and Their Families
June	Viktor Fainberg emigrates to Israel
September	Amnesty International formally recognizes Group 73 as a local chapter
December	The KGB begins arresting, blacklisting, and exiling members of Group 73
December	Vladimir Dremlyuga emigrates to the United States

1975

	Valentin Turchin circulates in samizdat his essay "What Is Impartiality?"
February	Viktor Krasin and Nadezhda Yemelkina emigrate to the United States
August	Thirty-five countries, including the USSR, sign the Helsinki Accords
October	Sakharov is awarded the Nobel Peace Prize
November	Amnesty International publishes *Prisoners of Conscience in the USSR: Their Treatment and Conditions*
November	KGB head Yuri Andropov submits a proposal to "resettle" Andrei Sakharov and Yelena Bonner at an undisclosed location away from Moscow
November	Vadim Deloné and Irina Belogorodskaya emigrate to France
December	Natalya Gorbanevskaya emigrates to France

1976

May	Yuri Orlov founds the Public Group to Assist in Implementing the Helsinki Accords in the USSR
June	The samizdat journal *Memory* begins to appear, focusing on Soviet history
July	Andrei Amalrik emigrates to the Netherlands

November	Founding of Helsinki Groups in the Ukrainian and Lithuanian Soviet Republics
December	Vladimir Bukovsky is expelled from the USSR

1977

January	Founding of the Working Group to Investigate the Abuse of Psychiatry for Political Purposes
January	Founding of the Helsinki Group in the Georgian Soviet Republic
February	Arrest of Alexander Ginzburg, Yuri Orlov, and Mykola Rudenko
April	Founding of the Helsinki Group in the Armenian Soviet Republic
March–December	Arrest of the majority of members of all the various Soviet Helsinki groups
October	Amnesty International is awarded the Nobel Peace Prize
November	Valentin Turchin and Petr Grigorenko emigrate to the United States

1978

July–August	The samizdat journal *Search* begins to appear
December	The samizdat literary journal *Metropole* begins to appear

1979

	Ivan Kovalev circulates in samizdat his manual *Some Thoughts on How to Help Political Prisoners*
April	Alexander Ginzburg is expelled from the Soviet Union
November	Arrest of Tatyana Velikanova and Gleb Yakunin
December	Soviet troops enter Afghanistan

1980

January	Andrei Sakharov condemns the Soviet invasion of Afghanistan, calls for a boycott of the 1980 Summer Olympics in Moscow, and is banished without trial to Gorky
July	The KGB orchestrates a campaign against Amnesty International in the Soviet press
July	Andrei Tverdokhlebov emigrates to the United States

1981

November	The Initiative Group for the Defense of Human Rights in the USSR issues its final appeal by members imprisoned in forced-labor camps

1982

September	The Moscow Helsinki Group announces its closure
December	The *Chronicle of Current Events* and several other samizdat periodicals cease circulating

1983

The Russian Social Fund to Aid Political Prisoners and Their Families ceases operations

Vladimir Albrekht emigrates to the United States

March   The KGB reports that "the so-called 'Russian section of Amnesty International'" has been liquidated

1985

March   Mikhail Gorbachev becomes general secretary of the Communist Party of the Soviet Union

1986

October   Yuri Orlov is expelled from the USSR

December   Anatoly Marchenko dies at age forty-eight in Chistopol Prison

ACKNOWLEDGMENTS

You know this moment. You've seen it a hundred times: the movie ends and the credits begin to scroll upward on the screen. They reveal that behind the dozen or so actors you've been watching stand not only an unseen director and producer and cinematographer, but a casting director, costume designer, line producer, visual effects person, storyboard artist, prop master, gaffer, best boy, key grip, swing gang, and dozens more, including my favorite, the continuity supervisor. You are reminded once again that not only is art itself a sovereign illusion; so is the process of making it.

The sovereign illusion behind a book is that a single person, "the author," made it. The reality is a long series of dialogues with colleagues, students, friends, and relatives, some explicitly about the subject of the book, others touching on it only obliquely, and still others apparently unrelated. Among the conversations that helped make this book, some of the most memorable have been with former Soviet dissidents whom it was my privilege to meet: Ludmilla Alexeyeva, Larisa Bogoraz, Yelena Bonner, Alexander Daniel, Alexander Gribanov, Sergei Kovalev, Pavel Litvinov, Yuri Orlov, Arseny Roginsky, Gabriel Superfin, and Alexander Volpin. Andrew Blane, Edward Kline, and Peter Reddaway, who for many years worked to support Soviet dissidents, generously shared their recollections and insights and were unfailingly encouraging of my interests, even when they disagreed with some of my interpretations. Ed and Peter also gave me access to their invaluable personal archives.

Helpful staff members and colleagues at various archives and repositories were instrumental to the research on which this book is based. They include Gennadii Kuzovkin, Aleksei Makarov, and Arseny Roginsky at the Memorial Society in Moscow; Bela Koval at the Andrei Sakharov

Archive in Moscow; Irina Flige at Memorial in St. Petersburg; Wolfgang Eichwede, Maria Klassen, Susanne Schattenberg, and Gabriel Superfin at the University of Bremen's Forschungsstelle Osteuropa; Stanley Rabinowitz at the Amherst Center for Russian Culture; Carol Leadenham, Anatol Shmelev, and Lora Soroka at the Hoover Institution Library and Archives; and, not least, the amazingly efficient staff at the Office of Interlibrary Loan at the University of Pennsylvania's Van Pelt Library.

Several of this book's chapters benefited from discussions hosted by various institutions. For their critical engagement, I thank colleagues at the universities of Bremen, California at Berkeley, Cambridge, Chicago, Delaware, Freiburg, Georgia, Illinois at Urbana-Champaign, Manchester, Michigan, Oxford, Potsdam, Tel Aviv, Tübingen, Turin, Virginia, and Wisconsin at Madison, as well as Arizona State University, the Carl Friedrich von Siemens Stiftung, City College of New York, Columbia, the CUNY Graduate Center, Duke, the École des Hautes Études en Sciences Sociales, the European University at St. Petersburg, the Fritt Ord Foundation, George Washington University, Georgetown, Harvard, the Hebrew University in Jerusalem, the Higher School of Economics in St. Petersburg, the Humboldt University, Indiana University in Bloomington, Johns Hopkins, Lasalle, the Massachusetts Institute of Technology, Miami University of Ohio, New York University, Ohio State University, my colleagues at the Penn *kruzhok*, Princeton, Rutgers University in New Brunswick, Saitama University in Tokyo, the University of Southern California Law School, Stanford, Temple University, the Van Leer Institute, and the Wissenschaftszentrum Berlin für Sozialforschung.

Undergraduate research assistants at Penn provided much-appreciated help: my thanks to Tatyana Levina, Pavel Suyumov, Paul Tylkin, Anya Vodopyanov, and Sarabeth Zielonka. A research workshop hosted by Paul Gregory at the Hoover Institution Library and Archives introduced me to some of the richest historical documents on which the book is based. Two glorious sabbaticals were critical for the writing phase: one at the Shelby Cullom Davis Center for Historical Studies at Princeton (special thanks to Angela Creager and Jennifer Houle), the other at the Freiburg Institute for Advanced Studies in Germany (special thanks to

Bernd Kortmann). For their comments and suggestions on various chapters, I am grateful to the late Andrew Blane, Laura Engelstein, Gregory Freidin, Slava Gerovich, Loren Graham, Stefan-Ludwig Hoffmann, Peter Holquist, Dani Holtz, the late Edward Kline, Stephen Kotkin, Bruce Kuklick, Damon Moglen, Samuel Moyn, my mother, Joanne Nathans, my cousin Sydney Nathans, Philip Nord, Serguei Oushakine, Kevin Platt, Sophie Rosenfeld, Kathleen Smith, the late Jonathan Steinberg, Barbara Walker, Jeffrey Wasserstrom, and Amir Weiner. In a category all their own—because they generously read and commented on the entire manuscript—are Warren Breckman, Michael Gordin, James Heinzen, Ann Komaromi, Stephen Lehmann, George Liber, my brother Eli Nathans, Joshua Rubenstein, Yuri Slezkine, and Judith Surkis.

A number of foundations and organizations provided essential financial support for the research and writing of this book. I am deeply grateful to the American Philosophical Society, the John Simon Guggenheim Memorial Foundation, the Andrew W. Mellon Foundation, the National Council for East European and Eurasian Research, and the University of Pennsylvania Research Foundation. For their unflagging encouragement of this project over the years, I thank my friends Stefan Creuzberger and Tatiana Yankelevich. Antonina Bouis, translator extraordinaire, helped me untangle some particularly knotty Russian expressions. My cousin Leah Spiro, in addition to cheering me on, put me in touch with the literary agent Wendy Strothman, who pushed me to think in new ways about reaching a wide readership.

There must have been times when my splendid editor at Princeton University Press, Priya Nelson, suspected that the dissidents' toast "To the success of our hopeless cause" doubled as commentary on my relationship to deadlines. Not only did she not lose hope, she provided invaluable suggestions for improving the book's style and architecture. I am lucky to have worked with her, along with editorial assistants Emma Wagh and Morgan Spehar and my omnicompetent copyeditor, Jane M. Lichty. Many thanks also to Lachlan Brooks for preparing the bibliography.

To the Success of Our Hopeless Cause is dedicated to my wife and children. They too never ran out of hope, or patience, or love.

NOTES

Endnotes use the short form of citation; for full citations see the bibliography. A number of samizdat documents lack pagination, whether hard copies housed in archives or digitized online versions. Endnotes follow the Library of Congress system of transliterating Russian words. In the main body of the text, I transliterate Russian names and words in accordance with conventional usage (Volpin, not Vol'pin; Alexeyeva, not Alekseeva; Sinyavsky, not Siniavskii; glasnost not *glasnost'*). Unless noted otherwise, all translations are my own.

Prologue

1. Shragin, *Mysl' i deistvie*, 365–67.
2. The 1,601 respondents to the Levada Center's poll were eighteen years or older and lived in forty-five different regions of Russia, rural as well as urban. See Levada-Tsentr, "Sovetskie dissidenty v pamiati rossiian," https://www.levada.ru/2013/08/20/sovetskie-dissidenty-v-pamyati-rossiyan/ (published August 20, 2013); Gessen, *Words Will Break Cement*, 287; Parisi, "Dissidents Reloaded? Anti-Putin Activists and the Soviet Legacy," in Beumers et al., *Cultural Forms of Protest in Russia*, 33-47.
3. Anton Troianovski, "Inside Aleksei Navalny's Final Months, in His Own Words," *New York Times* (February 20, 2024), 1; Bari Weiss, "Navalny's Letters from the Gulag," *Free Press* (February 19, 2024); Semichastnyi, *Bespokoinoe serdtse*, 255.
4. Lefebvre, *The Coming of the French Revolution*, 210.
5. Nathans, "Talking Fish: On Soviet Dissident Memoirs."
6. In her article on memoirs by activists in the incomparably larger, indeed mass-scale American civil rights movement, Kathryn L. Nasstrom records roughly two hundred works published since 1958. See Nasstrom, "Between Memory and History," 328, n. 12.
7. Birkbeck et al., *Boswell's Life of Johnson*, vol. 2, 231. I thank Leona Toker for bringing this passage to my attention.
8. Kind-Kovács, *Written Here, Published There*; Kind-Kovács and Labov, *Samizdat, Tamizdat, and Beyond*; Klots, *Tamizdat*. See also the Tamizdat Project website: https://tamizdatproject.org.
9. Abram Terts [Andrei Siniavskii], *Spokoinoi nochi*, 22 and 428.
10. Brodsky, "A Writer Is a Lonely Traveler."
11. Alexeyeva and Goldberg, *The Thaw Generation*, 4–5.
12. Kolakowski, *Main Currents of Marxism*, vol. 1, 1. For a parallel approach to Ukrainian cultural dissent in the 1960s, see Bellezza, *The Shore of Expectations*, xv.
13. Lenin, "Detskaia bolezn' 'levizny' v kommunizme," in idem, *Polnoe sobranie sochinenii*, vol. 41, 33.
14. Quoted in Hosking, *The First Socialist Society*, 404.

15. Neumann, *Behemoth*; Arendt, *The Origins of Totalitarianism* (1951); Friedrich and Brzezinski, *Totalitarian Dictatorship*.

16. On participatory dictatorship, see Fulbrook, *The People's State*. Beyond taking part in official campaigns, state-sponsored organizations, and single-candidate elections, millions of citizens participated in the Soviet dictatorship by reporting on and denouncing each other to the authorities, in what the historian Jan Gross describes as "us[ing] the state to settle private disputes" or "privatization of the public domain (i.e., legitimate coercion)." Gross, "A Note on the Nature of Soviet Totalitarianism," 374.

17. Suny, "Reading Russia and the Soviet Union in the Twentieth Century," in idem, *The Cambridge History of Russia*, vol. 3, 23.

18. Weiner, "Robust Revolution to Retiring Revolution," 214.

19. Baberowski, "Nikita Khrushchev and De-Stalinization," in Naimark et al., *The Cambridge History of Communism*, vol. 2, 121.

20. Mandel'shtam, *Sobranie sochinenii v dvukh tomakh*, vol. 1, 198. Mandelstam applied the term "vegetarian" to the period before the Great Terror of the 1930s; more recently, it has also been used to describe the decades after Stalin's death.

21. On the Bolsheviks' revolutionary pedigree, see Malia, *History's Locomotives*; and Slezkine, *The House of Government*.

22. Orlov, "Before and After Glasnost," 27.

23. Arendt, *The Origins of Totalitarianism* (1966 [3rd ed.]), ix, xx. See also Baehr, "Stalinism in Retrospect," 363.

24. Turchin, *Inertsiia strakha*, 14; Havel, "The Power of the Powerless," in idem, *Living in Truth*, 36.

25. Krylova, "Imagining Socialism."

26. See Arendt's comments on the trial of the writers Andrei Sinyavsky and Yuli Daniel, in *The Origins of Totalitarianism* (1966 [3rd ed.]), xxxvii.

27. Solzhenitsyn, "Na vozvrate dykhaniia i soznaniia," in idem, ed., *Iz-pod glyb*.

28. Turchin, *Inertsiia strakha*.

29. Havel, "The Power of the Powerless," in idem, *Living in Truth*, 36.

30. Shragin, *Mysl' i deistvie*, 185.

31. See Edward Kline's foreword to Sakharov, *Memoirs*, xiv (emphasis in original) and 361. Sakharov himself preferred the old Russian word *vol'nomysliashchii* ("freethinker"). Sakharov, "How I Became a Dissident."

32. Pipes, *Vixi*, 186.

33. Barkhudarov et al., *Slovar' russkogo iazyka XVIII veka*, vol. 6, 139. In the 1860s, the renowned lexicographer Vladimir Dal included the word "disident" [sic] in his dictionary, where it was identified as of Latin origin, signifying "a person of an alien faith or a schismatic; in general, someone who rejects the dominant confession in any given place." The non-standard orthography suggests the term's lack of settled usage at the time. See Dal', *Tolkovyi slovar'*, vol. 1, 437. Two decades later, I. Iuzov published his pioneering study *Russkie dissidenty: Starovery i dukhovnye khristiane*.

34. Pomerants, *Zapiski gadkogo utenka*, 322. A variant, even more difficult to render in English: "A question to Radio Armenia: into what categories can Soviet dissidents be divided? Answer: into *sidents* [those currently doing time], *dosidents* [those who have not yet done time], *otsidents* [those who served their full sentence], *peresidents* [those who did more time than they were supposed to], *ozhidants* [those who expect to do time], and *vnovsidents* [those who are doing time again]." Mel'nichenko, *Sovetskii anekdot*, 413. Both versions suggest a desire to domesticate (by Russifying) the foreign word *dissident*, and perhaps to parody the Western obsession with politically classifying Soviet dissidents as liberals, nationalists, technocrats, Marxists, neo-Slavophiles, or other categories. Rather than worldview or political affinity, it was one's location in the life cycle of imprisonment that most deeply shaped dissident identities.

35. On the history of the terms "dissent" and "dissident" in the English-language context, see Crowley, "'Dissident'—a Brief Note."
36. Quoted in Bassow, *The Moscow Correspondents*, 243.
37. Fainberg, *Cold War Correspondents*, 186–87.
38. Zinov'ev, *Russkaia sud'ba*, 422.
39. Aleksandr Ginzburg, letter to the editor, *Vechernaia Moskva* (June 3, 1965), reprinted in Orlov, *Aleksandr Ginzburg: Russkii roman*, 174.
40. Scammell, *Solzhenitsyn*, 795.
41. On the KGB's monitoring of the report, see Rubenstein and Gribanov, *The KGB File of Andrei Sakharov*, 93.
42. Thomas Dodd, preface to Yakobson and Allen, *Aspects of Intellectual Ferment and Dissent in the Soviet Union*, 4. A decade later, Winthrop Knowlton, chairman of Harper & Row publishing house, returned from his encounters with Moscow dissidents bearing similar impressions. "I had thought that my publishing colleagues and I were the ones bringing freedom—a sense of how it really works—with us from America. We have so much of it, after all, and we talk about it so readily, with our freedom of expression. But do we understand it? Do we know how precious it is? And knowing that, do we use it well? I had not dreamed that I would bring back from Russia a deeper, truer sense of what freedom can be." Knowlton, "A Day in the Country," 89.
43. Alexeyeva and Goldberg, *The Thaw Generation*, 35.
44. Tocqueville, *The Old Regime and the Revolution*, chap. 11, "Of the Kind of Liberty Enjoyed under the Old Regime, and of Its Influence upon the Revolution," 137.
45. Skinner, *Liberty before Liberalism*.
46. Millar, "The Little Deal."
47. Amal'rik, *Zapiski dissidenta*, 39.

Chapter One

1. Amal'rik, *Zapiski dissidenta*, 42. Volpin was hardly the first to demand that the authorities observe their own laws, a tradition going back to the tsarist era. But he does appear to have been the first to generalize this approach and apply it systematically, rather than to an individual case, vis-à-vis the Soviet government. See Nathans, "The Many Shades of Soviet Dissidence."
2. Alexeyeva, *Soviet Dissent*, 275; Alexeyeva and Goldberg, *The Thaw Generation*, 108. For additional descriptions of Volpin's influence, see Sakharov, *Memoirs*, 273, 314; and Chalidze, *Prava cheloveka i sovetskii soiuz*, 65.
3. Bukovskii, *"I vozvrashchaetsia veter . . . ,"* 144.
4. Madden, "Civil Disobedience," in *Dictionary of the History of Ideas*, vol. 1, 435.
5. Boym, *Another Freedom*, 237.
6. Aikhenvald in *Literaturnaia gazeta* (November 16, 1994), 5, reprinted in Esenin-Vol'pin, *Filosofiia, logika, poeziia, zashchita prav cheloveka: Izbrannoe*, 243–45 (henceforth cited as Vol'pin, *Izbrannoe*).
7. Archive of the Memorial Society, Moscow (henceforth Memorial-Moscow), fond (f.) 120, korobka (kor.) 1, papka (pap.) 1, delo (d.) 41; quotation on list (l.) 5. Founded in 1989, the Memorial Society became Russia's leading human rights organization and the most important and accessible repository of documents pertaining to the history of the dissident movement (among other topics). In 2014, the Russian government declared Memorial a "foreign agent." It was shut down in 2021 and officially liquidated in 2022. As of 2023, the fate of its archive remains uncertain. During the period when I conducted research there, with the invaluable assistance of Gennadii Kuzovkin, Aleksei Makarov, Alexander Daniel, and the late Arseny Roginsky, many

archival materials, including Volpin's papers, were not yet catalogued—hence the non-standard form of citation.

8. Ibid., ll. 4–13; see also Volpin's poem "Ne igral ia rebenkom s det'mi" [As a child I did not play with children] in Esenin-Vol'pin, *Vesennii list*, 44. By the late 1940s, Volpin had devised an elaborate calendrical system for dating entries in his diary, in which May 12, 1925 (his date of birth), became "quasi"-January 1 of the year zero. Entries were dated according to this calendar as well as by the exact number of days (reaching into the tens of thousands) that had passed since his date of birth.

9. Memorial-Moscow, f. 120, kor. 1, pap. 1, d. 41, l. 12. Before Volpin recorded this episode in detail on its tenth anniversary (April 1949), he alluded to it in his poem "Ot ottsa rodnogo li rozhden" (January 1946): "I disciplined my thinking at fifteen."

10. See Volpin's interview with Valery Chalidze in idem, *Otvetstvennost' pokoleniia*, 138.

11. Memorial-Moscow, f. 120, kor. 1, pap. 1, d. 1 (*anketa* dated July 12, 1953).

12. Memorial-Moscow, f. 120, kor. 1, pap. 5, d. 2, "Rossiia i ia," ll. 37–38; Esenin-Vol'pin, "Pamiatka dlia neozhidaiushchikh doprosa."

13. Sakharov, *Memoirs*, 42–49.

14. Alexeyeva and Goldberg, *The Thaw Generation*, 107.

15. Ibid. Volpin, it should be noted, claimed that neither of the preceding stories was accurate (interview conducted by the author, April 30, 2004). On Volpin's non-membership in the Communist Youth League, see his 1953 *anketa* in Memorial-Moscow, f. 120, kor. 1., pap. 1, d. 1, l. 1, as well as his interview with Chalidze, *Otvetstvennost' pokoleniia*, 137, where he states that he was a Pioneer for two years and then quit.

16. Alexeyeva and Goldberg, *The Thaw Generation*, 107.

17. Chalidze, *Otvetstvennost' pokoleniia*, 137–40.

18. Ibid., 137–41. On the "fig" gesture, see Arkhipova et al., *"Figa v karmane" i drugie teorii simvolicheskogo soprotivleniia*.

19. Volpin himself proposed the latter explanation (interview with the author, March 1, 2003). According to a Western scholar who spent time with him in Moscow in 1958, Volpin was accused of trying to escape to Romania, whose border lies twenty miles from Chernivtsi. See Comey, review of Yesenin-Volpin, *Vesennij list*, 150.

20. Memorial-Moscow, f. 120, kor. 1, pap. 6, l. 62.

21. From Esenin-Vol'pin, "Ia vchera rezvilsia na polianke" (July 1949–March 1951), in idem, *Vesennii list*, 72–74.

22. Kirk, *Profiles in Russian Resistance*, 119–21.

23. Esenin-Vol'pin, "Fronda," in idem, *Vesennii list*, 62.

24. Memorial-Moscow, f. 120, pap. 6, l. 61.

25. Lovell, *The Shadow of War*.

26. By way of comparison, roughly 410,000 U.S. citizens—1 in 320—were killed during the war, nearly all of them soldiers.

27. See Kozlov and Gilburd, "The Thaw as an Event in Russian History," in idem, *The Thaw*, 18–81, and the scholarship cited there.

28. Gilburd, *To See Paris and Die*, 1, 119–57; Feyginberg, *Glenn Gould: The Russian Journey*; Reid, "Who Will Beat Whom?"

29. Quoted in Reid, "(Socialist) Realism Unbound," in Bazin et al., *Art beyond Borders*, 282.

30. Volpin, private letter to Alexander Chizhevsky and Nina Chizhevskaya, date of receipt September 4, 1957, in Arkhiv Rossiiskoi akademii nauk (Moscow), f. 1703, opis' (op.) 1, d. 478, ll. 6–7. My thanks to Alexei Kojevnikov and Alexander Lokshin for helping to procure this document.

31. The original English edition appeared in 1952; Volpin's translation was published in Moscow in 1957.

NOTES TO CHAPTER 1   659

32. Gerovitch, *From Newspeak to Cyberspeak*, 1–2.
33. Graham, *Science, Philosophy, and Human Behavior in the Soviet Union*, 266.
34. Quoted in Slezkine, *The House of Government*, 203.
35. Ivanov and Kolmogorov quoted in Gerovitch, *From Newspeak to Cyberspeak*, 155, 232.
36. Tamiment Library and Robert F. Wagner Labor Archives, Sally Belfrage Papers, Series 1: Biographical Materials, Subseries D: Notebooks, Box 4, Folder 3, "Moscow and China, 1957–1958," 33–34.
37. Belfrage, *A Room in Moscow*, 152–59. This is Belfrage's published account of her experience at the youth festival and the year thereafter, including a series of meetings with Volpin beginning in December 1957. Volpin appears in the book under the pseudonym "Tolya," a fact that became known to Soviet officials soon after the book's publication (and was confirmed by Volpin in an interview with the author on March 1, 2003). See the evaluation of the book's "anti-Soviet slander" by the Council of Ministers in Rossiiskii gosudarstvennyi arkhiv noveishei istorii (henceforth RGANI), f. 5, op. 33, d. 121, rolik 4798, ll. 1–12. Volpin's Jewishness (on his mother's side) rarely surfaces in his diaries and notes, though it is prominent in Belfrage's account of her conversations with him.
38. Russell, *Why I Am Not a Christian*.
39. On the Yale Russian Chorus: Edward Kline, interview with the author, April 30, 2004; Wittgenstein, *Logiko-filosofskii traktat*. A possible explanation for the nominal switch from Russell to Wittgenstein was Russell's controversial call in the mid-1940s for the United States to use its monopoly on nuclear weapons to create a world government whose mission would include the destruction of any country that tried to create nuclear weapons of its own. For this Russell was repeatedly denounced as a warmonger in the Soviet press. See Ryan, *Bertrand Russell*, 177.
40. Wittgenstein, *Tractatus logico-philosophicus*, 36–37, entry no. 4.003. I have slightly altered the English translation given in this bilingual edition.
41. Only rarely does Wittgenstein actually argue a point; each assertion is delivered, as Russell archly put it, "as if it were a Tsar's *ukaz*." Quoted in Monk, *Wittgenstein*, 156.
42. Esenin-Vol'pin, "Svobodnyi filosofskii traktat, ili Mgnovennoe izlozhenie moikh filosofskikh vzgliadov," in idem, *Vesennii list*, 170.
43. Copies of the 1961 (New York) edition circulated within the Soviet Union; one was found on May 14, 1964, in a search of Aleksandr Ginzburg's apartment in Moscow. See Gosudarstvennyi arkhiv Rossiiskoi Federatsii (henceforth GARF), f. 8131 (1961 g.), op. 31, d. 89189a, l. 52, where it is listed as *Vesennii listok* (Spring leaflet). The Russian text was reprinted in *Sobranie dokumentov samizdata* (henceforth SDS) (New York and Munich, 1972–78), vol. 3, doc. 234.
44. Esenin-Vol'pin, *Vesennii list*, 170–72.
45. Emphasis in original. "One of the most interesting things about Volpin," Conquest goes on to say, "is that he shows how impossible it is for even the most efficient system of thought-control to prevent the spontaneous arising of the old questions and aspirations." Conquest's review is reproduced in his *Tyrants and Typewriters*, 79–81. The original venue of publication is not given. A recent assessment of Volpin's "Tractate" calls it "probably the first original philosophical treatise written in Soviet Russia after the 1930s." Epstein, *The Phoenix of Philosophy*, 156.
46. Esenin-Vol'pin, *Vesennii list*, 140–42, 128–30, 136.
47. Ibid., 116.
48. In an interview with the author (March 1, 2003), Volpin acknowledged that certain aspects of his thought were "removed from reality." "*Nu*," came the follow-up, "*tem khuzhe dlia deistvitel'nosti*" ("well, too bad for reality").
49. Volpin's simultaneous endorsement of "anarchy" and "rule of law" is puzzling. He once defined anarchy as the "absence of power" (*bezvlastie*), which could in theory leave room for law. Memorial-Moscow, f. 120, kor. 1, pap. 2, d. 3, l. 2. When asked, decades later, about his use

of the term "anarchy," he replied, "If we had had the term 'libertarianism,' I would have used that." Interview with the author, April 30, 2004.

50. Esenin-Vol'pin, *Vesennii list*, 134–36.

51. Clark, *Petersburg, Crucible of Cultural Revolution*, 201–23; Gorham, *Speaking in Soviet Tongues*, 8–10.

52. Such, at any rate, was the impression Volpin gave in conversations with psychiatrists at Moscow's Serbsky Institute, where he was involuntarily confined for several weeks in 1959. Memorial-Moscow, f. 120, kor. 1, pap. 1, d. 28, ll. 1–7.

53. Quoted in Gerovitch, *From Newspeak to Cyberspeak*, 235.

54. Memorial-Moscow, f. 120, kor. 3, pap. 1, d. 28, l. 3.

55. The Soviet use of psychiatric confinement to silence dissidents made the topic of their actual mental health all but taboo among Western supporters during the Cold War. The American publisher of Volpin's *Vesennii list*, for example, insisted that the charge of "mental instability" (*nevmeniaemost'*) was a "euphemism" meant to disguise the political motives behind state repression of Volpin and other freethinkers. A sympathetic visiting American mathematician, however, who interacted with Volpin over the course of a month in 1958, insisted that "the truth is not so simple, nor so easy to tell." He considered the charge of mental instability "not necessarily euphemistic." Volpin's friends in the USSR tended to consider him eccentric—a *chudak*, or oddball—and occasionally referred to him as "the computer" because of his insistence on procedures and tendency toward literalism. His adolescent obsession with numbers and calendars along with his difficult social relations might conceivably be situated on the autism spectrum. Or not. If it is unethical for a psychiatrist to offer a professional opinion about a person's mental health without conducting an examination, then it is even riskier for a historian to attempt to do so. Applying psychological labels to Volpin, moreover—whether "autistic," "neurodiverse," or any other—does not seem to me to add explanatory power. For the American publisher's comment, see the introduction to Esenin-Vol'pin, *Vesennii list*, 2–3; the visiting American mathematician, David Dinsmore Comey, offered his assessment in his review of Yesenin-Volpin, *Vesennij list*, in *Studies in Soviet Thought*, 152.

56. On Jewish lawyers in pre-revolutionary Russia, see Nathans, *Beyond the Pale*, 311–66.

57. Esenin-Vol'pin, "Pamiatka dlia neozhidaiushchikh doprosa"; Berman, *Justice in the USSR*, 66–96.

58. Memorial-Moscow, f. 120, kor. 3, pap. 6, d. 1, l. 64.

59. Finn and Couvée, *The Zhivago Affair*.

60. "Stenograma vstrechi rukovoditelei KPSS i sovetskogo pravitel'stva s deiateliami literatury i iskusstva, 17.XII.1962," in Tomilina, *Nikita Sergeevich Khrushchev: Dva tsveta vremeni*, vol. 2, 593.

61. Il'ichev, *Iskusstvo prinadlezhit narodu*, 11–12.

62. "Iz stenogrammy zasedaniia Ideologicheskoi komissii TsK KPSS s uchastiem molodykh pisatelei, khudozhnikov, kompozitorov, tvorcheskikh rabotnikov kino i teatrov Moskvy" [December 24, 1962], in Afanas'eva, Afiani, et al., *Ideologicheskie komissii TsK KPSS*, 310.

63. Article: *Pravda* (December 22, 1962); letters to the editor: *Pravda* (December 27, 1962).

64. Shatunovskii, "Iz biografii podletsa," 28.

65. From Volpin's 1972 testimony to the U.S. Senate. For the transcript, see Committee on the Judiciary, *Abuse of Psychiatry for Political Repression in the Soviet Union*, 15. Volpin had arrived in the United States ten days earlier.

66. Memorial-Moscow, f. 120, kor. 1, pap. 2, d. 10, ll. 2–4.

67. The Soviet Constitution of 1936, which was in effect until it was superseded in 1977, mentions the Communist Party twice (in Articles 126 and 141), but only in passing, as befits a

document devoted to the structure and functions of the Soviet state. It makes no mention of Marx, Lenin, Marxism, or Leninism.

68. Memorial-Moscow, f. 120, kor. 1, pap. 5, d. 2, l. 10. In his memoirs, Shatunovsky does not mention Volpin's lawsuit. He does note, however, that over the course of his (Shatunovsky's) career he was brought to court thirty-four times for unspecified journalistic infractions. Shatunovskii, *Zapiski strelianogo vorob'ia*, 207.

69. As paraphrased in Bukovskii, *"I vozvrashchaetsia veter...,"* 212.

70. Quoted in Taubman, *Khrushchev*, 620.

71. Volpin, "On the Constitution of the USSR: Report to the Moscow Human Rights Committee," reprinted in Chalidze and Lipson, *Papers on Soviet Law*, vol. 1, 66; see also the interview with Volpin in Kirk, *Profiles in Russian Resistance*, 115.

72. Volpin, unpublished memoir (typescript, no title, no date, but composed sometime after 1990), 45. My gratitude to Lowry Wyman for sharing this document with me.

73. Berdiaev, *Istoki i smysl russkogo kommunizma*, 93.

74. Memorial-Moscow, f. 120, kor. 2, pap. 4, d. 21, l. 47.

Chapter Two

1. See John Glad's 1983 interview with Sinyavsky in Dzhon Gled, *Besedy v izgnanii*, 177; Abram Tertz [Andrei Sinyavsky], *Goodnight!*, 190, 10. All passages quoted from *Goodnight!* (*Spokoinoi nochi*) are from Richard Lourie's splendid translation. The original, virtually untranslatable Russian phrase "ochnaia stavka so samim soboi" refers to a technique in judicial investigations in which a defendant and a witness are brought together to confront each other's version of events. On the ambiguous status of *Goodnight!* between memoir and fiction, see Matich, "Spokojnoj noči: Andrej Sinjavskij's Rebirth as Abram Terc"; and Markesinis, *Andrei Siniavskii*, 137–75.

2. "Socialist Surrealism," *Time*, vol. 76, no. 14 (October 3, 1960): 84. *The Court Is in Session* is my rendition of Sinyavsky's title *Sud idet*, which has appeared in English under the title *The Trial Begins*. "*Sud idet*" is the phrase pronounced by the bailiff to mark the beginning or resumption of proceedings in Russian courts.

3. "Pasternak's Way," *Time*, vol. 72, no. 18 (November 3, 1958): 26.

4. Siniavskii, *Chto takoe sotsialisticheskii realizm?*, 5.

5. Riurikov, "Sotsialisticheskii realizm i ego 'nizprovergateli,'" 197. On Sinyavsky's presence at the meeting, see Golomshtok, *"Zaniatie dlia starogo gorodovogo,"* 126.

6. For one example among many, see the section "Underground Literature" in Slonim, *Soviet Russian Literature*, 330–41.

7. For a summary of the KGB's investigation, see the Vasili Mitrokhin Archive, Woodrow Wilson International Center for Scholars, Cold War International History Project Digital Archive (henceforth Mitrokhin Archive), The Chekist Anthology, Folder 41 ("Pervoprokhodtsy").

8. O'Keeffe and Szamuely, *Samizdat*, 21. For a "lyrical digression" on typewriters and samizdat, see Aksel'rod-Rubina, *Zhizn' kak zhizn'*, vol. 2, 84–85.

9. Hoover Institution Library and Archives (henceforth Hoover Archive), Andrei Sinyavsky Papers, Box 5, Folder 3; Ginzburg, *Belaia kniga*, 205.

10. Pasternak, *Stikhotvoreniia i poemy*; this volume was subsequently reissued with a different introduction. For the leading theories of who helped the KGB identify Sinyavsky as Tertz, see Nepomnyashchy, *Abram Tertz and the Poetics of Crime*, 322–23.

11. Mitrokhin Archive, The Chekist Anthology, Folder 41 ("Pervoprokhodtsy"), p. 3.

12. Velikanova, *Tsena metafory*, 58–88, 146. The Cheka (Extraordinary Commission) and the NKVD (People's Commissariat of Internal Affairs) were predecessors of the KGB. According

to a written statement prepared by Daniel on November 17, 1965, while in KGB custody, he composed "This Is Moscow Speaking" in 1956 and 1957: Hoover Archive, Sinyavsky Papers, Box 4, Folder 14, p. 73. As with Sinyavsky, Daniel's identity as Arzhak was not definitively established until shortly before his arrest; his children's story "Flight" was about to be published under his real name in September 1965 when the KGB confiscated and destroyed the entire print-run. See Daniel', *"Ia vse sbivaius' na literaturu...,"* 727. As a poet and translator, including from French, Daniel no doubt savored the wordplay between "moi" ("L'état, c'est moi") and "my" (Russian for "we"; "gosudarstvo—eto my").

13. See the "Declaration" signed by the five writers in Labedz and Hayward, *On Trial*, unpaginated front matter. This is the British, and much fuller, edition of the work that appeared in the United States as Hayward, *On Trial: The Soviet State versus "Abram Tertz" and "Nikolai Arzhak"* (New York, 1966).

14. Labedz and Hayward, *On Trial*, 11.

15. "I have never belonged to any movement," Sinyavsky wrote, "or any dissident community." Siniavskii, "Dissidentstvo kak lichnyi opyt," 131. Yuli Daniel's son, one of the subtlest interpreters of the history of the dissident movement, notes that his father "did not count himself among the dissidents." Aleksandr Daniel', "Predislovie," in Iulii Daniel', *"Ia vse sbivaius' na literaturu...,"* 7.

16. "Trois écrivains auraient été arrêtés," *Le Monde* (October 20, 1965).

17. Hoover Archive, Sinyavsky Papers, Box 5, Folder 2, pp. 245–46, 251, 266. A French translation of "American Disturbers of the Russian Conscience" appeared in *Les lettres nouvelles* (April–May, 1964), 43–75.

18. Hoover Archive, Sinyavsky Papers, Box 4, Folder 12, pp. 214–15. Remizov traveled to France twice during the late 1950s and early 1960s as an interpreter with the Moscow circus and the Red Army band, thus heightening suspicion of his contacts with bourgeois enemies of the USSR. According to Sinyavsky, he also helped another writer, Yuri Pavlovsky, transmit work abroad; Pavlovsky was never caught. Nepomnyashchy, *Abram Tertz and the Poetics of Crime*, 323, n. 6.

19. "'Chast' rukopisei popadaet za rubezh,'" 144–48.

20. RGANI, f. 5, op. 30, d. 462, Semichastnyi report to the Central Committee, December 11, 1965, p. 255. This view echoes that expressed in a private exchange between two men who were instrumental in promoting the work of Sinyavsky and Daniel abroad. In 1961, the Oxford literary scholar Max Hayward wrote (in Russian) to Boris Filippov, a Soviet émigré in the United States: "It is extraordinarily difficult to interest publishers in translations of émigré writers, especially now, when so much material is arriving from [the Soviet Union]. Purely literary considerations are beside the point—it's all about provenance." Beinecke Library, Yale University, Boris Filippov Papers, General MSS 334, Box 3, Folder 85, Series 1, Correspondence from Hayward to Filippov, August 11, 1961, p. 8.

21. The saying is cited inter alia in Savitskii, "Kogda teoriia stanovitsia real'nost'iiu," in Savitskii and Larin, *Operedivshii vremia*, 15.

22. Hoover Archive, Sinyavsky Papers, Box 5, Folder 10, pp. 12–13. According to the KGB, the anti-Soviet émigré organization the People's Labor Alliance (NTS), based in West Germany, had sent unsolicited copies of Tarsis's autobiographical novel *Ward 7* to an unknown number of Soviet intellectuals, including Sinyavsky. Box 4, Folder 13, pp. 210–11.

23. Hoover Archive, Sinyavsky Papers, Box 5, Folder 2, pp. 238 and 248. A December 6, 1965, memorandum from Semichastnyi states: "Materials regarding Remizov have been sorted into a separate case file, and the question of his criminal prosecution will be decided after the completion of the Sinyavsky and Daniel case." RGANI, f. 5, op. 30, d. 462, l. 180. Semichastnyi's account nearly four decades later sheds no additional light on the decision to remove Remizov from the case: see Semichastnyi, *Bespokoinoe serdtse*, 253.

24. RGANI, f. 5, op. 30, d. 462, ll. 250–52. I find this explanation of the KGB's decision not to arrest and try Remizov more persuasive than the notion that his interrogators feared he might commit suicide in prison. For the latter, see Andrew and Mitrokhin, *The Sword and the Shield*, 209.

25. Remizov—who had known Sinyavsky since childhood—confirmed the KGB's sense of the latter's apoliticism: "As for A. D. Sinyavsky, it seems to me that his protest against the Party's politics in the ideological arena (between 1946 to 1964) turned into disgust toward politics in general." Hoover Archive, Sinyavsky Papers, Box 5, Folder 2, p. 252.

26. Daniel' and Roginskii, *Piatoe dekabria*, 19, 23, 25, and 48 (recollections by Iuliia Vishnevskaia, Dmitrii Zubarev, and Oleg Vorob'ev); Tvardovskii, *Novomirskii dnevnik*, vol. 1, 388 (entry for September 15, 1965); Levitin-Krasnov, *Rodnoi prostor*, 75.

27. KGB records of Western shortwave broadcasts indicated that the Russian-language text of *The Court Is in Session* was broadcast by Radio Liberty on May 25, 1963; *Lyubimov* was broadcast on April 3 and 10, 1965; and *This Is Moscow Speaking* was broadcast on May 3, 1965. Hoover Archive, Sinyavsky Papers, Box 5, Folder 4, p. 241; Box 5, Folder 7, pp. 76–79. *The Court Is in Session* contains several satirical scenes of Soviet citizens listening to or talking about Radio Liberty and other "émigré radio stations."

28. Hoover Archive, Sinyavsky Papers, Box 5, Folder 10, pp. 11–12. The memorandum also mentions the possibility of confiscating Solzhenitsyn's unpublished plays *The Republic of Labor* and *Feast of Victors*. On the actual confiscations, see Scammell, *Solzhenitsyn*, 217, 326, 347, 526–41. On responses to Sinyavsky's arrest and trial by his former co-workers at the Gorky Institute, see Bittner, *The Many Lives of Khrushchev's Thaw*, 174–210.

29. Hoover Archive, Sinyavsky Papers, Box 4, Folder 13, p. 147.

30. Ibid., Folder 12, pp. 23–49. One song in particular, according to the KGB's inventory of seized items, contained "uncensored expressions" that "slandered the names of V. I. Lenin and N. K. Krupskaya" (Lenin's wife). For a transcription of the offending song, see Vladimir Vysotskii, "Krokodily, medvedi i Reks: umneishaia sobaka," accessed November 30, 2023, http://www.pomnim-lyubim.ru/rex.php.

31. Biographical information on Sinyavsky's father is taken from Hoover Archive, Sinyavsky Papers, Box 5, Folder 4, Case R-7306: Donat Evgenievich Sinyavsky, 137–39; Tertz/Sinyavsky, *Goodnight!*, 188; Hélène Zamoyska [Peltier], "Sinyavsky, the Man and the Writer," in Labedz and Hayward, *On Trial*, 49. The characterization of Donat Sinyavsky as a repentant nobleman comes from Frank, "The Triumph of Abram Tertz," *New York Review of Books*, vol. 38, no. 12 (1991).

32. "When first questioned," Sinyavsky wrote of his father, "he said to his investigator, 'I give you my word as a revolutionary!' The investigator roared with laughter: A wooly mammoth! A mastodon! He would have been better off giving his word as a nobleman." Tertz/Sinyavsky, *Goodnight!*, 168, 188.

33. Hoover Archive, Sinyavsky Papers, Box 4, Folder 12, pp. 62–64. A day later, Daniel admitted that he was Arzhak. Kantov's full title was "Senior Investigator for Especially Important Cases, Investigative Division of the KGB, Council of Ministers of the USSR." Box 5, Folder 7, p. 83. In a commonly used technique, the KGB also planted an informant in Sinyavsky's cell, who eventually won his trust. Mitrokhin Archive, The Chekist Anthology, Folder 41 ("Pervoprokhodtsy"), p. 3.

34. Hoover Archive, Sinyavsky Papers, Box 4, Folder 12, pp. 97–98.

35. Tertz/Sinyavsky, *Goodnight!*, 8.

36. Hoover Archive, Sinyavsky Papers, Box 4, Folder 12, pp. 98 and 127–28; Folder 13, p. 191; Folder 14, pp. 41–45, 125. In both cases, the lists of listeners were far from complete; Sinyavsky insisted that he never revealed to his listeners the fact that he was the author of the

stories. On Yuli Daniel's father, see Aleksandr Daniel', "Predislovie," in Iulii Daniel', *"Ia vse sbivaius' na literaturu...,"* 8–9.

37. Azbel et al., "Dvadtsat' let spustia," 145. Azbel's vignette stemmed from a prior case unconnected to that of Sinyavsky and Daniel.

38. The interrogations occurred between September 8, 1965, and January 14, 1966. Sinyavsky's KGB dossier, a copy of which can be found at the Hoover Archive, contains transcripts of most but not all of those interrogations.

39. See text and commentary in Berman, *Soviet Criminal Law and Procedure*, 57, 81–83, and 153–54.

40. Hoover Archive, Sinyavsky Papers, Box 4, Folder 13, pp. 205–7.

41. Ibid., Folder 14, p. 88.

42. Siniavskii, *Chto takoe sotsialisticheskii realizm?*, 40. In a typically prescient comment, Sinyavsky noted that proponents of socialist realism would regard such a position as "blasphemy"—a word in fact used repeatedly by the prosecution in his trial.

43. Hoover Archive, Sinyavsky Papers, Box 4, Folder 14, pp. 182–83. The article in question is Lenin's "Partiinaia organizatsiia i partiinaia literatura," *Novaia zhizn'*, no. 12 (November 13, 1905), reprinted in Lenin, *Polnoe sobranie sochinenii*, vol. 12, 99–105. It seems likely that the interrogator's questions were inspired by having read not Lenin but the introduction to the 1962 Russian-language American edition of *This Is Moscow Speaking*, edited by the émigré Boris Filippov, who quotes Lenin's article and describes the work of Arzhak and Tertz as belonging to the "literature of non-heroes." Filippov, "Vmesto predisloviia," in Arzhak, *Govorit Moskva*, 6–7.

44. Hoover Archive, Sinyavsky Papers, Box 4, Folder 13, p. 199.

45. Schwartz, "Globov, Marina, and Seryozha," 4; "Notes from Underground," *Time*, vol. 86, no. 18 (1965): 47; Filippov, "Vmesto predisloviia," in Arzhak [Iulii Daniel'], *Ruki: Chelovek iz minapa: Rasskazy*, 5; Filippov, "Vmesto predisloviia," in Arzhak, *Govorit Moskva*, 7.

46. Hoover Archive, Sinyavsky Papers, Box 4, Folder 12, p. 100.

47. Miłosz in Tertz [Sinyavsky], *The Trial Begins*, 132; Field, "Abram Tertz's Ordeal by Mirror," 9, 15. An unpublished analysis of the Western reception of works by Tertz and Arzhak by Glavlit (short for the Main Administration for Literary and Publishing Affairs, the chief agency of Soviet censorship) tendentiously asserted that Andrew Field's review confirms those works' "hostile and slanderous" character. Hoover Archive, Sinyavsky Papers, Box 5, Folder 4, pp. 200–14.

48. Alexeyeva and Goldberg, *The Thaw Generation*, 113; Azbel, *Refusenik*, 199; Daniel' and Roginskii, *Piatoe dekabria*, 43.

49. Siniavskii, letter to the editor, *Kontinent*, vol. 49 (1986): 337–42; Hélène Zamoyska [Peltier], "Quelques souvenirs sur André Siniavski," 258. Sinyavsky's service as an informant is confirmed in the Mitrokhin Archive, The Chekist Anthology, Folder 41 ("Pervoprokhodtsy"), p. 2. It is also possible that the KGB knew that Sinyavsky had had a brief affair with Svetlana Alliluyeva. See Sullivan, *Stalin's Daughter*, 227–28.

50. Siniavskii, "Dissidentstvo kak lichnyi opyt"; Tertz/Sinyavsky, *Goodnight!*, 332.

51. My account draws on the following sources: Zamoyska, "Sinyavsky, the Man and the Writer"; idem, "Quelques souvenirs sur André Siniavski"; Terts/Siniavskii, *Spokoinoi nochi*, 339–445.

52. Khmel'nitskii, "Iz chreva kitova," 152, 162. The title of Khmelnitsky's autobiographical account, "From the Belly of the Whale," is taken directly from the title of the devastating chapter about him in Sinyavsky's *Goodnight!*

53. Kabo, *The Road to Australia*, 106; Tertz/Sinyavsky, *Goodnight!*, 321 and 324. Vladimir Kabo, a fellow Moscow State University student, was introduced to Peltier and Sinyavsky by Khmelnitsky.

54. Khmel'nitskaia, *Tak slozhilas' nasha zhizn'*, 115. Viktoria Khmelnitskaya (née Fainshtein) married Sergei Khmelnitsky in 1952 and became friendly with Sinyavsky thereafter.
55. Khmel'nitskii, "Iz chreva kitova," 162; Tertz/Sinyavsky, *Goodnight!*, 342.
56. Zamoyska, "Sinyavsky, the Man and the Writer," 49.
57. Ibid., 52.
58. Tertz/Sinyavsky, *Goodnight!*, 322.
59. Khmel'nitskii, "Iz chreva kitova," 162.
60. Zamoyska, "Sinyavsky, the Man and the Writer," 53–54.
61. Zamoyska, "Quelques souvenirs sur André Siniavski," 262; idem, "Sinyavsky, the Man and the Writer," 51.
62. Tertz/Sinyavsky, *Goodnight!*, 335–36.
63. Ibid., 339. Describing these circumstances decades later, Sinyavsky insisted that he did not "choose" to protect Peltier: "At the critical junctures, the soul has no more power of choice than we have a choice of our children or our parents. And I was not acting of my own free will when I informed you of the plot against you, even though that may have been the most serious crisis in my life."
64. Responding to the publication in 1984 of Sinyavsky's memoir *Goodnight!*, Sergei Khmelnitsky disputed many aspects of his extremely negative portrayal in that work (where he was easily recognizable as "Seryozha"). He also claimed that Sinyavsky had willingly served as an informant against other students, including some whose fates were less fortunate than Peltier's. Khmel'nitskii, "Iz chreva kitova."
65. Zamoyska, "Sinyavsky, the Man and the Writer," 51. Sinyavsky himself has written, "as a joke," that his differences with the Soviet regime were "basically aesthetic." Siniavskii, "Dissidentstvo kak lichnyi opyt," 132.
66. Hoover Archive, Sinyavsky Papers, Box 4, Folder 12, p. 214. "To tell the truth," Sinyavsky later wrote in his memoir, "they should have arrested me, not him. Without knowing anything [about the incident with Peltier], my mother suspected this was so and, like a she-wolf who has just whelped a cub, she used my father to shield me. She simply loved me more." Tertz/Sinyavsky, *Goodnight!*, 197.
67. Hoover Archive, Sinyavsky Papers, Box 4, Folder 12, p. 214. Apparently the KGB had not given up its ambition to gain influence over Peltier's father. When she revealed in a 1952 letter to Sinyavsky that she was planning a short trip to Vienna—still jointly occupied, like Berlin, by the four victorious powers—the KGB arranged to fly him there for a rendezvous. His imprisoned father served as de facto hostage against any attempt to defect while abroad. Furnished by his handler with an ill-fitting new suit and with only the foggiest sense of the purpose of his mission (another attempt at seduction, or perhaps kidnapping?), Sinyavsky again shielded Peltier from danger, "tricking the devil rather than cutting a deal with him." Tertz/Sinyavsky, *Goodnight!*, 348–62; Sinyavsky in *Kontinent*, vol. 49 (1986): 337. The passage about the devil is from Golomshtok, *"Zaniatie dlia starogo gorodovogo,"* 301, quoting Hélène Peltier.
68. Zamoyska, "Sinyavsky, the Man and the Writer," 56.
69. The Soviets themselves sometimes translated the phrase as "cult of the individual": see, for example, Khrushchev's speech in *The Road to Communism: Documents of the 22nd Congress*, 124.
70. Archive of the Forschungsstelle Osteuropa, University of Bremen (henceforth FSO Bremen), f. 50, Ernst Orlovskii Papers, uncatalogued folder marked "Samizdat 1968-ogo goda."
71. Zamoyska, "Sinyavsky, the Man and the Writer," 57.
72. Daniel', "Iskuplenie," in Velikanova, *Tsena metafory*, 139.
73. Hoover Archive, Sinyavsky Papers, Box 4, Folder 14, p. 180. A month later, Daniel reiterated this view at his trial, insisting that it was not anti-Soviet: "I consider every member of society

responsible for what happens in society. I do not exclude myself. I wrote 'everyone is to blame' insofar as there was no answer to the question 'who is to blame?'" Ginzburg, *Belaia kniga*, 201.

74. Library of Congress, Hannah Arendt Papers, Speeches and Writings File: 1923–1975, "Moral Responsibility under Totalitarian Dictatorships" (undated manuscript, written after 1953 and most likely before the 1961 Adolf Eichmann trial), https://www.loc.gov/item/mss1105601266: "Totalitarian tyranny is 'democratic': the citizens are deprived of all power, they are carefully atomized, but they constantly appear in public . . . they [are] implicated directly in all crimes. These crimes are not just committed in their name, but they themselves are asked to do it. They are participants—and this they never were in classic tyranny." This essay is not to be confused with Arendt's "Personal Responsibility under Dictatorship," in idem, *Responsibility and Judgment*, 17–48, written in 1964 and dealing largely with responses to her controversial book on the Eichmann trial, *The Banality of Evil*.

75. Quoted in Dobson, *Khrushchev's Cold Summer*, 50.

76. Etkind, "A Parable of Misrecognition."

77. Daniel', "Iskuplenie," in Velikanova, *Tsena metafory*, 146.

78. Siniavskii, *Chto takoe sotsialisticheskii realizm?*, 16.

79. Zamoyska, "Quelques souvenirs sur André Siniavski," 264.

80. During his trial, Sinyavsky described "An Experiment in Self-Analysis" as "an incomplete rough draft, a fragment of an article that I never finished and tossed aside. The first portion was written in 1953–54, another section in 1960." Ginzburg, *Belaia kniga*, 238.

81. Hoover Archive, Sinyavsky Papers, Box 5, Folder 4, pp. 148–58; here p. 148. The essay's title, "Opyt samo-analiza," can also be translated as "An Attempt at Self-Analysis."

82. Ibid., pp. 150–51.

83. Ibid.

Chapter Three

1. Hoover Archive, Sinyavsky Papers, Box 5, Folder 10, p. 11.

2. Azbel, *Refusenik*, 173; Nina Voronel', *Sodom tekh let*, 141; Golomshtok, "Zaniatie dlia starogo gorodovogo," 131.

3. De Boer et al., *Biographical Dictionary of Dissidents*, 270; Orlova and Kopelev, *My zhili v Moskve*, 184.

4. Azbel, *Refusenik*, 199.

5. Gurevich, *From Lenin to Lennon*, 3 (emphasis in original).

6. Kaminskaia, *Zapiski advokata*, 142. See also Vail' and Genis, *60-e: Mir sovetskogo cheloveka*, 181.

7. For a contemporary example of the term *kritika molchaniem*, see Tvardovskii, *Novomirskii dnevnik*, vol. 1, 429 (entry for February 17, 1966).

8. Azbel et al., "Dvadtsat' let spustia," 150–51. For similar impressions, see Kaminskaia, *Zapiski advokata*, 148.

9. Azbel et al., "Dvadtsat' let spustia," 162–63. On Alexeyeva and others, see Alexeyeva and Goldberg, *The Thaw Generation*, 130 and 169.

10. Azbel, *Refusenik*, 175. Azbel declared that "if Sinyavsky and Daniel hadn't been my friends, I would no doubt have remained on the sidelines." Azbel et al., "Dvadtsat' let spustia," 162–63.

11. Fleishman, *Boris Pasternak*, 273–300. On various individual attempts to defend Pasternak—mostly from outside the Soviet Union—see Barnes, *Boris Pasternak*, vol. 2, 350–51.

12. Bulat Okudzhava, "Soiuz druzei," Bard.ru, accessed December 12, 2023, http://www.bards.ru/archives/part.php?id=17922.

13. By his own account, Volpin had never met Sinyavsky and had only a superficial acquaintance with Daniel; of the latter's double life as Nikolai Arzhak, he was completely ignorant prior to the arrests. See Vol'pin, "O protsesse Siniavskogo i Danielia, ot"ezde Tarsisa i moikh besedakh s zapadnymi korrespondentami," in Ginzburg, *Belaia kniga*, 402.

14. English translations of various Soviet constitutions can be found in Unger, *Constitutional Development in the USSR*; quotation on 154. See Article 18 of the Criminal Code of the RSFSR. When Golomshtok—a mutual friend of Sinyavsky, Daniel, and Volpin—began to explain to Volpin that "these remarkable people have been arrested," he cut him off: "I don't care whether they're remarkable or not. What's the charge?" Golomshtok, *"Zaniatie dlia starogo gorodovogo,"* 128.

15. Daniel' and Roginskii, *Piatoe dekabria*, 27.

16. Both components of the term *miting glasnosti* require explanation. The Russian word *miting* (from the English "meeting") carries a distinctly political connotation, akin to "rally" or "demonstration." *Glasnost'* can signify transparency but also openness or publicity. A literal rendition of *miting glasnosti* might thus be "rally for openness," which struck me as potentially cumbersome, especially when used repeatedly. "Transparency meeting" nicely captures the strange novelty of the Russian phrase. My thanks to Michael Gordin for helping me think this through.

17. Volpin's brief, retrospective account of the drafting process can be found in Daniel' and Roginskii, *Piatoe dekabria*, 28. For original hand-written copies of various drafts, see Memorial-Moscow, f. 120, kor. 3, pap. 4, d. 4, ll. 3–12.

18. Memorial-Moscow, f. 120, kor. 3, pap. 4, d. 4, l. 3.

19. Volpin, unpublished memoir, 16; Daniel' and Roginskii, *Piatoe dekabria*, 27.

20. Volpin, unpublished memoir, 16. Shortly after Tarsis emigrated from the USSR, Volpin wrote that "he is too emotional, and elevates to the level of a cult his incapacity for (or extreme hostility to) systematic thought." Quoted in Ginzburg, *Belaia kniga*, 402.

21. Daniel' and Roginskii, *Piatoe dekabria*, 32.

22. Ibid., 29; Moscow-Memorial, f. 120, kor. 3, pap. 4, d. 4, ll. 3 and 11.

23. Daniel' and Roginskii, *Piatoe dekabra*, 18, 27.

24. Alexeyeva and Goldberg, *The Thaw Generation*, 122.

25. Memorial-Moscow, f. 120, kor. 3, pap. 4, d. 4, ll. 3 and 6. On the saying, see the recollections of Volpin's first wife, Victoria, in Vol'pin, *Izbrannoe*, 315. Volpin used the phrase again in an open letter to Solzhenitsyn dated July 20, 1970, but now with the following explanation: "I am not against the capacity for self-preservation and the concept of measuredness, but I prefer that they be the result of free reason, not of instinct or emotions, which can always be influenced." Vol'pin, "Vechnuiu ruchku Petru Grigor'evichu Grigorenko," in *SDS*, vol. 6, doc. 406.

26. *Grani*, no. 61 (1966): 14–15, quoted in Alexeyeva, *Soviet Dissent*, 273.

27. Batshev, *SMOG*, 155. For a complete list of members, see idem, 213–14.

28. Daniel' and Roginskii, *Piatoe dekabria*, 14 and 20, where Yuliya Vishnevskaya recalls that Volpin expressed a keen interest in the details of the April 1965 SMOG demonstration. On the latter, see Batshev, *SMOG*, 54–69; Polikovskaia, *My predchuvstvie*, 315–21.

29. The KGB's September 16, 1965, report to the Central Committee on the activities of the "Amazers" claimed that the majority "suffer from mental illness." RGANI, f. 5, op. 30, d. 462, rolik 4655, l. 190.

30. Bukovskii, *"I vozvrashchaetsia veter...,"* 208–15; Boobbyer, "Vladimir Bukovsky and Soviet Communism," 456.

31. Bukovsky was born in 1942, Batshev in 1947, and Vishnevskaya in 1949. In her poem "A ia zhid i lirichnyi tsinik," composed in 1965–66 and dedicated to Volpin, Vishnevskaya wrote, "Our world is like a carousel / (whose movements are general and therefore belong to no one) / And at the very center: Aleksandr Esenin [Volpin], / Around whom gather rabble-rousers and informers." Memorial-Moscow, f. 120, kor. 3, pap. 1, d. 40, l. 1. See also Batshev, *SMOG*, 174.

32. Daniel' and Roginskii, *Piatoe dekabria*, 22, 29. Later SMOG itself split into two factions, one supporting and the other keeping its distance from the dissident movement; see Batshev, *SMOG*, 155.

33. Memorial-Moscow, f. 120, kor. 3, pap. 4, d. 4, ll. 4–5.

34. Daniel' and Roginskii, *Piatoe dekabria*, 28; Memorial-Moscow, f. 120, kor. 3, pap. 4, d. 4, l. 4.

35. Volpin, unpublished memoir, 8; Daniel' and Roginskii, *Piatoe dekabria*, 27.

36. The practice of restricting knowledge of the chain of recruitment goes back to the nineteenth century and was designed to limit the damage tsarist police informers could do to underground organizations. As in pre-revolutionary Russia, anxiety about informers was ubiquitous in the Soviet dissident movement. "In those days," Andrei Amalrik recalled, "we all feared that people who were afraid of the regime would take us for provocateurs, and we [too] feared provocateurs." Amal'rik, *Zapiski dissidenta*, 11. To underscore the distinction between revolutionary conspiracy and the new kind of independent social action embodied in the transparency meeting, Volpin insisted that "under no circumstances" should the invitation be called a "leaflet" (*listovka*), the revolutionary movement's traditional mode of communication with the masses. Memorial-Moscow, f. 120, kor. 3, pap. 4, d. 4, l. 4.

37. Ginzburg, *Belaia kniga*, 61.

38. Gorbanevskaya, "Writing for 'Samizdat,'" 34.

39. Vol'pin, *Izbrannoe*, 399, n. 13; Daniel' and Roginskii, *Piatoe dekabria*, 31. A third unintended alteration found in most "final" versions of the Civic Appeal was accidental, a result of confusion between roman and arabic numerals: the relevant article of the Soviet Constitution was given as "3" instead of "111," an error that persists in most published versions of the text. Article 3 has nothing to do with judicial transparency. For examples, see *Grani*, no. 62 (1966): 19; Ginzburg, *Belaia kniga*, 61; Alexeyeva, *Soviet Dissent*, 275; Velikanova, *Tsena metafory*, 18; Vol'pin, *Izbrannoe*, 313; Daniel' and Roginskii, *Piatoe dekabria*, 17.

40. Daniel' and Roginskii, *Piatoe dekabria*, 19.

41. Ibid., 22–23; Batshev, *SMOG*, 176–82.

42. Tvardovskii, *Novomirskii dnevnik*, vol. 1, 403.

43. Bukovskii, "I vozvrashchaetsia veter...," 225. Alexeyeva, using a similar shorthand, reports that "just about all of Moscow" knew of the meeting. Alexeyeva and Goldberg, *The Thaw Generation*, 119.

44. The SMOGist Yevgeny Kushev recalls hearing the Civic Appeal on Western shortwave radio broadcasts, which, if true, would represent a communicative network of an entirely different scale than samizdat. See Kushev's account in Batshev, *SMOG*, 174–75.

45. In a report to the Central Committee, KGB chairman Semichastnyi concurred: "To a certain degree, the measures taken prevented further distribution of the leaflet and limited the number of participants in the meeting." Both reports were dated December 6, 1965. RGANI, f. 5, op. 30, d. 462, ll. 166–68 and 178–80.

46. Bukovskii, "I vozvrashchaetsia veter...," 225.

47. "Sometimes," wrote one participant, "it seemed to me there were about 500 people at the monument—sometimes, a thousand." See Gershuni, "Pushkinskaia vakhta svobody," pt. 2, 351.

48. Daniel' and Roginskii, *Piatoe dekabria*, 33 and 42.

49. Alexeyeva, *Soviet Dissent*, 276.

50. Ginzburg, *Belaia kniga*, 84. For an account by a visiting American of the twenty-year-old Ginzburg at the 1957 World Festival of Youth and Students in Moscow, see Garthoff, *A Journey through the Cold War*, 36, 53.

NOTES TO CHAPTER 3   669

51. Alexeyeva and Goldberg, *The Thaw Generation*, 121; Daniel' and Roginskii, *Piatoe dekabria*, 37. Lyudmila Polikovskaya, a student at Moscow State University, also recalled seeing a large number of unfamiliar faces, while the older and better-connected Alexeyeva, who came strictly to watch, recalled constantly bumping into groups of people she knew. Daniel' and Roginskii, *Piatoe dekabria*, 45, 42; Alexeyeva and Goldberg, *The Thaw Generation*, 122.

52. Daniel' and Roginskii, *Piatoe dekabria*, 39 and 50.

53. Volpin would have preferred the word "observe" rather than "respect" since the latter was too "emotional." Volpin, unpublished memoir, 2.

54. Gershuni, "Pushkinskaia vakhta svobody," pt. 2, 351.

55. Alexeyeva and Goldberg, *The Thaw Generation*, 123–24.

56. Daniel' and Roginskii, *Piatoe dekabria*, 45.

57. RGANI, f. 5, op. 30, d. 462, ll. 169–70.

58. Hoover Archive, Sinyavsky Papers, Box 5, Folder 10, p. 14; RGANI, f. 5, op. 30, d. 462, ll. 178–80.

59. Daniel' and Roginskii, *Piatoe dekabria*, 40–41.

60. Shalamov, "Pis'mo staromu drugu," in Ginzburg, *Belaia kniga*, 405–15. On Shalamov's eyewitnessing of the transparency meeting, see Thun-Hohenstein, *Das Leben schreiben*, 420. Shalamov was referring to the demonstration of so-called Oppositionists in Moscow on November 7, 1927 (Revolution Day), in which he himself had participated. In fact, there had been other unsanctioned demonstrations in the years since.

61. Volpin, unpublished memoir, 29–30. The *New York Times*, apparently unsure what to make of the event, reported on the transparency meeting twice (December 12 and 18, 1965), the second time on the front page.

62. Rubin, *Dnevniki*, vol. 1, 69.

63. Aksel'rod-Rubina, *Zhizn' kak zhizn'*, vol. 2, 96.

64. Alexeyeva and Goldberg, *The Thaw Generation*, 124.

65. RGANI, f. 5, op. 30, d. 462, ll. 181–89. Nearly four decades later, former KGB chairman Semichastnyi maintained this position, describing the crowd that gathered at the Pushkin monument as having simply migrated from its former gathering spot at the Mayakovsky monument. Semichastnyi, *Bespokoinoe serdtse*, 251.

66. Daniel' and Roginskii, *Piatoe dekabria*, 63. Interrogator #3 was twenty-three-year-old Ruslan Khasbulatov, who would go on to become a leading antagonist of Boris Yeltsin in the early post-Soviet era. "Organs" was a common shorthand expression for the KGB.

67. RGANI, f. 5, op. 30, d. 462, l. 180.

68. Daniel' and Roginskii, *Piatoe dekabria*, 48–49. These verses are excerpted from the fuller version of the poem.

69. One participant, the student Oleg Vorobev, reports that upon seeing Volpin's banner "Respect the Soviet Constitution!," his first instinct was to leave in disgust. Ibid., 50.

70. Granovetter, "The Strength of Weak Ties"; idem, "The Strength of Weak Ties: A Network Theory Revisited."

71. On the role of friendship in Soviet society, see Kon, *Druzhba*, 128–37; Shlapentokh, *Love, Marriage, and Friendship in the Soviet Union*, 213–45. Shlapentokh (213 and 244) characterizes friendship in the USSR—tendentiously, in my view—as "an institution against the state."

72. Hosking, "Trust and Distrust in the USSR."

73. Solzhenitsyn, *Odin den' Ivana Denisovicha*, 126.

74. See Ledeneva, "The Genealogy of *Krugovaia Poruka*," in Marková, *Trust and Democratic Transition*, 85–108; Gorlizki, "Structures of Trust after Stalin."

Chapter Four

1. Hoover Archive, Sinyavsky Papers, Box 5, Folder 10, p. 37.
2. Ibid.
3. Baron, *Bloody Saturday*, 97–103.
4. Hoover Archive, Sinyavsky Papers, Box 5, Folder 10, pp. 37–38.
5. Soviet courtrooms were not entirely unique in this respect: Western legal systems also recognize that, in addition to rendering justice vis-à-vis particular individuals, public trials serve as a source of legal and moral instruction for a broader population. On show trials as morality plays, see Labedz and Hayward, *On Trial*, 310.
6. For an evocative treatment of the trials, see Schlögel, *Traum und Terror*, 103–18, 174–97, 661–70.
7. Practices described by Khrushchev in his "Secret Speech" of February 24, 1956: Khrushchev, "O kul'te lichnosti i ego posledstviiakh," in Aimermakher et al., *Doklad N. S. Khrushcheva o kul'te lichnosti Stalina*.
8. Kozlov and Mironenko, *Kramola*, 26.
9. Hoover Archive, Sinyavsky Papers, Box 5, Folder 10, pp. 39–40; Semichastnyi, *Bespokoinoe serdtse*, 253; Feofanov and Barry, "The Siniavskii-Daniel Trial: A Thirty Year Perspective," 618–19. In their memoirs, neither Yakovlev nor Bobkov mentions this episode.
10. Dmitrii Eremin, "Perevertyshi," *Izvestiia*, no. 10 (January 12, 1966): 6. Eremin was secretary of the Moscow branch of the Union of Soviet Writers. Ginzburg's *Belaia kniga* and many works that cite it incorrectly date the article to January 13.
11. Tvardovskii, *Novomirskii dnevnik*, vol. 1, 421–22 (entry for January 18, 1966).
12. Quoted in Alexeyeva and Goldberg, *The Thaw Generation*, 129.
13. Orlova, "'Rodinu ne vybiraiut,'" 271 (diary entry for February 1, 1966).
14. Chukovsky, *Diary, 1901–1969*, 513 (entry for December 27, 1965).
15. Evtushenko, *Volchii pasport*, 375.
16. FSO Bremen, f. 23, unpaginated diary of Chingiz Guseinov. I thank Gasan Chingizovich Guseinov for granting me access to and permission to quote from his father's diary.
17. Voinovich, *Avtoportret*, 461.
18. Hoover Archive, Sinyavsky Papers, Box 5, Folder 10, pp. 39–40.
19. See the letter to Leonid Brezhnev from five Soviet scholars criticizing the conduct of the trial, in Hayward, *On Trial*, 293.
20. Kaminskaia, *Zapiski advokata*, 145–57. According to Kaminskaya (188), "An unspoken rule requires a defense attorney in a case involving ideology not merely to condemn [his or her client's] convictions which the state considers harmful, but to announce his or her own civic position." On Sinyavsky's dim view of his attorney's strategy, see Tertz/Sinyavsky, *Goodnight!*, 34–36.
21. On confession as the "queen of proofs" in Soviet jurisprudence, see Kozlov and Mironenko, *Kramola*, 17.
22. Ginzburg, *Belaia kniga*, 169.
23. Ibid., 184. As part of their preparation for the trial, the prosecution and the KGB prepared lists of questions to be asked of each defendant (seventy-two for Sinyavsky, sixty-four for Daniel), along with detailed summaries of dozens of interrogations, including key quotations and anticipated counter-arguments. Hoover Archive, Sinyavsky Papers, Box 5, Folder 8, pp. 1–100.
24. Ginzburg, *Belaia kniga*, 204.
25. Ibid., 244.
26. Ibid., 190.
27. See, for example, Sinyavsky's discussion of Andrew Field's review in the *New Leader* (July 19, 1965), in Ginzburg, *Belaia kniga*, 222, 229, and 244; and Daniel's mention of Vladimir Mayakovsky's *The Bathhouse* and *The Bedbug*, in *Belaia kniga*, 187.

28. Ginzburg, *Belaia kniga*, 185–86.
29. Ibid., 225–26.
30. Ibid., 301.
31. FSO Bremen, f. 23, unpaginated diary of Chingiz Guseinov, February 1966.
32. Ginzburg, *Belaia kniga*, 239.
33. FSO Bremen, f. 23, diary of Chingiz Guseinov, February 1966; Ginzburg, *Belaia kniga*, 305–6.
34. Shalamov, "Pis'mo staromu drugu," in Ginzburg, *Belaia kniga*, 405–15.
35. Tvardovskii, *Novomirskii dnevnik*, vol. 1, 428 (entry for February 16, 1966).
36. Tertz/Sinyavsky, *Goodnight!*, 11. The applause following the announcement of the sentences is confirmed in Guseinov's diary.
37. FSO Bremen, f. 23, diary of Chingiz Guseinov, February 1966.
38. Ibid.
39. Orlova, "'Rodinu ne vybiraiut,'" 271–72 (diary entries for February 22 and 24, 1966).
40. Hoover Archive, Sinyavsky Papers, Box 5, Folder 11, report by Semichastnyi to the Central Committee, February 19, 1966, unpaginated.
41. Orlova, "'Rodinu ne vybiraiut,'" 272. For a fuller account of the meeting, see Bittner, *The Many Lives of Khrushchev's Thaw*, 193–98.
42. FSO Bremen, f. 23, diary of Chingiz Guseinov, entry for March 2, 1966. See also the account by the writer Vladimir Voinovich, who submitted one of the questions, in idem, *Avtoportret*, 462–63.
43. Hoover Archive, Sinyavsky Papers, Box 5, Folder 11, report by Semichastnyi to the Central Committee, February 19, 1966, unpaginated.
44. Ginzburg, *Belaia kniga*, 385–87. Lydia Chukovskaya signed the petition despite rejecting the assumption that Sinyavsky and Daniel needed to be supervised: "This proves how close those who protest against the trial are to those who ran the trial." Quoted in Orlova, "'Rodinu ne vybiraiut,'" 276 (diary entry for March 22, 1966).
45. Tvardovskii, *Novomirskii dnevnik*, vol. 1, 433–34 (text of letter reproduced in entry for March 1, 1966). The idea of exile was also proposed by the writer Marietta Shaginian to the editor of *Izvestiia*, who forwarded the suggestion to the Central Committee. See Kozlov, *The Readers of "Novyi mir,"* 246. As an exchange student at Moscow State University in 1965–66, the future political scientist William Taubman was told by a Russian student that "most of us think that Sinyavsky and Daniel were wrong to send their work abroad" but that "the trial was unfair" because "the issue was moral not criminal." Taubman, *The View from Lenin Hills*, 192.
46. Hoover Archive, Sinyavsky Papers, Box 5, Folder 10, p. 12.
47. Comrades' courts were first established in the tsarist period by factory managers who believed that "the shame of peer-inflicted public punishment" would deter wayward workers more effectively than "penalties imposed by the management." See Andrle, *A Social History of Twentieth-Century Russia*, 101. On the use of collectives during the Khrushchev era for purposes of discipline and surveillance, see Kharkhordin, *The Collective and the Individual in Russia*, 279–328. Yoram Gorlizki, in "Delegalization in Russia" 404–5, argues that, beyond their ideological justification, the courts were driven by the practical needs of the judicial system, which was overwhelmed with petty complaints by private citizens against each other.
48. Hoover Archive, Sinyavsky Papers, Box 5, Folder 11, pp. 142–43.
49. Clowes et al., *Between Tsar and People*; Hagen, *Die Entfaltung politischer Öffentlichkeit in Rußland*; Riasanovsky, *A Parting of the Ways*.
50. Lietuvos ypatingasis archyvas (Lithuanian Special Archive, formerly the archive of the Lithuanian branch of the KGB, Vilnius) (henceforth LYA), fond (f.) K-1, apyrosa (ap.) 3, bylo (b.) 642, puslapį (p.) 51. Similar graffiti appeared on several buildings in Moscow in June; see

GARF, f. 8131, op. 31, d. 99976, l. 29. A report by the Lithuanian KGB (February 27, 1966) noted that many people wanted to know why Western authors of works criticizing capitalism, published in translation in the USSR, were not punished by their own governments. It also noted a "pathological reaction" on the part of certain writers of Jewish nationality who "consider the conviction of Sinyavsky and Daniel a 'new campaign of persecution against Jews.'" See LYA, f. K-1, ap. 3, b. 647, pp. 3–4.

51. Shalamov, "Pis'mo staromu drugu," in Ginzburg, *Belaia kniga*, 405.

52. For the most thorough analysis to date of letters responding to the Sinyavsky-Daniel affair, see Kozlov, *Readers of "Novyi mir,"* 239–62.

53. Hoover Archive, Sinyavsky Papers, Box 5, Folder 11, report by Semichastnyi to the Central Committee, March 22, 1966, unpaginated; GARF, f. r-8131, op. 31, d. 99563, l. 23, quoted in Putz, *Kulturraum Lager*, 139.

54. For video footage of Sholokhov's speech, see "Vystuplenie Sholokhova M.A. na 23 s"ezde KPSS (1966)," Net-Film.ru, accessed December 12, 2023, https://www.net-film.ru/film-58472/.

55. Ulanovskaia and Ulanovskaia, *Istoriia odnoi sem'i*, 271.

56. Vol'pin, "O protsesse Siniavskogo i Danielia, ot"ezde Tarsisa i moikh besedakh s zapadnymi korrespondentami," in Ginzburg, *Belaia kniga*, 401.

57. Ibid., 400.

58. Hoover Archive, Sinyavsky Papers, Box 5, Folder 10, p. 38.

59. Well before the trial, Bogoraz had attempted to send abroad an appeal on behalf of Daniel and Sinyavsky, together with supporting materials. A KGB agent posing as a helpful foreigner was given the documents, ensuring that they never left the country. Mitrokhin Archive, The Chekist Anthology, Folder 41 ("Pervoprokhodtsy"), p. 5. The pre-trial "texts" that scripted Stalin's show trials were not verbatim transcripts; they instructed defendants regarding what crimes to confess, what explanations to give, and what punishments to beg the court to deliver.

60. Azbel et al., "Dvadtsat' let spustia," 169–71; Nina Voronel', *Sodom tekh let*, 140.

61. In her 1990 memoirs, Ludmilla Alexeyeva claimed that the unofficial trial transcript was made using a tape recorder smuggled into the courtroom by a relative of Sinyavsky or Daniel. Bogoraz's account (published posthumously in 2009) mentions neither a tape recorder nor notes taken during the trial. Instead, she describes recounting, together with Sinyavsky's wife, Maria Rozanova, the trial proceedings to friends after each courtroom session. Alexeyeva and Goldberg, *The Thaw Generation*, 131; Bogoraz, *Sny pamiati*, 114.

62. Tsentral'nyi munitsipal'nyi arkhiv goroda Moskvy (Central Municipal Archive of the City of Moscow) (henceforth TsMAM) f. R-493, op. 1, d. 367, l. 105. For Rakhunov's views regarding judicial procedure, see his "Sovetskoe pravosudie i ego rol' v ukreplenii zakonnosti," 42–54.

63. Vigdorova, *Pravo zapisyvat'*. On Vigrodova's transcription practices, see "O 'literaturnosti,' 'dokumentalizme' i 'iuridicheskom iazyke' zapisei suda nad Brodskim," in Strukova and Belenkin, *Acta Samizdatica / Zapiski o samizdate: Al'manakh*, vol. 4, 249–78; Reich, "Words on Trial."

64. "'Chast' rukopisei popadaet za rubezh,'" 146. Vigdorova's was not the only unofficial transcript of Brodsky's trial: the geneticist Raissa Berg, also present, created but did not disseminate her own. See Berg, *Sukhovei*, 230.

65. Quoted in Naiman, *Rasskazy o Anne Akhmatovoi*, 17. The literary critic and children's book author Kornei Chukovsky agreed: "By persecuting Pasternak and putting Brodsky on public trial," he confided to his diary, "our idiots made them famous on five continents." Chukovsky, *Diary, 1901–1969*, 505 (entry for June 8, 1965).

66. Golomshtok, *"Zaniatie dlia starogo gorodovogo,"* 135; Tvardovskii, *Novomirskii dnevnik*, vol. 1, 427; FSO Bremen, f. 23, diary of Chingiz Guseinov, annotation to entry for February 12, 1966.

67. Ginzburg, *Belaia kniga*, 4.

68. The official transcript can be found in the Hoover Archive, Sinyavsky Papers, Box 6, Folder 1, pp. 74–131, and Folder 2, pp. 132–61. I am grateful to my former student and research assistant Sarabeth Zielonka for her line-by-line comparison of the two texts. Apart from occasional variations in wording, the only substantive difference between the two documents is the absence in the official version of a handful of passages that make the prosecution look incompetent. If we assume that they were intentionally excised for purposes of image management, then they too are useful, suggesting by their absence which subjects were cause for official anxiety. The missing passages include the previously quoted exchange between the prosecutor and Daniel in which the latter asks, "To whom are you speaking? To me or to my protagonist?" and Daniel's sharp rejoinder to Judge Smirnov's folksy but unsuccessful analogy to the argument between two women in a communal apartment. Also absent from the official transcript is an attempt by the prosecution to portray certain passages in *This Is Moscow Speaking* as anti-semitic, along with Sinyavsky's wry observation on the implausibility of that interpretation given Daniel's Jewish background, as well as the prosecution's claim that the statement "the entire Soviet people is responsible for the personality cult" constitutes slander against the entire Soviet people. As if to diminish the impression of their unbroken spirit, the official transcript drastically abbreviated the final statements by the two defendants and eliminated all expressions of humor on their part—as when Daniel, having just been told by the prosecutor "You're slandering again, Sinyavsky!," bows and says, "My name is Daniel" (Ginzburg, *Belaia kniga*, 193). Indications of various sounds from the audience, which enhance the sense of presence in Bogoraz's transcript, are absent from the official text, as is a comment by Sinyavsky about his apartment being bugged. In the one significant passage present in the official transcript but missing from its unofficial counterpart, Daniel mentions the Kalmyks, Ingushetians, Balkartsy, Chechens, and other Soviet ethnic groups exiled under Stalin. Hoover Archive, Sinyavsky Papers, Box 6, Folder 1, p. 88.

69. *Grani*, no. 62 (1966); Filippov, "Predislovie," in *Siniavskii i Daniel' na skam'e podsudimykh*, 13–16.

70. On Ginzburg's background and motives for compiling the *Belaia kniga*, see Orlov, *Aleksandr Ginzburg*, 178–241; Litvinov, *Protsess chetyrekh*, 7–11; Amal'rik, *Nezhelannoe puteshestvie*, 86–90; interview with Ginzburg, "Dvadtsat' let tomu nazad," in *Russkaia mysl'* (February 14, 1986).

71. Van het Reve, *Dear Comrade*, 16–17.

72. Alexeyeva and Goldberg, *The Thaw Generation*, 141; Litvinov, *Protsess chetyrekh*, 195. Ehrenburg had previously helped Ginzburg obtain a Moscow residence permit following his release from his first term of imprisonment. Vigdorova, too, had sent a copy of her transcript of Brodsky's trial to Soviet officials, in her case, to Rudenko, the procurator-general. Rozenblium, "Frida Vigdorova's Transcript of Joseph Brodsky's Trial."

73. Ginzburg, "Dvadtsat' let tomu nazad," 10.

74. For the most recent incarnation of this legend, see the Russian-language Wikipedia entry "Гинзбург, Александр Ильич," last modified November 1, 2023, http://ru.wikipedia.org/wiki/Гинзбург,_Александр_Ильич. Like Volpin, Ginzburg was the son of a Russian father and a Jewish mother and took his mother's last name and nationality in his passport. Unlike Volpin, Ginzburg was also a practicing Russian Orthodox Christian.

75. Amal'rik, *Nezhelannoe puteshestvie*, 88.

76. At his own subsequent trial, Ginzburg denied that he had given his samizdat collection the title *White Book* (Litvinov, *Protsess chetyrekh*, 134). It remains unclear who did. Both the French and German editions of the collection retained the title: *Le livre blanc de l'affaire Siniavsky/Daniel*

(Paris, 1967); *Weissbuch in Sachen Sinjawskij-Daniel* (Frankfurt am Main, 1967). According to Natalya Gorbanevskaya, it was Western publishers who came up with the name *White Book*; see Ulitskaia, *Poetka*, 264.

77. Ginzburg, *Belaia kniga*, 4; Hoover Archive, Sinyavsky Papers, Box 5, Folder 7, p. 74. The English, French, German, and Italian editions do not preserve Ginzburg's chronological arrangement of the documents, instead segregating them into Soviet and Western sections.

78. According to the 1968 indictment against Ginzburg, as reproduced in Litvinov, *Protsess chetyrekh*, 60.

79. Korotkov et al., *Kremlevskii samosud*, 55.

80. Amal'rik, "Dvizhenie za prava cheloveka v SSSR," 3.

81. Daniel', *"Ia vse sbivaius' na literaturu..."* 25 (letter dated March 2, 1966); see also Anatoly Marchenko's account of Daniel's arrival in Marchenko, *My Testimony*, 367–72. For Sinyavsky's account of his hero's welcome, see Terts/Siniavskii, *Spokoinoi nochi*, 102–6.

82. Korotkov et al., *Kremlevskii samosud*, 171.

83. Bogoraz, *Sny pamiati*, 119.

84. GARF, f. 8131, op. 36, d. 4999, l. 286.

85. Testimony of Aida Topeshkina, quoted in Litvinov, *Protsess chetyrekh*, 182. For additional examples, see Blium, *Tsenzura v Sovetskom Soiuze*, 429 and 453; Bukovskii, *"I vozvrashchaetsia veter...,"* 233.

86. Roi Medvedev, "K sudebnomu protsessu nad Danielem i Siniavskim," *Politicheskii dnevnik* (February 1966). My thanks to Barbara Martin for generously sharing this document, housed in the Memorial-Moscow archive, f. 128, kor. 1.

87. Azbel et al., "Dvadtsat' let spustia," 165.

88. For a detailed analysis of Sinyavsky's strategy in this regard, see Murav, *Russia's Legal Fictions*, 193–232.

Chapter Five

1. "The term Soviet Law," wrote one of its leading interpreters, "will at first seem to many people to be a self-contradiction." See Berman, *Justice in the USSR*, 7. Another leading scholar of Soviet law, John N. Hazard of Columbia Law School, was asked "again and again" by those who learned of his expertise: "'Soviet law? Is there any?'" Hazard, *Recollections of a Pioneering Sovietologist*, 161. See also Brunner, *Die Grundrechte im Sowjetsystem*; Engelstein, "Combined Underdevelopment: Discipline and Law in Imperial and Soviet Russia"; Wortman, "Russian Monarchy and the Rule of Law."

2. A bracing version of this claim can be found in Martin Malia's foreword to Courtois et al., *The Black Book of Communism*, xvii: "Communist regimes did not just commit criminal acts (all states do so on occasion); they were criminal enterprises in their very essence: on principle, so to speak, they all ruled lawlessly." It is unclear what "criminal" means in a context in which there are allegedly no laws.

3. Personal remark to the author, Berkeley, California, early 1990s. On Malia's activities as conduit of forbidden texts from the émigré scholar Gleb Struve to his Soviet colleague (and Gulag survivor) Yulian Oksman, see the June 24, 1965, KGB report in RGANI, f. 5, op. 33, d. 210, l. 152. See also Malia's account of the operation in Malia, "Historian of Russian and European Intellectual History," an oral history conducted in 2003 by David Engerman, Regional Oral History Office, Bancroft Library, University of California, Berkeley, 2005, 116–23.

4. Weber, *Russlands Übergang zum Scheinkonstitutionalismus*. For a recent analysis of prerevolutionary ideas of constitutionalism, see Khristoforov, "'Oznachaiushchee i oznachaemoe.'"

NOTES TO CHAPTER 5   675

5. The saying has been attributed to many figures, including the historian Nikolai Karamzin, the statesman Mikhail Speransky, and the writer Mikhail Saltykov-Shchedrin—in each case without firm textual evidence.

6. Shklar, *Legalism*, 144.

7. For the preeminent refutation of the latter notion, see Kornai, *The Socialist System*.

8. Pravilova, *Zakonnost' i prava lichnosti*; Pomerantz, *Law and the Russian State*.

9. Hooper, "A Darker 'Big Deal,'" in Fürst, *Late Stalinist Russia*, 142–60.

10. Kotkin, *Magnetic Mountain*, especially chap. 5, "Speaking Bolshevik." Kotkin's analysis has inspired a rich literature on language and belief in the Soviet Union, including influential works by Jochen Hellbeck, Igal Halfin, Anna Krylova, and Alexei Yurchak.

11. On the various historical forms of the Russian service state, see Hellie, "The Structure of Modern Russian History." The rote performance of ideological slogans and the resulting hollowing out of their semantic content are not unique to the post-Stalin era. In fact, they predate Stalin's rise to power. Acute observers began to notice this trend shortly after the October Revolution. In his 1923 article "Revoliutsionnaia frazeologiia," for example, the Formalist Grigory Vinokur noted that constantly repeated Party slogans such as "Long live the victory of the working class and its vanguard, the Russian Communist Party!" had become "hackneyed clichés . . . worthless tokens . . . a collection of sounds to which our ear has become so accustomed that it is utterly impossible to react." Applying the theory of defamiliarization (*ostranenie*), Vinokur argued that constant repetition had made it impossible for the masses to register the meaning of revolutionary language: "Strike once," he urged Party spokespersons, "strike twice, but don't beat them unconscious!" Quoted in Groys, *The Total Art of Stalinism*, 44–45. Following a 1931 visit to the USSR, the American sociologist David Riesman noted the tendency of Soviet citizens to "develop ritualized ways of handling their political exhortations without inner conviction." Quoted in Baehr, *Hannah Arendt*, 48. By the 1960s, as the anthropologist Alexei Yurchak has shown, the hollowing out of Party slogans was virtually complete, thanks not only to constant repetition by Party leaders, but to the now ubiquitous performance of those slogans by ordinary (and overwhelmingly literate) Soviet people. Yurchak, *Everything Was Forever*, 10–16, 47–54, 93–98.

12. Maza, *The Myth of the French Bourgeoisie*, 145.

13. Lenin, *Polnoe sobranie sochinenii*, vol. 20, 11; vol. 17, 345.

14. See Marx's essay "On the Jewish Question," in Tucker, *The Marx-Engels Reader*, 26–46.

15. Bukharin and Preobrazhenskii, *Azbuka kommunizma*, chap. 3, article 21.

16. Quoted in Berman, *Justice in the USSR*, 26.

17. Nathans, "Soviet Rights-Talk," in Hoffmann, *Human Rights in the Twentieth Century*, 166–90.

18. Electronic Library of the History Department at Moscow State University, "Konstitutsiia (Osnovnoi Zakon) Rossiiskoi sotsialisticheskoi federativnoi sovetskoi respubliki," accessed December 12, 2023, http://www.hist.msu.ru/ER/Etext/cnst1918.htm#13, Article 64.

19. For the classic statement, see Berlin, *Two Concepts of Liberty*.

20. Lenin, "Doklad o peresmotre programmy i izmenenii nazvaniia partii 8 marta [1918]," in idem, *Polnoe sobranie sochinenii*, vol. 36, 52–53.

21. Electronic Library of the History Department at Moscow State University, "Konstitutsiia (Osnovnoi Zakon) Soiuza Sovetskikh Sotsialisticheskikh Respublik," accessed December 12, 2023, http://www.hist.msu.ru/ER/Etext/cnst1936.htm#10.

22. On the materialization of rights in the context of Soviet legal thought, see Brunner, *Die Grundrechte im Sowjetsystem*, 43–52.

23. Taubman, *The View from Lenin Hills*, 192.

24. Nor was the linking of rights and duties unique to the Soviet Constitution. The "Rules of the First International" drafted by Marx in 1864 acknowledged "no rights without duties, no

duties without rights." Similar arrangements may be found in the constitutions of Weimar Germany and the Fourth French Republic (as well as post–World War II Japan and Italy). See Unger, *Constitutional Development in the USSR*, 136–37, nn. 51 and 52.

25. On the "all-people's discussion" and what it reveals about popular attitudes toward rights and other topics, see Lomb, *Stalin's Constitution*, 19–33; Getty, "State and Society under Stalin"; Wimberg, "Socialism, Democratism and Criticism"; Davies, *Popular Opinion in Stalin's Russia*, chap. 6; Velikanova, *Mass Political Culture under Stalinism*. The comparison to the American Constitution dates from 1963: RGANI, f. 5, op. 30, d. 385, l. 204.

26. The constitution of the Weimar Republic (1919) pledged a broad program of social welfare for German citizens, including universal health and unemployment insurance and care for the elderly, but these were formulated as goals, not rights, and were eviscerated by the hyperinflation of the early 1920s. See Peukert, *Die Weimarer Republik*, 132–37. On Soviet social rights, see Smith, "Social Rights in the Soviet Dictatorship."

27. See, for example, the high school report cards of Andrei Amalrik from the mid-1950s, in which "Constitution of the USSR" is listed as a required subject: Houghton Library, Harvard University (henceforth Houghton Library), Andrei Sakharov Archives, Andrei Amalrik Papers, Box 3, Folder 27. On the influence more generally of the rights language in the 1936 constitution, see Davies, *Popular Opinion in Stalin's Russia*, 102–8.

28. Sergej Kowaljow [Sergei Kovalev], *Der Flug des weißen Raben*, 19–22. The standard interpretation of Article 125 held that the subordinate clauses "in conformity with the interests of the toilers" and "for the purpose of strengthening the socialist order" were meant to impose limits on the content of the various enumerated rights. In Kovalev's creative (mis)reading, however, the two clauses are semantically linked not to the rights, but to the phrase "shall be guaranteed by law," suggesting that it is the singular act of granting civil rights to Soviet citizens—rather than the ongoing interpretation of their content—that is supposed to conform to the interest of toilers and strengthen the socialist order.

29. The number of letters received by Khrushchev's constitutional subcommission in the early 1960s pales in comparison to the "all-people's discussion" that preceded the 1936 constitution. The Khrushchev-era letters themselves have not (to my knowledge) been preserved; one has to rely on the extensive quotations from them in reports prepared by the constitutional subcommission.

30. RGANI, f. 5, op. 30, d. 385, l. 109.

31. RGANI, f. 5, op. 30, d. 384, l. 93.

32. RGANI, f. 5, op. 30, d. 471, l. 7.

33. RGANI, f. 5, op. 30, d. 444, l. 11.

34. RGANI, f. 5, op. 30, d. 385, l. 112.

35. Numerous letters call for a return to the "Leninist principle" of the Soviet state as a "unified power." "The strength of the Soviet system consists of the unity of legislative and executive powers," wrote N. V. Fedorenko of Frunze. See RGANI, f .5, op. 30, d. 470, l. 233. I. G. Mrevlishvili (residence unknown) criticized Andrei Vyshinsky for seeking to give the judiciary a monopoly over enforcement of law. "The theoretical source of this thesis," Mrevlishvili explained, "is the arch-bourgeois conception of Montesquieu regarding the separation of judicial power from legislative and executive power. This proposition contradicts the Leninist teaching about the unity of organs in states of the new type." RGANI, f. 5, op. 30, d. 384, l. 8.

36. RGANI, f. 5, op. 30, d. 471, l. 4.

37. Abraham, "Liberty without Equality," 44. The Czech jurist Karel Vasak is generally credited with introducing the notion of "generations" in the history of human rights. See Weston, "Human Rights."

38. Berman, *Justice in the USSR*, 66–96.

39. The right to legal representation paid for by the government—paralleling the logic of Soviet civil rights—was established by the United States Supreme Court at roughly the same time (1963) in the landmark case *Gideon v. Wainwright*. The right to self-financed legal representation dated back to the Sixth Amendment, ratified as part of the Bill of Rights in 1791.

40. Zemskov, "Politicheskie repressii v SSSR." My thanks to Steven Barnes for bringing this article to my attention.

41. On debates over the presumption of innocence, the adversarial system, and other legal principles, see Savitskii, "Kogda teoriia stanovitsia real'nost'iiu," in Savitskii and Larin, *Operedivshii vremia*, 15–40; Strogovich, *Osnovnye voprosy sovetskoi sotsialisticheskoi zakonnosti*, 5–6; Savitskii, "Sostiazatel'nost' i sud v sovetskom ugolovnom protsesse," 13. The Rudenko memorandum (responding inter alia to Strogovich's arguments in favor of the presumption of innocence) is in RGANI, f. 5, op. 35, d. 197, ll. 12–27.

42. For the text of the Moral Code, see De George, *Soviet Ethics and Morality*, 83. The Code's religious overtones are not accidental. Fedor Burlatsky, one of its authors, recounts how biblical injunctions were deliberately incorporated. See the interview with Burlatsky in *Obshchestvenno-pravovoi zhurnal "Rossiiskii advokat,"* no. 5 (2007), 4.

43. Quoted in Field, "Irreconcilable Differences," 602. Already in the 1960s—that is, well before the anticipated arrival of communism—the Moral Code began to serve as the basis for verdicts rendered by comrades' courts, local tribunals run by lay citizens. See De George, *Soviet Ethics and Morality*, 84.

44. Taubman, *Khrushchev*, 620–45.

45. Lushin, *Sovetskoe gosudarstvo i oppozitsiia*, 112–13.

46. Memorial-Moscow, f. 120, kor. 3, pap. 4, d. 4, l. 6. During the drafting of the Civic Appeal, Volpin repeatedly referred to "transparency meetings" in the plural.

47. *SDS*, vol. 2, doc. 203-A. The leaflet is undated.

48. Lushin, *Sovetskoe gosudarstvo i oppozitsiia*, 113.

49. Two years earlier, in its pivotal 1964 ruling *New York Times Co. v. Sullivan*, the U.S. Supreme Court *added* the element of deliberate intent ("actual malice") to the constitutional definition of libel against public officials, thereby radically limiting the range of statements that could qualify as defamatory.

50. Medvedev, *Andropov*, 106.

Chapter Six

1. The letter is reproduced in Litvinov, *Pravosudie ili rasprava?*, 6–9.

2. Ibid., 8.

3. Bukovskii, *"I vozvrashchaetsia veter . . . ,"* 250.

4. Tarrow, *Power in Movement*, especially 195–214. Tarrow prefers "cycles of contention."

5. Bruderer, *Protsess tsepnoi reaktsii*; Bukovskii, *"I vozvrashchaetsia veter . . . ,"* 303; A. Mikhailov [Malinovskii], "Soobrazheniia po povodu liberal'noi kampanii 1968 g.," Memorial-Moscow, f. 156, pap. "Mikhailov."

6. Golomshtok, *"Zaniatie dlia starogo gorodovogo,"* 144.

7. Bogoraz, *Sny pamiati*, 114, 117.

8. Van het Reve, introduction to Litvinov, *Demonstration in Pushkin Square*, 8.

9. Mandelstam, *Hope Against Hope*, 333.

10. Moroz, *A Chronicle of Resistance in Ukraine*, 17.

11. Bruderer, *Protsess tsepnoi reaktsii*, 7.

12. Kaminskaia, *Zapiski advokata*, 11.

13. Ibid., 14; these are the opening lines of Kogan's poem "A Lyrical Digression."

14. Ibid., 147.
15. Bukovskii, "I vozvrashchaetsia veter...," 264.
16. Kaminskaya notes that in certain areas of the Soviet Union "a judge who didn't accept bribes was not only an exceptional phenomenon but practically unbelievable." Kaminskaia, *Zapiski advokata*, 50. On the logistics of bribery, see Heinzen, *The Art of the Bribe: Corruption under Stalin*.
17. Kaminskaia, *Zapiski advokata*, 162.
18. Bukovskii, "I vozvrashchaetsia veter...," 260-63.
19. Kaminskaia, *Zapiski advokata*, 172.
20. Vadim Deloné, "Otkrytoe pis'mo o demonstratsii na ploshchadi Pushkina," in *SDS*, vol. 1, doc. 49.
21. Bukovskii, "I vozvrashchaetsia veter...," 260, 265. On Volpin's "apostles," see Alexeyeva and Goldberg, *The Thaw Generation*, 121.
22. Kaminskaia, *Zapiski advokata*, 172-74.
23. Litvinov, *Pravosudie ili rasprava?*, 28.
24. Ibid., 98.
25. Contemporary commentary on the samizdat transcript can be found in Medvedev, *Politicheskii dnevnik*, no. 31 (April 1967): 156-58.
26. Litvinov, *Pravosudie ili rasprava?*, 76-78.
27. Bukovskii, "I vozvrashchaetsia veter...," 270.
28. Kaminskaia, *Zapiski advokata*, 175.
29. Litvinov, *Pravosudie ili rasprava?*, 30.
30. Ibid., 108-9.
31. Ibid., 115-17.
32. Kaminskaia, *Zapiski advokata*, 175.
33. Mitrokhin Archive, The Chekist Anthology, Folder 25 ("Muchenik"), p. 1. Quotation from Bukovskii, "I vozvrashchaetsia veter...," 248-51.
34. Bukovskii, "I vozvrashchaetsia veter...," 248-49 (emphasis in original). Anatoly Marchenko sounded a similar theme in his 1968 samizdat memoir: "It is essential that these things be known by everyone—by those who want to know the truth, but who instead are fed on slick, mendacious newspaper articles; and by those who do not want to know, who close their eyes and ears so that someday, after emerging cleansed from this filth, they may again justify themselves: 'My God, but we knew nothing about it.'" Marchenko, *Moi pokazaniia*, 7.
35. "Ofitsial'nyi sait poeta Vadima Nikolaevicha Delone (1947–1983)," accessed December 12, 2023, http://vadim-delaunay.org/poetry?t=n#1.
36. On the Soviet debate, see Jones, *Myth, Memory, Trauma*.
37. Weiner, *Making Sense of War*.
38. Kozlov and Gilburd, "The Thaw as an Event in Russian History," in idem, *The Thaw*, 30-36.
39. Harris, *Fatherland*.
40. Bukovskii, "I vozvrashchaetsia veter...," 142. Others testified to the difficult and despotic side of Konstantin Bukovsky: see Voinovich, *Avtoportret*, 352.
41. Konstantin Bukovskii, "Otvet na lestnitse," 199-201. On the reception history of Ehrenburg's memoirs, as well as efforts to censor passages dealing with Stalin's repressions, see Rubenstein, *Tangled Loyalties*, 334-51.
42. Bukovskii, "I vozvrashchaetsia veter...," 90.
43. Ibid., 103.
44. FSO Bremen, f. 23, diary of Chingiz Guseinov, February 1966, as discussed in chapter 4.

45. *Vechernaia Moskva* (September 4, 1967), cited in Chukovsky, *Diary, 1901–1969*, 527, n. 350.

46. Litvinov's account of this visit can be found in the Hoover Archive, Pavel Litvinov Papers, Box 3, Folder 16, pp. 36–39.

47. On Litvinov's activities and the preliminary title, see the International Institute for Social History, Amsterdam (henceforth IISH), Alexander Herzen Foundation Collection, Correspondence of Karel and Jozien van het Reve, Folder 24, p. 2. See also the May 21, 2015, interview with Litvinov in Morev, *Dissidenty*, 115–35.

48. Morev, *Dissidenty*, 116. See also David Remnick's portrait of the Litvinov family in *Lenin's Tomb*, 10–21.

49. Van het Reve, *Dear Comrade*, 2–16.

50. *International Herald Tribune* (December 29, 1967), 1, 4, 5. Aksel'rod-Rubina, *Zhizn' kak zhizn'*, vol. 2, 99. On Litvinov's professional elevation by the Voice of America, see Alexeyeva and Goldberg, *The Thaw Generation*, 164.

51. See chapter 14, "The Fifth Directorate."

52. Amal'rik, *Zapiski dissidenta*, 36.

53. Karel van het Reve, "Waarde kameraad? Spontane reacties van Russische burgers," *Het Parool* (April 4, 1969), reprinted in his *Verzameld werk*, vol. 3, 611–17. I thank Richard Calis for his translation of this and other Dutch-language documents.

54. Amal'rik, *Zapiski dissidenta*, 67.

55. Van het Reve, *Dear Comrade*, 30–33.

56. Amal'rik, *Zapiski dissidenta*, 14. One retiree who purportedly got his news from the BBC was Khrushchev. See Bukovskii, *"I vozvrashchaetsia veter . . . ,"* 123. Soviet citizens were also known to engage in what the KGB called "collective listening" to foreign radio broadcasts. LYA, f. K-1, ap. 10, b. 130, p. 118.

57. Alexeyeva and Goldberg, *The Thaw Generation*, 170.

58. Hoover Archive, Pavel Litvinov Papers, Box 3, Folder 16, Document 2, p. 59.

59. Author's interview with Pavel Litvinov, New York City, January 20, 2017.

60. Chukovsky, *Diary, 1901–1969*, 519–29.

61. GARF, f. 8313, op. 31, d. 89189a, ll. 99–101. Additional letters requesting that the trial be open to the public came from individuals in Riga, Ivanovo, and other Soviet cities.

62. Litvinov, *Protsess chetyrekh*, 48–53, 55, 290. In a March 29, 1999, interview at the Memorial Society, Moscow, Litvinov noted that his transcripts of political trials were reconstructed from accounts by multiple people who had attended such trials. Hoover Archive, Pavel Litvinov Papers, Box 3, Folder 15, Document 2, p. 15.

63. Meerson-Aksenov and Shragin, *The Political, Social, and Religious Thought of Russian "Samizdat,"* 227–28, n. 3.

64. Stephan, *Von der Küche auf den Roten Platz*, 312. The signer was the ethnographer Natalya Sadomskaya, wife of Boris Shragin.

65. Turchin, *Inertsiia strakha*, 76; Glazov, *V kraiu ottsov*, 75 and 81–82.

66. Amal'rik, *Zapiski dissidenta*, 39.

67. Alexeyeva and Goldberg, *The Thaw Generation*, 167–69; Chuprinin, "U cherty."

68. Litvinov, *Protsess chetyrekh*, 50.

69. Alexeyeva and Goldberg, *The Thaw Generation*, 166.

70. Letter from O. Tymchuk, quoted in Litvinov, *Protsess chetyrekh*, 484. The quoted passage comes from Pierce, *An Introduction to Information Theory*, 2. Pierce, it should be noted, expressed skepticism about whether such mastery is actually possible.

71. Volkov et al., *Pressa v obshchestve (1959–2000)*, 545–47.

72. Versions of these anecdotes can be found inter alia in Mel'nichenko, *Sovetskii anekdot*, 625. The earliest written version of the *War and Peace* anecdote I have found is in *SDS*, vol. 5, doc. 331, dated March 21, 1969.

73. Binge reading of unpublished texts was not limited to the dissident milieu. In 1968, members of the Politburo were given a 450-page typed transcript of ex–general secretary Khrushchev's memoirs (first published two years later in the West as *Khrushchev Remembers*), with instructions to return it in forty-eight hours. See Shelest, *Da ne sudimy budete*, 317.

74. Quoted in von Zitzewitz, *The Culture of Samizdat*, 68.

75. Komaromi, *Soviet Samizdat*, 81.

76. Hoover Archive, Pavel Litvinov Papers, Box 3, Folder 15, Document 1, pp. 11, 13.

77. Alexeyeva and Goldberg, *The Thaw Generation*, 130.

78. Shub, *The New Russian Tragedy*, 53; van het Reve, *Verzameld werk*, vol. 2, 797–98 (originally published in *Het Parool* [January 31, 1968]). See also the transcript of the CBS News interview with Petr Yakir, in Cole, "Three Voices of Soviet Dissent," 137. On the lack of KGB interference: Litvinov, *Protsess chetyrekh*, 257.

79. *SDS*, vol. 1, doc. 17. On the genesis of the appeal, see the 2018 interview by Olga Rozenblium with Pavel Litvinov and Alexander Daniel, posted by the Moscow Memorial Society, accessed December 12, 2023, https://www.memo.ru/ru-ru/biblioteka/pavel-litvinov-i-aleksandr-daniel-ob-istorii-otkrytogo-pisma-k-m/.

80. On the system of popular election of Soviet judges, see the entry on "Courts" in Feldbrugge et al., *The Encyclopedia of Soviet Law*, 212.

81. Rolf, *Das sowjetische Massenfest*, 337–46; Yurchak, *Everything Was Forever*, 74–76, 93–102.

82. In Russian, *mirovaia obshchestvennost'*.

83. "V tsentre vnimaniia obshchestvennosti vsego mira," *Pravda* (January 29, 1959), 1. See also Matsui, "*Obshchestvennost'* across Borders," in idem, *Obshchestvennost' and Civic Agency*, 198–218.

84. Haslam, *Russia's Cold War*, ix; Shubin, *Tainy sovetskoi epokhi*, 214.

85. Van het Reve, *Verzameld werk*, vol. 3, 696–703.

86. Bukovskii, *"I vozvrashchaetsia veter...,"* 218.

87. Litvinov, *Protsess chetyrekh*, 266.

88. Alexeyeva and Goldberg, *The Thaw Generation*, 196.

89. Quoted in von Zitzewitz, *The Culture of Samizdat*, 53.

90. Litvinov, *Protsess chetyrekh*, 404 and 395.

91. Ibid., 367–70, 373, 394, and 413.

92. Ulanovskaia and Ulanovskaia, *Istoriia odnoi sem'i*, 271.

93. Berg, *Sukhovei*, 267.

94. The 1967 attack on censorship was subsequently published in Solzhenitsyn, *Bodalsia telenok*, 486–92 (quotation: 486). The attack on Soviet law appeared as "Zakon segodnia," the final chapter in idem, *Arkhipelag Gulag 1918–1956*, vol. 3, 478–97 (quotation: 491).

95. Litvinov, *Protsess chetyrekh*, 266, 289–93, 301–9, 316–17, 333–37.

96. Bellezza, *The Shore of Expectations*, 141–56; Alexeyeva, *Soviet Dissent*, 37–38.

97. Daniel', *"Ia vse sbivaius' na literaturu...,"* 19.

98. On dissidents in the post-Stalin Gulag, see Putz, *Kulturraum Lager*; Nathans, "The Many Shades of Soviet Dissidence; Finkelman, "Ghetto, Gulag, *Geulah*," especially chap. 5.

99. Marchenko, *Zhivi kak vse*, 6–7.

100. Marchenko, *Moi pokazaniia*, 336.

101. Esenin-Vol'pin, *Vesennii list*, 78.

102. Bukovskii, *"I vozvrashchaetsia veter...,"* 64.

103. Recounted by Alexander Gribanov in a conversation with the author in 2007.

104. On the return of Gulag prisoners to Soviet society, see Dobson, *Khrushchev's Cold Summer*; Adler, *The Gulag Survivor*; Cohen, *The Victims Return*, and Applebaum, *Gulag*, 506-51. Alexeyeva cites a remark by Khrushchev, not known for understatement, to the effect that "ten thousand memoirs by former political prisoners" were submitted to Soviet publishing houses. Alexeyeva and Goldberg, *The Thaw Generation*, 99.

105. *Pravda*, no. 28 (January 28, 1959). In fact, political arrests spiked precisely during the two years preceding this statement. In 1957 and 1958, 2,380 Soviet citizens were sentenced for "anti-Soviet" activity under Article 58-10. The total number of political prisoners, including those already in jails, camps, or exile, was considerably higher. See Hornsby, *Protest, Reform and Repression*, 108-34.

106. Bogoraz, *Sny pamiati*, 118.

107. Pliushch, *Na karnivale istorii*, 220. On the composition and publishing history of Marchenko's *My Testimony* and other dissident memoirs, see Nathans, "Talking Fish: On Soviet Dissident Memoirs."

108. *Khronika tekushchikh sobytii*, no. 1 (April 30, 1968); Medvedev, "'Chernye spiski' pisatelei," 340. Blacklisted writers included Bella Akhmadulina, Lev Kopelev, Raisa Orlova, Vasily Aksenov, and Vladimir Voinovich. According to Medvedev, Voinovich was told that two of his plays, blocked from performance, would be restored if he publicly repudiated the protest letter he had signed.

109. The Novosibirsk case is, so far, the most thoroughly documented episode of extrajudicial retaliation against signers of open letters in 1968. See Kuznetsov, *Novosibirskii Akademgorodok v 1968 godu*; Berg, *Sukhovei*.

110. For Gerlin's biographical details and tributes from her former students, see Frierson and Vilensky, *Children of the Gulag*, 78 and 162-65; and Margolina, *Obraz zhizni*, 10-107.

111. *SDS*, vol. 1, doc. 42. Transcripts of the two meetings were created by Gerlin herself, from memory. Both circulated in samizdat and were first published abroad in the NTS journal *Posev*, no. 10 (1968).

112. The school director, reluctant to deliver anything less than a unanimous resolution to his superiors, advised the five teachers who voted against firing Gerlin (and the two who abstained) to "think about yourself." "The collective," he warned, "won't want to work with those who defend her." Under considerable pressure, two later changed their votes. *SDS*, vol. 1, doc. 42.

113. Ibid., vol. 1, doc. 42, and vol. 9, doc. 698.

114. Ibid., vol. 9, doc. 698. The quotation, surely familiar to everyone in the auditorium, is from Pushkin's poem "Ia pamiatnik sebe vozdvig nerukotvornyi." The Third Section of His Imperial Majesty's Own Chancellery was established in 1825 to serve as Tsar Nicholas I's secret police.

115. Ibid., vol. 1, doc. 42. Raissa Berg similarly interpreted her disciplinary hearing in Novosibirsk as symptomatic of a slide back toward the Stalinist order. See Berg, *Sukhovei*, 322.

116. De Boer et al., *Biographical Dictionary of Dissidents*, 157.

117. Bukovskii, *"I vozvrashchaetsia veter..."*, 273; Pavel Litvinov, quoted in Hopkins, *Russia's Underground Press*, 8.

118. Ulitskaia, *Poetka*, 277.

119. LYA, f. K-1, ap. 58, (investigative file on Sergei Kovalev), t. 11, p. 167 (interrogation of Viktor Krasin on January 19, 1973). The thirty volumes of Kovalev's investigative file are a treasure trove of information about his case and those of many other dissidents interrogated by the KGB as part of its investigation of the *Chronicle of Current Events*. They include materials from KGB archives in Moscow, Kyiv, and other cities. I am grateful to Ivan Sergeyevich Kovalev for sharing a digitized copy of his father's entire file. Microfilmed copies can be found at Harvard University's Houghton Library (in the Sakharov Archive) and at the Hoover Institution Library

and Archives. A digitized version has been posted online by the Moscow Memorial Society, accessed December 12, 2023, https://www.memo.ru/ru-ru/biblioteka/delo-kovaleva-podokumentnaya-opis/.

120. Krasin, *Poedinok*, 124.

121. Author's interview with Pavel Litvinov, January 20, 2017. For examples of KGB informant reports on gatherings at Litvinov's apartment, as summarized by KGB chairman Yuri Andropov, see Hoover Archive, Pavel Litvinov Papers, Box 1, Folder 4; Rubinstein and Gribanov, *The KGB File of Andrei Sakharov*, 90.

122. LYA, f. K-1, ap. 58, t. 11, p. 167 (interrogation of Viktor Krasin on January 19, 1973); Hoover Archive, Pavel Litvinov Papers, Box 3, Folder 16, Document 2, p. 41.

123. LYA, f. K-1, ap. 58, t. 14, pp. 89–90 (interrogation of Vera Gromichenko-Modzelevskaia on March 23, 1970).

124. Ibid., t. 13, p. 342 (interrogation of Viktor Krasin on January 18, 1973).

125. Krasin, *Poedinok*, 167; LYA, f. K-1, ap. 58, t. 14, p. 78 (interrogation of Petr Yakir on June 28, 1973).

126. A. Antipov [pseudonym], "Ot brozheniia umov—k umstvennomu dvizheniiu" (January 1969), in *SDS*, vol. 6, doc. 388. "Unbounded sea of talk" was how the Hungarian dissident writer Georgi Konrad described analogous gatherings in Budapest. See Konrad, *Anti-Politics*, 206. On Yakir's apartment, see Ulanovskaia and Ulanovskaia, *Istoriia odnoi sem'i*, 306.

127. Alexeyeva and Goldberg, *The Thaw Generation*, 84.

128. Hoover Archive, Pavel Litvinov Papers, Box 1, Folder 4, p. 161.

129. Kuznetsov, "O 40-kh i 50-kh," 790–91.

130. Ulitskaia, *Poetka*, 14.

131. Gorbanevskaia, "Eti 'slovno' i 'budto' i 'kak'," *Zvezda* no. 10 (2010), 3. My thanks to Kevin Platt for his advice on this translation.

132. See Linor Goralik's 2011 interview with Gorbanevskaya on the OpenSpace.ru platform, accessed December 12, 2023, http://os.colta.ru/literature/events/details/32573/page1/; Gorbanevskaya's September 1968 psychiatric evaluation in LYA, f. K-1, ap. 58, t. 14, pp. 94–96; Reid, "Ot Bartoka do Butyrki: O konflikte grazhdanskogo i liricheskogo v rannei poezii Natal'i Gorbanevskoi."

133. Ulitskaia, *Poetka*, 278. Andrei Amalrik referred to Gorbanevskaya in private as the "godmother" of the *Chronicle of Current Events*. See IISH, Alexander Herzen Foundation, Folder 24, letter from Amalrik to Karel van het Reve (January 23, 1969, mistakenly dated 1968), p. 1.

134. Zhelvakova, *Gertsen*, 448.

135. Bourdeaux, *Religious Ferment in Russia*; Baran, *Dissent on the Margins*. For an example of a trial transcript, see Bourdeaux and Howard-Johnston, *Aida of Leningrad*, 45–104.

136. Ulitskaia, *Poetka*, 279.

137. The original Russian is even subtler: "Kuda nam do istiny, nam by khot' do pravdy." Alexander Pyatigorsky, quoted in Golomshtok, *"Zaniatie dlia starogo gorodovogo,"* 149.

138. Medvedev started his unofficial journal in 1964 in the wake of Khrushchev's ouster, with the goal of fostering "political creativity." Monthly issues appeared continuously until 1971. Among its contributors was the physicist Valery Pavlinchuk, who in early 1968 befriended Pavel Litvinov and Petr Grigorenko and who, according to Medvedev, participated in discussions that led to the launching of the *Chronicle of Current Events* later that year. Only when published abroad was Medvedev's journal given the name *Political Diary*. See Medvedev, "Kak sozdavalsia 'Politicheskii dnevnik.'"

139. This is my translation of the Russian text, which varies slightly from the official English-language version in the Universal Declaration of Human Rights. See Rubenstein, *Soviet Dissidents*, 100.

NOTES TO CHAPTER 7   683

140. Schattenberg, *Leonid Breschnew: Staatsmann und Schauspieler*, 428.

141. Ulitskaia, *Poetka*, 278.

142. Hoover Archive, Pavel Litvinov Papers, Box 3, Folder 15, Document 1, p. 11.

143. Ulitskaia, *Poetka*, 278.

144. See the prefatory material posted by the Moscow Memorial Society, accessed December 12, 2023, http://www.memo.ru/history/diss/chr/about.htm.

145. Already in August 1968—following the first two issues—Pavel Litvinov anticipated that the bulletin would be "continued under a different name." IISH, Alexander Herzen Foundation, Folder 56, letter from Litvinov to Stephen Spender (August 8, 1968), p. 2.

146. Ulitskaia, *Poetka*, 281.

147. Gorbanevskaya, "Writing for 'Samizdat,'" 34. The phrase "No one is forgotten, nothing is forgotten" originated in a poem by Olga Berggolts about the Nazi siege of Leningrad.

148. *Prestuplenie i nakazanie*, no. 2 (1969), reproduced in *SDS*, vol. 7, doc. 490. The April 1969 issue of Medvedev's samizdat journal contained a similar exposé of the Bolshevik leader Yakov Sverdlov's son Andrei Sverdlov, a notoriously sadistic NKVD interrogator who sent hundreds of victims to the Gulag. Medvedev shared with his readers the younger Sverdlov's address (apartment 319 in the famous House of Government) as well as his home and office telephone numbers, noting bitterly that "it is only for *fascist* murderers that the statute of limitations has been lifted." The exposé was reproduced in *Khronika tekushchikh sobytii*, no. 7 (April 30, 1969). Andrei Sverdlov died several months later at the age of fifty-eight. On Sverdlov's sadism, see Slezkine, *The House of Government*, 882–86 and 931–32.

149. Valerii Chalidze, "Ot sostavitelia," *Obshchestvennye problemy*, vol. 1 (1969), reproduced in *SDS*, vol. 16, doc. 479-a.

150. The essential bibliographical resource for the study of samizdat periodicals is Ann Komaromi's website and the accompanying database: Soviet Samizdat Periodicals, University of Toronto Libraries, http://samizdat.library.utoronto.ca/content/about-samizdat. See also Komaromi and Kuzovkin, *Katalog periodiki samizdata, 1956–1986*.

151. As noted in chapter 3, Volpin's Civic Appeal, inviting Soviet citizens to the original transparency meeting on December 5, 1965, had referred to "transparency meetings" in the plural. An early draft of the Civic Appeal had stated that "in cases of extra-judicial repressions against participants in a transparency meeting, new meetings will be organized," including in cities other than Moscow. From the outset, therefore, Volpin had in mind a technique that could be repeated, whether in response to specific reprisals—that is, as part of the chain reaction—or as an annual event. See Memorial-Moscow, f. 120, kor. 3, pap. 4, d. 4, l. 6.

152. See note 114 in this chapter.

Chapter Seven

1. Apart from in the USSR's western borderlands, where sporadic anti-Communist and anti-Russian guerrilla warfare persisted into the 1950s.

2. Kozlov and Mironenko, *Kramola*; Kazakov, "Dissens und Untergrund," in Belge and Deuerlein, *Goldenes Zeitalter der Stagnation?*, 75–93; Budraitskis, *Dissidenty sredi dissidentov*, 30–87.

3. Igrunov, "O Glebe Pavlovskom, i o tom, kak poiavilas' stat'ia 'K problematike obshchestvennogo dvizheniia.'"

4. "Anekdoty pro Vovochku," Informatsionno-istoricheskii portal vash 1922-91 god rozhdeniia, accessed December 12, 2023, http://22-91.ru/sovetskii-folklor/2/anekdoty-pro-vovochku.

5. On doctrine-based reformism, see Tromly, "Intelligentsia Self-Fashioning in the Postwar Soviet Union," 164.

6. Kozlov and Mironenko, *Kramola*, 350–52.

7. Bukovskii, "I vozvrashchaetsia veter...," 130.

8. Scott, *Domination and the Arts of Resistance*, 100.

9. Amherst Center for Russian Culture (henceforth ACRC), Amherst College, Grigorenko Family Papers, Series 2, Box 3, letter from Petr Grigorenko to Leonid Plyushch (March 24, 1976), p. 3.

10. Quoted in Goldberg, *The Final Act*, 125.

11. Alexeyeva and Goldberg, *The Thaw Generation*, 65–76.

12. Ibid., 65–76.

13. Grigorenko, *V podpol'e*, 490–92.

14. On dissident readings of Lenin, see Nathans, "Talking Fish: On Soviet Dissident Memoirs." Quotation from Voinovich, *Avtoportret*, 289. For additional accounts of disillusioning encounters with Lenin's works by future dissidents, see Marchenko, *Moi pokazaniia*, 210–14; and Turchin, *Inertsiia strakha*, 66.

15. Gurevich, *Istoriia istorika*, 70–71.

16. Cherniaev, *Sovmestnyi iskhod*, 80 and 88.

17. Recent works by Kazakov and Budraitskis (see note 2 in this chapter) make the case for a revival of socialist opposition in the USSR in the late Brezhnev era while confirming the eclipse of neo-Leninism in the 1960s. The left-wing groups that sprouted in the late 1970s and early 1980s could hardly be called Leninist or Marxist-Leninist, however; they were oriented toward the New Left in the West, or Eurocommunism, or socialist democracy.

18. Bergman, "Soviet Dissidents on the Russian Intelligentsia."

19. Prominent examples: Shatz, *Soviet Dissent in Historical Perspective*; Beyrau, *Intelligenz und Dissens*; Bergman, *Meeting the Demands of Reason*. For exceptions, see Siniavskii, "Dissidentstvo kak lichnyi opyt"; Daniel', "Wie freie Menschen," in Eichwede, *Samizdat: Alternative Kultur*, 38–50; Vaissié, *Pour votre liberté et pour la nôtre*, 345-49.

20. Shalamov, "Pis'mo staromu drugu," in Ginzburg, *Belaia kniga*, 410.

21. Trigos, *The Decembrist Myth in Russian Culture*, 145–48.

22. Alexeyeva and Goldberg, *The Thaw Generation*, 34–35.

23. Berlin, *Affirming: Letters 1975–1997*, 81–82. Berlin again likened Sakharov to Herzen in a letter to the historian Andrzej Walicki (316). The historian Sydney Monas compared Andrei Amalrik to both Chaadaev and Herzen: see Monas's postscript, "Amalrik's Vision of the End," in Amalrik, *Will the Soviet Union Survive until 1984?* (1970), 86–88.

24. Chalidze, *Prava cheloveka i Sovetskii Soiuz*, 72.

25. Ibid., 70, 76.

26. Walicki, *A History of Russian Thought*, 3, 88, 166, 200.

27. By the 1930s, Steffens too had changed his mind. See Hartshorn, *I Have Seen the Future*, chap. 24, "The Future Revisited."

28. Pipes, "Russia's Exigent Intellectuals," 80–82.

29. Sakharov, *Vospominaniia*, vol. 1, 19 and 854. The volume Sakharov had in mind was M. N. Gernet, O. B. Gol'dovskii, and I. N. Sakharov, eds., *Protiv smertnoi kazni* (Moscow, 1906), which included texts by prominent religious thinkers such as Nikolai Berdyaev, Sergei Bulgakov, Vasily Rozanov, and Vladimir Solovyov—but not Tolstoy. Tolstoy's essay "I Cannot Be Silent" was composed in 1908.

30. Tolstoi, "Ne mogu molchat'," in idem, *Polnoe sobranie sochinenii*, vol. 37, 85.

31. This is my rendering of the original Russian-language text and capitalization; Sakharov, "Nobel Lecture." Tolstoy refers to *um* and Sakharov to *Razum*, but the two words are essentially synonymous.

32. Among Tilly's many influential works on this subject, see especially *The Contentious French* and *Regimes and Repertoires*.

33. Insightful discussions of the study of social movements can be found in Calhoun, "'New Social Movements' of the Early Nineteenth Century"; Buechler, "New Social Movement Theories"; and Pichardo, "New Social Movements." See also Snow et al., *The Wiley-Blackwell Encyclopedia of Social and Political Movements*.

34. To her credit, the keenest scholar of totalitarianism, Hannah Arendt, recognized the extreme fluidity of such regimes, their perpetual renaming, shape-shifting, and repositioning of the institutions of power. The totalitarian model itself, however, remained static; the only source of systemic change it recognized was external, in the form of a cataclysmic military defeat.

35. Notable exceptions: Weiner, *A Little Corner of Freedom*; Osa, *Solidarity and Contention*; Stephan, "Von 'Dissidenten,' 'Ehefrauen,' und 'Sympathisantinnen,'" in Roth, *Soziale Netzwerke und soziales Vertrauen*, 122–33; Edele, *Soviet Veterans of the Second World War*.

36. Grigorenko, *V podpol'e*, 600.

37. Hoover Archive, Pavel Litvinov Papers, Box 3, Folder 15, Document 1, p. 12.

38. LYA, f. K-1, ap. 58, t. 11, pp. 166–71 (interrogation of Viktor Krasin on January 19, 1973).

39. Hoover Archive, Pavel Litvinov Papers, Box 3, Folder 15, Document 1, pp. 3–7. About his close friend Litvinov, Amalrik wrote: "He was a lousy organizer—in general, I don't know whether he was capable of finishing things." Amal'rik, *Zapiski dissidenta*, 56.

40. LYA, f. K-1, ap. 58, t. 11, pp. 166–71 (interrogation of Viktor Krasin on January 18, 1973).

41. Ibid., p. 167.

42. Khrushchev, *Vospominaniia*, vol. 1, 182–83.

43. See Khrushchev's speech in *The Road to Communism: Documents of the 22nd Congress*, 347–48; Robert Conquest's introduction to Yakir, *A Childhood in Prison*, 13; Reddaway, "Pamiati Petra Iakira," in *SSSR: Vnutrennie protivorechiia*, vol. 7, 193.

44. Iakir and Geller, *Komandarm Iakir*; Roitman and Tseitlin, *Polkovodets-kommunist*; Esaulenko and Kozhukhar', *Iona Emmanuilovich Iakir*. For the commemorative stamp, see Wikipedia, s.v. "Iona Yakir," last modified August 9, 2023, https://en.wikipedia.org/wiki/Iona_Yakir#/media/File:1966_CPA_3342.jpg.

45. See "Forty-Three Children of Murdered Bolsheviks Protest Rehabilitation of Stalin," in Saunders, *Samizdat: Voices of the Soviet Opposition*, 248–50. Petr Yakir's name stands at the head of the list of signatories.

46. Bonavia, *Fat Sasha*, 26 and 33. David Bonavia was the Moscow correspondent for *The Times* of London. *Fat Sasha* uses the pseudonym "Pavel" for Petr Yakir; the nameless "Westerner" was almost certainly Bonavia himself. The Russian visitor was Amalrik. Amal'rik, *Zapiski dissidenta*, 114. See also Krasin, *Poedinok*, 131–32.

47. LYA, f. K-1, ap. 58, t. 11, p. 167 (interrogation of Viktor Krasin on January 18, 1973). Krasin paraphrased Litvinov and Bogoraz's position as follows: "The entire moral strength of protests consists precisely in the fact that people protest voluntarily, unconnected to any organizational principles, when they want, and against whatever they are against; the basis of protests is free will, and there's no need to introduce any elements that in any way oblige or coerce people to take actions they don't wish to take."

48. Hoover Archive, Pavel Litvinov Papers, Box 3, Folder 16, Document 2, p. 52. Litvinov indicated that Gabai shared his and Bogoraz's views on avoiding hierarchy and organization.

49. Author's interview with Pavel Litvinov, New York City, January 20, 2017.

50. Hoover Archive, Pavel Litvinov Papers, Box 3, Folder 15, Document 1, p. 3.

51. Alexeyeva and Goldberg, *The Thaw Generation*, 250.

52. Orlova, "'Rodinu ne vybiraiut,'" 286.

53. Chalidze, *Prava cheloveka i Sovetskii Soiuz*, 67–68.

54. On Spender's disillusionment with communism and his decision to offer support to Soviet dissidents, see Matsui, "Forming a Transnational Moral Community."

55. Sutherland, *Stephen Spender*, 458.

56. Litvinov, "Waiting for the New Dictators," 2. See also the Hoover Archive, Pavel Litvinov Papers, Box 3, Folder 15, Document 1, p. 1.

57. IISH, Alexander Herzen Foundation, Correspondence of K. and J. van het Reve, Folder 56, letter from Pavel Litvinov to Stephen Spender (August 8, 1968), p. 1.

58. Ibid., pp. 2–3. On samizdat typists, see von Zitzewitz, *The Culture of Samizdat*, 65–92.

59. Reddaway, *The Dissidents: A Memoir*, 216–19.

60. For an early mission statement, see IISH, Alexander Herzen Foundation, Folder 9; for a notice informing samizdat readers of the Foundation's existence, see *Khronika tekushchikh sobytii*, no. 11 (December 31, 1969). Before leaving his journalism job in Moscow and returning to his life as a professor of Russian literature in the Netherlands, van het Reve was issued a playful certificate, signed by Amalrik, Bogoraz, and Litvinov, declaring him "henceforth and forevermore to be considered an honorable and indispensable participant in all protests, collective and individual letters, appeals, demonstrations, proclamations, mass meetings, and other types of actions occurring in the USSR" and stating that "this document is to be presented to organs of the police, procuracy, and the KGB, as well as to the administration of the University of Leiden." IISH, Alexander Herzen Foundation, Folder 24. The absurdity of such a certificate, needless to say, made it funny. But its humor also lay in the incongruity between its pretentious formalism and the dissidents' aversion to formality and structure of any kind. It is telling that the certificate was issued to a foreigner—as if to emphasize that formality itself was foreign to the dissident milieu.

61. *Index on Censorship*, vol. 4, no. 1 (1975): 8–10. The journal's sponsoring organization was Writers and Scholars International, established by Spender, the Soviet expert Edward Crankshaw, and other prominent British intellectuals.

62. See, for example, Gilcher-Holtey, *Die 68er Bewegung*; Klimke and Scharloth, *1968 in Europe*; Davis et al., *Changing the World, Changing Oneself*; Klimke et al., *Between Prague Spring and French May*.

63. Bukovskii, "I vozvrashchaetsia veter...," 248–49.

64. In a 2014 interview, Vera Lashkova characterized Bukovsky as a "leader by instinct, an authentic leader. He was born that way." See Morev, *Dissidenty*, 101.

65. See the several thousand samizdat authors listed in the SDS index, *Polnyi spisok dokumentov s podrobnym nomernym ukazatelem* (1977); on typists, see von Zitzewitz, *The Culture of Samizdat*, 66.

66. Stephan, *Von der Küche auf den Roten Platz*, 402–8. Stephan notes that arrests of female dissidents increased sharply after 1980; see ibid., 355

67. Babitskii and Makarov, "Kto delal 'Khroniku tekushchikh sobytii,'" 9.

68. These calculations are based on 911 signatures published in Litvinov, *Protsess chetyrekh*, of which 440 belong to men, 253 to women, and 218 do not reveal the gender of the signer.

69. Stephan, *Von der Küche auf den Roten Platz*; Brier, "Gendering Dissent," 18.

70. Boobbyer, *Conscience, Dissent and Reform*, 87.

71. The sociologist Oleg Kharkhordin asserts that "the political dissident movement... was more like a secret society than a network," a claim difficult to reconcile with a movement whose open letters to Soviet authorities included signatories' names, addresses, and occasionally telephone numbers, and which featured weekly "visiting days" at activists' apartments,

where little effort was made to screen guests. See Kharkhordin, *The Collective and the Individual in Russia*, 314. For criticism of Soviet dissidents for refusing to employ underground conspiratorial techniques, see Ushakov, *In the Gunsight of the KGB*, 87–90.

72. Grigorenko, *V podpol'e*, 218–22, 262–68.
73. Ibid., 184.
74. Ibid., 304.
75. Ibid., 502–4.
76. Ibid., 509–10.
77. Ibid., 577.
78. Amal'rik, *Zapiski dissidenta*, 41.
79. Alexeyeva and Goldberg, *The Thaw Generation*, 250.
80. Grigorenko, *V podpol'e*, 575.
81. ACRC, Grigorenko Family Papers, Series 2, Box 3, letter from Petr Grigorenko to Pavel Litvinov (April 23, 1969), p. 1.
82. Grigorenko, *V podpol'e*, 523–24 (emphasis in original).
83. Ibid., 575–76 (emphasis in original).
84. Bukovskii, "I vozvrashchaetsia veter...," 248–49.
85. Grigorenko, *V podpol'e*, 577.
86. ACRC, Grigorenko Family Papers, Series 2, Box 3, letter from Petr Grigorenko to Pavel Litvinov (January 23, 1969), p. 2.
87. Ibid., letter from Petr Grigorenko to Pavel Litvinov (April 23, 1969), p. 1 (emphasis in original).
88. Azbel et al., "Dvadtsat' let spustia," 171.
89. Ibid., 172–73. Strangely, Azbel placed Alexeyeva among the "new people," when in fact she was among those standing vigil outside the courthouse in 1966.
90. ACRC, Grigorenko Family Papers, Series 2, Box 3, letter from Petr Grigorenko to Pavel Litvinov (April 23, 1969), p. 2.
91. Pliushch, *Na karnivale istorii*, 7.
92. Amal'rik, "Dvizhenie za prava cheloveka v SSSR," 3.
93. Grigorenko, *V podpol'e*, 400.
94. Ibid., 519–20.
95. Chalidze often encountered this view in the USSR, whether during interrogations by the KGB or with friends of the dissident movement. "In the West, too," he noted with irritation after emigrating in 1972, "people who apparently understand my position better than I do tell me that my strictly legal approach is the result of tactical considerations." Chalidze, *Prava cheloveka i Sovetskii Soiuz*, 71.
96. Litvinov, "O dvizhenii za prava cheloveka v SSSR," in Litvinov et al., *Samosoznanie*, 79 (emphasis in original).
97. Quotations from, respectively, Vadim Deloné, Liudmila Kats, and Evgenii Kushev, in Litvinov, *Pravosudie ili rasprava?*, 74, 87, 125; and Bukovskii, "I vozvrashchaetsia veter...," 249.
98. Alexeyeva and Goldberg, *The Thaw Generation*, 169.
99. Chebrikov et al., *Istoriia sovetskikh organov gosudarstvennoi bezopasnosti*, 543. Some leaflets arrived in the USSR via air balloons sent from abroad.
100. Litvinov associated leaflets with "Bolshevik" techniques. Hoover Archive, Pavel Litvinov Papers, Box 3, Folder 15, Document 1, p. 12.
101. Kozlov and Mironenko, *Kramola*, 252, 259–60, 267. See Olga Edelman's discussion of anonymous leaflets and letters, ibid., 224–37.
102. Ibid., 227, table 4. See also Bobkov, *Kak gotovili predatelei*, 174–75.

103. Volpin, unpublished memoir, 27.

104. Litvinov, *Pravosudie ili rasprava?*, 96. Evidently Tatyana Litvinova considered silent demonstrations with no demands, such as those held annually on December 5 on Pushkin Square, to belong to a different category; the KGB recorded her presence at such a gathering in 1975. RGANI, f. 89, op. 37, d. 14, l. 1.

105. Litvinov, *Pravosudie ili rasprava?*, 120.

106. Alexeyeva and Goldberg, *The Thaw Generation*, 167.

107. Sakharov, *Memoirs*, 476. For an eye-witness account of Sakharov's harassment, see Podrabinek, *Dissidenty*, 15.

108. Daucé, "Lettres de protestation dans l'U.R.S.S. post-stalinienne"; Rozenblium, "'Diskussii ne bylo . . .': Otkrytye pis'ma kontsa 1960-kh godov," in Atnashev et al., *Nesovershennaia publichnaia sfera*, 416–46.

109. Polikovskaia, *My predchuvstvie*, 153.

110. Karel van het Reve, "Russen horen van vreemde zenders wat niet in hun 'Pravda' staat," *Het Parool* (March 30, 1968), reprinted in idem, *Verzameld Werk*, vol. 2, 815–17.

111. Orlova, "'Rodinu ne vybiraiut,'" 277. Orlova mistakenly wrote Ginzburg's patronymic as Semyonovna; it was Solomonovna.

112. Berg, *Sukhovei*, 263.

113. Sakharov, *Memoirs*, 271. "Private interventions," Sakharov went on to say, "are sometimes useful as a supplement—not a replacement—for public actions."

114. Kowaljow, *Der Flug des weißen Raben*, 76.

115. See, for example, the interview with Fedor Burlatsky in Cohen and vanden Heuvel, *Voices of Glasnost*, 179.

116. *SDS*, vol. 5, doc. 360.

117. Rubenstein and Gribanov, *The KGB File of Andrei Sakharov*, 98–99.

118. See Dzhemilev's account of the meeting in Allworth, *The Tatars of Crimea*, 164–65.

119. ACRC, Grigorenko Family Papers, Series 2, Box 3, letter from Petr Grigorenko to Reshat Dzhemilev (undated but likely spring of 1968), pp. 1–3.

120. On the Stockholm Appeal, also known as the World Peace Appeal, see Wittner, *The Struggle against the Bomb*, vol. 1, 182–90.

121. *Pravda* (March 20, 1950).

122. Krasin, *Poedinok*, 173–75.

123. Zisserman-Brodsky, *Constructing Ethnopolitics in the Soviet Union*, 132. Other examples include a 1963 letter with 2,500 signatures, addressed to Khrushchev from Armenian residents of Nagorno-Karabakh, demanding that their enclave be detached from the Azerbaijan Soviet Socialist Republic (119), and a 1973 appeal by Lithuanian Catholics demanding the right to religious education, which gained over 16,000 signatures (129).

124. ACRC, Grigorenko Family Papers, Series 2, Box 3, letter from Petr Grigorenko to Reshat Dzhemilev (undated but likely spring of 1968), p. 2.

125. Alexeyeva and Goldberg, *The Thaw Generation*, 108.

126. Iulii Kim, "Moskovskii sud v ianvare 1968 goda," Mezhdunarodnyi portal avtorskoi pesni, accessed December 12, 2023, http://www.bards.ru/archives/part.php?id=6178.

127. Gabowitsch, "Gewaltfreier Widerstand," 61.

128. On this claim, see Lovell, *The Russian Reading Revolution*, 21–22.

129. On the extra-legal status of strikes, see Berman, *Justice in the USSR*, 358.

130. Feldbrugge, *Samizdat and Political Dissent in the Soviet Union*, 300.

131. Liebling, "The Wayward Press," 109.

132. Chalidze, *Prava cheloveka i Sovetskii Soiuz*, 68.

133. See the classic analysis in Clark, *The Soviet Novel: History as Ritual*. For elaboration and critique, see Krylova, "Beyond the Spontaneity-Consciousness Paradigm"; and Zelnik, "A Paradigm Lost? Response to Anna Krylova."

134. ACRC, Grigorenko Family Papers, Series 2, Box 3, letter from Petr Grigorenko to Pavel Litvinov (April 23, 1969), p. 2. Grigorenko described stationary photons as being "without rest mass" (*bez massy pokoia*); according to the theory of relativity, photons cease to have mass when still.

Chapter Eight

1. Bukovskii, "*I vozvrashchaetsia veter* . . . ," 289.
2. Navrátil, *The Prague Spring 1968*, xxviii–xxix.
3. *Khronika tekushchikh sobytii*, no. 18 (March 5, 1971); Hoover Archive, Pavel Litvinov Papers, Box 3, Folder 16, Document 2, p. 52.
4. Quoted in Kramer, "The Czechoslovak Crisis and the Brezhnev Doctrine," in Fink et al., *1968: The World Transformed*, 143.
5. Those themes would gain prominence only later, after the crushing of the Prague Spring and the emergence of a strikingly Soviet-style dissident movement in Czechoslovakia in the 1970s. See Bolton, *Worlds of Dissent*.
6. Hoover Archive, Pavel Litvinov Papers, Box 3, Folder 16, Document 2, p. 58.
7. Alexeyeva and Goldberg, *The Thaw Generation*, 210.
8. Grigorenko, *V podpol'e*, 628.
9. Hoover Archive, Pavel Litvinov Papers, Box 3, Folder 16, Document 2, p. 57.
10. *SDS*, vol. 1, doc. 36.
11. Ibid.
12. Alexeyeva and Goldberg, *The Thaw Generation*, 211.
13. *SDS*, vol. 1, doc. 78.
14. Alexeyeva and Goldberg, *The Thaw Generation*, 212–13.
15. Bogoraz, *Sny pamiati*, 127; Litvinov remembered a call coming at 6 A.M. and not from Gorbanevskaya but from Petr Yakir. Hoover Archive, Pavel Litvinov Papers, Box 3, Folder 16, p. 59.
16. IISH, Alexander Herzen Foundation, Folder 24, letter from Andrei Amalrik to Karel van het Reve (September 26, 1968), p. 1.
17. Hoover Archive, Pavel Litvinov Papers, Box 3, Folder 16, p. 59.
18. Tvardovskii, *Novomirskii dnevnik*, vol. 2, 219.
19. Evtushenko, *Sobranie sochinenii*, vol. 5, 365–66.
20. Orlov, *Opasnye mysli*, 133.
21. Bogoraz, *Sny pamiati*, 197.
22. De Boer et al., *Biographical Dictionary of Dissidents*, 347–48; Amal'rik, *Zapiski dissidenta*, 72.
23. A second layer of the joke, involving accents on the wrong syllables of the three nouns, mocked the unrefined speech characteristic of Brezhnev and other Soviet leaders. See Komaromi, "Samizdat as Extra-Gutenberg Phenomenon," 663.
24. Bogoraz, *Sny pamiati*, 129.
25. From the front pages of Soviet newspapers reproduced in Gorbanevskaia, *Polden'*, 24, 27, 31.
26. Quoted in Firsov, *Raznomyslie v SSSR*, 357.
27. Alexeyeva and Goldberg, *The Thaw Generation*, 217.

28. ACRC, Grigorenko Family Papers, Box 3, Folder 2, letter from Petr Grigorenko to Pavel Litvinov (February 11, 1969).

29. Within twenty-four hours of the invasion, the Moscow City Committee of the Communist Party claimed, over nine thousand meetings had been held to explain events in Czechoslovakia to roughly 885,000 workers and technicians in the Soviet capital. The recorded responses overwhelmingly expressed "full support for the foreign and domestic policies of the Soviet government" and "firm conviction" in the necessity of strengthening the alliance of socialist countries worldwide. Nonetheless, a dozen instances of public criticism were also noted. See Tomilina et al., *Chekhoslovatskii krizis*, 866–70. A more limited survey of the public mood five days later, on August 27, reported similar results (ibid., 873–76). The historian Vladislav Zubok correctly notes that, in comparison with the Soviet crushing of the uprising in Budapest in 1956, the invasion of Czechoslovakia in 1968 produced relatively little unrest inside the USSR, but mistakenly claims that there was only a single protest against the invasion. Zubok, "Die sowjetrussische Gesellschaft in den sechziger Jahren," in Karner et al., *Prager Frühling*, vol. 1, 844. In fact, there were several dozen small-scale protests, in Moscow and other cities. See Weiner, "Déjà Vu All Over Again," 181–83. See also the interviews and other texts collected in Pazderka, *The Soviet Invasion of Czechoslovakia*. Alexander Daniel perceptively observes that the invasion of Czechoslovakia "produced a tremendous psychological fissure in the souls of several generations of the Soviet intelligentsia. Many years later a group of young people conducted a kind of sociological survey on the subject 'What does August 21 mean to you?' They received a wide variety of responses, but there was nonetheless one feature common to all: all of the respondents were able to remember precisely where and how they spent every minute and hour of that day. This rare phenomenon of collective individual memory occurs only at the turning point of an era. In Russia people remember only three other 20th-century dates in this way: 22 June 1941 (the beginning of the war); 9 May 1945 (Victory Day) and 5 March 1953 (Stalin's death). It is interesting that the public did not perceive the much bloodier suppression of the revolution in Budapest in November 1956 with that level of tragic intensity. That event was not perceived as the end of an era. This fact testifies to the remarkable evolution undergone by the civic mentality between 1956 and 1968." Daniel, "1968 in Moscow: A Beginning," in Farik, *1968 Revisited*, 32.

30. Among those patriotic citizens was the thirty-seven-year-old Mikhail Gorbachev, whose father was wounded in 1944 during the liberation of Slovakia. Gorbachev, *Zhizn' i reformy*, vol. 1, 49–50. Not all Soviet citizens followed this logic, however. Andrei Amalrik overheard the following exchange between two working-class prisoners in a corrective labor camp, apropos the Soviet invasion of Czechoslovakia: "One of them said, 'We saved them from the Germans, and now they tried to reject us!' The other retorted, 'So, if you rescued a girl who was drowning, would that give you the right to fuck her for the rest of her life?'" Amal'rik, *Zapiski dissidenta*, 72. For a similar exchange on the streets of Yerevan, see Orlov, *Opasnye mysli*, 134.

31. Gorbanevskaia, *Polden'*, 7–8, 316; Boobbyer, *Conscience, Dissent and Reform*, 100.

32. Gorbanevskaia, "'Mozhesh' vyiti na ploshchad'"; idem, *Polden'*, 7. Philip Boobbyer, in his pioneering book *Conscience, Dissent and Reform*, writes (83) that in her 1983 essay in *Russkaia mysl'*, Gorbanevskaya "argued that it was a deliberately unselfish act in the sense that the demonstrators were sacrificing their freedom in defense of others." In fact, Gorbanevskaya argued that the demonstration was "perceived" this way—a perception she described as "a misunderstanding, or more precisely, a half-truth."

33. IISH, Alexander Herzen Foundation, letter from Andrei Amalrik to Karel van het Reve (September 26, 1968). Amalrik subsequently conceded that some Czechs resisted the Soviet occupation: Amal'rik, *Zapiski dissidenta*, 86.

34. Hoover Archive, Pavel Litvinov Papers, Box 3, Folder 15, p. 18; Gorbanevskaia, *Polden'*, 315.

35. Gorbanevskaia, "'Mozhesh' vyiti na ploshchad'."
36. Hoover Archive, Pavel Litvinov Papers, Box 3, Folder 15, p. 17.
37. Ibid., Folder 15, p. 18.
38. Galich, *Sochineniia*, vol. 1, 59. Galich deliberately uses the verb "to go out" (*vyiti*) to convey the sense of leaving the private space of one's apartment for the public space of the square.
39. Alexeyeva and Goldberg, *The Thaw Generation*, 218–19.
40. For a useful overview of the participants in the demonstration and their backgrounds, see Holler, "'Für eure Freiheit und unsere!': Die Demonstration sowjetischer Dissidenten auf dem Roten Platz," in Karner, *Prager Frühling*, vol. 1, 849–68.
41. The term "sit-down demonstration" (*sidiashchaia demonstratsiia*) appears in several accounts: *Khronika tekushchikh sobytii*, no. 3 (August 31, 1968); SDS, vol. 1, doc. 67; Hoover Archive, Pavel Litvinov Papers, Box 3, Folder 16, doc. 3, p. 61.
42. See the KGB report reproduced in Karner, *Prager Frühling*, vol. 2, 1184.
43. IISH, Alexander Herzen Foundation, Folder 24, letter from Andrei Amalrik to Karel van het Reve (September 26, 1968), p. 1. On Wolzak, see van Voren, *On Dissidents and Madness*, 42–43.
44. Hoover Archive, Pavel Litvinov Papers, Box 3, Folder 15, doc. 1, p. 19.
45. Gorbanevskaia, *Polden'*, 39. Prominent among those Russian democrats was Alexander Herzen, an outspoken supporter of the movement for Polish independence from imperial Russia.
46. Ibid., 158.
47. Ibid., 42–43.
48. Ibid., 37–38.
49. Ibid., 63.
50. Thomas Dodd, preface to Yakobson and Allen, *Aspects of Intellectual Ferment and Dissent in the Soviet Union*, 4.
51. Carswell, *The Exile: A Life of Ivy Litvinov*, 182.
52. Mitrokhin Archive, The Chekist Anthology, Folder 51 ("O pravakh cheloveka"), p. 3.
53. Orlova, *Vospominaniia o neproshedshem vremeni*, 365.
54. Amal'rik, *Zapiski dissidenta*, 75.
55. A handful of other protests were reported in the *Chronicle of Current Events*, no. 3 (August 30, 1968), including anonymous leaflets distributed across Moscow with the heading "Let's Think for Ourselves." For the fullest survey to date of protests in the Soviet Union against the invasion of Czechoslovakia, see Kuzovkin, Makarov, et al., "Spisok grazhdan, vyrazivshikh protest ili nesoglasie s vtorzheniem v Chekhoslovakiiu," posted on Polit.ru on September 2, 2008, under the title "Liudi avgusta 1968 . . . ," https://polit.ru/article/2008/09/02/people68/.
56. Amal'rik, *Zapiski dissidenta*, 75; IISH, Alexander Herzen Foundation, Folder 24, letter from Andrei Amalrik to Karel van het Reve (September 26, 1968), p. 2.
57. Alexeyeva and Goldberg, *The Thaw Generation*, 122; Gorbanevskaia, *Polden'*, 315.
58. ACRC, Grigorenko Family Papers, Box 3, Folder 2, letter from Petr Grigorenko to Pavel Litvinov (February 11, 1969); Chalidze, *Prava cheloveka i Sovetskii Soiuz*, 67.
59. Gorbanevskaia, *Polden'*, 320; Krasin, *Poedinok*, 142. Other dissidents shared Krasin's view that Litvinov's significance for the burgeoning world of samizdat should have trumped his desire to take part in the demonstration; see Pliushch, *Na karnivale istorii*, 247. Krasin denied telling Gorbanevskaya that he would have canceled the demonstration; of the two, however, she is the more credible source. According to Roy Medvedev, Grigorenko's immediate reaction to the demonstration was similar to Krasin's. Grigorenko claimed that, had he been in town,

"the demonstration would not have happened." Roi Medvedev, "12 oktiabria. O novom sudebnom protsesse," *Politicheskii dnevnik* (October 1968), Memorial-Moscow, f. 128, kor. 1–2, l. 43. My thanks to Barbara Martin for sharing Medvedev's text with me.

60. Chukovskii, *Dnevnik 1901–1969*, vol. 2, 545 (entry for October 13, 1968).

61. Glazov, *V kraiu ottsov*, 75.

62. Jacobs, *Cold War Mandarin*, 149. The global dissemination of photographs of Thích Quang Duc's ritual suicide stands in stark contrast to the absence of visual images of the 1968 protest in Red Square. Over the next several years, Quang Duc's self-immolation was copied by other monks in South Vietnam as well as by several anti-war protesters in the United States. On "self-imprisonment," see Anatolii Iakobson, "Pis'mo v zashchitu demonstrantov na Krasnoi ploshchadi, 18 sept. 1968 g.," in *SDS*, vol. 2, doc. 146. On the history of self-immolation as a technique of political protest, see the website devoted to the memory of Jan Palach, accessed on December 12, 2023, http://www.janpalach.cz/en/default/zive-pochodne. In 1969, there were at least two acts of self-immolation in protest against the invasion of Czechoslovakia: by Ryszard Siwiec in Warsaw and Jan Palach in Prague.

63. Pliushch, *Na karnivale istorii*, 463.

64. *SDS*, vol. 2, doc. 146 (emphasis in original). It may have seemed to Yakobson and others during the chain reaction that every "act of arbitrariness and violence by the government" provoked "public protest and rebuke," but evidence from the archives of the Soviet Procuracy shows that many political trials received no public notice at all.

65. Gorbanevskaia, *Polden'*, 175.

66. Ibid., 100.

67. Ibid., 113 and 169.

68. Ibid., 211–12.

69. Ibid., 205. On her refusal of a defense attorney, see Bogoraz's *Sny pamiati*, 130.

70. Chalidze, *Otvetstvennost' pokoleniia*, 10 (interview with Tatyana Litvinova).

71. My summary of Bogoraz's childhood draws on her posthumously published memoir, *Sny pamiati*, 6–75; quotations on 71 and 75. On her youthful oscillations among Jewish, Ukrainian, and Russian identities, see Bogoraz, "Kto ia?"

72. Bogoraz, *Sny pamiati*, 117.

73. As late as 1965, when her husband, Yuli Daniel, was arrested, Bogoraz was unaware that there were political prisoners in the USSR. Ibid., 118.

74. Library of Congress, Hannah Arendt Papers, Speeches and Writings File: 1923–1975, "Moral Responsibility under Totalitarian Dictatorships" (undated typescript composed sometime after 1953), accessed on December 14, 2023, https://www.loc.gov/item/mss11056012666.

75. Ginzburg, *Chelovek za pis'mennym stolom*, 196.

Chapter Nine

1. Bogoraz, *Sny pamiati*, 138.

2. Vysotskii, "Nam ni k chemu siuzhety i intrigi."

3. Kim, "Monolog p'ianogo genseka."

4. Bogoraz, *Sny pamiati*, 147 and 164. It was not unusual for prison and camp guards to keep inmates informed about their celebrity—telling them, for example, when their names were mentioned in broadcasts by Western radio stations. See Goldberg, *The Final Act*, 100. Vail' and Genis, in *60-e: Mir sovetskogo cheloveka* (183), describe prison guards sneaking glimpses of Petr Yakir, Pavel Litvinov, and Petr Grigorenko.

5. Pliushch, *Na karnivale istorii*, 225.

6. See, for example, "Vnesudebnye politicheskie repressii 1968 goda," *Khronika tekushchikh sobytii*, no. 2 (June 30, 1968), which lists ninety-one cases, including names and punishments, of people who had signed various open letters and petitions during the previous six months.

7. Igrunov, "K problematike obshchestvennogo dvizheniia."

8. On Brezhnev's attitude toward Sakharov, see Schattenberg, *Leonid Breschnew*, 433–40.

9. Sakharov, *Memoirs*, 275.

10. On Sakharov's role in the development of Soviet nuclear weapons, see Holloway, *Stalin and the Bomb*, 312–19.

11. Sakharov, *Memoirs*, 97.

12. Schattenberg, *Leonid Breschnew*, 434; Khrushchev, *Vospominaniia*, vol. 4, 212.

13. Lourie, *Sakharov*, 113–15, 146–48, 243; Solzhenitsyn, *Bodalsia telenok*, 397.

14. Several members of Sakharov's extended family, however, were arrested in the 1930s. See Bergman, *Meeting the Demands of Reason*, 14–15.

15. A KGB personality profile prepared in 1971 stated that "having made a great contribution to the creation of thermonuclear weapons, Sakharov felt his 'guilt' before mankind, and, because of that, he has set himself the task of fighting for peace and preventing thermonuclear war." Rubenstein and Gribanov, *The KGB File of Andrei Sakharov*, 115–16.

16. For these and other approaches to Sakharov, see Bergman, *Meeting the Demands of Reason*; Gorelik, *Andrei Sakharov: Nauka i svoboda*; Lourie, *Sakharov*.

17. Sakharov, *Memoirs*, 96–97.

18. Ibid., 150.

19. Sakharov, *Vospominaniia*, vol. 1, 651.

20. Sakharov, *Trevoga i nadezhda*, 38ff.; Gribanov, "Ob izuchenii istochnikov pervoi 'gumanitarnoi' raboty Sakharova," in *30 let "Razmyshlenii..." Andreia Sakharova*, 14–22.

21. Memorial-Moscow, f. 102, op. 1, d. 5, Turchin, "Inertsiia strakha" (samizdat, 1969), 15–19. I thank Gennadii Kuzovkin for making a copy of Turchin's essay available to me.

22. Amal'rik, *Prosushchestvuet li Sovetskii Soiuz do 1984 goda?*, 12–14; Tromly, *Making the Soviet Intelligentsia*; Shlapentokh, *Soviet Intellectuals and Political Power*.

23. Sakharov, *Vospominaniia*, vol. 1, 429–30.

24. Sakharov, *Memoirs*, 218 and 225.

25. Ibid., 278.

26. Sakharov, *Vospominaniia*, vol. 1, 477–78. Khrushchev's account of his interactions with Sakharov, composed after his fall from power, do not include this episode. See Khrushchev, *Vospominaniia*, vol. 4, 212–13.

27. Sakharov, *Memoirs*, 217.

28. Ibid., 233–34.

29. Ibid., 273.

30. Ibid., 273.

31. Ibid., 276–77.

32. Ibid., 299.

33. Sakharov, "Simmetriia vselennoi."

34. Ibid., 74.

35. Sakharov, *Memoirs*, 275 and 96.

36. Ibid., 276, 281–82.

37. Sakharov, *Vospominaniia*, vol. 1, 620.

38. Rubenstein and Gribanov, *The KGB File of Andrei Sakharov*, 87.

39. This was a common impression of Sakharov. Fedor Burlatsky, an adviser to Khrushchev and later to Gorbachev, wrote that Sakharov's modest demeanor and halting speech "deceived me at first. I did not think it would be difficult to influence him, to persuade him to change the

wording of his ideas a bit or even to express them in a different way. I soon realized that this was a hopeless endeavor." Burlatsky, *Khrushchev and the First Russian Spring*, 258.

40. The term Sakharov used for "science"—*nauka*—refers to scholarly expertise across disciplines, not only in the natural sciences.

41. Gribanov, "Ob izuchenii istochnikov pervoi 'gumanitarnoi' raboty Sakharova."

42. The appeal is reproduced in Solzhenitsyn, *Bodalsia telenok*, 486–92.

43. Solzhenitsyn, *The First Circle*, 415. Percy Bysshe Shelley made a similar claim, in his "The Defense of Poetry" (1821): "Poets are the unacknowledged legislators of the world." The difference, of course, is that at the respective times, the world did not have an actual legislature, whereas the Soviet Union had an actual government.

44. Sakharov, "Razmyshleniia o progresse," in idem, *Trevoga i nadezhda*, 12.

45. Ibid., 46.

46. Ibid., 27 (quoting Chekhov), 33, 36, 47; Gessen, "Fifty Years Later, Andrei Sakharov's Seminal Essay Is a Powerful Model of Writing for Social Change."

47. Sakharov, *Memoirs*, 285.

48. *New York Times* (July 22, 1968), 1.

49. For detailed accounts of the transmission of *Reflections* to the West, see van het Reve, "Hoe Sacharovs tekst in Holland kwam," *Het Parool* (October 21, 1975), reprinted in idem, *Verzameld werk*, vol. 4, 651–55; Sakharov, *Memoirs*, 286–88; Anderson, "Vystuplenie," in *30 let "Razmyshlenii . . ." Andreia Sakharova*, 93–96.

50. See the National Security Council memorandum dated December 9, 1969, reproduced in Mahan, *Foreign Relations of the United States*, vol. 12, *Soviet Union, January 1969–October 1970*, doc. 103, "For the 303 Committee," p. 312. A *New York Times* correspondent in Moscow was told that "tens of thousands" of Soviet scientists read Sakharov's *Reflections*. See Smith, *The Russians*, 451.

51. Bergman, *Meeting the Demands of Reason*, 135; Taubman, *Gorbachev*, 122–23; Sakharov, *Memoirs*, 286–87.

52. Rubenstein and Gribanov, *The KGB File of Andrei Sakharov*, 94.

53. Bergman, *Meeting the Demands of Reason*, 150.

54. Rubenstein and Gribanov, *The KGB File of Andrei Sakharov*, 89.

55. Sakharov, *Memoirs*, 287.

56. On Slavsky's biography, see the Russian-language Wikipedia page "Славский, Ефим Павлович," last modified November 7, 2023, https://ru.wikipedia.org/wiki/Славский,_Ефим_Павлович; and Sakharov, *Memoirs*, 367.

57. Sakharov, *Memoirs*, 287.

58. Amal'rik, *Zapiski dissidenta*, 69.

59. Solzhenitsyn, interview with the Associated Press news agency and *Le Monde* (Moscow, August 23, 1973), reprinted in idem, *Bodalsia telenok*, 596. Solzhenitsyn gave this interview at the height of a vicious campaign against Sakharov in the Soviet press. After his exile to the West, he published a memoir in which his assessment of Sakharov struck a rather different tone. Like Amalrik, he referred to Sakharov as "a miracle"—but now sarcastically, in order to dismiss those in the dissident movement who sought to elevate themselves by association with him, "hitching [themselves] to this strange, enormous, conspicuous hot-air balloon, which was soaring to the heights without engine or fuel." Ibid., 399.

60. Ludmilla Alexeyeva represents a rare exception to these two dominant stances vis-à-vis veterans of the Gulag. Of her second husband and other survivors (including Petr Yakir), she wrote, "People who ended up in the camps early in life grew up only physically; their personality was preserved as it was before the arrest. Even later, they matured slowly or not at all. Many remain infantile into their old age." Alexeyeva and Goldberg, *The Thaw Generation*, 88.

61. Andrei Tverdokhlebov, "V zashchitu pis'ma Sakharova 'Razmyshleniia o progresse . . . ,'" *Obshchestvennye problemy*, vol. 1 (July 1969), reproduced in *SDS*, vol. 16, doc. 479-a.
62. Pliushch, *Na karnivale istorii*, 246–47.
63. Solzhenitsyn, "Nobel Lecture." This passage was directed explicitly against the idea of "the levelling of nations, of the disappearance of different races in the melting-pot of contemporary civilization."
64. Sakharov, "Razmyshleniia o progresse," 38.

Chapter Ten

1. Alexeyeva and Goldberg, *The Thaw Generation*, 166; for the joke, see Susan Jacoby's introduction to Amalrik, *Notes of a Revolutionary*, xv.
2. Amal'rik, "O sebe kak pisatele p'es," in idem, *P'esy*, 5.
3. Amal'rik, *Nezhelannoe puteshestvie*, 205–6.
4. Houghton Library, Sakharov Archives, Amalrik Papers, Box 1, Folder 10.
5. Ibid., Box 3, Folder 27; Amal'rik, "O sebe kak pisatele p'es," in idem, *P'esy*, 5.
6. Ibid., Box 3, Folder 27.
7. Ibid., Box 6, Folder 91, pp. 2–5.
8. Ibid., Box 6, Folder 91, pp. 6–7, 19.
9. Riasanovsky, "The Norman Theory of the Origin of the Russian State," 96; Iakovlev, "O prepodavanii otechestvennoi istorii," 27.
10. Houghton Library, Sakharov Archives, Amalrik Papers, Box 3, Folder 28, letter from Andrei Amalrik to Leonid Ilichev (December 19, 1961); Folder 30, letter from Andrei Amalrik to professor Adolf Stender-Petersen (May 5, 1961).
11. Amal'rik, *Nezhelannoe puteshestvie*, 32.
12. Houghton Library, Sakharov Archives, Amalrik Papers, Box 6, Folder 91, pp. 28–30; Box 3, Folder 30, letter from Andrei Amalrik to Adolf Stender-Petersen (March 24, 1961), letter from Stender-Petersen to Amalrik (April 17, 1961).
13. Half a century after its composition, Amalrik's work was published: Amal'rik, *Normanny i Kievskaia Rus'* (Moscow, 2018).
14. Amal'rik, *Nezhelannoe puteshestvie*, 62.
15. Ibid., 206–7.
16. Houghton Library, Sakharov Archives, Amalrik Papers, Box 6, Folder 91, letter from Andrei Amalrik to the Central Committee (July 17, 1961), p. 30.
17. Houghton Library, Sakharov Archives, Amalrik Papers, Box 13, Folder 167, letters from Richard Wortman to Andrei Amalrik (October 31, 1962, and March 17, 1963). I thank Richard Wortman for sharing with me Amalrik's responses (June 7 and November 20, 1963). See also Pipes, *Vixi: Memoirs of a Non-belonger*, 120–21. At Moscow State University, Amalrik also became friends with Gilbert Osay, an exchange student from Ghana, whose souring on the USSR and early departure university officials blamed in part on Amalrik's corrosive influence (Osay blamed it on the racial prejudice he encountered). According to Amalrik, Osay's treatment by the history department mirrored his own: "Those who turn friends into near-enemies are that much less capable of turning enemies into friends." Houghton Library, Sakharov Archives, Amalrik Papers, Box 6, Folder 91, pp. 9–11.
18. Krasin, *Poedinok*, 125.
19. Amal'rik, *Zapiski dissidenta*, 11, quoting Alexander Ginzburg; Iar. Iasnyi [pseudonym of Boris Shragin], "Andrei Amal'rik kak publitsist," in *SDS*, vol. 6, doc. 408, p. 4.
20. Amal'rik, *Nezhelannoe puteshestvie*, 11; idem, *Zapiski dissidenta*, 18–19.

21. Houghton Library, Sakharov Archives, Amalrik Papers, Box 14, Folder 179; Amal'rik, *Nezhelannoe puteshestvie*, 33.

22. Amal'rik, *Nezhelannoe puteshestvie*, 14–15.

23. Houghton Library, Sakharov Archives, Amalrik Papers, Box 6, Folder 91, p. 14.

24. Blending the absurdist style of Eugène Ionesco and Samuel Beckett with the humor of Soviet anecdotes, the plays were eventually published by the Alexander Herzen Foundation as *P'esy* (Amsterdam, 1970) and then translated into English: Amalrik, *Nose! Nose? No-se! and Other Plays* (New York, 1973).

25. Houghton Library, Sakharov Archives, Amalrik Papers, Box 14, Folder 179; de Boer et al., *Biographical Dictionary of Dissidents*, 17; Amal'rik, *Nezhelannoe puteshestvie*, 206–7. The Supreme Court's annulment of Amalrik's conviction cited the weakness of the evidence presented against him as well as his heart condition, which rendered him "unable to perform labor that involves the lifting of heavy objects."

26. Aksel'rod-Rubina, *Zhizn' kak zhizn'*, vol. 2, 118.

27. Alexeyeva and Goldberg, *The Thaw Generation*, 166; Aksel'rod-Rubina, *Zhizn' kak zhizn'*, vol. 2, 118; Amal'rik, *Zapiski dissidenta*, 38.

28. Amalrik, preface to *Will the Soviet Union Survive until 1984?* (1981 [2nd ed.]), 12; Amal'rik, *Zapiski dissidenta*, 97.

29. Shub, "'Will the USSR Survive . . . ?' A Personal Comment," 88.

30. For this way of deploying Amalrik's book, see, for example, Malia, *The Soviet Tragedy*, 351, 384, 405, 493.

31. Amalrik had originally put the year 1980 in the book's title, perhaps to mock Nikita Khrushchev's incautious claim that the USSR would complete its transition from socialism to communism by that year. His friend Vitaly Rubin persuaded him instead to invoke the Orwellian year 1984. Amal'rik, *Prosushchestvuet li Sovetskii Soiuz do 1984 goda?*, 2.

32. See the epigraph to chapter 7.

33. Yurchak, *Everything Was Forever*, 1.

34. Amal'rik, *Prosushchestvuet li Sovetskii Soiuz do 1984 goda?*, 7–11.

35. Ibid., 13.

36. Ibid., 15–16.

37. Ibid., 18–19.

38. Ibid., 22–23, 42.

39. Ibid., 27.

40. Ibid., 25–26, 54.

41. Ibid., 30–31, 39–40.

42. Ibid., 32.

43. Ibid., 33–34.

44. Ibid., 56. In this and certain other respects, *Will the Soviet Union Survive until 1984?* reads like an updated version of Petr Chaadaev's *Philosophical Letters*, which similarly circulated in manuscript in the 1830s under Tsar Nicholas I, who had Chaadaev declared insane and placed under house arrest. In the preface to the second English edition of his book, Amalrik confessed that the phrase "a country without beliefs, without tradition, without culture" was "written in the heat of the moment. Russia has always had, and has, traditions, faith, and culture; it is just that, for some strange reason, she tries either to repudiate them utterly or, on the contrary, to fence herself off with them from the rest of the world." Amalrik, *Will the Soviet Union Survive until 1984?*, 8.

45. Ibid., 28, 35–36.

46. Ibid., 1, 64. In his preface to the second English edition (1981), Amalrik specifically—and, in my view, unconvincingly—denied any *Schadenfreude* with respect to the USSR.

47. Verblovskaia, *Moi prekrasnyi strashnyi vek*, 316.

48. IISH, Alexander Herzen Foundation, Folder 24, letter from Andrei Amalrik to Karel van het Reve (undated but probably from 1970), p. 1.

49. Solzhenitsyn, *Bodalsia telenok*, 596; Sergei Bondarenko, "Chelovek, kotoryi otmenil Olimpiiskie igry," interview conducted on December 29, 2013, quoted in Doikov, *Ob Andree Amal'rike*, 14; Houghton Library, Sakharov Archives, Amalrik Papers, Box 8, Folder 118, Document 1.

50. Shub, *The New Russian Tragedy*, 38. Shub grew up in a Bundist family in Brooklyn, where guests included Leon Trotsky and Alexander Kerensky.

51. Ian Iasnyi [Shragin], "Andrei Amal'rik kak publitsist," 4.

52. "This is my personal promise, a matter of honor," Zheludkov entreated, explaining his desire to prevent his essay from going abroad. "They already understand everything well enough 'over there.' It is not for them to fix our grief or release us from our Russian sadness." Entreaty notwithstanding, the essay was published in the bulletin of the émigré Russian Student Christian Movement: *Vestnik RSKhD*, vol. 94 (1969): 46–57.

53. IISH, Alexander Herzen Foundation, Folder 24, letters from Andrei Amalrik to Karel van het Reve (March 4 and July 4, 1969; January 26, 1970).

54. Ibid., Folder 24, letters from Andrei Amalrik to Karel van het Reve (March 4, 1969; January 26 and February 28, 1970; undated but probably from 1970); letters from van het Reve to Amalrik (February 10 and April 14, 1970); Amal'rik, *Zapiski dissidenta*, 101; Kissinger, *Years of Renewal*, 100.

55. Amal'rik, *Zapiski dissidenta*, 36; Shafarevich, "Est' li u Rossii budushchee?," in Solzhenitsyn, *Iz-pod glyb*, 262–63; Sokirko, "Otklik na stat'iu A. A. Amal'rika"; IISH, Alexander Herzen Foundation, Folder 24, letter from Andrei Amalrik to Karel van het Reve (undated but probably early 1970).

56. Al'brekht, *Kak vesti sebia na obyske*, 12.

57. Jacoby, *Moscow Conversations*, 117.

58. IISH, Alexander Herzen Foundation, Folder 24, letter from Andrei Amalrik to Karel van het Reve (undated but probably from early 1970); Folder 75, letter from Amalrik to the editors of various Western newspapers (undated).

59. Ibid., Folder 75, letter from Amalrik to the editors of various Western newspapers (undated).

60. Nikolai Bokov, "Predislovie k reprintu 'Vani Chmotanova,'" quoted in Doikov, *Ob Andree Amal'rike*, 13.

61. For this latter reading, see "Ob Amal'rike," *Novoe russkoe slovo* (January 30, 1971), 2. Rumors of Amalrik's collaboration with the KGB persisted even after his arrest, conviction, and sentencing to three years in the Gulag.

62. IISH, Alexander Herzen Foundation, Folder 32, letters from Gleb Struve to Karel van het Reve (October 10 and 26, 1968).

63. Alexeyeva and Goldberg, *The Thaw Generation*, 166.

64. *Khronika tekushchikh sobytii*, no. 13 (April 30, 1970), 34–35; *SDS*, vol. 5, doc. 370.

65. Quoted in Taubman, *The View from Lenin Hills*, 192.

66. Amal'rik, *Prosushchestvuet li Sovetskii Soiuz do 1984 goda?*, 1.

67. Fitzgerald, "The Crack-Up," in idem, *My Lost City: Personal Essays*, 139.

68. Siniavskii, "Dissidentstvo kak lichnyi opyt," 142.

69. Quoted in Gabowitsch, *Protest in Putin's Russia*, 73.

Chapter Eleven

1. Sakharov's *Reflections* wasn't mentioned until five years after it appeared (A. Chakovskii, "Chto zhe dal'she?," *Literaturnaia gazeta*, no. 7 [February 14, 1973], 14); Amalrik's *Will the Soviet Union Survive* wasn't mentioned in the Soviet press until eight years after it appeared (*Nedelia* [June 20, 1977]).

2. The Alexander Herzen Foundation published the manifesto as *Programma Demokraticheskogo dvizheniia Sovetskogo Soiuza* (Amsterdam, 1970). Its anonymous author was Sergei

Soldatov; see his *Zarnitsy vozrozhdeniia*, 166–69, 201–2. After reading the manifesto, which called for "limitations on the authority of the Communist Party" and multi-party elections, Politburo member Petro Shelest described it in his diary as "destructive, counter-revolutionary, and chauvinistic." "By interpreting 'democratism' to mean freedom of speech, the press, and assembly," the manifesto promoted "anarchism." Shelest, *Da ne sudimy budete*, 499.

3. "Pamiati Liudmily Polikovskoi i Aleksandra Malinovskogo" (March 1, 2017).

4. A. Mikhailov [Aleksandr Malinovskii], "Soobrazheniia po povodu liberal'noi kampanii 1968 goda," Memorial-Moscow, f. 156, pap. "A. Mikhailov." All subsequent quotations and paraphrases from Malinovsky derive from this unpaginated text. To the best of my knowledge, Malinovsky's essay has not appeared in any published collection of samizdat. A three-page summary appeared in *Khronika tekushchikh sobytii*, no. 17 (December 31, 1970), followed by brief critical responses.

5. Malinovsky approvingly cites Sakharov's *Reflections on Progress, Peaceful Coexistence, and Intellectual Freedom*, in particular its appeal for scientific governance and "democratization," but makes no mention of Sakharov's emphasis on human rights and convergence.

6. In this regard, Malinovsky's argument anticipates more recent analyses of dissident rhetoric as mimicry of official Soviet ideology and dissidents as mirror-images of Party activists. See Oushakine, "The Terrifying Mimicry of Samizdat"; and Yurchak, *Everything Was Forever*, 102–8 and 129–31. For a critique, see Platt and Nathans, "Socialist in Form, Indeterminate in Content."

7. Here and elsewhere in his essay, Malinovsky used the old-fashioned term *soslovnoe*— meaning "of or pertaining to a particular social estate"—to suggest the narrow, archaic nature of the dissidents' focus on the interests of the intelligentsia.

8. Pliushch, *Na karnivale istorii*, 461.

9. Igrunov, "K problematike obshchestvennogo dvizheniia."

10. As recounted in Igrunov, "O biblioteke Samizdata." On Marchenko's memoir, see chapter 6.

11. Andrei Slavin [Venyamin Kozharinov], "Nekotorye zametki o sovetskom demokraticheskom dvizhenii," Memorial-Moscow, f. 103, Kollektsiia Kronida Liubarskogo, uncatalogued. I am grateful to Aleksei Makarov of the Memorial Society for sharing this document with me. Remarkably, Kozharinov mentioned several such "princes" by name, including Vladimir Bukovsky, Ilya Gabai, Vladimir Gershuni, and Julius Telesin, all of whom were already arrested or exiled.

12. Memorial-Moscow, f. 120, pap. 2, doc. 3, Aleksandr Vol'pin, "Vse my zashchitniki svobody, zakonnosti i glasnosti," 18–20.

13. A. Strikh [Anatolii Iakobson], "Otvet Mikhailovu," *Khronika tekushchikh sobytii*, no. 17 (December 31, 1970), in the section "Novosti samizdata."

14. On the biography of Vladimir Zelinsky (not to be confused with Ukrainian president Volodymyr Zelensky), see his Russian-language Wikipedia page, Зелинский, Владимир Корнельевич, accessed on December 12, 2023, https://ru.wikipedia.org/wiki/Зелинский_Владимир_Корнельевич.

15. Dmitrii Nelidov [Vladimir Zelinskii], "Ideokraticheskoe soznanie i lichnost'," 204–5.

16. Ibid., 206–7.

17. Shklovsky, "Art as Device," in Berlina, *Viktor Shklovsky: A Reader*, 73–96; Nelidov/Zelinskii, "Ideokraticheskoe soznanie i lichnost'," 207 and 213.

18. Nelidov/Zelinskii, "Ideokraticheskoe soznanie i lichnost'," 213. The Russian word *litso*, which I am rendering here as "identity," can also mean "face" or "person."

19. Ginzburg, *Chelovek za pis'mennym stolom*, 196.

20. Nelidov/Zelinskii, "Ideokraticheskoe soznanie i lichnost'," 213.

21. See chapter 8.
22. *SDS*, vol. 8, doc. 607.
23. Grigorenko, *V podpol'e*, 519–20; Igrunov, "K problematike obshchestvennogo dvizheniia"; Amal'rik, *Zapiski dissidenta*, 39. On the concept of virtuosi, see Goldman and Pfaff, "Reconsidering Virtuosity: Religious Innovation and Spiritual Privilege."
24. Paperno, *Chernyshevsky and the Age of Realism*, 11.
25. Cole, "Three Voices of Dissent," 145. The text quoted here is the published translation of Cole's interview with Bukovsky (along with Amalrik and Yakir) filmed in Moscow in the spring of 1970.
26. Bukovskii, *"I vozvrashchaetsia veter...,"* 223–24. This passage made a deep impression on Grigorenko. See idem, *V podpol'e*, 576.
27. The analogy to the biblical Sodom is explicit in Ternovskii, *Vospominaniia i stat'i*, 272.
28. For classic treatments of the peasant moral economy, see Thompson, "The Moral Economy of the English Crowd in the Eighteenth Century"; and Scott, *The Moral Economy of the Peasant*.
29. A. Antipov [pseudonym], "Ot brozheniia umov—k umstvennomu dvizheniiu" (January 1969), in *SDS*, vol. 6, doc. 388.
30. Igrunov, "K problematike obshchestvennogo dvizheniia"; Strukova, "Odesskaia biblioteka samizdata Viacheslava Igrunova," in *Dissidenty SSSR, vostochnoi i tsentral'noi Evropy*, 167–73.
31. Medvedev, "K voprosu o formakh i metodakh bor'by"; Valentin Turchin, "Inertsiia strakha" (samizdat, 1969), 39–43. "I want to caution our intelligentsia," Turchin wrote, "and especially the youth, against creating any sort of 'opposition.' There is nothing more fatal for society than when, instead of thinking in terms of truth, duty, and freedom, people start thinking in terms of 'us' and 'them.'"
32. Roi Medvedev, "12 oktiabria. O novom sudebnom protsesse," *Politicheskii dnevnik* (October 1968), Memorial-Moscow, f. 128, kor. 1–2, l. 43.
33. Mikhailov/Malinovskii, "Soobrazheniia po povodu liberal'noi kampanii 1968 goda."
34. Slavin/Kozharinov, "Nekotorye zametki o sovetskom demokraticheskom dvizhenii."

Chapter Twelve

1. Hoover Archive, Jahimovics Papers, Box 2, "Delo Ivana Iakhimovicha," pp. 217–23; de Boer et al., *Biographical Dictionary of Dissidents*, 202–3; Ivan Iakhimovich, "Vmesto poslednogo slova" (March 24, 1969), in *SDS*, vol. 1, doc. 102.
2. Amal'rik, *Zapiski dissidenta*, 86.
3. Hoover Archive, Jahimovics Papers, Box 1, Diary #12, November 1965–February 1966 (unpaginated section).
4. Ibid., Box 1, Diary #12, p. 187.
5. Hoover Archive, Jahimovics Papers, Box 2, "Statsionarnaia sudebno-psikhiatricheskaia ekspertiza Iakhimovicha I. A.," pp. 158–62. Remarkably, the version in this file of Yakhimovich's letter to Suslov was a KGB transcript created from a March 10, 1968, broadcast by the Voice of America, rather than the copy Yakhimovich sent to Suslov at the Central Committee. Neither Andrei Vyshinsky nor any other Soviet official signed the Universal Declaration of Human Rights, as discussed later in this chapter.
6. Ibid., pp. 157–62.
7. Hoover Archive, Jahimovics Papers, Box 2, "Delo Ivana Iakhimovicha," p. 218.
8. Ibid., p. 217.
9. *SDS*, vol. 1, doc. 102.

10. The original lyrics of "The Internationale" were composed in 1871 by the French anarchist Eugene Pottier. The 1902 Russian translation by Arkady Yakovlevich Kots was the official anthem of the Soviet Union from 1918 to 1944, at which point it was replaced by the State Anthem of the USSR. The text quoted here is my translation of Kots's Russian translation—the version that would have been familiar to Yakhimovich's daughters.

11. *SDS*, vol. 1, doc. 102; "Arest Ivana Iakhimovicha," *Khronika tekushchikh sobytii*, no. 7 (April 30, 1969).

12. Hoover Archive, Jahimovics Papers, Box 2, "Statsionarnaia sudebno-psikhiatricheskaia ekspertiza Iakhimovicha I. A.," pp. 57–58.

13. ACRC, Grigorenko Family Papers, Series 2, Box 3, letter from Petr Grigorenko to Pavel Litvinov (April 23, 1969), p. 2.

14. Electronic Library of the History Department at Moscow State University, "Konstitutsiia (Osnovnoi Zakon) Soiuza Sovetskikh Sotsialisticheskikh Respublik," accessed December 12, 2023, http://www.hist.msu.ru/ER/Etext/cnst1936.htm#10.

15. Gorlizki, "Delegalization in Russia."

16. Amal'rik, *Zapiski dissidenta*, 63. One such person was Maya Ulanovskaya, who was present at the meeting at Grigorenko's apartment and who was "terrified by the word 'organization.'" Ulanovskaia and Ulanovskaia, *Istoriia odnoi sem'i*, 302.

17. For a sampling of Tsukerman's briefs, see Memorial-Moscow, f. 102, op. 1, d. 8, ll. 1–76.

18. Amal'rik, *Zapiski dissidenta*, 88.

19. Ibid., 61; Horvath, "Breaking the Totalitarian Ice," 154. Horvath's article is the most complete account in English of the events leading up to the founding of the Initiative Group.

20. Grigorenko, *V podpol'e*, 675; Kowaljow, *Der Flug des weißen Raben*, 57. Petr Yakir gave a similar account of Ulanovskaya's intervention in a 1972 KGB interrogation. See LYA, f. K-1, ap. 58, t. 11, p. 89 (interrogation of Petr Yakir on November 17, 1972).

21. Grigorenko, *V podpol'e*, 675; Kowaljow, *Der Flug des weißen Raben*, 58.

22. Krasin, *Poedinok*, 169. Several leading dissidents were at various times thought (by other dissidents) to have been cooperating with the KGB, including Amalrik, Valery Chalidze, Sinyavsky, and Yakir.

23. Bukovsky Archives, Document 0039, Memorandum 887-A from Yuri Andropov to the Central Committee (April 16, 1969), pp. 3–5, http://www.bukovsky-archives.net/pdfs/dis60/dis60-r.html.

24. Grigorenko raised the possibility of self-immolation in a February 1969 letter to Solzhenitsyn, shortly after Palach set himself on fire in Prague. But he went on to reject that option "not out of fear, but because acts of despair are not part of my character. What in Czechoslovakia was a heroic deed would look like a farce [in the USSR]." ACRC, Grigorenko Family Papers, Series 2, Box 3, letter from Petr Grigorenko to Alexander Solzhenitsyn (February 2, 1969), p. 1.

25. Ibid.

26. Horvath, "Breaking the Totalitarian Ice," 156.

27. LYA, f. K-1, ap. 58, t. 11, pp. 114–17 (interrogation of Viktor Krasin on December 21, 1972); p. 264 (interrogation of Aleksandr Ziuzikov on October 27, 1972).

28. Kuzovkin and Makarov, *Dokumenty Initsiativnoi gruppy*, Document 1.

29. LYA, f. K-1, ap. 58, t. 11, pp. 27–28 (interrogation of Petr Yakir on August 29, 1972); pp. 114–17 (interrogation of Viktor Krasin on December 21, 1972).

30. The term can be found as early as 1917: see Holquist, *Making War, Forging Revolution*, 66. "Initiative groups" were often sponsored and supervised by the Soviet government.

31. Krasin, *Sud*, 39; Kovalev puts the number at a dozen: Kowaljow, *Der Flug des weißen Raben*, 60.

32. Ternovskii, "Taina IG," in idem, *Vospominaniia i stat'i*, 192–94.

33. Kowaljow, *Der Flug des weißen Raben*, 60–61; Krasin, *Poedinok*, 197; LYA, f. K-1, ap. 58, t. 11, pp. 121–22 (interrogation of Viktor Krasin on December 22, 1972).
34. "I had to lie," was how Krasin put it some four decades later. Krasin, *Poedinok*, 197.
35. LYA, f. K-1, ap. 58, t. 11, p. 122 (interrogation of Viktor Krasin on December 22, 1972); p. 234 (interrogation of Venyamin Kozharinov on March 1, 1973); Krasin, *Poedinok*, 97.
36. *Delovoi chelovek*: Vyacheslav Bakhmin's description of Krasin during an interview in the documentary film by Andrei Loshak, *Anatomiia protsessa* (2013).
37. Natalya Gorbanevskaya, interviewed in Loshak, *Anatomiia protsessa*.
38. LYA, f. K-1, ap. 58, t. 11, p. 206 (interrogation of Venyamin Kozharinov on January 23, 1973).
39. LYA f. K-1, ap. 58, t. 11, pp. 118–19 (interrogation of Viktor Krasin on December 21, 1972). The tape recorder and cassettes were a gift from Henk Wolzak, the Dutch journalist who misunderstood the start time of the demonstration on Red Square (see chapter 8); ibid., t. 11, p. 141 (interrogation of Viktor Krasin on January 17, 1973); t. 13, p. 348 (interrogation of Viktor Krasin on June 28, 1973).
40. Pavel Litvinov, interviewed in Loshak, *Anatomiia protsessa*.
41. Krasin, *Sud*, 39.
42. Kowaljow, *Der Flug des weißen Raben*, 59.
43. Frank Starr, "Pyotr Yakir Is New Russian Voice of Intellectual Dissent," *Chicago Tribune* (May 25, 1969), 1. The KGB kept track of the Western periodicals that printed the Initiative Group's inaugural appeal: see LYA, f. K-1, ap. 58, t. 14, p. 111 (interrogation of Abel Aganbegyan on September 29, 1969). The invisibility of dissident women in accounts by Western journalists occurred elsewhere in the Soviet bloc: see Szulecki, *Dissidents in Communist Central Europe*, 155–58. See also Bolton, *Worlds of Dissent*, 43.
44. Shub, "'Will the USSR Survive . . . ?' A Personal Comment," 91.
45. *Khronika tekushchikh sobytii*, no. 10 (October 31, 1969).
46. Alexeyeva and Goldberg, *The Thaw Generation*, 251 and 262. Kovalev came to a similar judgment of Yakir's Western-media-driven elevation to the status of leader: "It seems to me that this flattered him and in the depths of his soul terrified him." Kowaljow, *Der Flug des weißen Raben*, 59. On letters from Soviet citizens to Yakir, see Horvath, "Breaking the Totalitarian Ice," 164.
47. Kuzovkin and Makarov, *Dokumenty Initsiativnoi gruppy*, 271.
48. Krasin, *Sud*, 40.
49. LYA, f. K-1, ap. 58, t. 11, p. 242 (interrogation of Anatolii Levitin on April 3, 1973). Levitin, the son of a Jewish convert to Russian Orthodoxy, initially wrote under the pseudonym "Krasnov" before deciding to merge his two identities with the hyphenated last name Levitin-Krasnov.
50. Ternovskii, *Vospominaniia i stat'i*, 196. The Russian term used by Ternovsky, *nezakonnorozhdennaia*, literally "illegally born," captures the irony of an organization devoted to the rule of law that operates outside the law.
51. Krasin, *Sud*, 40–41. Unlike most dissidents, Krasin was an outspoken anti-communist.
52. Amal'rik, *Zapiski dissidenta*, 114–15; Alexeyeva and Goldberg, *The Thaw Generation*, 252; Bonavia, *Fat Sasha*, 25–37.
53. Alexeyeva and Goldberg, *The Thaw Generation*, 252–53.
54. Barsukova and Ledeneva, "Are Some Countries More Informal Than Others? The Case of Russia," in Ledeneva, *The Global Encyclopaedia of Informality*, vol. 2, 487. On the general absence of charters, programs, formal membership, and dues in dissident organizations, see Stephan, "Von 'Dissidenten,' 'Ehefrauen,' und 'Sympathisantinnen,'" in Roth, *Soziale Netzwerke und soziales Vertrauen*, 132.

55. Kuzovkin and Makarov, *Dokumenty Initsiativnoi gruppy*, Document 6.

56. Not surprisingly, the statement makes no mention of the ethical lapses that accompanied the Initiative Group's founding, presenting that event as a collective, democratic decision.

57. Kuzovkin and Makarov, *Dokumenty Initsiativnoi gruppy*, Document 6.

58. The first member of the Initiative Group to be arrested was Vladimir Borisov, who was seized from his apartment in Leningrad on June 12, 1969, and spent the next five years confined, without trial, in various psychiatric hospitals and prisons. He had met Grigorenko in a psychiatric prison in Leningrad in 1965. De Boer et al., *Biographical Dictionary of Dissidents*, 63–64.

59. Amal'rik, *Zapiski dissidenta*, 89; Horvath, "Breaking the Totalitarian Ice," 164.

60. The Initiative Group was not the first to address appeals to entities outside the USSR, including the United Nations. As Manuela Putz has shown, political prisoners from the Ukrainian and Baltic Soviet republics had sent protests to émigré organizations in the West starting in the 1950s, including references to the Universal Declaration of Human Rights. Similar appeals went to the United Nations. See Putz, *Kulturraum Lager*, 104–7.

61. Moyn, *The Last Utopia*, 135.

62. See the preamble to the Universal Declaration of Human Rights, United Nations, accessed December 12, 2023, https://www.un.org/en/about-us/universal-declaration-of-human-rights.

63. Moyn, *The Last Utopia*, 134.

64. Kuzovkin and Makarov, *Dokumenty Initsiativnoi gruppy*, Document 2.

65. Among the group's official "supporters," one—Leonid Vasilyev—was identified as a "jurist." Vasilyev's name, however, appeared only on the first and third of the group's several dozen appeals; he seems to have dropped out by 1970.

66. Kuzovkin and Makarov, *Dokumenty Initsiativnoi gruppy*, Document 5 (January 17, 1970). The original phrase is "*zashchita grazhdanskikh prav cheloveka.*" In Russian, the standard rendition of "human rights" is *prava cheloveka*, literally "rights of the person" or "rights of the human being," harking back to the idea's eighteenth-century formulation as the "rights of man." For purposes of readability, I generally translate *prava cheloveka* as "human rights," except in cases such as this, where "civil human rights" would be awkward. The Initiative Group's first letter (Document 1, May 20, 1969) similarly requested that the United Nations's "Committee for the Defense of Human Rights" (whose actual name was Commission on Human Rights) investigate "violations of fundamental civil rights in the Soviet Union." Horvath ("Breaking the Totalitarian Ice," 175) attributes the confusion to problems of translation of Initiative Group documents by Western journalists and commentators. The confusion already existed, however, within the Russian-language texts produced by the Initiative Group itself and persisted for years thereafter. A 1979 document produced by the Moscow Helsinki Group (see chapter 22) also contains the phrase "civil rights of man": Zubarev and Kuzovkin, *Dokumenty Moskovskoi Khel'sinkskoi gruppy*, 326.

67. LYA, f. K-1, ap. 58, t. 13, p. 285 (interrogation of Viktor Krasin on October 6, 1972); vol. 11, pp. 167–68 (interrogation of Viktor Krasin on January 18, 1973).

68. For a prominent example, see Sakharov, "Razmyshleniia o progresse," in idem, *Trevoga i nadezhda*, 19.

69. This misconception persists even today in some of the scholarly literature: see, for example, Beyrau, "Im Kreislauf der Geschichte: Die Macht und ihre Widersacher in Russland," 117; von Zitzewitz, *The Culture of Samizdat*, 41.

70. In a review of the Soviet position vis-à-vis the Universal Declaration on the occasion of its twentieth anniversary in 1968, the Soviet Ministry of Foreign Affairs noted that Vyshinsky had been unable to insert language into the Declaration regarding "the material and social conditions necessary for realizing rights"—language at the heart of the distinctive Soviet logic of rights. The Soviet delegation had also unsuccessfully attempted to limit the purview of Article

19, on freedom of opinion and expression, to "democratic views" and to explicitly prohibit "fascist propaganda." GARF, f. 89, op. 19, d. 47.

71. Morsink, *The Universal Declaration of Human Rights*, 21–28. At the time, "Soviet-bloc" meant the USSR, the Belorussian and the Ukrainian Soviet Socialist Republics, Poland, Czechoslovakia, and Yugoslavia. Bulgaria, Hungary, and Romania did not join the United Nations until 1955.

72. Dudziak, *Cold War Civil Rights*.

73. Amos, "Embracing and Contesting," in Hoffmann, *Human Rights in the Twentieth Century*, 153–64.

74. Nathans, "Dictatorship of Reason," 660. The journalist David Bonavia, who covered the dissident scene closely as Moscow correspondent for the *Times* of London from 1969 to 1972, reported that the Universal Declaration "is difficult to get hold of in the Soviet Union, and is regarded as an article of faith by the adherents of the Democratic Movement." Bonavia, *Fat Sasha*, 77. For Volpin's analysis of the UN pact and its relation to Soviet law, see *SDS*, vol. 16, doc. 530.

75. Eckel and Moyn, *The Breakthrough*.

76. Moyn, *The Last Utopia*, 9.

77. Exceptions include Krasin, Solzhenitsyn, and above all Viktor Sokirko, whose samizdat essays often appeared under the pseudonym "K. Burzhuademov" ("K. Bourgeois-Democrat"). For examples of Sokirko's championing of private enterprise and private ownership, complete with references to Adam Smith and Max Weber, see Sokirko, *Zhit' ne po lzhi (sbornik otklikov-sporov na stat'iu A. I. Solzhenitsyna)*.

78. Orlov, "Vozmozhen li sotsializm ne-totalitarnogo tipa?" (samizdat, 1975), published in *Materialy samizdata*, vol. 11 (1976), doc. 2425. A decade and a half later, Orlov claimed that by the time of the Prague Spring, "I already preferred the reliable and tested Scandinavian model of capitalism with a human face. I doubted whether the Czechs' experiment was realistic." Idem, *Opasnye mysli*, 134.

79. Alexeyeva and Goldberg, *The Thaw Generation*, 5.

80. Nathans, "The Disenchantment of Socialism," in Eckel and Moyn, *The Breakthrough*, 33–48.

81. Moyn, *The Last Utopia*, 5.

82. Kuzovkin and Makarov, *Dokumenty Initsiativnoi gruppy*, Document 1. This explanation appeared again a year later (Document 6): "People ask: why do you appeal to the UN instead of to your government? It was not the Initiative Group that started the conversation about violations of civil rights in the USSR. During the past several years, various individuals and groups of people have produced hundreds if not thousands of statements condemning arbitrariness, sending them to every Party, governmental, and public entity in the Soviet Union. And all in vain." Krasin reiterated precisely this explanation while under interrogation: LYA, f. K-1, ap. 58, t. 11, p. 116 (interrogation of Viktor Krasin on December 21, 1972).

83. Alexeyeva and Goldberg, *The Thaw Generation*, 252. One can imagine, counter-factually, a less ham-fisted and savvier Soviet government successfully entangling dissidents in a sham dialogue and thereby preempting or at least postponing their turn to foreign entities. In reality, the only actual conversations between dissidents and representatives of the Soviet government took place during interrogations by the KGB.

84. Solzhenitsyn, "Vot kak my zhivem," in idem, *Bodalsia telenok*, 542.

85. Krasin, *Poedinok*, 145.

86. For a sampling of relevant works, see Mozokhina, *Svoboda lichnosti i osnovnye prava grazhdan v sotsialistichekikh stranakh Evropy*; Mal'tsev, *Sotsialisticheskoe pravo i svoboda lichnosti*; Mel'nikov, *Pravovoe polozhenie lichnosti v sovetskom grazhdanskom protsesse*; Kuchinskii,

Lichnost', svoboda, pravo; Kuznetsov, *Ugolovno-pravovaia okhrana interesov lichnosti v SSSR*; Savaneli, *Iuridicheskie formy polozheniia lichnosti v sovetskom obshchestve*.

87. Kharkhordin, *The Collective and the Individual in Russia*, 184–200.

88. Kuzovkin and Makarov, *Dokumenty Initsiativnoi gruppy*, Document 6.

89. Ibid., Document 1. Within this text, there are several variants of the name for the "Committee for the Defense of Human Rights"—an indication, perhaps, of the haste with which the text was prepared and the scant information available to Soviet citizens about the inner workings of the United Nations. The name Initiative Group for the Defense of Civil Rights (later Initiative Group for the Defense of Human Rights) may well have been an attempt to mirror what members mistakenly took to be the name of the corresponding UN entity.

90. John Humphrey, quoted in Horvath, "Breaking the Totalitarian Ice," 159.

91. Kuzovkin and Makarov, *Dokumenty Initsiativnoi gruppy*, Documents 3, 6, and 13.

92. Alexeyeva, *Soviet Dissent*, 292.

93. Horvath, "Breaking the Totalitarian Ice," 169.

94. Evangelista, *Unarmed Forces*, 7–8.

95. Eckel, *Die Ambivalenz des Guten*, 207–22.

96. In 1968, the prosecutor in the "Trial of the Four" accused Yuri Galanskov and other defendants of maintaining contacts with the émigré organization NTS, a charge they denied. Information that came to light later, following Galanskov's death in a Soviet labor camp, suggests that he had indeed maintained contacts with and possibly joined NTS as early as 1966. See Galanskov, *Iurii Galanskov*, 129–35. A more recent work elides the issue: Kaganovskii, *Khronika Iuriia Galanskova*.

97. Information on activities by Scandinavian supporters of SMOG can be found in the papers of the Russian émigré organization NTS: FSO Bremen, f. 98, Bookcase 9, Shelf 2, Box 36, Item 30, pp. 1–16; and in a report by KGB head Yuri Andropov dated 1970: RGANI, f. 89, op. 37, d. 17, l. 1.

98. *Khronika tekushchikh sobytii*, no. 9 (August 31, 1969).

99. Arkhiv upravleniia KGB po Orlovskoi oblasti, Delo 27 (po obvineniiu Superfina G. G.), t. 3, l. 198 (interrogation of Gabriel Superfin on August 14, 1973). My thanks to Gabriel Superfin for providing me with a scanned copy of his five-volume KGB dossier.

100. FSO Bremen, f. 98, Bookcase 9, Shelf 2, Box 36, Item 30, pp. 9–13; LYA, f. K-1, ap. 58, t. 11, pp. 8–9 (interrogation of Petr Yakir on August 18, 1972). On the four thousand rubles—equivalent to roughly two years' salary—see Krasin, *Poedinok*, 221. For an example of coverage by the *Chronicle*, see "Inostrantsy—v zashchitu sovetskikh politzakliuchennykh," *Khronika tekushchikh sobytii*, no. 12 (February 28, 1970).

101. LYA, f. K-1, ap. 58, t. 11, pp. 10–11 (interrogation of Petr Yakir on August 18, 1972).

102. FSO Bremen, f. 98, Bookcase 9, Shelf 2, Box 36, Item 28, p. 1.

103. FSO Bremen, f. 98, Bookcase 9, Shelf 2, Box 36, Item 28, Bukovsky leaflet (undated, but not earlier than January 1972).

104. Bukovskii, "I vozvrashchaetsia veter...," 223–24.

105. LYA, f. K-1, ap. 10, b. 130, pp. 145 and 253.

106. LYA, f. K-1, ap. 58, t. 11, pp. 131–34 (interrogation of Viktor Krasin on January 17, 1973). Four decades later, in his memoirs, Krasin gave a slightly different version of this story. Krasin, *Poedinok*, 219–21.

107. On NTS, see Tromly, *Cold War Exiles and the CIA*, 174–91.

108. On Europa Civiltà, see Dogliotti, "L'eversione nera negli anni settanta." My thanks to Ben Mercer and the late Jonathan Steinberg for summarizing the content of this article for me.

109. Under interrogation by the KGB, Krasin suggested that many of the young tourists who smuggled leaflets and brochures in support of Soviet dissidents, and who attempted to foster

links with the Initiative Group, were sponsored by NTS. See LYA, f. K-1, ap. 58, t. 11, pp. 134–36 (interrogation of Viktor Krasin on January 17, 1973). Krasin repeated this claim in his memoirs: *Poedinok,* 220.

110. LYA, f. K-1, ap. 58, t. 11, pp. 135–36 (interrogation of Viktor Krasin on January 17, 1973).

111. Ibid., p. 138.

112. Andrei Slavin [Venyamin Kozharinov], "Nekotorye zametki o sovetskom demokraticheskom dvizhenii," Memorial-Moscow, f. 103, Kollektsiia Kronida Liubarskogo, uncatalogued.

113. By the 1980s, critics outside the USSR had come up with a name for these and other entities that mimic the constituent elements of civil society: GONGOs—government-organized non-governmental organizations.

114. Beliakov, *Mezhdunarodnye nepravitel'stvennye organizatsiia,* 16–17.

Chapter Thirteen

1. Chalidze's wife, Vera Ilinichna Slonim (b. 1948), is not to be confused with Vera Evseevna Slonim (1902–1991), whose husband was the writer Vladimir Nabokov.

2. Chalidze, *Prava cheloveka i Sovetskii Soiuz,* 69.

3. Bonavia, *Fat Sasha,* 14 and 24.

4. Sakharov, *Memoirs,* 314.

5. Bonavia, *Fat Sasha,* 18; Sakharov, *Memoirs,* 314.

6. All fifteen issues of *Problems of Society* were reprinted in *SDS,* vol. 16.

7. Chalidze interview with Jay Axelbank, *Newsweek,* vol. 76, no. 25 (December 21, 1970); David Bonavia, "Four Men in Moscow Talk to Russians about Their Rights," *Times* (London) (November 5, 1971), 1 and 6.

8. *Grazhdanskii kodeks RSFSR ot 11.06.1964.* For Chalidze's deployment of Article 482, see *Dokumenty Komiteta prav cheloveka,* 246. There were two obvious problems with Chalidze's reading of the Civil Code. First, Article 482 appears in a section devoted to intellectual property, not associations; the latter were treated separately and in a manner not particularly conducive to his argument. Second, the copyright status of samizdat texts was uncertain, even in international law, let alone in Soviet law at the time.

9. *Dokumenty Komiteta prav cheloveka,* 14.

10. New York Public Library, Manuscripts and Archives Division (henceforth NYPL-MAD), International League for Human Rights, Group D, Box 61, Folder: General Correspondence, USSR, September–December 1971, Valery Chalidze, "Human Rights in Russia," p. 1.

11. Bonavia, "Four Men in Moscow," 1 and 6; Harrison Salisbury, "Struggling Now for Human Rights," *New York Times* (March 4, 1973), 56.

12. Sakharov, *Memoirs,* 320.

13. Pliushch, *Na karnivale istorii,* 425; Rubenstein and Gribanov, *The KGB File of Andrei Sakharov,* 103 and 114–15; memorandum from Yuri Andropov to the Central Committee (March 18, 1971), from the Archive of the President of the Russian Federation (henceforth APRF), f. 3, op. 80, d. 638, ll. 78–79. This document was not included in *The KGB File of Andrei Sakharov* but is posted on the accompanying website: Yale University Press, Annals of Communism: The Andrei Sakharov KGB File, accessed December 12, 2023, http://yupnet.org/annals/sakharov/documents_frames/Sakharov_026.htm. I thank the late Edward Kline for bringing it to my attention.

14. Rubenstein and Gribanov, *The KGB File of Andrei Sakharov,* 21.

15. Pliushch, *Na karnivale istorii,* 425–26 ("Zakony—nashe oruzhie, no ne nasha illiuziia").

16. Sakharov, *Vospominaniia,* vol. 1, 697.

17. Alexeyeva and Goldberg, *The Thaw Generation*, 254–55.
18. Ibid., 256–57.
19. Smith, *The Russians*, 442.
20. Sakharov, *Vospominaniia*, vol. 1, 696; Scammell, *Solzhenitsyn*, 795. Solzhenitsyn characteristically found nothing funny about the Committee's by-laws, seeing them instead as "cunningly devised" to allow Chalidze "to thwart the will of any other member." Solzhenitsyn, *Bodalsia telenok*, 399.
21. Quoted in Shelest, *Da ne sudimy budete*, 507.
22. Chalidze's work on gay rights, which many dissidents considered inappropriate, included challenging a Soviet law that punished sodomy with up to five years in prison. See Sophia Kishkovsky, "Valery Chalidze, Soviet Dissident Exile and Advocate for Justice, Dies at 79," *New York Times* (January 22, 2018), B17.
23. After contacting the Institute of State and Law (part of the Soviet Academy of Sciences), Chalidze was visited by someone claiming to be a staff member but who wished to remain anonymous and declined further communication. Chalidze, "Zapiska o tvorcheskikh sviazakh Komiteta prava cheloveka," in Memorial-Moscow, f. 102, d. 47, l. 33. On the lack of response of Soviet lawyers, see NYPL-MAD, International League for Human Rights, Group D, Box 61, Folder: General Correspondence, USSR, September–December 1971, letter from Leonid Rigerman to Roger Baldwin (September 23, 1971), 2.
24. Andrei Sakharov Archive, Moscow, f. 1, razdel 3.1.1 and 8.1. My thanks to Bela Koval for making these and other letters available to me.
25. Rubenstein and Gribanov, *The KGB File of Andrei Sakharov*, 115 and 124–25. *The Insulted and the Injured* is the title of an 1861 novel by Dostoevsky. Andropov memorandum to the Central Committee (March 18, 1971), APRF, f. 3, op. 80, d. 638, ll. 78–79.
26. Sakharov, *Vospominaniia*, vol. 1, 693 and 696.
27. On the Human Rights Committee as a "legal aid bureau," see Beckerman, *When They Come for Us, We'll Be Gone*, 256.
28. Sakharov, *Memoirs*, 501–6.
29. Memorial-Moscow, f. 102, d. 47, l. 37.
30. Beliakov, *Mezhdunarodnye nepravitel'stvennye organizatsiia*, 16–17.
31. Chalidze interview, *Newsweek* (December 21, 1970), 57.
32. *Khronika tekushchikh sobytii*, no. 18 (March 5, 1971).
33. NYPL-MAD, International League for Human Rights, Group D, Box 61, Folders: General Correspondence, USSR, January–August 1971 and General Correspondence, USSR, September–December 1971.
34. *Dokumenty Komiteta prav cheloveka*.
35. Rubenstein and Gribanov, *The KGB File of Andrei Sakharov*, 127–28.
36. Yale University Press, Annals of Communism: The Andrei Sakharov KGB File, accessed December 12, 2023, http://yupnet.org/annals/sakharov/documents_frames/Sakharov_034.htm.
37. NYPL-MAD, International League for Human Rights, Group D, Box 61, Folder: General Correspondence, USSR, September–December 1971 (message sent on December 2, 1971, by the Human Rights Committee).
38. See, for example, Daniel', "Pochemu ne 'perestroilis'' dissidenty?," 15.
39. Hedrick Smith, "Soviet Repression Leaves the Dissidents in Disarray," *New York Times* (December 11, 1972), 26.
40. Krasin, *Poedinok*, 145.
41. Bonavia, *Fat Sasha*, 168. Solzhenitsyn would soon start referring to the dissident "movement" (using quotation marks). Solzhenitsyn, *Bodalsia telenok*, 402.

42. Arkhiv upravleniia KGB po Orlovskoi oblasti, Delo 27 (po obvineniiu Superfina G. G.), t. 4, l. 118 (interrogation of Gabriel Superfin on December 11, 1973).

43. Sakharov, *Vospominaniia*, vol. 2, 59. Imshenetsky was not against signing collective letters: when Sakharov was awarded the Nobel Peace Prize in 1975, he added his name to a statement by Soviet Academicians condemning the award.

44. Ibid.

45. Igrunov, "O biblioteke Samizdata."

46. Andrei Slavin [Venyamin Kozharinov], "Nekotorye zametki o sovetskom demokraticheskom dvizhenii," Memorial-Moscow, f. 103, Kollektsiia Kronida Liubarskogo, uncatalogued.

47. LYA, f. K-1, ap. 58, t. 13, pp. 316–18 (interrogation of Viktor Krasin on December 26, 1972).

48. Eckel, *Die Ambivalenz des Guten*, 208.

49. Igrunov, "K problematike obshchestvennogo dvizheniia."

Chapter Fourteen

1. Weiner, "The Empires Pay a Visit," 376.

2. Chebrikov et al., *Istoriia sovetskikh organov gosudarstvennoi bezopasnosti*, 542. On the covert book program, see Reisch, *Hot Books in the Cold War*.

3. Zubok, *Zhivago's Children*, 141.

4. Mitrokhin, *KGB Lexicon: The Soviet Intelligence Officer's Handbook*, 45.

5. LYA, f. K-1, ap. 10, b. 325, pp. 25–36, F. D. Bobkov, "Ideologicheskaia diversiia imperializma protiv SSSR i deiatel'nosti organov KGB po bor'be s nei" (report presented at a January 1964 conference at the Dzerzhinsky KGB Higher School in Moscow).

6. Mitrokhin Archive, The Chekist Anthology, Folder 9 ("Inakomyslie"), p. 2.

7. Bobkov, *Agenty*, 98.

8. LYA, f. K-1, ap. 10, b. 325, pp. 38–40, V. N. Strunnikov, "Sovremennyi antikommunizm—ideinnoe oruzhie imperializma" (report presented at a January 1964 conference at the Dzerzhinsky KGB Higher School in Moscow).

9. Bobkov, *Agenty*, 100–101.

10. Fedor, "Chekists Look Back on the Cold War," 852.

11. No entity in the world paid more sustained attention to the Soviet dissident movement than the KGB. Nobody amassed a larger archive of samizdat texts, criminal investigation records, intercepted letters, and transcripts of bugged phone calls; nobody spent as many hours eavesdropping, tailing, filming, and interrogating its participants. The KGB, and the Fifth Directorate in particular, is thus a crucial repository of evidence for the subject of this book. It is also mostly inaccessible to historians. Its post-Soviet successor, Russia's Federal Security Service (FSB), like most intelligence organizations, has jealously guarded its historical records, and for obvious reasons: in addition to documenting the unsavory and sometimes criminal aspects of its work as well as the identities of its agents at home and abroad, KGB files from the post-Stalin era implicate hundreds of thousands of Soviet citizens who informed on their colleagues, friends, lovers, and relatives, more than a few of whom are still alive. As long as Russia's political elite, including President Vladimir Putin and many of his closest associates, are drawn disproportionately from the ranks of the security services (the so-called *siloviki*, or "enforcers"), these circumstances are unlikely to change. There have, however, been significant cracks in the wall of secrecy. When the Soviet Union splintered in 1991, so did the KGB, leaving substantial chunks of its archive stranded in fifteen former Soviet republics, now independent states. In Estonia, Latvia, Lithuania, and Ukraine, former branch offices of the KGB have been transformed into archives open to researchers. A handful of far-flung provincial KGB archives within Russia

itself, moreover, have until recently been more hospitable to scholars than the central repository in Moscow. Copies of certain KGB documents circulated through the Central Committee of the Communist Party or through one or more Soviet government ministries and are now preserved in their archives, with varying degrees of accessibility. In addition, during the 1990s a handful of Soviet dissidents (or their relatives) took advantage of a law allowing them to request copies of their KGB dossiers, and in several cases those dossiers, although filtered by the FSB prior to their release, are now accessible in Western archives. Finally, a handful of defectors from the Soviet secret services, preeminent among them the former KGB archivist Vasiliy Mitrokhin, have smuggled valuable materials to the West concerning covert activities at home and abroad.

12. According to Bobkov, "Andropov did not hide the fact that his sense of the need for the [Fifth] Directorate derived from the events in Hungary in 1956." Bobkov, *Agenty*, 115. See also Andrew and Mitrokhin, *The Sword and the Shield*, 5.

13. Bobkov joined the Fifth Directorate as deputy director when it was created in July 1967; two years later he was promoted to director. For a lightly sourced account, see Makarevich, *Filipp Bobkov i Piatoe upravlenie KGB*.

14. Mitrokhin Archive, The Chekist Anthology, Folder 51 ("O pravakh cheloveka"), p. 1.

15. Bobkov, *Agenty*, 115; idem, *Kak gotovili predatelei*, 154–56; Petrov, "Podrazdeleniia KGB," in Behrends et al., *Povsednevnaia zhizn' pri sotsializme*, 163; Smykalin, "Ideologicheskii kontrol' i Piatoe upravlenie." As Smykalin notes, the figure of twenty-five thousand is based on averages per district; this was the number of individuals working exclusively for the Fifth Directorate and therefore represents a fraction of the KGB's total pool of domestic informants. On Putin's service in the Fifth Directorate, see Short, *Putin*, 72–73.

16. Chebrikov et al., *Istoriia sovetskikh organov gosudarstvennoi bezopasnosti*, 552.

17. Andrew, "Intelligence in the Cold War," in Leffler and Westad, *The Cambridge History of the Cold War*, vol. 2, 436.

18. For Brezhnev's views on dissidents, see Schattenberg, *Leonid Breschnew*, 423–40.

19. After the Fifth Directorate found it expedient to change its name in 1989 to Directorate Z (Directorate for the Defense of the Constitutional Order), Bobkov maintained—improbably—that the West's "psychological warfare" aimed to "change the constitutional socialist order in the USSR" by manipulating "those who were prepared to fight against the constitutional order of their own country, from within." Bobkov, *Kak gotovili predatelei*, 139. He also maintained—even more improbably—that under Andropov, the KGB itself was a paragon of transparency and the rule of law: "The activities of KGB officers had to be known and understood by the population. This is how *glasnost* was not just nurtured but acquired real substance." The KGB's work thus involved "constant communication with the population, always keeping society informed by means of the mass media." Andropov, moreover, "was uncompromising in his insistence on observing the rule of law: KGB officers were required to be models of fidelity to the law." Bobkov, *Agenty*, 111–12. In this retrospective fantasy, the KGB successfully appropriated the central values of the dissident movement: constitutionalism, transparency, and the rule of law.

20. KGB data for the year 1977 indicate that the majority of defendants in cases involving "anti-Soviet agitation and propaganda" refused to plead guilty. Nearly a fifth of defendants in such cases pleaded guilty but refused to repent for their actions. See Mitrokhin Archive, The Chekist Anthology, Folder 21 ("Priznaki anti-sovietizma"), p. 5.

21. Cherniaev, *Sovmestnyi iskhod*, 108. The precipitating event in this case was the bulldozing of an unofficial art exhibition on the outskirts of Moscow in September 1974.

22. *Vlast' i dissidenty*, 20–21.

23. Bredikhin, *Lubianka—staraia ploshchad'*, 68–72.

24. Vol'pin, "Iuridicheskaia pamiatka dlia tekh, komu predstoiat doprosy," in idem, *Izbrannoe*, 369.

25. RGANI, f. 5, op. 30, d. 462, l. 250.

26. Andropov, *Leninizm—neischerpaemyi istochnik revoliutsionnoi energii i tvorchestva mass*, 100, quoted in Petrov, "Podrazdeleniia KGB," 158.

27. Bukovsky Archives, Document no. 181-A, memorandum from Yuri Andropov to the Central Committee (January 26, 1968), https://bukovsky-archive.com/1937-1969/1968-1969/.

28. Chebrikov et al., *Istoriia sovetskikh organov gosudarstvennoi bezopasnosti*, 545. For additional examples of this concern, see the April 1975 memorandum by Andropov and Soviet procurator-general Roman Rudenko in RGANI, f. 89, op. 37, d. 11, l. 1; and Mitrokhin Archive, The Chekist Anthology, Folder 9 ("Inakomyslie"), pp. 2–3.

29. Rubenstein and Gribanov, *The KGB File of Andrei Sakharov*, 155.

30. Shelest, *Da ne sudimy budete*, 415.

31. The four identifiable locations of copies of *Kolokol* mentioned in figure 14.1 are Leningrad, Morozova (on Lake Ladoga), Voronezh, and the Saratov region. On the Soviet Union–wide distribution of the *Chronicle of Current Events*, see GARF, f. 8131, op. 36, d. 4999 and d. 5000.

32. Butov, "Vospominaniia ob odesskikh dissidentakh," 10. Another rumor claimed that the KGB added radioactive material to certain samizdat texts in order to track their movement. See Martin, *Dissident Histories in the Soviet Union*, 113.

33. On Case 24, see Hopkins, *Russia's Underground Press*, 47–65; and Alexeyeva, *Soviet Dissent*, 310–17.

34. On the origins of *profilaktika*, see Fedor, *Russia and the Cult of State Security*, 52–55; Hornsby, *Protest, Reform and Repression*, 211–21; Cohn, "Coercion, Reeducation, and the Prophylactic Chat."

35. Cohn, "Coercion, Reeducation, and the Prophylactic Chat," 274.

36. See the 2014 interview with Lashkova, in Morev, *Dissidenty*, 103.

37. Ibid. For another semi-humorous account of a prophylactic conversation, see Voinovich, "Est' zhelanie vstretit'sia," in idem, *Antisovetskii Sovetskii Soiuz*, 306–13. Voinovich was probably too talented a satirist to be fully trusted as a reporter.

38. Cohn, "Coercion, Reeducation, and the Prophylactic Chat," 282; Petrov, "Podrazdeleniia KGB," 164.

39. On the early 1960s, see Hornsby, *Protest, Reform and Repression*, 220; for the period 1967–74, see Andropov's memorandum to the Central Committee (October 31, 1975), "O nekotorykh itogakh predupreditel'no-profilakticheskoi raboty organov gosbezopasnosti," in *Vlast' i dissidenty*, 61–62.

40. Petrov, "Podrazdeleniia KGB," 165.

41. Kozlov and Mironenko, *Kramola*, 23.

42. Smykalin, "Ideologicheskii kontrol' i Piatoe upravlenie," 34; Mitrokhin Archive, The Chekist Anthology, Folder 21 ("Priznaki antisovetizma"), p. 1.

43. Mitrokhin Archive, The Chekist Anthology, Folder 21 ("Priznaki antisovetizma"), p. 3.

44. "Russian metaphysical police," the protagonist of Nabokov's novel *Pnin* (1957) remarks, "can break physical bones also very well" (28).

45. Halfin, *Stalinist Confessions*, 145–46.

46. One of the foremost historians of the Soviet security services, Nikita Petrov, claims that "the KGB researched and analyzed the factors that led citizens who were dissatisfied with the Soviet order to form groups and organizations." But he cites only a single unpublished study, from 1975, by a graduate student in criminology, focusing on statistical data. To my mind, this is the exception that proves the rule, namely, the KGB's conspicuous lack of curiosity about the etiology of dissent. Petrov, "Podrazdeleniia KGB," 163–64.

47. Halfin, *Stalinist Confessions*, 39 and 119.
48. Gluzman, *Risunki*, 24.
49. Kozlov and Mironenko, *Kramola*, 24–25.
50. Fedor, "Chekists Look Back on the Cold War," 846 and 860.
51. Ibid., 856.
52. Bobkov, *Kak gotovili predatelei*, 16.
53. LYA, f. K-1, ap. 3, b. 700, pp. 3–7. "Radio Liberty Committee" refers to the American Committee for the Liberation of the Peoples of Russia (also known as Amcomlib), the organization that launched Radio Liberation in 1953.
54. Tumanov, *Podlinnaia "Sud'ba rezidenta."* The quotation is from a report by Tumanov (under the code name "Neris") from 1972. LYA, f. K-1, ap. 3, b. 700, pp. 104–5. See also Sosin, *Sparks of Liberty*, 128–29.
55. LYA, f. K-1, ap. 3, b. 700, p. 7.
56. Iurii Andropov, "Kommunisticheskaia ubezhdennost'—velikaia sila stroitelei novogo mira," *Pravda*, no. 253 (September 10, 1977): 1–2. The nearly exclusive focus on Western sources of Soviet dissent can be found as early as 1960, for example, in the candidate dissertation by V. V. Nazarov, "Antisovetskaia agitatsiia i propaganda kak osobo opasno gosudarstvennoe prestuplenie" (Vysshaia krasnoznamennaia shkola KGB SSSR, 1960). See the detailed summary in GARF, f. 8131 op. 31 d. 90081, ll. 35–42.
57. See the discussion of Article 190 at the conclusion of chapter 5.
58. Malia, *The Soviet Tragedy*, 167; Halfin, *Intimate Enemies*, 53.
59. Beer, *The House of the Dead*, 6.
60. Among the aforementioned names, only Sakharov was exiled without a trial. De Boer et al., *Biographical Dictionary of Dissidents*, mentions 348 cases of internal exile from 1956 to 1975 involving people who had engaged in dissenting activities such as attending demonstrations or writing, compiling, and disseminating samizdat.
61. Ro'i, *The Jewish Movement in the Soviet Union*; Fleischhauer, *The Soviet Germans*.
62. The historian Karl Schlögel estimates that roughly half (400–500) of the roughly 1,000 "signers" of the late 1960s left the Soviet Union in the 1970s. See Schlögel, *Der renitente Held*, 186. I have identified by name 103 dissidents who were forced to emigrate from the USSR between 1970 and 1987. This list is incomplete, possibly by a significant margin. It intentionally excludes anyone who had already applied to leave, for example, the considerable number of Jewish activists seeking to move to Israel.
63. Golomshtok, *"Zaniatie dlia starogo gorodovogo,"* 150.
64. In 1960, Volpin was subjected to forced injections of the tranquilizer reserpine. See Bloch and Reddaway, *Psychiatric Terror*, 71.
65. Smith and Oleszczuk, *No Asylum*, 36.
66. For an insider account, see Korotenko and Alikina, *Sovetskaia psikhiatriia*, 9–88.
67. Mel'nichenko, *Sovetskii anekdot*, 182.
68. Bukovsky, preface to Bloch and Reddaway, *Psychiatric Terror*, 15.
69. This number should be taken as an approximation of the actual total. On the one hand, it includes individuals who did not necessarily fit the category of dissident as used in the present study, for example, people who were caught attempting to cross the Soviet border without permission or who engaged in various religious practices. On the other hand, it almost certainly does not capture all the relevant cases, especially in far-off provinces. For details on the data set, see Smith and Oleszczuk, *No Asylum*, 49–57. On the use of punitive psychiatry against non-dissidents, see Werkmeister, "Wahnsinn mit System: Psychiatrische Anstalten in der späten Sowjetunion."
70. Reddaway, "Soviet Policies toward Dissent, 1953–1986," 69–71.

71. "Slow progressive schizophrenia" (*vialotekushchaia* or *vialo protekaiushchaia shizofreniia*) has often been translated as "sluggish schizophrenia," which potentially creates the misimpression that the patient, rather than the alleged illness, was sluggish. See Jargin, "Some Aspects of Psychiatry in Russia."

72. Bloch and Reddaway, *Psychiatric Terror*, 250. Yofe spent the next six months in various psychiatric hospitals against her will. See de Boer et al., *Biographical Dictionary of Dissidents*, 192.

73. For Great Britain, see Hurst, *British Human Rights Organizations*, 43–78.

74. Bloch and Reddaway, *Psychiatric Terror*, 24–27; Reich, *State of Madness*, 64.

75. Bloch and Reddaway, *Psychiatric Terror*, 254.

76. Ibid., 251; "Shizofreniia," in Snezhnevskii, *Spravochnik po psikhiatrii*, 84.

77. Or, rather, she was re-diagnosed: Gorbanevskaya had been similarly labeled nearly a decade earlier, when she had voluntarily checked herself into the same psychiatric hospital in connection with the severe depression and guilt she felt after cooperating with the KGB's investigation of two fellow students who had criticized the Soviet invasion of Hungary in 1956 (see chapter 6). This KGB-induced episode—with the KGB's role conveniently absent from Gorbanevskaya's psychiatric case file—now served as evidence for long-gestating mental illness.

78. Bloch and Reddaway, *Psychiatric Terror*, 132–35.

79. Andropov's memorandum about the trial, sent to the Central Committee on July 10, 1970, noted cryptically that the KGB had taken steps to influence "Western public opinion by means of information favorable to us." RGANI, f. 89, op. 55, d. 3, ll. 1–3. For the samizdat transcript of the trial, see *Khronika tekushchikh sobytii*, no. 15 (August 31, 1970).

80. Bloch and Reddaway, *Psychiatric Terror*, 143–44.

81. Ibid., 145–46.

82. For the English translation of the interview transcripts, see Cole, "Three Voices of Dissent." For the KGB's report to the Central Committee on Cole's interviews, see *Vlast' i dissidenty*, 24–25.

83. Edvard Klain [Edward Kline], *Moskovskii komitet prav cheloveka*, 107–8.

84. For textual variants, see https://cyclowiki.org/wiki/Обменяли_хулигана_на_Луиса_Корвалана (accessed December 12, 2023). For background, see Ulianova, "Corvalán for Bukovsky: A Real Exchange of Prisoners during an Imaginary War."

85. Medvedev, *Opasnaia professiia*, 169–74. On the Lysenko case, see Gordin, "Lysenko Unemployed: Soviet Genetics after the Aftermath."

86. Medvedev, *Opasnaia professiia*, 227.

87. Solzhenitsyn, *Bodalsia telenok*, 542–43 (emphasis in original).

88. Ibid., 322.

89. Medvedev, *Opasnaia professiia*, 237.

90. Scammell, *Solzhenitsyn*, 721–22. There may have been another option: in August 1971, the KGB allegedly poisoned Solzhenitsyn with ricin while he was waiting in line to buy food in Novocherkassk. Bedridden for three months with a mysterious malady, he narrowly survived. Details of the plot, including who authorized it, remain uncertain. See David Remnick, "KGB Plot to Assassinate Solzhenitsyn Reported," *Washington Post* (April 21, 1992).

91. Korotkov et al., *Kremlevskii samosud*, 198–201.

92. Ibid., 204–5.

93. Chuev, *Molotov: Poluderzhavnyi vlastelin*, 491–92. On the reliability of Feliks Chuev's transcripts, see von Hagen, review of *Molotov Remembers*, an English translation of an earlier edition of Chuev's conversations with Molotov.

94. Schattenberg, *Leonid Breschnew*, 433.

95. Korotkov et al., *Kremlevskii samosud*, 208–11.
96. Ibid., 209–10.
97. *Pravda*, no. 28 (January 28, 1959).
98. Korotkov et al., *Kremlevskii samosud*, 210.
99. Ibid., 210–16.
100. Scammell, *Solzhenitsyn*, 813–28.
101. Korotkov et al., *Kremlevskii samosud*, 353–63. It is indeed very cold in Verkhoyansk, which lies north of the Arctic Circle. But the reason foreign correspondents would not go there is that they were barred from traveling outside Moscow without the Soviet government's permission.
102. Ibid., 353–63.
103. Scammell, *Solzhenitsyn*, 831.
104. Korotkov et al., *Kremlevskii samosud*, 438–40.
105. For the list of charges, see ibid., 362.
106. Solzhenitsyn, *Bodalsia telenok*, 621 (emphasis in original).
107. Scammell, *Solzhenitsyn*, 829–46.
108. Salisbury, *Sakharov Speaks*, 186.
109. *Pravda* (September 5, 1973).
110. Arbatov, *The System*, 138 and 224.

Chapter Fifteen

1. Ulanovskaia and Ulanovskaia, *Istoriia odnoi sem'i*, 273; Vyacheslav Bakhmin, speaking in the documentary film by Andrei Loshak, *Anatomiia protsessa* (2013). The linguist Yuri Glazov called Yakir "one of the leaders of the democratic movement." Glazov, *V kraiu ottsov*, 73.
2. Bukovskii, *"I vozvrashchaetsia veter . . . ,"* 123; Bonavia, *Fat Sasha*, 33–5; LYA, f. K-1, ap. 58, t. 11, p. 138 (interrogation of Viktor Krasin on January 17, 1973).
3. Bonavia, *Fat Sasha*, 28.
4. LYA, f. K-1, ap. 58, t. 13, p. 329 (interrogation of Viktor Krasin on January 4, 1973).
5. Cole, "The Russian Underground," broadcast on *60 Minutes* on July 28, 1970. My thanks to CBS producer Jill Rosenbaum Meyer for helping me procure video footage of the original 1970 interviews with Yakir, Vladimir Bukovsky, and Andrei Amalrik.
6. Bonavia, *Fat Sasha*, 25. Numerous KGB documents describe Yakir as a key figure in the dissident movement, and at least one refers to him as "heading the 'democratic movement,'" but none, to my knowledge, use the term *vozhd*. Bonavia appears to have derived this claim from an incident in which a KGB agent who was tailing a friend of Yakir called out, "Where is your *vozhd*?," in reference to Yakir. Ibid., 35. For the "heading" reference, see Rubenstein and Gribanov, *The KGB File of Andrei Sakharov*, 108.
7. Bonavia, *Fat Sasha*, 28. On relations between dissidents and foreign journalists, see Walker, "Moscow Human Rights Defenders Look West"; idem, "The Moscow Correspondents," in Chatterjee and Holmgren, *Americans Experience Russia*, 139–57; Julia Metger, "Writing the Papers: How Western Correspondents Reported the First Dissident Trials in Moscow, 1965–72," in Brier, *Entangled Protest*, 87–108; Metger, *Studio Moskau. Westdeutsche Korrespondenten im Kalten Krieg*, 140–54; Fainberg, *Cold War Correspondents*, 206–20.
8. *Khronika tekushchikh sobytii*, no. 13 (April 30, 1970).
9. Amal'rik, *Zapiski dissidenta*, 114; Krasin *Poedinok*, 131. Bonavia similarly described Yakir as "no more than a steady drinker, not an alcoholic, because he can do without drink when he has to." Bonavia, *Fat Sasha*, 27. Yakir's arrest would test that claim.
10. Alexeyeva, *Soviet Dissent*, 312.

11. Yakir, *A Childhood in Prison*.
12. David Bonavia, "Message to the West by Man KGB Seized," *Times* (London) (June 23, 1972), 7. Bonavia was expelled from the Soviet Union shortly after Yakir made this recording.
13. Aleksandrovskii and Kislykh, "O nekotorykh takticheskikh priemakh," 71. My thanks to Mark Kramer, director of the Cold War History Project at Harvard University, for generously sharing a copy of this article.
14. Glazov, *V kraiu ottsov*, 81.
15. Krasin, *Sud*, 19 and 85.
16. Aleksandrovskii and Kislykh, "O nekotorykh takticheskikh priemakh," 74; de Boer et al., *Biographical Dictionary of Dissidents*, 126. Krasin's first wife and their three sons emigrated to Israel in 1972.
17. Krasin, *Sud*, 28 and 32.
18. Ibid., 25.
19. Ibid., 37–38.
20. Alexandrovsky as paraphrased, ibid., 41–42.
21. Krasin, *Poedinok*, 117.
22. Krasin, *Sud*, 50.
23. For a partial list, see LYA, f. K-1, ap. 58, t. 12, pp. 113–14.
24. Ibid., t. 12, p. 116.
25. According to Krasin, the "overwhelming majority" of those who were interrogated refused to cooperate. Krasin, *Sud*, 67.
26. *Khronika tekushchikh sobytii*, no. 29 (July 31, 1973) (capitalizations in original). It should be noted that, for reasons explained later in this chapter, issue no. 29 did not actually appear until May 1974; it was backdated by the editors. But the information on witness confrontations had already begun to circulate in real time: see, for example, *SDS*, vol. 25, docs. 1423 and 1424.
27. Alexeyeva, *Soviet Dissent*, 316.
28. Krasin, *Sud*, 50.
29. Sakharov, *Memoirs*, 369.
30. Amal'rik, *Zapiski dissidenta*, 250; *Khronika tekushchikh sobytii*, no. 29 (July 31, 1973).
31. Krasin, *Sud*, 16–18. Yemelkina's point was that perhaps her husband was using the threat of capital punishment to justify—to himself and to others—his capitulation to the KGB.
32. *Khronika tekushchikh sobytii*, no. 30 (December 31, 1973).
33. Krasin, *Sud*, 71.
34. Ibid., 73.
35. Ibid.
36. Ibid., 77–80.
37. *News Conference at the Journalists' Club, Moscow*, 16–17 and 24–25.
38. *Khronika tekushchikh sobytii*, no. 30 (December 31, 1973); *News Conference at the Journalists' Club, Moscow*, 19 and 30. For filmed excerpts from the press conference, see "Press Conference in the Central House of Journalists (1973)," British Pathé, https://www.britishpathe.com/asset/15939/.
39. Mitrokhin Archive, The Chekist Anthology, Folder 9 ("Inakomyslie"), p. 12.
40. Aleksandrovskii and Kislykh, "O nekotorykh takticheskikh priemakh." The article is notable for its silence regarding the use of intimidation and blackmail.
41. GARF, f. R-8131, op. 32, d. 8533, l. 112, quoted in Kozlov and Mironenko, *Kramola*, 52–53.
42. Loshak, *Anatomiia protsessa*.
43. Krasin, *Sud*, 56–57; *Khronika tekushchikh sobytii*, no. 30 (December 31, 1973).
44. Amal'rik, *Zapiski dissidenta*, 250.

45. Mitrokhin Archive, The Chekist Anthology, Folder 12 ("OSOT—Dremliuga"), pp. 1–5; de Boer et al., *Biographical Dictionary of Dissidents*, 112.
46. *Chronicle of Human Rights in the USSR*, no. 27 (July–September 1977): 8–9.
47. Mitrokhin Archive, The Chekist Anthology, Folder 9 ("Inakomyslie"), p. 3.
48. *News Conference at the Journalists' Club, Moscow*, 9–10.
49. Alexeyeva and Goldberg, *The Thaw Generation*, 269.
50. Babitskii and Makarov, "Kto delal 'Khroniku tekushchikh sobytii,'" 9.
51. *Khronika tekushchikh sobytii*, no. 28 (December 31, 1972); Alexeyeva and Goldberg, *The Thaw Generation*, 262–63. There is some confusion on the latter point: elsewhere in the same issue of the *Chronicle*, the editors indicate that the threat of new arrests related to individuals connected, or previously connected, to the *Chronicle*.
52. Kowaljow, *Der Flug des weißen Raben*, 64–65, 83.
53. Alexeyeva and Goldberg, *The Thaw Generation*, 275.
54. One participant in this debate (on Kovalev's side) attributed the eventual victory of those in favor of resuming publication not to superior arguments, but to attrition of their opponents via arrests and emigration. See the testimony by Yuri Gastev in Hopkins, *Russia's Underground Press*, 89.
55. Robert G. Kaiser, "Banned Clandestine Journal of Soviet Dissidents Reappears," *Washington Post* (May 13, 1974); Hopkins, *Russia's Underground Press*, 85–90.
56. "Chitateliam 'Khroniki,'" *Khronika tekushchikh sobytii*, no. 28 (December 31, 1972 [1974]).
57. Krasin, *Sud*, 84.
58. LYA, f. K-1, ap. 58, t. 7, p. 200.
59. Ibid., t. 1, pp. 419–28.

Chapter Sixteen

1. Memorial-Moscow, f. 120, pap. 1, doc. 24, l. 3.
2. Vol'pin, *Izbrannoe*, 11.
3. Article 13 of the Universal Declaration of Human Rights reads: "Everyone has the right to leave any country, including his own, and to return to his country."
4. Chuev, *Molotov: Poluderzhavnyi vlastelin*, 492.
5. Esenin-Vol'pin, *Iuridicheskaia pamiatka dlia tekh, komu predstoiat doprosy*.
6. See the interview with Volpin: Esenin-Vol'pin, "Pamiatka dlia ne ozhidaiushchikh doprosa."
7. Krastev, *Eksperimental'naia rodina*, 123. Leonid Plyushch, describing one of his early interrogations, noted that he "replied eagerly" to the questions put to him, "like most novices with the KGB." "It is hard to believe that the person smiling at you is a dimwit or a scoundrel. It seems like you can convince him of the truth of your views, or at least that you're sincere and not anti-Soviet." Pliushch, *Na karnivale istorii*, 111.
8. Solzhenitsyn, *Bodalsia telenok*, 464–65.
9. For some readers, the interrogations of Decembrist revolutionaries depicted in Natan Eidelman's popular novel *Lunin* (1970) served as a guide for how to approach their own encounters with the KGB. See Rogov, *Semidesiatye kak predmet istorii russkoi kul'tury*, 16. The fascination with interrogations seems to have endured well into the 1980s, as evidenced in the pages of the samizdat journal *Poiski*; see, for example, vol. 4 (1982): 260–72.
10. Full title: *A Billion Years before the End of the World* (*A Manuscript Discovered under Strange Circumstances*) (*Za milliard let do kontsa sveta. Rukopis', obnaruzhennaia pri strannykh obstoiatel'stvakh*). The Russian edition of the novel was serialized in the journal *Znanie-sila* in

1976–77 before appearing as a book in 1984. An English translation appeared in 1978 under the title *Definitely Maybe*.

11. Boris Strugatskii, interview, *Smena*; B[oris] Strugatskii, "Kommentarii k proidennomu."

12. Strugatskii and Strugatskii, *Za milliard let do kontsa sveta*, 69.

13. Al'brekht, *Svidetel' po sobstvennomu delu*, 47.

14. See chapter 6.

15. Podrabinek, *Dissidenty*, 117–23.

16. See chapter 1.

17. All quotations are from the corrected version of the memorandum, in Vol'pin, *Izbrannoe*, 356–72.

18. Ibid. 369 and 405, n. 47.

19. Zitzewitz, *The Culture of Samizdat*, 73; Al'brekht, *Kak byt' svidetelem*, 4.

20. Kelly, *Refining Russia: Advice Literature, Polite Culture, and Gender*.

21. Amal'rik, *Zapiski dissidenta*, 117.

22. Krastev, *Eksperimental'naia rodina*, 142–43.

23. Roginskii, "Dlia nas, eto bylo rubezhnyi den'," in Pazderka, *Vtorzhenie*, 213.

24. Krasin, *Sud*, 85; Medvedev, *Opasnaia professiia*, 143.

25. Quoted in Venclova, *Forms of Hope*, 180.

26. Amal'rik, *Zapiski dissidenta*, 39.

27. On Zelinsky/Nelidov, see chapter 11; Halfin and Hellbeck, "Rethinking the Stalinist Subject"; Hellbeck, *Revolution on My Mind*.

28. Igrunov, "K problematike obshchestvennogo dvizheniia."

29. Aksel'rod-Rubina, *Zhizn' kak zhizn'*, vol. 2, 96. For another example of Volpin's principle of transparency in everyday life, see Reddaway, *The Dissidents: A Memoir*, 172.

30. See chapter 11.

31. Dzhon Gled [John Glad], *Besedy v izgnanii*, 215.

32. Krastev, *Eksperimental'naia rodina*, 135.

33. Al'brekht, *Kak vesti sebia na obyske*; idem, *Kak byt' svidetelem*, 43.

34. See the entry for Jan Albrekht in the Last Address database, accessed December 12, 2023, https://www.poslednyadres.ru/news/news781.htm.

35. Al'brekht, *Kak byt' svidetelem*, 6.

36. Rubin, *Dnevniki. Pis'ma*, vol. 2, 162 (diary entry for July 9, 1975). During Albrekht's interrogations—or at least in his retrospective accounts of them, according to Rubin—he was "like a fish in water. [He was] in his element."

37. Extra-textual in the sense of not present in the original record of the interrogation or apartment search. Soviet law itself was, of course, textual.

38. Al'brekht, *Svidetel' po sobstvennomu delu*, 22–23.

39. Ibid., 23–24.

40. In at least one of his works on the experience of interrogation, Albrekht apparently went wholly over to the side of fiction, imagining a session in which he persuades a KGB officer to let him both pose and respond to the questions. Al'brekht, "190 voprosov po 'delu' Tverdokhlebova" (samizdat, 1976), reprinted in *Materialy samizdata*, vol. 4 (1976), doc. 2501.

41. Al'brekht, *Kak vesti sebia na obyske*, 14 (Question #13).

42. Al'brekht, *Svidetel' po sobstvennomu delu*, 27–28.

43. Al'brekht, *Kak vesti sebia na obyske*, 3.

44. Al'brekht, *Kak byt' svidetelem*, 16.

45. Ibid., 19, 48–49.

46. Krastev, *Eksperimental'naia rodina*, 125–68.

47. Al'brekht, "Vladimir Al'brekht—beseda s Glebom Pavlovskim."
48. Ibid.
49. See chapter 14, note 11.
50. While one cannot rule out KGB redaction of transcripts in order to eliminate passages that reflected poorly on interrogators or cast dissidents in a favorable light, I am unaware of any such cases or even of allegations thereof. On the contrary, among the limited pool of available transcripts there are more than a few moments that cast dissident witnesses and defendants in a far more favorable light than their interrogators.
51. Butov, "Vospominaniia ob odesskikh dissidentakh," 9.
52. Al'brekht, *Kak byt' svidetelem*, 28. At the periphery of the genre of dissident conduct manuals, and considerably less well known than the works analyzed in this chapter, were the following guides: Valerii Chalidze, "Ko mne prishel inostranets" (samizdat, 1971), on how to deal with the KGB in cases involving contact with foreigners; Valerii Senderov, "Pamiatka abiturientu, kotorogo priemnaia komissiia mozhet schest' evreem" (samizdat, 1979), on how university applicants can manage an admissions office that may consider them Jewish; and Ivan Kovalev, "Nekotorye soobrazheniia o pomoshchi politzakliuchennym" (samizdat, 1979), on the most effective techniques—especially for people outside the USSR—for helping Soviet political prisoners. See Aleksei Makarov's introduction to the reprint of Ivan Kovalev's guide in Strukova and Belenkin, *Acta samizdatica / Zapiski o samizdate*, 101–2.
53. Al'brekht, *Kak vesti sebia na obyske*, 2.
54. Grigorenko, *Mysli sumasshedshego*, 293.
55. Gluzman, *Risunki*, 481–82. In order to make his report conform to Soviet professional standards, which required the endorsement of two fellow psychiatrists in addition to the author, Gluzman invented two colleagues whose names he added to his own. Tat'iana Cherbova, interview with Gluzman, *Bul'var Gordona* (August 18, 2011), http://www.bulvar.com.ua/gazeta/archive/s33_64856/7017.html.
56. Gluzman, *Risunki*, 479.
57. Ibid., 476 and 483.
58. Cherbova, interview with Gluzman, *Bul'var Gordona*.
59. Gluzman, "Pis'mo roditeliam," in idem, *Risunki*, 20–21.
60. Ibid., 23–25. The *oprichnina* refers to the special police force established by Tsar Ivan the Terrible in the mid-sixteenth century, which he used to crush the aristocratic boyars and other perceived political opponents.
61. Ibid., 35.
62. Ibid., 210–13.
63. Vol'pin, "Iuridicheskaia pamiatka dlia tekh, komu predstoiat doprosy," in idem, *Izbrannoe*, 370.
64. Bukovskii and Gluzman, "Posobie po psikhiatrii dlia inakomysliashchikh," in Gluzman, *Risunki*, 28, 37, and 47.
65. Ibid., 39.
66. Reich, *State of Madness*, 10 and 60–100.
67. Gluzman, *Risunki*, 36–42, 49.
68. Ibid., 42–43. "Worldly" and "grounded" are my translation of *zazemlennaia*. The disturbing analogy to Hottentots aimed to expose Marxist "class morality" as little more than tribalism, which does nothing to diminish the analogy's racist character.
69. Lenin, *Imperializm, kak vysshaia stadiia kapitalizma*, in idem, *Polnoe sobranie sochinenii*, vol. 27, 397.
70. "The concept of the *rentier* is not scientific," they added. "One will not find it anywhere in Soviet psychiatric literature. But practicing physicians use it in their daily activity, though not

always consciously and of course not as a firm, unchanging template." Gluzman, *Risunki*, 33–34.

71. Ibid., 43–45. Amalrik took this approach even further, boasting that Soviet psychiatrists would never find him mentally ill: "Their basic test involves assessing your relationship to personal gain. If a person does something selflessly, on behalf of an idea, of truth, for freedom and the motherland, that means he's nuts. But I always tell them, 'I, Mr. Investigator, do not work for free, but only for money, because I love money, and my money is busy earning very high interest in a Swiss bank account." Quoted in Voinovich, *Antisovetskii Sovetskii Soiuz*, 355.

72. Reich, *State of Madness*, 89.

73. Gluzman, *Risunki*, 28 and 399.

74. Solzhenitsyn's *Live Not by the Lie* was originally meant to be paired with his *Letter to the Soviet Leaders*: "two sides of a single unity," through which he could "induce both the people [*narod*] and the government to forsake the very same abomination." Solzhenitsyn, *Bodalsia telenok*, 415.

75. *Zhit' ne po lzhi*, 196. Solzhenitsyn invokes a "we" that "writes and reads samizdat," implying that his message was intended for a broad swathe of the intelligentsia. In an interview conducted one month prior to the statement's release (in January 1974), however, when asked what his "fellow countrymen and the youth"—a much larger constituency—could do to support him, he delivered the same message: "to refuse to collaborate in lying," whether in speaking, writing, quoting, signing, voting, or reading. Ibid., 41.

76. Ibid., 195–96 (emphasis in original).

77. Ibid., 196–98.

78. In an otherwise insightful article, the anthropologist Alexei Yurchak mischaracterizes Solzhenitsyn as calling on Soviet citizens "to expose 'the official Lie' at every opportunity." In fact, Solzhenitsyn's essay urged silent non-participation, which in the context of a participatory dictatorship nonetheless involved considerable risk. As Philip Boobbyer notes, *Live Not by the Lie* "was apophatic in the sense that it stressed what should not be done rather than what should be done." Yurchak, "The Cynical Reason of Late Socialism," 169; Boobbyer, *Conscience, Dissent and Reform*, 90. The phrase "speaking Bolshevik" is from Stephen Kotkin, not Solzhenitsyn.

79. Solzhenitsyn, *Zhit' ne po lzhi*, 196–98.

80. According to Ludmilla Alexeyeva, *Live Not by the Lie* "passed almost unnoticed": idem, *Soviet Dissent*, 443. In his authoritative biography of Solzhenitsyn, Michael Scammell offers an uncertain assessment of the reception of Solzhenitsyn's essay. It "circulat[ed] widely in dissident circles," he writes, and became "the chief topic of conversation in the samizdat collection" that bore the same name as the essay. He also describes it, however, as having been met without "much enthusiasm" (Scammell, *Solzhenitsyn*, 862, 896). *Live Not by the Lie* did indeed become the subject of samizdat commentary, but not in the collection cited by Scammell (*Zhit' ne po lzhi: Sbornik materialov avgust 1973–fevral' 1974*). That collection includes a copy of Solzhenitsyn's essay but not responses to it; the volume is devoted rather to events leading up to Solzhenitsyn's expulsion from the USSR in February 1974. It was not until three years later that a different eponymous collection (*Zhit' ne po lzhi [sbornik otklikov-sporov na stat'iu A. I. Solzhenitsyna]* [samizdat, 1977–78]) appeared in three consecutive samizdat volumes. The editor of those volumes, Viktor Sokirko, noted in his introduction that "although there were no immediate discussions or arguments in samizdat regarding [Solzhenitsyn's] testament, it continues to live and exercise its influence on us." See also Boobbyer, *Conscience, Dissent and Reform*, 90–93. I return to the puzzle of the delayed and idiosyncratic reception of *Live Not by the Lie* in chapter 22.

81. In the short run, *Live Not by the Lie* seems to have found its deepest resonance abroad, and in particular in the work of the Czech playwright Václav Havel, who cited Solzhenitsyn's example as he developed the concept of "living in truth." Havel, "The Power of the Powerless," 59.

82. Pliushch, *Na karnivale istorii*, 125; Amal'rik, *Zapiski dissidenta*, 295–99; Litvinov, "Otkrytoe pis'mo A. Solzhenitsynu." On Litvinov's critique (and Solzhenitsyn's reply), see the excellent analysis in Boobbyer, *Conscience, Dissent and Reform*, 90–93.

83. Amal'rik, *Zapiski dissidenta*, 33.

Chapter Seventeen

1. See Svetlana Boym's persuasive reading of one of Pushkin's best-known poems, "From Pindemonte" (1836), as expressing his critique of American political institutions as portrayed in Alexis de Tocqueville's *Democracy in America*, which Pushkin was reading at the time. Boym, *Another Freedom*, 31, 82–88.

Officials at the American embassy in Moscow estimated attendance on Pushkin Square on December 5, 1976, at 150. See Ronald Reagan Presidential Library, Norman Bailey Collection, Folder: Soviet Policy (05/17/1983–05/30/1983), RAC Box 3, "Chronology of Soviet Dissidence: January 1970 through December 1982," p. 20. My thanks to Marc Trachtenberg for sharing this document with me.

2. Alexeyeva, *Soviet Dissent*, 290 and 342.

3. In Russian, these silent allies were called *polu-dissidentskii* or *okolodissidentskii*.

4. *Pravda* (March 22, 1977), quoted in Parchomenko, *Soviet Images of Dissidents and Nonconformists*, 143.

5. *Pravda* (February 12, 1977), quoted in Parchomenko, *Soviet Images of Dissidents and Nonconformists*, 141.

6. *Literaturnaya gazeta* (February 2, 1977), 14, quoted in Parchomenko, *Soviet Images of Dissidents and Nonconformists*, 140; Amal'rik, *Zapiski dissidenta*, 57.

7. Igrunov, "K problematike obshchestvennogo dvizheniia."

8. Igrunov was unaware at the time that Bogoraz and Daniel had separated years before and that Bogoraz was now involved with Anatoly Marchenko. Igrunov, "O biblioteke Samizdata."

9. Ibid.

10. Marynovych, *The Universe behind Barbed Wire*, 45–48.

11. On relations between rights-defenders in the Ural region and their counterparts in Moscow, see the doctoral dissertation by Aleksandr Prishchepa, "Inakomyslie na Urale," 285–342 and 442. My thanks to Amir Weiner for helping to procure a copy of Prishchepa's dissertation.

12. Cherniaev, *Sovmestnyi iskhod*, 89 (diary entry for February 21, 1974).

13. Jallot, *Viktor Orekhov: Un dissident au KGB*, 51–52, 60–61.

14. Ibid., 61–66.

15. Orlov, *Opasnye mysli*, 208–9; Podrabinek, *Dissidenty*, 221–25.

16. Kowaljow, *Der Flug des weißen Raben*, 66–71.

17. Ibid., 68.

18. Ibid., 66–69.

19. Paperno, *Stories of the Soviet Experience*.

20. Verblovskaia, *Moi prekrasnyi strashnyi vek*, 143.

21. Yuri Slezkine, "'He Loved Handing Out Decorations," *New York Review of Books*, vol. 64, no. 10 (June 9, 2022): 35.

22. Millar, "The Little Deal," 697.

23. Kim, "Informal Economy Activities of Soviet Households"; Alexeev and Pyle, "Measuring the Unofficial Economy in the Former Soviet Republics." My thanks to James Heinzen for bringing these articles to my attention.

24. Millar, "The Little Deal," 697; on the legal status of samizdat, see Loeber, "Samizdat under Soviet Law," in Barry et al., *Contemporary Soviet Law*, 84–101.

25. Quoted in Kadarkay, *Human Rights in American and Russian Political Thought*, 109.
26. Shlapentokh, "The Justification of Political Conformism," 114.
27. See chapter 6.
28. Harris, *Lying*, 4.
29. Iurchak, *Eto bylo navsegda*, 222–23.
30. FSO Bremen, f. 98, Bookcase 10, Shelf 2, Box 1, Item 186, pp. 1–10. The author of this 1978 report considered it "extremely likely" that these sentiments were spread by the Communist Party in an attempt "to damage the dissident movement." I do not share this assessment.
31. The phrase comes from Reinhart Koselleck's classic study of the origins of the Enlightenment critique of absolutism: *Kritik und Krise*, 60.
32. Iurchak, *Eto bylo navsegda*, 259. One of the chapters of Valentin Turchin's *Inertsiia strakha* is titled "Everybody Knows Everything." To which Turchin adds, "And they admit that they know everything." *Inertsiia strakha*, 36.
33. Kuzovkin and Makarov, *Dokumenty Initsiativnoi gruppy*, Document 17.
34. Pliushch, *Na karnivale istorii*, 453–54.
35. Versions of this anecdote appear at least three times in publications by Alexei Yurchak, each slightly different and without a cited source. In the first ("The Cynical Reason of Late Socialism," 1997), the person who falls into the pond is identified as a "newcomer"; in the second (*Everything Was Forever*, 2006), he or she is identified as a "dissident." In the expanded Russian edition of Yurchak's book (*Eto bylo navsegda*, 2014, 548), the term is *dissident*. I have used my own translation of Yurchak's Russian-language version. On decoding late Soviet humor, see Graham, *Resonant Dissonance*; and Klumbyte, *Authoritarian Laughter*.
36. Reemstma, *"Wie hätte ich mich verhalten?,"* 9–29.
37. See, for example, Dershowitz, *The Best Defense*, 238: "Some of my own family—like many American-Jewish families—had emigrated from what is currently the Soviet Union. I wondered what I would now be doing if they had not had the foresight to leave. Would I be a dissident follower of Andrei Sakharov? Would I be a 'refusenik,' seeking to emigrate to Israel or the United States? Or would I be one of the silent millions? As I thought about these alternatives I began to feel a bond with Soviet dissidents."
38. Private letter from Sophia Dubnov-Erlikh to Efim Etkind, December 29, 1977, reproduced in Etkind, *Perepiska za chetvert' veka*, 111–12. Dubnov-Erlikh left Russia in 1918.
39. Gurevich, *Istoriia istorika*, 212; idem, "'Put' priamoi, kak Nevskii prospekt,' ili Ispoved' istorika," in Gurevich, *Istoriia—neskonchaemyi spor*, 461.
40. For a third-person example, see the following obituary from 1996: "In the 73rd year of his life, the writer and publisher Anatoly Zlobin has died—a person of the front generation who was never a dissident in the literal sense but who for years wrote for the desk drawer." *Kommersant* (March 16, 1996), 13.
41. Navrozov, "Getting out of Russia," 48–52; idem, *The Education of Lev Navrozov*, 22.
42. Navrozov, "Getting out of Russia," 48–52. See also Bergman, "Reading Fiction to Understand the Soviet Union: Soviet Dissidents on Orwell's *1984*"; Masha Karp, *George Orwell and Russia*, 224–48.
43. Quoted in Komaromi, *Soviet Samizdat*, 76–77. I have slightly amended the translation.
44. Quoted from an interview by Sofia Tchouikina, in idem, "Anti-Soviet Biographies: The Dissident Milieu and Its Neighboring Milieux," in Humphrey et al., *Biographical Research in Eastern Europe*, 133–34.
45. Dissident networks had been assisting the families of political prisoners ever since the 1966 trial of Andrei Sinyavsky and Yuli Daniel. Altruism on a larger scale began to materialize several years later. When Sakharov was awarded the 1974 Prix mondial Cino Del Duca for

humanitarian service, he and Yelena Bonner used the prize money to establish a fund for the children of Soviet political prisoners. In the same year, the exiled Solzhenitsyn used the considerable royalties from *The Gulag Archipelago* to establish the Russian Social Fund to Aid Political Prisoners and Their Families. By 1977, the Solzhenitsyn Fund had dispensed roughly 200,000 rubles and raised an additional 70,000 rubles from anonymous donors inside the USSR. The number of prisoners and families assisted by the Fund ranged from 120 to 720 per year. See Sakharov, *Memoirs*, 414; Scammell, *Solzhenitsyn*, 874; Goldberg, *The Final Act*, 216. On the culture of altruism within the dissident movement, see Walker, "Pollution and Purification in the Moscow Human Rights Networks."

46. Sakharov, *Memoirs*, 361–62. The dissident Yuri Glazov noted, "Not infrequently, we hear that educated Soviet citizens blame the participants of the human rights movement for being possessed by devils." Glazov, "*The Devils* by Dostoevsky and the Russian Intelligentsia," 327.

47. Arbatov, *The System*, 225. On the public campaigns against Sakharov and Solzhenitsyn, see Sakharov, *Memoirs*, 385–91; Bergman, *Meeting the Demands of Reason*, 199–206; and Scammell, *Solzhenitsyn*, 807–8 and 829–31.

48. Krastev, *Eksperimental'naia rodina*, 154.

49. Alexeyeva and Goldberg, *The Thaw Generation*, 244.

50. Zinov'ev, *Bez illiuzii*, 91.

51. Goritschewa, *Von Gott zu reden ist gefährlich*, 47, 57–58, 122.

52. Shimanov, *Zapiski iz krasnogo doma*, 226–27. For the broader evolution of Shimanov's thought and activity, see Suslov, "Utopicheskii proekt G. M. Shimanova v kontekste pravogo dissidentskogo dvizheniia v SSSR"; and Mitrokhin, *Russkaia partiia*, 516–26.

53. Shimanov, *Zapiski iz krasnogo doma*, 223, 228.

54. Ibid., 228; see also the 2004 transcript of Shimanov's memoirs in "Russkaia narodnaia liniia: Pravoslavie, samoderzhavie, narodnost'"; and the interview with Shimanov in *Evrei v SSSR*, no. 13 (1977), reprinted in *Evreiskii samizdat*, vol. 13, 177. My thanks to Samuel Finkelman for bringing the interview to my attention.

55. Solzhenitsyn, "Nashi pliuralisty," 133 and 151–52.

56. Ibid., 149 and 154.

57. Tromly, *Making the Soviet Intelligentsia*; see also Joseph Berliner's foreword to Millar, *Politics, Work, and Daily Life in the USSR*, especially xi–xii.

58. Vaissié, *Pour votre liberté et pour la nôtre*, 343.

59. Amalrik, preface to *Will the Soviet Union Survive until 1984?* (1981 [2nd ed.]), 10; Gluzman, *Risunki*, 413; Al'brekht, *Kak vesti sebia na obyske*; Solzhenitsyn, *Zhit' ne po lzhi*, 196. Writing under the pseudonym "Iar. Iasnyi," Boris Shragin compared the medium of samizdat itself to the boy who cried out about the emperor's lack of clothing. See *SDS*, doc. 408, p. 2.

60. For a different "Soviet" adaptation of the tale, see Vladimir Voinovich, "Novaia skazka o golom korole," *Literaturnaia Gazeta* no. 22 (June 5, 1991), 12.

Chapter Eighteen

1. Krastev, *Eksperimental'naia rodina*, 123.

2. Volpin, unpublished memoir, 16.

3. Signers of protest letters often indicated their occupation or workplace affiliation, making it relatively easy to classify them as intellectuals, workers, students, and so on. To ascertain their nationality would have required additional research, using their last names and patronymics (hardly a foolproof method) or making inquiries among acquaintances, which would have been potentially awkward, even for someone as brash as Amalrik. The point, however, is that it seems not to have occurred to him to investigate the national composition or gender of the signers.

NOTES TO CHAPTER 18   721

4. Remeikis, *Opposition to Soviet Rule in Lithuania*, 100.

5. Vardys, *The Catholic Church, Dissent and Nationality in Soviet Lithuania*, 145–46; Remeikis, *Opposition to Soviet Rule in Lithuania*, 114–21.

6. Kowaljow, *Der Flug des weißen Raben*, 76; Alexeyeva and Goldberg, *The Thaw Generation*, 288.

7. Vardys, *The Catholic Church, Dissent and Nationality in Soviet Lithuania*, 148–51.

8. Spengla, *The Church, the "Kronika," and the KGB Web*, 44; Kowaljow, *Der Flug des weißen Raben*, 75–78.

9. Spengla, *The Church, the "Kronika," and the KGB Web*, 48 and 75.

10. *Chronicle of the Lithuanian Catholic Church*, no. 21 (January 25, 1976), accessed at https://lkbkronika.lt/en/.

11. Quoted in Remeikis, *Opposition to Soviet Rule in Lithuania*, 162.

12. Electronic Library of the History Department at Moscow State University, "Konstitutsiia (Osnovnoi Zakon) Rossiiskoi sotsialisticheskoi federativnoi sovetskoi respubliki," accessed December 12, 2023, http://www.hist.msu.ru/ER/Etext/cnst1936.htm#1, Article 17. The absence of any guidelines regarding the procedure by which a republic might secede from the USSR suggests the shallowness of the commitment to this particular right.

13. Bilocerkowycz, *Soviet Ukrainian Dissent*, 191–93.

14. Liber, "Ivan Dzyuba, Cultural Cringe, and the Origins of Ukraine's Revolution of Dignity," 57–58.

15. See in this regard Astrouskaya, *Cultural Dissent in Soviet Belarus*.

16. *Ukrainsky visnyk*, no. 5 (1971), reprinted in *Khronika tekushchikh sobytii*, no. 22 (November 10, 1971).

17. Alexeyeva, *Soviet Dissent*, 52; Komaromi, *Soviet Samizdat*, 116–17. On the whole, Alexeyeva's description is accurate. But as Samuel Finkelman has shown, several prominent Ukrainian nationalists publicly expressed support for Ukrainian Jews. See Finkelman, "Ghetto, Gulag, Geulah," chaps. 4 and 5.

18. My treatment of relations between Soviet dissidents and the Jewish national movement in the USSR builds on Rubenstein, *Soviet Dissidents*, chap. 5, "Zionists and Democrats," and Fürst, "Born under the Same Star," in Ro'i, *The Jewish Movement in the Soviet Union*, 137–65. For a fuller version of my thinking on this topic, see Nathans, "Refuseniks and Rights Defenders," in Moss et al., *From Europe's East to the Middle East*, 362–75.

19. This is a (very) partial list of dissidents whose mother or father (or both) were Jewish according to Soviet nationality categories used in passports and other forms of personal identification.

20. Quoted in Amal'rik, *Zapiski dissidenta*, 88. This passage was left out of the English translation of Amalrik's memoir.

21. Quoted in Azbel et al., "Dvadtsat' let spustia," 175. Maya Ulanovskaya recounted a meeting between dissidents and Zionists in which only Jews were present. Ulanovskaia and Ulanovskaia, *Istoriia odnoi sem'i*, 273.

22. Zinov'ev, *Sobranie sochinenii*, vol. 5, *Gomo sovetikus*, 76.

23. Rubenstein, *Soviet Dissidents*, 183; Scammell, *Solzhenitsyn*, 664 (the variant "Solzhenitser" was also deployed).

24. Amal'rik, *Zapiski dissidenta*, 101.

25. See the interview with Shimanov in *Evrei v SSSR*, no. 13 (1977), reprinted in Ingerman, *Evreiskii samizdat*, vol. 13, 175–77. Speculation regarding Shimanov's allegedly Jewish roots centered on the similarity of his last name to the Hebrew word *shem* (name) as well as the shape of his nose.

26. See the report to the Central Committee by Filipp Bobkov (May 10, 1972), in Morozov, *Evreiskaia emigratsiia v svete novykh dokumentov*, 139.

27. On pre-revolutionary Jewish integration into Russian society, see Nathans, *Beyond the Pale*.

28. Valery Chalidze participated in the drafting of the statement. Quoted in Rubenstein, *Soviet Dissidents*, 166. I have slightly modified the translation for purposes of readability.

29. See Beckerman, *When They Come for Us, We'll Be Gone* and the literature cited there.

30. Chalidze, *Prava cheloveka i Sovetskii Soiuz*, 115.

31. Voronel', "O vrednoi funktsii slov i probleme assimiliatsii evreev," 10. Reprinted in *Evreiskii samizdat*, vol. 10.

32. Weinberg, *Stalin's Forgotten Zion*; Gessen, *Where the Jews Aren't*.

33. See the online discussion of James Loeffler's *Rooted Cosmopolitans: Jews and Human Rights in the Twentieth Century* in H-Diplo, vol. 20, no. 31 (April 1, 2019), accessed December 12, 2023, https://networks.h-net.org/node/28443/discussions/3947096/roundtable-xx-31-rooted-cosmopolitans-jews-and-human-rights.

34. The dwindling population of Soviet Jews who were familiar with the Talmud or other traditional texts, moreover, remained entirely aloof from the dissident movement.

35. On appeals to Jewish history and values by Jewish participants in the African-American civil rights movement, see Greenberg, *Troubling the Waters*, 4–5. I know of only a single Soviet analogy: the Ukrainian-Jewish psychiatrist Semyon Gluzman, whose 1974 letter to his parents explaining his decision to protest against Soviet punitive psychiatry (discussed in chapter 16) invoked the Holocaust and the founding of the state of Israel (though not by name).

36. Bogoraz, "Kto ia?"

37. In the opening of *The Eighteenth Brumaire of Louis Bonaparte* (1852), Marx wrote, "People make their own history, but they do not make it as they please; they do not make it under self-selected circumstances, but under circumstances existing already, given and transmitted from the past," 15.

38. Belfrage, *A Room in Moscow*, 154–55.

39. Slezkine, *The Jewish Century*, 342–46; Fürst, "Born under the Same Star."

40. Quoted in Zubok, *Zhivago's Children*, 305.

41. Gluzman, *Risunki po pamiati*, 417.

42. I am using "libertarian" in the sense proposed by Volpin, namely, seeking to maximize human freedom within a framework of law and rights, but without any assumptions about a market economy as a necessary component of that framework.

43. Kowaljow, *Der Flug des weißen Raben*, 76–77. In Russian, as in French, the term "human rights" has traditionally emphasized the singular individual as bearer of those rights: *prava cheloveka* (*droits de l'homme*), that is, "rights of the person" or "rights of man."

44. Sakharov, "Razmyshleniia o progresse," in idem, *Trevoga i nadezhda*, 38.

45. Zubok, *Zhivago's Children*, 304–5.

Chapter Nineteen

1. For helpful background on Kochetov's novel, see Glad, "Vsevolod Kochetov: An Overview"; and the Wikipedia page "Кочетов, Всеволод Анисимович," accessed on December 12, 2023, https://ru.wikipedia.org/wiki/Кочетов,_Всеволод_Анисимович.

2. Kochetov, "Chego zhe ty khochesh'?," *Oktiabr'*, no. 9, 116–17.

3. Ibid., 117–18.

4. Ibid., 118.

5. Ibid., 119.

6. Sokirko, "'Krivoe zerkalo.'"

7. Memorial Society, "Grigorii Pod"iapol'skii," accessed on December 12, 2023, http://old.memo.ru/history/podjap/stih.htm, "Kochetiana."

8. "Vokrug romana Kochetova 'Chego zhe ty khochesh'?,'" *Khronika tekushchikh sobytii*, no. 12 (February 28, 1970); no. 14 (June 30, 1970).

9. For writing and disseminating *What Is It, Rooster?*, Z. Papernyi, a senior researcher at the Institute of World Literature in Moscow, was expelled from the Communist Party. *Khronika tekushchikh sobytii*, no. 14 (June 30, 1970).

10. Anatoly Sofronov was the long-serving editor of the journal *Ogonek* (Spark) and, according to the literary historian Evgeny Dobrenko, "one of the most feared literary hangmen of the Stalinist era." Dobrenko, "Stalinskaia kul'tura: Skromnoe obaianie antisemitizma," 52.

11. For a thorough analysis of the tangled relations between Kochetov and Blake, see Kukulin, "A Bizarre Encounter."

12. Pliushch, *Na karnivale istorii*, 384. Plyushch's diagnosis of Kochetov's alleged pathologies may itself have been an act of revenge for Soviet psychiatrists' diagnosis of Plyushch as suffering from "sluggish schizophrenia" and "reformist delusions," for which he was confined in a psychiatric hospital and drugged for three years.

13. Kochetov, "Chego zhe ty khochesh'?," *Oktiabr'*, no. 11, 123.

14. Sokirko, "Krivoe zerkalo."

15. Ibid. The phrase "cultural revolution" gestures to contemporaneous upheavals in Mao Zedong's China. The phrase "Red Hundreds" merges the activists of that revolution, the so-called Red Guards (Hong Weibing), with the pro-monarchist, ultra-nationalist Black Hundreds of pre-revolutionary Russia.

16. On Soviet dissidents and the *Landmarks* (*Vekhi*) collection, see Bergman, "Soviet Dissidents on the Russian Intelligentsia." Fear of the *demos* by self-styled democrats, it should be noted, is by no means unique to the Soviet setting.

17. Sokirko, "Krivoe zerkalo."

18. Wikipedia, "Кочетов, Всеволод Анисимович."

19. Alexeyeva and Goldberg, *The Thaw Generation*, 96; Litvinov, *Protsess chetyrekh*, 48. According to Nina Voronel, Zlobina chose the pseudonym "Anna Gerts" to link her work to that of Abram Terts, that is, Andrei Sinyavsky, with whom she—according to Voronel—had an affair before his arrest in 1965. "A. Gerts" may also be a nod toward Alexander Herzen, or Gertsen in Russian.

20. Anna Gerts [Maia Zlobina], "K vol'noi vole zapovednye puti . . . ," *Novyi zhurnal*, no. 121, 33.

21. Ibid., 40–41.

22. Ibid., 40–45, 56.

23. Ibid., 31–32. Marishka, it should be noted, is Tatar, not Russian.

24. Ibid., no. 123, 15–16; no. 122, 73–74.

25. Ibid., no. 122, 69.

26. *Khronika tekushchikh sobytii*, no. 42 (October 8, 1976).

27. On the movement's "home front," see Chuikina, "The Role of Dissident Women in Creating the Milieu," 190.

28. Bogoraz, "Melkie besy," 213–15. The title of Bogoraz's review refers to Fedor Sologub's 1907 novel *Petty Demon*, which explores the banality and malevolence of Russian provincial life. A slightly shorter version of her review, under the same title, appeared in 1976 in New York as a pamphlet published by Valery Chalidze. In both versions, Bogoraz refers to the author of *Sacred Paths to Willful Freedom* exclusively by her pseudonym, Gerts. Bogoraz was almost certainly aware that Gerts was in fact Maya Zlobina, whom she knew personally. Despite her sharply negative review, she did not violate the unspoken dissident rule against unmasking pseudonymous authors. I use the name Zlobina except when quoting Bogoraz.

29. Ibid., 216.

30. Ibid., 220.

31. Other contemporary fictional treatments of the dissident milieu can be found in Feliks Roziner, *Nekto Finkel'maier: Roman* (London, 1981); Aleksandr Zinov'ev, *Gomo sovetikus* (Lausanne, 1981); Evgenii Kozlovskii, *Krasnaia ploshchad': Dissident i chinovnitsa* (Paris, 1982); and Vladimir Voinovich, *Shapka* (London, 1988). To the best of my knowledge, none of these works elicited reactions from readers in the USSR as vigorous as those provoked by Kochetov and Zlobina. More recently, Soviet dissident characters have appeared in novels by Lyudmila Ulitskaya (*Zelenyi shater*, 2010), David Bezmozgis (*The Betrayers*, 2014), Paul Goldberg (*The Dissident: A Novel*, 2023), and other post-Soviet writers.

32. The title was lifted without attribution. The poem appears within Yuli Daniel's story "Atonement."

33. Komaromi, *Soviet Samizdat*, 140.

34. Bogoraz, "Melkie besy," 221.

35. Ibid.

36. Gorbanevskaia, "Neskol'ko slov v 'poslesloviie,'" 222–23.

37. Ibid., 224–25.

38. Gerts, "K vol'noi vole," *Novyi zhurnal*, no. 124, 58–59.

39. Turchin, *Inertsiia strakha*, 35.

40. Bonavia, *Fat Sasha*, 13; Alexeyeva and Goldberg, *The Thaw Generation*, 275.

41. Alexeyeva and Goldberg, *The Thaw Generation*, 244; Bonner, *Alone Together*.

Chapter Twenty

1. Hughes, *Sophisticated Rebels*, 13. Hughes makes the case for a shared culture of protest after 1968 that was "sophisticated in the sense of recognizing realistic limits and frequently defying conventional classification as right or left." The title of Hughes's book was meant to highlight the contrast to earlier modes of protest analyzed in Hobsbawm, *Primitive Rebels*. "Global lingua franca of protest": see Klimke and Scharloth, *1968 in Europe*, 106. "International dissident culture": see Suri, *Power and Protest*, 94. More broadly, see della Porta, "'1968'—Zwischen nationale Diffusion und transnationale Strukturen," in Gilcher-Holtey, *1968: Vom Ereignis zum Gegenstand der Geschichtswissenschaft*, 131–50. See also Fürst, "From the Maiak to the Psichodrom: How Sixties Global Counterculture Came to Moscow," in Jian et al., *The Routledge Handbook of the Global Sixties*, 180–92; and Müller, *Contesting Democracy*, 171–242.

2. Jeremi Suri, in *Power and Protest*, offers thought-provoking but empirically unsustainable arguments regarding a language and culture of dissent shared by student radicals in the West and their alleged counterparts in the Soviet Union and China (see especially 3, 94, and 164). Claims about the radicalizing effects of an expanding postwar university population do not hold for the Soviet case, where, in any event, the majority of dissidents were well beyond their university years. Suri's reading of Solzhenitsyn's *One Day in the Life of Ivan Denisovich* and his use of Solzhenitsyn as representative of the Soviet dissident movement as a whole are wide of the mark. *One Day* does not "single out Stalin for criticism," nor does it offer a "condemnation of the communist political project" (Suri, 105), which, needless to say, would have torpedoed its publication, not to mention Khrushchev's personal approval thereof. Suri's central claim that a "social crisis of the nation-state" (211) helped usher in détente doesn't apply to Brezhnev's Soviet Union. The USSR's shift toward détente was driven by the desire for geopolitical, technological, and economic gains; détente itself was associated in the USSR with heightened risk of ideological contamination from the West, and thus the possibility of social unrest.

3. See the classic study by Gitlin, *The Whole World Is Watching: Mass Media in the Making and Unmaking of the New Left*.

4. Dovlatov, *Kompromiss*, 44.

5. These included professional organizations of Western scientists, mathematicians, writers, and psychiatrists. On their mobilization on behalf of Soviet dissident colleagues, see Rhéaume, *Sakharov: Science, morale et politique*, which shows that Soviet dissidents had less in common with Berkeley students in the 1960s than with their politically engaged physics professors in the 1970s.

6. On human rights language as "thin description," see Cmiel, "The Emergence of Human Rights Politics in the United States," 1248–49. Human rights language is "thin," or minimalist, in the sense that claims of a *right* to speak freely, associate freely, or worship freely have nothing to say about the *content* of that speech, association, or worship, or whether it is wise or necessary to exercise such rights in a particular way or in a particular setting. The right to vote similarly tells us nothing about how to vote.

7. Amnesty International, *Annual Report*, 1961–62 (London, 1962), 1. On Amnesty's founding, see Buchanan, "'The Truth Will Set You Free.'" For histories of Amnesty, see Eckel, *Die Ambivalenz des Guten*, 347–434; Clark, *Diplomacy of Conscience*; Hopgood, *Keepers of the Flame*.

8. Peter Benenson, "The Forgotten Prisoners," *Observer* (London) (May 28, 1961).

9. See, for example, Amnesty International, *Annual Report*, 1962–63 (London, 1963), 12.

10. Ibid., 3.

11. Hunt, *Inventing Human Rights*, 35–69.

12. Amnesty International, *Annual Report*, 1962–63, 1.

13. Amnesty International, *Personal Freedom in the Marxist-Leninist Countries*, 5.

14. On Benenson's biography, see Loeffler, *Rooted Cosmopolitans*, 202–29. Benenson's maternal aunt, Manya Harari, was a prominent translator from Russian, including works by Andrei Amalrik, Yevgenia Ginzburg, Pavel Litvinov, Boris Pasternak, Andrei Sinyavsky (his fiction as well as the transcript of his trial), and Alexander Solzhenitsyn.

15. International Institute for Social History, Amnesty International Archive, Index Documents on the USSR (EUR 46) (henceforth IISH AI-Index), Folder 438, "Report on political prisoners in USSR" (not later than 1969), p. 1.

16. Thakur, "Human Rights: Amnesty International and the United Nations."

17. International Institute for Social History, Amnesty International Archive, International Executive Committee (EUR 46) (henceforth IISH AI-IEC), Microfilm 243, "International Executive Committee meeting, February 17/18, 1968," frames 330 and 387.

18. Amnesty International, *Annual Report*, 1967–68 (London, 1968), 7. In October 1968, Peter Archer, a British Member of Parliament and Amnesty supporter, applied for permission to attend the trial of Larisa Bogoraz and the other Red Square demonstrators, but was refused a visa by the Soviet embassy in London. A year later, Amnesty's Swedish section applied for permission to send an observer to the trial of Ivan Yakhimovich, to which it received no response. An internal Amnesty report on the Soviet Union noted in 1973: "It has not been possible for an Amnesty observer to attend a single trial in the Soviet Union.... Applications to attend trials, even by foreign journalists, are considered by the authorities as an intervention in their internal affairs." IISH AI-Index, Folder 446, "Situation Paper on the USSR, November 1973," p. 6.

19. Amnesty International, *Annual Report*, 1967–68, 7.

20. IISH AI-Index, Folder 446, "Situation Paper on the USSR, November 1973," pp. 5–6.

21. For a sample form letter, together with instructions on how to address Soviet officials, see IISH AI-Index, Folder 433, pp. 1–3. On the epigraph's widespread but incorrect attribution to Voltaire, see Kinne, "Voltaire Never Said It!" My thanks to Fara Dabhoiwala for this reference.

22. S. S. Smirnov, "Moguchii golos solidarnosti," *Pravda* (July 17, 1968), 4. Smirnov was chairman of the Soviet Committee for Solidarity with Greek Democrats and a member of the

Moscow branch of the Union of Soviet Writers. He had chaired the session that demanded the expulsion of Boris Pasternak from the Soviet Union. My thanks to Alexander Gribanov for this information.

23. Beliakov, *Mezhdunarodnye nepravitel'stvennye organizatsiia*, 23–24.

24. Rhodesia's white minority, determined to avoid majority rule, had unilaterally declared independence from Great Britain a year before, to London's great displeasure. See Buchanan, "Amnesty International in Crisis, 1966–7"; and Tolley, *The International Commission of Jurists*, 125–27. The International Commission of Jurists, with headquarters in Geneva, was covertly funded by the CIA via its U.S. affiliate, the American Fund for Free Jurists, established as a counterweight to the International Association of Democratic Lawyers, a Soviet-bloc organization. Not all of the relevant information about Amnesty International emerged during the 1966–67 crisis. As Buchanan shows, Benenson had worked for the British intelligence services during World War II; in addition, he received covert assistance from the Foreign Office's Information Research Department when he established Amnesty and continued to do so for several years thereafter.

25. They did not, however, go unnoticed. For examples of letters from local Amnesty adoption groups in England, Holland, Sweden, and Australia, preserved in the Soviet Procuracy's case files on various dissidents, see GARF, f. 8131, op. 36, d. 5459; op. 31, d. 99561; and op. 36, d. 2752. In none of the examples I examined did the case files include comments on or responses to the letters. On occasion, local Amnesty groups who sent letters to the families of political prisoners did receive responses. My thanks to Peter Reddaway for this information.

26. The prisoners and their families belonged to a group that in 1961 had broken from the All-Union Council of Evangelical Christian Baptists, the state-authorized organization designed to maintain the Kremlin's grip on the USSR's various Baptist communities, whose members numbered in the hundreds of thousands.

27. *SDS*, vol. 14, doc. 770; Sawatsky, *Soviet Evangelicals since World War II*, 139–41.

28. IISH AI-Index, Folder 437, p. 35; Amnesty International, *Annual Report*, 1966–67 (London, 1967), section on USSR (no pagination).

29. Columbia University, Rare Book and Manuscript Library (henceforth Columbia RBML), Amnesty International USA (AI-USA), Record Group II, Series 5, Box 9, Folder 22, Document 1. Although the Lviv trial was held in May 1961, the prisoners' petition may have been sent as late as 1964. It was published, with accompanying documents, in Maistrenko, *Ukrainski iurysty pid sudom KGB* (Munich, 1968).

30. On contemporary techniques of information-gathering by human rights NGOs, see Weissbrodt and McCarthy, "Fact-Finding by International Nongovernmental Human Rights Organizations," which notes that during the period 1971–78, Amnesty sponsored 111 inquiry missions and 76 trial observer missions, none of them in the USSR. Amnesty's relations with the Soviet Union's East European allies were, if anything, even chillier: see, for example, Mihr, *Amnesty International in der DDR*, especially 311–14; Miedema, *Not a Movement of Dissidents*.

31. IISH, AI-Index, Folder 446, "Situation Paper on the USSR, November 1973", p. 2; Folder 455, "Report #2 on meeting of USSR coordination groups, London, June 14–15, 1975," p. 3; Columbia RBML, AI-USA, Record Group II, Series 5, Box 9, Folder 23, "Prisoners of Conscience in Eastern Europe: A Report published to mark the centenary of Lenin's birth (April 22, 1870)."

32. Amnesty International, *Annual Report*, 1970–71 (London, 1971), 70, describes the *Chronicle* as "Amnesty's main source of information on Soviet prisoners of conscience and new arrests." See also Reddaway, *The Dissidents: A Memoir*, 120–24.

33. O'Keeffe and Szamuely, *Samizdat*, 24.

34. IISH AI-Index, Folder 455, "Report on meeting of USSR coordination groups, London, June 14–15, 1975," pp. 13–14. On the strain created by increasing flows of samizdat to Amnesty's researchers, see Hurst, *British Human Rights Organizations*, 150–60.

35. IISH AI-IEC, Microfilm 243, "Development of research on the USSR, March 21–22, 1970," p. 708; Amnesty International, *Annual Report*, 1971–72 (London, 1972), 43.

36. Columbia RBML, AI-USA, Record Group IV, Series 1, Subseries 3, Box 11 (Soviet Coordination Group), Folder 9, Document 1, p. 1.

37. Data for 1968: IISH AI-IEC, Microfilm 243, "Currently adopted prisoners as of August 1968," p. 386. Data for 1982: Columbia RBML, AI-USA, Record Group IV, Series 1, Subseries 3, Box 11, Folder 12, Document 1, "Minutes of international meeting (Paris, March 20–21, 1982) of USSR Coordination Groups," p. 4; Record Group II, Series 2, Box 11, Folder 15, Document 1, "International Meeting of USSR Coordinators, March 20–21, 1982," pp. 1–8. "The balance of [our] work as a whole" refers to Amnesty's commitment to proportionality in its advocacy on behalf of prisoners in the capitalist, communist, and developing worlds.

38. The *Chronicle*'s coverage of the persecution of religious believers led some of its readers inside the USSR to accuse the journal of "conducting religious propaganda," a charge its editors vigorously denied. *Khronika tekushchikh sobytii*, no. 10 (October 31, 1969), "Otvety chitateliam."

39. Columbia RBML, AI-USA, Record Group II, Series 5, Box 9, Folder 23, "Monthly Newsletter from AI: Postcards for Prisoners Campaign" (issues from April and June 1971); *SDS*, vol. 14, doc. 770.

40. IISH AI-IEC, Microfilm 243, "International Executive Committee meeting, February 17–18, 1968," p. 330.

41. Arendt, *The Origins of Totalitarianism* (1950), 290–302.

42. In the Russian Orthodox Bible, the first quotation is located in Psalm 118. These are among the dozens of biblical quotations found, for example, in the dissenting Baptists' public appeals from the first half of the 1960s. See *SDS*, vol. 14, docs. 673, 770, and 771.

43. These distinctions are not absolute: Soviet Baptists were open to newcomers and united by faith, although in practice they tended to be organized in extended family units. For their part, the metropolitan dissidents, while highly individualized and brought together primarily by commitment to civil liberties, were not infrequently linked via marriage and family. Thus, for example, in addition to the married couples Bonner/Sakharov, Bogoraz/Marchenko, Krasin/Yemelkina, Shragin/Sadomskaya, and Volpin/Kristi (his second wife, the mathematician Irina Kristi), Pavel Litvinov's wife, Maya, was the daughter of Raisa Orlova and Lev Kopelev; Valery Chalidze was married to Litvinov's cousin Vera Litvinova; Petr Yakir's daughter, Irina, was married to Yuli Kim; Larisa Bogoraz and Irina Belogorodskaya were cousins; the dissident sisters Tatyana, Zoya, Ksenya, Yekaterina, and Maria Velikanova were married to, respectively, the dissidents Konstantin Babitsky, Nikolai Yarnykh, Sergei Myuge, Alexander Daniel (son of Yuli Daniel and Larisa Bogoraz; Yekaterina Velikanova subsequently married the Leningrad dissident Arseny Roginsky), and Andrei Grigorenko (son of Petr Grigorenko).

44. IISH AI-IEC, Microfilm 243, "Development of research on the USSR, March 21–22, 1970," p. 708; Columbia RBML, AI-USA, Record Group IV, Series 1, Subseries 3, Box 11, Folder 9, Document 3, "Notes from April 20–21, 1974 meeting of national sections on the USSR."

45. Tolley, *The International Commission of Jurists*, 100–102 and 120.

46. IISH AI-Index, Folder 446, "Situation Paper on the USSR, November 1973," p. 7.

47. "Ne po adresu, gospoda!," *Pravda Ukrainy*, no. 256 (November 1, 1970): 2.

48. K. Briantsev, "Lzheradeteli v triasine klevety," *Izvestiia* (October 23, 1971), 4.

49. IISH AI-IEC, Microfilm 244, "Report from Nigel Rodley on visit to Moscow, August 24–29, 1973," pp. 763–64.

50. Quoted in Christopher Wren, "Amnesty Group in Private Soviet Talks," *New York Times* (November 2, 1973), A2.

51. Author's interview with Yadja Zeltman, October 9, 2012. Group 11's unofficial name came from the Madison Avenue address of Zeltman, its chair. See also the papers of the late

Edward Kline, Folder "Soviet Human Rights Movement—General," transcript of telephone conversation with A. Tverdokhlebov, 10 P.M. London time (September 5, 1973); letter from Andrew Blane to Ivan Morris, professor of Japanese literature at Columbia University and chair of Amnesty International in the United States (January 31, 1974), p. 1. One of the most important American supporters of the Soviet dissident movement and an engaged member of Amnesty International, Kline graciously allowed me to conduct research in his personal archive, then stored in his New York City apartment. The bulk of the collection can now be accessed in Columbia University's Rare Book and Manuscript Library, where as of 2023 it is in the process of being catalogued. A smaller portion of the Kline papers is housed in the Andrei Sakharov Archives at Harvard University's Houghton Library. All citations here are to the original collection (henceforth Kline Archive) with Kline's own subject headings and labels.

52. For the tables of contents of the first three samizdat collections of Amnesty materials, see Chalidze, *Andrei Tverdokhlebov—v zashchitu prav cheloveka*, 115–17.

53. In an increasingly common ritual among Soviet rights activists, several days later Tverdokhlebov circulated in samizdat a brief description of the search and a list of seized items. *SDS*, vol. 25, doc. 1478.

54. Group 73's charter invoked "international experience in helping prisoners and their families, and in particular the experience of Amnesty International, which sets as its goal the provision of assistance to prisoners of conscience and political prisoners." Tverdokhlebov et al., "Printsipy Gruppy-73," in *SDS*, vol. 25, doc. 1486.

55. Kline Archive, Folder "Soviet Human Rights Movement—General," Otchet Arkhangel'skogo sovetu uchreditelei "Gruppy-73" (September 1, 1973).

56. Columbia RBML, AI-USA, Record Group II: Executive Director Files, 1967–97, Series 5: National Section Memos, Box 9, Folder 23, Document 3. The quoted passage is my translation of the original Russian text in *SDS*, vol. 25, doc. 1487.

57. For a contrasting analysis of the "vernacularization" of human rights language in the United States during the same period, in part under the influence of the Soviet dissident movement, see Bradley, "American Vernaculars: The United States and the Global Human Rights Imagination."

58. Nathans, "The Disenchantment of Socialism," in Eckel and Moyn, *The Breakthrough*, 33–48.

59. Turchin, *Inertsiia strakha*, 288 (emphasis in original). Idem, *Fiziki shutiat*.

60. Orlov, *Opasnye mysli*, 149.

61. Mikola Rudenko, "Otkrytoe pis'mo L. I. Brezhnevu," *Materialy samizdata*, vol. 29 (1975), doc. 2215.

62. Turchin, *Inertsiia strakha*, 195 and 202.

63. Ibid., 288–89.

64. The five others were the mathematicians Nikolai Beloozerov and Boris Landa, the biologist Sergei Kovalev, the engineer Ernst Orlovsky, and the Russian Orthodox priest Sergei Zheludkov. For the application to join Amnesty, see Kline Archive, Folder "Soviet Human Rights Movement—General," "Zaiavlenie [October 6, 1973]." Identical copies can be found in IISH AI-IEC, Microfilm 244, pp. 1955–57; and *SDS*, vol. 28, doc. 1501.

65. For these and other concerns, see Kline Archive, Folder "Soviet Human Rights Movement—General," letter from Jane Ward of Amnesty International, International Secretariat Research Department, to Yadja Zeltman (August 16, 1973).

66. Kline Archive, letter from Andrew Blane to Ivan Morris (January 31, 1974), p. 4.

67. In an October 21, 1973, memorandum to the Central Committee, Yuri Andropov wrote: "The Committee for State Security [KGB] realizes that the irresponsible demagogic statements

and provocations of foreign delegates to the Congress may evoke increased activity on the part of antisocial and hostile elements among Soviet citizens. Consequently, it intends to take steps to contain the harmful influence of foreigners. At the same time, Soviet citizens known for their antisocial and hostile orientation are being kept under close surveillance in order to prevent possible undesirable excesses on their part." Rubenstein and Gribanov, *The KGB File of Andrei Sakharov*, 170–71.

68. Kline Archive, Folder "Soviet Human Rights Movement—General," letter from Andrew Blane to Ivan Morris (January 31, 1974), p. 4.

69. On calls for MacBride's resignation, see Kline Archive, Folder "Soviet Human Rights Movement—General," letter from Herbart Ruitenberg [secretary of Amnesty's Dutch section] to Yadja Zeltman (December 31, 1973). Copies of this letter went to Peter Reddaway, a professor of Soviet politics at the London School of Economics and one of Amnesty's most forceful (and knowledgeable) critics of Soviet policies, as well as to Leonid Rigerman. On the spreading of rumors that Reddaway and others were CIA moles: personal email to the author from Peter Reddaway, March 12, 2011.

70. Kline Archive, Folder "Soviet Human Rights Movement—General," letter from Jane Ward to Peter Reddaway (October 31, 1973), p. 1. One of the eleven applicants, Ernst Orlovsky, lived in Leningrad.

71. Ibid.

72. Kline Archive, Folder "Soviet Human Rights Movement—General," letter from Leonid Rigerman to Valery Chalidze (November 6, 1973), pp. 1–5.

73. Personal communication from Edward Kline to the author, September 28, 2012.

Chapter Twenty-One

1. Amnesty International, *Annual Report*, 1968–69 (London, 1969), 3.

2. Kline Archive, Folder "Amnesty International Historical Documents," background paper for working party #2, AI-USA Annual Meeting, March 1977: "Perception vs. Effectiveness: How Amnesty International Is Viewed by the Non-Western World," pp. 1–5.

3. By the late 1960s, however, there were a handful of individual members from socialist countries, including the Leningrad engineer Ernst Orlovsky and the Moscow biologist Zhores Medvedev.

4. On the distinction between self-defense and moral intervention, see Eckel, "Utopie der Moral, Kalkül der Macht," 462.

5. IISH AI-IEC, Microfilm 244, "Amnesty International—Five Years Hence: Report and Recommendation by the Long-Range Planning Committee, May 31, 1972," pp. 96–117.

6. IISH AI-IEC, Microfilm 244, letter from Lothar Belck to Eric Baker, June 20, 1974, pp. 1332–33.

7. Kline Archive, Folder "Soviet Human Rights Movement—General," letter from Andrew Blane to Ivan Morris (January 31, 1974), p. 5. MacBride's position is summarized in this letter.

8. Amnesty's expansion into the United States also raised questions, although of a very different kind, about the transferability beyond Europe of Amnesty's model of citizen activism. See Snyder, "Exporting Amnesty International to the United States."

9. Kline Archive, Folder "Soviet Human Rights Movement—General," letter from Yadja Zeltman to Herbart Ruitenberg, secretary of Amnesty's Dutch section (December 21, 1973), p. 2.

10. Kline Archive, Folder "Soviet Human Rights Movement—General," letter from Andrew Blane to Ivan Morris and the Madison Avenue Group (January 31, 1974), p. 6.

11. Columbia RBML, AI-USA, Record Group II: Executive Director Files, 1967–97, Series 5: National Section Memos, Box 9, Folder 23, Document 5, letter from Peter Reddaway to Martin Ennals and fifteen other Amnesty leaders (December 27, 1973), p. 2.

12. Ibid., p. 3 (emphasis in original).

13. Kline Archive, Folder "Soviet Human Rights Movement—General," letter from Irmgard Hutter to Amnesty's International Executive Committee (January 7, 1974). The Soviet biologist Zhores Medvedev, who may have been the first Soviet citizen to join Amnesty as an individual member (in 1969, four years before he was stripped of his Soviet citizenship), remarked to Reddaway that Soviet dissidents would regard a rejection by Amnesty of their bid for affiliation as "deplorable cowardice." See ibid., letter from Peter Reddaway to Martin Ennals (February 7, 1974). There was ample precedent for Medvedev's warning, given the combustible mix of admiration, dependency, and disenchantment exhibited by many dissidents vis-à-vis their putative Western allies. In a letter to the London-based journal *New Scientist*, Tverdokhlebov expressed bitter disappointment at the lack of solidarity on the part of Western scientists, and in particular psychiatrists, who had failed to respond to evidence of Soviet psychiatric abuse of dissident scientists: "Unfortunately we are compelled to recognize yet again the correctness of Soviet propaganda when it claims that in the West 'man is wolf to man.'" For Tverdokhlebov's letter (dated September 17, 1973), see Kline Archive, Folder "Soviet Human Rights Movement—General."

14. IISH AI-IEC, Microfilm 244, letter from Dirk Börner to Martin Ennals and the International Executive Committee (August 13, 1974), p. 1511. Arrowsmith was imprisoned for, among other things, distributing leaflets at a British army base urging soldiers to quit the army or refuse to serve in Northern Ireland.

15. Kline Archive, Folder "Soviet Human Rights Movement—General," letter from Herbart Ruitenberg to Yadja Zeltman (December 31, 1973), p. 2; letter from Leonid Rigerman to Peter Reddaway (August 29, 1973), p. 2.

16. Amnesty International, *Prisoners of Conscience in the USSR*, 5; IISH AI-IEC, Microfilm 244, letter from Lothar Belck to Martin Ennals (May 20, 1974), p. 1334. Belck refers to discussions with "our Russian Friend," who in all probability was Zivs. See Zivs's reference to various meetings with Amnesty officials in Zivs, *The Anatomy of Lies*, 6.

17. Orlov, *Opasnye mysli*, 156.

18. Rubenstein and Gribanov, *The KGB File of Andrei Sakharov*, 141.

19. RGANI, f. 89, op. 37, d. 11, l. 1.

20. IISH AI-Index, Folder 455, Clayton Yeo [Research Department, Amnesty International], "Recent Events in the USSR," p. 2. The *New York Times* quoted an anonymous Soviet source several days later as claiming that the authorities "feel the dissidents are using Amnesty as a sort of shield." Christopher Wren, "Soviet Dissidents Fear Crackdown," *New York Times* (April 20, 1975), 8. The charge appears again in David K. Shipler, "A Small Chapter of Amnesty International, Begun in a Moscow Apartment, Survives Its First Anxious Year," *New York Times* (September 7, 1975), A1.

21. Buchanan, "'The Truth Will Set You Free,'" 589. The Soviet government occasionally granted amnesty (the Russian term, *amnistiia*, is the same as that in the name Amnesty International) to those it had convicted of various crimes, going as far back as the participants in the anti-Bolshevik Kronstadt rebellion in 1921.

22. Kline Archive, Folder "Amnesty International—1970s," "AI and the charge of interference in the internal affairs of states" (May 20, 1976), p. 1.

23. On the circumstances of MacBride's Nobel Prize, see Keane, *An Irish Statesman and Revolutionary*, 187–91. Curiously, in his own memoirs, MacBride passes over in silence the

entire history of Amnesty's engagement with the Soviet government and Soviet dissidents: see MacBride, *L'exigence de la liberté: Amnesty International*.

24. Amnesty International, *Annual Report*, 1974–75 (London, 1975), 119–20; Kline Archive, Folder "Soviet Human Rights Movement—General," letter from Masha Vorobiev [AI-USA] to Clayton Yeo (November 11, 1975), p. 1; Shipler, "A Small Chapter of Amnesty International," A1; Yadja Zeltman, private communication to the author, October 9, 2012. Sastre's release, it should be noted, came after broad international protests; in this, as in many other cases, it was impossible to gauge the specific effect of Amnesty's intervention. The fate of the Moscow group's Yugoslav adoptee, a woman arrested for inciting ethnic nationalism, is unknown.

25. Kline Archive, Folder "Amnesty International Moscow Group," petition to the president of Uruguay from Moscow Amnesty members and other Soviet citizens protesting torture (March 23, 1976).

26. Markarian, *Left in Transformation*, 88.

27. Kline Archive, Folder "Amnesty International Moscow Group," Amnesty press release (May 10, 1976).

28. Leeuwenberg, "Is Amnesty Impartial Enough?," *Statute of Amnesty International, as Amended by the Eighth International Council Meeting in St. Gallen, Switzerland, 12–14 September 1975*.

29. The phrase has been attributed to former Senegalese president Léopold Senghor. For an example of its invocation in Amnesty's internal deliberations, see Kline Archive, Folder "Amnesty International Historical Documents," background paper for working party #2, AI-USA Annual Meeting, March 1977: "Perception vs. Effectiveness: How Amnesty International Is Viewed by the Non-Western World," p. 3.

30. V. F. Turchin, "Chto takoe bespristrastnost'?," *Materialy samizdata*, no. 276 (January 23, 1976), doc. 2401, p. 4, reprinted in Litvinov et al., *Samosoznanie*, 305–26; an English translation appeared as "What Is Impartiality?," *A Chronicle of Human Rights in the USSR*, no. 19 (1976): 29–37.

31. Turchin, "Chto takoe bespristrastnost'?," 9–10.

32. IISH AI-Index, Folder 424, memorandum from the Research Department to the International Executive Committee (August 16, 1976), p. 7.

33. RGANI, f. 89, op. 37, d. 11, l. 3. The memorandum, written prior to Tverdokhlebov's trial, recommended that he be deprived of his right to reside in Moscow and "exiled to one of the distant regions" of the USSR. A Soviet court dutifully delivered precisely this sentence.

34. Amnesty International, *Prisoners of Conscience in the USSR*, 5 and 148.

35. Nathans, "Soviet Rights-Talk," in Hoffmann, *Human Rights in the Twentieth Century*, 166–90. The "Asian values" argument was expressed most famously in the 1993 Bangkok Declaration. See Burke, *Decolonization and the Evolution of International Human Rights*, 112–44.

36. Gushchin, "Human Rights in the Soviet Union," 18–23. Sukharev was correct in the narrow sense that Soviet jurisprudence did not employ the term "political prisoner" (*politicheskii zakliuchennyi*). But the term was sufficiently embedded in vernacular usage as to have acquired an abbreviated form: *politzek*. In 1974, Gabriel Superfin and other prisoners in the Perm-35 labor camp petitioned for official recognition as political prisoners.

37. A. Amal'rik, "Est' li politzakliuchennye v SSSR?," *Materialy samizdata*, vol. 23 (1976), doc. 2547; republished in Amal'rik, *SSSR i zapad v odnoi lodke*, 120–26. I have used the translation by Hilary Sternberg in Amalrik, *Will the Soviet Union Survive until 1984?* (1981 [2nd ed.]), 135–36.

38. Amal'rik, "Est' li politzakliuchennye v SSSR?," *Materialy samizdata*, vol. 23 (1976), doc. 2547. In the 1950s, the Soviet playwright Leonid Zorin recorded in his notebook an aphorism

attributed to the eighteenth-century German scientist George Christoph Lichtenberg: "The first book that must be banned is the catalogue of banned books." See Zorin, *Zelenye tetradi*, 31.

39. Amnesty International, *Prisoners of Conscience in the USSR*, 53.

40. Samuil Siws [Zivs], "Spekulationen über UdSSR," 2.

41. Kline Archive, Folder "Amnesty International Historical Documents," undated letter from Thomas Hammarberg.

42. Yuri Orlov and other dissidents who formed the Moscow Helsinki Group in 1976 (see chapter 22) estimated the number of political prisoners in the USSR as "many hundreds." See Zubarev and Kuzovkin, *Dokumenty Moskovskoi Khel'sinkskoi gruppy*, 66. Solzhenitsyn, by contrast, repeated Amnesty's numbers ("there are tens of thousands of political prisoners in our country") in speeches he delivered at the time in the West. Solzhenitsyn, *Warning to the West*, 44.

43. "Soviet, at U.N., Attacks Human Rights Groups," *New York Times* (May 3, 1977), A2. The delegate's threat to cancel Amnesty's consultative status at the United Nations was never carried out.

44. RGANI, f. 89, op. 25, d. 51, ll. 2–4.

45. Aase Lionæs, "Award Ceremony Speech," The Nobel Prize, 1977, https://www.nobelprize.org/prizes/peace/1977/ceremony-speech/. Solzhenitsyn was sharply critical of Amnesty's policy of balancing attention to prisoners of conscience in capitalist, socialist, and developing countries. See Scammell, *Solzhenitsyn*, 900.

46. For budget data from 1961 to 1979, see Amnesty International, *Annual Report*, 1979 (London, 1979), 182.

47. Hopgood, *Keepers of the Flame*, 85.

48. Amnesty International, *Annual Report*, 1978 (London 1978), 242–44.

49. Columbia RBML, AI-USA, Record Group V, Series V.1, Box 264, Folder 17, letter to all AI-USA groups that had adopted Soviet prisoners (November 29, 1978).

50. RGANI, f. 89, op. 25, d. 52.2, ll. 35–36.

51. V. Barsov and M. Mikhailov, "Kogo zashchishchaete, gospoda! Ob odnom iz filialov imperialisticheskikh sekretnykh sluzhb," *Izvestiia*, no. 200 (August 25, 1980): 5.

52. The latter point was developed at length in K. Kirillov, "'Amnistiia'—sluzhanka spetssluzhb," *Izvestiia*, no. 67 (March 19, 1981), 5. Kirillov insisted that Amnesty knew about cases of punitive psychiatric imprisonment in Italy but did nothing about them. See also V. Skosyrev, "Bespristrastnost's dvoinym dnom," *Izvestiia*, no. 64 (March 16, 1981), 5.

53. Zivs, *Human Rights*, 93.

54. Other works in the same genre, targeted primarily at foreign audiences, include Yakovlev, *Solzhenitsyn's Archipelago of Lies*, and idem, *CIA Target: The USSR*, both of which were published in multiple European languages (in Moscow) as well as in Russian, and both of which deal extensively with Soviet dissidents. The campaign to discredit Amnesty continued well into the Gorbachev era with the publication of Boris Antonov's *Prisoners of Conscience in the USSR and Their Patrons*. It gained further momentum in the Putin era, as evidenced by the Russian government's raid on Amnesty's Moscow office in March 2013, part of a broader offensive against Western-sponsored NGOs in the Russian Federation. See David Herszenhorn and Andrew Roth, "Russian Authorities Raid Amnesty International Office," *New York Times* (March 26, 2013), A5. In April 2022, Amnesty's operations in Russia were liquidated.

55. Readers who bothered to compare the analysis of sources in Amnesty's report with Zivs's description of that analysis would have quickly grasped his method. At one point he quotes the report claiming as "absolutely reliable" the figure of ten thousand total prisoners of conscience. The report claimed no such degree of certainty. On the contrary, it repeatedly notes the difficulty of calculating the number of prisoners in the absence of published statistics and the wide

range of existing estimates. See Amnesty International, *Prisoners of Conscience in the USSR*, 48–53.

56. Zivs, *Anatomy of Lies*, 9–16, 25–34, 47; idem, *Human Rights*, 12, 97, 104.

57. Zivs, *Human Rights*, 89 and 100.

58. Andrei Sakharov, "A Reply to Slander," *New York Review of Books* (July 21, 1983).

59. Ibid. The joke plays on the fact that the Russian rendition of Amnesty's name never became standardized in the Soviet press, so that both versions—one translated, with the adjective preceding the noun, the other transliterated from English and therefore foreign-sounding—continued to be used.

60. Shaposhnikov, *Mezhdunarodnye nepravitel'stvennye organizatsiia i uchrezhdeniia*, 159.

61. Kline Archive, Folder "Soviet Human Rights Movement—General," letter from Valentin Turchin and Vladimir Albrekht to the meeting of Amnesty's International Council in Strasbourg (September 3, 1976), pp. 1–2.

62. On Benenson's vision, expressed in an extraordinary memorandum not meant for publication, see Buchanan, "'The Truth Will Set You Free,'" 595.

63. Harrison, *Secret Leviathan*.

64. Peter Benenson, "The Forgotten Prisoners," *Observer* (London) (May 28, 1961).

Chapter Twenty-Two

1. Morgan, *The Final Act*.

2. For an explicit comparison to the Munich agreement, see Orlov, *Opasnye mysli*, 172.

3. From his exile in the United States, Alexander Solzhenitsyn issued the sharpest denunciation of the Helsinki Final Act, describing it as "the funeral of Eastern Europe." Solzhenitsyn, *Warning to the West*, 41.

4. According to KGB surveillance records, the writer Vladimir Voinovich, a member of the Moscow chapter of Amnesty International, planned in 1974 to found a Moscow chapter of PEN, the international association of writers. See Yale University Press, Annals of Communism: The Andrei Sakharov KGB File, accessed December 12, 2023, http://yupnet.org/annals/sakharov/documents_frames/Sakharov_090.htm.

5. Amal'rik, *Zapiski dissidenta*, 330–31.

6. Ibid., 331–32; Orlov, *Opasnye mysli*, 167–68.

7. Amal'rik, *Zapiski dissidenta*, 331.

8. Eichwede, "'Entspannung mit menschlichem Antlitz,'" 62.

9. Goldberg, *The Final Act*, 36–40; see also Kuzovkin et al., *K istorii Moskovskoi Khel'sinkskoi gruppy*.

10. Orlov, *Opasnye mysli*, 168.

11. Alexeyeva and Goldberg, *The Thaw Generation*, 279–80.

12. Orlov, *Opasnye mysli*, 168; Goldberg, *The Final Act*, 39.

13. De Boer et al., *Biographical Dictionary of Dissidents*, 51.

14. Orlov, *Opasnye mysli*, 170; Sakharov, *Memoirs*, 456.

15. Goldberg, *The Final Act*, 19.

16. Ibid., 48. For the complete text of the announcement, see Zubarev and Kuzovkin, *Dokumenty Moskovskoi Khel'sinkskoi gruppy*, 23–24.

17. The TASS communiqué was re-translated into Russian and appeared in the *Chronicle of Current Events*, no. 40 (May 20, 1976). TASS rendered what it called the group's "pretentious and provocative name" as "the Organization for Monitoring Compliance by the Soviet Union with the Conditions of the Final Act."

18. Zubarev and Kuzovkin, *Dokumenty Moskovskoi Khel'sinkskoi gruppy*, 5.

19. Ibid., 239.

20. Alexeyeva, *Soviet Dissent*, 339. For memoirs by participants in the Ukrainian Helsinki Group, see Marynovych, *The Universe behind Barbed Wire*, 85–133; Rudenko, *Naibil'she dyvo—zhyttia*; and Zisel's, *Esli ne seichas*, 8–66.

21. Morgan, *The Final Act*, 95.

22. Kennan, "Morality and Foreign Policy," 207.

23. The phone call, recorded by Kissinger, is dated September 14, 1973. Both Kissinger and Kennan refer specifically to Sakharov and Solzhenitsyn. For the transcript, see "The Kissinger Telephone Conversations: A Verbatim Record of U.S. Diplomacy, 1969–1977, Digital National Security Archive, ProQuest.

24. On Fenwick, see Snyder, *Human Rights Activism*, 40–43. On Fenwick's meeting with dissidents, see Orlov, *Dangerous Thoughts*, 188.

25. Snyder, *Human Rights Activism*, 40–41; Orlov, *Opasnye mysli*, 279.

26. Quoted in Snyder, *Human Rights Activism*, 48.

27. Ibid., 115–28.

28. Zubarev and Kuzovkin, *Dokumenty Moskovskoi Khel'sinkskoi gruppy*, 67.

29. Based on my analysis of 169 dossiers prepared by the Moscow Helsinki Group, reproduced in Zubarev and Kuzovkin, *Dokumenty Moskovskoi Khel'sinkskoi gruppy*. Among those dossiers are also several dealing with repression of activists on behalf of various non-Russian national movements. For the most part, however, such cases were handled by the Helsinki Groups in the relevant Soviet republics.

30. Ibid., 259.

31. Rubenstein and Gribanov, *The KGB File of Andrei Sakharov*, 218. I have slightly amended the translation in accordance with the original Russian text at Yale University Press, Annals of Communism: The Andrei Sakharov KGB File, accessed December 12, 2023, http://yupnet.org/annals/sakharov/documents_frames/Sakharov_118.htm.

32. Mitrokhin Archive, The Chekist Anthology, Folder 42 ("Vokrug nominatsii"), pp. 1–2.

33. Selvage, "From Helsinki to 'Mars.'"

34. Mitrokhin Archive, The Chekist Anthology, Folder 51 ("O pravakh cheloveka"), pp. 1–2. It is uncertain whether the Foundation was actually established under this or a similar name.

35. Selvage, "From Helsinki to 'Mars,'" 92.

36. Sakharov, "Nobel Lecture."

37. Sakharov does not mention this message in his memoirs. The KGB reported it to the Central Committee on September 19, 1976. See Yale University Press, Annals of Communism: The Andrei Sakharov KGB File, accessed December 12, 2023, http://yupnet.org/annals/sakharov/documents_frames/Sakharov_117.htm.

38. On the slightly awkward circumstances that led to the exchange of letters, see Sakharov, *Memoirs*, 464–66.

39. Zubarev and Kuzovkin, *Dokumenty Moskovskoi Khel'sinkskoi gruppy*, 337. For sentencing data, see individuals' Wikipedia pages; the entry on Oleksa Tykhy in Canadian Institute of Ukrainian Studies, Internet Encyclopedia of Ukraine, accessed December 12, 2023, https://www.encyclopediaofukraine.com/display.asp?linkpath=pages%5CT%5CY%5CTykhyOleksa.htm; Ainars Bruvelis and Aivars Borovkovs, Timenote, accessed December 12, 2023, https://timenote.info/en/Viktoras-Petkus. For a complete list of members of the Ukrainian Helsinki Group and their punishments, see Marynovych, *The Universe behind Barbed Wire*, 96–103.

40. See note 29 above. Of 169 dossiers sent by the Moscow Helsinki Group to representatives of states that had signed the Final Act, 53 concerned arrested members of the various Soviet Helsinki Groups, more than twice as many as any other category.

41. Zubarev and Kuzovkin, *Dokumenty Moskovskoi Khel'sinkskoi gruppy*, 534–35.

42. Quoted in Goldberg, *The Final Act*, 25.
43. Mitrokhin Archive, The Chekist Anthology, Folder 42 ("Vokrug nominatsii"), pp. 2–4. The 1978 Nobel Peace Prize was shared by Menachem Begin and Anwar Sadat.
44. National Security Archive Electronic Briefing Book No. 391, "Soviet Dissidents and Jimmy Carter," Document 18, "Memorandum of Conversation, 'The President's Meeting with USSR Foreign Minister A. A. Gromyko,' Secret, September 23, 1977," p. V-40, National Security Archive, https://nsarchive2.gwu.edu/NSAEBB/NSAEBB391/.
45. Selvage, "From Helsinki to 'Mars,'" 67–69; Snyder, *Human Rights Activism*, 81–114.
46. Selvage, "From Helsinki to 'Mars,'" 75–84.
47. Thomas, *The Helsinki Effect*; Morgan, *The Final Act*; Nathans, "Helsinki Syndrome: Human Rights and International Diplomacy."
48. Dobrynin, *Sugubo doveritel'no*, 397 and 499.
49. Reddaway, "Soviet Policies toward Dissent, 1953–1986."
50. Solzhenitsyn, *Warning to the West*, 48.
51. Dobrynin, *Sugubo doveritel'no*, 592.
52. See chapter 14 at note 94.
53. Rubenstein and Gribanov, *The KGB File of Andrei Sakharov*, 111, 115–16, 122.
54. Ibid., 201–4; Selvage, "KGB, MfS, und Andrej Sacharow," in Selvage and Herbstritt, *Der "grosse Bruder,"* 248.
55. Rubenstein and Gribanov, *The KGB File of Andrei Sakharov*, 243–50.
56. Sakharov, *Memoirs*, 510–11.
57. Zubarev and Kuzovkin, *Dokumenty Moskovskoi Khel'sinkskoi gruppy*, 443. Sakharov's American biographers correctly emphasize that the manner of his exile violated Soviet law, but then erroneously describe him as having been "arrested." Not even that formality of criminal procedure was observed. See Bergman, *Meeting the Demands of Reason*, chap. 19, "Arrested but Still Defiant"; and Lourie, *Sakharov*, 303.
58. Sakharov, *Memoirs*, 515–16.
59. Zubarev and Kuzovkin, *Dokumenty Moskovskoi Khel'sinkskoi gruppy*, 487–88; "Мейланов, Васиф Сиражутдинович," Wikipedia, accessed December 12, 2023, https://ru.wikipedia.org/wiki/Мейланов_Васиф_Сиражутдинович. When he arrived at the Perm-35 labor camp, Meilanov refused to perform coerced labor and spent over a year in solitary confinement, during which he managed to write and smuggle out several essays that circulated in samizdat. He also maintained contact with Amnesty International. He was released in 1988.
60. Zubarev and Kuzovkin, *Dokumenty Moskovskoi Khel'sinkskoi gruppy*, 555.
61. *Vlast' i dissidenty*, 256.
62. Alexeyeva and Goldberg, *The Thaw Generation*, 295.
63. *Khronika tekushchikh sobytii*, no. 65 (December 31, 1982).
64. A handful of dissidents insisted that the post-Helsinki crackdown marked not the end of the movement but the beginning of another rebirth. See Abovin-Egides and Podrabinek, "Nekotorye aktual'nye problemy dissidentskogo dvizheniia," 8.
65. Vol'pin, *Izbrannoe*, 351.
66. See chapter 16.
67. K. Burzhuademov [V. Sokirko], "Aktivno dumat', uspeshno rabotat', smelo zhit'," in Sokirko, *Zhit' ne po lzhi*, vol. 1 (samizdat, 1977).
68. T. M. Velikanova, "Vozrazheniia," and V. Nikol'skii, "Mysli po povodu . . . ," in Sokirko, *Zhit' ne po lzhi*, vol. 1.
69. A. Grineva [Sonia Sorokina], "Uchitel' i stroptivyi uchenik-priverzhenets," in Sokirko, *Zhit' ne po lzhi*, vol. 2 (samizdat, 1978). On Sorokina's exit from the Communist Party, see Sorokina, "Personal'noe delo S. Sorokinoi."

70. Grigorii Pomerants, "Pis'ma o nravstvennom vybore," in Sokirko, *Zhit' ne po lzhi*, vol. 3 (samizdat, 1978).

71. Osnos, "Soviet Dissidents and the American Press," 32–36. From his exile in Europe, Amalrik accused Osnos of disregarding "the moral aspects of the situation in which a free journalist finds himself in a country where freedom—in particular, freedom of the press—does not exist. Faith in the supra-individual organization and lack of faith in the power of the individual are . . . at the base of everything [Osnos] has written about the Soviet Union." See Amalrik, "Soviet Dissidents and the American Press: A Reply," 63–64. Osnos's rejoinder in the same issue similarly sheds more heat than light: "American reporters—in contrast to Amalrik and other dissidents—have a responsibility to reflect the Soviet Union as it is and not as we wish it to be." Decades later, Osnos's memoirs revealed the personal antipathy between the two: Osnos, *An Especially Good View*, 135 and 144. See also Fainberg, *Cold War Correspondents*, 219–20.

72. Bordewich, "The Press Harmonizes on a Presidential Theme," 36–37.

73. Komaromi, *Soviet Samizdat*, 18.

74. See Ann Komaromi's online database of samizdat periodicals: Soviet Samizdat Periodicals, University of Toronto Libraries, https://samizdat.library.utoronto.ca/. On samizdat readerships and publics, see Komaromi, *Soviet Samizdat*.

75. For examples of samizdat crossword puzzles, see Nauchno-informatsionnyi tsentr Memorial, St. Petersburg, papers of Revol't Pimenov (uncatalogued).

76. "Игрунов, Вячеслав Владимирович," Wikipedia, last modified October 9, 2023, https://ru.wikipedia.org/wiki/Игрунов_Вячеслав_Владимирович.

77. On laws of history and positive law, see Nathans, "Soviet Rights Talk," 190.

Epilogue

1. Amal'rik, "Konformist li diadia Dzhek?," in idem, *P'esy*, 197–232.

2. Al'brekht, *Kak vesti sebia na obyske*, 2.

3. Iar. Iasnyi [Boris Shragin], "Andrei Amal'rik kak publitsist," in *SDS*, vol. 6, doc. 408; Solzhenitsyn, *Pis'mo vozhdiam*, 40; Dmitrii Nelidov [Vladimir Zelinskii], "Ideokraticheskoe soznanie i lichnost'," 206–7.

4. Igrunov, "K problematike obshchestvennogo dvizheniia"; Amal'rik, *Zapiski dissidenta*, 39.

5. Siniavskii, "Dissidentstvo kak lichnyi opyt," 133.

6. Zinov'ev, *Bez illiuzii*, 59.

7. Korotkov et al., *Kremlevskii samosud*, 359.

8. Orlov, *Opasnye mysli*, 151.

9. For a polemical look at Soviet dissident émigrés' views of the West, see Jelen and Wolton, *L'Occident des dissidents*, especially 11–88.

10. Quoted in Mazour, *The First Russian Revolution*, 154.

11. A. I. Gertsen [Herzen], "Le peuple russe et le socialisme," in idem, *Sobranie sochinenii*, vol. 7, 299.

12. Karel van het Reve, "Russen horen van vreemde zenders wat niet in hun 'Pravda' staat," *Het Parool* (March 30, 1968), reprinted in idem, *Verzameld werk*, vol. 2, 815–17.

13. Columbia RBML, AI-USA, Record Group II, Series 5, Box 9, Folder 23, Document 2, "Report from the AI Research Department on Prisoners of Conscience in the USSR" (November 9, 1971).

14. Kochanski, *Resistance: The Underground War against Hitler*, xx.

15. Hans Mommsen, "Gesellschaftsbild und Verfassungspläne des deutschen Widerstandes," in Schmitthenner and Buchheim, eds., *Der deutsche Widerstand gegen Hitler*, 76.

16. It is telling that none of the major scholarly attempts at systematic comparison of Nazi Germany and the Soviet Union take up the comparative analysis of resistance under the two

regimes. See Geyer and Fitzpatrick, *Beyond Totalitarianism: Stalinism and Nazism Compared*; Kershaw and Lewin, *Stalinism and Nazism: Dictatorships in Comparison*; Rousso, *Stalinisme et nazisme: Histoire et mémoire comparées*. A notable exception is Overy, *The Dictators: Hitler's Germany and Stalin's Russia*, chap. 8, "Friend and Foe: Popular Responses to Dictatorship." Comparisons between Nazi Germany and the USSR invariably limit themselves (on the Soviet side) to the period before 1953. The Soviet dissident movement thus appears outside the chronological limits of the comparison.

17. Von Klemperer, "'What Is the Law That Lies behind These Words?'"
18. See Judt, "The Dilemmas of Dissidence"; and Brier, *Entangled Protest*, 11–42.
19. Judt, "The Dilemmas of Dissidence," 193.
20. Eichwede, *Samizdat: Alternative Kultur*.
21. Quoted in Judt, "The Dilemmas of Dissidence," 195.
22. Ibid., 191.
23. Pils, *China's Human Rights Lawyers*, 64.
24. According to Ian Johnson, a significant proportion of the Weiquan movement's lawyers are also practicing Christians. See idem, "When the Law Meets the Party," *New York Review of Books*, vol. 64, no. 13 (August 17, 2017): 58–60.
25. Anderson, *Eyes off the Prize*.
26. Reflecting on her defense of Vladimir Bukovsky at his trial in 1967, the lawyer Dina Kaminskaya noted that "Soviet courts are not authorized to declare this or that law unconstitutional." See Kaminskaia, *Zapiski advokata*, 43.
27. Alexeyeva and Goldberg, *The Thaw Generation*, 7.
28. Gorbachev, *Zhizn' i reformy*, vol. 1, 527.
29. Zubok, *Collapse*, 427.
30. Natal'ia Gorbanevskaia, "Dvoinoe vos'mistishie," in idem, *Goroda i dorogi*, 370.
31. Horvath, "The Putin Regime and the Heritage of Dissidence," 28.
32. Bogoraz, *Sny pamiati*, 253; Alexeyeva and Goldberg, *The Thaw Generation*, 222, 317.
33. McVicker, "The Creation and Transformation of a Cultural Icon," 317–18; Horvath, "Apologist of Putinism?"
34. Smith, *Remembering Stalin's Victims*.
35. On Strategy-31 and other protest movements of the Putin era, see Horvath, "'Sakharov Would Be with Us'"; Gilligan, "Refashioning the Dissident Past"; Gabowitsch, "Are Copycats Subversive?" More generally, see Horvath, *The Legacy of Soviet Dissent*.
36. For one example, see Tat'iana Uskova, "Chto delat', esli vas vyzyvaiut na dopros," MBKh Media (July 27, 2019), https://mbk-news.appspot.com/practica/vyzyvayut-na-dopros/.
37. Quoted in *Vedemosti* (January 22, 2008), 1.
38. Filippov et al., *Noveishaia istoriia Rossii, 1945–2006 gg.*, 256–57. On the controversies surrounding the handbook's ambiguous treatment of Stalinism, see Brandenburger, "A New Short Course?"
39. Quoted in Michael Schwirtz, "On Russian TV News, in Startling Shift, Straightforward Account of Day's Events," *New York Times* (December 11, 2011), A12.
40. See Benjamin Nathans, "Profiles in Decency," *New York Review of Books*, vol. 67, no. 7 (April 23, 2020): 50–52.
41. On the creeping militarization of Russian society in the decade leading up to 2022, see Dietmar Neutatz, "Die Verteidigung im Westen stärken," *Frankfurter Allgemeine Zeitung* (May 9, 2022).
42. See the virtual exhibition curated by Alexandra Arkhipova and Yuri Lapshin, *"No wobble/Net voble": Russian Anonymous Street Art against War 2022/23*, accessed January 12, 2024, https://www.nowobble.net/.
43. Medinskii and Torkunov, *Istoriia Rossii: 1945 god–nachalo XXI veka*, 138, 208.

BIBLIOGRAPHY

Only works mentioned in the text or notes are listed here. With a few exceptions, specific articles in the contemporary press are not included in the bibliography.

Archival Sources

Germany
    Archiv der Forschungsstelle Osteuropa (FSO), University of Bremen, Bremen
    Gabriel Superfin, private collection

Lithuania
    Lietuvos ypatingasis archyvas (LYA), Vilnius

The Netherlands
    International Institute for Social History (IISH), Amsterdam

Russian Federation
    Arkhiv Obshchestva "Memorial" (Memorial-Moscow), Moscow
    Arkhiv Prezidenta Rossiiskoi Federatsii (APRF), Moscow
    Arkhiv Rossiiskoi akademii nauk (ARAN), Moscow
    Arkhiv Sakharova, Moscow
    Gosudarstvennyi arkhiv Rossiiskoi Federatsii (GARF), Moscow
    Nauchno-informatsionnyi tsentr Memorial, St. Petersburg
    Rossiiskii gosudarstvennyi arkhiv noveishei istorii (RGANI), Moscow
    Tsentral'nyi munitsipal'nyi arkhiv goroda Moskvy (TsMAM), Moscow

United States
    Amherst Center for Russian Culture (ACRC), Amherst College, Amherst, Massachusetts
    Beinecke Library, Yale University, New Haven, Connecticut
    Edward Kline, private collection, New York
    Hoover Institution Library and Archives, Stanford University, Stanford, California
    Houghton Library, Harvard University, Cambridge, Massachusetts
    New York Public Library, Manuscripts and Archives Division (NYPL-MAD), New York
    Rare Book and Manuscript Library, Columbia University (Columbia RBML), New York
    Tamiment Library and Robert F. Wagner Labor Archives, New York University, New York
    Vasili Mitrokhin Archive, Woodrow Wilson International Center for Scholars, Cold War International History Project Digital Archive (Mitrokhin Archive)

Periodicals

Bolshevik
Chicago Tribune
Chronicle of Human Rights in the USSR
Chronicle of the Lithuanian Catholic Church
Commentary
Encounter
Esprit
Evrei v SSSR
Frankfurter Allgemeine Zeitung
Frankfurter Rundschau
Izvestiia
Khronika tekushchikh sobytii
Kommersant
Le Monde
Literaturnaia gazeta
Nedelia
New Leader
Newsweek
New Times
New Yorker
New York Review of Books
New York Times
Novoe russkoe slovo
Novoe vremia
Novyi mir
Novyi zhurnal
Observer (London)
Obshchestvennye problemy
Ogonek
Oktiabr'
Pamiat'
Poiski
Politicheskii dnevnik
Pravda
Pravda Ukrainy
Russkaia mysl'
Time
Times (London)
Vedemosti
Vestnik Russkogo studencheskogo khristianskogo dvizheniia (*Vestnik RSKhD*)
Washington Post
Znamia

Other Published Sources

Abovin-Egides, Petr, and Pinkhos Podrabinek. "Nekotorye aktual'nye problemy dissidentskogo dvizheniia v nashei strane." *Poiski: Svobodnyi moskovskii zhurnal*, no. 2 (1980). Moscow, samizdat.

Abraham, David. "Liberty without Equality: The Property-Rights Connection in a Negative Citizenship Regime." *Law and Social Inquiry*, vol. 21, no. 1 (1996): 1–65.
Adler, Nanci. *The Gulag Survivor: Beyond the Soviet System*. New Brunswick, NJ: Transaction Publishers, 2002.
Aksel'rod-Rubina, I. M. *Zhizn' kak zhizn': Vospominaniia*. 2 vols. Jerusalem: [s.n.], 2006.
Al'brekht, Vladimir. *Kak byt' svidetelem*. Paris: Izd. zhurnala "A–Ia," 1983.
Al'brekht, Vladimir. *Kak vesti sebia na obyske*. Samizdat, 1976.
Al'brekht, Vladimir. "190 voprosov po 'delu' Tverdokhlebova." Samizdat, 1976.
Al'brekht, Vladimir. *Svidetel' po sobstvennomu delu i drugie dokumenty*. Frankfurt am Main: Posev, 1974.
Al'brekht, Vladimir. "Vladimir Al'brekht—beseda s Glebom Pavlovskim 3 aprelia 1987 g." *Russkii zhurnal* (November 5, 1997 [sic]). http://old.russ.ru/journal/travmp/97-11-05/alpavl.htm.
Aleksandrovskii, P., and G. Kislykh. "O nekotorykh takticheskikh priemakh doprosa lits, obviniaemikh v antisovetskoi agitatsii i propagande." *KGB Sbornik*, vol. 57, no. 2 (1973): 70–75.
Alexeev, Michael, and William Pyle. "A Note on Measuring the Unofficial Economy in the Former Soviet Republics." *Economics of Transition*, vol. 11, no. 1 (2003): 153–75.
Alexeyeva, Ludmilla. *Soviet Dissent: Contemporary Movements for National, Religious, and Human Rights*. Middletown, CT: Wesleyan University Press, 1985.
Alexeyeva, Ludmilla, and Paul Goldberg. *The Thaw Generation: Coming of Age in the Post-Stalin Era*. Pittsburgh, PA: University of Pittsburgh Press, 1990.
Allworth, Edward A., ed. *The Tatars of Crimea: Return to the Homeland*. Durham, NC: Duke University Press, 1998.
Amal'rik, Andrei. "Dvizhenie za prava cheloveka v SSSR." *Russkaia mysl'*, no. 3158 (June 30, 1977).
Amal'rik, Andrei. *Nezhelannoe puteshestvie v Sibir'*. New York: Harcourt Brace Jovanovich, 1970.
Amal'rik, Andrei. *Normanny i Kievskaia Rus'*. Moscow: Novoe literaturnoe obozrenie, 2018.
Amalrik, Andrei. *Nose! Nose? No-se! and Other Plays*. New York: Harcourt, 1973.
Amal'rik, Andrei. *P'esy*. Amsterdam: Fond imeni Gertsena, 1970.
Amal'rik, Andrei. *Prosushchestvuet li Sovetskii Soiuz do 1984 goda?* Amsterdam: Fond imeni Gertsena, 1970.
Amalrik, Andrei. "Soviet Dissidents and the American Press: A Reply." *Columbia Journalism Review*, vol. 16, no. 6 (1978): 63–64.
Amal'rik, Andrei. *SSSR i zapad v odnoi lodke*. London: Overseas Publications Interchange, 1978.
Amalrik, Andrei. *Will the Soviet Union Survive until 1984?* 2nd ed. New York: Harper & Row, 1981.
Amal'rik, Andrei. *Zapiski dissidenta*. Ann Arbor, MI: Ardis, 1982.
Amnesty International. *Annual Report*. London: Amnesty International, various years.
Amnesty International. *Personal Freedom in the Marxist-Leninist Countries. Report of Conference, June 16th 1962*. London: Amnesty International Publications, n.d.
Amnesty International. *Prisoners of Conscience in the USSR: Their Treatment and Conditions*. London: Amnesty International Publications, 1975.
Amos, Jennifer. "Embracing and Contesting: The Soviet Union and the Universal Declaration of Human Rights, 1948–58." In *Human Rights in the Twentieth Century*, edited by Stefan-Ludwig Hoffmann, 153–64. New York: Cambridge University Press, 2011.
Anderson, Carol. *Eyes off the Prize: The United Nations and the African American Struggle for Human Rights, 1944–1955*. New York: Cambridge University Press, 2003.
Anderson, Raymond. "Vystuplenie." In *30 let "Razmyshlenii..." Andreia Sakharova*, 93–96. Moscow: Prava cheloveka, 1998.

Andrew, Christopher. "Intelligence in the Cold War." In *The Cambridge History of the Cold War*, vol. 2, *Crises and Détente*, edited by Melvyn Leffler and Odd Arne Westad, 417–37. Cambridge: Cambridge University Press, 2010.

Andrew, Christopher, and Vasili Mitrokhin. *The Sword and the Shield: The Mitrokhin Archive and the Secret History of the KGB*. New York: Basic Books, 1999.

Andrle, Vladimir. *A Social History of Twentieth-Century Russia*. London: Edward Arnold, 1994.

Andropov, Iu. V. *Leninizm—neischerpaemyi istochnik revoliutsionnoi energii i tvorchestva mass: Izbrannye rechi i stat'i*. Moscow: Izdatel'stvo politicheskoi literatury, 1984.

Antonov, Boris. *Prisoners of Conscience in the USSR and Their Patrons*. Moscow: Novosti, 1988.

Applebaum, Anne. *Gulag: A History*. New York: Doubleday, 2003.

Arbatov, Georgi. *The System: An Insider's Life in Soviet Politics*. New York: Times Books, 1993.

Arendt, Hannah. *The Origins of Totalitarianism*. New York: Harcourt Brace, 1951.

Arendt, Hannah. *The Origins of Totalitarianism*. 3rd ed. New York: Harcourt Brace, 1966.

Arendt, Hannah. "Personal Responsibility under Dictatorship." In *Responsibility and Judgment*, by Hannah Arendt, 17–48. New York: Schocken Books, 2003.

Arkhipova, Aleksandra, et al., eds. *"Figa v karmane" i drugie teorii simvolicheskogo soprotivleniia*. Moscow: Delo, 2016.

Arzhak, Nikolai. *Govorit Moskva: Povest'*. Washington, DC: n.p., 1962.

Arzhak, Nikolai. *Ruki: Chelovek iz MINAPa: Rasskazy*. Washington, DC: n.p., 1963.

Astrouskaya, Tatsiana. *Cultural Dissent in Soviet Belarus (1968–1988): Intelligentsia, Samizdat, and Nonconformist Discourses*. Wiesbaden: Harrassowitz Verlag, 2019.

Azbel, Mark. *Refusenik: Trapped in the Soviet Union*. Boston: Paragon House, 1987.

Azbel, Mark, Nina Voronel', and Aleksandr Voronel'. "Dvadtsat' let spustia (vospominaniia o protsesse Siniavskogo-Danielia)." *Dvadtsat' dva*, no. 46 (January–March 1986): 132–77.

Babenko, P. I. *E. Iakir: Ocherk boevogo puti*. Moscow: Gosudarstvennoe izdatel'stvo politicheskoi literatury, 1964.

Baberowski, Jörg. "Nikita Khrushchev and De-Stalinization in the Soviet Union, 1953–1964." In *The Cambridge History of Communism*, edited by Norman Naimark, Silvio Pons, and Sophie Quinn-Judge, vol. 2, 113–38. Cambridge: Cambridge University Press, 2017.

Babitskii, Andrei, and Aleksei Makarov. "Kto delal 'Khroniku tekushchikh sobytii.'" *Zhurnal Kommersant Weekend*, April 26, 2013.

Baehr, Peter. *Hannah Arendt, Totalitarianism, and the Social Sciences*. Stanford, CA: Stanford University Press, 2010.

Baehr, Peter. "Stalinism in Retrospect: Hannah Arendt." *History and Theory*, vol. 54, no. 3 (October 2015): 353–66.

Baran, Emily. *Dissent on the Margins: How Soviet Jehovah's Witnesses Defied Communism and Lived to Preach About It*. New York: Oxford University Press, 2014.

Barkhudarov, S. G., et al., eds. *Slovar' russkogo iazyka XVIII veka*. Vol. 6. Leningrad: Nauka, 1984.

Barnes, Christopher. *Boris Pasternak: A Literary Biography*. Vol. 2. Cambridge: Cambridge University Press, 1998.

Baron, Samuel H. *Bloody Saturday in the Soviet Union: Novocherkassk, 1962*. Stanford, CA: Stanford University Press, 2001.

Barsukova, Svetlana, and Alena Ledeneva. "Are Some Countries More Informal Than Others? The Case of Russia." In *The Global Encyclopaedia of Informality*, edited by Alena Ledeneva, vol. 2, 487–92. London: UCL Press, 2018.

Bassow, Whitman. *The Moscow Correspondents: Reporting on Russia from the Revolution to Glasnost*. New York: William Morrow, 1988.

Batshev, Vladimir. *SMOG: Pokolenie s perebitymi nogami*. [USA]: Franc-Tireur, 2009.

Beckerman, Gal. *When They Come for Us, We'll Be Gone: The Epic Struggle to Save Soviet Jewry.* New York: Houghton Mifflin Harcourt, 2010.
Beer, Daniel. *The House of the Dead: Siberian Exile under the Tsars.* New York: Knopf, 2017.
Belfrage, Sally. *A Room in Moscow.* New York: Reynal, 1959.
Beliakov, A. S., ed. *Mezhdunarodnye nepravitel'stvennye organizatsiia: Spravochnik.* Moscow: Nauka, 1967.
Bellezza, Simone Attilio. *The Shore of Expectations: A Cultural Study of the Shistdesiatnyky.* Toronto: Canadian Institute of Ukrainian Studies Press, 2019.
Berdiaev, Nikolai. *Istoki i smysl russkogo kommunizma.* Paris: YMCA Press, 1955 [1937].
Berg, Raissa. *Sukhovei: Vospominaniia genetika.* New York: Chalidze Publications, 1983.
Bergman, Jay. *Meeting the Demands of Reason: The Life and Thought of Andrei Sakharov.* Ithaca, NY: Cornell University Press, 2009.
Bergman, Jay. "Reading Fiction to Understand the Soviet Union: Soviet Dissidents on Orwell's *1984*." *History of European Ideas*, vol. 23, nos. 5–6 (1997): 173–92.
Bergman, Jay. "Soviet Dissidents on the Russian Intelligentsia, 1956–1985: The Search for a Usable Past." *Russian Review*, vol. 51 (January 1992): 16–35.
Berlin, Isaiah. *Affirming: Letters 1975–1997.* London: Chatto & Windus, 2015.
Berlin, Isaiah. *Two Concepts of Liberty: An Inaugural Lecture Delivered before the University of Oxford on 31 October 1958.* Oxford: Clarendon Press, 1958.
Berliner, Joseph. Foreword to *Politics, Work, and Daily Life in the USSR*, edited by James Millar, vii–xii. New York: Cambridge University Press, 1987.
Berman, Harold J. *Justice in the USSR: An Interpretation of Soviet Law.* Cambridge, MA: Harvard University Press, 1963.
Berman, Harold J. *Soviet Criminal Law and Procedure: The RSFSR Codes.* Cambridge, MA: Harvard University Press, 1972.
Beumers, Birgit, et al., eds. *Cultural Forms of Protest in Russia.* London: Routledge, 2018.
Beyrau, Dietrich. "Im Kreislauf der Geschichte: Die Macht und ihre Widersacher in Russland." *Osteuropa*, vol. 71, no. 8/9 (2021): 113–37.
Beyrau, Dietrich. *Intelligenz und Dissens: Die russischen Bildungsschichten in der Sowjetunion 1917 bis 1985.* Göttingen: Vandenhoeck & Ruprecht, 1993.
Bezmozgis, David. *The Betrayers.* Toronto: HarperCollins, 2014.
Bilocerkowycz, Jaroslaw. *Soviet Ukrainian Dissent: A Study of Political Alienation.* Boulder, CO: Westview Press, 1988.
Birkbeck, George, Norman Hill, and L. F. Powell, eds. *Boswell's Life of Johnson.* Vol. 2. Oxford: Clarendon Press, 1934.
Bittner, Stephen. *The Many Lives of Khrushchev's Thaw: Experience and Memory in Moscow's Arbat.* Ithaca, NY: Cornell University Press, 2008.
Blium, A. V. *Tsenzura v Sovetskom Soiuze 1917–1991: Dokumenty.* Moscow: ROSSPEN, 2004.
Bloch, Sydney, and Peter Reddaway. *Psychiatric Terror: How Soviet Psychiatry Is Used to Suppress Dissent.* New York: Basic Books, 1977.
Bobkov, Filipp. *Agenty: Opyt bor'by v "Smershe" i "Piatke."* Moscow: Algoritm, 2012.
Bobkov, Filipp. *Kak gotovili predatelei: Nachal'nik politicheskoi kontrrazvedki svidetel'stvuet.* Moscow: Algoritm, 2011.
Bogoraz, Larisa. "Kto ia?" *Evrei v SSSR*, no. 1 (1972). Samizdat.
Bogoraz, Larisa. "Melkie besy." *Kontinent*, no. 12 (1977): 213–21.
Bogoraz, Larisa. *Sny pamiati.* Kharkiv: Prava liudini, 2009.
Bolton, Jonathan. *Worlds of Dissent: Charter 77, the Plastic People of the Universe, and Czech Culture under Communism.* Cambridge, MA: Harvard University Press, 2012.

Bonavia, David. *Fat Sasha and the Urban Guerilla: Protest and Conformism in the Soviet Union.* New York: Atheneum, 1973.

Bonner, Elena. *Alone Together.* New York: Knopf, 1986.

Boobbyer, Philip. *Conscience, Dissent and Reform in Soviet Russia.* New York: Routledge, 2005.

Boobbyer, Philip. "Vladimir Bukovsky and Soviet Communism." *Slavonic and East European Review,* vol. 87, no. 3 (July 2009): 452–87.

Bordewich, Fergus. "The Press Harmonizes on a Presidential Theme." *Columbia Journalism Review,* vol. 16, no. 4 (1977): 36–37.

Bourdeaux, Michael. *Religious Ferment in Russia: Protestant Opposition to Soviet Religious Policy.* London: Macmillan, 1968.

Bourdeaux, Michael, and Xenia Howard-Johnston, eds. *Aida of Leningrad: The Story of Aida Skripnikova.* London: Mowbrays, 1972.

Boym, Svetlana. *Another Freedom: The Alternative History of an Idea.* Chicago: University of Chicago Press, 2010.

Bradley, Mark Philip. "American Vernaculars: The United States and the Global Human Rights Imagination." *Diplomatic History,* vol. 38, no. 1 (January 2014): 1–21.

Brandenburger, David. "A New *Short Course*? A. V. Filippov and the Russian State's Search for a 'Usable Past.'" *Kritika: Explorations in Russian and Eurasian History,* vol. 10, no. 4 (Fall 2009): 825–33.

Bredikhin, V. N. *Lubianka—staraia ploshchad': Sekretnye dokumenty TsK KPSS i KGB o repressiiakh 1937–1990 gg.* Moscow: Posev, 2005.

Brier, Robert, ed. *Entangled Protest: Transnational Approaches to the History of Dissent in Eastern Europe and the Soviet Union.* Osnabrück: Fibre, 2013.

Brier, Robert. "Gendering Dissent: Human Rights, Gender History, and the Road to 1989." *L'Homme: European Journal of Feminist History,* vol. 28, no. 1 (2019): 15–32.

Brodsky, Joseph. "A Writer Is a Lonely Traveler, and No One Is His Helper." *New York Times Magazine,* October 1, 1972.

Bruderer, Georg, ed. *Protsess tsepnoi reaktsii: Sbornik dokumentov po delu Iu. T. Galanskova, A. I. Ginzburga, A. A. Dobrovol'skogo, V. I. Lashkovoi.* Frankfurt am Main: Posev, 1971.

Brunner, Georg. *Die Grundrechte im Sowjetsystem.* Cologne: Verlag Wissenschaft und Politik, 1963.

Buchanan, Tom. "Amnesty International in Crisis, 1966–7." *Twentieth-Century British History,* vol. 15, no. 3 (2004): 267–89.

Buchanan, Tom. "'The Truth Will Set You Free': The Making of Amnesty International." *Journal of Contemporary History,* vol. 37, no. 4 (October 2002): 575–97.

Budraitskis, Il'ia. *Dissidenty sredi dissidentov.* Moscow: Svobodnoe marksistskoe izdatel'stvo, 2017.

Buechler, Steven. "New Social Movement Theories." *Sociological Quarterly,* vol. 36, no. 3 (Summer 1995): 441–64.

Bukharin, Nikolai, and Evgenii Preobrazhenskii. *Azbuka kommunizma.* St. Petersburg [Petrograd]: Gosudarstvennoe izdatel'stvo, 1920.

Bukovskii, Konstantin. "Otvet na lestnitse." *Oktiabr',* no. 9 (September 1966): 199–201.

Bukovskii, Vladimir. *"I vozvrashchaetsia veter..."* New York: Khronika, 1979.

Bukovskii, Vladimir, and Semen Gluzman. "Posobie po psikhiatrii dlia inakomysliashchikh." In *Risunki po pamiati, ili Vospominaniia otsidenta,* by S. F. Gluzman, 27–53. Kyiv: Izdatel'skii dom Dmitriia Burago, 2012.

Burke, Roland. *Decolonization and the Evolution of International Human Rights.* Philadelphia: University of Pennsylvania Press, 2010.

Burlatsky, Fedor. Interview. *Obshchestvenno-pravovoi zhurnal "Rossiiskii advokat,"* no. 5 (2007): 4–9.

Burlatsky, Fedor. *Khrushchev and the First Russian Spring: The Era of Khrushchev through the Eyes of His Advisor.* New York: Scribner's, 1991.
Butov, Petr. "Vospominaniia ob odesskikh dissidentakh." Igrunov.ru, accessed December 12, 2023. http://igrunov.ru/cv/odessa/dissident_od/samizdat/1109017845.html.
Calhoun, Craig. "'New Social Movements' of the Early Nineteenth Century." *Social Science History*, vol. 17 (1993): 385–427.
Carswell, John. *The Exile: A Life of Ivy Litvinov.* London: Faber and Faber, 1983.
Chalidze, Valerii, comp. *Andrei Tverdokhlebov—v zashchitu prav cheloveka.* New York: Khronika, 1975.
Chalidze, Valerii. "Ko mne prishel inostranets." Samizdat, 1971.
Chalidze, Valerii. "Ot sostavitelia." *Obshchestvennye problemy*, vol. 1 (1969). Samizdat.
Chalidze, Valerii. *Otvetstvennost' pokoleniia: Interv'iu Valeriia Chalidze.* New York: Chalidze Publications, 1981.
Chalidze, Valerii. *Prava cheloveka i Sovetskii Soiuz.* New York: Khronika, 1974.
Chalidze, Valery, and Leon Lipson, eds. *Papers on Soviet Law.* Vol. 1. New York: Institute on Socialist Law, 1977.
"'Chast' rukopisei popadaet za rubezh': O nastroenii tvorcheskoi intelligentsia." *Istochnik: Dokumenty russkoi istorii*, vol. 29, no. 4 (1997): 143–48.
Chebrikov, V. M., et al., eds. *Istoriia sovetskikh organov gosudarstvennoi bezopasnosti: Uchebnik.* Moscow: Vysshaia krasnoznamennaia shkola Komiteta gosudarstvennoi bezopasnosti pri Sovete ministerstva SSSR imeni F. E. Dzerzhinskogo, 1977.
Cherniaev, Anatolii. *Sovmestnyi iskhod: Dnevnik dvukh epokh, 1972–1991 gody.* Moscow: ROSSPEN, 2008.
Chuev, Feliks. *Molotov: Poluderzhavnyi vlastelin.* Moscow: Olma-Press, 2000.
Chuikina [Tchouikina], Sof'ia. "Otkrytyi dom i ego khoziaika (iz istorii uchastnits dissidentskogo dvizheniia)." In *Feministskaia teoriia i praktiki: Vostok—Zapad. Materialy mezhdunarodnoi nauchno-prakticheskoi konferentsii*, 287–94. St. Petersburg: PTsGI, 1996.
Chuikina [Tchouikina], Sof'ia. "Uchastie zhenshchin v dissidentskom dvizhenii. Sluchai Leningrada (1956–1986)." Feministskaia biblioteka, accessed May 31, 2017. https://genderlibrary.wordpress.com/2015/10/13/участие-женщин-в-диссидентском-движе/.
Chuikina [Tchouikina], Sofya. "The Role of Dissident Women in Creating the Milieu." In *Women's Voices in Russia Today*, edited by Anna Rotkirch and Elina Haavio-Mannila, 189–205. Brookfield, VT: Dartmouth Publishing Company, 1996.
Chukovskii, Kornei. *Dnevnik 1901–1969.* 2 vols. Moscow: OLMA Press, 2003.
Chukovsky, Kornei. *Diary, 1901–1969.* New Haven, CT: Yale University Press, 2005.
Chuprinin, S. I. "U cherty: Opyt i uroki rannego 'podpisantstva.'" *Znamia*, no. 4 (2019): 160–70.
Clark, Ann Marie. *Diplomacy of Conscience—Amnesty International and Changing Human Rights Norms.* Princeton, NJ: Princeton University Press, 2001.
Clark, Katerina. *Petersburg, Crucible of Cultural Revolution.* Cambridge, MA: Harvard University Press, 1995.
Clark, Katerina. *The Soviet Novel: History as Ritual.* Chicago: University of Chicago Press, 1981.
Clowes, Edith, Samuel Kassow, and James West, eds. *Between Tsar and People: Educated Society and the Quest for Public Identity in Late Imperial Russia.* Princeton, NJ: Princeton University Press, 1991.
Cmiel, Kenneth. "The Emergence of Human Rights Politics in the United States." *Journal of American History*, vol. 86, no. 3 (December 1999): 1231–50.
Cohen, Stephen F. *The Victims Return: Survivors of the Gulag after Stalin.* Exeter, NH: Publishing Works, 2010.

Cohen, Stephen F., and Katrina vanden Heuvel. *Voices of Glasnost: Interviews with Gorbachev's Reformers*. New York: W. W. Norton, 1989.

Cohn, Edward D. "Coercion, Reeducation, and the Prophylactic Chat: *Profilaktika* and the KGB's Struggle with Political Unrest in Lithuania, 1953–64." *Russian Review*, vol. 76 (April 2017): 272–93.

Cole, William. "Three Voices of Dissent." *Survey*, no. 77 (Autumn 1970): 128–45.

Comey, David Dinsmore. Review of *Vesennij list*, by Aleksandr Sergeyevich Yesenin-Volpin. *Studies in Soviet Thought*, vol. 2, no. 2 (June 1962): 148–57.

Committee on the Judiciary. *Abuse of Psychiatry for Political Repression in the Soviet Union: Hearing before the Subcommittee to Investigate the Administration of the Internal Security Act and Other Internal Security Laws of the Committee of the Judiciary, United States Senate, Ninety-Second Congress, Second Session, September 26, 1972*. Washington, DC: Government Printing Office, 1972.

Conquest, Robert. Introduction to *A Childhood in Prison*, by Pyotr [sic] Yakir. New York: Coward, McCann & Geoghegan, 1973.

Conquest, Robert. *Tyrants and Typewriters: Communiqués from the Struggle for Truth*. Lexington, MA: Lexington Books, 1989.

Courtois, Stéphane, et al. *The Black Book of Communism: Crimes, Terror, Repression*. Cambridge, MA: Harvard University Press, 1999.

Crowley, Tony. "'Dissident'—a Brief Note." *Critical Quarterly*, vol. 53, no. 2 (2011): 1–11.

Dal', V. I. *Tolkovyi slovar' zhivogo velikorusskogo iazyka*. 4 vols. Moscow: Russkii iazyk, 1981 [1862].

Daniel', A. Iu., and A. B. Roginskii, eds. *Piatoe dekabria 1965 goda v vospominaniiakh uchastnikov sobytii, materialakh samizdata, dokumentakh partiinykh i komsomol'skikh organizatsii i v zapiskakh Komiteta gosudarstvennoi bezopasnosti v TsK KPSS*. Moscow: Memorial, 1995.

Daniel', Aleksandr. "Pochemu ne 'perestroilis' dissidenty?" *Novoe vremia*, no. 15 (1995): 15.

Daniel', Aleksandr. "Wie freie Menschen: Ursprung und Wurzeln des Dissens in der Sowjetunion." In *Samizdat: Alternative Kultur in Zentral- und Osteuropa: Die 60er bis 80er Jahre*, edited by Wolfgang Eichwede, 38–50. Bremen: Edition Temmen, 2000.

Daniel, Alexander. "1968 in Moscow: A Beginning." In *1968 Revisited: 40 Years of Protest Movements*, edited by Nora Farik, 27–32. Brussels: Heinrich Böll Foundation, 2008.

Daniel', Iulii. *Govorit Moskva: Povesti i rasskazy*. [New York]: Inter-Language Literary Associates, 1966.

Daniel', Iulii. *"Ia vse sbivaius' na literaturu…": Pis'ma iz zakliucheniia: Stikhi*. Moscow: Zven'ia, 2000.

Daucé, Françoise. "Lettres de protestation dans l'U.R.S.S. post-stalinienne: Du choix des justifications." *Revue russe*, no. 32 (2009): 143–52.

Davies, Sarah. *Popular Opinion in Stalin's Russia: Terror, Propaganda and Dissent, 1934–1941*. Cambridge: Cambridge University Press, 1997.

Davis, Belinda, Wilfried Mausbach, Martin Klimke, and Carla MacDougall, eds. *Changing the World, Changing Oneself: Political Protest and Collective Identities in West Germany and the U.S. in the 1960s and 1970s*. New York: Berghahn Books, 2010.

De Boer, S. P., et al., eds. *Biographical Dictionary of Dissidents in the Soviet Union, 1956–1975*. The Hague: M. Nijhoff, 1982.

De George, Richard T. *Soviet Ethics and Morality*. Ann Arbor: University of Michigan Press, 1969.

Della Porta, Donatella. "'1968'—Zwischen nationale Diffusion und transnationale Strukturen." In *1968: Vom Ereignis zum Gegenstand der Geschichtswissenschaft*, edited by Ingrid Gilcher-Holtey, 131–50. Göttingen: Vandenhoeck & Ruprecht, 1998.

Dershowitz, Alan. *The Best Defense*. New York: Vintage Books, 1983.
Dobrenko, Evgenii. "Stalinskaia kul'tura: Skromnoe obaianie antisemitizma." *Novoe literaturnoe obozrenie*, no. 1 (2010): 52–74.
Dobrynin, Anatolii. *Sugubo doveritel'no: Posol v Vashingtone pri shesti prezidentakh SShA (1962–1986 gg.)*. Moscow: Avtor, 1996.
Dobson, Miriam. *Khrushchev's Cold Summer: Gulag Returnees, Crime, and the Fate of Reform after Stalin*. Ithaca, NY: Cornell University Press, 2009.
Dogliotti, Chiara. "L'eversione nera negli anni settanta." *Asti contemporanea*, no. 10 (2004): 203–61.
Doikov, Iurii. *Ob Andree Amal'rike: Bibliograficheskii ukazatel'*. Arkhangel'sk: self-published, 2017.
Dokumenty Komiteta prav cheloveka. New York: International League for the Rights of Man, 1972.
Dovlatov, Sergei. *Kompromiss*. New York: Serebrianyi vek, 1981.
Dudziak, Mary. *Cold War Civil Rights: Race and the Image of American Democracy*. Princeton, NJ: Princeton University Press, 2000.
Eckel, Jan. *Die Ambivalenz des Guten: Menschenrechte in der internationalen Politik seit den 1940ern*. Göttingen: Vandenhoeck & Ruprecht, 2014.
Eckel, Jan. "Utopie der Moral, Kalkül der Macht: Menschenrechte in der globalen Politik seit 1945." *Archiv für Sozialgeschichte*, vol. 49 (2009): 437–84.
Eckel, Jan, and Samuel Moyn, eds. *The Breakthrough: Human Rights in the 1970s*. Philadelphia: University of Pennsylvania Press, 2014.
Edele, Mark. *Soviet Veterans of the Second World War: A Popular Movement in an Authoritarian Society, 1941–1991*. New York: Oxford University Press, 2008.
Eichwede, Wolfgang. "'Entspannung mit menschlichem Antlitz': Die KSZE, die Menschenrechte und der Samizdat." *Osteuropa*, vol. 60, no. 11 (2010): 58–83.
Eichwede, Wolfgang, ed. *Samizdat: Alternative Kultur in Zentral- und Osteuropa: Die 60er bis 80er Jahre*. Bremen: Edition Temmen, 2000.
Engelstein, Laura. "Combined Underdevelopment: Discipline and Law in Imperial and Soviet Russia." *American Historical Review*, vol. 98, no. 2 (April 1993): 338–53.
Epstein, Mikhail. *The Phoenix of Philosophy: Russian Thought of the Late Soviet Period (1953–1991)*. London: Bloomsbury, 2019.
Esaulenko, A. S., and P. M. Kozhukhar'. *Iona Emmanuilovich Iakir (1896–1937): Biobibliograficheskii ukazatel'*. Kishinev: Kartia moldoveniaske, 1967.
Esenin-Vol'pin, Aleksandr. *Iuridicheskaia pamiatka dlia tekh, komu predstoiat doprosy*. Samizdat, 1969.
Esenin-Vol'pin, Aleksandr Sergeevich. *Filosofiia, logika, poeziia, zashchita prav cheloveka: Izbrannoe*. Compiled by A. Iu. Daniel' et al. Moscow: RGGU, 1999.
Esenin-Vol'pin, Aleksandr Sergeevich. "Pamiatka dlia ne ozhidaiushchikh doprosa: Beseda s Aleksandrom Eseninym-Vol'pinym." *Neprikosnovennyi zapas*, vol. 1, no. 21 (2002).
Esenin-Vol'pin, Aleksandr Sergeevich. *Vesennii list / A Leaf of Spring*. New York: Praeger, 1961.
Etkind, Alexander. "A Parable of Misrecognition: Anagnorisis and the Return of the Repressed from the Gulag." *Russian Review*, vol. 68 (October 2009): 623–40.
Etkind, Efim. *Perepiska za chetvert' veka*. St. Petersburg: European University at St. Petersburg, 2012.
Evangelista, Matthew. *Unarmed Forces: The Transnational Movement to End the Cold War*. Ithaca, NY: Cornell University Press, 1999.
Evreiskii samizdat. 27 vols. Jerusalem: Center for Documentation of East European Jewry, Hebrew University, 1971–92.

Evtushenko, Evgenii. *Volchii pasport*. Moscow: Vagrius, 1998.
Evtushenko, Evgenii. *Sobranie sochinenii*. Vol. 5. Moscow: Eksmo, 2014.
Fainberg, Dina. *Cold War Correspondents: Soviet and American Reporters on the Ideological Frontlines*. Baltimore: Johns Hopkins University Press, 2021.
Fedor, Julie. "Chekists Look Back on the Cold War: The Polemical Literature." *Intelligence and National Security*, vol. 26, no. 6 (2011): 842–63.
Fedor, Julie. *Russia and the Cult of State Security: The Chekist Tradition, from Lenin to Putin*. New York: Routledge, 2011.
Feldbrugge, Ferdinand Joseph Maria. *Samizdat and Political Dissent in the Soviet Union*. Leiden: A. W. Sijthoff, 1975.
Feldbrugge, Ferdinand Joseph Maria, G. P. van den Berg, and William B. Simons, eds. *The Encyclopedia of Soviet Law*. Dordrecht: M. Nijhoff Publishers, 1985.
Feofanov, Yuri, and Donald Barry. "The Siniavskii-Daniel Trial: A Thirty Year Perspective." *Review of Central & East European Law*, vol. 22, no. 6 (1996): 603–20.
Feyginberg, Yosef, dir. *Glenn Gould: The Russian Journey*. CBC Home Video, 2003. DVD.
Field, Andrew. "Abram Tertz's Ordeal by Mirror." *New Leader*, July 19, 1965, 9–15.
Field, Deborah A. "Irreconcilable Differences: Divorce and Conceptions of Private Life in the Khrushchev Era." *Russian Review*, vol. 57, no. 4 (October 1998): 599–613.
Filippov, A. V., A. I. Utkin, and S. V. Sergeev, eds. *Noveishaia istoriia Rossii, 1945–2006 gg.: Kniga dlia uchitelia*. Moscow: Prosveshchenie, 2007.
Finkelman, Samuel. "Ghetto, Gulag, *Geulah*: Jewish National Revival, Inter-ethnic Encounters, and Collective Memory of Catastrophe in the Post-Stalin Soviet Union, 1954–1991." PhD diss., University of Pennsylvania, 2023.
Finn, Peter, and Petra Couvée. *The Zhivago Affair: The Kremlin, the CIA, and the Battle over a Forbidden Book*. New York: Pantheon Books, 2014.
Firsov, Boris. *Raznomyslie v SSSR 1940–1960-e gody: Istoriia, teoriia i praktika*. St. Petersburg: Evropeiskii dom, 2008.
Fitzgerald, F. Scott. "The Crack-Up." In *My Lost City: Personal Essays, 1920–1940*, 139–44. Cambridge: Cambridge University Press, 2005.
Fleischhauer, Ingeborg. *The Soviet Germans: Past and Present*. London: Hurst, 1986.
Fleishman, Lazar. *Boris Pasternak: The Poet and His Politics*. Cambridge, MA: Harvard University Press, 1990.
Friedrich, Carl J., and Zbigniew K. Brzezinski. *Totalitarian Dictatorship and Autocracy*. Cambridge, MA: Harvard University Press, 1956.
Frierson, Cathy, and Semyon Vilensky. *Children of the Gulag*. New Haven, CT: Yale University Press, 2010.
Fulbrook, Mary. *The People's State: East German Society from Hitler to Honecker*. New Haven, CT: Yale University Press, 2006.
Fürst, Juliane. "Born under the Same Star: Refuseniks, Dissidents and Late Socialist Society." In *The Jewish Movement in the Soviet Union*, edited by Yaacov Ro'i, 137–65. Baltimore: Johns Hopkins University Press, 2012.
Fürst, Juliane. "From the Maiak to the Psichodrom: How Sixties Global Counterculture Came to Moscow." In *The Routledge Handbook of the Global Sixties: Between Protest and Nation-Building*, edited by Chen Jian et al., 180–92. New York: Routledge, 2018.
Gabowitsch, Mischa. "Are Copycats Subversive? Strategy-31, the Russian Runs, the Immortal Regiment, and the Transformative Potential of Non-hierarchical Movements." *Problems of Post-Communism*, vol. 65, no. 5 (2018): 297–314.
Gabowitsch, Mischa. "Gewaltfreier Widerstand: Vergleichende Betrachtungen zu Dynamik und Erfolgsbedingungen." Supplement, *Mittelweg 36*, August/September 2012, 61–67.

Gabowitsch, Mischa. *Protest in Putin's Russia*. Cambridge, UK: Polity Press, 2017.
Galanskov, Iurii. *Iurii Galanskov*. Frankfurt am Main: Posev, 1980.
Galich, Aleksandr. *Sochineniia v dvukh tomakh*. Moscow: Lokid, 1999.
Garthoff, Raymond. *A Journey through the Cold War: A Memoir of Containment and Coexistence*. Washington, DC: Brookings Institution Press, 2001.
Gernet, M. N., O. B. Gol'dovskii, and I. N. Sakharov, eds. *Protiv smertnoi kazni*. Moscow: Tipografiia I. D. Sytina, 1906.
Gerovitch, Slava. *From Newspeak to Cyberspeak: A History of Soviet Cybernetics*. Cambridge, MA: MIT Press, 2002.
Gershuni, V. "Pushkinskaia vakhta svobody." *Poiski: Svobodnyi moskovskii zhurnal*, no. 5–6 (1983). Moscow, samizdat.
Gerts, Anna [Maia Zlobina]. "K vol'noi vole zapovednye puti ..." *Novyi zhurnal*, no. 120 (1975): 31–77; no. 121 (1975): 25–70; no. 122 (1976): 27–77; no. 123 (1976): 15–38; no. 124 (1976): 45–72.
Gertsen [Herzen], A. I. "Le peuple russe et le socialisme." In *Sobranie sochinenii v tridtsati tomakh*, vol. 7, 279–314. Moscow: Izdatel'stvo Akademii nauk SSSR, 1956.
Gessen, Masha. "Fifty Years Later, Andrei Sakharov's Seminal Essay Is a Powerful Model of Writing for Social Change." *New Yorker*, July 25, 2018. https://www.newyorker.com/news/our-columnists/fifty-years-later-andrei-sakharovs-most-famous-essay-is-a-powerful-model-of-writing-for-social-change.
Gessen, Masha. *Where the Jews Aren't: The Sad and Absurd Story of Birobidzhan, Russia's Jewish Autonomous Region*. New York: Nextbook/Schocken, 2016.
Gessen, Masha. *Words Will Break Cement: The Passion of Pussy Riot*. New York: Riverhead Books, 2014.
Getty, Arch. "State and Society under Stalin: Constitutions and Elections in the 1930s." *Slavic Review*, vol. 50, no. 1 (1991): 18–35.
Geyer, Michael, and Sheila Fitzpatrick, eds. *Beyond Totalitarianism: Stalinism and Nazism Compared*. New York: Cambridge University Press, 2009.
Gilburd, Eleanory. *To See Paris and Die: The Soviet Lives of Western Culture*. Cambridge, MA: Belknap Press of Harvard University Press, 2018.
Gilcher-Holtey, Ingrid. *Die 68er Bewegung: Deutschland, Westeuropa, USA*. Munich: Beck, 2001.
Gilligan, Emma. *Defending Human Rights in Russia: Sergei Kovalyov, Dissident and Human Rights Commissioner, 1969–2003*. London and New York: RoutledgeCurzon, 2004.
Gilligan, Emma. "Refashioning the Dissident Past: Politics and Resistance in the Putin Era." *Russian Review*, vol. 74, no. 4 (October 2015): 559–62.
Ginzburg, Aleksandr. *Belaia kniga: Sbornik dokumentov po delu A. Siniavskogo i Iu. Danielia*. Frankfurt am Main: Posev, 1967.
Ginzburg, Aleksandr. "Dvadtsat' let tomu nazad." *Russkaia mysl'*, no. 3608 (February 14, 1986).
Ginzburg, Lidiia. *Chelovek za pis'mennym stolom*. Leningrad: Sovetskii pisatel', Leningradskoe otd-nie, 1989.
Gitlin, Todd. *The Whole World Is Watching: Mass Media in the Making and Unmaking of the New Left*. Berkeley: University of California Press, 1980.
Glad, John. "Vsevolod Kochetov: An Overview." *Russian Language Journal / Russkii iazyk*, vol. 32, no. 113 (Fall 1978): 95–102.
Glazov, Iurii. *V kraiu ottsov: Khronika nedavnego proshlogo*. Moscow: Khronika i zhizn', 1998.
Glazov, Yuri. "*The Devils* by Dostoevsky and the Russian Intelligentsia." *Studies in Soviet Thought*, vol. 17, no. 4 (December 1977): 309–30.
Gled, Dzhon [John Glad]. *Besedy v izgnanii: Russkoe literaturnoe zarubezh'e*. Moscow: Knizhnaia palata, 1991.

Gluzman, S. F. *Risunki po pamiati, ili Vospominaniia otsidenta.* Kyiv: Izdatel'skii dom Dmitriia Burago, 2012.

Goldberg, Paul. *The Dissident: A Novel.* New York: Farrar, Straus and Giroux, 2023.

Goldberg, Paul. *The Final Act: The Dramatic, Revealing Story of the Moscow Helsinki Watch Group.* New York: William Morrow, 1988.

Goldman, Marion, and Steven Pfaff. "Reconsidering Virtuosity: Religious Innovation and Spiritual Privilege." *Sociological Theory*, vol. 32, no. 2 (June 2014): 128–46.

Golomshtok, Igor'. *"Zaniatie dlia starogo gorodovogo": Memuary pessimista.* Moscow: Redaktsiia Eleny Shubinoi, 2015.

Gorbachev, Mikhail. *Zhizn' i reformy.* 2 vols. Moscow: Novosti, 1995.

Gorbanevskaia, Natal'ia. *Goroda i dorogi: izbrannye stikhotvoreniia 1956-2011.* Moscow: Russkii Gulliver, 2013.

Gorbanevskaia, Natal'ia. "'Mozhesh' vyiti na ploshchad', smeesh' vyiti na ploshchad.'" *Russkaia mysl'*, no. 3479 (August 25, 1983).

Gorbanevskaia, Natal'ia. "Neskol'ko slov v 'posleslovie.'" *Kontinent*, no. 12 (1977): 222–25.

Gorbanevskaia, Natal'ia. *Polden': Delo o demonstratsii na Krasnoi ploshchadi 25 avgusta 1968 goda.* Moscow: Novoe izdatel'stvo, 2007 [1970].

Gorbanevskaya, Natalya. "Writing for 'Samizdat.'" Interview by Michael Scammell. *Index on Censorship*, vol. 6, no. 1 (1977): 29–36.

Gordin, Michael. "Lysenko Unemployed: Soviet Genetics after the Aftermath." *ISIS*, vol. 109, no. 1 (2018): 56–78.

Gorelik, Gennadii. *Andrei Sakharov: Nauka i svoboda.* Moscow: R&C Dynamics Izhevsk, 2000.

Gorham, Michael. *Speaking in Soviet Tongues: Language Culture and the Politics of Voice in Revolutionary Russia.* DeKalb: Northern Illinois University Press, 2003.

Goritschewa, Tatjana. *Von Gott zu reden ist gefährlich. Meine Erfahrungen im Osten und im Westen.* Freiburg: Herder, 1984.

Gorlizki, Yoram. "Delegalization in Russia: Soviet Comrades' Courts in Retrospect." *American Journal of Comparative Law*, vol. 46, no. 3 (Summer 1998): 403–25.

Gorlizki, Yoram. "Structures of Trust after Stalin." *Slavonic and East European Review*, vol. 91, no. 1 (January 2013): 119–46.

Graham, Loren. *Science, Philosophy, and Human Behavior in the Soviet Union.* New York: Columbia University Press, 1987.

Graham, Seth. *Resonant Dissonance: The Russian Joke in Cultural Context.* Evanston, IL: Northwestern University Press, 2009.

Granovetter, Mark. "The Strength of Weak Ties." *American Journal of Sociology*, vol. 78, no. 6 (1973): 1360–80.

Granovetter, Mark. "The Strength of Weak Ties: A Network Theory Revisited." *Sociological Theory*, vol. 1 (1983): 201–33.

Grazhdanskii kodeks RSFSR ot 11.06.1964. http://www.kremlin.ru/acts/bank/3/print.

Greenberg, Cheryl Lynn. *Troubling the Waters: Black-Jewish Relations in the American Century.* Princeton, NJ: Princeton University Press, 2006.

Gribanov, Aleksandr. "Ob izuchenii istochnikov pervoi 'gumanitarnoi' raboty Sakharova." In *30 let "Razmyshlenii..." Andreia Sakharova*, 14–22. Moscow: Prava cheloveka, 1998.

Grigorenko, Petr. *Mysli sumasshedshego: Izbrannye pis'ma i vystupleniia Petra Grigor'evicha Grigorenko.* Amsterdam: Fond imeni Gertsena, 1973.

Grigorenko, Petr. *V podpol'e mozhno vstretit' tol'ko krys.* New York: Detinets, 1981.

Gross, Jan. "A Note on the Nature of Soviet Totalitarianism." *Soviet Studies*, vol. 34, no. 3 (July 1982): 367–76.

Groys, Boris. *The Total Art of Stalinism: Avant-Garde, Aesthetic Dictatorship, and Beyond.* Princeton, NJ: Princeton University Press, 1992.

Gurevich, Aron. *Istoriia istorika.* Moscow: ROSSPEN, 2004.

Gurevich, Aron. *Istoriia—neskonchaemyi spor.* Moscow: RGGU, 2005.

Gurevich, David. *From Lenin to Lennon: A Memoir of Russia in the Sixties.* New York: Harcourt Brace, 1991.

Gushchin, Viktor. "Human Rights in the Soviet Union: Putting the Record Straight." *New Times,* January 1976.

Hagen, Manfred. *Die Entfaltung politischer Öffentlichkeit in Rußland, 1906–1914.* Wiesbaden: F. Steiner, 1982.

Hagen, Mark von. Review of *Molotov Remembers: Inside Kremlin Politics; Conversations with Felix Chuev,* edited by Albert Resis. *Political Science Quarterly,* vol. 109, no. 5 (Winter 1994–95): 926–28.

Halfin, Igal. *Intimate Enemies: Demonizing the Bolshevik Opposition, 1918–1928.* Pittsburgh, PA: University of Pittsburgh Press, 2007.

Halfin, Igal. *Stalinist Confessions: Messianism and Terror at the Leningrad Communist University.* Pittsburgh, PA: University of Pittsburgh Press, 2009.

Halfin, Igal, and Jochen Hellbeck. "Rethinking the Stalinist Subject: Stephen Kotkin's 'Magnetic Mountain' and the State of Soviet Historical Studies." *Jahrbücher für Geschichte Osteuropas,* vol. 44 (1996): 456–63.

Harris, Robert. *Fatherland.* New York: Random House, 1992.

Harris, Sam. *Lying.* New York: Four Elephants Press, 2013.

Harrison, Mark. *Secret Leviathan: Secrecy and State Capacity under Soviet Communism.* Stanford, CA: Stanford University Press, 2023.

Hartshorn, Peter. *I Have Seen the Future: A Life of Lincoln Steffens.* Berkeley, CA: Counterpoint, 2011.

Haslam, Jonathan. *Russia's Cold War: From the October Revolution to the Fall of the Wall.* New Haven, CT: Yale University Press, 2011.

Havel, Václav. "The Power of the Powerless." In *Václav Havel, or Living in Truth,* edited by Jan Vladislav, 36–122. London: Meulenhoff Amsterdam in association with Faber and Faber, 1986.

Hayward, Max, ed. *On Trial: The Soviet State versus "Abram Tertz" and "Nikolai Arzhak."* New York: Harper & Row, 1966.

Hazard, John N. *Recollections of a Pioneering Sovietologist.* New York: Oceana Publications, 1987.

Heinzen, James. *The Art of the Bribe: Corruption under Stalin, 1943–1953.* New Haven, CT: Yale University Press, 2016.

Hellbeck, Jochen. *Revolution on My Mind: Writing a Diary under Stalin.* Cambridge, MA: Harvard University Press, 2006.

Hellie, Richard. "The Structure of Modern Russian History: Towards a Dynamic Model." *Russian History,* vol. 4, no. 1 (1977): 1–22.

Hobsbawm, Eric. *Primitive Rebels: Studies in Archaic Forms of Social Movement in the 19th and 20th Centuries.* New York: W. W. Norton, 1965.

Holloway, David. *Stalin and the Bomb: The Soviet Union and Atomic Energy, 1939–1956.* New Haven, CT: Yale University Press, 1994.

Holler, Markus. "'Für eure Freiheit und unsere!': Die Demonstration sowjetischer Dissidenten auf dem Roten Platz." In *Prager Frühling: Das internationale Krisenjahr 1968,* edited by Stefan Karner et al., vol. 1, 849–68. Cologne: Verein zur Förderung der Forschung von Folgen nach Konflikten und Kriegen, 2008.

Holquist, Peter. *Making War, Forging Revolution: Russia's Continuum of Crisis*. Cambridge, MA: Harvard University Press, 2002.

Hooper, Cynthia. "A Darker 'Big Deal': Concealing Party Crimes in the Post–Second World War Era." In *Late Stalinist Russia: Society between Reconstruction and Reinvention*, edited by Juliane Fürst, 142–60. New York: Routledge, 2006.

Hopgood, Stephen. *Keepers of the Flame: Understanding Amnesty International*. Ithaca, NY: Cornell University Press, 2006.

Hopkins, Mark. *Russia's Underground Press: "The Chronicle of Current Events."* New York: Praeger, 1983.

Hornsby, Robert. *Protest, Reform and Repression in Khrushchev's Soviet Union*. Cambridge: Cambridge University Press, 2013.

Horvath, Robert. "Apologist of Putinism? Solzhenitsyn, the Oligarchs, and the Specter of Orange Revolution." *Russian Review*, vol. 70, no. 2 (April 2011): 300–318.

Horvath, Robert. "Breaking the Totalitarian Ice: The Initiative Group for the Defense of Human Rights in the USSR." *Human Rights Quarterly*, vol. 36, no. 1 (February 2014): 147–75.

Horvath, Robert. *The Legacy of Soviet Dissent: Dissidents, Democratisation and Radical Nationalism in Russia*. New York: RoutledgeCurzon, 2005.

Horvath, Robert. "The Putin Regime and the Heritage of Dissidence." *Telos*, no. 145 (Winter 2008): 7–30.

Horvath, Robert. "'Sakharov Would Be with Us': Limonov, Strategy-31, and the Dissident Legacy." *Russian Review*, vol. 74, no. 4 (October 2015): 581–98.

Hosking, Geoffrey. *The First Socialist Society: A History of the Soviet Union from Within*. Cambridge, MA: Harvard University Press, 1992.

Hosking, Geoffrey. "Trust and Distrust in the USSR: An Overview." *Slavonic and East European Review*, vol. 91, no. 1 (January 2013): 1–25.

Hughes, H. Stuart. *Sophisticated Rebels: The Political Culture of European Dissent, 1968–1987*. Cambridge, MA: Harvard University Press, 1990.

Hunt, Lynn. *Inventing Human Rights: A History*. New York: W. W. Norton, 2007.

Hurst, Mark. *British Human Rights Organizations and Soviet Dissent, 1965–85*. London: Bloomsbury, 2016.

Iakir, Petr, and Iurii Geller, eds. *Komandarm Iakir: Vospominaniia druzei i soratnikov*. Moscow: Voen. Izd-vo, 1963.

Iakovlev, N. "O prepodavanii otechestvennoi istorii." *Bolshevik*, November 30, 1947, 25–32.

Igrunov, Viacheslav. "K problematike obshchestvennogo dvizheniia." Igrunov.ru, accessed December 12, 2023. http://igrunov.ru/cv/vchk-cv-chosenpubl/vchk-cv-chosenpubl-ego.html.

Igrunov, Viacheslav. "O biblioteke Samizdata, o Gruppe sodeistviia kul'turnomu obmenu i o Larise Bogoraz-Brukhman." Igrunov.ru, accessed December 12, 2023. http://igrunov.ru/cv/vchk-cv-memotalks/talks/vchk-cv-memotalks-talks-bogoraz.html.

Igrunov, Viacheslav. "O Glebe Pavlovskom, i o tom, kak poiavilas' stat'ia 'K problematike obshchestvennogo dvizheniia.'" Igrunov.ru, accessed December 12, 2023. http://igrunov.ru/cv/vchk-cv-memotalks/memories/vchk-cv-memotalks-memo-pavlovsk.html.

Il'ichev, L. F. *Iskusstvo prinadlezhit narodu*. Moscow: Gosudarstvennoe izdatel'stvo politicheskoi literatury, 1963.

Iurchak, Aleksei. *Eto bylo navsegda, poka ne konchilos': Poslednee sovetskoe pokolenie*. Moscow: Novoe literaturnoe obozrenie, 2014.

Iuzov, I. *Russkie dissidenty: Starovery i dukhovnye khristiane*. St. Petersburg: Tipografiia A. M. Kotomina, 1881.

"Iz stenogrammy zasedaniia Ideologicheskoi komissii TsK KPSS s uchastiem molodykh pisatelei, khudozhnikov, kompozitorov, tvorcheskikh rabotnikov kino i teatrov Moskvy"

[December 24, 1962]. In *Ideologicheskie komissii TsK KPSS 1958–1964: Dokumenty*, edited by E. S. Afanas'eva, V. Iu. Afiani, et al., 293–339. Moscow, 2000.
Jacobs, Seth. *Cold War Mandarin: Ngo Dinh Diem and the Origins of America's War in Vietnam, 1950–1963*. Lanham, MD: Rowman & Littlefield, 2006.
Jacoby, Susan. Introduction to *Notes of a Revolutionary*, by Andrei Amalrik, xi–xv. New York: Knopf, 1982.
Jacoby, Susan. *Moscow Conversations*. New York: Coward, McCann & Geoghegan, 1972.
Jallot, Nicolas. *Viktor Orekhov: Un dissident au KGB*. Paris: Stock, 2011.
Jargin, Sergei. "Some Aspects of Psychiatry in Russia." *International Journal of Culture and Mental Health*, vol. 4, no. 2 (2011): 116–20.
Jelen, Christian, and Thierry Wolton. *L'Occident des dissidents*. Paris: Stock, 1979.
Jones, Polly. *Myth, Memory, Trauma: Rethinking the Stalinist Past in the Soviet Union, 1953–70*. New Haven, CT: Yale University Press, 2013.
Judt, Tony. "The Dilemmas of Dissidence: The Politics of Opposition in East-Central Europe." *East European Politics and Societies*, vol. 2, no. 2 (1988): 185–240.
Kabo, Vladimir. *The Road to Australia: Memoirs*. Canberra: Aboriginal Studies Press, 1998.
Kadarkay, Arpad. *Human Rights in American and Russian Political Thought*. Washington, DC: University Press of America, 1982.
Kaganovskii, Gennadii, ed. *Khronika Iuriia Galanskova*. Moscow: Agraf, 2006.
Kaminskaia, Dina. *Zapiski advokata*. Kharkiv: Prava liudini, 2000 [1984].
Karner, Stefan, et al. *Prager Frühling: Das internationale Krisenjahr 1968*. 2 vols. Cologne: Verein zur Förderung der Forschung von Folgen nach Konflikten und Kriegen, 2008.
Karp, Masha. *George Orwell and Russia*. London: Bloomsbury Academic, 2023.
Kazakov, Evgenii. "Dissens und Untergrund: Das Wiederaufkommen der linken oppositionellen Gruppen in der späten Brežnev-Zeit." In *Goldenes Zeitalter der Stagnation? Perspektiven auf die sowjetische Ordnung der Brežnev Ära*, edited by Boris Belge and Martin Deuerlein, 75–93. Tübingen: Mohr Siebeck, 2014.
Keane, Elizabeth. *An Irish Statesman and Revolutionary: The Nationalist and Internationalist Politics of Seán MacBride*. New York: Tauris Academic Studies, 2006.
Kelly, Catriona. *Refining Russia: Advice Literature, Polite Culture, and Gender from Catherine to Yeltsin*. New York: Oxford University Press, 2001.
Kennan, George F. "Morality and Foreign Policy." *Foreign Affairs*, vol. 64, no. 2 (Winter 1985–86): 205–18.
Kenney, Padraic. "The Gender of Resistance in Communist Poland." *American Historical Review*, vol. 104, no. 2. (April 1999): 399–425.
Kershaw, Ian, and Moshe Lewin, eds. *Stalinism and Nazism: Dictatorships in Comparison*. New York: Cambridge University Press, 1997.
Kharkhordin, Oleg. *The Collective and the Individual in Russia: A Study of Practices*. Berkeley: University of California Press, 1999.
Khmel'nitskaia, Viktoriia. *Tak slozhilas' nasha zhizn'*. St. Petersburg: Aleteiia, 2011.
Khmel'nitskii, Sergei. "Iz chreva kitova." *Dvadtsat' dva*, no. 48 (June–July 1986): 151–80.
Khristoforov, Igor' A. "'Oznachaiushchee i oznachaemoe': Poniatie 'konstitutsiia' v rossiiskom politicheskom diskurse do i posle 17 oktiabria 1905 goda." *Cahiers du monde russe*, vol. 48, no. 2/3 (2007): 173–84.
Khrushchev, N. S. "O kul'te lichnosti i ego posledstviiakh." In *Doklad N. S. Khrushcheva o kul'te lichnosti Stalina na XX s"ezde KPSS: Dokumenty*, edited by K. Aimermakher et al., 51–119. Moscow: ROSSPEN, 2002.
Khrushchev, N. S. *Vospominaniia: Vremia, liudi, vlast'*. 4 vols. Moscow: Moskovskie novosti, 1999.

Kim, Byung-Yeon. "Informal Economy Activities of Soviet Households: Size and Dynamics." *Journal of Comparative Economics*, vol. 31, no. 3 (September 2003): 532–51.

Kim, Iulii. "Monolog p'ianogo genseka," Bards, 1968. http://www.bards.ru/archives/part.php?id=6188.

Kind-Kovács, Friederike. *Written Here, Published There: How Underground Literature Crossed the Iron Curtain*. Budapest: Central European University Press, 2014.

Kind-Kovács, Friederike, and Jessie Labov, eds. *Samizdat, Tamizdat, and Beyond: Transnational Media during and after Socialism*. New York: Berghahn Books, 2013.

Kinne, Burdette. "Voltaire Never Said It!" *Modern Language Notes*, vol. 58, no. 7 (November 1943): 534–35.

Kirk, Irina. *Profiles in Russian Resistance*. New York: Quadrangle, 1975.

Kissinger, Henry. *Years of Renewal*. New York: Simon & Schuster, 1999.

"The Kissinger Telephone Conversations: A Verbatim Record of U.S. Diplomacy, 1969–1977." Digital National Security Archive, 2014. ProQuest. https://www.proquest.com/dnsa_ka/docview/1679070314/fulltextPDF/578C089F7B4582PQ/1?.

Klain, Edvard [Edward Kline]. *Moskovskii komitet prav cheloveka*. Moscow: Prava cheloveka, 2004.

Klemperer, Klemens von. "'What Is the Law That Lies behind These Words?' Antigone's Question and the German Resistance against Hitler." In "Resistance against the Third Reich." Supplement, *Journal of Modern History*, vol. 64 (December 1992): S102–11.

Klimke, Martin, Jacco Pekelder, and Joachim Scharloth, eds. *Between Prague Spring and French May: Opposition and Revolt in Europe, 1960–1980*. New York: Berghahn Books, 2011.

Klimke, Martin, and Joachim Scharloth, eds. *1968 in Europe: A History of Protest and Activism, 1956–77*. New York: Palgrave Macmillan, 2008.

Klots, Yasha. *Tamizdat: Contraband Russian Literature in the Cold War Era*. Ithaca, NY: Northern Illinois University Press, an imprint of Cornell University Press, 2023.

Klumbyte, Neringa. *Authoritarian Laughter: Political Humor and Soviet Dystopia in Lithuania*. Ithaca, NY: Cornell University Press, 2022.

Knowlton, Winthrop. "A Day in the Country." *Atlantic Monthly*, vol. 241, no. 2 (February 1978): 87–89.

Kochanski, Halik. *Resistance: The Underground War against Hitler, 1939–1945*. New York: Liveright, 2022.

Kochetov, Vsevolod. "Chego zhe ty khochesh'?" *Oktiabr'*, no. 9 (September 1969): 11–136; no. 10 (October 1969): 41–138; no. 11 (November 1969): 107–72.

Kolakowski, Leszek. *Main Currents of Marxism*. 3 vols. Oxford: Clarendon Press, 1978.

Komaromi, Ann. "Samizdat as Extra-Gutenberg Phenomenon." *Poetics Today*, vol. 29, no. 4 (Winter 2008): 629–67.

Komaromi, Ann. *Soviet Samizdat: Imagining a New Society*. Ithaca, NY: Northern Illinois University Press, an imprint of Cornell University Press, 2022.

Komaromi, Ann, and Gennadii Kuzovkin. *Katalog periodiki samizdata, 1956–1986*. Moscow: Mezhdunarodnyi Memorial, 2018.

Kon, Igor. *Druzhba*. 4th ed. St. Petersburg: Piter, 2005.

Konrad, Georgi. *Anti-Politics*. San Diego, CA: Harcourt Brace Jovanovich, 1984.

Kornai, János. *The Socialist System: The Political Economy of Socialism*. Oxford: Clarendon Press, 1992.

Korotenko, A., and N. Alikina. *Sovetskaia psikhiatriia: Zabluzhdeniia i umysel*. Kyiv: Sfera, 2002.

Korotkov, A., et al., comps. *Kremlevskii samosud: Sekretnye dokumenty Politbiuro o pisatele A. Solzhenitsyne*. Moscow: Rodina, 1994.

Koselleck, Reinhart. *Kritik und Krise: Eine Studie zur Pathogenese der bürgerlichen Welt.* Frankfurt am Main: Suhrkamp Verlag, 1973 [1959].
Kotkin, Stephen. *Magnetic Mountain: Stalinism as a Civilization.* Berkeley: University of California Press, 1997.
Kovalev, Ivan. "Nekotorye soobrazheniia o pomoshchi politzakliuchennym." Samizdat, 1979.
Kowaljow, Sergej [Sergei Kovalev]. *Der Flug des weißen Raben: Von Sibirien nach Tschetschenien: Eine Lebensreise.* Berlin: Rowohlt, 1997.
Kozlov, Denis. *The Readers of "Novyi mir": Coming to Terms with the Stalinist Past.* Cambridge, MA: Harvard University Press, 2013.
Kozlov, Denis, and Eleonory Gilburd. "The Thaw as an Event in Russian History." In *The Thaw: Soviet Society and Culture during the 1950s and 1960s,* edited by Denis Kozlov and Eleonory Gilburd, 18–81. Toronto: University of Toronto Press, 2021.
Kozlov, V. A., and S. V. Mironenko, eds. *Kramola: Inakomyslie v SSSR pri Khrushcheve i Brezhneve 1953–1982 gg.* Moscow: Materik, 2005.
Kozlovskii, Evgenii. *Krasnaia ploshchad': Dissident i chinovnitsa.* Paris: Tret'ia volna, 1982.
Kramer, Mark. "The Czechoslovak Crisis and the Brezhnev Doctrine." In *1968: The World Transformed,* edited by Carole Fink, Philipp Gassert, and Detlef Junker, 111–72. New York: Cambridge University Press, 1998.
Krasin, Viktor. *Poedinok: Zapiski antikommunista.* Surrey, UK: Hodgson Press, 2012.
Krasin, Viktor. *Sud.* New York: Chalidze Publications, 1983.
Krastev, Ivan. *Eksperimental'naia rodina: Razgovor s Glebom Pavlovskim.* Moscow: Evropa, 2018.
Krylova, Anna. "Beyond the Spontaneity-Consciousness Paradigm: 'Class Instinct' as a Promising Category of Historical Analysis." *Slavic Review,* vol. 62, no. 1 (Spring 2003): 1–23.
Krylova, Anna. "Imagining Socialism in the Soviet Century." *Social History,* vol. 42, no. 3 (2017): 315–41.
Kuchinskii, V. A. *Lichnost', svoboda, pravo.* Minsk: Nauka i tekhnika, 1969.
Kukulin, Ilya. "A Bizarre Encounter: O vliianii prototipa Portsii Braun na roman Vs. Kochetova 'Chego zhe ty khochesh'?'" *Novoe literaturnoe obozrenie,* no. 4 (2018): 146–58.
Kuznetsov, A. V. *Ugolovno-pravovaia okhrana interesov lichnosti v SSSR.* Moscow: Iuridicheskaia literatura, 1969.
Kuznetsov, I. S. *Novosibirskii Akademgorodok v 1968 godu: "Pis'mo soroka shesti." Dokumental'noe izdanie.* Novosibirsk: Klio, 2007.
Kuznetsov, Vladimir. "O 40-kh i 50-kh: Dom, shkola, filfak MGU." *Tynianovskii sbornik,* no. 10 (1998): 761–93.
Kuzovkin, Gennadii, and Aleksei Makarov, eds. *Dokumenty Initsiativnoi gruppy po zashchite prav cheloveka v SSSR.* Moscow: n.p., 2009. http://old.memo.ru/history/diss/ig/docs/igdocs.html.
Kuzovkin, Gennadii, Aleksei Makarov, et al. "Spisok grazhdan, vyrazivshikh protest ili nesoglasie s vtorzheniem v Chekhoslovakiiu." *Polit,* September 2, 2008. https://polit.ru/article/2008/09/02/people68/.
Kuzovkin, Gennadii, et al., eds. *K istorii Moskovskoi Khel'sinkskoi gruppy: Vospominaniia, dokumenty TSK KPSS, KGB i drugie materialy.* Moscow: Zatsepa, 2001.
Labedz, Leopold, and Max Hayward, eds. *On Trial: The Case of Sinyavsky (Tertz) and Daniel (Arzhak).* London: Collins and Harvill Press, 1967.
Ledeneva, Alena. "The Genealogy of *Krugovaia Poruka*: Forced Trust as a Feature of Russian Political Culture." In *Trust and Democratic Transition in Post-Communist Europe,* edited by Ivana Marková, 85–108. Oxford: Oxford University Press, 2004.

Leeuwenberg, Huib. "Is Amnesty Impartial Enough?" In *Statute of Amnesty International, as Amended by the Eighth International Council Meeting in St. Gallen, Switzerland, 12–14 September 1975*. London: Amnesty International Publications, 1975. Unpaginated.

Lefebvre, George. *The Coming of the French Revolution*. Princeton, NJ: Princeton University Press, 2019 [1939].

Lenin, V. I. *Polnoe sobranie sochinenii*. 55 vols. 5th ed. Moscow: Gosudarstvennoe izdatel'stvo politicheskoi literatury, 1958–65.

Levitin-Krasnov, Anatolii. *Rodnoi prostor: Demokraticheskoe dvizhenie*. Frankfurt am Main: Posev, 1981.

Liber, George. "Ivan Dzyuba, Cultural Cringe, and the Origins of Ukraine's Revolution of Dignity." *New Zealand Slavonic Journal*, vol. 49/50 (2015–16): 51–66.

Liebling, A. J. "The Wayward Press: Do You Belong in Journalism?" *New Yorker*, May 14, 1960.

Litvinov, Pavel, ed. *The Demonstration in Pushkin Square*. Boston: Gambit, 1969.

Litvinov, Pavel. "O dvizhenii za prava cheloveka v SSSR." In *Samosoznanie: Sbornik statei*, edited by Pavel Litvinov, Mikhail Meerson-Aksenov, and Boris Shragin, 63–88. New York: Khronika, 1976.

Litvinov, Pavel. "Otkrytoe pis'mo A. Solzhenitsynu." *Vestnik RSKhD*, vol. 114, no. 4 (1974): 258–60.

Litvinov, Pavel. "Pavel Litvinov and Index on Censorship." *Index on Censorship*, vol. 4, no. 1 (1975): 6–10.

Litvinov, Pavel, ed. *Pravosudie ili rasprava? Delo o demonstratsii na Pushkinskoi ploshchadi 22 ianvaria 1967 goda. Sbornik dokumentov*. London: Overseas Publications Interchange, 1968.

Litvinov, Pavel, ed. *Protsess chetyrekh: Sbornik materialov po delu Galanskova, Ginzburga, Dobrovol'skogo i Lashkova*. Amsterdam: Alexander Herzen Foundation, 1971.

Litvinov, Pavel. "Waiting for the New Dictators." *Index on Censorship*, vol. 20, no. 1 (1992): 2–4.

Loeber, D. A. "Samizdat under Soviet Law." In *Contemporary Soviet Law: Essays in Honor of John N. Hazard*, edited by Donald Barry, William Butler, and George Ginsburgs, 84–123. The Hague: Martinus Nijhoff, 1974.

Loeffler, James. *Rooted Cosmopolitans: Jews and Human Rights in the Twentieth Century*. New Haven, CT: Yale University Press, 2018.

Lomb, Samantha. *Stalin's Constitution: Soviet Participatory Politics and the Discussion of the 1936 Draft Constitution*. New York: Routledge, 2017.

Loshak, Andrei, dir. *Anatomiia protsessa*. Blinibioscoop, December 20, 2014. https://www.youtube.com/watch?v=XpkixqlpN9w.

Lourie, Richard. *Sakharov: A Biography*. Hanover, NH: Brandeis University Press, 2002.

Lovell, Stephen. *The Russian Reading Revolution: Print Culture in the Soviet and Post-Soviet Eras*. New York: St. Martin's Press, 2000.

Lovell, Stephen. *The Shadow of War: Russia and the USSR, 1941 to the Present*. Malden, MA: Wiley-Blackwell, 2010.

Lushin, A. I. *Sovetskoe gosudarstvo i oppozitsiia v seredine 1950–1980-kh godov*. St. Petersburg: Izdatel'stvo Severo-Zapadnoi akademii gosudarstvennoi sluzhby, 2011.

MacBride, Seán. *L'Exigence de la liberté: Amnesty International*. With Éric Laurent. Paris: Stock, 1981.

Madden, Edward. "Civil Disobedience." In *Dictionary of the History of Ideas*, vol. 1, 434–41. New York: Scriber, 1973.

Mahan, Erin R., ed. *Foreign Relations of the United States, 1969–1976*. Vol. 12, *Soviet Union, January 1969–October 1970*. Washington, DC: Government Printing Office, 2006.

Maistrenko, Ivan, ed. *Ukrainski iurysty pid sudom KGB*. Munich: Suchasnist', 1968.

Makarevich, Eduard. *Filipp Bobkov i Piatoe upravlenie KGB: Sled v istorii*. Moscow: Eduard Fedorovich, 2015.

Malia, Martin. *History's Locomotives: Revolutions and the Making of the Modern World*. New Haven, CT: Yale University Press, 2006.
Malia, Martin. *The Soviet Tragedy: A History of Socialism in Russia, 1917–1991*. New York: Free Press, 1994.
Mal'tsev, G. V. *Sotsialisticheskoe pravo i svoboda lichnosti*. Moscow: Iuridicheskaia literatura, 1968.
Mandel'shtam, Nadezhda. *Sobranie sochinenii v dvukh tomakh*. Ekaterinburg, 2014.
Mandelstam, Nadezhda. *Hope Against Hope: A Memoir*. New York: Atheneum, 1983.
Marchenko, Anatolii. *Moi pokazaniia*. Paris: Presse Libre, 1969 [samizdat, 1968].
Marchenko, Anatolii. *Zhivi kak vse*. New York: Problemy Vostochnoi Evropy, 1987.
Marchenko, Anatoly. *My Testimony*. New York: E. P. Dutton, 1969.
Margolina, T. A., comp. *Obraz zhizni: Ob uchiteliakh Iu. A. Aikhenval'de i V. M. Gerlin*. Moscow: Vozvrashchenie, 2014.
Markarian, Vania. *Left in Transformation: Uruguayan Exiles and the Latin American Human Rights Networks, 1967–1984*. New York: Routledge, 2005.
Markesinis, Eugenie. *Andrei Siniavskii: A Hero of His Time?* Boston: Academic Studies Press, 2013.
Martin, Barbara. *Dissident Histories in the Soviet Union: From De-Stalinization to Perestroika*. London: Bloomsbury, 2019.
Martin, Barbara. *Roy and Zhores Medvedev: Loyal Dissent in the Soviet Union*. Boston: Academic Studies Press, 2023.
Marx, Karl. *The Eighteenth Brumaire of Louis Bonaparte*. New York: International Publishers, 1963.
Marx, Karl. "On the Jewish Question." In *The Marx-Engels Reader*, edited by Robert Tucker, 26–46. New York: W. W. Norton, 1978.
Marynovych, Myroslav. *The Universe behind Barbed Wire: Memoirs of a Ukrainian Soviet Dissident*. Rochester, NY: University of Rochester Press, 2021.
Matich, Olga. "*Spokojnoj noči*: Andrej Sinjavskij's Rebirth as Abram Terc." *Slavic and East European Journal*, vol. 33, no. 1 (Spring 1989): 50–63.
Matsui, Yasuhiro. "Forming a Transnational Moral Community between Soviet Dissidents and Ex-Communist Western Supporters: The Case of Pavel Litvinov, Karel van het Reve, and Stephen Spender." *Contemporary European History*, vol. 29, no. 1 (February 2020): 77–89.
Matsui, Yasuhiro. "*Obshchestvennost'* across Borders: Soviet Dissidents as a Hub of Transnational Agency." In *Obshchestvennost' and Civic Agency in Late Imperial and Soviet Russia: Interface between State and Society*, edited by Matsui Yasuhiro, 198–218. London: Springer, 2015.
Maza, Sarah. *The Myth of the French Bourgeoisie: An Essay on the Social Imaginary, 1750–1850*. Cambridge, MA: Harvard University Press, 2003.
Mazour, Anatole. *The First Russian Revolution, 1825*. Stanford, CA: Stanford University Press, 1964.
McVicker, Ben. "The Creation and Transformation of a Cultural Icon: Aleksandr Solzhenitsyn in Post-Soviet Russia, 1994–2008." *Canadian Slavonic Papers*, vol. 53, nos. 2–4 (June-September-December 2011): 305–33.
Medinskii, V. R., and A. V. Torkunov. *Istoriia Rossii: 1945 god–nachalo XXI veka. Uchebnik*. Moscow: Prosveshchenie, 2023.
Medvedev, Roi. *Andropov*. Moscow: Molodaia gvardiia, 2006.
Medvedev, Roi. "'Chernye spiski' pisatelei." *Politicheskii dnevnik*, no. 43 (April 1968). Moscow, samizdat.
Medvedev, Roi. "Kak sozdavalsia 'Politicheskii dnevnik.'" *Vnutrennie protivorechiia*, vol. 6 (1982): 147–54.
Medvedev, Roi. "K voprosu o formakh i metodakh bor'by za demokratizatsiiu v strane i v partii." *Politicheskii dnevnik* (June 1968). Moscow, samizdat.

Medvedev, Zhores. *Opasnaia professiia*. Moscow: Vremia, 2019.
Meerson-Aksenov, Michael, and Boris Shragin, eds. *The Political, Social, and Religious Thought of Russian "Samizdat": An Anthology*. Belmont, MA: Nordland, 1977.
Mel'nichenko, Mikhail, comp. *Sovetskii anekdot (ukazatel' siuzhetov)*. Moscow: Novoe literaturnoe obozrenie, 2014.
Mel'nikov, A. A. *Pravovoe polozhenie lichnosti v sovetskom grazhdanskom protsesse*. Moscow: Nauka, 1969.
Metger, Julia. *Studio Moskau. Westdeutsche Korrespondenten im Kalten Krieg*. Paderborn: Schöningh, 2016.
Miedema, Christie. *Not a Movement of Dissidents: Amnesty International beyond the Iron Curtain*. Göttingen: Wallstein Verlag, 2019.
Mihr, Anja. *Amnesty International in der DDR: Der Einsatz für Menschenrechte im Visier der Stasi*. Berlin: Ch. Links, 2002.
Millar, James. "The Little Deal: Brezhnev's Contribution to Acquisitive Socialism." *Slavic Review*, vol. 44, no. 4 (Winter 1985): 694–706.
Mitrokhin, Nikolai. *Russkaia partiia: Dvizhenie russkikh natsionalistov v SSSR, 1953–1985*. Moscow: Novoe literaturnoe obozrenie, 2003.
Mitrokhin, Vasiliy, ed. *KGB Lexicon: The Soviet Intelligence Officer's Handbook*. London: Frank Cass, 2002.
Monas, Sydney. "Amalrik's Vision of the End." In *Will the Soviet Union Survive until 1984?*, by Andrei Amalrik, 69–90. New York: Harper & Row, 1970.
Monk, Ray. *Wittgenstein: The Duty of Genius*. New York: Free Press, 1990.
Morev, Gleb, ed. *Dissidenty: Dvadtsat' razgovorov*. Moscow: Izdatel'stvo AST, 2017.
Morgan, Michael C. *The Final Act: The Helsinki Accords and the Transformation of the Cold War*. Princeton, NJ: Princeton University Press, 2018.
Moroz, Valentyn. *A Chronicle of Resistance in Ukraine*. Baltimore: Smoloskyp, 1970.
Morozov, Boris, ed. *Evreiskaia emigratsiia v svete novykh dokumentov*. Tel Aviv: Ivrus, 1998.
Morsink, Johannes. *The Universal Declaration of Human Rights: Origins, Drafting, and Intent*. Philadelphia: University of Pennsylvania Press, 1999.
Moyn, Samuel. *The Last Utopia: Human Rights in History*. Cambridge, MA: Belknap Press of Harvard University Press, 2010.
Mozokhina, A. G. *Svoboda lichnosti i osnovnye prava grazhdan v sotsialistichekikh stranakh Evropy*. Moscow: Nauka, 1965.
Müller, Jan-Werner. *Contesting Democracy: Political Ideas in Twentieth-Century Europe*. New Haven, CT: Yale University Press, 2011.
Murav, Harriet. *Russia's Legal Fictions*. Ann Arbor: University of Michigan Press, 1998.
Nabokov, Vladimir. *Pnin*. New York: Vintage, 1989 [1957].
Naiman, Anatolii. *Rasskazy o Anne Akhmatovoi*. 2nd ed. Moscow: Vagrius, 1999.
Nasstrom, Kathryn L. "Between Memory and History: Autobiographies of the Civil Rights Movement." *Journal of Southern History*, vol. 74 (2008): 325–64.
Nathans, Benjamin. *Beyond the Pale: The Jewish Encounter with Late Imperial Russia*. Berkeley: University of California Press, 2002.
Nathans, Benjamin. "Dictatorship of Reason: Aleksandr Vol'pin and the Idea of Rights under 'Developed Socialism.'" *Slavic Review*, vol. 66, no. 4 (Winter 2007): 630–63.
Nathans, Benjamin. "The Disenchantment of Socialism: Soviet Dissidents, Human Rights, and the New Global Morality." In *The Breakthrough: Human Rights in the 1970s*, edited by Jan Eckel and Samuel Moyn, 33–48. Philadelphia: University of Pennsylvania Press, 2014.
Nathans, Benjamin. "Helsinki Syndrome: Human Rights and International Diplomacy." *Times Literary Supplement*, no. 6038/9 (December 2018): 6–7.

Nathans, Benjamin. "The Many Shades of Soviet Dissidence." *Kritika: Explorations in Russian and Eurasian History*, vol. 23, no. 1 (Winter 2022): 185–96.
Nathans, Benjamin. "Refuseniks and Rights Defenders: Jews and the Soviet Dissident Movement." In *From Europe's East to the Middle East: Israel's Russian and Polish Lineages*, edited by Kenneth B. Moss, Benjamin Nathans, and Taro Tsurumi, 362–75. Philadelphia: University of Pennsylvania Press, 2021.
Nathans, Benjamin. "Soviet Rights-Talk in the Post-Stalin Era." In *Human Rights in the Twentieth Century*, edited by Stefan-Ludwig Hoffmann, 166–90. New York: Cambridge University Press, 2011.
Nathans, Benjamin. "Talking Fish: On Soviet Dissident Memoirs." *Journal of Modern History*, vol. 87, no. 3 (September 2015): 579–614.
Navrátil, Jaromíl, ed. *The Prague Spring 1968: A National Security Archive Documents Reader*. New York: Central European University Press, 1998.
Navrozov, Lev. *The Education of Lev Navrozov: A Life in the Closed World Once Called Russia*. New York: Harper's Magazine Press, 1975.
Navrozov, Lev. "Getting out of Russia." *Commentary*, vol. 54, no. 4 (October 1972): 45–53.
Nazarov, V. V. "Antisovetskaia agitatsiia i propaganda kak osobo opasno gosudarstvennoe prestuplenie." PhD diss., Vysshaia krasnoznamennaia shkola KGB SSSR, 1960.
Nelidov, Dmitrii [Vladimir Zelinskii]. "Ideokraticheskoe soznanie i lichnost." *Vestnik RSKhD*, vol. 111 (1974): 185–214.
Nepomnyashchy, Catherine Theimer. *Abram Tertz and the Poetics of Crime*. New Haven, CT: Yale University Press, 1995.
Neumann, Franz. *Behemoth: The Structure and Practice of National Socialism*. New York: Octagon Books, 1944.
News Conference at the Journalists' Club, Moscow, September 5, 1973 (Concerning the Trial of Yakir and Krasin). Moscow: Novosti Press Agency Publishing House, 1973.
O'Keeffe, Dennis, and Helen Szamuely. *Samizdat, Based on a Discussion at the CRCE Led by Dennis O'Keeffe and Helen Szamuely*. London: Centre for Research into Post-Communist Economies, 2004.
Oleszczuk, Thomas A. *Political Justice in the USSR: Dissent and Repression in Lithuania, 1969–1987*. New York: East European Monographs, 1988.
"O 'literaturnosti,' 'dokumentalizme' i 'iuridicheskom iazyke' zapisei suda nad Brodskim (stenogramma obsuzhdeniia doklada O. Rozenbliuma)." In *Acta Samizdatica / Zapiski o samizdate: Al'manakh*, edited by Elena Nikolaevna Strukova and Boris Isaevich Belenkin, vol. 4, 249–78. Moscow: GPIB Rossii, 2018.
Orlov, Aleksandr. *Aleksandr Ginzburg: Russkii roman*. Moscow: Russkii put', 2017.
Orlov, Iurii. *Opasnye mysli: Memuary iz russkoi zhizni*. Moscow: Moskovskaia Khel'sinskaia gruppa, 2006.
Orlov, Iurii. "Vozmozhen li sotsializm ne-totalitarnogo tipa?" Samizdat, 1975.
Orlov, Yuri. "Before and After Glasnost." *Commentary*, vol. 86, no. 4 (October 1988): 24–34.
Orlov, Yuri. *Dangerous Thoughts: Memoirs of a Russian Life*. Translated by Thomas P. Whitney. New York: William Morrow, 1991.
Orlova, Raisa. "'Rodinu ne vybiraiut. Ia vernus'...': Iz dnevnikov i pisem 1964–1988 godov." *Voprosy literatury*, no. 5 (2010): 261–302.
Orlova, Raisa. *Vospominaniia o neproshedshem vremeni*. Ann Arbor, MI: Ardis, 1983.
Orlova, Raisa, and Lev Kopelev. *My zhili v Moskve, 1956–1980*. Ann Arbor, MI: Ardis, 1988.
Osa, Maryjane. *Solidarity and Contention: The Networks of Polish Opposition, 1954–81*. Minneapolis: University of Minnesota Press, 2003.
Osnos, Peter. *An Especially Good View: Watching History Happen*. New York: Platform Books, 2021.

Osnos, Peter. "Soviet Dissidents and the American Press." *Columbia Journalism Review*, vol. 16, no. 4 (1977): 32–37.

Oushakine, Serguei. "The Terrifying Mimicry of Samizdat." *Public Culture*, vol. 13, no. 2 (2001): 191–214.

Overy, Richard. *The Dictators: Hitler's Germany and Stalin's Russia*. New York: W. W. Norton, 2004.

"Pamiati Liudmily Polikovskoi i Aleksandra Malinovskogo." *Polit*, March 1, 2017. https://polit.ru/article/2017/03/01/in_memoriam/.

Paperno, Irina. *Chernyshevsky and the Age of Realism: A Study in the Semiotics of Behavior*. Stanford, CA: Stanford University Press, 1988.

Paperno, Irina. *Stories of the Soviet Experience: Memoirs, Diaries, Dreams*. Ithaca, NY: Cornell University Press, 2009.

Parchomenko, Walter. *Soviet Images of Dissidents and Nonconformists*. New York: Praeger, 1986.

Parisi, Valentina, ed. *Samizdat: Between Practices and Representations*. Budapest: CEU Institute for Advanced Study, 2015.

Pasternak, Boris. *Stikhotvoreniia i poemy*. Moscow: Sovetskii pisatel', 1965.

Pazderka, Josef, ed. *The Soviet Invasion of Czechoslovakia in 1968: The Russian Perspective*. Lanham, MD: Lexington Books, 2019.

Petrov, Nikita. "Podrazdeleniia KGB SSSR po bor'be s inakomysliem 1967–1991 godov." In *Povsednevnaia zhizn' pri sotsializme: Nemetskie i rossiiskie podkhody*, edited by Jan Behrends et al., 158–84. Moscow: ROSSPEN, 2015.

Peukert, Detlev. *Die Weimarer Republik: Krisenjahre der klassischen Moderne*. Frankfurt am Main: Suhrkamp, 1987.

Pichardo, Nelson. "New Social Movements: A Critical Review." *Annual Review of Sociology*, vol. 23 (1997): 411–30.

Pierce, John Robinson. *An Introduction to Information Theory: Symbols, Signals and Noise*. New York: Dover, 1980 [1961].

Pils, Eva. *China's Human Rights Lawyers: Advocacy and Resistance*. New York: Routledge, 2015.

Pipes, Richard. "Russia's Exigent Intellectuals." *Encounter*, vol. 22 (January 1964): 79–84.

Pipes, Richard. *Vixi: Memoirs of a Non-belonger*. New Haven, CT: Yale University Press, 2006.

Platt, Kevin, and Benjamin Nathans. "Socialist in Form, Indeterminate in Content: The Ins and Outs of Late Soviet Culture." *Ab Imperio*, no. 2 (2011): 301–24.

Pliushch, Leonid. *Na karnivale istorii*. London: Overseas Publications Interchange, 1979.

Podrabinek, Aleksandr. *Dissidenty*. Moscow: AST, 2014.

Polikovskaia, Liudmila, ed. *My predchuvstvie . . . predtecha . . . : Ploshchad' maiakovskogo 1958–1965*. Moscow: Zven'ia, 1997.

Pomerants, Grigorii. *Zapiski gadkogo utenka*. Moscow: Moskovskii rabochii, 1998.

Pomeranz, William. *Law and the Russian State: Russia's Legal Evolution from Peter the Great to Vladimir Putin*. New York: Bloomsbury Academic, 2019.

Pravilova, Ekaterina A. *Zakonnost' i prava lichnosti: Administrativnaia iustitsiia v Rossii (vtoraia polovina XIX v.–oktiabr' 1917 g.)*. St. Petersburg: Obrazovanie-kul'tura, 2000.

Prishchepa, Aleksandr. "Inakomyslie na Urale (ser. 1940-x–ser. 1980-x gg.)." PhD diss., Surgut State University, 1999.

Putz, Manuela. *Kulturraum Lager: Politische Haft und dissidentisches Selbstverständnis in der Sowjetunion nach Stalin*. Wiesbaden: Harrassowitz Verlag, 2019.

Rakhunov, R. "Sovetskoe pravosudie i ego rol' v ukreplenii zakonnosti." *Kommunist*, no. 7 (1956): 42–54.

Reddaway, Peter. *The Dissidents: A Memoir of Working with the Resistance in Russia, 1960–1990*. Washington, DC: Brookings Institution Press, 2020.

Reddaway, Peter. "Pamiati Petra Iakira." In *SSSR: Vnutrennie protivorechiia*, vol. 7, 193–96. New York: Chalidze Publications, 1982.
Reddaway, Peter. "Soviet Policies toward Dissent, 1953–1986." *Journal of Interdisciplinary Studies*, vol. 24, no. 1/2 (2012): 57–82.
Reemstma, Jan Philipp. *"Wie hätte ich mich verhalten?" und andere nicht nur deutsche Fragen: Reden und Aufsätze*. Munich: Beck, 2001.
Reich, Rebecca. *State of Madness: Psychiatry, Literature, and Dissent after Stalin*. DeKalb: Northern Illinois University Press, 2018.
Reich, Rebecca. "Words on Trial: Morality and Legality in Frida Vigdorova's Journalism." *Slavic Review*, vol. 81, no. 2 (Summer 2022): 349–69.
Reid, Allan. "Ot Bartoka do Butyrki: O konflikte grazhdanskogo i liricheskogo v rannei poezii Natal'i Gorbanevskoi." *Novoe literaturnoe obozrenie*, no. 3 (2008): 251–62.
Reid, Susan. "(Socialist) Realism Unbound: The Effects of International Encounters on Soviet Art Practice and Discourse in the Khrushchev Thaw." In *Art beyond Borders: Artistic Exchange in Communist Europe (1945–1989)*, edited by Jérôme Bazin, Pascal Dubourg Glatigny, and Piotr Piotrowski, 267–95. Budapest: Central European University Press, 2016.
Reid, Susan. "Who Will Beat Whom? Soviet Popular Reception of the American National Exhibition in Moscow, 1959." *Kritika: Explorations in Russian and Eurasian History*, vol. 9, no. 4 (Fall 2008): 855–904.
Reisch, Alfred. *Hot Books in the Cold War: The CIA-Funded Secret Book Distribution Program behind the Iron Curtain*. Budapest: Central European University Press, 2013.
Remeikis, Thomas. *Opposition to Soviet Rule in Lithuania, 1945–1980*. Chicago: Institute of Lithuanian Studies Press, 1980.
Remnick, David. *Lenin's Tomb: The Last Days of the Soviet Empire*. New York: Vintage, 1994.
Reve, Karel van het, ed. *Dear Comrade: Pavel Litvinov and the Voices of Soviet Citizens in Dissent*. New York: Pitman Publishing, 1969.
Reve, Karel van het. *Verzameld werk*. 7 vols. Amsterdam: Van Oorschot, 2008–11.
Rhéaume, Charles. *Sakharov: Science, morale et politique*. Laval, Canada: Les Presses de l'Université Laval, 2004.
Riasanovsky, Nicholas. "The Norman Theory of the Origin of the Russian State." *Russian Review*, vol. 7, no. 1 (Autumn 1947): 96–110.
Riasanovsky, Nicholas. *A Parting of the Ways: Government and the Educated Public in Russia, 1801–1855*. Oxford: Clarendon Press, 1976.
Riurikov, Boris. "Sotsialisticheskii realizm i ego 'nizprovergateli.'" *Inostrannaia literatura*, vol. 1, no. 1 (1961): 191–200.
The Road to Communism: Documents of the 22nd Congress of the Communist Party of the Soviet Union, October 17–31, 1961. Moscow: Foreign Languages Publishing House, 1962.
Roginskii, Arsenii. "Dlia nas, eto bylo rubezhnyi den'." In *Vtorzhenie: Vzgliad iz Rossii. Chekhoslovakiia, avgust 1968*, edited by Iosef Pazderka, 209–19. Moscow: Novoe literaturnoe obozrenie, 2016.
Rogov, K. Iu., ed. *Semidesiatye kak predmet istorii russkoi kul'tury*. Moscow: OGI, 1998.
Ro'i, Yaacov, ed. *The Jewish Movement in the Soviet Union*. Baltimore: Johns Hopkins University Press, 2012.
Roitman, N. D., and V. L. Tseitlin. *Polkovodets-kommunist*. Kishinev: Kartia moldoveniaske, 1967.
Rolf, Malte. *Das sowjetische Massenfest*. Hamburg: Hamburger Edition, 2006.
Rousso, Henry, ed. *Stalinisme et nazisme: Histoire et mémoire comparées*. Brussels: Complexes, 1999.
Rozenblium, Ol'ga. "'Diskussii ne bylo…': Otkrytye pis'ma kontsa 1960-kh godov kak pole obshchestvennoi refleksii." In *Nesovershennaia publichnaia sfera. Istoriia rezhimov publichnosti*

v Rossii. Sbornik stat'ei, edited by T. Atnashev, T. Vaizer, and M. Velizhev, 416–46. Moscow: Novoe literaturnoe obozrenie, 2021.

Rozenblium, Olga. "Frida Vigdorova's Transcript of Joseph Brodsky's Trial: Literary Tradition vs. Legal Aspects." Paper presented at the Annual Conference of the Association for Slavic, East European, and Eurasian Studies, 2015.

Roziner, Feliks. *Nekto Finkel'maier: Roman*. London: Overseas Publications Interchange, 1981.

Rubenstein, Joshua. *Soviet Dissidents: Their Struggle for Human Rights*. 2nd ed. Boston: Beacon Press, 1985.

Rubenstein, Joshua. *Tangled Loyalties: The Life and Times of Ilya Ehrenburg*. New York: Basic Books, 1996.

Rubenstein, Joshua, and Alexander Gribanov, eds. *The KGB File of Andrei Sakharov*. New Haven, CT: Yale University Press, 2005.

Rubin, Vitalii. *Dnevniki. Pis'ma*. 2 vols. Jerusalem: Biblioteka Aliia, 1989.

Rudenko, Mykola. *Naibil'she dyvo—zhyttia: Spohady*. Kyiv: Takson, 1998.

Russell, Bertrand. *Why I Am Not a Christian and Other Essays on Religion and Related Subjects*. New York: Simon and Schuster, 1957.

Ryan, Alan. *Bertrand Russell: A Political Life*. New York: Hill and Wang, 1988.

Sakharov, Andrei. *Memoirs*. New York: Knopf, 1992.

Sakharov, Andrei. "Nobel Lecture." The Nobel Prize, 1976. https://www.nobelprize.org/prizes/peace/1975/sakharov/26035-andrei-sakharov-nobel-lecture-1975/.

Sakharov, Andrei. "Razmyshleniia o progresse, mirnom sosushchestvovanii i intellektual'noi svobode." In *Trevoga i nadezhda*, by Andrei Sakharov, 11–47. Moscow: Inter-Verso, 1990.

Sakharov, Andrei. "Simmetriia vselennoi." *Budushchee nauki*, vol. 2 (1967): 74–96.

Sakharov, Andrei. *Vospominaniia*. 3 vols. Sobranie sochinenii. Moscow: Vremia, 2006.

Salisbury, Harrison, ed. *Sakharov Speaks*. New York: Knopf, 1974.

Samizdat Archive Association. *Materialy samizdata*. 91 vols. Columbus: Ohio State University, Center for Slavic and East European Studies, 1975–91.

Saunders, George ed. *Samizdat: Voices of the Soviet Opposition*. New York: Monad Press, 1974.

Savaneli, B. V. *Iuridicheskie formy polozheniia lichnosti v sovetskom obshchestve*. Tbilisi: Metsniereba, 1969.

Savitskii, V. M. "Kogda teoriia stanovitsia real'nost'iiu." In *Operedivshii vremia. K stoletiiu s dnia rozhdeniia M. S. Strogovicha*, edited by V. M. Savitskii and A. M. Larin, 6–40. Moscow: Serial, 1994.

Sawatsky, Walter. *Soviet Evangelicals since World War II*. Scottdale, PA: Herald Press, 1981.

Scammell, Michael. *Solzhenitsyn: A Biography*. New York: W. W. Norton, 1984.

Schattenberg, Susanne. *Leonid Breschnew: Staatsmann und Schauspieler im Schatten Stalins. Eine Biographie*. Cologne: Böhlau Verlag, 2017.

Schlögel, Karl. *Der renitente Held: Arbeiterprotest in der Sowjetunion, 1953–1983*. Hamburg: Junius, 1984.

Schlögel, Karl. *Traum und Terror: Moskau 1937*. Munich: Carl Hanser Verlag, 2008.

Schmitthenner, Walter, and Hans Buchheim, eds. *Der deutsche Widerstand gegen Hitler*. Köln-Berlin: Kiepenheuer & Witsch, 1966.

Schwartz, Harry. "Globov, Marina, and Seryozha of the Soviet New Class." *New York Times Book Review*, October 16, 1960.

Scott, James C. *Domination and the Arts of Resistance: Hidden Transcripts*. New Haven, CT: Yale University Press, 1990.

Scott, James C. *The Moral Economy of the Peasant: Rebellion and Subsistence in Southeast Asia*. New Haven, CT: Yale University Press, 1976.

Selvage, Douglas. "From Helsinki to 'Mars': Soviet-Bloc Active Measures and the Struggle over Détente in Europe, 1975–1983." *Journal of Cold War Studies*, vol. 23, no. 4 (Fall 2021): 34–94.
Selvage, Douglas. "KGB, MfS, und Andrej Sacharow 1975 bis 1980." In *Der "grosse Bruder": Studien zum Verhältnis von KGB und MfS 1958–1989*, edited by Douglas Selvage and Georg Herbstritt, 243–86. Göttingen: Vandenhoeck & Ruprecht, 2022.
Semichastnyi, Vladimir. *Bespokoinoe serdtse*. Moscow: Vagrius, 2002.
Senderov, Valerii. "Pamiatka abiturientu, kotorogo priemnaia komissiia mozhet schest' evreem." Samizdat, 1979.
Shafarevich, Igor. "Est' li u Rossii budushchee?" In *Iz-pod glyb: Sbornik stat'ei*, edited by Aleksandr Solzhenitsyn, 261–76. Paris: YMCA Press, 1974.
Shaposhnikov, V. S., ed. *Mezhdunarodnye nepravitel'stvennye organizatsiia i uchrezhdeniia: Spravochnik*. Moscow: Mezhdunarodnye otnosheniia, 1982.
Shatunovskii, Il'ia. "Iz biografii podletsa." *Ogonek*, January 1963.
Shatunovskii, Il'ia. *Zapiski strelianogo vorob'ia*. Moscow: Voskresen'e, 2003.
Shatz, Marshall. *Soviet Dissent in Historical Perspective*. Cambridge: Cambridge University Press, 1980.
Shelest, Petro. *Da ne sudimy budete: Dnevnikovye zapisi, vospominaniia chlena Politbiuro TsK KPSS*. Moscow: Edition Q, 1995.
Shelley, Percy Bysshe. "A Defense of Poetry." In *Shelley's Poetry and Prose*, edited by Donald H. Reiman, 478–508. New York: W. W. Norton, 1977.
Shimanov, Gennady. "Russkaia narodnaia liniia: Pravoslavie, samoderzhavie, narodnost'." *Ruskline*, October 16, 2013. https://ruskline.ru/analitika/2013/10/17/vospominaniya.
Shimanov, Gennady. *Zapiski iz krasnogo doma*. Moscow: Institut russkoi tsivilizatsii, 2013.
Shklar, Judith N. *Legalism: Law, Morals, and Political Trials*. Cambridge, MA: Harvard University Press, 1986 [1964].
Shklovsky, Viktor. "Art as Device." In *Viktor Shklovsky: A Reader*, edited by Alexandra Berlina, 73–96. London: Bloomsbury, 2016.
Shlapentokh, Vladimir. "The Justification of Political Conformism: The Mythology of Soviet Intellectuals." *Studies in Soviet Thought*, vol. 39, no. 2 (March 1990): 111–35.
Shlapentokh, Vladimir. *Love, Marriage, and Friendship in the Soviet Union: Ideals and Practices*. New York: Praeger, 1984.
Shlapentokh, Vladimir. *Soviet Intellectuals and Political Power: The Post-Stalin Era*. Princeton, NJ: Princeton University Press, 1990.
Short, Philip. *Putin: His Life and Times*. London: Bodley Head, 2022.
Shragin, Boris. *Mysl' i deistvie*. Moscow: RGGU, 2000.
Shub, Anatole. *The New Russian Tragedy*. New York: W. W. Norton, 1969.
Shub, Anatole. "'Will the USSR Survive . . . ?' A Personal Comment." *Survey*, no. 74/75 (Winter–Spring 1970): 87–94.
Shubin, Aleksandr. *Tainy sovetskoi epokhi: Dissidenty, neformaly i svoboda v SSSR*. Moscow: Veche, 2008.
Siniavskii, Andrei. *Chto takoe sotsialisticheskii realizm?* Paris: Sintaksis, 1988 [1959].
Siniavskii, Andrei. "Dissidentstvo kak lichnyi opyt." *Sintaksis*, vol. 15 (1985): 131–47.
Siniavskii, Andrei. Letter to the editor. *Kontinent*, vol. 49 (1986): 337–42.
Siniavskii i Daniel' na skam'e podsudimykh. New York: Mezhdunarodnoe literaturnoe sodruzhestvo, 1966.
Siws, Samuil [Samuil Zivs]. "Spekulationen über UdSSR." *Frankfurter Rundschau*, January 12, 1976.
Skinner, Quentin. *Liberty before Liberalism*. New York: Cambridge University Press, 1998.
Slezkine, Yuri. *The House of Government: A Saga of the Russian Revolution*. Princeton, NJ: Princeton University Press, 2017.

Slezkine, Yuri. *The Jewish Century.* Princeton, NJ: Princeton University Press, 2004.
Slonim, Mark, ed. *Soviet Russian Literature: Writers and Problems.* New York: Oxford University Press, 1964.
Smith, Hedrick. *The Russians.* New York: Quadrangle, 1976.
Smith, Kathleen. *Remembering Stalin's Victims: Popular Memory and the End of the USSR.* Ithaca, NY: Cornell University Press, 1996.
Smith, Mark B. "Social Rights in the Soviet Dictatorship: The Constitutional Right to Welfare from Stalin to Brezhnev." *Humanity,* vol. 3, no. 3 (Winter 2012): 385–406.
Smith, Theresa C., and Thomas A. Oleszczuk. *No Asylum: State Psychiatric Repression in the Former USSR.* New York: New York University Press, 1996.
Smykalin, A. S. "Ideologicheskii kontrol' i Piatoe upravlenie KGB SSSR v 1967–1989 gg." *Voprosy istorii,* no. 8 (August 2011): 30–40.
Snezhnevskii, A. V., ed. *Spravochnik po psikhiatrii.* Moscow: Meditsina, 1974.
Snow, David A., et al., eds. *The Wiley-Blackwell Encyclopedia of Social and Political Movements.* 3 vols. Malden, MA: Wiley, 2013.
Snyder, Sarah B. "Exporting Amnesty International to the United States: Transatlantic Human Rights Activism in the 1960s." *Human Rights Quarterly,* vol. 34, no. 3 (August 2012): 779–99.
Snyder, Sarah B. *Human Rights Activism and the End of the Cold War: A Transnational History of the Helsinki Network.* New York: Cambridge University Press, 2011.
Sobranie dokumentov samizdata. 30 vols. Munich: Samizdat Archive Association, 1972–78.
Sokirko, Viktor. "'Krivoe zerkalo'—retsenziia na roman V. Kochetova 'Chego zhe ty khochesh'?" Samizdat, 1970. Sokirko, accessed December 12, 2023. https://sokirko.info/ideology/pv1/33.htm.
Sokirko, Viktor. "Otklik na stat'iu A. A. Amal'rika 'Prosushchestvuet li Sovetskii Soiuz do 1984 g.?'" Sokirko, accessed December 12, 2023. https://sokirko.info/kb#_Toc391668290.
Sokirko, V[iktor]. *Zhit' ne po lzhi (sbornik otklikov-sporov na stat'iu A. I. Solzhenitsyna).* 3 vols. Samizdat, 1977–78. Sokirko, accessed December 12, 2023. https://sokirko.info/ideology/gnl/index.html.
[Soldatov, Sergei]. *Programma Demokraticheskogo dvizheniia Sovetskogo Soiuza.* Amsterdam: Fond imeni Gertsena, 1970.
Soldatov, Sergei. *Zarnitsy vozrozhdeniia.* London: Overseas Publications Interchange, 1984.
Solzhenitsyn, Aleksandr. *Arkhipelag Gulag 1918–1956: Opyt khudozhestvennogo issledovaniia.* 3 vols. Ekaterinburg: U-Faktoriia, 2006 [1973].
Solzhenitsyn, Aleksandr. *Bodalsia telenok s dubom: Ocherki literaturnoi zhizni.* Paris: YMCA Press, 1975.
Solzhenitsyn, Alexander. *The First Circle.* New York: Harper & Row, 1968.
Solzhenitsyn, Aleksandr. "Nashi pliuralisty." *Vestnik Russkogo khristianskogo dvizheniia,* no. 139 (1983): 133–60.
Solzhenitsyn, Aleksandr. "Na vozvrate dykhaniia i soznaniia," In *Iz-pod glyb: Sbornik stat'ei,* edited by Aleksandr Solzhenitsyn, 7–28. Paris: YMCA Press, 1974.
Solzhenitsyn, Alexander. "Nobel Lecture." The Nobel Prize, 1970. https://www.nobelprize.org/prizes/literature/1970/solzhenitsyn/lecture/.
Solzhenitsyn, Aleksandr. *Odin den' Ivana Denisovicha.* Moscow: Sovetskii pisatel', 1963.
Solzhenitsyn, Aleksandr. *Pis'mo vozhdiam Sovetskogo Soiuza.* Paris: YMCA-Press, 1974.
Solzhenitsyn, Alexander. *Warning to the West.* New York: Farrar, Straus and Giroux, 1976.
Sorokina, Sonia. "Personal'noe delo S. Sorokinoi." *Poiski: Svobodnyi moskovskii zhurnal,* vol. 5–6, no. 2 (1983): 367–78. Moscow, samizdat.
Sosin, Gene. *Sparks of Liberty: An Insider's Memoir of Radio Liberty.* University Park: Pennsylvania State University Press, 1999.

Spengla, Vidas. *The Church, the "Kronika," and the KGB Web*. Vilnius: Kataliku akademija, 2002.
"Stenograma vstrechi rukovoditelei KPSS i sovetskogo pravitel'stva s deiateliami literatury i iskusstva, 17.XII.1962." In *Nikita Sergeevich Khrushchev: Dva tsveta vremeni. Dokumenty iz lichnogo fonda N. S. Khrushcheva*, edited by N. G. Tomilina, vol. 2, 533–601. Moscow: Mezhdunarodnyi fond Demokratiia, 2009.
Stephan, Anke. *Von der Küche auf den Roten Platz: Lebenswege sowjetischer Dissidentinnen*. Zurich: Pano, 2005.
Stephan, Anke. "Von 'Dissidenten,' 'Ehefrauen,' und 'Sympathisantinnen': Das Entstehen und Funktionieren dissidentischer Netzwerken in der Sowjetunion der 1960er bis 1980er Jahre." In *Soziale Netzwerke und soziales Vertrauen in den Transformationsländern*, edited by Klaus Roth, 122–33. Zurich: LIT, 2007.
Strogovich, M. S. *Osnovnye voprosy sovetskoi sotsialisticheskoi zakonnosti*. Moscow: Iuridicheskaia literatura, 1959.
Strogovich, M. S. "Sostiazatel'nost' i sud v sovetskom ugolovnom protsesse." *Izvestiia vysshykh uchebnykh zavedenii: Pravovedenie*, no. 2 (1962).
Strugatskii, Arkadii, and Boris Strugatskii. *Za milliard let do kontsa sveta. Rukopis', obnaruzhennaia pri strannykh obstoiatel'stvakh*. Moscow: Sovetskii pisatel', 1984.
Strugatskii, Boris. Interview. *Smena*, July 5, 1990. http://www.rusf.ru/abs/books/publ34.htm.
Strugatskii, B[oris]. "Kommentarii k proidennomu." *Russian Science Fiction and Fantasy* (blog), accessed December 12, 2023. http://www.rusf.ru/abs/books/bns-07.htm.
Strukova, Elena. "Odesskaia biblioteka samizdata Viacheslava Igrunova i masterskaia narodnykh promyslov Avraama Shifrina." In *Dissidenty SSSR, vostochnoi i tsentral'noi Evropy: Epokha i nasledie*, 167–73. Moscow: Mezhdunarodnyi Memorial, 2020.
Strukova, Elena Nikolaevna, and Boris Isaevich Belenkin, eds. *Acta samizdatica / Zapiski o samizdate (pilotnyi vypusk)*. Moscow: Mezhdunarodnyi Memorial, 2012.
Sullivan, Rosemary. *Stalin's Daughter: The Extraordinary and Tumultuous Life of Svetlana Alliluyeva*. New York: HarperCollins, 2015.
Suny, Ronald Grigor. "Reading Russia and the Soviet Union in the Twentieth Century: How the 'West' Wrote Its History of the USSR." In *The Cambridge History of Russia*, edited by Ronald Grigor Suny, vol. 3, 5–64. Cambridge: Cambridge University Press, 2006.
Suri, Jeremi. *Power and Protest: Global Revolution and the Rise of Détente*. Cambridge, MA: Harvard University Press, 2003.
Suslov, Mikhail. "Utopicheskii proekt G. M. Shimanova v kontekste pravogo dissidentskogo dvizheniia v SSSR v 1960–1980-kh gg." *Forum noveishei vostochnoevropeiskoi istorii i kul'tury*, vol. 2 (2008): 73–98.
Sutherland, John. *Stephen Spender: A Literary Life*. New York: Oxford University Press, 2005.
Szulecki, Kacper. *Dissidents in Communist Central Europe: Human Rights and the Emergence of New Transnational Actors*. Cham, Switzerland: Palgrave Macmillan, 2019.
Tarrow, Sydney. *Power in Movement: Social Movements and Contentious Politics*. New York: Cambridge University Press, 1998.
Taubman, William. *Gorbachev: His Life and Times*. New York: W. W. Norton, 2017.
Taubman, William. *Khrushchev: The Man and His Era*. New York: W. W. Norton, 2003.
Taubman, William. *The View from Lenin Hills: Soviet Youth in Ferment*. London: H. Hamilton, 1968.
Tchouikina, Sofia. "Anti-Soviet Biographies: The Dissident Milieu and Its Neighboring Milieux." In *Biographical Research in Eastern Europe: Altered Lives and Broken Biographies*, edited by Robin Humphrey, Robert Miller, and Elena Zdravomyslova, 129–39. Aldershot, UK: Ashgate, 2003.
Ternovskii, Leonard. *Vospominaniia i stat'i*. Moscow: Vozvrashchenie, 2006.

Tertz, Abram [Andrei Sinyavsky]. *Goodnight!* Translated by Richard Lourie. New York: Viking, 1989.
Terts, Abram [Andrei Siniavskii]. *Spokoinoi nochi: Roman*. Paris: Sintaksis, 1984.
Tertz, Abram [Andrei Sinyavsky]. *The Trial Begins and On Socialist Realism*. Berkeley: University of California Press, 1982.
Thakur, Ramesh. "Human Rights: Amnesty International and the United Nations." *Journal of Peace Research*, vol. 31, no. 2 (May 1994): 143–60.
Thomas, Daniel. *The Helsinki Effect: International Norms, Human Rights, and the Demise of Communism*. Princeton, NJ: Princeton University Press, 2001.
Thompson, E. P. "The Moral Economy of the English Crowd in the Eighteenth Century." *Past & Present*, no. 50 (1971): 76–136.
Thun-Hohenstein, Franziska. *Das Leben schreiben: Warlam Schalamow: Biographie und Poetik*. Berlin: Matthes & Seitz, 2022.
Tilly, Charles. *The Contentious French*. Cambridge, MA: Belknap Press of Harvard University Press, 1986.
Tilly, Charles. *Regimes and Repertoires*. Chicago: University of Chicago Press, 2006.
Tocqueville, Alexis de. *The Old Regime and the Revolution*. Translated by John Bonner. New York: Harper and Brothers, 1856.
Tolley, Howard B., Jr. *The International Commission of Jurists: Global Advocates for Human Rights*. Philadelphia: University of Pennsylvania Press, 1994.
Tolstoi, L. N. "Ne mogu molchat'." In *Polnoe sobranie sochinenii*, by L. N. Tolstoi, vol. 37, 83–96. Moscow: Gosudarstvennoe izdatel'stvo khudozhestvennoi literatury, 1956.
Tomilina, N. G., et al., eds. *Chekhoslovatskii krizis 1967–1969 gg. v dokumentakh TsK KPSS*. Moscow: ROSSPEN, 2010.
Trigos, Ludmilla. *The Decembrist Myth in Russian Culture*. New York: Palgrave Macmillan, 2009.
Tromly, Benjamin. *Cold War Exiles and the CIA: Plotting to Free Russia*. New York: Oxford University Press, 2019.
Tromly, Benjamin. "Intelligentsia Self-Fashioning in the Postwar Soviet Union: Revol't Pimenov's Political Struggle, 1949–57." *Kritika: Explorations in Russian and Eurasian History*, vol. 13, no. 1 (Winter 2012): 151–76.
Tromly, Benjamin. *Making the Soviet Intelligentsia: Universities and Intellectual Life under Stalin and Khrushchev*. New York: Cambridge University Press, 2013.
Tumanov, Oleg. *Podlinnaia "Sud'ba rezidenta": Dolgii put' na rodinu*. Moscow: Algoritm, 2017.
Turchin, V. F. "Chto takoe bespristrastnost'?" In *Samosoznanie: Sbornik statei*, edited by Pavel Litvinov, Mikhail Meerson-Aksenov, and Boris Shragin, 305–26. New York: Khronika, 1976.
Turchin, V. F. *Fiziki shutiat: Sbornik perevodov*. Moscow: Mir, 1966.
Turchin, Valentin. *Inertsiia strakha: Sotsializm i totalitarizm*. New York: Khronika, 1978.
Tvardovskii, Aleksandr. *Novomirskii dnevnik*. 2 vols. Moscow: Prozaik, 2009.
Tverdokhlebov, Andrei. "V zashchitu pis'ma [sic] Sakharova 'Razmyshleniia o progresse...'" *Obshchestvennye problemy*, vol. 1 (July 1969).
Ulanovskaia, Nadezhda, and Maia Ulanovskaia. *Istoriia odnoi sem'i*. St. Petersburg: INAPRESS, 2003.
Ulianova, Olga. "Corvalán for Bukovsky: A Real Exchange of Prisoners during an Imaginary War. The Chilean Dictatorship, the Soviet Union, and US Mediation, 1973–1976." *Cold War History*, vol. 14, no. 3 (August 2014): 315–36.
Ulitskaia, Liudmila. *Poetka: Kniga o pamiati: Natal'ia Gorbanevskaia*. Moscow: ACT, 2014.
Ulitskaia, Liudmila. *Zelenyi shater*. Moscow: Eksmo, 2010.
Unger, Aryeh. *Constitutional Development in the USSR: A Guide to the Soviet Constitutions*. New York: Pica Press, 1981.

Ushakov, Alexander A. *In the Gunsight of the KGB*. New York: Knopf, 1989.
Vail', Petr, and Aleksandr Genis. *60-e: Mir sovetskogo cheloveka*. Moscow: Novoe literaturnoe obozrenie, 2001.
Vaissié, Cécile. *Pour votre liberté et pour la nôtre: Le combat des dissidents de Russie*. Paris: Robert Laffont, 1999.
Vardys, V. Stanley. *The Catholic Church, Dissent and Nationality in Soviet Lithuania*. Boulder, CO: East European Quarterly, 1978.
Velikanova, E., comp. *Tsena metafory ili prestuplenie i nakazanie Siniavskogo i Danielia*. Moscow: Kniga, 1989.
Velikanova, Olga. *Mass Political Culture under Stalinism: Popular Discussion of the Soviet Constitution of 1936*. Cham: Palgrave Macmillan, 2018.
Venclova, Tomas. *Forms of Hope*. Riverdale-on-Hudson, NY: Sheep Meadow Press, 1999.
Verblovskaia, Irina. *Moi prekrasnyi strashnyi vek*. St. Petersburg: Zvezda, 2011.
Vigdorova, Frida. *Pravo zapisyvat': Sbornik*. Moscow: Izdatel'stvo AST, 2017.
Vlast' i dissidenty: Iz dokumentov KGB i TsK KPSS. Moscow: Moskovskaia Khel'sinskaia gruppa, 2006.
Voinovich, Vladimir. *Antisovetskii Sovetskii Soiuz*. Moscow: Materik, 2002.
Voinovich, Vladimir. *Avtoportret: Roman moei zhizni*. Moscow: Eksmo, 2010.
Voinovich, Vladimir. *Shapka*. London: Overseas Publications Interchange, 1988.
Volkov, A., M. Pugacheva, and S. Iarmoliuk, eds. *Pressa v obshchestve (1959–2000): Otsenki zhurnalistov i sotsiologov: Dokumenty*. Moscow: Moskovskaia shkola politicheskikh issledovanii, 2000.
Voren, Robert van. *On Dissidents and Madness: From the Soviet Union of Leonid Brezhnev to the "Soviet Union" of Vladimir Putin*. Amsterdam: Editions Rodopi B.V., 2009.
Voronel', Aleksandr. "O vrednoi funktsii slov i probleme assimiliatsii evreev." *Evrei v SSSR*, no. 7 (1974).
Voronel', Nina. *Sodom tekh let*. Rostov-na-Donu: Feniks, 2006.
Vysotskii, Vladimir. "Nam ni k chemu siuzhety i intrigi." Bards, 1964. http://www.bards.ru/archives/part.php?id=15321.
Walicki, Andrzej. *A History of Russian Thought: From the Enlightenment to Marxism*. Stanford, CA: Stanford University Press, 1979.
Walker, Barbara. "The Moscow Correspondents, Soviet Human Rights Activists, and the Problem of the Western Gift." In *Americans Experience Russia: Encountering the Enigma, 1917 to the Present*, edited by Choi Chatterjee and Beth Holmgren, 139–57. New York: Routledge, 2012.
Walker, Barbara. "Moscow Human Rights Defenders Look West: Attitudes toward U.S. Journalists in the 1960s and 1970s." *Kritika: Explorations in Russian and Eurasian History*, vol. 9, no. 4 (Fall 2008): 905–27.
Walker, Barbara. "Pollution and Purification in the Moscow Human Rights Networks of the 1960s and 1970s." *Slavic Review*, vol. 68, no. 2 (Summer 2009): 376–95.
Weber, Max. "Russlands Übergang zum Scheinkonstitutionalismus." Supplement, *Archiv für Sozialwissenschaft und Sozialpolitik*, vol. 23, no. 1 (1906): 165–401.
Weinberg, Robert. *Stalin's Forgotten Zion: Birobidzhan and the Making of a Soviet Jewish Homeland*. Berkeley: University of California Press, 1998.
Weiner, Amir. "Déjà Vu All Over Again: Prague Spring, Romanian Summer and Soviet Autumn on the Soviet Western Frontier." *Contemporary European History*, vol. 15, no. 2 (2006): 159–94.
Weiner, Amir. "The Empires Pay a Visit: Gulag Returnees, East European Rebellions, and Soviet Frontier Politics." *Journal of Modern History*, vol. 78, no. 2 (June 2006): 333–76.

Weiner, Amir. *Making Sense of War: The Second World War and the Fate of the Bolshevik Revolution*. Princeton, NJ: Princeton University Press, 2001.

Weiner, Amir. "Robust Revolution to Retiring Revolution: The Life Cycle of the Soviet Revolution, 1945–1968." *Slavonic and East European Review*, vol. 86, no. 2 (April 2008): 208–31.

Weiner, Douglas. *A Little Corner of Freedom: Russian Nature Protection from Stalin to Gorbachev*. Berkeley: University of California Press, 1999.

Weissbrodt, David, and James McCarthy. "Fact-Finding by International Nongovernmental Human Rights Organizations." *Virginia Journal of International Law*, vol. 22 (1981–82): 1–91.

Werkmeister, Christian. "Wahnsinn mit System: Psychiatrische Anstalten in der späten Sowjetunion." *Osteuropa*, vol. 62, no. 11–12 (November/December 2014): 133–52.

Weston, Burns. "Human Rights." In *Encyclopaedia Britannica Online*. Accessed December 12, 2023. http://www.search.eb.com/eb/article-9106289.

Wimberg, Ellen. "Socialism, Democratism and Criticism: The Soviet Press and the National Discussion of the 1936 Draft Constitution." *Soviet Studies*, vol. 44, no. 2 (1992): 313–32.

Wittgenstein, Ludwig. *Logiko-filosofskii traktat*. Moscow: Izdatel'stvo inostrannoi literatury, 1958.

Wittgenstein, Ludwig. *Tractatus logico-philosophicus*. London: Routledge & Kegan Paul, 1966 [1921].

Wittner, Lawrence S. *The Struggle against the Bomb*. Vol. 1, *One World or None*. Stanford, CA: Stanford University Press, 1993.

Wortman, Richard. "Russian Monarchy and the Rule of Law: New Considerations of the Court Reform of 1864." *Kritika: Explorations in Russian and Eurasian History*, vol. 6, no. 1 (Winter 2005): 145–70.

Yakir, Pyotr. *A Childhood in Prison*. London: Macmillan, 1972.

Yakobson, Sergius, and Robert V. Allen. *Aspects of Intellectual Ferment and Dissent in the Soviet Union*. 2nd ed. Washington, DC: Government Printing Office, 1968.

Yakovlev, Nikolai. *CIA Target: The USSR*. Moscow: Progress Publishers, 1980.

Yakovlev, Nikolai. *Solzhenitsyn's Archipelago of Lies*. Moscow: Novosti Publishing House, 1974.

Yurchak, Alexei. "The Cynical Reason of Late Socialism: Power, Pretense, and the *Anekdot*." *Public Culture*, vol. 9, no. 2 (1997): 161–88.

Yurchak, Alexei. *Everything Was Forever, Until It Was No More: The Last Soviet Generation*. Princeton, NJ: Princeton University Press, 2006.

Zakharova, Svetlana. "Gendering Soviet Dissent: How and Why the Woman Question was Excluded from the Agenda of Soviet Dissidents (1964–1982)." Master's thesis, Central European University, 2013.

Zamoyska [Peltier], Hélène. "Quelques souvenirs sur André Siniavski." *Esprit*, February 1957, 258–70.

Zamoyska [Peltier], Hélène. "Sinyavsky, the Man and the Writer." In *On Trial: The Case of Sinyavsky (Tertz) and Daniel (Arzhak)*, edited by Leopold Labedz and Max Hayward, 46–69. London: Collins and Harvill Press, 1967.

Zelnik, Reginald E. "A Paradigm Lost? Response to Anna Krylova." *Slavic Review*, vol. 62, no. 1 (Spring 2003): 24–33.

Zemskov, Viktor. "Politicheskie repressii v SSSR (1917–1990 gg.)." *Rossiia XXI*, no. 1–2 (1994): 107–25.

Zhelvakova, Irena. *Gertsen*. Moscow: Molodaia gvardiia, 2010.

Zhit' ne po lzhi: Sbornik materialov avgust 1973–fevral' 1974. Paris: YMCA Press, 1975.

Zinov'ev, Aleksandr. *Bez illiuzii*. Lausanne, Switzerland: Éditions L'Âge d'Homme, 1979.

Zinov'ev, Aleksandr. *Gomo sovetikus*. Lausanne, Switzerland: Éditions L'Âge d'Homme, 1981.

Zinov'ev, Aleksandr. *Russkaia sud'ba: Ispoved' otshchepentsa*. Moscow: Tsentrpoligraf, 1999.
Zinov'ev, Aleksandr. *Sobranie sochinenii v desiati tomakh*. Moscow: Tsentrpoligraf, 2000.
Zisel's, Iosif. *Esli ne seichas: Stat'i, interv'iu, vystupleniia, 1989–2006 gg*. Kyiv: Dukh i litera, 2006.
Zisserman-Brodsky, Dina. *Constructing Ethnopolitics in the Soviet Union: Samizdat, Deprivation, and the Rise of Ethnic Nationalism*. New York: Palgrave Macmillan, 2003.
Zitzewitz, Josephine von. *The Culture of Samizdat: Literature and Underground Networks in the Late Soviet Union*. London: Bloomsbury, 2020.
Zivs, Samuil. *The Anatomy of Lies*. Moscow: Progress Publishers, 1984.
Zivs, Samuil. *Human Rights: Continuing the Discussion*. Moscow: Progress Publishers, 1980.
Zorin, Leonid. *Zelenye tetradi*. Moscow: Novoe literaturnoe obozrenie, 1999.
Zubarev, Dmitrii, and Gennadii Kuzovkin, eds. *Dokumenty Moskovskoi Khel'sinkskoi gruppy, 1976–1982*. Moscow: Moskovskaia Khel'sinskaia gruppa, 2006.
Zubok, Vladislav. *Collapse: The Fall of the Soviet Union*. New Haven, CT: Yale University Press, 2021.
Zubok, Vladislav. "Die sowjetrussische Gesellschaft in den sechziger Jahren." In *Prager Frühling: Das internationale Krisenjahr 1968*, edited by Stefan Karner et al., vol. 1, 823–48. Cologne: Verein zur Förderung der Forschung von Folgen nach Konflikten und Kriegen, 2008.
Zubok, Vladislav. *Zhivago's Children: The Last Russian Intelligentsia*. Cambridge, MA: Belknap Press of Harvard University Press, 2009.

ILLUSTRATION CREDITS

1.1 Obshchestvo Memorial, Moscow

2.1 Hoover Institution Library and Archives, Andrei Siniavskii Papers, Box 121, Folder 5

2.2 Hoover Institution Library and Archives, Andrei Siniavskii Papers, Box 5, Folder 3

2.3 Obshchestvo Memorial, Moscow

2.4 Obshchestvo Memorial, Moscow

2.5 Inside the Zhivago Storm, https://zhivagostorm.org/2015/03/07/helene-peltier-boris-pasternak-and-giangiacomo-feltrinelli/

3.1 Obshchestvo Memorial, Moscow

3.2 Archiv der Forschungsstelle Osteuropa - FSO 01-121

3.3 Undated advertisement in *Izvestiia*

3.4 Obshchestvo Memorial, Moscow

4.1 Obshchestvo Memorial, Moscow

4.2 Archiv der Forschungsstelle Osteuropa - FSO 01-023

4.3 Hoover Institution Library and Archives, Andrei Siniavskii Papers, Box 6, Folder 1

5.1 Wikimedia Commons

6.1 Houghton Library, Harvard University, Peter Reddaway Collection, Box 1, Folder 1, Document P31

6.2 Houghton Library, Harvard University, Peter Reddaway Collection, Box 5, Folder 41, Document G39

6.3 Obshchestvo Memorial, Moscow

6.4 Obshchestvo Memorial, Moscow

6.5 Houghton Library, Harvard University, Peter Reddaway Collection, Box 1, Folder 9, Document P461

6.6 Ann Komaromi, Soviet Samizdat Periodicals, University of Toronto Libraries, https://samizdat.library.utoronto.ca/

7.1 Houghton Library, Harvard University, Peter Reddaway Collection, Box 4, Folder 30, Document P1749

7.2 Houghton Library, Harvard University, Peter Reddaway Collection, Box 1, Folder 9, Document P477

7.3 Houghton Library, Harvard University, Peter Reddaway Collection, Box 1, Folder 9, Document P479

7.4 Lietuvos Ypatingasis Archyvas f. K-1, ap. 3, b. 642, p.4

8.1 Aleksandr Zinov'ev, *Russkaia sud'ba: Ispoved' otshchepentsa* (Moscow: Tsentrpoligraf, 1999)

8.2 Obshchestvo Memorial, Moscow

8.3 Wikimedia Commons

8.4 Houghton Library, Harvard University, Peter Reddaway Collection, Box 4, Folder 30, Document P1756

8.5 Houghton Library, Harvard University, Peter Reddaway Collection, Box 5, Folder 37, Document P2107

8.6 Obshchestvo Memorial, Moscow

9.1 Houghton Library, Harvard University, Peter Reddaway Collection, Box 3, Folder 24, Document P1324

10.1 Houghton Library, Harvard University, Peter Reddaway Collection, Box 5, Folder 42, Document G90

10.2 Houghton Library, Harvard University, Peter Reddaway Collection, Box 5, Folder 37, Documents P2099 and P2102

11.1 Obshchestvo Memorial, Moscow

12.1 Houghton Library, Harvard University, Peter Reddaway Collection, Box 5, Folder 42, Document G111

12.2 Obshchestvo Memorial, Moscow

12.3 Houghton Library, Harvard University, Peter Reddaway Collection, Box 4, Folder 27, Document P1599

12.4 Houghton Library, Harvard University, Peter Reddaway Collection, Box 5, Folder 41, Document G74

12.5 Archiv der Forschungsstelle Osteuropa - FSO 01-098

13.1 Obshchestvo Memorial, Moscow

14.1 Nauchno-informatsionnyi tsentr Memorial, St. Petersburg

14.2 Archiv der Forschungsstelle Osteuropa - FSO 01-024B

14.3 Graph adapted from Theresa C. Smith and Thomas A. Oleszczuk, *No Asylum: State Psychiatric Repression in the Former USSR* (New York: New York University Press, 1996), 93.

ILLUSTRATION CREDITS 773

14.4 Archiv der Forschungsstelle Osteuropa - FSO 01-024B

14.5 Houghton Library, Harvard University, Peter Reddaway Collection, Box 4, Folder 25, Document P1504

15.1 Obshchestvo Memorial, Moscow

15.2 Houghton Library, Harvard University, Peter Reddaway Collection, Box 4, Folder 30, Document P1750

15.3 British Pathé, Press Conference in the Central House of Journalists (1973), https://www.britishpathe.com/asset/15939/

15.4 Houghton Library, Harvard University, Peter Reddaway Collection, Box 1, Folder 5, Document P260

15.5 Obshchestvo Memorial, Moscow

16.1 Houghton Library, Harvard University, Peter Reddaway Collection, Box 3, Folder 23, Document P1268

16.2 Houghton Library, Harvard University, Peter Reddaway Collection, Box 4, Folder 25, Document P1467

16.3 Houghton Library, Harvard University, Peter Reddaway Collection, Box 1, Folder 1, Document P30

16.4 Author's private collection

16.5 Lietuvos Ypatingasis Archyvas, f. K-1, ap. 58, t.1, p.234

16.6 Houghton Library, Harvard University, Peter Reddaway Collection, Box 1, Folder 9, Document P447

16.7 Houghton Library, Harvard University, Peter Reddaway Collection, Box 1, Folder 9, Document P442

20.1 Houghton Library, Harvard University, Peter Reddaway Collection, Box 4, Folder 27, Document P1638

20.2 Hoover Institution Library and Archives, Turchin Papers, Box 1

21.1 Viacheslav Sysoev, *Khodite tikho, govorite tikho: zapiski iz podpol'ia* (Moscow: Novoe literaturnoe obozrenie, 2004), 171

22.1 Courtesy of Ivan Kovalev and Tatyana Osipova

E.1 Wikimedia Commons

E.2 *Moscow Times* (May 11, 2021)

INDEX

Page numbers in *italics* refer to illustrative material.

ABC of Communism, The (Bukharin & Preobrazhensky), 133
Abramov, Ivan, 402–4
Abramovich, I. M., 140
Acheson, Dean, 301
African-American civil rights movement, 626–29
Aikhenvald, Yuri, 25, 81, 84–85, 182, 220
Akhmatova, Anna, 72, 121, 125, 408, 481
Akselrod-Rubina, Ina, 93
Aksenov, Vasily, 103
Albrekht, Jan, 463
Albrekht, Vladimir, 455, 463–70, *464*, 509, 562, 591; emigration, 632, 649; Jewish background, 517; manual for dissidents, 616, 646, 647
Alexander Herzen Foundation, 6, 212, 311, 645, 686n60
Alexander II (tsar), 189
Alexandrov, Anatoly, 503
Alexandrovsky, Pavel, 436–38, 440; interrogation techniques, 444; press conference coaching, 442–43; strategy of, 447, 448–49
Alexeyeva, Ludmilla: on Decembrists, 200; on dissidence movement, 7, 445, 504, 510, 542–43; dissident circles/apartment gatherings, 75, 492; emigration, 632, 633; on hierarchy in Communist Party, 210; on human rights, 18; Human Rights Committee, reaction to, 378–79; and Initiative Group, 354; interrogations at Lefortovo, 446; Marxism-Leninism, 197–98; open letter on Marchenko's arrest, 242; on *perestroika*, 629; photo, *170*; Public Group, 597; reaction to military action in Prague, 246; reaction to ordinary citizens, 330; on support for socialism, 361; transparency meetings, 90, 255; Trial of the Four, 167–69; and Volpin, 23–24, 82; on Western journalism, 352, 547
Aliger, Margarita, 103
Alliluyeva, Svetlana, 64, 113, 427, 664n49
All-Union Council of Evangelical Christian Baptists, 553
Amalrik, Aleksei, 295, 297, 301
Amalrik, Andrei: Amnesty International, 580; apartment gatherings (1968), 206, 224; arrest (1965), 301; contact with non-Russians, 299–301; dissident, formation as, 294–303; on dissident ideology, 305–6, 485; on dissident organization, 343; exile of, 410, 647; Helsinki Group, 605; individualism vs. collective action, 303; on Initiative Group, 356; on invasion of Czechoslovakia, 248, 254–55; *Is Uncle Jack a Conformist?*, 615–17; on KGB techniques, 458; and Natan Sharansky, 594; on open petitions, 168–69; on ordinary citizens, 331; photos, *300*, *303*; on political prisoners, 586; on prophylactic conversations, 165; rumors of Jewish identity, 518; Sakharov's *Reflections*, 288, 291; samizdat essays, 2–3, 19; on Soviet leadership, 306–7; suspected of KGB involvement, 314; Tarasov, exchange with, 432; on Volpin, 23; *Will the Soviet Union Survive until 1984?*, 304, 309–15, 508, 511, 611, 622, 645, 696n44
Amalrik, Zoya, 295, 297

"Amazers" (student association), 83, 667n29
American Civil Liberties Union, 382
"American Disturbers of the Russian Conscience" (Remizov), 51
American Federation of Labor and Congress of Industrial Organizations (AFL-CIO), 633
American Relief Administration, 57
Amnesty International: and Association of Soviet Lawyers, 576–78; Baptists in Soviet Union, 553–54; *Chronicle of Current Events*, 646; creation and background, 549–50, 642; efforts to open Moscow branch, 568–70; empathy, distanced, 575; Group 73, 562–63, 565, 567–68, 646, 647; and International League, 384, 388; language used, 564–65; Madison Avenue Group, 561–62, 562, 575; Moscow Group, 573, 579–83; Nobel Peace Prize, 587–88, 648; non-intervention, principles of, 572; official Soviet stance regarding, 551–52; Prisoner of the Year (1968), 556–58, 644; *Prisoners of Conscience in the USSR*, 584–87, 647; publicity, use of, 570–71; relationships with Soviet dissidents, 551–52, 590–92; samizdat and dissemination of information, 554–56; Seán MacBride, 558–61; Sixth International Council, 563–64; Soviet counter-attack on, 588–90; Turchin's views of valence communities, 565–67; universalism and impartiality, 570, 572–76, 578–79, 581–83
Amnesty International (samizdat periodical), 562
Anatomy of Lies, The (Zivs), 589
Andersen, Hans Christian, 508–9
Anderson, Carol, 628
Anderson, Raymond, 288–89
Andrei Sakharov Center, 4, 634, 637
Andrew, Christopher, 395
Andropov, Yuri: alternatives to legal system, 486; on Amnesty International, 578; on Fifth Directorate, 394, 395–97, 430; Gorbanevskaya's arrest/forced emigration, 417, 418; on Helsinki Accords, 601–2; on International League, 384; as KGB chairman, 232, 253–54, 643; and

Krasin, 442; practice of involuntary psychiatric hospitalization, 413; reaction to Amnesty International, 587, 588–89; on resistance to Western capitalism, 393–94; on Sakharov, 275, 283, 381, 607; on samizdat, 169; on Solzhenitsyn, 421–22, 426, 427–28, 429; on treatment of dissidents, 397–99; on Yakhimovich's defense, 344–45
Animal Farm (Orwell), 76
anonimki, 227–29
anti-Semitism, 51, 265, 506, 519
Antokolsky, Pavel, 114
Arbatov, Georgi, 503
Arendt, Hannah, 8–9, 11, 72, 266–67, 557, 685n34
Arkhangelsky, Vladimir, 561–63, 564, 567
Armenia, 516, 598; Helsinki Group, 609, 648
arms control. *See* nuclear weapons
Arrowsmith, Pat, 576, 730n14
Arzhak, Nikolai, 48. *See also* Daniel, Yuli
Ashkenazy, Vladimir, 189
Aspects of Intellectual Ferment and Dissent in the Soviet Union, 16–17
Association for International Law, 384
Association of Soviet Jurists, 100
Association of Soviet Lawyers, 559, 574, 583
Astrov, Valentin, 192–93
Atonement (Arzhak/Daniel), 48–50, 71–72, 78, 173
Auden, W. H., 210
author's methodology, 5–7
Avgustovsky, A. I., 139–40
Axelbank, Jay, 383
Aygi, Gennady, 506
Azbel, Mark, 45, 58, 75, 76–78, 103, 119
Azbel, Naya, 75

Baberowski, Jörg, 10
Babitsky, Konstantin, 249, 251, 644
Baez, Joan, 253
Baldwin, Roger, 382–83
"Ballad on Disbelief" (Deloné), 159
Baptists, 151, 180, 189, 511, 553, 557–58, 601, 642, 727n43
Barsukova, Svetlana, 354
Batshev, Vladimir, 83, 88, 667n31
Bayeva, Tatyana, 251, 254, 259, *261*, 644

BBC news radio: "English by Radio," 349; on Red Square demonstrations, 338; Soviet listenership, 211, 392; and tamizdat, 6, 55, 351, 628; "To the Global Public," 337; and transparency meetings, 93; and Trial of the Four, 164, 175, 176
Belck, Lothar, 573
Belfrage, Sally, 34–35, 523, 659n37
Bell (Kolokol) (journal), 189, 399–400, *400*
Belogorodskaya, Irina, 419, 438–39, 466; emigration, 632, 647
Beloozerov, Nikolai, 728n64
Benedict, Ruth, 8
Benenson, Peter, 549, 550, 552, 591, 592, 726n24
Bentham, Jeremy, 25
Berdyaev, Nikolai, 43, 535
Berg, Raissa, 182, 231, 517
Bergman, Jay, 277
Beria, Lavrenti, 144, 159, 278, 288
Berlin, Isaiah, 134, 141, 200, 202
Berlinguer, Enrico, 199
Bernshtam, Mikhail, 596, 597
Bernstein, Robert, 600
Bezemer, Jan Willem, 212, 645
Billion Years before the End of the World, A (Strugatsky & Strugatsky), 454
Blake, Patricia, 527, 530, 531
Blane, Andrew, 575
Blishchenko, Igor, 560, 574
Bobkov, Filipp, 102, 393, 407, 416; alternatives to legal system, 486; Fifth Directorate, 394–95, 492, 607; Gorbanevskaya's arrest, 417; memoir, 408, 409
Bogdanov, Alexander, 317
Bogoraz, Larisa Yosifovna: Amnesty International, 580; "Appeal to the Global Public," 172, 174–75, 186–87, 210, 337, 362, 644; criticism of, 325; and dissident strategy, 206; emigration, 633; exile of, 410; family background, 264–66, *266*; on Gulag, 180; Jewish background/identity, 517, 523; labor camps, 162; Lefortovo jail, 271–73; meeting with Igrunov, 490–91; open letter on Marchenko's arrest, 242–43; photos of, *76*, *173*, *181*, *339*; reaction to military action in Prague, 248–49; reaction to ordinary citizens, 330; Red Square sit-down demonstration (1968), 250–51, 260–64; on *Sacred Paths to Willful Freedom*, 536–39; on Soviet military in Prague: 244–46; Trial of the Four, 119–20, 171–74, *173*; Yuli Daniel, 48, 75, 149, 151–52, 179, 490
Bogoraz, Natan, 265
Bogoraz, Yosif Aronovich, *261*, 264–65
Böll, Heinrich, 116
Bolshevik (journal), 297
Bolshevik party, 214; struggle against dissidents, 422; Tenth Party Congress (1921), 410
Bolshevik Revolution: and building of Soviet state, 8; Constituent Assembly, 56; and historical evolution, 15; law and rights, 133–34
Bolshoi Ballet, 492
Bonavia, David, 433–34, 435, 542–43, 685n46, 703n74, 712nn6–7
Bonch-Osmolovsky, Gleb, 406
Bondarenko, Sergei, 310
Bonner, Yelena: Amnesty International, 580; on "dissident" label, 14; Helsinki Accords, 601; human rights advocate, 633; Jewish background, 517, 518, 523; Nobel Peace Prize lecture, 203; Public Group, 597; and Sakharov, 13, 378, 596–97
Bordaberry, Juan María, 580
Bordewich, Fergus, 613
Börner, Dirk, 573, 576, 577
Brandys, Kazimierz, 625
Brezhnev, Leonid, 176; Amnesty International, 570; approach to dissidents, 395–96, 430–31; bilateral talks with Czech leaders, 240; caricature of, *247*; handling of dissidents, 422–24, 488; handling of Solzhenitsyn, 426–29; Helsinki Accords, 593–94, 599; on Human Rights Committee, 381; interrogation techniques, 407; leadership of, 147, 495; legal system under, 142–43; on "mature socialism," 11, 199; open letters to, 232; Politburo, 638; and Sakharov, 274, 607; stability, focus on, 334; Universal Declaration of Human Rights, 359–60; in Yuli Kim song, 272
Brodsky, Joseph, 7, 90, 120–21, 154, 301, 410, 455, 501, 642

Brothers Karamazov, The (Dostoevsky), 454
Brownell, Sonia, 211
Bruderer, Georg, 147, 150
Brukhman, Maria Samuilovna, 264–65
Brzezinski, Zbigniew K., 9
Bubnys, Prosperas, 513, 646
Bukharin, Nikolai, 133, 192, 208
Bukovskaya, Nina, 152, 157
Bukovsky, Konstantin, 160–61
Bukovsky, Vladimir: Amnesty International, 556; arrest and trial, 148–57, 396, 646; CBS interview, 418; deportation, 419, 648; on dissident structures/hierarchies, 213, 331–32; with Gluzman in labor camp, 475–76, 478; goal of dissent, 161; influence of, 505; internal exile, 410, 413; International Committee publicity, 367–68; *Manual on Psychiatry for Other-Thinkers,* 478–82; Mayakovsky Square poetry readings, 196–97; and punitive psychiatry, 473, 561, 643; on rights, 238; and samizdat, 88–89; at Serbsky Institute, 220; in U.S., 602; and Volpin, 24, 83–84
Burzhuademov (pseudonym), 611. *See also* Sokirko, Viktor

Calas, Jean, 50
Campaign for Nuclear Disarmament, 576
Cancer Ward (Solzhenitsyn), 420
capitalism: analysis of, 132, 437–38; convergence with socialism, 289–90, 292, 293; *Imperialism, the Highest Stage of Capitalism* (Lenin), 480; in Japan, 492; Krasin's dissertation, 437–38; transition from, 133–34, 481
capital punishment, 142, 202–3, 386
Carey, John, 383
Carson, Rachel, 285
Carter, Jimmy, 600, 602, 604
Case 24, 400, 439, 448
Cassin, René, 382, 383
censorship, 62, 177, 235, 286, 408, 495, 569, 624. *See also samizdat*
Central House of Writers, 82–83, 113
Chaadaev, Petr, 200
Chalidze, Valery: on dissidence in Soviet Union, 201, 210, 236, 248, 263; emigration, 452, 632, 646; on fidelity to Soviet law, 374; formalism, advocacy for, 373–74; on freedom of movement, 521; Gorbanevskaya's trial, 417; Human Rights Committee, 374–85, 609; Khronika Press, 568–69; photo, 375; *Problems of Society,* 193, 292, 374, 377–78, 380, 645; reaction to Red Square demonstration, 255; on revolution, 623; techniques used during interrogation, 469; and Western NGOs, 590–91
Charter 77 (Czechoslovakia), 625
Chebrikov, Viktor, 609
Chechen people, 178, 206
Chernyaev, Anatoly, 199, 396, 491
Chernyshevsky, Nikolai, 200, 331
Chicago Tribune (newspaper), 15, 347, 351
Childhood in Prison, A (Yakir), 435
China, People's Republic of, 309, 310, 585, 625–26
Chișinău (Kishinev), 352
Chistopol Prison, 631
Chornovil, Viacheslav, 178, 491, 516, 633
Christianity, 67–68, 305, 309; anti-Christian campaigns, 189–90; Protestant ethic, 611. *See also* Baptists; Russian Orthodox Church
Chronicle of Current Events: Amnesty International, 555; background/overview, 191–92, 214, 231, 644; cessation of, 614, 648; Chalidze in, 383; crackdown on, 490; dissident movement, dissolution of, 609; Gorbanevskaya, 191–92, 416, 418; and information dissemination, 400; KGB investigation, 463; on Kochetov's writings, 529; name, origin of, 191, 621; news from Lefortovo, 439; "News from the Jewish Community," 520; and petition campaigns, 513; on protest letters, 273; *Sacred Paths to Willful Freedom,* 536; threats to/disruption of/resumption, 445–49; on Ukrainian nationalism, 516
Chronicle of the Lithuanian Catholic Church, 513–14, 646
Chukovskaya, Lydia, 526
Chukovsky, Kornei, 103, 114, 255, 391
CIA (Central Intelligence Agency): Amnesty International, 568, 589; covert operations in Soviet Union, 289, 316, 392, 601; International Commission of Jurists, 552, 726n24; People's Labor Alliance (NTS), 150, 370; Radio Liberty

Committee, 409; satire of, 176; in *What Is It You Want?*, 527–28, 532
Civic Appeal, Sinyavsky-Daniel affair: distribution of, 84–85, 88–89, 668n44; early drafts, 78–84, 683n151; impact of, 78, 96–97, 122, 145, 186, 227, 334; text of, 85–88, 668n39; transparency meetings, call for, 79, 80, 84, 145, 643, 677n46, 683n151
civil disobedience, 24, 25, 234, 485
civil obedience, 24, 25, 44, 485, 547, 592, 610, 618
civil rights movement, U.S., 626–29, 722n35
class hierarchies, 132–33, 323, 511–12; class antagonisms, 179, 304, 371, 392–93, 406–7; class morality, 479–80; illusion of disappearance of, 2, 115, 143, 263; middle class, 306, 307, 309; working class, 138, 152, 181, 214, 228, 336
Code of Criminal Procedure. *See* Criminal Code
Cold War, 16–17, 408; and emergence of Soviet dissidence, 2, 4, 235, 415, 489; reduction of tension, 605; Western views of, 52, 104, 141, 312, 359, 392, 414–15, 557, 561, 660n55
Cole, William, 331, 418–19, 434, 444, 645
collective action: citizenship and weak ties, 97; contrasted with individuality, 209–10, 224–25, 233–34, 252–53, 303; in dissident movement, 75–78, 205–6, 209–10, 224–25, 233–34, 263–64, 434, 524
collective farms, 301, 308, 337–38, 496, 601
Collings, Anthony, 347
color revolutions, 636
Committee to Defend Civil Rights in the USSR, 206, 213
Communist Manifesto (Marx & Engels), 48
Communist Party: Central Committee, 43, 102, 139, 298; and elections, 174; expulsion of writers, 239; and literacy levels, 13, 101, 280; monopoly on power, 195, 202, 219, 340, 342, 349, 371, 385, 620; Moscow City Committee, 423, 690n29; post-Stalin, 80; Twentieth Congress, 129; Twenty-Fourth Congress, 381; Twenty-Second Congress, 143; Twenty-Third Congress, 233; under Stalin, 71; workers' protests, 216

Communist Party of Youth, 195
Communist Youth League (Komsomol), 27, 64, 82, 89–90, 336, 395, 453; disciplinary hearings after 1965 transparency meeting, 94
Compromise, The (Dovlatov), 548
Comte, August, 337
Conference on Security and Cooperation in Europe, 426, 599–600
Conquest, Robert, 36, 659n45
corps intermédiaires (Montesquieu), 385
corpus delicti, denial of, 397
Corvalán, Luis, 419
Council of Orthodox Churches, 364
Council of Prisoners' Relatives, 553
Court Is in Session, The (Tertz/Sinyavsky), 45–46, 54, 55, 109–10
Crimean peninsula, annexation of (2014), 636–37
Crimean Tatars: and chain reaction, 178, 179; deportations, 190; in exile, 344, 474, 511, 643; petitions by, 232, 233; protests by, 151; as Soviet minority, 206, 225
Crimean War (1856), 630
Criminal Code, 59, 79, 177, 197, 272, 400–401, 456, 618; Article 64, 429, 441; Article 70, 118, 145, 441; Article 190, 145–49, 152–54, 158, 163, 400, 417, 643
cybernetics, 33–34, 39, 43, 216
"cycles of protest," 149
Czechoslovakia, 232, 624–25, 626; Czechoslovak Communist Party, 239–40; protest (1968), 391; Soviet military action, 243–44, 690n29; Soviet occupation, protests of, 340, 691n55; Václav Havel, 3. *See also* Prague Spring; Red Square sit-down demonstration (1968)
Czechoslovak Writers' Union, 239

Daniel, Mark, 58
Daniel, Yuli: Jewish background, 517; KGB case against, 47, 48–50, 50; Larisa Bogoraz, 490; literary fiction, 71–72, 129–30, 526; and Marchenko, 220; photos, 52, 76, 105; political camps, 178–80; and samizdat, 478. *See also* Sinyavsky-Daniel affair
Darkness at Noon (Koestler), 56, 76, 260, 454
Day-Lewis, Cecil, 210

death penalty, 142, 202–3, 386, 441
Decembrists, 95–96, 200, 249, 526–27, 622, 714n9
Degree of Trust, A (Voinovich), 526
Deloné, Vadim, 148, 153, 159, 161, 230, 251, 262, 419, 632, 647
DemDvizhenie (Democratic Movement of the Soviet Union), 223, 317, 330, 378, 386, 434, 443, 518, 539
Department of Agitation and Propaganda, 102
détente, 305, 599, 606, 724n2
de Tocqueville, Alexis, 18
Deutsche Welle, 6, 55, 392
Devils, The (Dostoevsky), 214, 537–38
dialectical materialism *(diamat)*, 33
dictatorships, nature of, 9, 72, 77, 198.
 See also totalitarianism
Dictionary of the History of Ideas, 24
disenchantment (Weber), 564
Disraeli, Benjamin, 337
dissident movement: autobiographies/first-person accounts, 5, 235; avoidance of politics, 225–26; belief systems, 409–10; "camp fear" of Gulag survivors, 460; celebrity culture of, 316; as chain reaction, 150, 236, 238–39, 274, 327, 619; collective action, 75–78, 263–64, 434, 524; communication methods, 367–68, 388; conduct manuals, 457–60, 463–71; Constitutional rights, 238–39, 396; contact persons, 517; critiques of, 318–19; defamiliarization/estrangement, 329, 675n11; dialogue with Soviet government, 385–86, 485–86; ethical example of, 491–92; explanations for origin of, 430–31; formal organization vs. conscience-driven approach, 341–42, 373, 446, 619; freedom vs. unfreedom, framework of, 16–18; friendship as organizational structure, 226–27, 565–66; global, in 1960s, 547–48, 724n1; goals and methods, 331–33, 334–35; hierarchy and divisions of labor, 236–37; and human rights, 287–88; impact of, 489–90, 497–98, 631–39; increasing despair, early 1970s, 445–46; individual vs. collective action, 209–10, 224–25, 233–34, 252–53, 303; and intelligentsia, 199–204, 234–35, 305–6; interrogation, approaches to, 455–60, 470–73, 472; Jewish involvement, 517–24; law-based dissent, 342, 374–76, 379–80, 397, 437, 495–96, 617; leaflets, use of, 227–29, 230; liberty/inner freedom, 18–19, 461–62; mindfulness and self-unmasking, 329; minority communities, 515; and moral issues, 257–58, 510–11; 1968 as year of failure, 317, 325–26, 333; and non-dissident Soviet citizens, 488, 493–94, 500–504, 524–25, 541–42; and non-participation in Soviet life, 483–84; open/protest letters, 273, 327; persecution of, 3, 394–96; professional dissidents, 503; recantations, 273; rights-defense strategy, 327, 334, 610–14; and scientific worldview, 276–78; shift to NGOs, 388; slogans, use of, 230; socialism vs. Soviet dogma, 361; social movements, models of, 204; transnational language of, 548; voluntary citizens' associations, 385; West, relationship with, 384–85; women's participation, 213–14; words/terms/labels, origins and use, 13–16; youth involvement, 196–97
Dobrovolsky, Aleksei, 148, 167, 192, 220, 338, 403. *See also* "Trial of the Four"
Dobrynin, Anatoly, 606–7
Doctor Zhivago (Pasternak), 40, 46
Dodd, Thomas, 17, 253, 657n42
Domshlak-Gerchuk, Marina, 76
Don Quixote on Russian Soil (Aikhenvald), 25
Don Quixotism, 32, 542, 612–13
Dostoevsky, Fedor, 214, 454
double-speak, 183, 209, 497, 498
Dovlatov, Sergei, 487, 548
Dremlyuga, Vladimir, 251, 262, 445, 446, 644, 647
"Drunken General Secretary's Monologue, The" (Kim), 272
druzhinniki (anti-crime squads), 89
Dubček, Alexander, 239, 240–41
Dubin, Boris, 82
Dubnov-Erlikh, Sophia, 501
Dudko, Dmitry, 505
Dzhemilev, Reshat, 232–33
Dziuba, Ivan, 515–16, 633, 643

INDEX 781

Ehrenburg, Ilya, 114, 115, 124, 160–61
Eidelman, Natan, 200, 526, 714n9
Einstein, Albert, 285
Elvin, Lionel, 550
emigration: forced, 410–11, 413, 416, 429; inner, 461, 462; Jewish, 517, 520–21, 523. *See also* exile, practice of
empathy: distanced, 549–50, 558, 575, 577, 591–92; and human rights, 550
"Emperor's New Clothes, The" (Andersen), 508–9
Encounters with Solzhenitsyn (television program), 633–34
Engels, Friedrich, 12, 48; on freedom, 36
Ennals, Martin, 568–69, 573, 576, 577, 578
Erofeev, Venedikt, 526
Esenin, Georgy Sergeyevich, 26
Esenin, Sergei, 25, 41
Estonia, 189, 368, 392, 512
Etkind, Alexander, 72
Etkind, Efim, 501
Europa Civiltà (European Civilization, NGO), 365–66, 370, 372
exile, practice of, 3, 29, 410–11, 603. *See also* emigration
Exodus (samizdat journal), 645
"Experiment in Self-Analysis, An" (Sinyavsky), 73

Fainberg, Viktor, 251–52, 259–60, 644; emigration, 632, 647
Fantastic Stories (Tertz/Sinyavsky), 46, 60
Feast of Victors (Solzhenitsyn), 421
Fedin, Konstantin, 503
Fenwick, Millicent, 599
fiction, influence of, 526–27, 541–43; Anna Gerts, 532–41; literary evidence in Sinyavsky-Daniel affair, 106; literary texts, export of, 107; Vsevolod Kochetov, 527–32
Fifth Directorate: conversations, 453; creation of, 430, 643; and dissemination of samizdat, 400; dissident critiques, 407–8; and forced emigration, 411; formation of, 394–95; interrogation transcripts, 439; organizational structure, 395; prophylactic measures, 401–2, 404, 486; punishment, 405; use of Yakir and Krasin, 438. *See also* KGB

Filippov, Boris, 62, 107–9, 122, 662n20
Final Act. *See* Helsinki Accords
First Circle, The (Solzhenitsyn), 55, 286, 421
Fitzgerald, F. Scott, 315
Flemish Committee for Eastern Europe, 365–66, 367
"For a Sister" (Rich), 253
Ford, Gerald, 599, 602
forensic linguistics, 47–48
forensic psychiatry, 29, 220, 414, 480
"forest brothers," 391–92, 512
Foucault, Michel, 415
Foundation for Spiritual Freedom, 365
Foundation for Victims of Imperialism and Colonialism, 602
Franco, Francisco, 579
Frankfurter Allgemeine Zeitung, 351
Frankfurter Rundschau, 586, 589
Fraternal Leaflet (Bratskii listok), 189, 642
freedom: contrasted with transparency, 81; inner freedom, 18, 311, 326–28, 334, 461–62, 498, 504; liberty, concepts of, 17, 18–19; negative freedoms, 134, 140; of the press, 235; public conformism to preserve private freedom, 44; and unfreedom during Cold War, 16–17; Volpin's concept of, 36–37
Freud, Sigmund, 25
Friedrich, Carl J., 9
Frunze Military Academy, 216
Future of Science (journal), 284
futurology, 283–84, 287, 292, 304

Gabai, Ilya, 149, 206, 208, 303, 345, 346, 445, 517, 645
Gagarin, Yuri, 31
Galanskov, Yuri, 91, 148, 158, 167, 274, 338, 631, 642, 646, 704n96; Amnesty International, 556; arrest, 403. *See also* "Trial of the Four"
Galich, Alexander, 1, 249, 272, 376, 632
Gamsakhurdia, Zviad, 3, 603
Gandhi, Mahatma, 24
Gannushkin Psychiatric Hospital, 42
Garnett, Constance, 255
Gasilov, Georgy Vasilievich, 185
Geneva Convention, 358
Georgia: color revolution, 636; Helsinki Group, 598, 609; post-Soviet, 3

Gerasimov, Alexander, 1–2
Gerchuk, Yuri, 76, 242
Gerlin, Valeria, 182–86, *183*, 497, 681nn111–12
Germany: invasion of Soviet Union, 30–31; parallels with Soviet Union, 624, 736–37n16; Third Reich, 159, 160. *See also* Nazi Party, Germany
Gerts, Anna (Maya Lazarevna Zlobina), 532–36, 611, 723n19
Gessen, Masha, 288
Gilman, Inessa Markovna, 64
Ginzburg, Alexander: Amnesty International, 556, 580; apartment searches, 148; arrest and trial, 167, 177, 403, 602; defense of, 396; and Dobrovolsky, 220; emigration, 632; Jewish background, 517; *Journey into the Whirlwind*, 231; KGB surveillance, 124; on Leninism, 197; letter-writing campaign for, 274; Public Group, 597; and samizdat, 90, 122, 162, 642; transparency meetings, 322; *White Book*, 124, 148, 162, 178, 197, 556, 643. *See also* "Trial of the Four"
Ginzburg, Lidiya, 267, 329
Ginzburg, Lyudmila, 366
Ginzburg, Vitaly, 147
Ginzburg, Yevgenia, 180, 530
glasnost (transparency), 3, 25, 122, 364–65, 564, 667n16
Glazov, Yuri, 168, 517
Gluzman, Fishel, 474–77
Gluzman, Galina, 474–77
Gluzman, Semyon, 407, 473–78, 475, 483, 508–9; Jewish background, 517; *Manual on Psychiatry for Other-Thinkers*, 478–82; and punitive psychiatry, 561
God That Failed, The (manifesto), 210
Goffman, Erving, 415
Goldberg, Arthur, 605
Golomshtok, Igor, 75–76, 121, 149, 411, 667n14
Gorbachev, Mikhail, 3, 10, 25, 102, 289, 629–30, 649, 690n30
Gorbanevskaya, Natalya: activist meetings, 206; background, 187–92; *Chronicle of Current Events*, 210, 214, 273; forced emigration, 418–19, 647; on informal organization of dissidence movement, 325, 352; and Initiative Group, 354; on inner freedom, 462; Jewish background, 517; open letters, 242; photos, *190, 419*; pressure from Soviet power, 510, 641; psychiatric hospitalization, 416–18, 473, 644, 711n77; on Quixotism, 631–32; reaction to military action in Prague, 246–48; reaction to Red Square demonstration, 255; Red Square demonstrations/trial, 251–53, 260; *Sacred Paths to Willful Freedom*, 539–40; on samizdat, 86, 240
Goricheva, Tatyana, 504–5, 536
Gorky Institute of World Literature, 46–47, 55, 113
Gostev (KGB agent), 163–64, 167, 455, 643
Grachev, Andrei, 289
Granovetter, Mark, 97
Grass, Günther, 50
gray economy, 612
Great Soviet Encyclopedia, 288
Greene, Graham, 50
Grigorenko, Petr: Amnesty International, 556, 580; background and arrest, 214–20, 645; and Bukovsky, 220–21; civil rights organization in Moscow, 220–23; Committee in Defense of Grigorenko, 345–46; Crimean Tatars, 517; defense of Yakhimovich, 342–45; emigration, 632; on Leninism, 198; and Litvinov, 223–24; open letters, 230–33, 242; organization vs. individualism, 325, 342; photos, *215, 221, 339*; psychiatric hospitalization, 473–74, 642; Public Group, 597; reaction to military action in Prague, 246, 248; reaction to ordinary citizens, 330–31; reaction to Red Square demonstration, 255; Serbsky Institute, 220, 224–25; on Soviet Constitution, 197; Union for the Struggle for the Rebirth of Leninism, 216–17, 642; use of leaflets, 227–29; and Viktor Krasin, 187
Grishin, Viktor, 423
Gromyko, Andrei, 359, 587, 604
Group 73, 562–63, 565, 567–68, 646, 647, 728n54
Guevara, Ernesto "Che," 337
Gul, Roman, 536
Gulag Archipelago, The (Solzhenitsyn), 425, 428, 484, 492, 585, 634, 647

Gulag system, 178; "camp fear," 460; credibility of survivors, 291; memoirs/accounts of, 180, 231; persistence of, 180–81; Solzhenitsyn's history of, 421; songs and language, 208; Vorkuta, 265. *See also* labor camps
Gurevich, Aron, 198, 501–2, 504, 536
Guseinov, Chingiz, 103, 110, *111*, 112–13, 161

Halfin, Igal, 461
Hammarberg, Thomas, 577, 578, 589
Hampshire, Stuart, 210
Handbook for Parents (Makarenko), 295
Handbook of Psychiatry (Snezhnevsky), 416
Harris, Robert, 160
Havel, Václav, 3, 11, 625; "The Power of the Powerless," 12–13
Hegel, Georg Wilhelm Friedrich, 36, 245, 337
Heidegger, Martin, 328
Hellbeck, Jochen, 461
Helsinki Accords, 426, 593–94, 647, 648; Helsinki effect and collapse of communism, 605–7; Helsinki Groups, dismantling of, 608–10; Helsinki Groups/arrests, 602–5; human rights articles, 595–96; Public Group to Assist . . . Helsinki Accords, 596–98; Sakharov case, 608; Western foreign policy and control, 599–602
Helsinki Watch, 600
Henri, Ernst, 284, 288, 395
Herzen, Alexander, 189, 200, 258, 622
Hess, Rudolf, 160
Het Parool (newspaper), 172, 288, 313, 623
Hitler, Adolf, 9, 159, 160
Hitler-Stalin Mutual Non-Aggression Pact, 336
Homo Sovieticus (Zinoviev), 518
Hoover, Herbert, 57
Hosking, Geoffrey, 97
How to Be a Witness (Albrekht), 463–70, *465*, 646
How to Conduct Yourself at an Interrogation (Volpin), 456–57, 478, 645
How to Conduct Yourself During a Search (Albrekht), 647
"How Would I Have Conducted Myself?" (Reemtsma), 500
Hryhorenko, Petro. *See* Grigorenko, Petr

hub-and-spoke model for samizdat dissemination, 399–400, *400*, 446
Humanist Manifesto (Galanskov), 91
Human Rights Committee (Valery Chalidze), 374–85, 645; and "creative partnerships," 382–33; and Sakharov's 1972 petitions, 386; Soviet reactions to, 378–80
Human Rights: Continuing the Discussion (Zivs), 589
Human Rights Movement, Soviet Union, 594–95, 727–28n51, 730n13
Human Rights Year in the USSR, 191–92
Hungary, 611, 624–25, 626; uprisings, 188, 248, 391, 394, 641, 711n77
Hunt, Lynn, 550
Hutter, Irmgard, 576, 730n13

"I Cannot Be Silent" (Tolstoy), 202–3
"Ideocratic Consciousness and the Individual Person" (Zelinsky), 328, 329–30, 461, 498
"Idiots" literary club, 90
Igrunov, Vyacheslav: on Human Rights Committee, 387; on Marxism-Leninism, 196; meeting with Larisa Bogoraz, 490–91; on morality of dissident movement, 316, 331, 489–90, 536; photo, *326*; samizdat, 325–26, 333–34, 461, 614
Ilichev, Leonid, 41, 51, 61, 120
Imperialism, the Highest Stage of Capitalism (Lenin), 480
Imshenetsky, Alexander, 386, 707n43
inakomysliashchie (other-thinkers), 13, 371, 523–24, 531, 542
Index on Censorship (journal), 212, 646, 686n61
Initiative Group for the Defense of Civil Rights, 347, 351–60, 591, 603, 609, 618; achievements of, 356; appeals to United Nations, 360–61, 362–63; arrest of members, 364, 645; civil rights vs. human rights, 357–58; formal structure, lack of, 354–55; formation of, 417, 645; international appeals, 357; and international organizations, 370–71; Jewish members, 517–18; lack of broader support, 387; reaction to Human Rights Committee, 378; turn to human rights, 361–62; Western NGOs, 364–66; Western support, 366–68

"inner emigration," 461–62
inner freedom, 18, 311, 326–28, 334, 461–62, 498, 504
Institute of Red Professors, 264–65
International Commission of Jurists, 552, 726n24
International Committee for the Defense of the Rights of Man, 365, 366–67, 370–72
International Covenant on Civil and Political Rights (1966), 628, 643
International Federation for Human Rights, 384
International Herald Tribune (newspaper), 164
International Institute of Human Rights, 382, 388, 646
Internationalism or Russification? (Dziuba), 515–16
International League for the Rights of Man, 364, 382–83, 388, 570, 646
International Peace Bureau, 559
interrogation: KGB techniques, 39–40, 56–61, 406–7, 409–10, 436–39, 453–55; transcripts of, 455
Introduction to Metamathematics (Kleene), 33
Iron Curtain, 5, 122, 166, 213, 289, 384, 392, 484, 548, 593, 613, 628–29
Iskra/Spark (newspaper), 445
Israel, 411, 447, 452, 518, 520, 594, 631, 632, 722n35
"Is There Life on Mars?" (Remizov), 51
Is Uncle Jack a Conformist? (Amalrik), 615–17
Italy and Fascism, 9–10, 204
Ivanov, Vyacheslav, 33, 39
Izvestiia (newspaper), 87, 88, 93, 99, 163, 289, 560, 588–89; Sinyavsky-Daniel trial coverage, 102, 103

Jackson, James, 15
Jahimowicz, Jan, 336, 339. *See also* Yakhimovich, Ivan
Jaunā Gvarde collective farm, 337
Jaurès, Jean, 420
Jewish Anti-Fascist Committee, 519
Jewish Autonomous Region of Birobidzhan, 521–22
Jews, 179, 225; and dissident involvement, 517–24; and emigration, 192, 411, 422, 512, 517, 521–22; Pale of Jewish Settlement, 39

Jokes Physicists Tell (Turchin), 565
Joliot-Curie, Frédéric, 233
Journey into the Whirlwind (Ginzburg), 180, 231
judicial transparency, 24, 80–81, 93, 122, 169, 175

Kafka, Franz, 76
Kaganovich, Lazar, 208
Kagarlitsky, Boris, 315
Kallistratova, Sofia, 260, 382, 397, 417, 474, 603, 608
Kalygina, Anna, 337
Kaminskaya, Dina, 104, 151–54, 157–58, 243, 260, 396, 397, 517
Kaminsky, Isaak, 151
Kantov, Lieutenant-Colonel Georgy, 57, 218
Kapitonov, Ivan, 426
Kapitsa, Petr, 386, 503
Kataev, Valentin, 103
Kaverin, Venyamin, 114, 147
Kazakhstan, 29, 178, 182, 336
Kedrina, Zoia, 100
Keldysh, Mstislav, 503
Kennan, George F., 312, 599
Kennedy, John F., 256, 281, 337
Kesey, Ken, 415
KGB (Committee for State Security), 2; apartment searches, 458; approach to constitutionalists, 396; archival materials and case files, 6, 707–8n11; arrests, increasing rate of (1969), 346; arrest of Andropov, 345; classified history of itself, 398; control of attorneys, 397; control strategies, 416; "conversations," 181–82, 402–4; crackdown on dissidents, 606–7; data collection, 227; dissident attitudes toward, 326–27; forced emigrations, 418–20; ideological commitment, 408–9; Initiative Group, 356–57; interrogation techniques, 39–40, 56–61, 406–7, 409–10, 436–39, 453–55; involuntary psychiatric hospitalization, 411–18, 414, 420; memoirs of officials, 408; mole at Litvinov's apartment, 187–88; network theory, 399–400; Pavel Litvinov and "prophylactic conversation," 162–65; press conferences, 442–44; "prophylactic measures," 402–5; "psychological warfare," 392–94; repertoire of techniques,

429–30; retaliatory gestures, 182, 327, 385–86; Sinyavsky-Daniel interrogations, 56–61; Sinyavsky-Daniel transparency meeting, 89, 91, 93–94; Sinyavsky-Daniel trial, 99–111, *105*, 121, 124; surveillance of Amalrik, 298–99; surveillance of Chalidze, 384; surveillance of Krasin, 349–51; surveillance of Sakharov, 283, 288, 381; terror and social control, 52–53; threats to *Chronicle of Current Events*, 447–49; Vladimir Bukovsky's trial, 152–53. *See also* Fifth Directorate

Khachaturian, Aram, 503

Khakhayev, Sergei, 399

Khariton, Yuli, 288, 503

Kharkhordin, Oleg, 686–87n71

Khaustov, Viktor, 148, 155, 161

Kheifets, Mikhail, 454

Khentov, Leonid Abramovich (pseudonym), 395

Khlebnikov, Velimir, 38

Khmelnitsky, Sergei, 66–69, 664n52, 665n64

Khodorovich, Tatyana, 440, 447, 448, 596

Khronika Press, 568–69

Khrushchev, Nikita, 10, 12; anti-religious campaign, 505; on arms control, 281; cult of personality, 51; Gulag mass amnesties, 180; impact of, 423–24; legal system under, 142, 144; on Petr Yakir, 207–8; removal from power (1964), 43, 142, 144, 337, 642; and Sakharov, 278; Secret Speech (1956), 31, 72–73, 83–84, 151, 187, 641; Stalin's cult of personality, unmasking of, 70–71, 117, 138; Thaw, 31–32, 33, 55, 160–61, 362, 393, 495, 618; Twentieth Party Congress speech, 111, 129, 159; on Volpin, 41

Kim, Yuli, 195, 233–34, 272, 304, 609

Kislykh, Gennady, 436, 444, 447

Kissinger, Henry, 312, 599, 606

Kleene, Stephen, 33

Kline, Edward, 569, 632

Kochetov, Vsevolod, 527–32, 536, 539, 541–42, 611, 645

Koestler, Arthur, 56, 76, 260, 454

Kogan, Pavel, 151

Kohout, Pavel, 232

Kolmogorov, Andrei, 33–34

Kolokol (Bell), 189, 399–400, *400*

Kol Yisrael (Voice of Israel), 392

Kolyma Tales (Shalamov), 180

Komaromi, Ann, 538

Komsomolskaya pravda (newspaper), 176, 337

Kopelev, Lev, 76, 243, 517, 631

Kopeleva, Maya, 243, 249, *261*

Korchak, Alexander, 597

Korneyev, Ilya, 562, 567

Kornienko, Georgy, 588

Korzhavin, Naum, 518

Kostava, Merab, 603

Kosygin, Alexei: and Andropov, 424, 607; Chairman of Council of Ministers, 43, 147, 231; on Czechoslovakia, 240; open letters to, 232; on Solzhenitsyn, 426; and Soviet scientists, 283

Kotkin, Stephen, 131

Kovalev, Ivan, 603, *604*; *Some Thoughts on How to Help Political Prisoners*, 648

Kovalev, Sergei: arrest and trial, 447, 448, 449–50, 583, 728n64; background in Soviet constitution, 138–39; dissident activities, impact of, 493–94; individual vs. group approach, 524; interrogation of, 471–72; Lithuanian Catholics, 517; on Litvinov's interrogation, 440; on open letters, 231; petitions in Moscow, 513; photo of, *450*; service to Yeltsin, 633; on Yakir/Krasin and Initiative Group, 348–49, 350–51

Kozharinov, Venyamin, 326–27, 335, 349, 371, 372, 387

Kozlov, Vladimir, 407

Krasin, Viktor: arrest and imprisonment, 436–37; Committee to Defend Civil Rights, 213; dissident movement, organization of, 187, 206–7, 325; emigration, 450, 632, 647; and foreign journalists, 433–34; as Gulag survivor, 291; on human rights, 358; Initiative Group, 348–49, 352, 353–54; Jewish background, 517; KGB, 349–50, 436–41; open letters, 242, 345–46; on partnership with international organizations, 371; photos, *351, 433*; possession of samizdat, 370; press conference, 442–44, *444*, 448; on protest letters, 385–86; public disavowal, 540–41; reaction to military action in Prague, 248; reaction to Red Square demonstration, 255; and rights-defenders, 236, 343; trial, 432, 441–42

Kravchenko, Natalya, 439
krugovaya poruka, 72
Krupskaya, Nadezhda, 463
Kushev, Yevgeny, 148–49, 161, 668n44
Kuznetsov, Anatoly, 421
Kuznetsov, Viktor, 345, 346
Kyrgyzstan, 636

labor camps, 3, 412; connections forged in, 125, 162, 178–79; German POWs, 66; Kolyma, 110, 206; *My Testimony* (Marchenko), 241; Perm-35, 475; sentences during 1970s, 602–3. *See also* Gulag system
Landa, Boris, 728n64
Landa, Malva, 597
Landmarks essay collection, 532, 723n16
language: Amnesty International's language of protest, 564–65, 580; ideal, 25, 27; of law, 43–44, 60, 79, 379, 625; literary, 107, 110, 117–18; of mathematics, 33; scientific, modal or ideal, 34, 37–39
Lashkova, Vera, 148, 167, 177, 228, 229, 322, 338, 402–4, 403, 551, 643. *See also* "Trial of the Four"
Latvia, 336–37, 341, 392, 512
Latvian Agricultural Academy, 337
law. *See* rule of law strategies
Leaf of Spring, A (Volpin), 40–42, 53, 412
Ledeneva, Alena, 354
Lefebvre, George, 5
Lefortovo Prison, 149, 152, 271, 429, 435, 438, 442, 453–54
legalism, 35, 198–99, 205, 214, 234, 235, 319, 378, 547, 625, 626. *See also* rights-defenders
Legal Memorandum for Those Facing Interrogations (Volpin), 453, 456
Le Monde (newspaper), 51, 313
Lenin, Vladimir: on building socialism, 8; *Imperialism, the Highest Stage of Capitalism*, 480; on law and rights, 132–33; on peasant rebellions, 25; Seventh Congress (1918), 134
Leningrad Psychiatric Prison Hospital, 29
Leninism, 196, 197–99, 219; neo-Leninism, 214, 216–17, 229, 548, 684n17. *See also* Marxism-Leninism
Leontovich, Mikhail, 148

Let History Judge (Medvedev), 646
Letter of the 46, 182
Letter to the Soviet Leaders (Solzhenitsyn), 611, 717n74
Levitin-Krasnov, Anatoly, 353, 378, 518, 701n49
L'Humanité (newspaper), 163
liberty. *See* freedom
lichnost, 362
Liebling, A. J., 235
Lietuvos katalikų bažnyčios kronika (Chronicle of the Lithuanian Catholic Church), 513
Lionæs, Aase, 588
literature. *See* fiction, influence of
Literaturnaya gazeta (journal), 284, 489
Lithuanian Soviet Socialist Republic, 115–16, 392, 450, 512–15, 646; Catholics and religious freedom, 512–15; Helsinki monitoring groups, 598, 609
Litvinov, Ivy, 253
Litvinov, Maxim, 113, 162, 167, 253, 374
Litvinov, Mikhail Maximovich, *261*
Litvinov, Pavel, 206; "Appeal to the Global Public," 172, 174–75, 186–87, 210, 337, 362, 644; exile of, 410, 646; and Grigorenko, 223–24; Jewish background, 517, 523; KGB interrogation/surveillance, 161–67, 440, 455; open letter on Marchenko's arrest, 242–43; photos, *300, 339, 351*; on Prague Spring, 241; protest and free will, 208–10; on protest movements, 211–13; public reaction to, 165–66; publishing house, 632; reaction to military action in Prague, 248–49; Red Square sit-down demonstration (1968), 250–52, 253; samizdat publication, 311–12, 644; on Solzhenitsyn, 485; and Stephen Spender, 384, 595; Trial of the Four, 171–74, *173*; "visiting days," 186–87
Litvinova, Tatyana, 113, 175, 229–30, 577, 688n104
Live Not by the Lie (Solzhenitsyn), 483–85, 611, 612, 647, 717nn80–81
Lobnoe Mesto, Red Square, 250
Loginov, Gleb Pavlovich, 611
Lotman, Yuri, 189
Lubyanka Prison, 28–29, 48, 56–57, 58, 99, 106, 188, 218; executions, 207

Lunacharsky, Anatoly, 317–18
Lunin (Eidelman), 526, 714n9
L'Unità (newspaper), 163
Lunts, Aleksandr, 580
Lunts, Daniil, 418, 473, 474, 476, 479
Lysenko, Trofim, 282
Lyubimov (Tertz/Sinyavsky), 55, 530

MacBride, Seán, 552, 558–61, 567–69, 571, 573–77, 579, 621
Mackay, Charles, 510
Madison Avenue Group, 561, 568, 569, 574, 575. *See also* Amnesty International
Makarenko, Anton, 295
Makhno, Nestor, 207
Makudinova, Gyuzel, 299, *300,* 439
Malia, Martin, 129–30
Malinovsky, Alexander, 317–18, 524, 645; critique of dissident movement, 318–24, 329–30; moral revolution, 324; reactions to Malinovsky's criticism, 325; views on the West, 335
Malyarov, Mikhail, 429, 443, 445–46
Mandelstam, Nadezhda, 10, 150, 312
Mandelstam, Osip, 533
Manual on Psychiatry for Other-Thinkers (Bukovsky & Gluzman), 478–82, 647
Marchenko, Anatoly, 3, 178–80, *181,* 220, 354, 597, 631, 643; arrest of, 242; death, 649; on Igrunov's methods, 326; on Prague Spring, 241–42; trial, 243, 248
Markov, D. K., 140
Marr, Nikolai, 38
Marx, Karl, 12, 48, 360, 494; on freedom, 36; law and rights, 132–34
Marxism-Leninism, 31, 34, 47, 195–96, 197, 305, 408, 512, 565. *See also* Leninism
Marynovych, Myroslav, 491, 536, 603
Mauriac, François, 50
McCarthy, Mary, 210
Medinsky, Vladimir, 637–38
Medvedev, Dmitry, 632
Medvedev, Roy, 126, 191, 232, 284, 288, 335, 420, 517, 610, 636, 642; *Let History Judge,* 646; samizdat journal, 191, 682n138, 683n148
Medvedev, Zhores, 420, 517, 645, 730n13
Meerson-Aksenov, Mikhail, 518
Meilanov, Vazif, 608, 735n59

Meiman, Naum, 603
Meirovich, Mark Mendelevich, 58
Memorial Society, Moscow, 4, 634, 657–58n7; dissolution of, 637; Nobel Peace Prize (2022), 637
Memory (samizdat journal), 647
Men, Alexander, 518
Mensheviks, 57, 236
Menuhin, Yehudi, 210
Metropole (samizdat journal), 648
Michelet, Jules, 622
Mikhailov, A., 318. *See also* Malinovsky, Alexander
Miller, Arthur, 50
Miłosz, Czesław, 63
Misik, Olga, 634
Mitrokhin, Vasily, 405
Molotov, Vyacheslav, 208, 422, 453
Mommsen, Hans, 624
Montesquieu, 385
Months (samizdat journal, also known as *Political Diary*), 191, 642
Moore, Henry, 210
Moral Code of the Builder of Communism, 143
Mordovian Autonomous Region, labor camps, 125, 179, 609, 631
Morning Star (newspaper), 163
Moroz, Valentyn, 150
Moscow Helsinki Group, 4, 634; dissolution of, 637
Moscow State University: Amalrik as student, 296–98; "Idiots," 90; Sakharov's *Reflections,* 289; Sinyavsky-Daniel social circle, 75–76
Movement, Democratic, 223, 317, 330, 378, 386, 434, 443, 518, 539. *See also* dissident movement
Moyn, Samuel, 357, 360, 361
Mussolini, Benito, 9
"mutual assured destruction" (MAD), 9. *See also* nuclear weapons
My Testimony (Marchenko), 3, 181, 326, 643

Nabokov, Vladimir, 406, 705n1
"Natalya" (Baez), 253
nationalism, 9–10, 298, 309, 423, 516, 612; Ukrainian, 516–17
Navalny, Aleksei, 4

Navrozov, Lev, 502, 536
Nazi Party, Germany, 9–10, 159–60, 500, 623–24; resistance to, 266, 504; Soviet Union, comparisons with, 736–47n16
Nechaev, Sergei, 538
Nedelin, Mitrofan, 278–79, 281–82
"negative freedoms," 134, 140
Neizvestny, Ernst, 496
Nekrasov, Viktor, 103, 147
Nelidov, Dmitry (pseudonym), 328. *See also* Zelinsky, Vladimir
Nelson, Lars, 347
networks: Amnesty International, 565–66; KGB theory of, 399–400, *400*; of micro-communities, 97; repression of, 149; and samizdat, 194, 229, 305, 332, 387, 388, 540, 619; and social movements, 149, 179, 204–5, 228–29, 490, 600–601
Neumann, Franz, 8
Nevler, Leonid, 119
New Leader, The (journal), 63
Newsweek (journal), 383
New Times (journal), 585
New York Times (newspaper), 62, 288–89, 313, 351, 613
Ngo Dinh Diem, 256
NGOs (non-governmental organizations), 364–72, 382, 384–85, 388, 409, 548, 602, 619, 645. *See also* Amnesty International
Nicholas I (tsar), 96, 200
Nicholas II (tsar), 24, 96, 130
Nikolskaya, Ada, 90
Nikolsky, Valery, 79, 85, 90, 611–12
1984 (Orwell), 3, 502
Nixon, Richard, 312, 606
NKVD (People's Commissariat of Internal Affairs), 182, 216, 395, 406, 463, 467
Norman Theory of Russian state formation, 297
Notes from Underground (Dostoevsky), 214
Novocherkassk, workers' protest, 100, 216
Novosti Press Agency, 104
Novotný, Antonín, 239
Novozhilova, A. V., 184
Novyi mir (journal), 88, 102, 160, 475, 533; publication of Solzhenitsyn's works, 180
Novyi zhurnal (New Journal), 536
nuclear weapons, 9, 274–75, 391; moratorium on, 280–81; radioactive fallout, 280
Nuzhdin, Nikolai, 282

Observer, London, 351, 549
ochnaya stavka, 438
Ogarev, Nikolai, 189
Ogibalov, Pyotr, 27–28
Ogonek/Spark (journal), 41, 42
Oistrakh, David, 503
Oktiabr/October (journal), 160, 527, 529
Okudzhava, Bulat, 78, 90
One Day in the Life of Ivan Denisovich (Solzhenitsyn), 180, 642, 724n2
"On the Choice of Terms" (Shragin), 13
open/protest letters, 214, 230–32, 334, 364, 453, 570, 720n3; KGB reprisal, 327, 385–86, 404
Operation Barbarossa, 304, 391
Oppenheimer, Robert, 276
Orekhov, Viktor, 492–93
Origins of Totalitarianism, The (Arendt), 11
Orlov, Yuri: Amnesty International, 566, 577–78, 580; arrest of, 492–93, 603; emigration, 632, 649; Helsinki Accords, 599–600; internal exile, 410; on international support, 620; KGB, seizure by, 597–98; and Natan Sharansky, 594; Nobel Peace Prize nomination, 603–4; on post-Stalin Soviet Union, 11, 244; Public Group, 596–97; samizdat, 360–61
Orlova, Raisa, 103, 113, 210, 231, 235, 243, 254, 517, 631
Orlovsky, Ernst, 728n64
Orwell, George, 76, 211, 502
Osipova, Tatyana, 603, *604*, 609
Osnos, Peter, 613, 736n71
"other-thinkers," 13, 371, 523–24, 531, 542

Palach, Jan, 340, 645
parasitism, charges of, 60, 120, 314, 583, 642, 643
Park Chung-hee, 568
Parks, Rosa, 24
Pashko, Atena, 491
Pasternak, Boris, 40, 46, 48, 52, 55, 78, 151, 642
Patočka, Jan, 625
Paul VI (pope), 364
Paustovsky, Konstantin, 103
Pavlovsky, Gleb, 453, 458, 503–4, 633; on inner freedom, 462; on interrogation techniques, 470; order to shave beard, 510

Peltier [Zamoyska], Hélène, 51, 63–74, *64*, 80, 122, 126, 205, 641, 665n67
People, Years, Life (Ehrenburg), 160
People's Commissariat of Internal Affairs (NKVD), 182, 216, 395, 406, 463, 467
People's Labor Alliance (NTS), 150, 368–70, 497–98
perestroika, 3, 629
"Petersburg Romance" (Galich), 249
petition campaigns, 103, 114, 150, 167–69, 177, 233, 513–14, 554, 643
Petkus, Viktoras, 603
Petrov, Nikita, 404, 709n46
Petty Demon (Sologub), 723n28
Philosophical Letters (Chaadaev), 200
Phoenix (samizdat journal), 642
Pierce, John Robinson, 169
Pimenov, Revolt, 3
Pinochet, Augusto, 419
Pipes, Richard, 202, 299
Plehve, Vyacheslav, 96
Plyushch, Leonid, 256, 292, 325, 378, 485; criticism of nationalism, 517; on cynicism, 499; emigration, 632; on Kochetov, 530; psychiatric hospitalization, 473
Podgorny, Nikolai, 232, 425–26, 607, 620
podpisant, 169
Podrabinek, Alexander, 455, 492
Podyapolsky, Grigory, 529
pokazatel'nyi protsess, 100–101
Poland, 3, 624–25, 626; uprisings, 188
Polikovskaya, Lyudmila, 230–31, 669n51
Political Diary (samizdat journal), 191, 642, 682n138
Poltorak, Arkady, 560
Poluektov, A. S., 140
Pomerants, Grigory, 517, 593, 612–13
Ponomarev, Boris, 424
Potapov, Ivan, 196
Power and Protest (Suri), 724n2
"Power of the Powerless, The" (Havel), 12–13
Prague Spring, 240–42, 292, 317, 644. See also Czechoslovakia; Red Square sit-down demonstration (1968)
Pravda (newspaper), 41, 233, 289, 409, 488; Ukrainian edition, 559
pravozashchitniki (rights-defenders), 13. See also rights-defenders
Preobrazhensky, Evgeny, 133
Principia Mathematica (Whitehead), 27

Prisoners of Conscience in the USSR (Amnesty International), 584–87
Problems of Society (journal), 193, 292, 374, 377–78, 380, 645
profilaktika, 401–2
propaganda, 100–101, 161, 246, 370; Anti-Soviet Agitation and Propaganda, 32, 51, 57, 59, 142, 145, 404; Department of Agitation and Propaganda, 102; KGB programs, 427, 435; literature as propaganda, 110; concerning the West, 371–72, 559
"prophylactic conversations," 89, 163, 165, 181–82, 401, 404–5, 453, 486, 531, 601
pseudonyms, use of, 53, 105, 175, 194, 326
psychiatry, punitive, 411–18, *414*, 420, 455, 473–78, 561, 648, 660n55; *Manual on Psychiatry for Other-Thinkers*, 478–82. See also Serbsky Institute of Forensic Psychiatry
"psychological warfare," 392–93, 394, 415. See also KGB
Public Group to Assist in Implementing the Helsinki Accords, 596–98, 601, 608–9, 647
Pushkin, Alexander, 25, 87, 275
Pushkin Square, 87, 89, 96, 152, 157, 193–94, 229, 230, 283, 374, 487, 643
Pussy Riot, 4
Putin, Vladimir, 4, 395, 633, 634, 636, 707–8n11

quantum mechanics, 276–277
Quiet Flows the Don (Sholokhov), 117
Quixotism, 32, 318, 542, 612–13

Radek, Karl, 208
Radio Armenia jokes, 14
radio broadcasts, 6, 165, 392, 548. See also shortwave radio broadcasts
Radio Liberty, 6, 55, 124, 168, 310, 392, 409, 628, 663n27, 710n53
Radzievsky, Pavel, 148
Rakhunov, R. D., 120
Ramparts (magazine), 409
Raškinis, Arimantas, 513–14, 517
Razumovsky, Andrei, 82
Reagan, Ronald, 13
Reddaway, Peter, 212, 574, 575–76, 578, 589, 645, 729n69, 730n13

Red Square at Noon (samizdat transcript), 260
Red Square sit-down demonstration (1968), 250–53, 644; criticism of, 319; impact of, 273–74; Lobnoe Mesto, 250; reactions to, 253–59, 267; trial, 259–64, 261, 302, 303
Red Virgin Soil (literary journal), 33
Reemtsma, Jan Philipp, 500–501
REFAL (Recursive Functions Algorithmic Language), 565
Reflections on Progress, Peaceful Coexistence, and Intellectual Freedom (Sakharov), 286–93, 367, 644
refuseniks, 411, 520
Reich, Rebecca, 479
Remizov, Andrei, 51, 54, 71, 126, 662n18, 662n23, 663nn24–25
rentiers, 480–81, 716–17n70
Reuters (news agency), 347
Riasanovsky, Nicholas, 297
Rich, Adrienne, 253
Rigerman, Leonid, 562–63, 569
rights: civil rights vs. human rights, 358, 702n66; as dissident tool, 361, 385, 505, 550–51, 605–6; doctrine of universal human rights, 17–18; empathy and universal rights, 550; freedom of movement, 521; human rights activism, 388; human rights NGOs, 364, 372, 382, 384–85, 409, 557; human rights and Soviet Constitution, 357–61; and pre-Soviet peasantry, 25; public attention and human rights advocacy, 550–51; rights of the person, 362; Soviet citizens' reluctance to claim, 24. *See also* Amnesty International; Human Rights Committee (Valery Chalidze), Initiative Group for the Defense of Civil Rights; Universal Declaration of Human Rights
rights-defenders, 13, 197, 504; allies of, 489, 493, 515–17; failure of, 610–14; and marginalized groups, 178, 512, 515–17; and Jewish dissidents, 518, 521–22; and Old Regime intelligentsia, 200, 202; resentment of, 510–11, 542; strategies and structure, 205, 234, 236, 327, 434, 505–7, 554–55; women as, 538
Rimsky-Korsakov, Nikolai, 211

Rise and Fall of T. D. Lysenko, The (Medvedev), 420
Rodley, Nigel, 560–61, 568
Roginsky, Arseny, 458, 459, 651
Romanov dynasty, 630
Romm, Mikhail, 147
Ronkin, Valery, 220, 399
Rostovsky, Semyon Nikoleavich (pseudonym), 395
Rostropovich, Mstislav, 310
Roy, Manabendra, 420
Rozanova, Maria, 66, 75, 76, 105, 461, 647
Rubin, Vitaly, 93, 517, 520, 580, 597, 632
Rudakov, Ivan, 439
Rudenko, Mykola, 566–67, 583, 603, 632
Rudenko, Roman, 53, 99, 100, 102, 143, 231, 570, 578, 583, 607; on Criminal Code, 144–45; Solzhenitsyn, 421–22
Rudé právo (newspaper), 240
rule of law strategies, 23–24, 327, 342, 374–76, 437, 495–96
Rumyantsev, Valery, 478
Russell, Bertrand, 27, 35, 211, 285
Russia: current high school history textbooks, 637–38; intelligentsia, 622; literary tradition, 176; post-Soviet, 3–5; pre-Bolshevik Revolution, 131–32, 199–200; protests post-2011 parliamentary elections, 636; Romanov dynasty, 630; Russian culture, 236–37; serfdom, 630
Russian Code of Criminal Procedure (1960). *See* Criminal Code
Russian Orthodox Church, 225, 317, 328, 504, 505–6, 518, 727n42
Russian Social Fund to Aid Political Prisoners, 647, 649, 729–30n45
Russian Supreme Soviet, 145
Ryleev, Kondraty, 622

Sacred Paths to Willful Freedom (Gerts/Zlobina), 533–36, 611, 646; criticism of, 536–41
Sadomskaya, Natalya, 90
Saigon, Vietnam, 256, 257
Sakharov, Anatoly, 297
Sakharov, Andrei, 3, 147, 202, 271; Amnesty International, 580, 590; Andrei Sakharov Center, 634, 637; Brezhnev's handling of, 422–23; Chalidze and Human Rights

Committee, 374–85; dissidence, journey to, 274–83, 282; exile of, 410; expulsion from Moscow, 607–8; Human Rights Committee, 380; on intellectual freedom, 285–87; moral questions, 510; at Moscow State University, 27; 1972 petitions, 386–87; Nobel Peace Prize (1975), 602, 647; open letters on public issues, 231; Public Group, 596–97; *Reflections*, 286–90, 367, 621, 644; *Reflections*, criticism of, 290–91; *Reflections*, distribution of, 303–4; *Reflections*, public opinion of, 292–93; on science and freedom of opinion, 284–85; Soviet media campaigns against, 503; "The Symmetry of the Universe," 283–84
Sakharov, Ivan, 203
"Sakharov Committee". *See* Human Rights Committee (Valery Chalidze)
Sakharov Speaks (Sakharov), 3
Salisbury, Harrison, 377
samizdat: Amnesty International and information dissemination, 554–55; and Baptist resisters, 553; *Chronicle of Current Events*, 490; distribution in the West, 311; duplication and distribution, 399–400, *400*, 482, 540, 598; expansion of (1980s), 613–14; hub-and-spoke model for samizdat dissemination, 399–400, *400*, 446; impact of for dissident movement, 234–35; KGB charges against (1969), 169–70; Krasin's collection, 434; lending libraries, 387, 614; literary journals, 90; modes of reading of, 170–71; open letters, 231–32; periodicals/journals, 189, 191–93, *193*; poetry, 240–41; production of, 212, 234, 235–36; *samizdatchiki*, 169, 175; smuggled from camps, 478; social networks, 229; text-centrism, 7; translation of, 312; vitality of post-1968, 333–34. See also *tamizdat* (publishing over there)
samosazhanie, 256
Samsonov, Vasily, 104
Sartre, Jean-Paul, 116
Sastre, Alfonso, 579
Savenkova, Valentina, 207
Sazonov, Yegor, 96
Scammell, Michael, 16, 212, 646

Schapiro, Leonard, 550
schizophrenia, 27, 413–14, 416, 417, 711n71. *See also* psychiatry, punitive
Schlesinger, Arthur, 200
science: as method of governance, 286–87; postwar Soviet achievements, 31; scientific models and social activism, 323, 522; scientists in Soviet intelligentsia, 277–78
Scopes, John, 50
Scott, James C., 197
Search (samizdat journal), 648
self-immolation/self-denunciation, 256–58, 340, 645, 692n62, 700n24
self-liberation, 73, 78, 258, 315, 328, 434
self-perfection, 362
self-scrutiny, 500, 502, 541
self-unmasking, 329
Semerova, L. P., 184
Semichastnyi, Vladimir, 4, 52–53, 75–76, 94–96, 99, 102, 393–94; on Criminal Code, 144–45; questioning of Grigorenko, 218–19; Sinyavsky-Daniel trial verdict, 113, 115
Semyonovna, Yevgenia, 231
Serbsky Institute of Forensic Psychiatry, 29, 220, 224–25, 414, 418, 473–74. *See also* psychiatry, punitive
Šeškevičius, Antanas, 513, 646
Shafarevich, Igor, 312
Shalamov, Varlam, 93, 111, 116, 180, 200, 208, 669n60
Sharansky, Anatoly (Natan), 4, 594–95; arrest of, 603; emigration, 632; Jewish background, 517; Public Group, 597
Shatunovsky, Ilya, 41, 42, 59, 79, 642
Shchelokov, Nikolai, 125
Shelest, Petro, 399, 697–98n2
Shikhanovich, Yuri, 439
Shimanov, Gennady, 505–6, 518, 522, 536
Shklar, Judith, 130
Shklovsky, Viktor, 114, 329
Sholokhov, Mikhail, 117, 532
shortwave radio broadcasts, 6, 55, 165, 175, 194, 235, 240, 367, 392, 514, 548, 623. *See also* "Voices, the"
Shostakovich, Dmitry, 147, 503
Shragin, Boris, 1–2, 13, 299, 311, 517, 616, 632
Shub, Anatole, 172, 311, 352

Silone, Ignazio, 50
Simes, Konstantin, 104
Simonov, Konstantin, 103, 114, 115, 503
Sintaksis (samizdat journal), 642
Sinyavsky, Andrei Donatovich, 7, 46, 90, 105, 188; as Abram Tertz, 45–48, 56; arrest, 45–46; "Dissent as a Personal Experience," 617–18; emigration, 632, 647; *Goodnight!* (memoir), 661n1, 665n64; Gorky Institute, 46–47; as KGB informant, 64–65, 510; literary fiction, 129–30, 526; Lubyanka prison, 56–58; on mental state after Gulag, 460; at Moscow State University, 65–66; photos, 46, 52, 76, 461; political camps, 178. *See also* Sinyavsky-Daniel affair
Sinyavsky, Donat Evgenievich, 56–57, 69
Sinyavsky-Daniel affair: charges and trial, 59, 61, 643; Civic Appeal, Alexander Volpin's, 78–89, 96–97; Civic Appeal, text of, 85–86; collective psyche and responsibility, 72–74; defense of by Moscow intelligentsia, 75–78; and freedom of expression, 140; Hélène Peltier, 63–74, 64; impact and aftermath, 45–51, 205; KGB and threats of terror, 51–55; KGB interrogations, 56–61; Lubyanka prison, 99; Moscow State University social circle, 75–76, 76; public opinion of verdict, 115–17; transparency meeting (December 5, 1965), 89–98; trial, 142, 407–8, 616; trial, documentation of, 122–24, 123; trial, groundwork for, 99; trial, literary language and elements, 118; trial, reactions to, 337; trial, transcriptions of, 119–21, 672n61, 673n68; trial, verdict and aftermath, 111, 125–26; Western reception of defendants' writing, 62–63
skaz literary style, 464
Skinner, Quentin, 18
Skripnikova, Aida, 556
Slavin, Andrei (pseudonym), 326. *See also* Kozharinov, Venyamin
Slavsky, Yefim, 274, 289–91
Slepak, Vladimir, 580
Slonim, Vera, 374, 705n1
Slovakia, 239
Slutsky, Boris, 103

SMERSH (Death to Spies), 394
Smirnov, Lev, 100, 106, 109, 253, 365, 583, 584–86
Smith, Adam, 611
SMOG (Boldness, Thought, Form, Depth), 82, 83–84, 94, 96, 365, 643, 668n32
Snezhnevsky, Andrei, 414, 416
socialism: "developed/mature socialism" and rights, 11, 142, 144, 362, 626; European-style, 132, 239–41, 305; Soviet system, 8, 12, 141, 174, 290, 292, 360–62, 460, 495, 498; transition to communism, 115, 199, 292
Socialist Revolutionary (SR) Party, 56, 182
Sokirko, Viktor, 312, 531–32, 542, 611
Sokolov, Sasha, 82
Sologub, Fedor, 723n28
Solomentsev, Mikhail, 423
Solovyev, Vladimir, 636
Solzhenitsyn, Alexander, 4, 11, 13, 55, 176, 177, 208, 286; on Amalrik's writing, 310; on appeals to authority, 336, 361–62; call for honesty, 610–11; criticism of dissidents, 506–8; on convergence of capitalism/socialism, 292; deportation to Germany, 453–54; emigration, 429, 632; on Helsinki effect, 606; Human Rights Committee, 376–78, 380; KGB handling of, 424–29; literary fiction, 526; *Live Not by the Lie*, 483–85; Nobel Prize and increasing pressure on, 420–21, 645; open letters of, 420, 643; photo, 425; return to Russia (1994), 633–34; and Sakharov, 291, 386, 694n59; Soviet media campaigns against, 503; Universal Copyright Convention violations, 428–29
Some Thoughts on How to Help Political Prisoners (Ivan Kovalev), 648
Sörheim, Ingjald Orbeck, 551
Sorokin, Viktor, 612
Sorokina, Sonya, 612
Sotin, Boris, 91–92
Soviet Academy of Sciences, 147, 182, 208, 249, 275, 282, 304, 328, 367, 381, 608
Soviet Constitution (1918), 134
Soviet Constitution (1936), 23–24, 135–38, 136, 137, 641; Article 111, 79; Article 125, 134–35, 138–39; Article 126, 342;

Article 160, 608; free association, 355; free speech rights, 333–34; planning for new (1960s), 139, 142; Union republics' rights to secede, 515
Soviet Constitution Day, 193–94; transparency meetings, 487–88
Soviet Union: Afghanistan, invasion of, 648; agricultural policies, 337; anti-Soviet agitation and propaganda, charges of, 32, 57, 59, 142, 167, 404, 435; Central Committee Ideological Commission, 41, 120; Civil Code (1964), 375–76, 379, 705n8; collective responsibility for state action, 174; Communism, as final stage, 12; Communist Party, 42, 620; court system, 155; dissolution of, 2–3, 9–10, 629–31; early disobedience and uprisings, 24–25; education and literacy, 101, 279–80; elections, 174, 410; ethnic minorities and marginalized groups, 179, 356, 357, 380, 555–56, 601; exile, practice of, 410–11; guerrilla resistance to, 392–92; informality, embeddedness of, 354–55; intelligentsia, 305–6; liberalization in 1960s, 307–9; Marxism-Leninism, 195–97; mass media, control of, 14, 190–91, 488–89; micro-communities, 97; nationalist movements, 214; Olympics (1980), 607–8; ordinary citizens, 322, 330; parental role of, 279–80; post-Stalin leadership, 10–12; rural/urban divides, 308–9; Russian Supreme Soviet, 145; "self-criticism" ritual, 41, 406; shadow economy, 495–96; show trials, 100–101, 104; slander, legal stance, 108–9; Soviet Code of Criminal Procedure, 39–40, 177, 453, 509; Soviet Constitution Day, 85, 89, 93, 193, 487; Soviet values, 276–77; state functions vs. citizen-volunteers, 342; state goals/juridical means, 144–46; terrorism, 10–11, 26, 51–53, 80, 125, 144, 205, 216, 280, 323, 395, 430, 460, 637–38; Thaw period, 31–32, 33, 55, 160–61, 362, 393, 495, 618; Twentieth Party Congress, 111, 159, 228; Western journalists in, 15–16
Soviet Uzbekistan, 345
Spark/Ogonek (journal), 41, 42
Speer, Albert, 160

Spender, Stephen, 210, 211, 311, 384, 595, 644, 646
Sputnik, 31, 275
Stalin, Joseph, 495; cult of personality, 70–71; de-Stalinization, 159–60; legal system under, 142; opponents, treatment of, 97; political trials during rule, 101; secret police practices, 406, 407; Stalin-era criminal procedure, 149–50, 528–29
Stalin Constitution (1936). *See* Soviet Constitution (1936)
Starr, Frank, 347, 349, 350–51, 352, 701n43
Steffens, Lincoln, 201
Stender-Petersen, Adolf, 298
Stockholm Appeal (1950), 232–33
Stolypin, Petr, 203
Strada, Vittorio, 527
Strategy-31, 634, 635
Stravinsky, Igor, 211
Stroeva, Yelena, 79, 88
Strugatsky, Arkady, 452, 454–55
Strugatsky, Boris, 452, 454–55
Strunnikov, Vladimir, 393
Struve, Gleb, 530
Stuchka, Petr, 133
Stus, Vasyl, 631
Sukharev, Alexander, 585–87, 590
Sunday Times, London (newspaper), 344
Superfin, Gabriel, 439, 651, 652, 731n36
Suri, Jeremy, 724n2
Suslov, Mikhail, 278, 279, 284, 337–38, 340, 380, 384, 423, 699n5
Svetlova, Natalya, 518
Svoboda, Ludvík, 239, 244–45, 251, 253
Sysoyev, Vyacheslav, 584
Szasz, Thomas, 415

tamizdat (publishing over there), 6, 190, 212, 312, 471, 456, 526, 530, 646. *See also* samizdat
Tamm, Igor, 147
Tan-Bogoraz, Vladimir, 265
Tarsis, Valery, 53, 55, 80–81, 114, 421, 662n22, 667n20
TASS (Soviet news agency), 103–4, 597
telefonnoe pravo (law by telephone), 130
Temushkin, Oleg, 100, 106, 107, 110, 119
Tereshkova, Valentina, 31
Ternovsky, Leonard, 347–49, *348*, 353

Tertz, Abram. *See* Sinyavsky, Andrei Donatovich
Theodorakis, Mikis, 212
Thích Quang Duc, 256, 257, 692n62
Thirty-Seven (journal), 502
This Is Moscow Speaking (Arzhak/Daniel), 48, 54, 55, 61, 77, 106–7, 673n68
Thoreau, Henry David, 24
Tiagai, Ida, 463
Tilly, Charles, 204
Time (magazine), 46, 62, 63
Times, London (newspaper), 253, 313, 315, 433, 542; *Sunday Times,* London (newspaper), 344
Titov, Yuri, 79
Tolstoy, Lev, 24, 26, 200; "I Cannot Be Silent," 202–3
totalitarianism, 8–9, 685n34; collective/mass responsibility, 72–74; collective psyche and face-to-face community, 80; and cults of personality, 140–41; dissent, treatment of, 9, 11–12; law, role of, 129–32; modern contrasted with classic tyrannies, 72; totalitarian responsibility, 263–64; Western ways of regarding Soviet Union, 204. *See also* dictatorships, nature of
"To the Global Public" (Bogoraz & Litvinov), 172, 186, 210, 262, 337, 362, 644
Tractatus Logico-Philosophicus (Wittgenstein), 35
transparency: *glasnost,* 3, 25, 122, 364–65, 564, 667n16; judicial, 24, 80–81, 93, 122, 169, 175; laws regarding, 118, 145, 167; as tactic, 485–86, 512–13, 624, 629; transparency from below, 231–32
transparency meetings, 138, 145, 193, 250, 283, 322, 487–88, 492, 612, 645; Volpin's, in Sinyavsky-Daniel affair, 79–98, 148, 193, 229, 255, 365, 397, 418
Trepov, Fedor, 96
"Trial of the Four," 404, 643; in *Chronicle of Current Events,* 192; impact of, 397; Letter of the 46, 182; petitions to attend, 167–69, 533, 551; and protest letters, 206, 232; public reaction to, 174–78, 186; Western coverage of, 171–72, 174–75

Trofimov, Viktor, 196
Trotsky, Leon, 133, 495
Tsukerman, Boris, 343, 374, 376, 380, 452, 517, 518, 632
Tumanov, Oleg, 409
Turchin, Valentin: on Amnesty International, 565–67, 570, 577, 579–80, 591; emigration, 632; on fear, 11, 460; open letters, 168, 232; photo, 566; on quantum mechanics, 277; on reformation of Communist Party, 335; samizdat essay, 581–83, 647; on Soviet society, 542; on totalitarianism, 11; on valence bonds, 565–66, 567, 597
Tvardovsky, Alexander, 88, 102–3, 112, 114, 115, 121, 125, 243–44
Tverdokhlebov, Andrei, 292, 376–78, 469, 561–63, 568–69, 574, 577, 579, 646; arrest of, 583; emigration, 632, 648
"Two Thousand Words" (Vaculík), 240
Tykhy, Oleksa, 602–3

Ukraine, 392, 512; annexation, 636–37; color revolution, 636; demonstrations against war in, 637; Helsinki monitoring groups, 598, 609; independence movement, 553; nationalists, 557–58; People's Movement of Ukraine, 633; resistance to Soviet policies, 515–17; Ukrainian intellectuals, 491; Ukrainian Soviet Socialist Republic, 178, 264–65
Ukrainsky visnyk/Ukrainian Herald (journal), 516, 645
Ulanovskaya, Maya, 256–57, 343–44, 632
Ulitskaya, Ludmila, 188
Union for the Struggle for the Rebirth of Leninism, 216–18, 642
Union of Communards, 195
"Union of Friends" (Okudzhava), 78
Union of Leninist Communists, 196
Union of Patriots, 195
Union of Soviet Writers, 40, 55, 100, 105, 114, 160, 176, 286, 421, 643, 645
Union of Struggle for the Revolutionary Cause, 195
United Nations: and Amnesty International, 587; Commission on Human Rights, 362–63; Convention against Discrimination in Education, 380;

Initiative Group contacts, 357, 360, 362–64, 366, 383, 387; International Year of Human Rights, 191. *See also* Universal Declaration of Human Rights
United States: Amnesty International involvement, 588, 590, 729n8; civil rights movement, 522, 626–29, 722n35; exile in, 427, 450, 452, 632–33, 646, 647, 648, 649; human rights violations, 359; and Puerto Rico, 602; Soviet views of, 613; Vietnam War and American imperialism, 43, 559, 692n62. *See also* Cold War; West
Universal Declaration of Human Rights, 191, 232, 287, 346, 357–60, 452–53, 570, 626–28, 641, 702–3n70; and Amnesty International, 553; free association, 355; Soviet obligations, 360
U Thant, 346, 363
utopianism, 38, 43–44, 360
Uzbekistan, 345

Vaculík, Ludvík, 240, 625
Vakhtin, Boris, 121
valence bonds, among rights-defenders, 565–67, 597
van het Reve, Karel, 172, 211, 212, 231, 288, 311–12, 313, 623, 645
Vasilev, Arkady, 100
Vatican Radio, 514
Vavilov, Nikolai, 208
Vechernaya Moskva/Evening Moscow (newspaper), 161
Velikanova, Tatyana, 440, 447, 448, 609, 648
Velikanova, Yekaterina, 261, 727n43
Venclova, Tomas, 179, 181, 632
Verblovskaya, Irina, 310
Vigdorova, Frida, 120–21, 122, 124, 642
Vikhireva, Klavdia, 378
Vinogradov, Igor, 121
Vins, Georgi, 556
Virgin Lands campaign, 336
Vishnevskaya, Yuliya, 83, 89
Vladimov, Georgi, 607
Voice of America, 6, 55, 93, 164, 166, 182, 240, 351, 392, 628
"Voices, the" (radio programs), 6, 55, 234, 274, 289, 312, 352, 386, 597, 610, 617

Voinovich, Vladimir, 198, 517, 526, 733n4
Volnyi, K. (pseudonym), 330
Volpin, Alexander, 24, 200; and Aikhenvald, 220; Amnesty International, 556; anarchism vs. rule of law, 237, 659n49; anti-Soviet agitation charges (1957), 32; arrest (1949), 28–29, 641; in Bukovsky's trial, 155; childhood and education, 25–27; Civic Appeal, 145, 334, 643, 683n151; Civic Appeal, text of, 85–86; Communist Youth League, 27; constitutionalism, 150; cybernetics, 33–34; departure from Soviet Union, 609; emigration, 452–53, 632, 646; exposure to Western media during Thaw, 33; on fidelity to Soviet law, 374; "A Free Philosophical Tractate," 35–37, 40; *glasnost* as term, 564; impact of, 23–25, 505; interrogation, strategy to cope with, 456–59, 645; involuntary psychiatric hospitalization, 39, 41, 42, 412, 473, 642; Jewish identity, 517, 523; language, philosophy of, 27, 35, 37–39; language of Soviet law, 43–44; law-based approach, 23–24, 462–63, 657n1; *A Leaf of Spring*, 40–42, 53, 412, 642; meta-revolution, 44, 324, 334, 511, 521, 610, 619; Moscow State University, 27–28; photo, 30; poetry, 28–29; rights-defense strategy, 334; rule-of-law strategy, 327, 437, 659–660n49; Sinyavsky-Daniel case, 78–89, 96–97, 117–19; on Soviet law and rights, 39–40, 132; Soviet values, 276–77; transparency meeting, 90–91, 91, 92–93, 96–97, 193
Volpina, Nadezhda, 25, 27
Volpina, Victoria, 82
von Clausewitz, Carl, 392
Vorobev, Oleg, 90
Voronel, Alexander, 75, 78, 119, 223–24, 521
Voronel, Nina, 75, 77–78, 119
Voronsky, Alexander, 33
Voronyanskaya, Elizaveta, 425, 492
Voroshilov, Kliment, 1, 208
Voznesensky, Andrei, 529
Vyshinsky, Andrei, 101, 129, 216, 358–59, 702–3n70
Vysotsky, Vladimir, 56, 208, 271

Waldheim, Kurt, 363, 513, 570
Wałęsa, Lech, 3
Washington Post (newspaper), 172, 311, 313, 344, 352, 613
weak ties, strength of, 97, 255, 399
Weber, Max, 130, 564, 611
Weiner, Amir, 392
Weiquan (Rights-Defense) movement, 625–26, 737n24
West: criticism of Soviet Union, 431, 605–6; and Human Rights Committee, 377; journalism, 6, 15–16, 418, 434, 500–501, 599, 610, 736n71; Soviet communication with foreigners, 63, 172–73, 349, 352, 433–34, 489, 554–55; dissident movement, relationship with, 14, 384, 393–94, 443, 500–502, 620–22; and human rights, 17–18; protest movements in, 548, 552; views of Cold War, 2–3, 414–15; views of Soviet Union, 107; Western gaze, power of, 14–15, 316, 610, 620–21. *See also* samizdat; shortwave radio broadcasts; *tamizdat* (publishing over there)
What Is It, Rooster?, 529
What Is It You're Laughing At?, 529
What Is It You Want? (Kochetov), 527–32, 536, 539, 542, 611, 645
What Is Socialist Realism? (Tertz/Sinyavsky), 47, 61, 63
White Book (Ginzburg), 124, 148, 162, 167, 178, 197, 556, 643
Whitehead, Alfred North, 27
Williams, Nikolai, 93
Will the Soviet Union Survive until 1984? (Amalrik), 304, 309–15, 508, 511, 518, 611, 622, 645, 696n44
Wilson, Woodrow, 360
Wittgenstein, Ludwig, 35, 38, 43, 659n39
Wolzak, Henk, 250
World Congress of Peace Forces, 559, 560, 568
World Festival of Youth and Students (Moscow, 1957), 1–2, 32, 34, 412, 641
World Health Organization, 363–64, 590
World Peace Council, 233
World Psychiatric Association, 363–64
World War II, 30–31, 51, 336; European resistance during, 623; German assault on the Soviet Union, 295; Helsinki Accords, 593–94; Nazi Party, 27, 30–31, 204, 336, 500, 460; SMERSH counterintelligence unit, 394
Wortman, Richard, 299

Yakhimovich, Ivan (Jan Jahimowicz), 336–40, 339, 517; arrest of, 340–41, 645; defense of, 342–45; and formation of Initiative Group, 347–49; psychiatric hospitalization, 473
Yakir, Irina, 251, 435, 439, 447
Yakir, Morris, 207
Yakir, Petr: apartment gatherings, 345, 347–49; arrest, 435–36; death, 609; family background, 206–8; and foreign journalists, 433–34; as Gulag survivor, 291; impact of, 450–51; and Initiative Group, 352, 354; Jewish background, 517; KGB interrogation, 436–41; leadership of, 352; on leadership structures, 236, 314–15, 325; photos, *209, 351, 433, 435*; press conference, 442–44, *444*, 448; public disavowal, 540–41; removal from Moscow, 423–24; status as dissident, 350–51; trial, 432, 441–42; visiting days, 187
Yakir, Sarra, 207, *209*
Yakir, Yona, 207, 290
Yakobson, Anatoly, 223, 242, 256–59, *259, 261, 262*, 327, 343, 439, 447, 632
Yakovlev, Alexander, 102
Yakovlevich, Valery, 81
Yakunin, Gleb, 648
Yale Russian Chorus, 35, 659n39
Yeltsin, Boris, 633
Yemelkina, Nadezhda, 350, 410, 436–37, 632, 647
Yevtushenko, Yevgeny, 41, 103, 244
Yofe, Olga, 414
Yofe, Venyamin, *419*
Young Pioneers, 318, 320
Yudenich, Nikolai, 207
Yugoslavia, 240, 579, 605, 611
Yurchak, Alexei, 498–99, 717n78, 719n35

zakonnost (rule of law), 25. *See also* rule of law strategies
Zasulich, Vera, 95, 96

Zdebskis, Juozas, 513, 646
Zeldovich, Yakov, 147
Zelinsky, Vladimir (Dmitry Nelidov), 328–30, 461, 462, 498, 506, 616
Zeltman, Yadja, 575
Zhdanov, Andrei, 68
Zheludkov, Sergei, 311, 697n52, 728n64
Zhivlyuk, Yuri, 285
Zhuchkov, Anatoly Gavrilovich, 466–68
Zinoviev, Alexander, 8, 15, 247, 502, 504, 518, 619, 631
Zivs, Samuil, 560, 576–77, 578, 586, 589–90
Zlobina, Maya Lazarevna (Anna Gerts), 532–36, 611, 646, 723n19
Zolotukhin, Boris, 182, 396

GPSR Authorized Representative: Easy Access System Europe - Mustamäe tee 50, 10621 Tallinn, Estonia, gpsr.requests@easproject.com

www.ingramcontent.com/pod-product-compliance
Lightning Source LLC
Jackson TN
JSHW020911090525
84110JS00001B/1/J